Governance of Marine Fisheries and Biodiversity Conservation

Governance of Marine Fisheries and Biodiversity Conservation: Interaction and Coevolution

Edited by

Serge M. Garcia
IUCN/CEM/FEG, Gland, Switzerland

Jake Rice
Fisheries and Oceans Canada, Ottawa, Canada
IUCN/CEM/FEG, Gland, Switzerland

Anthony Charles
Sobey School of Business and the Environmental Science Department, Saint Mary's University, Halifax, Nova Scotia, Canada
IUCN/CEM/FEG, Gland, Switzerland

IUCN CEM FEG | ebcd EUROPEAN BUREAU FOR CONSERVATION & DEVELOPMENT

WILEY Blackwell

Registered Office
John Wiley & Sons, Ltd, The Atrium, Southern Gate, Chichester, West Sussex, PO19 8SQ, UK

Editorial Offices
9600 Garsington Road, Oxford, OX4 2DQ, UK
The Atrium, Southern Gate, Chichester, West Sussex, PO19 8SQ, UK
111 River Street, Hoboken, NJ 07030-5774, USA

For details of our global editorial offices, for customer services and for information about how to apply for permission to reuse the copyright material in this book please see our website at www.wiley.com/wiley-blackwell.

Library of Congress Cataloging-in-Publication Data

Governance of marine fisheries and biodiversity conservation : interaction and coevolution/edited by Serge M. Garcia, Jake Rice, and Anthony Charles.
pages cm
Includes bibliographical references and index.
ISBN 978-1-118-39264-5 (cloth)
1. Fishery management. 2. Marine biodiversity. I. Garcia, Serge, editor of compilation. II. Rice, J. C. (Jake C.) III. Charles, Anthony Trevor, 1956– editor of compilation.
QH91.G68 2014
333.95′6–dc23

2014016945

A catalogue record for this book is available from the British Library.

Wiley also publishes its books in a variety of electronic formats. Some content that appears in print may not be available in electronic books.

Cover image: Left panel: School of cod near an invertebrate-covered shipwreck.
NOAA Stellwagen Bank National Marine sanctuary. Courtesy of Matthew Lawrence.
Top right panel: Trawler in the port of Tromsø. Courtesy of Dr. David E. David Johnson.
Bottom right panel: Small-scale fishermen pulling a beach seine; Zanzibar. Courtesy of Rob Tarr.

Set in 9/11.5pt Myriad by SPi Publishers Services, Pondicherry, India
Printed and bound in Singapore by Markono Print Media Pte Ltd

1 2014

Contents

Notes on contributors

Armada, N. MSc in Hydrobiology and Fisheries Science, Hamburg University. Fisheries biologist with over 30 years of experience in fisheries resources assessment and management. Former faculty member of the College of Fisheries and Ocean Sciences of the University of the Philippines. He is currently the fisheries management advisor of the project Ecosystems Improved for Sustainable Fisheries (ECOFISH) Project of USAID, a project designed to improve management of important fisheries resources of the eight key biodiversity areas in the Philippines. *nbarmada@gmail.com*

Augustyn, J. PhD, Zoology (Marine biology). Chief Director: Fisheries Research and Development, Department of Agriculture Forestry and Fisheries, South Africa. Chair of the Ecosystem Advisory Committee of the Benguela Current Commission and International Commission for the Conservation of Atlantic Tunas (ICCAT) Panel 3 on Temperate Tunas. He recently chaired the FAO Technical Consultation on Flag State Performance. He has broad experience of regional organizations in Southern Africa and has authored and co-authored several papers and presentations in international fora on the Ecosystem Approach to Fisheries. Member of the Fisheries Expert Group of the IUCN Commission on Ecosystem Management (IUCN/CEM/FEG). *cjaugustyn@yahoo.com*

Bavinck, M. PhD. Specialized in coastal and marine geography and in legal pluralism, and focuses largely on South Asia. He has authored two monographs and two edited volumes on fisheries governance, in addition to a large number of academic articles on related topics. He is presently an Associate professor in the Department of Human Geography, Planning and International Development Studies, University of Amsterdam. He is also the co-founder and co-director of the Centre for Maritime Research (MARE). *j.m.bavinck@uva.nl*

Béné, C. PhD Department of Ecology, University of Paris VI, France. Research Fellow in the Vulnerability and Poverty Reduction team at the Institute of Development Studies in UK. He is a socio-economist and policy analyst by training with 15 years of experience in sub-Sahara Africa, South and Southeast Asia, Caribbean, and South America. His research focuses on poverty, vulnerability and governance in relation to the socio-political economy of natural resources-dependent communities in developing countries. *c.bene@ids.ac.uk*

Beslier, S. Master in Public Law and Political Science (Paris, France) and graduated from the Navy School for the Administration of Maritimes Affairs (Bordeaux, France). 39 years of practical and senior management in maritime and fisheries administration, both in the French Maritime Affairs Administration and the European Commission services. Former chair of the board of the European Fisheries Control Agency. Member of the Fisheries Expert Group of the IUCN Commission on Ecosystem Management (IUCN/CEM/FEG) as a legal expert. *serge.beslier@yahoo.fr*

Cabanban, A.S. PhD. Marine ichthyologist based in Manila, Philippines. She was a Lecturer at Silliman University, Philippines and Associate Professor at the Universiti Malaysia Sabah, Malaysia from 1979 to 2006. She has assisted in coordinating activities in the East Asian Seas region at the United Nations Environment Programme in 1994-1996 and 2002 and managed for 2. 5 years a conservation program of the World Wide Fund for Nature, Malaysia. She is involved intermittently in the past 17 years in designing, implementing, executing, and evaluating of projects of the Global Environment Facility. Member of the Fisheries Expert Group of the IUCN Commission on Ecosystem Management (IUCN/CEM/FEG). *annadel.cabanban@gmail.com*

Camilleri, M. PhD. Fishery Liaison Officer within the FAO Fisheries and Aquaculture Policy and Economics Division.

Graduated in Fisheries and Ocean Science from the University of Plymouth (UK) in 1995 where he went on to obtain a PhD in fisheries management. He is a member and chartered biologist of the Society of Biology (UK). Served as Consultant to the Maltese Government and Head of the Malta Centre for Fisheries Sciences between 1998 and 2007. Joined FAO in 2007 taking up a post within the Secretariat of the General Fisheries Commission for the Mediterranean before moving to his current position in Italy. *matthew.camilleri@fao.org*

Campbell, L.M. PhD in Geography (University of Cambridge, UK). She has studied on marine fisheries, protected species, coastal development, and conservation in Latin America, the Caribbean, and North Carolina, USA. She is a Rachel Carson Associate Professor, Marine Affairs and Policy, Nicholas School of Environment and Duke University Marine Laboratory. Member of the IUCN's Marine Turtle Specialist Group and Commission on Environment, Economic, and Social Policy and Editor of the journal Conservation and Society. *lisa.m.campbell@duke.edu*

Charles, A. PhD. Natural science and Engineering. At Saint Mary's University for over 25 years, holding a dual appointment as Director of the School of the Environment at Saint Mary's University (Canada), and Professor in the Sobey School of Business and the Environmental Science Department. Lead researcher of the Community Conservation Research Network. Pew Fellow in Marine Conservation, Senior Research Fellow of the International Ocean Institute and Past President of the International Institute of Fisheries Economics and Trade. Within Canada, he was formerly a research scientist with the Department of Fisheries and Oceans (DFO), later served on the DFO Science Advisory Council and was founding Director of Canada's Ocean Management Research Network. Author/co-author of books on sustainable fishery systems, integrated fish farming, coastal fisheries of Latin America and the Caribbean, Canadian marine fisheries, community fisheries management and human dimensions of the ecosystem approach to fisheries. Member of the Fisheries Expert Group of the IUCN Commission on Ecosystem Management (IUCN/CEM/FEG). *Tony.Charles@smu.ca; http://smu-facweb.smu.ca/vcharles*

Christie, P. PhD. Environmental sociology and marine policy. Professor, School of Marine and Environmental Affairs and Jackson School of International Studies, University of Washington. His research focuses on the social ecological dimensions of marine and coastal policy in the Southeast Asia, Latin America, and the United States. Prior to his academic work, he was involved in the implementation of a community-based marine protected area in the Philippines as a Peace Corps Volunteer. His Pew Fellowship in Marine Conservation focused on governance factors affecting the success of Philippine MPAs and translating this research into practical solutions. He is Editor-in-Chief of the journal Coastal Management. *patrickc@uw.edu*

Deupmann, P. Masters Degree in Marine and Environmental Law of the University of Cape Town, Masters Degree in Dutch Law of the University of Amsterdam, where he also obtained an M.A. Legal officer, Development Law Service, Food and Agriculture Organization of the United Nations. With the FAO since 2007. Presently in the Development Law Service. Peter is specialized in marine and inland fisheries and aquaculture as well as in biological diversity related issues. *Peter.Deupmann@fao.org*

De Young, C. MS on environmental and natural resource economics from West Virginia University, USA (1997). Environmental and natural resource economist, Food and Agriculture Organization of the United Nations (FAO). She has a particular interest in promoting discussions on the social and economic contributions of the aquatic systems as part of integrated management efforts. Most of her 11-year tenure at the FAO has related to the development of fisheries management plans and strategies and supporting the ecosystem approach to fisheries. Understanding human choices and drivers of change, the benefits derived from the sector as well as the sector's impacts on the aquatic systems have always been central to her work. *cassandra.deyoung@fao.org*

Drobenko, B. Doctor in Public Law (1990). University Professor and Director of the "Campus de la mer ». His area of competence covers the Law of the Sea, environment, and coastal urban development law. He is presently in the French Université du littoral de la Côte d'Opale Lille, France). *bernard.drobenko@univ-littoral.fr*

Freestone, D. Doctor of Laws, LL.D University of Hull (1999). Professor. Executive Director of the Sargasso Sea Alliance led by the Government of Bermuda, and Visiting Scholar at The George Washington University Law School. He is former Head of the International Environment Law Group and Deputy General Counsel at The World Bank. He is the founding Editor of the International Journal of Marine and Coastal Law, and has published some 20 books and more than 150 articles on the law of the sea and international and marine environmental law issues He is the winner of the 2007 Haub Prize for Environmental Law. *dfreestone@sargassoalliance.org*

Garcia, S.M. Doctorate in Science of the University of Marseille (France, 1976). Specialized in population dynamics and management (Penaeid shrimps; demersal resources). Long research and management experience in the tropics. Former research Director in IRD (France) before moving to FAO in 1979 where he held the positions of Chief Marine Resources Service and Director Fisheries and Aquaculture Management Division before retiring in 2007. Chairs the Fisheries Expert Group of the IUCN Commission on Ecosystem Management (IUCN/CEM/FEG). *garcia.sergemichel@gmail.com*

Gjerde, K.M. JD. Graduated from New York University School of Law in 1984. Is a Pew Fellow in Marine Conservation and past fellow at the Woods Hole Oceanographic Institution and the Law School of the University of Hull. Specializes in international marine environmental law including law of the sea, fisheries, shipping, seabed mining and marine biodiversity conservation, with a focus on areas beyond national jurisdiction. Currently Senior High Seas Advisor to IUCN's Global Marine and Polar Programme and Chair, Specialist Group on High Seas, IUCN World Commission on Protected Areas. *kgjerde@eip.com.pl*

Gullestad, P. Candidatus Oeconomices 1976. Norwegian Delegate to the International Council of the Exploration of the Seas (ICES). Graduated from the University of Oslo with a master of economics (Cand. oecon.) in 1976. His professional career since 1977 has been with the Norwegian Directorate of Fisheries and for 12 years (1996-2008) he was the Director General of Fisheries of Norway, managing the Norwegian fisheries and aquaculture sectors. Most of his career has been dedicated to the development of a sustainable management regime for the living marine resources. He I now Specialist Director at the Norwegian Directorate of Fisheries. As Norway is sharing most of its resources with neighboring countries, these efforts have been directed both at the national as well as the international level. *Peter.Gullestad@fiskeridir.no.*

Hamukuaya, H. MSc in Marine Environmental Sciences from the State University of New York at Stony Brook, USA (1991) and a PhD degree in zoology from the University of Port Elizabeth in South Africa (2000). Between 1991 and 1998 he worked as a fishery biologist in the Ministry of Fisheries and Marine Resources (MFMR), Namibia responsible for the Demersal Research Section and in 1999 he was promoted as Deputy Director of Resources Management, responsible for fisheries research and international fisheries initiatives including SADC, CCAMLR and ICCAT. During 2003-2005, he was a Director of the Benguela Current Large Marine Ecosystem Program, coordinating dozens of research projects on living marine resources in Angola, Namibia and South Africa. From 2005-2008, he worked as the first Executive Secretary of South East Atlantic Fisheries Organisation (SEAFO) before he joined the Benguela Current Commission as its first Executive Secretary of Benguela Current Commission (BCC). *hashali@benguelacc.org*

Hanna, S. PhD. Natural Resource Economics (Oregon State University). Professor emeritus of marine economics at Oregon State University, USA. Scientific advisor to multiple regional, national and international resource agencies. Recipient of the Distinguished Service Award of the International Institute of Fisheries Economics and Trade. Research areas include marine economics and policy, economic performance of institutions, ecologic-economic linkages. *hannasu@onid.orst.edu; susan.hanna7@gmail.com*

Hersoug, B. Master in sociology from the University of Tromsø in 1976.Worked as planner for the county for three years before returning to the university, where he has had various positions, including two periods as rector of the Norwegian College of Fishery Science. Later he took his Dr. philos. degree on fisheries development projects. Since 2001 he has been professor at NCFS within fisheries and aquaculture management. He has been a consultant to Norad for 30 years and published extensively on fisheries in Norway as well as in various other countries (South Africa, New Zealand, Nambia, and Chile). At present he is responsible for the Master study in International Fisheries Management. *bjorn.hersoug@gmail.com*

Hoel, A.H. Candidatus Politicarum. Research Director of the Institute of Marine Research. He lives and works in Tromsø, Norway at 70 degrees North, well beyond the Arctic Circle. A political scientist by training, his research centers on the institutional issues in management of the marine environment and its natural resources, in particular in the polar areas. Hoel is the author of numerous academic publications. Recent works addresses ecosystem-based management, international cooperation in marine scientific research, and implications of climate change for the management of the marine environment. Hoel has also participated in numerous assessment of marine ecosystems, such as the Arctic Climate Impact Assessment and the Arctic Biodiversity Assessment. *alf.haakon.hoel@imr.no*

Hoydal, K. Candidatus scientarum (Ms) in Marine Ecology from the Copenhagen University. His field of competence is fishery biology and fisheries management, as well as regional fisheries management organisation and governance. Former Secretary of the Northeast Atlantic Fisheries

Commission (NEAFC) from March 2011 to June 2011. He works as Director for "Skrivarastova – Fish and Film", Consultants. Ltd. Board member of Nordic Marine Think Tank, NMTT since 2012. Member of the Fisheries Expert Group of the IUCN Commission on Ecosystem Management (IUCN/CEM/FEG) (2012). *kjartanhoy@gmail.com*

Jennings, S. PhD in fish population biology from the University of Wales. Lead Adviser, Centre for Environment, Fisheries and Aquaculture Science, Lowestoft, UK. Current role preceded by other roles as advisor and research scientist, working nationally and internationally on marine environmental management, biodiversity and fisheries issues. Former Chair of the International Council for the Exploration of the Sea, Advisory Committee on Ecosystems. Honorary Professor at the University of East Anglia and Visiting Professor at the University of Newcastle, UK. *simon.jennings@cefas.co.uk*

Jentoft, S. PhD. Sociologist. Professor at the Norwegian College of Fishery Science, Faculty of Bioscience, Fisheries and Economics, University of Tromsø, Norway. He is specializing on the governance of fisheries and has written and edited many books and journal articles this issue, most recently on small-scale fisheries, poverty alleviation and marine protected areas. He has also written extensively on issues such as fisheries co-management, fisheries organizations and coastal community development in Norway and in a number of other countries in the north as well as in the south. *svein.jentoft@uit.no*

Johnson, D. PhD in intertidal conservation. Director of Seascape Consultants Ltd and Emeritus Professor of Coastal Management at the UK Southampton Solent University. He was previously Executive Secretary to the OSPAR Commission (2006-2012) and his career has included work in practical conservation, environmental consultancy, and higher education as well as intergovernmental marine environmental protection. In 2002 David held a Caird Fellowship at the UK National Maritime Museum. He is a member of the GESAMP pool of experts and a Visiting Professor at the World Maritime University in Malmo Sweden. *david.johnson@seascapeconsultants.co.uk*

Kenchington, R. MSc Marine Biology, University of Wales Thesis, 1968. D.Sc (Hon) Marine and Coastal Environmental, 2001. Management. Professorial Fellow at the Australian National Centre for Ocean Resources and Security of the University of Wollongong, Australia. Consultant and advisor to governments and international agencies. As marine ecologist, he studied crown of thorns starfish and large scale surveys of the Great Barrier Reef and was a foundation member of the planning and research staff of the Great Barrier Reef Marine Park Authority in 1978, retiring in 1999 as Executive Director of the Authority. He specialized in ecosystem based management of coral reef, coastal and marine ecosystems. He is a member of Fisheries Expert Group of the IUCN Commission on Ecosystem Management (IUCN/CEM/FEG). *rkenchin@uow.edu.au*

Khan, Z. Master in Environment Management and Development. The Australian National University, Canberra, 2009. Foundation for the Peoples of the South Pacific International (FSPI), Suva, Fiji. Marine biologist previously engaged as a Marine Protected Area specialist with Centre for Resource Management and Environmental Studies (CERMES), The University of the West Indies (UWI), under marine protected governance project. Her work has been largely with civil society non-government organization in implementing locally managed marine area program in the Pacific Region. Her recent work has been in Bahamas and in Grenada in the Caribbean region on marine protected areas. Her area of professional interest is in building community resilience for sustainable livelihoods under the changing climate and ecosystems. *zaidy.khan@gmail.com*

Kolding, J. *Doctor Scientiarum* (PhD) Degree in Fisheries from University of Bergen (1994), Associate Professor, Department of Biology, University of Bergen, Norway. specializing on the ecology and exploitation of fish in fluctuating tropical freshwater systems. Employed by Department of Biology of the University of Bergen, Norway from 1987-2000 and from 2002 to present. In 2000-2001 he was employed by the Institute of Marine Research (IMR, Norway), Centre for Development Cooperation in Fisheries where he worked with the R/V Dr. Fridtjof Nansen programme. His primary research interest is fish stock assessment, ecology and management of small-scale tropical fisheries, with particular experience from inland fisheries. Member of the Fisheries Expert Group of the IUCN Commission on Ecosystem Management (IUCN/CEM/FEG). *jeppe.kolding@bio.uib.no*

Kuemlangan, B. LLB from the University of Papua New Guinea (1986). LLM in International and Comparative Law from the Chicago-Kent College of Law of the Illinois Institute of Technology (1995). Chief of the Development Law Service (LEGN) in the Legal Office of the Food and Agriculture Organization of the United Nations (FAO). Specialized in international Law of the Sea and fisheries and aquaculture law, he manages FAO's legal assistance work programme food and agriculture including the natural resource sectors of land, fisheries, forestry, and water and has provided legal assistance in developing national legal frameworks in Africa, Asia and the Pacific, Central Asia and the Caucasus, Latin America and the Caribbean, and the Near East. He was a technical and legal adviser in the process of establishing two regional fisheries management organizations and recently provided secretarial and legal support to the development of the Agreement on Port State Measures to Prevent, Deter and Eliminate IUU Fishing. Prior to joining FAO, he was a senior legal officer with the Department of Justice and Attorney General of Papua New Guinea and advised on foreign affairs, trade and fisheries related treaties and conventions including the negotiations of the Convention on Biological Diversity and the 1995 UN Fish Stocks Agreement. *Blaise.kuemlangan@fao.org*

Lee, J. PhD in Coastal Oceanography (State Unviersity of New York, 1998). M.A. in International Relations (Tufts University, 2010). Environmental Affairs Officer for marine and coastal biodiversity within the Secretariat of the Convention on Biological Diversity/the United Nations Environment Programme. She covers various issues related to the implementation of Jakarta Mandate, such as integrated marine and coastal management, marine and coastal protected areas, conservation of deep sea biodiversity, marine invasive alien species, etc. In particular, she coordinates the organization of a series of CBD regional workshops to facilitate the description of ecologically or biologically significant marine areas (EBSAs), including marine areas beyond national jurisdiction (ABNJ). She also coordinates the implementation of the CBD's Sustainable Ocean Initiative (SOI) aimed at capacity development toward achieving Aichi Biodiversity Targets. In addition, she currently focuses on the implementation of the Secretariat's work on ocean acidification, anthropogenic underwater noise, marine debris, addressing biodiversity concerns on sustainable fisheries, application of area-based management tools, including marine protected areas and marine spatial planning, and the use of voluntary guidelines on impact assessments in marine and coastal areas. *jihyun.lee@cbd.int*

Mace, P. PhD (1983) from the University of British Columbia. Currently the Principal Advisor Fisheries Science at the New Zealand Ministry for Primary Industries. Her key responsibilities are to ensure the integrity of the Ministry's fisheries research, stock assessment and environmental assessment programs. She has worked extensively in the United States, Canada, and New Zealand, as well as in Europe and Australia. Her research interests include the national and international development of biological reference points and harvest control rules for fisheries, ecosystem approaches, development of criteria for defining species at risk, and fish stock assessments. *Pamela.Mace@fish.govt.nz*

Makino, M. M.Phil. (University of Cambridge) and MA and PhD (University of Kyoto). Professor. Fishery research Agency, Japan. He is also the Head of Fisheries Management Group of the Fisheries Research Agency, Japan. He is the Chair of PICES Section on Human Dimensions (SG-HD) and the and the member of IMBER Working Group on Human Dimensions (IMBER-HD). He is also a member of the Editorial Board of ICES Journal of Marine Science. He is also involved in international scholarly programs such as IMBER, etc. He is specializing in the fisheries and ecosystem-based management policy analysis. Member of the Fisheries Expert Group of the IUCN Commission on Ecosystem Management (IUCN/CEM/FEG). *mmakino@affrc.go.jp*

McCay, B.J. PhD in Anthropology (Columbia University). Anthropologist and Human Ecologist. Professor, Department of Human Ecology, School of Environmental and Behavioral Sciences, Rutgers the State University, New Brunswick, New Jersey, USA. Member of the U.S. National Academy of Sciences Specialized in research on coastal fisheries and fisheries management as problems in managing "the commons." Her work has largely been with commercial fishing communities and marine fisheries management institutions in North America and Mexico, with a focus on intersections of property rights, knowledge systems, and community. Member of the Fisheries Expert Group of the IUCN Commission on Ecosystem Management (IUCN/CEM/FEG). *mccay@rutgers.edu*

McConney, P. PhD. Resource Management, University of British Columbia. 1995, CanadaSenior Lecturer in Marine Resource Management Planning at the Centre for Resource Management and Environmental Studies (CERMES), The University of the West Indies (UWI) Cave Hill Campus in Barbados. His work covers many aspects of small-scale fisheries and marine protected areas, but especially governance, livelihoods, socio-economics and building capacity. *patrick.mcconney@cavehill.uwi.edu*

Miller, D. PhD in marine biology from the University of Cape Town, and Professorial Fellow at two Australian universities. Semi-retired. He was CCAMLR Executive Secretary from 2002 to 2010, and received the South African Antarctic Medal in 1995 and Duke of Edinburgh Conservation Medal in 2007. He became an Honorary Member of the Order of Australia (AM) in 2011 for his Antarctic conservation and management work. He served on the NAFO and WCPFC review panels, and various FAO bodies. He has published widely on marine resource, biology, policy, management and conservation issues. He is currently Director: Antarctic

Tasmanian and Science Research Development - a branch of the Tasmanian Government. *denzilgmiller@gmail.com*

O'Criodain, C. BSc and PhD in botany from National University of Ireland. He is a Policy Analyst on matters relating to international wildlife trade in the Global Species Programme of WWF International. *cocriodain@wwfint.org*

Petersen, S. PhD in Ecosystem-based Fisheries Management with a focus on addressing bycatch in southern African trawl and longline fisheries at the University of Cape Town in 2008. Her degree was recognised by FAO as contributing significantly to conservation of migratory species. She is currently, Senior Manager of the WWF Marine Programme, WWF-South Africa. This programme focuses on the implementation of an EAF in southern African fisheries. Highlights of this programme include the development of an EAF Tracking tool, Responsible Fisheries training, bycatch mitigation and the formation of the Responsible Fisheries Alliance, a pioneering alliance specifically aimed to facilitate the implementation of an EAF in South Africa. *spetersen@wwf.org.za*

Pomeroy, R. PhD, Cornell University, Resource Economics, Marketing, International Agriculture. 1989. Professor in the Department of Agricultural and Resource Economics and Connecticut Sea Grant College Fisheries Extension Specialist at the University of Connecticut – Avery Point in Groton, CT. USA. He is a marine resource economist whose areas of professional interest are small-scale fisheries management and development, aquaculture economics, international development, policy analysis, and seafood marketing. Dr. Pomeroy has worked on research and development projects in over 60 countries in Asia, Africa, the Caribbean and Latin America. *robert.pomeroy@uconn.edu; R. Pomeroy@cgiar.org*

Rajagopalan, R. Masters in Marine Affairs (2008). Has an inter-disciplinary academic background in ecology and marine affairs and has been working with the International Collective in Support of Fishworkers (ICSF) since 2000. Her main interests are related to governance of marine protected areas (MPAs), highlighting social dimensions of such conservation interventions and issues of community participation and equity. *ramya.rajagopalan@gmail.com*

Rice, J. BSc from Cornell (1970 - Conservation) and PhD from University of Toronto (1974 - Ornithology). Chief Scientist - Department of Fisheries and Oceans Canada, Canada. He has held faculty positions at Memorial University, Arizona State University, and University of Copenhagen. He had senior positions in Pacific Region and Newfoundland Region. He has been the Director of Peer Review and Science Advice for the Department of Fisheries and Oceans (Canada). He is also Vice Chair of the Fisheries Expert Group of the IUCN Commission on Ecosystem Management (IUCN/CEM/FEG). *Jake.Rice@dfo-mpo.gc.ca*

Ridgeway, L. MA Economics, University of Alberta. 1981. 32 years in policy and governance in Canadian government and at OECD. Formerly with Fisheries and Oceans Canada (DFO) since 1999. Now retired. Senior Advisor to Associate Deputy Minister; Exec Dir. Arctic; Director General (DG) International Policy and Integration (Fisheries, Oceans and Biodiversity, International Trade, Business Development); DG Economic and Policy Analysis. Also Privy Council Office and Finance Canada. Experience across international oceans/fisheries fora: inter alia, Chair, OECD Fisheries Committee; also OECD Sustainable Development (subsidy reform); Chair APEC Fisheries Committee; Co-Chair, UN Informal Consultative Process on Oceans and the Law of the Sea (UNICPOLOS); expert and Head of Delegation roles at FAO, other UN, UNEP, WTO and international processes. Editorial Board, ICES-JMS. *Lori.Ridgeway@dfo-mpo.gc.ca*

Sainsbury, K. PhD. Population dynamics and marine ecology, University of Canterbury, New Zealand (1977). Vice-chair of the Board of Trustees of the Marine Stewardship Council, Fishery Commissioner with the Australian Fisheries Management Authority and Professor of Marine System Management in the Institute for Marine and Antarctic Studies at the University of Tasmania. Former senior scientist in the Australian federal research organisation CSIRO for over 20y up until 2006 where he directed and conducted research on: (i) ecologically and economically sustainable fisheries, and (ii) assessment and management of the cumulative impacts of human use on regional marine ecosystems. In 2004, he was awarded the Japan Prize for his contribution to understanding and sustainable management of continental shelf ecosystems. In 2011 he was awarded the Swedish Seafood Award for the development and practical application of the ecosystem approach to fishery management. *ksainsbury@netspace.net.au*

Salas, S. PhD from the University of Bristish Columbia, Canada. Professor at Marine Resources Department at Cinvestav Unidad Merida, Mexico. She has competence in fisheries bioeconomics and management, fleet dynamics and fishing strategies, vulnerability and risk perception and her work is oriented with emphasis in small scale fisheries. She has participated as a consultant in several FAO workshops and has collaborated as a member of the advisory committee of the WECAF Region (FAO). Currently she is the regional coordinator of Thhe Latin America and the Caribbean region in the world project "To big to Ignore: Global Parthership for small-scale fisheries". *ssalas@mda.cinvestav.mx; marquezs.silvia@gmail.com*

Sanders, J. Master Degree in International Environmental Policy of the Monterey Institute of International Studies and a BA in Environmental Studies of Scripps College. Fishery analyst with the Policy,Eeconomics and Institutions Service in the Fisheries and Aquaculture Department of the FAO. She is specialized in policy issues related to fisheries and aquaculture. *Jessica.Sanders@fao.org*

Sant, G. BSc (Hons). He is the Fisheries Trade Programme Leader in TRAFFIC (ENGO) and Senior Research Fellow, ANCORS (Australian National Centre for Ocean Resources & Security, University of Wollongong). Has worked for TRAFFIC for twenty years specialising in research and management of trade in marine species, publishing extensively in the area, particularly in regard to sharks, RFMO's and trade related measures, CITES and most recently risk assessment work of aquatic species in trade. Vice Chair IUCN Shark Specialist Group. Prior to working for TRAFFIC worked in fisheries research within Australia. *glenn.sant@traffic.org*

Sbai, L. Professor. Graduated from the University of Grenoble in France and the University of Casablanca on the law of marine fisheries in Morocco, he has published 5 books and over than 50 articles on the Law of the Sea and the Law of the Marine Environment. Legal Adviser of the Secretary-General of the Department of Marine Fisheries and Chairman of the Compliance Committee of the MAP. Expert of the Commission on Environmental Law of the IUCN with which he participated to the publication of several Mediterranean studies. *larbisbai@yahoo.fr*

Seijo, J.C. PhD (1986) from Michigan State University in Resource Economics and Systems Science. He is currently a Professor of fisheries bioeconomics, and natural resource economics at Marist University of Mérida, (founding President of this University, 1996-2004), and FAO consultant for 23 years. He is author/co-author of 53 scientific papers and 4 books. He is currently President of the North American Association of Fisheries Economists (NAAFE). His academic interests and publications deal with fisheries bioeconomics, decision theory and risk analysis in fisheries, and spatial modeling of fisheries. *jseijo@marista.edu.mx*

Shannon, L. PhD, University of Cape Town, 2001 University of Cape Town. Her competencies include: ecological research and modelling in support of the Ecosystem Approach to Fisheries; ecosystem (trophic) modelling; ecological indicators of fishing; fisheries management; ecosystem changes; She co-chairs the IndiSeas Working Group (Indicators for the Seas, www.indiseas.org), which aims to evaluate the effects of fishing and natural variability on marine ecosystems by means of a suite of ecological, environmental, biodiversity and human dimension indicators using a comparative approach across a broad range of ecosystem types. *lynne.shannon@uct.ac.za*

Sharma, C. She has been working with the International Collective in Support of Fishworkers (ICSF), a non-governmental organization that supports small-sale fisheries, since 1996. Prior to ICSF she had worked for several years on issues related to community-based forest management among tribal communities in northern India. With a multi-disciplinary academic background in the social sciences, she is committed to issues of social and environmental justice, in particular gender justice. *chandmegh@gmail.com*

Shelley, P. BA in Economics and English from Hobart College (1969) and law degree with honors from Suffolk University Law School (1978).Vice President and Senior Counsel with Conservation Law Foundation in Boston, MA. His focus is on implementing the Magnuson-Stevens Fishery Conservation and Management Act, advancing marine public trust law, developing special marine area protections, and coastal pollution and restoration. He was awarded a Pew Marine Fellowship in 1996 and the David B. Stone Medal by the New England Aquarium in 2003. As a private consultant, he is part of the effort to create the Phoenix Islands Protected Area in Kiribati. *Pshelley@clf.org; pshelley@me.com*

Simard, F. Master degree in marine ecology. Worked as at the Tokyo University of Fisheries during 8 years and, after 1990, worked with the Oceanographic Museum of Monaco as Deputy Director for 12 years. Presently Deputy Director of IUCN Global Marine and Polar Programme and Senior Advisor for Fisheries. His current responsibilities bring him on diverse subjects such as Marine Protected Areas and Ocean governance. He worked for IUCN's regional Centre for Mediterranean in Spain, from 2003 to 2009 as Marine Programme Coordinator. He is a Member of the Fisheries Expert Group of the IUCN Commission on Ecosystem Management (IUCN/CEM/FEG). *francois.simard@iucn.org*

Slicer, N.M. Former CCAMLR Compliance Officer, who now works as an independent consultant specialising in catch and trade documentation. She has worked for organisations including the Association of Responsible Krill Harvesting Companies (ARK), the Coalition of Legal

Toothfish Operators (COLTO), FAO, MRAG Ltd, Pew and national governments on issues ranging from catch documentation and trade traceability schemes, surveillance, vessel monitoring systems and IUU fishing. Currently manages the ARK Secretariat and has co-authored a number of articles and chapters on MCS and catch documentation schemes. *natashaexel@gmail.com*

Sowman, M.R. PhD in integrated coastal management. University of Cape Town (1994). Associate Professor in the Department of Environmental and Geographical Sciences at UCT. Director of the Environmental Evaluation Unit of UCT. She has been involved in research, consulting and teaching in the general field of integrated coastal and fisheries management with particular interest in small-scale fisheries governance, human dimensions of MPAs, and policy development and analysis. Contributed to the FAO technical guidelines on human dimensions of EAF and the development of the new small-scale fisheries policy for South Africa. *Merle.Sowman@uct.ac.za*

Sunde, J. MA (Women's Studies) from the University of York, UK. Social researcher. Currently conducting her PhD at the University of Cape Town on customary marine governance and small-scale fisheries. Her interest is in activist scholarship on issues of marine and coastal governance, social justice and development. She is a member of the International Collective in Support of Fishworkers. *jsunde@telkomsa.net*

Tandstad, M. MSc in Aquatic Resource Management. Fishery scientist. Experience in marine biology and fisheries, stock assessment, scientific surveys and elaboration of management advice. Presently Fishery Resources Officer in the Fisheries and Aquaculture Department of Food and Agriculture Organization of the United Nations (FAO). Main areas of work include the ecosystem approach to fisheries and science based advice for fisheries management related to coastal as well as deep-sea fisheries. For deep-sea fisheries she is working on the follow- up on the implementation of the Guidelines on the Management of Deep-sea Fisheries in the High seas, including in relation to Vulnerable Marine Ecosystems. *Merete.Tandstad@fao.org*

VanderZwaag, D. PhD (University of Wales, Cardiff, 1994). Professor of Law and Canada Research Chair in Ocean Law and Governance at the Marine & Environmental Law Institute, Dalhousie University. He teaches in the areas of International Ocean Law and Governance and International Environmental Law. He co-chairs the IUCN World Commission on Environmental Law's Specialist Group on Oceans, Coasts and Coral Reefs. *david.vanderzwaag@dal.ca*

van Rijn, T. Master of Laws (Nijmegen, NL). Special Master in European Law (Brussels B). PhD (Leiden, NL) Working on many areas of European law, e.g. fisheries, agriculture, banking, fundamental freedoms, intellectual property, transport, media, telecom, internet. Principal Legal Advisor of the European Commission. *van.rijn.thomas@gmail.com; thomas.van-rijn@ec.europa.eu*

Vestergaard, O. Masters in aquatic ecology (University of Copenhagen). Specialized in marine and coastal ecosystem-based management and adaptation with focus on decision-support for spatial planning and management and policy development. Programme Officer with UNEP's Freshwater & Marine Ecosystems Branch, Division for Environmental Policy Implementation, United Nations Environment Programme, HQ, Nairobi, Kenya. *ole.vestergaard@unep.org*

Warner, R. PhD. University of Sydney, 2006. My current position is Associate Professor Australian national Centre for Ocean Resources and Security (ANCORS) University of Wollongong, NSW Australia.Associate Professor at the Australian National Centre for Ocean Resources and Security (ANCORS), University of Wollongong. NSW, Australia.. Her current research interests include law of the sea, oceans governance, marine environmental law and climate. She is the author of Protecting the Oceans Beyond National Jurisdiction: Strengthening the International Law Framework (Martinus Nijhoff, Leiden, 2009), editor (with Simon Marsden) of Transboundary Environmental Governance: Inland Coastal and Marine Perspectives (Ashgate Publishers, Farnham, UK, 2012) and editor (with Clive Schofield) of Climate Change and the Oceans: Gauging the Legal and Policy Currents in the Asia Pacific (Edward Elgar, UK, 2012). *rwarner@uow.edu.au*

Foreword

Bonnie J. McCay

Professor, Rutgers State University, USA

The world's oceans constitute a vast 'commons', that is, places and resources that are shared by many different peoples and institutions. Discussions about the fate of the world's oceans and fisheries are thus discussions about the global commons, about whether and how human beings can improve our ways of interacting with the oceans for the sake of present and future generations.

The past 50 years have seen the evolution of more complex and diverse governance of the ocean commons. The Law of the Sea is one important step in this regard, designating greater responsibility to coastal nation-states for parts of the oceans and, through the United Nations, we have come up with many regional organizations to help govern the high seas and transboundary dimensions of the global marine commons. Nation-states have multiple and complex ways of assigning authority and responsibility for commons governance, sometimes top-down but also devolved to smaller units and user-groups. Increasingly, non-governmental organizations are also engaged in efforts to improve commons governance, ranging from the piracy of anti-whaling groups such as Sea Shepherd to the eco-certification work of the Marine Stewardship Council and industry associations such as the World Ocean Council. Parallel efforts on an international, regional and local scale are being made to grapple with challenges to sustainable development and protection of biodiversity. This book is exemplary in tracing such paths and addressing how marine fisheries, sustainable development and biodiversity protection come together in oceans governance now and in the future.

The language of commons governance is usually linked to 'tragedies of the commons', an idea popularized by Oxford economist William Forster Lloyd on a lecture series in 1832 (Lloyd, 1833). Thomas Malthus was on the same lecture circuit, and the two gentlemen hoped to explain a major issue of the time: the role of population in causing poverty. Lloyd used the idea of a village common pasture with too many animals out to graze to exemplify the problem of too many people for the labour market. His image was picked up in the 20th century by biologist Garrett Hardin, who published *The Tragedy of the Commons* in Science magazine in 1968. He added the dramatic notion of 'tragedy' to underscore the need for extreme measures on the part of government to avert over-population in the world.

The 'tragedy of the commons' idea is that rational individuals could have incentives to have too many children, or put too many cows on the pasture, or take too many fish from the sea in 'open access' or 'free access' situations. The argument that followed – central to modern oceans governance – underscored both the value of privatized property rights and the need for strong government where privatization is not feasible. A conceptual error (later recognized in Hardin, 1994) was equating 'commons' with 'open access' rather than recognizing that people can and sometimes do find ways to retain the conditions of sharing and local governance in managing the commons. Much scholarship has followed to explore those conditions and their implications (McCay and Acheson, 1987; Ostrom, 1990, Ostrom et al., 2002), leaving open the opportunity to re-title the challenge for oceans governance as 'comedies' or 'dramas' of the commons (Ostrom et al., 2002; Rose, 1994).

Following Greek dramaturgy, 'comedy' means dramas in which people come together to try to address problems, in sharp contrast to tragedy which features individuals destined to behave in ways that lead to tragic outcomes (http://www.britannica.com/EBchecked/topic/127459/comedy 2011). From another perspective, in the tragedy of the commons the efforts of each person to maximize their own interests result in everyone losing because limited resources are over-utilized; in the comedy of the commons, people contribute to the common good while getting something for themselves.

The 'comedy of the commons' is therefore highly relevant to modern oceans governance, reflected in this volume. The topics covered reflect important new and old directions and innovations in fisheries and biodiversity governance: innovations that improve the transparency and participatory nature of decision-making; innovations involving fisheries and marine biodiversity programs that more fully engage resource users and their knowledge and expertise; and innovations that rely upon economic incentives for short-term gains but are balanced with larger and longer concerns for social justice and biological conservation. In a sense, this volume itself is an outcome of comedies of the commons, in which scholars and practitioners coming from the very different perspectives of biodiversity conservation, social justice and fisheries management have sought and found ways to work together on oceans governance.

References

Hardin, G. (1968) The tragedy of the commons. *Science* **162**, 1243–1248.

Hardin, G. (1994) The tragedy of the unmanaged commons. *Trends in Ecology and Evolution* **9**, 199.

Lloyd, W.F. (1833) *Two Lectures on the Checks to Population Delivered before the University of Oxford in Michaelmas Term 1832*. Oxford, London.

McCay, B.J. and Acheson, J.M. (1987) *The Question of the Commons: The Culture and Ecology of Communal Resources*. University of Arizona Press, Tucson.

Ostrom, E. (1990) *Governing the Commons : The Evolution of Institutions for Collective Action*. Cambridge University Press, Cambridge, New York.

Ostrom, E., Dietz, T., Dolsak, N., Stern, P.C., Stonich, S. and Weber, E.U. (2002) *The Drama of the Commons*. Division of Behvioral and Social Sciences and Education, National Research Council. National Academy Press, Washington, DC.

Rose, C.M. (ed.) (1994) The comedy of the commons: custom, commerce, and inherently public property. In: *Property & Persuasion: Essays on the History, Theory, and Rhetoric of Ownership*. Westview Press, Boulder, CO, pp. 105–162.

Foreword

Árni M. Mathiesen

Assistant Director-General, FAO Fisheries and Aquaculture Department

Modern fishery governance is a systemic concept relating to the exercise of economic, political and administrative authority. It establishes the overriding principles and objectives of the sector. It develops the policy and regulatory frameworks. It connects government with civil society, harmonizing individual, sectoral and societal perspectives and maintaining social order and productive socio-ecological systems. It legitimates and balances stakeholders' interaction, enforces decisions and regulations and maintains coherence across jurisdictional, space and time scales. Finally, it conditions the allocation of power, resources and benefits and maintains the governance system capacity to learn and change.

FAO has been dealing with fisheries governance for sustainability since its creation in 1946, struggling with legal and institutional frameworks, policies, technological developments, managing capacity and measures and global monitoring. The organization works with its members to enhance the effectiveness of governance systems at the national, regional and global levels. Progress has been made but a lot remains to be done, in Exclusive Economic Zones – where implementation is still a problem – and in the high seas where frameworks are still developing, posing important challenges to both conservation and governance.

One area of our work focuses on the marine Areas Beyond National Jurisdiction (ABNJ), commonly called the high seas. These are the areas of ocean for which no one nation has the specific or sole responsibility for management. They include the water column of the high seas and the seabed falling within. These areas make up 40% of the surface of our planet, comprising 64% of the surface of the oceans and nearly 95% of its volume.

Achieving sustainable management of the fisheries resources and biodiversity conservation in the ABNJ is extremely difficult given the complexity of the ecosystems, including their great depths and distances from the coasts as well as the large number and wide diversity of all the public and private actors involved. The obligation made by UNCLOS to States to collaborate for decision-making, and the difficulties of enforcement, complicates the fight against free-riders and illegal fishing (IUU). Only limited progress has therefore been made so far in meeting the already-agreed global targets from international forums. For example, there has been very little application of an ecosystem approach across the ABNJ despite the target from the 2002 Johannesburg World Summit on Sustainable Development (WSSD) which reads 'to encourage the application by 2010 of the ecosystem approach to ensure sustainable utilization of the ocean'. Similarly, notwithstanding the healthy state of some species targeted by fisheries, there is considerable doubt that the intention underlying the target of 'maintaining or restoring stocks to levels that can produce the maximum sustainable yield where possible and not later than 2015' will be fully realized. Greater progress is also required on the need to 'encourage relevant Regional Fisheries Management Organization and Arrangements (RFMO/As) to give due consideration to the rights, duties and interests of costal States and the special requirements of developing States'. Also, the Millennium Development Report of 2010 indicates that 'the world has missed the 2010 target for biodiversity conservation, with potentially grave consequences'.

The ABNJ poses serious and distinctive governance and conservation challenges. The high-value fisheries and associated biodiversity in the ABNJ are now subject to mounting threats from fisheries such as overcapacity, IUU fishing, increased hardship for fishers, foregone wealth, threatened food supplies, inappropriate fishing practices and inadequate protection of the related ecosystems. Additional environmental problems are mounting on the horizon related to deep-sea mining as well as oil and gas exploitation on deep slopes. These problems call for an increased focus on reforms of systems. The seriousness of the threats on the health of the oceans' resources and biodiversity is compounded by the weak implementation of regulatory frameworks that exist and management options that have been agreed but call for more coordinated, cohesive and integrated interventions. An important part of the reform concerns the definition of user rights systems for sustainable and economically rational fisheries in ABNJ.

Information flow, outreach and awareness raising of the critical management issues are seen as fundamental in engaging the general public. FAO continues to play a key role in building and improving cross-sectoral dialogue, engaging high-level decision-makers and contributing to information sharing and public awareness of ABNJ issues, bringing together governments, UN-agencies, NGOs, regional organizations, academia and the private sector to establish linkages for information-sharing on ABNJ across sectors and between global and regional levels.

In the end, our work must lead to changes in fishing behaviour on the water. We need to bring about transformational changes such as moving away from the race to fish, increasing capacity to protect fragile ecosystems, reducing or removing barriers to international and cross-sectoral sharing of knowledge and experiences and moving towards an ecosystem approach. If we do not achieve these changes, we can anticipate a future in which major fisheries in ABNJ will face serious declines, with resulting negative impacts on national economies and the fishing industry. We may see negative impacts on livelihoods of fishing populations in many countries, rich and poor, as well as threats to global food security and unacceptable consequences for global biodiversity priorities: the species and habitats that depend on good ABNJ management.

ABNJ fisheries are critically important for a host of biodiversity and public welfare outcomes which extend far beyond the limits of ABNJs. Putting it simply, we cannot achieve sustainability of global fisheries and their ecosystems without addressing ABNJ management.

The management of fisheries and conservation in ABNJ concerns the largest part of the world ocean and of Earth. It is necessarily interconnected and interdependent with their governance in Exclusive Economic Zones and on land, as the ocean is the final receptacle of all human pollutions. In this respect, the efforts made in this book to highlight the history, common points, contradictions, progress and hiccups of conservation and fisheries, identifying gaps, further synergies and some ways forward, is invaluable.

Foreword

Braulio Ferreira de Souza Dias

Executive Secretary of the Convention on Biological Diversity

The Convention on Biological Diversity (CBD) was adopted more than two decades ago, with a three-fold mission: (1) the conservation of biological diversity; (2) the sustainable use of biological diversity; and (3) the fair and equitable sharing of benefits arising from the use of genetic resources. Its thematic programmes of work correspond to the major biomes on the planet, with the work devoted to marine and coastal biodiversity having increased its breadth and depth in recent years.

The Secretariat of the Convention on Biological Diversity (SCBD) is increasingly turning its attention to the implementation of the decisions adopted by the Conference of the Parties (COP), using the guidelines, tools and approaches developed over the years. In order to strengthen international commitment in this regard, in 2010 in Nagoya, Japan the tenth meeting of the Conference of the Parties (CoP) to the Convention adopted a revised and updated Strategic Plan for Biodiversity for 2011–2020, including a global biodiversity agenda with a balanced set of 20 targets: the Aichi Biodiversity Targets (decision X/2). Target 6 is devoted to sustainable fisheries:

> 'By 2020 all fish and invertebrate stocks and aquatic plants are managed and harvested sustainably, legally and applying ecosystem based approaches, so that overfishing is avoided, recovery plans and measures are in place for all depleted species, fisheries have no significant adverse impacts on threatened species and vulnerable ecosystems and the impacts of fisheries on stocks, species and ecosystems are within safe ecological limits.'

However, as the Convention's 193 Parties move forward in their pursuit to achieve the Aichi Targets by the year 2020, the trend of biodiversity loss continues unabated. The sustainable use of biodiversity remains a notable exception rather than the rule. If we look jointly at the trends in biodiversity loss and poverty, it seems obvious that the benefits of conservation and sustainable use are yet to be fairly shared. There are important exceptions that indicate such progress is possible, but the mismatch between the commitments we made and the actual implementation we are achieving is much too obvious.

It is generally recognized that the main reason for insufficient progress is inadequate implementation. This has been a central and recurrent issue, most recently pointed out in the outcome document of the United Nations Conference on Sustainable Development (Rio + 20) *The Future We Want*, which specified the problems and identified solutions. The implementation strategies at local, national and regional levels need to be pragmatic and multi-faceted, facing the dual challenges of achieving both biodiversity conservation and sustainable use.

Nowhere are these trends more apparent than in the oceans and coasts. Rio + 20 confirmed the central role of the oceans in most of the important challenges being faced today in sustainable development: food security, poverty eradication, sustainable livelihoods, overall human wellbeing and more. While issues such as data availability and technology transfer have been identified and still require substantial attention, the main cause of inefficiency remains in governance capacity and performance. This issue was central in the Rio + 20 Summit, particularly with regard to oceans and coasts. The way forward implies improving coherence in the initiatives and providing the means corresponding to the commitments. Improved coherence is needed at a national level to reduce cross-sectoral externalities and improve synergy. Looking at governance with finer lenses, we need a more systematic application of the principles of good governance including legitimacy, democracy, participation and transparency.

Simultaneously achieving biodiversity conservation, food and livelihood security and poverty eradication requires increased coherence and integration in international and national policies and programmes. It also requires stronger engagement of the coastal communities and the industry. Indeed, the CoP to the CBD noted the need for further improvement and implementation of the ecosystem approach in fisheries management by enhancing the capacity of these fisheries management organizations, constructive interagency collaboration and full and meaningful participation by a wide range of experts on biodiversity, indigenous and local communities (decision XI/18).

The editors and authors of this timely book should be commended for having undertaken this historical, multidisciplinary and multi-scale analysis of ocean governance for fisheries and biodiversity conservation. Bringing conservation and fisheries experts to write together on a comprehensive range of issues and across scales – to find common points, draw common lessons and to identify remaining impediments to further convergence and integration – was a major challenge and one that was met admirably. It is refreshing to see experts from the International Union for Conservation of Nature (IUCN), the United Nations Food and Agriculture Organization (FAO), the Convention on International Trade in Endangered Species of Wild Fauna and Flora (CITES), the CBD Secretariat and regional and national centres of excellence collaborating in this endeavour. Indeed, this book responds to the call from the CoP to the CBD which, in its decision XI/18, called for constructive collaboration between biodiversity and fisheries bodies.

The overall conclusion, among many others, is that the last decades have seen an important convergence in the objectives, principles and approaches of fisheries and conservation of biodiversity is encouraging. The ecosystem approach adopted by the CBD has been an important factor of convergence in the last two decades. However, the book also identifies important remaining tensions and real differences in the preferred paths to reach our shared future: differences in risk perceptions, in appreciation of where we stand compared to where we should be and in allocation of the costs and benefits of action (the central equity issue) all highlight the amount of work still required. The specific attention in the book to small-scale fisheries and local biodiversity conservation stresses that vulnerable small-scale coastal communities worldwide critically depend on the ocean and on the conservation of its biodiversity. It is probably at that level that the synergy is most obvious in the oceans.

There is no single best path forward. It has to be determined regionally, nationally and locally with involvement of the actors concerned. However, the various strategies that may be developed will most probably have in common a number of key ingredients: strengthening both the capacity and the role of coastal communities in decision-making and implementation, testing decentralized governance strategies, removing historical impediments, providing incentives and promoting multi-stakeholder governance systems involving both powerful and marginalized actors. Many of these elements are described in the book, which should be of great interest to both researchers and practitioners of fisheries, oceans governance and biodiversity conservation.

Preface

Serge M. Garcia, Jake Rice and Anthony Charles

This book examines the interface between two streams of ocean governance: one dealing with the development and management of fisheries, and the other with the conservation of biodiversity and the ecosystem. The key goal of the book is to improve understanding of both governance systems and the ways in which they do or might interact, thereby facilitating improved coherence between them.

The book examines the interactions of the two streams at different scales and across a range of thematic areas and institutions. Their respective principles, objectives, policies, plans, processes and instruments are all examined, focusing on conceptual and operational synergies and conflicts, duplications and gaps. The distinctive characteristics of the two streams of governance are considered, along with the involvement of different ministries, institutions and legislative instruments at all levels, from global bodies to local communities. A wide variety of concrete case studies are provided, at a variety of spatial and temporal scales, offering perspectives on their co-evolution in time and space and exploring impediments to faster and more effective implementation of coherent policies.

As editors, we conceived the overall structure of the book and for each chapter we identified a small team of authors with competence in both streams. This was done to promote interaction, bridge across streams and provide an objective range of past, present and future perspectives on fisheries and biodiversity conservation. This approach to producing the book is certainly not the norm for an edited volume but, in a world in which the two governance streams operate largely independently, this 'structured' interaction proved immensely enriching.

Before embarking on this project, we knew broadly that the two streams compete for social legitimacy, political influence and funding. The systematic comparison contained in this book provides the depth of understanding behind this knowledge, showing in detail the many ways in which the streams differ: (1) their different but interconnected histories and scientific perspectives; (2) the different mixes of stakeholders to which they are accountable; (3) their partially overlapping but not totally identical goals; (4) the relative importance given to human benefits vs. ecosystem protection and to fishery resources and socio-economic processes vs. biodiversity and natural processes; (5) the role of humans (e.g. as a source of problems for nature vs. recipients of its benefits); (6) the priority given to timeframes (e.g. long vs. short term) and operational objectives; (7) the tolerance for different types of risks (e.g. misses and false alarms); (8) the favourite management approaches (e.g. resource vs. space-based); and (9) the perceptions of key knowledge gaps (ecological vs. socio-economic) and methods used for diagnoses.

The comparison also shows fundamental similarities that have attracted less attention despite their potential to improve coherence and overall performance of governance. The two governance systems recognize that: (1) the conservation of a functioning ecosystem is a *sine qua non* condition for the provision of key services (e.g. food, livelihood) through a responsible fishery sector; (2) there is a need for simultaneous satisfaction of human and ecosystem wellbeing; (3) uncertainty generates risks (for people and nature) and calls for precaution; (4) an ecosystem approach to fisheries and biodiversity conservation is essential; and (5) 'good governance' principles play a central role in ensuring legitimacy and compliance. Both streams agree also that the status quo is not acceptable, but differences remain in terms of the acceptable balance of outcomes and the acceptable pathways to and pace of change.

The book illustrates that valuable efforts are being made to create institutional bridges at global, regional and national levels. The present situation reflects a progressive yet incomplete closure of the historical dichotomy between nature conservation and human development, in a difficult search for balance between two complex and partly overlapping sets of goals, priorities and constraints. Many points of friction remain, and without a better integration of assessment, decision-making and performance evaluation processes, both streams are likely to fail to achieve their main goals. The book offers insights as to how this better integration could be achieved.

Terminology

The book contains many concepts and terms that cut across the chapters. As editors we have striven in the opening chapters to provide clear explanations of how terms that thread through the book are to be interpreted in order to reduce uncertainty for the reader and improve coherence between chapters. We have also sought to have all chapters use these terms in consistent ways, although this was not always possible. In fact, the challenge we met in trying to achieve consistent and constructive dialogue within chapters and within the book reflected, in a microcosm, the challenge of achieving coherent governance in the oceans.

The reality is that many of the concepts underpinning the themes of the book – focusing on fishery resources and biodiversity – have been described, defined and explained in many different fora, resulting in an often confusing range of literature reflecting idiosyncrasies, 'turf demarcation' of different school of thoughts and ideological slants, as well as real differences. The simpler expressions of concepts are often the easiest to agree with, but also most vulnerable to different interpretations in different contexts. The term 'conservation' itself is an excellent example of the problem with its different interpretations in ecology, fishery science and social sciences, ranging from total protection from human activities to management and sustainable use by humans. The terms 'biodiversity' and 'governance' may also mean different things to different users. We did not consider it useful or even advisable to prescribe common definitions throughout the book, considering that a variety of viewpoints might be more important and informative.

Structure of the book

The book is divided into six parts. Part I describes **challenges and trends** in both governance streams, using a broad brush. Chapter 1 presents the history of the two streams in terms of a timeline of relevant international events subdivided in phases. It is supported by a large annex listing key events from 1850 to 2012. Chapter 2 contains an analysis of that timeline to identify the evolutionary strands running through the phases and characterizing the changes, posing hypotheses regarding both convergence and coevolution, concepts that are explored throughout the book. Chapter 3 offers a reflection on the challenges faced when aiming at a more effective integration between the two streams.

Part II offers a description of the different **dimensions** common to the two governance streams: biological (Chapter 4); economic (Chapter 5); social (Chapter 6); legal (Chapter 7); spatial (Chapter 8) and scientific (Chapter 9). Together, they illustrate the multidisciplinary nature of the issues grappled with by both governance streams, the roots of many of these tensions at play between them and the opportunities for increased integration.

Part III describes **global governance** issues, looking at policies and the management instruments developed for their implementation (Chapter 10); global governance processes, particularly those at the UN level (Chapter 11); and the process of translation of global instruments into implementation approaches and plans at the national level (Chapter 12). It also describes, as examples of global-level interactions, the issues and processes at play in both streams when dealing with the important issue of species at risk (Chapter 13); and the modern questions regarding Ecologically and Biologically Significant Areas (EBSAs) and Vulnerable Marine Ecosystems (VMEs) (Chapter 14).

Part IV looks at **regional governance**. The regional level is often fundamental for the translation of high-level international norms and policies into coherent national implementation approaches and plans. The overall regional framework and interplay of instruments and institutions is given in Chapter 15. Specific cases of regional collaboration are then examined for: the North East Atlantic via OSPAR-NEAFC interaction (Chapter 16); the Mediterranean Sea (Chapter 17); the Southern Ocean and CCAMLR (Chapter 18); the Benguela Current Large Marine Ecosystem (Chapter 19); and European waters (Chapter 20).

Part V explores **governance issues at national and sub-national scales**, beginning with a description of the role played by integrative national cross-sectoral frameworks (Chapter 21). The next several chapters focus on small-scale fisheries (SSFs), particularly those in the developing world, examining various aspects of their interaction with broad fishery governance trends and with biodiversity conservation. This focus on SSFs reflects the importance of these fisheries worldwide and their vulnerability to impacts of broad trends in fisheries as well as those arising as nature conservation practices move from land to the coastal and marine area. The chapters cover several main themes: the relationship between SSFs and conservation from a governance perspective (Chapter 22); the role that SSFs can play in marine conservation stewardship in tropical SSFs (Chapter 23); the impacts of spatial management measures, especially marine protected areas, on SSFs in the developing world (Chapter 24); and the important but complex role of ENGOs in small islands developing states (SIDS, Chapter 25). Part V ends with chapters on the importance of capacity-building in moving to an ecosystem approach to fisheries governance (Chapter 26); the important potential and actual role of fishers' organizations in governance and in marine conservation (Chapter 27) and the role of courts in translating sometimes fuzzy governance principles and policies into practice, comparing the systems in USA and EU (Chapter 28).

Part VI offers a single **concluding chapter** to the book (Chapter 29), presenting a brief synthesis on the structure, evolution and present state of governance of the two streams and their interaction. We base this on an evaluation of a set of hypotheses posed in Chapter 2, drawing on the rich contents of the subsequent chapters. This analysis leads finally to a range of conclusions and suggestions for improving the integration and performance of the two governance streams.

With the exceptions of the logical flow through the initial Chapters 1–3, and a summing up and reference back to the book's chapters in the concluding Chapter 29, other chapters can be read independently if so desired. The reader may choose to focus on one of the various scales at which the two streams interact (international, regional, national or local), or delve into specific dimensions or thematic topics. There are rich insights to be gained from each chapter. However, the book's chapters together constitute a coherent package, and we feel confident that those who read through the entirety of the book will emerge as we did, with a richer sense of the pervasiveness of the two streams of governance and both the importance of and the potential for further convergence and coevolution.

Acknowledgements

The authors of this book are extremely grateful to the many competent and patient individuals and institutions that have assisted them during the writing and publishing process. This book has been published at the initiative of the Fisheries Expert Group of the IUCN Commission on Ecosystem Management (IUCN/CEM/FEG) under the coordination of the European Board of Conservation and Development (EBCD).

The initial scientific workshop of the main authors, hosted by the Institute of Marine Research (IMR) in Bergen (Norway, 25–27 March 2012), was supported by the Norwegian Agency for Development Cooperation (NORAD) and the Norwegian Ministry of Fisheries and Coastal affairs with contributions from Fisheries and Oceans (Canada), the Global Guardian Trust (Japan), the National Committee of Fisheries (France) and the IUCN Commission on Ecosystem Management. Despina Symons, EBCD Director, coordinated the process and the workshop (assisted by Kathleen Laissy) and took care of fund raising. Research and editorial support was provided by Ashley Shelton at Saint Mary's University. Financial support for research activities was provided by the Natural Sciences and Engineering Research Council and the Social Sciences and Humanities Research Council (Canada).

Finally, we thank the numerous scientific reviewers whose input contributed to improvements in the original versions of the text.

List of selected acronyms

ABNJ	Areas beyond national jurisdiction
ACAP	Agreement on the Conservation of Albatrosses and Petrels
ACCOBAMS	Agreement on the Conservation of Cetaceans in the Black Sea Mediterranean Sea and Contiguous Atlantic Area
ACFM	Advisory Committee on Fisheries Management
ACME	Advisory Committee on the Marine Environment
AOSIS	Alliance of Small Island States
APA	Administrative Procedures Act (USA)
APF	Antarctic Polar Front (i.e. the 'Antarctic Convergence')
ASEAN	Association of South East Asian Nations
ASOC	Antarctic and Southern Ocean Coalition (ASOC)
ATCM	Antarctic Treaty Consultative Meeting
ATCP	Antarctic Treaty Consultative Party
ATS	Antarctic Treaty System
BBNJ	UN Ad Hoc Open-ended Informal Working Group to Study Issues Relating to the Conservation and Sustainable Use of Marine Biological Diversity Beyond Areas of National Jurisdiction
BCC	Benguela Current Commission
BCLME	Benguela Current Large Marine Ecosystem
BENEFIT	Benguela Environment Fisheries Interaction and Training Programme
BFCA	Belize Fishermen Cooperative Association
BIOMASS	Biological Investigations of Marine Antarctic Systems and Stocks Program
BNP	Big Numbers Project
CAF	GFCM Committee on Administration and Finance
CAMLR	Convention for the Conservation of Antarctic Marine Living Resources
CANARI	Caribbean Natural Resources Institute
CAQ	GFCM Committee on Aquaculture
CBC	Community-based conservation
CBD	Convention on Biological Diversity
CCAMLR	Commission for the Conservation of Antarctic Marine Living Resources
CCAS	Convention for the Conservation of Antarctic Seals
CCRF	FAO Code of Conduct for Responsible Fisheries
CCSBT	Commission for the Conservation of Southern Bluefin Tuna
CDS	Catch Documentation Scheme
CECAF	Commission for Eastern Central Atlantic Fisheries
CEMP	CCAMLR Ecosystem Monitoring Programme
CEP	Committee for Environmental Protection (under Article 11 of the Madrid Protocol)
CFP	Common Fishery Policy
CI	Conservation International
CIESM	Commission Intergouvernementale pour l'Exploration Scientifique de la Méditerranée
CITES	Convention on International Trade in Endangered Species of Wild Fauna and Flora
CLEC	Coastal Law Enforcement Council (Philippines)
CMs	CCAMLR Conservation Measures agreed under Article IX of the CAMLR Convention
CMS	Conservation of Migratory Species of Wild Animals (Bonn Convention)
CNFO	Caribbean Network of Fisherfolk Organizations
CoC	GFCM Compliance Committee
COFI	FAO Committee on Fisheries

COLTO	Coalition of Legal Toothfish Operators
CoML	Census of Marine Life project
COMNAP	Council of Managers of National Antarctic Programs
CoP	Conference of the Parties
COSEWIC	Committee on the Status of Endangered Wildlife in Canada
CP	Contracting Party
CPUE	Catch per Unit Effort
CRFM	Caribbean Regional Fisheries Mechanism
CRM	Coastal Resource Management
CSD	Commission on Sustainable Development
DEMA	Turks and Caicos Islands Department of Environment and Maritime Affairs
DWFN	Distant Water Fishing Nations
DFO	Department of Fisheries and Oceans (Canada)
EAF	Ecosystem Approach to Fisheries
EAFM	Ecosystem Approach to Fishery Management
EBFM	Ecosystem-Based Fishery Management
EBM	Ecosystem-Based Management
EBSA	Ecologically and Biologically Significant Areas
ECA	Court of Auditors (of the European Union)
EcAp	MAP Ecosystem Approach Coordination Group
ECJ	Court of Justice of the European Union
ECOSOC	UN Economic and Social Council
EEZ	Exclusive Economic Zone
EIA	Environmental Impact Assessment
ENGO	Environmental Non-Governmental Organization
EP	European Parliament
EPBC	Environment Protection and Biodiversity Conservation Act (Australia)
ESA	Endangered Species Act (USA)
ESSCP	Ecologically Significant Species and Community Property
ERA	Ecological Risk Assessment
ESD	Ecologically Sustainable Development
EU	European Union
FAO	Food and Agriculture Organization of the United Nations
FFA	Forum Fisheries Agency
FIBEX	First International BIOMASS Experiment
FISH	Fisheries Improved for Sustainable Harvest project (Philippines)
FRA	GFCM fisheries restricted areas
FSPI	Foundation of the Peoples of the South Pacific International
G-77	Group of 77
GATT	Global Agreement on Tariffs and Trade
GBRMP	Great Barrier Reef Marine Park
GCFI	Gulf and Caribbean Fisheries Institute
GEF	Global Environment Facility
GFCM	General Fisheries Commission for the Mediterranean
GIS	Geographical Information System
GPO	Global Partnership for Oceans
GYM	General Yield Model (used in CCAMLR)
HCR	Harvest Control Rules
HSP	Harvest Strategy Policy
HSTF	Inter-Ministerial High Seas Task Force
IATTC	Inter-American Tropical Tuna Commission
IBP	International Biological programme
ICAM	Integrated Coastal Areas Management
ICCAT	International Commission for the Conservation of Atlantic Tuna
ICDP	Integrated Conservation and Development Project
ICES	International Council for the Exploration of the Sea
ICM	Integrated Coastal Management
ICP	UN Informal Consultative Process (also called UNICPOLOS)
ICPM	International Commission on National Parks
ICRI	International Coral Reef Initiative
ICRW	International Convention for the Regulation of Whaling
ICSP	Informal Consultations of State Parties to the UNFSA
ICZM	Integrated Coastal Zone Management
IEC	Information, Education and Communication
IFMP	Integrated Fisheries Management Plan
IFSD	Institutional Framework for Sustainable Development
IGO	Inter-Governmental Organization
ILO	International Labour Organization
IM	Integrated Model (CCAMLR)
IMCAM	Integrated Marine and Coastal Integrated Management
IMF	International Monetary Fund
IMO	International Maritime Organization
IOM	Integrated Ocean Management
IOTC	Indian Ocean Tuna Commission
IPBES	Intergovernmental Science–Policy platform on Biodiversity and Ecosystem Services
IPHC	International Pacific Halibut Commission
IPOA	International Plan of Action

IPOA-IUU	IPOA to Prevent, Deter and Eliminate Illegal, Unreported and Unregulated Fishing
IPOA-Seabirds	IPOA for Reducing Incidental Catch of Seabirds in Longline Fisheries
IPOA-Sharks	IPOA for the Conservation and Management of Sharks
ITLOS	International Tribunal for the Law of the Sea
ITQ	Individual Transferable Quota
IUCN	International Union for the Conservation of Nature
IUPN	International Union for the Protection of Nature
IUU	Illegal, Unreported and Unregulated (fishing and related activities)
IWC	International Whaling Commission (ICRW's Executive Arm)
JAG	Joint Assessment Group of SC-CAMLR-SCIC
JGOFS	Joint Ocean Global Flux Study
JPOI	Johannesburg Plan of Implementation
KYM	Krill Yield Model (CCAMLR)
LBS	Land-Based Sources (of pollution) Protocol
LDCs	Least-Developed Countries
LGU	Local Government Unit
LME	Large Marine Ecosystem
LMMA	Locally Managed Marine Area
LON	League of Nations
LOSC	UN Law of the Sea Convention
LTL	Low Trophic Level (species or fisheries on these species)
MAB	Man and Biosphere
Madrid Protocol	Madrid Protocol on Environmental Protection to the Antarctic Treaty
MAP	Barcelona Convention/Mediterranean Action Plan
MARPOL	Protocol for the Prevention of Pollution from Ships
MCS	Monitoring, Control and Surveillance
MCSD	Mediterranean Commission on Sustainable Development
MDG	Millennium Development Goal
MEA	Multi-lateral Environmental Agreement
MedPAN	Mediterranean Protected Areas Network
MedPol	Programme for the Assessment and Control of Marine Pollution in the Mediterranean Region
MedU	MAP Coordinating Unit (also referred to asMed/RCU Mediterranean Regional Coordination Unit)
MGR	Marine Genetic Resource
MMPA	Marine Mammal Protection Act (USA 1972)
MOU	Memorandum of Understanding
MP	Management Procedure
MPA	Marine Protected Area
MRA	Marine Resources Act (Norway)
MS	Member State (of the European Union)
MSC	Marine Stewardship Council
MSFD	Marine Strategy Framework Directive (of the EU)
MSP	Marine Spatial Planning
MSSD	Mediterranean Strategy on Sustainable Development
MSY	Maximum Sustainable Yield
NDA	Nature Diversity Act (Norway)
NAFO	Northwest Atlantic Fisheries Organization
NASCO	North Atlantic Salmon Conservation Organization
NCP	Non-Contracting party
NEAFC	North East Atlantic Fisheries Commission
NEPA	National Environmental Policy Act (in USA)
NGO	Non-Governmental Organization
NRSMPA	National Representative System of Marine Protected Areas (Australia)
OBIS	Ocean Biogeographic Information System
ODA	Official Development Assistance
OECD	Organization for Economic Cooperation and Development
OM	Operating Model
OPRT	Organization for the Promotion of Responsible Tuna Fisheries
OSPAR	Convention for the Protection of the Marine Environment of the North-East Atlantic. Oslo and Paris Commission
PA	Precautionary Approach
PAF	Precautionary Approach to Fisheries
PICES	North Pacific Marine Science Organization
PICTs	Pacific Island Countries and Territories
PITIA	Pacific Islands Tuna Industry Association
POA	Programme of Action
PSMA	FAO Agreement on Port State Measures to Prevent, Deter and Eliminate Illegal, Unreported and Unregulated Fishing ('Port State Measures Agreement')
PSSAs	Particularly Sensitive Sea Areas
RAC/SPA	Regional Activity Centre for Specially Protected Areas
RAC-MED	European Union Regional Advisory Council for the Mediterranean
RACs	Mediterranean Action Plan Regional Activity Centres
RFA	Responsible Fisheries Alliance

RFB	Regional Fishery Body
RFMO	Regional Fishery Management Organization
RFMO/A	Regional Fisheries Management Organization and Arrangement
RFSN	FAO Regional Fisheries Secretariats Network
Rio+20	UN Conference on Sustainable Development
RSA	Regional Seas Arrangement
SARA	Species at Risk Act (Canada)
SASSI	Southern African Sustainable Seafood Initiative
SAC GFCM	Scientific Advisory Committee
SBSTTA	CBD Subsidiary Body on Technical and Technological Advice
SCAR	Scientific Committee on Antarctic Research
SC-CAMLR	Scientific Committee for the Conservation of Antarctic Marine Living Resources
SCIC	Standing Committee on Implementation and Compliance (CCAMLR)
SCOR	Scientific Committee on Oceanic Research
SD	Sustainable Development
SEA	Strategic Environmental Assessment
SFF	Sustainable Fisheries Framework (Canada)
SEAFO	Southeast Atlantic Fisheries Organization
SIDS	Small Island Developing States
SOFA	State of Food and Agriculture (FAO)
SOFIA	State of Fisheries and Aquaculture (FAO)
SOFIA	State of World Fisheries and Aquaculture (biennial FAO report)
SO-GLOBEC	Southern Ocean Global Ocean Ecosystem Dynamics Program
SPA	Special Protected Area
SPA/BD	Specially Protected Areas and Biodiversity Protocol
SPAMIs	Specially Protected Areas of Mediterranean Importance
SPC	Secretariat of the Pacific Community
SPREP	South Pacific Regional Environment Program
SSF	Small-Scale Fisheries
SUSG	Sustainable Use Specialist Group
TAC	Total Allowable Catch
TEU	Treaty on European Union
TFEU	Treaty on the Functioning of the European Union
TIDE	Toledo Institute for Development and Environment
TNC	The Nature Conservancy
TRIPs	Trade-Related Aspects of Intellectual Property Rights
TURF	Territorial Use Rights in Fisheries
UN	United Nations
UNCED	UN Conference on Environment and Development
UNCHE	UN Conference on Human Environment
UNCLOS III	Third UN Conference on the Law of the Sea
UNCLOS	UN Conference (or Convention) on the Law of the Sea
UNCTAD	UN Commission on Trade and Development
UNEP	UN Environment Programme
UNESCO	UN Educational, Scientific and Cultural Organization
UNFCCC	UN Framework Convention on Climate Change
UNFSA	UN Fish Stock Agreement (UN Agreement for the Implementation of the Provisions of the United Nations Convention on the Law of the Sea of 10 December 1982 relating to the Conservation and Management of Straddling Fish Stocks and Highly Migratory Fish Stocks)
UNGA	UN General Assembly
UNICPOLOS	UN Informal open-ended Consultative Process on Oceans and the Law of the Sea
UNCSD	UN Conference on Sustainable Development (Rio+20)
VME	Vulnerable Marine Ecosystem
VMS	Vessel Monitoring System
WCED	World Commission on Environment and Development
WCMC	World Conservation Monitoring Centre (of UNEP)
WCPA	World Commission on Protected Areas (of IUCN)
WCPFC	Western and Central Pacific Fisheries Commission
WCPN	World Commission for the Protection of Nature
WCS	World Conservation Strategy
WECAFC	Western Central Atlantic Fisheries Commission
WFC	World Fish Center
WIDECAST	Wider Caribbean Sea Turtle Conservation Network
WSSD	World Summit on Sustainable Development
WTO	World Trade Organization
WWF	Worldwide Fund for Nature
WWF-SASSI	WWF Southern African Sustainable Seafood Initiative

Glossary

In cases where multiple meanings of a given term exist (e.g. governance, precautionary principle/approach) no guidance was given to authors about distinctions to apply between the different meanings, so the intent of the term in each chapter has to be inferred from how the authors use the term.

Adaptive management An iterative management approach that stimulates social and institutional learning and emphasizes the importance of feedbacks in shaping policy (Folke et al., 2000). It explicitly recognizes the occurrence and potential consequences of uncertainties resulting from incomplete knowledge and adopts strategies and methods aimed explicitly at both managing risks under current uncertainty and progressively reducing uncertainties and risk through 'learning by management' (Garcia et al., 2008).

Advocacy Actions to influence public and political opinion to gain support for a particular change (Graham 1971, p. 124). In this book, frequently associated with an explicit or implied willingness to selectively use information to promote a pre-selected viewpoint.

Allocation Refers both to a share and the process of sharing. In fisheries management and coastal planning, the term is often used narrowly to refer to a share (a portion, a right, of the allowable catch, effort or area attributed to a person, a community, a vessel or a company) and the process of distributing shares. Beneficiaries could be contemporary (intra-generational allocation) or belong to successive generations (inter-generational allocation) (FAO Glossary; adapted from Garcia and Boncoeur, 2007).

Artisanal fisheries A fishery based on an individual (self-employed) or family type of enterprise (as opposed to an industrial company), most often operated by the owner (even though the vessels may sometimes belong to the fishmonger or some external investor), with the support of the household. The term has no obvious reference to size but tends to have a connotation of relatively low levels of technology although this may not always be the case. In practice, the definition varies between countries, e.g. from gleaning or a one-person canoe in poor developing countries, to more than 20 m trawlers, seiners or longliners in developed countries (e.g. in Europe). Artisanal fisheries can be subsistence or commercial fisheries, providing for local consumption or export (based on FAO Glossary). See also **small-scale fisheries.**

Benthic Living on or in the bottom of the ocean.

Biodiversity In this book 'biodiversity' and 'biological diversity' are not differentiated in usage. For the purposes of this book, biodiversity is the variability among living organisms from all sources, including terrestrial, marine and other aquatic ecosystems and the ecological complexes of which they are part (Convention on Biological Diversity, http://www.biodiv.org/convention/articles.asp).

Bycatch Part of a catch of a fishing unit taken incidentally in addition to the target species towards which fishing effort is directed. It may be retained for human use or some or all of it may be returned to the sea as discards, usually dead or dying (modified from FAO, 1998). Also written as **by-catch**.

Capacity building Enhancing the skills of people and the ability of institutions to participate in resources management through education and training (Fisheries and Oceans Canada, 2002). A process of strengthening or developing human resources, institutions or organizations

(Alcamo et al., 2003). The sum of efforts needed to nurture, enhance and utilize the skills and capabilities of people and institutions at all levels, towards a particular goal, e.g. participatory management (Berkes et al., 2001)

Co-adaptation: See **Coevolution**

Coevolution The process by which species interactively adapt and evolve in a changing physical, socio-economic environment. By extension, the term is also used to describe the process by which science and society interactively adapt in a changing socio-economic and institutional environment. In this process, the actions of each component of the system affect the actions of all other components in the same physical environment. Also referred to as **co-adaptation.**

Co-management A process of management in which government shares power with resource users, with each given specific rights and responsibilities relating to information and decision-making (OECD, 1996). A partnership arrangement in which government, the community of local resource users (fishers), external agents (non-governmental organizations, research institutions), and sometimes other fisheries and coastal resource stakeholders (boat owners, fish traders, credit agencies or money lenders, tourism industry, etc.) share the responsibility and authority for decision making over the management of a fishery (Berkes et al. 2001).

Common pool resources Also referred to as '**commons**'. A class of resource systems in which: (1) exclusion of any beneficiaries through physical or institutional means is especially costly if not impossible, even if they do not contribute to its management and undermines the capacity for others to do so; and (2) exploitation by one user reduces resource availability for others (subtractability). While a common pool resource system (a fishing ground) can be jointly appropriated (e.g. under a common property regime), the resource units withdrawn from it (e.g. the catch) are private (Berkes et al., 1989; Ostrom, 1990; Ostrom et al., 1999).

Conservation The management of human use of the biosphere so that it may yield the greatest sustainable benefit to present generations while maintaining its potential to meet the needs and aspirations of future generations. Conservation is positive, embracing preservation, maintenance, sustainable utilization, restoration and enhancement of the natural environment (IUCN/UNEP/WWF World Conservation Strategy 1980).

Conservation of biodiversity The management of human interactions with genes, species and ecosystems in order to provide the maximum benefit to the present generation while maintaining their potential to meet the needs and aspirations of future generations; encompasses elements of saving, studying and using biodiversity (Hesselink et al., 2007).

Critical habitat Specific area within the geographical range occupied by the species, that contain physical or biological features crucial to the survival of the species and hence essential to its conservation, and may require special management considerations or protection to maintain such features (FAO glossary).

Deep-sea fisheries Fisheries carried out in waters deeper than about 400 m are generally considered to be deep-sea fisheries (ICES). **Deep-sea fisheries in the high seas** are tentatively defined as fisheries that: (1) occur in areas beyond the limits of national jurisdiction; (2) take a catch (intended as everything brought up by the gear) that includes species that can only sustain low exploitation rates; and (3) use fishing gear that is likely to contact the seafloor during the normal course of fishing operations (FAO, 2008).

Discard To release or return fish to the sea, dead or alive, whether or not such fish are brought fully on board a fishing vessel (FAO glossary).

Diversity A measure of the complexity of an ecosystem. It is measured in terms of the number of different plant and animal species (Scialabba, 1998). A numerical measure combining the number of species in an area with their relative abundance (Nybakken, 1982). Often called *species richness*, the diversity of a community is represented by a mathematical expression relating species richness to the relative abundance of a species. For a fixed species richness, the diversity is maximum when the abundance of each species is the same.

Eco-label A seal of approval (or certification) of a product, process or service complying with a particular set of agreed environmental criteria usually awarded by an impartial third party (certification company). In fisheries, the label informs the buyer of the quality of the product as well as on the production and management processes.

Ecosystem approach An approach to management that recognizes the complexity of ecosystems and the interconnections among component parts (Fisheries and Oceans Canada, 2002). A strategy for the integrated management of land, water and living resources that promotes conservation and sustainable use in an equitable way. The Ecosystem Approach places human needs at the centre of biodiversity management. It aims to manage the ecosystem, based on the multiple functions that ecosystems perform and the multiple uses that are made of these functions. The ecosystem approach does not aim for short-term economic gains, but aims to optimize the use of an ecosystem without damaging it (Hesselink et al., 2007).

Ecosystem approach to fisheries An approach to fisheries management and development that strives to balance diverse societal objectives by taking into account the knowledge and uncertainties about biotic, abiotic and human components of ecosystems and their interactions and applying an integrated approach to fisheries within ecologically meaningful boundaries. The purpose of EAF is to plan, develop and manage fisheries in a manner that addresses the multiple needs and desires of societies, without jeopardizing the options for future generations to benefit from the full range of goods and services provided by marine ecosystems (FAO, 2003).

Ecosystem goods and services Material (goods) and non-material (services) commodities produced by the ecosystem that yield positive utility (modified from Sutinen, 2000).

Ecosystem objective A narrative or numeric statement on the desired condition of an ecosystem, or of one of its constituents (Fisheries and Oceans Canada, 2002).

Ecosystem The system of interactive relationships among organisms (e.g. energy transfer) and between organisms and their physical environment (e.g. habitat) in a given geographical unit (Fisheries and Oceans Canada, 2002). The ecosystem is conceived as comprising a natural and a human subsystem and is often referred to as socio-ecological system to stress this dual structure (Folke et al., 2000).

Ecosystem-based management The management of human activities so that ecosystems and their structure, function and composition are maintained at appropriate temporal and spatial scales (Fisheries and Oceans Canada, 2002).

Endangered species Species threatened by extinction as a direct or indirect result of human activities, e.g. through habitat degradation, overexploitation or competition with introduced species.

Equity A principle of stewardship that, in fisheries and environmental management, relates to fairness, justice, impartiality and freedom from bias or favouritism (e.g. in the allocation of rights or determination of claims). It requires that similar options be available to all parties. It is an important factor of compliance. See **inter-generational equity** (Garcia and Boncoeur, 2007).

Exclusive economic zone (EEZ) Area adjacent to a coastal state which encompasses all waters between: (1) the seaward boundary of that state; (2) a line on which each point is 200 nautical miles (370.40 km) from the baseline from which the territorial sea of the coastal state is measured (except when other international boundaries need to be accommodated); and (3) the maritime boundaries agreed between that state and the neighbouring states (FAO glossary).

Extinction The total disappearance of a species. The risk of extinction of a species depends on its population demographics, biological characteristics (such as body size, trophic level, life cycle, breeding structure or social structure requirements for successful reproduction) and vulnerability due to aggregating habits, natural fluctuations in population size (dimensions of time and magnitude) and residency/migratory patterns (FAO glossary).

Fishery management The integrated process of information gathering, analysis, planning, decision-making, allocation of resources and formulation and enforcement of fishery regulations by which the fishery management authority controls the present and future behaviour of interested parties in the fisheries, in order to ensure the continued productivity of the living resources (FAO, 1995).

Ghost fishing The capture of aquatic organisms by lost or abandoned gear (Bjorndal, 2009).

Good governance Governance that is participatory, transparent and accountable. Here transparency refers to openness, the free availability of information, decisions and plans, and the use of language that stakeholders can understand. Accountability means that people who make the decisions should be available to answer to the people who are affected by the decisions (United Nations, 2008).

Governance The formal and informal arrangements, institutions and mores which determine how resources or an environment are utilized, how problems and opportunities are evaluated and analysed, what behaviour is deemed acceptable or forbidden and what rules and sanctions are applied to affect the pattern of resource and environmental use (Sutinen, 2000).

High seas Waters beyond the areas of national jurisdiction (which can be 200 nautical miles or less) excluding species fixed on the continental shelf, which remain under the sovereign rights of the coastal States (FAO glossary).

Illegal fishing Refers to activities: (1) conducted by national or foreign vessels in waters under the jurisdiction of a State, without the permission of that State or in contravention of its laws and regulations; (2) conducted by vessels flying the flag of States that are parties to a relevant regional fisheries management organization but operate in contravention of the conservation and management measures adopted by that organization and by which the States are bound, or relevant provisions of the applicable international law; or (3) in violation of national laws or international obligations, including those undertaken by cooperating States to a relevant regional fisheries management organization (FAO glossary).

Incentives Something which encourages a person, corporation or organization to do something. Any factor (financial or non-financial) that provides a motive for a particular course of action, or counts as a reason for preferring one choice to the alternatives. Depending on the different ways in which they motivate agents to take a particular course of action, incentives may be classified as remunerative (or financial), moral or coercive.

Incidental catch: See **bycatch**

Input controls Management instruments used to control the time and place as well as type and/or amount of fishing with the view to limit yields and fishing mortality; e.g. restrictions on type and quantity of gear, effort and capacity, or closed seasons (FAO glossary).

Integrated assessment (IA) An interdisciplinary process of synthesizing, interpreting and communicating knowledge from diverse scientific disciplines in order to provide relevant information to policy makers on a specific decision problem (Toth, 2001). The process of assessing whole and dynamic complex fishery systems in their environment using quantitative and qualitative methods to advise centralized and decentralized government bureaucracies as well as self-governing communities. See Garcia (2008).

Integrated management (IM) A continuous process through which decisions are made for the sustainable use, development and protection of areas and resources. IM acknowledges the interrelationships that exist among different users and the environments they potentially affect. It is designed to overcome the fragmentation inherent in a sectoral management approach, analyses the implications of development, conflicting uses and promotes linkages and harmonization among various activities (Fisheries and Oceans Canada, 2002).

Integration The process of bringing together separate components as a functional whole that involves coordination of interventions. The act of combining into an integral whole. The more integrated the representation, the closer to a system representation (Garcia et al., 2008).

Inter-generational equity Requires that future generations be given the same opportunity as the present generation to decide on how to use the resources. It can be sought through avoidance of actions that are not potentially reversible on some agreed time scale (e.g. a human generation), consideration of long-term consequences in decision-making and rehabilitation of degraded physical and biological environments (Garcia and Boncoeur, 2007).

Livelihood A means of securing the necessities of life. A livelihood comprises the capabilities, assets (including both material and social resources) and activities required for a means of living. A livelihood is sustainable when it can cope with and recover from stresses and shocks and maintain or enhance its capabilities and assets both now and in the future, while not undermining the natural resource base (adapted from Chambers and Conway, 1992).

Management plan An explicit arrangement (contract) between the interested parties and the fisheries management authority which makes explicit the objectives and means of management, the nature of the management authority, its powers and responsibilities, its working and consultation procedures and the rights and responsibilities of the interested parties in the fishery (FAO, 1995).

Marine protected area (MPA) Any area of intertidal or subtidal terrain, together with its overlying water and associated flora, fauna, historical and cultural features, which has been reserved by law or other effective means to protect part or all of the enclosed environment (Kelleher and Kenchington, 1992; IUCN 1994). Any marine geographical area that is afforded greater protection than the surrounding waters for biodiversity conservation or fisheries management purposes (FAO, 2011).

Marine reserve Marine areas in which some or all extractive activities are prohibited for a lengthy period of time for the purpose of conservation of the area with its biodiversity (modified from FAO glossary and US Pacific Management Council). Often referred to also as **no-take area**.

Maximum sustainable yield (MSY) The highest theoretical equilibrium yield that can be continuously taken (on average) from a stock under existing (average) environmental conditions without significantly affecting the reproduction process (FAO glossary). Also referred to sometimes as **potential yield**.

Monitoring, control and surveillance (MCS) An integrated set of functions of a technical (data collection), legislative (enacting of instruments) and policing (enforcement) nature, essential for the effective implementation of fishery development and management plans. *Monitoring* includes the collection, measurement and analysis of fishing activity including, but not limited to: catch, species composition, fishing effort, bycatch, discards, area of operations, etc. *Control* involves the specification of the terms and conditions under which resources can be harvested. These specifications are normally contained in national fisheries legislation and other arrangements that might be nationally, subregionally or regionally agreed. *Surveillance* involves the regulation and supervision of fishing activity to ensure that national legislation and terms, conditions of access and management measures are observed (Flewwelling et al., 2003).

No-take MPA A type of marine protected area in which any extractive activity is strictly prohibited. Often considered equivalent to a **marine reserve** (modified from the FAO glossary).

Overexploitation: See **Overfishing**

Overfishing Exerting fishing pressure (fishing intensity) beyond an agreed optimum level (FAO glossary). The symptoms of ecosystem overfishing include: reduction in diversity, reduction in aggregate production of exploitable resources, decline in mean trophic level, increase in bycatch, greater variability in abundance of species and greater anthropogenic habitat modification (Hall, 1999).

Participatory management Any form of management involving a degree of stakeholder participation. Co-management is a specific form of participatory management in which there is a sharing of decision-making power between the State and the stakeholders (FAO glossary).

Precautionary Term used when referring to the **precautionary principle** and **precautionary approach.** No single definition of either term is applied consistently within or across disciplines, and both are used in international policy instruments and scientific debate. Further blurring the distinction, the precautionary approach is presented in Principle 15 in UNCED as 'in order to protect the environment, the precautionary approach shall be widely applied by States according to their capabilities (United Nations, 1992). Where there are threats of serious or irreversible damage, lack of full scientific certainty shall not be used as a reason for postponing cost-effective measures to prevent environmental degradation'. The **precautionary principle** is often used prescriptively, in contexts such as addressing severe and potentially irreversible risks for human health, species extinction or ecosystem degradation. The **precautionary approach** is encountered more commonly overall, particularly in fisheries agreements. It is typically viewed as being more flexible than the 'principle' with greater leeway for considering cost-effectiveness of options.

Productivity For a fish stock, relates to the birth, growth and death rates of a stock. A highly productive stock is characterized by high birth, growth and mortality rates and, as a consequence, a high turnover and production to biomass ratio (P/B). Such stocks can usually sustain higher exploitation rates and, if depleted, could recover more rapidly than comparatively less-productive stocks (FAO glossary).

Property right A type of resource *ownership* by an individual, a group (communal rights, common property) or the State (State property, public property). Property refers to a 'bundle' of rights including: the rights of access, withdrawal, management, exclusions and alienation. These rights might be separated. The granting or acquisition of all rights characterizes full property or *ownership* (Ostrom, 2000).

Protected areas Designated areas of the ecosystem which are protected by law and regulations with the view to conserving/rehabilitating them and the biodiversity they contain. Protection includes regulation or outright prohibition of human uses.

Quota A share of the total allowable catch (TAC) allocated to an operating unit such as a country, a community, a vessel, a company or an individual fisherman (individual quota) depending on the system of allocation. Quotas may or may not be transferable, inheritable and tradable. While generally used to allocate total allowable catch, quotas could be used also to allocate fishing effort or biomass (FAO glossary).

Risk A term used with many different meanings: (1) the magnitude of a hazard and the probability of its occurrence (Choudhury and Jansen, 1999); and (2) a measure of the probability that damage to life, health, property and/or the environment will occur as a result of a given hazard, and the magnitude of the consequences of the effect occurring (Choudhury and Jansen, 1999).

Seamount Seamounts are undersea mountains whose summits lie beneath the ocean waves. They are usually volcanic in origin and are generally defined as having an elevation of greater than 1000 m from the seabed (FAO glossary).

Small-scale fisheries A fishery that typically uses a relatively small-sized gear and vessel. The term sometimes has the added connotation of low levels of technology and capital investment per fisher, although that may not always be the case (Garcia et al., 2008). See: **artisanal fisheries**

Socio-ecological system A term used to emphasize the point that social and ecological systems are in fact linked, and that the delineation between social and ecological (and between nature and culture) is artificial and arbitrary. The integrated concept of humans-in-nature (Berkes et al., 2001).

Stewardship The careful and responsible management of something entrusted to one's care: stewardship of our natural resources (Webster Dictionary).

Stock (1) The part of a fish population which is under consideration from the point of view of actual or potential utilization (Ricker, 1975). (2) A group of individuals in a species occupying a well-defined spatial range independent of other stocks of the same species. Random dispersal and directed migrations due to seasonal or reproductive activity can occur. Such a group can be regarded as an entity for management or assessment purposes. Some species form a single stock (e.g. southern bluefin tuna) while others are composed of several stocks (e.g. albacore tuna in the Pacific Ocean comprises separate northern and southern stocks). The impact of fishing on a species cannot be determined without knowledge of this stock structure (FAO glossary).

Sustainable development Management and conservation of the natural resource base and the orientation of technological and institutional change in such a manner as to ensure the attainment of continued satisfaction of human needs for present and future generations. Such sustainable development conserves (land) water, plants and (animal) genetic resources is environmentally non-degrading, technologically appropriate, economically viable and socially acceptable (FAO, 1989). Development that meets the needs of the present generation without compromising the ability of future generations to meet their own needs (WCED, 1987).

Sustainable livelihoods A livelihood that can cope with and recover from stresses and shocks and maintain or enhance its capabilities, assets and entitlements, while not undermining the natural resource base (Chambers and Conway, 1992).

Sustainable use The use of components of biological diversity in a way and at a rate that does not lead to the long-term decline of biological diversity, thereby maintaining its potential to meet the needs and aspirations of present and future generations (Hesselink et al., 2007).

Target species Those species that are primarily sought by the fishermen in a particular fishery. The subject of directed fishing effort in a fishery. There may be primary as well as secondary target species (FAO glossary).

Technical measures Contain two categories of measures for regulating fisheries: *prohibitions*, which are used in case of high risk, and *regulated measures*. Use of toxic substances, dynamite and other destructive methods will usually be prohibited. Authorized uses are regulated through regulations, decrees, by laws and administrative orders. Area and time restrictions are technical measures. Technical measures**,** as opposed to effort or catch quotas or property rights, are often assumed (often wrongly) to have no distributional effect (FAO glossary).

Tenure Socially defined agreements held by individuals or groups (either recognized by law or customary norms) on the rights of access and the rules for use of either a land area or associated resources such as individual trees, plant species, water or animals (CIFOR, 1999).

Threatened species The IUCN Red List of Threatened Species defines a number of categories of threat based on the risk of extinction in the wild, for example: critically endangered (extremely high risk); endangered (very high risk); vulnerable (high risk); near-threatened (if management measures are not introduced); least concern (no imminent risk); data deficient (impossible to assess); and not evaluated. The vulnerability of a species to threats of extinction depends on its population demographics, biological characteristics (such as body size, trophic level, life cycle, breeding structure or social structure requirements for successful reproduction) and vulnerability due to aggregating habits, natural fluctuations in population size (dimensions of time and magnitude) and residency/migratory patterns (IUCN, FAO glossary).

Uncertainty The degree to which a present or future condition (e.g. of an ecosystem) is unknown. Uncertainty can result from the existence of natural fluctuations over time (e.g. in year-to-year survival rates), from a lack of information or from disagreement about what is known or even knowable. It may have many types of sources, from quantifiable errors in the data to ambiguously defined terminology or uncertain projections of human behaviour (Alcamo et al., 2003). In adaptive co-management, uncertainty refers to the extent to which actors are unable to understand, predict or control how system components, relationships and processes will interact, and what outcomes will result (Armitage et al. 2007). In statistics and risk analysis, the estimated amount (or percentage) by which an observed or calculated value may differ from the true value (FAO, 1995).

Unregulated fishing Refers to fishing activities: (1) in the area of application of a relevant regional fisheries management organization that are conducted by vessels without nationality, or by those flying the flag of a State not party to that organization, or by a fishing entity, in a manner that is not consistent with or contravenes the conservation and management measures of that organization; or (2) in areas or for fish stocks in relation to which there are no applicable conservation or management measures and where such fishing activities are conducted in a manner inconsistent with State responsibilities for the conservation of living marine resources under international law (FAO glossary).

Unreported fishing Refers to fishing activities: (1) which have not been reported, or have been misreported, to the relevant national authority, in contravention of national laws and regulations; or (2) undertaken in the area of competence of a relevant regional fisheries management organization which have not been reported or have been misreported, in contravention of the reporting procedures of that organization (FAO glossary).

Use rights The rights held by individual fishers, fishing groups, fishing communities or companies to use the fishery resources. These may be in the form of rights to an amount of fishing effort (effort rights) or catch that can be taken in the fishery (harvest rights or harvest quotas). They can be defined by local custom, mutual agreements or prescribed by other entities holding access rights. They may restrict the use of particular harvesting techniques (FAO glossary).

Vessel monitoring system (VMS) As part of modern monitoring, control and surveillance systems (MCS) the VMS is a vessel tracking system (usually satellite-based) which provides management authorities with accurate information on fishing vessels position, course and speed at time intervals. Detail of VMS approved equipment and operational use will vary with the requirements of the nation of the vessel's registry and the regional or national water in which the vessel is operating.

Vulnerability The degree to which a system is likely to experience harm due to exposure to a hazard which can either be either a perturbation (disturbance or shock) or a stress. Vulnerability is registered not by exposure to hazards alone; it also resides in the resilience of the system experiencing the hazard (Turner et al., 2003). A condition arising from the interaction of three factors, namely: (1) *risk exposure*, or the nature and degree to which a household (or community) is exposed to a certain risk such as a natural disaster, conflicts, macro-economic changes, etc.; (2) *sensitivity* to this risk, measured for instance through the dependence of the household (or community) on fishing activity for food security or income generation; and (3) *adaptive capacity* of the household (or community) to deal with risk, that is, its ability to cope with changes (FAO, 2005).

Vulnerable species Taxa of various types, including: (1) taxa believed likely to move into the 'endangered' category in the near future if the relevant causal factors continue to operate; these factors may include overexploitation, extensive destruction of habitat and other environmental disturbances; (2) taxa with populations that have been seriously depleted and whose ultimate security has not yet been assured; and (3) taxa with populations that are still abundant but are under threat from severe adverse factors throughout their range (United Nations, 1997).

References

Alcamo, J. et al. (eds) (2003) *Ecosystem and Human Well-Being. A Framework for Assessment. Millennium Ecosystem Assessment*. Island Press, Washington DC.

Armitage, D., Berkes, F. and Doubleday, N. (eds) (2007) *Adaptive Co-Management: Collaboration, Learning, and Multi-Level Governance*. University of British Columbia Press, Vancouver.

Berkes, F., Feeny, F., McCay, B.J. and Acheson, J.M. (1989) The benefits of the commons. *Nature* **340**, 91–93.

Berkes, F., Mahon, R., McConney, P., Pollnac, R.C. and Pomeroy, R.S. (2001) *Managing Small-Scale Fisheries: Alternative Directions and Methods*. International Development Research Centre, Ottawa.

Bjorndal, A. (2009) Regulation of fishing gears and methods. In: *A Fishery Managers' Guidebook* (eds K.L. Cochrane and S.M. Garcia), pp. 167–194. FAO, Rome and Wiley-Blackwell, Chichester.

Chambers, R. and Conway, G. (1992) Sustainable rural livelihoods: Practical concepts for the 21st century. IDS Discussion Paper 296. IDS, Brighton.

Choudhury, K. and Jansen, L.J.M. (1999) *Terminology for Integrated Resources Planning and Management*. FAO, Rome.

CIFOR (1999) The CIFOR criteria and indicators generic template. Center for International Forestry Research, Bogor, Indonesia.

FAO (nd) Glossary on responsible fisheries. Available at http://www.fao.org/fi/glossary/default.asp (accessed March 2014).

FAO (1989) Sustainable development and natural resources management. Conference of the Food and Agriculture Organization of the United Nations 26th session, 11–30 November 1989. FAO, Rome.

FAO (1995) Guidelines for responsible management of fisheries. In *Report of the Expert Consultation on Guidelines for Responsible Fisheries Management*. FAO Fisheries Report. Wellington, New Zealand, 23–27 January 1995.

FAO (1998) Guidelines for the routine collection of capture fishery data. FAO Fisheries Technical Paper No. 382. FAO, Rome.

FAO (2003) The Ecosystem Approach to Fisheries. FAO Technical Guidelines for Responsible Fisheries, No. 4 (Suppl. 2). FAO, Rome.

FAO (2005) Increasing the contribution of mall-scale fisheries to poverty alleviation and food security. FAO Technical Guidelines for Responsible Fisheries, No. 10. FAO, Rome.

FAO (2008) Draft international guidelines for the management of deep-sea fisheries in the high seas. FAO meeting document TC/DSF/2008.2. Document presented at the *FAO Technical Consultation on the International Guidelines for the Management of Deep-sea Fisheries in the High Seas*, Rome, 4–8 February 2008.

FAO (2011) Fisheries management. 4-Marine protected areas and fisheries. FAO Technical Guidelines for Responsible Fisheries, No. 4(suppl. 4). FAO, Rome.

Fisheries and Oceans Canada (2002) Canada's ocean strategy. Policy and operational framework for integrated management of estuarine, coastal and marine environments in Canada.

Flewwelling, P., Cullinan, C., Balton, D., Sautter, R.P. and Reynolds, J.E. (2003) Recent trends in monitoring, control and surveillance systems for capture fisheries. FAO Fisheries Technical Paper, No. 415. FAO, Rome.

Folke, C., Berkes, F. and Colding, J. (2000) Ecological practices and social mechanisms for building resilience and sustainability. In: *Linking Social and Ecological Systems. Management Practices and Social Mechanisms for Building Resilience* (eds F. Berkes, C. Folke and J. Colding), pp. 414–435. Cambridge University Press, Cambridge.

Garcia, S.M. (compiler) (2008) Glossary. In: *A Fishery Managers' Guidebook* (eds K.L. Cohrane and S.M. Garcia), pp. 473–507. FAO, Rome and Wiley-Blackwell, Chichester.

Garcia, S.M. and Boncoeur, J. (2007) Allocation and conservation of ocean fishery resources: Connecting rights and responsibilities. American Fisheries Society Symposium, 2007.

Garcia, S.M., Allison, E.H., Andrew, N.J., Bené, C., Bianchi, G., de Graaf, G.J., Kalikoski, D., Mahon, R. and Orensanz, J.M. (2008) Towards integrated assessment and advice in small scale fisheries: principles and processes. FAO Fisheries Technical Paper No. 515. FAO, Rome.

Graham, J. (1971) *Community Fisheries Management Handbook*. With A. Charles and A. Bull. Gorsebrool Research Institute, St Mary's University.

Hall, S.J. (1999) *The Effects of Fishing on Marine Ecosystems and Communities*. Blackwell Scientific Publications, London.

Hesselink, F., Goldstein, W., van Kempen, P.P., Garnett, T. and Dela, J. (2007) Communication, Education and Public Awareness (CEPA). A Toolkit for National Focal Points and NBSAP Coordinators. CBD/IUCN. Available at http://www.cbd.int/cepa/toolkit/2008/doc/CBD-Toolkit-Glossaries.pdf (accessed March 2014).

IUCN (1994) Resolution 19.46 of the 19th session of the General Assembly of the World Conservation Union (IUCN). Buenos Aires, Argentina

Kelleher, G. and Kenchington, R. (1992) Guidelines for Establishing Marine Protected Areas. A Marine Conservation Development Report. IUCN, Gland, Switzerland.

Nybakken, J.W. (1982) *Marine Biology. An Ecological Approach*. Harper and Row, New York.

OECD (1996) Synthesis report for the study on the economic aspects of the management of marine living resources. AGR/FI (96) 12, Paris.

Ostrom, E. (1990) *Governing the Commons. The Evolution of Institutions for Collective Action*. Cambridge University Press, Cambridge, UK.

Ostrom, E. (2000) Private and common-property rights. II. Civil law and economics. In: *Encyclopedia of Law and Economics* (eds B. Bouchaert and G. De Geest), pp. 332–379. Edward Elgar Publisher, Cheltenham, England.

Ostrom, E., Burger, J., Field, C.B., Norgaard, R.B. and Policansky, D. (1999) Revisiting the commons: Local lessons, global challenges. *Science* **284**, 278–282.

Ricker, W.E. (1975) Computation and interpretation of biological statistics of fish populations. *Bulletin of the Fisheries Research Board of Canada* **191**, 2–6.

Scialabba N. (ed.) (1998) *Integrated Coastal Area Management and Agriculture, Forestry and Fisheries*. FAO Guidelines. FAO, Rome.

Sutinen, J.G. (ed.) (2000) A framework for monitoring and assessing socioeconomics and governance of large marine ecosystems. NOAA Technical Memorandum NMFS-NE-158. NOAA, Maryland, US.

Toth, F.L. (2001) Participatory integrated assessment methods. An assessment of their usefulness to the European Environmental Agency (EEA). Copenhagen, EEA Technical Report 64.

Turner, B.L. II, Kasperson, R.E., Matson, P.A. et al. (2003) A framework for vulnerability analysis in sustainability science. *Proceedings of the National Academy of Sciences of the United States* **100**, 8074–8079.

United Nations (1992) Earth Summit Agenda 21. The United Nations programme of action from Rio: 294 p.

United Nations (1997) Glossary of Environment Statistics. Studies in Methods, Series F, No. 67.

United Nations (2008) Good governance. Available at http://www.unescap.org/pdd/prs/ProjectActivities/Ongoing/gg/governance.asp (accessed March 2014).

WCED (1987) *Our common future. World Commission on Environment and Development*. Oxford University Press, Oxford.

Part I

Governance trends and challenges

Chapter 1

Governance of marine fisheries and biodiversity conservation: A history

S.M. Garcia[1], J. Rice[1,2] and A. Charles[1,3]

[1]IUCN Commission on Ecosystem Management (IUCN/CEM/FEG), Gland, Switzerland
[2]Fisheries and Oceans Canada, Ottawa, Canada
[3]Sobey School of Business and the Environmental Science Department, Saint Mary's University, Halifax, Canada

Abstract: The governance of marine fisheries and biodiversity conservation has evolved significantly since its origin. The chapter describes its evolution, at a global scale, using the international events compiled from the literature as data. The history of the two streams of governance is rich and reflects a constant quest to improve performance with mixed results. These streams have important common roots in pre-capitalistic and pre-colonial communities. They tend to emerge more clearly during the 19th century as a utilitarian and an aesthetic branch of conservation with tumultuous relationships. In fisheries, centralized forms of governance already existed in the 13th century. In marine conservation, they were practically absent until the 1960s and have emerged forcefully since the mid-20th century. From 1970 onwards, the emergence of the Law of the Sea and a number of cross-sectoral summits established institutional bridges between the two streams, accelerating conceptual convergence. The trend today towards creating new incentive structures and market-based approaches in both streams has the potential to accelerate their convergence and produce better integration in their implementation.

Keywords: governance; biodiversity; fisheries; history; coevolution

Introduction

The governance of fisheries and of nature conservation has developed within two parallel and interacting streams of events and decisions. Their present characteristics arise from this complicated history, which must be understood to improve their performance. Towards that end, this chapter presents the political, legal and institutional international events that have shaped the evolution of the two streams of governance for centuries. The key events have been assembled in a timeline (Annex 1) that is the backbone of Chapters 1 and 2. Despite our efforts, the list of events in Annex 1 cannot claim to be exhaustive. In the early years, governance has focused necessarily on the northern hemisphere where modern conservation, fishery science and international governance originated. Being based on formal literature, the list is bound to be deficient with regard to evolution of customary practices. Even though the two streams seem to have developed with little direct interaction until recently, the processes of economic development, colonization, globalization and international collaboration as well as the Internet have progressively synchronized their evolution to some extent. Regarding conservation of biodiversity, the chapter starts necessarily on terrestrial systems where most of the concepts used in the marine environment originated. The phases and milestones of the timeline contained in Annex 1 reflect the purposes of this book, but we are confident that key events would be considered benchmarks in almost any reasonable timeline.

Governance of Marine Fisheries and Biodiversity Conservation: Interaction and Coevolution, First Edition.
Edited by Serge M. Garcia, Jake Rice and Anthony Charles.

This chapter describes the chronological phases of the historical development of the two streams of governance. The evolutionary strands contained in the two streams are identified and analysed in detail in Chapter 2 and the central issue of their integration is addressed in Chapter 3. More information on the underlying global frameworks, processes and instruments is provided in Chapters 7, 10 and 11.

Historical developments in fishery governance

The history of fisheries policy, development and management since the Second World War proposed by Garcia (1992), Garcia and Newton (1994, 1997), Caddy and Cochrane (2001), Roberts (2007) and Bolster (2012) have been consulted, completed and extended below to cover the period from 1850 to 2012. The succession of events (Annex 1) has been subdivided into phases (Figure 1.1) marked by major milestones such as wars, conferences, treaties and agreements. The phases reflect policy trajectories, from the traditionally integrated concern for 'wise use' to its modern avatars of sustainable development, responsible fishing and sustainable use. It presents them against an evolving background of industrial growth and of colonization, management modernization, trade globalization and responses to growing environmental concerns.

Before 1850: Early concerns

Total removals from the ocean are not known and, for many decades, scientists and managers have implicitly or explicitly considered that, in these early days, the fisheries impact on ocean resources and biodiversity was minimal. However, even if relative fishing pressures cannot be precisely measured, pre-industrial technologies were sufficient to put marine animal populations under severe stress; by the late 19th century, extractions in Europe, North America and the Caribbean had reached levels that would be equivalent to today's Total Allowable Catches (Holm et al., 2010, p. 19). Bolster (2012) confirms the severe declines in some marine birds and mammals in Europe as early as the 9th century, the progressive degradation of estuarine, anadromous and coastal marine resources in the Northeast Atlantic since the 13th century and the decline of Northwest Atlantic coastal resources from the middle of the 17th century.

The role of fishing and the environment on the degradation of coastal fisheries was noted as early as in 14th century England (Alward, 1932, in Roberts, 2007), 15th century France (Typhaigne de la Roche, 1760) and 17th century USA (Bolster, 2012). Indeed, since the second half of medieval

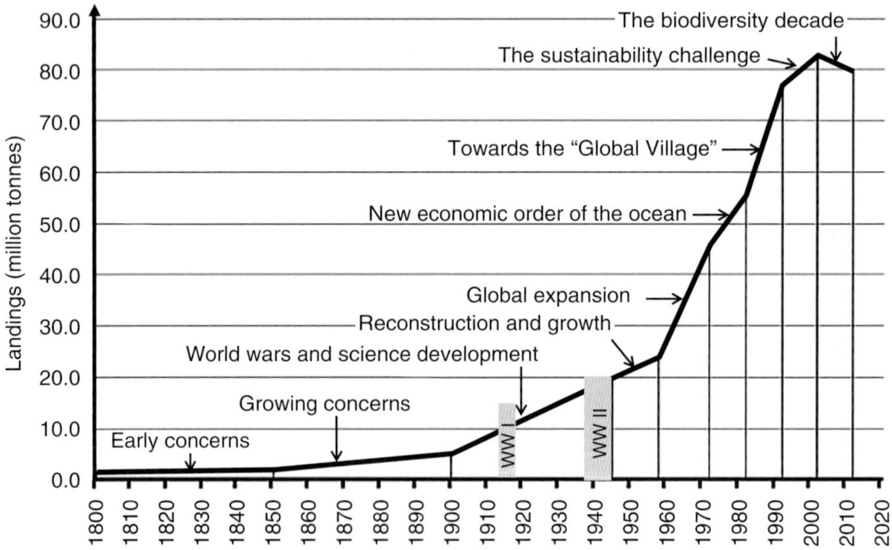

Figure 1.1 Landing trends: 1850–2012. The line joins the landings at the beginning and end of the different phases (FAO statistics).

times, some European marine fisheries have been regulated by laws and by-laws, community practices, inter-community agreements, town and Guilds' regulations and state intervention on gear and restrictions on period and place of the fishing including rotational harvest schemes, access to fishing and trade, etc. (Alward, 1932; Bolster, 2012; Caddy and Cochrane, 2001; Holm, 2010; Roberts, 2007). Fishing activities have been regulated through professional organizations with conditional decision power or community-based groups in France (*prud'homies*), Spain (*cofradias*) and Japan since the 17th century and in the Philippines, Oceania, Pacific coast of North America and Mexico since the 19th century (Garcia, 1989; McGoodwin in Caddy and Cochrane, 2001). In less-developed areas, traditional governance of fishing and conservation activities through local use rights and regulations as well as religious taboos (cf. Johannes, 2003 in the South Pacific) probably emerged as soon as growing human aggregations visibly started impacting local resources.[1]

During the second half of the 18th century however, developed states increasingly reduced the power of non-state institutions because of the rising concept of national sovereignty (in the coastal 3 miles), the need to make room for industrialization and the need to start subsidising economically failing industrial fleets (e.g. in USA, England and France; Miller, 1989). To allow for new developments, colonizing states neglected or depressed local traditions and marine tenure systems and modified access rules and the distribution of benefits (cf. Kittinger et al., 2011 in Hawai).

An epochal swing in fishery objectives and social relations resulted from the shift from pre-capitalistic (pre-colonial and pre-industrial) community-based management to capitalistic (colonial and industrial) and market-driven centralized management. It was accompanied by exponentially increasing landings (Figure 1.1).

From 1850 to 1900: Growing concerns

The period starts with accelerating industrial development and ends before the birth of the International Council for the Exploration of the Sea (ICES). Reported landings increased from 1.5–2 million tonnes in 1850 to about 5 million tonnes in 1900 (Garcia and Grainger, 2005).

Coastal states had exclusive jurisdiction only in the first 3 miles of the ocean. The repeated concerns of traditional fishers facing local depletions were disregarded (Bolster, 2012). The few warnings about the limitations of Earth productivity by T.R. Malthus (1766–1834), the vulnerability of ocean resources by Lankester (1884, 1890) and the early description of overfishing by J. Cleghorn of Wick in 1854 (Roswadowki, 2002) were silenced by the rumble of the fast industrialization process, the growth of trawl fishing industries and the call for deregulation by T.H. Huxley (1825–1895).[2] Some countries did impose limitations on trawling, fleet size, closed areas, mesh size, etc. (Anderson, 1998; Bolster, 2012) or established fishermen's organizations to resolve conflicts (Troadec, 1989, p. 45). The establishment of the first marine research stations (created in Naples in 1872 and in Woods Hole in 1875) and the Stockholm Conference on Fisheries (1899) signalled a rising interest in rational exploitation resting on scientific enquiry (Roswadowski, 2002). The period ended with growing scientific concern on the impoverishment of the sea (Garstang, 1900) and growing overfishing in the North Sea (Heinke in Roswadowski, 2002).

From 1901 to 1945: World Wars and science development

This period starts with the Kristiana Conference (in 1901) and establishment of ICES (in 1902) and ends with the end of the Second World War. Estimated landings grew rapidly from 5 to 15 million tonnes. With the exception of the North Atlantic (particularly the North Sea) and a few resources in the North Pacific, fishing impact was limited to coastal areas. The pressure on whale stocks grew rapidly and their demise was regularly forecast from the 1930s onwards.[3]

Fisheries development continued, and its heavy impact was confirmed by the significant rebuilding that took place when fishing was curtailed during the two World Wars (Atkinson, 1988; Gulland and Carroz, 1968). ICES established an Overfishing Committee in 1903, the first large-scale science-based international management experiment. The effects of natural oscillations and fishing mortality were identified through the work of Hjort, Russel, Graham, Baranov, Ottestad and others, but it took until mid-century for the fishing industry to recognize the necessity of scientific advice (Roswadowski, 2002, p. 85). The revolutionary theory of bio-economic overfishing elaborated by Warming (1911) in Danish went unnoticed.[4] The Great Law of Fishing of Graham (1935) foresaw that 'all fisheries that are unregulated become unprofitable' and that management measures elaborated without social and economic consideration were likely to be ineffective. The concept of 'optimum yield'[5] (a precursor of maximum sustainable yield or MSY, adopted in the 1982 United Nations Law of the Sea Convention) was proposed by Hjort, Jahn and Ottestad in 1933 (Roswadowski, 2002, p. 91). The conceptual biological foundations for modern fisheries management were therefore laid, but fishery regulations remained focused on mesh sizes and minimum landing sizes (Saetersdal, 1992).[6] The establishment of the first regional commissions on seals (in 1911), halibut (1932) and salmon (1937) signalled rising international awareness and institutional development

From 1946 to 1958: Reconstruction and growth

The period starts at the end of the Second World War and ends with the first United Nations Convention on the Law of the Sea (UNCLOS I). From 1950 to 1958, reported 'fish' landings rose from 15 to 24 million tonnes. North Atlantic fisheries expanded towards the Arctic Ocean and deep-sea stocks were discovered.

Countries focused on post-war reconstruction of the fishing capacity and development of food production. The Food and Agriculture Organization (FAO), created in 1946 to support implementation of the Marshall Plan, established a fisheries technical committee. Overfishing of high-value species was internationally recognized in the North Atlantic and North Pacific oceans (Gulland and Carroz, 1968). However, 'fishery regulations [were] still largely designed to promote orderly fishing and marketing and not really for the purpose of conservation' (Royce, 1988). Concerns increased further in ICES as marine biology moved more decisively towards providing advice for orderly fisheries development first, and management later (Roswadowski, 2002; Royce, 1988). Questions of open access and effort control were raised at the 2nd London Conference on Overfishing (in 1946) with little support and impact on regulations (Roswadowski, 2002, p. 107). New fishery bodies were created such as IWC in 1946, IPFC in 1948 and ICNAF, GFCM and IATTC in 1949. In the 1950s, the theory of economic overfishing was redeveloped by Gordon, Scott and Crutchfield. Simultaneously, Schaefer and Beverton and Holt turned the biological foundations established by their predecessors (e.g. Volterra, Gause, Baranov, Russel, Graham, etc.) into easy mathematics. These developments opened the possibility of scientific predictions that were subsequently pushed beyond their limits and the original intentions of their authors.

Irritated by the impact of foreign fleets on their resources, Peru and Chile (in 1947) and Ecuador (in 1952) unilaterally extended their jurisdiction to 200 miles and signed the 1952 Santiago Declaration, putting *de facto* the UNCLOS process on its starting blocks. After years of heated debate, the 1st UN Conference on the Law of the Sea (UNCLOS I in 1958) adopted a Convention on Fishing and Conservation of the Living Resources of the High Seas, which remained unsigned by some of the main actors concerned. The 1st 'Cod War' between Iceland and UK (19581961) started as Iceland extended its jurisdiction on fisheries to 12 miles.

From 1959 to 1972: Global expansion

The period starts with UNCLOS II (in 1960) and ends with the 1972 UN Conference on Human Environment (in Stockholm). Reported landings rose from 26 to 46 million tonnes as industrial fisheries spread through the world oceans.

During the 1960s, the priority of governments was discovering new grounds and resources with new technologies (e.g. tropical shrimp, tunas, small pelagics). Long-range fleets of developed nations (Europe, Japan, USSR) and developing nations (Cuba, Korea, Ghana, Thailand) expanded, supported by large subsidies and research programmes. The fishing capacity of developing countries (e.g. Peru, Chile, Morocco, Senegal, Côte d'Ivoire, Ghana, Angola and Thailand) grew rapidly with the support of former colonial powers. Knowledge of world resources potential increased (Gulland, 1971). Concerns increased in the North Atlantic regarding overcapacity and the inefficiency of conventional technical measures (Roswadowski, 2002, p. 189). Whales were sequentially depleted (Gulland and Carroz, 1968). The idea of imposing catch limits progressed[7] but direct regulation of fishing capacity was considered technically unfeasible (Garcia, 1992; Roswadowski, 2002). The ICES symposium on marine food chains (1968) opened a period of intense work for multispecies fishery research. The focus of the FAO Committee on Fisheries (COFI) moved from development to management. National jurisdiction became a major issue as the UNCLOS II process unfolded and, following the example of Iceland, many states extended their jurisdiction from 3–6 miles to 12 miles and even 100 miles in some Asian States.

The period closed in 1972 with: (1) the UN Conference on the Human Environment (UNCHE) in Stockholm and the formal birth of the sustainable development concept; (2) the spectacular collapse of the huge Peru–Chile anchoveta stock; (3) the recommendation of a 10 year moratorium on whaling; and (4) the 2nd Cod War between UK and Iceland as the latter claimed a 50 miles exclusive fishing zone. The time of uncontrolled geographical and technological expansion of fisheries seemed over.

From 1973 to 1982: A new economic order of the ocean

This period started with the FAO Technical Conference on Fishery Management and Development (in Vancouver in 1973) and the beginning of the UNCLOS III process. It ended with the adoption of the 1982 Law of the Sea Convention (LOSC) and the establishment of a new economic order in the oceans. Reported landings increased from 46 to 56 million tonnes, hiding growing sequential overfishing (Garcia, 1992) and discarding.

The 1973 FAO Technical Conference on Fishery Management and Development warned again about the fishing impact on resources, overcapitalization, incomplete science, free and open access and pollution. It called for precautionary, adaptive and anticipatory management, 9 years before adoption of the precautionary principle at the UN Conference on Environment and Development (UNCED) and 22 years before the publication of FAO guidelines on the subject. Many national jurisdictions were unilaterally extended in 1976–1979, and long-range fleets concentrated on straddling stocks and the few shelves that remained open (e.g. off Western Sahara, Namibia and in the Southwest Atlantic). New high-seas resources were targeted: tropical tunas, oceanic squids, sharks and horse mackerel and krill (Garcia, 1992). The third Cod War exploded in 1975 as Iceland extended its jurisdiction to 200 miles. The movement was followed by USA in 1976 and many countries followed suit under an emergent international customary law of the sea. The formal adoption of the UN Convention on the Law of the Sea in 1982 legalized the new claims, and established the International Tribunal for the Law of the Sea (ITLOS). Although it was not until 1994 that UNCLOS came into force, the battle over access to fishing grounds was over and a new battle for access to markets began.

The governance of inshore and coastal fisheries by communities emerges as a focus in the context of demographic increases, economic globalization, coastal developments, pollution and resilient pro-industrial national policy developments.

From 1983 to 1992: Towards the 'global village'

This period starts with the FAO World Conference on Fishery Management and Development (in 1984) and ends with the UN Conference on Environment and Development (in 1992). Reported marine 'fish' landings grew from 56 to 77 million tonnes.

Overfishing and overcapacity became global concerns together with pollution and natural fluctuations. The new Exclusive Economic Zone (EEZ) owners experimented with fishery development and management, limited entry, Individual Transferable Quotas (ITQs) and control and surveillance. On multispecies fisheries management, ICES concluded that mixed-species fisheries were a more serious issue than predator–prey relationships. Annual economic losses in the North Atlantic reached US$ 1 billion, and the poor state of many resources called for improved management (FAO, 1981; Garcia, 1992). With its fishery and environmental mandates, the Convention for the Conservation of Antarctic Marine Living Resources (CCAMLR; cf. Chapter 17) became a precursor of an ecological approach to fisheries. A moratorium on whaling was endorsed in 1982. The FAO (1984) Conference on Fisheries Management and Development focused on the new order of ocean governance, poor management, access to foreign markets and fish trade,[8] and social and economic issues. The 1984 International Conference of Fishworkers and Their Supporters, organized by the International Collective in Support of Fishworkers (ICSF), drew attention to the importance and problems of the neglected small-scale fisheries.

International concerns regarding the collateral effects of fishing grew rapidly. The 1989 Wellington Convention in the South Pacific and the United Nations General Assembly (UNGA) Resolution 44(225) banned large drift nets in the high seas from 1st July 1991 over concerns with bycatch, signalling a new importance of UNGA in cross-sectorial ocean matters. ICES established a Working Group on Ecosystem Effects of Fishing in 1990, incorporating ecosystem considerations within European fishery management advice. In 1991, the Australian Fisheries Management Act introduced the concept of Ecologically Sustainable Development (ESD) in fisheries (McLoughlin et al., 2008), strongly influencing the development of the precautionary and ecosystem approaches in FAO.

This period of rising concerns ended in 1992 with numerous significant events for fisheries and conservation. The Rio Declaration, its Principle 15 on the Precautionary Approach and its Agenda 21 provided guidance for fisheries and marine conservation for decades. The newly established the Commission on Biological Diversity (CBD) played a growing role at the interface between fisheries and conservation and enhanced the concept of sustainable use of biodiversity (hence fishery resources). Cod fishing in Canada became an emblematic example of a non-sustainable fishery, leading to a moratorium. For the first time, a fishery resource (tuna) was proposed for listing by the Convention on International Trade in Endangered Species (CITES).[9] The International Conference on Responsible Fishing (Cancun, Mexico) elaborated on the responsible fishing concept and called on FAO to develop a Code of Conduct.

From 1993 to 2002: Sustainability challenge

This active decade of implementation of UNCED Agenda 21 ended with the UN World Summit on Sustainable Development (WSSD). Reported marine landings grew from 78 to 83 million tonnes.

Overfishing was widespread (Garcia, 1992; Garcia and Grainger, 2005). The 1996 report of FAO on the State of Fisheries and Aquaculture (SOFA; FAO, 1996) indicated that the majority of the world's fish resources were in need of urgent management, that 35% were overfished and that 20 million tons more fish could be landed through adequate management and development efforts.

The institutional framework for fisheries was significantly reinforced. The 1993 FAO Compliance Agreement[10] and the 1995 UNFSA strengthened the arm of Regional Fisheries Management Organizations (RFMOs) and Port States in control and surveillance. The UNFSA also called for a precautionary approach and changed the role of MSY from a target to a limit reference value.

The 1995 FAO Code of Conduct for Responsible Fisheries provided a comprehensive international code of ethics for fishing in all aquatic systems under all jurisdictions. Its implementation was facilitated by International Plans of Action (IPOAs) on seabirds bycatch, sharks, fishing capacity and illegal, unreported and unregulated (IUU) fishing (others came later). The framework was further strengthened by the establishment of more RFMOs: the Commission for the Conservation of Southern Bluefin Tuna (CCSBT in 1993); the Central Bering Sea Donut Hole Agreement (in 1994); the Indian Ocean Tuna Commission (IOTC in 1995); the Western and Central Pacific Fisheries Commission (WCPFC in 2000); and Southeast Atlantic Fisheries Organization (SEAFO in 2001). The establishment of the Marine Stewardship Council (MSC) by WWF and Unilever in 1994 confirmed the increasing role of non-state market-driven instruments and consumers in fisheries policy. A collision between fisheries and conservation governance occurred in 1996 with the first International Union for the Conservation of Nature (IUCN) 'Red' listing of a dozen commercially exploited species (including Atlantic cod and haddock), triggering a long-standing debate on the listing of commercial species (cf. 'From 1950 to 1970-80' section; Chapter 12).

In 1999, the United Nations open-ended Informal Consultative Process on Oceans and the Law of the Sea (UNICPOLOS or the Informal Consultative Process, ICP) was established under the UNGA to facilitate an annual review of developments in ocean affairs and the law of the sea (resolution 54/33). Science advisory bodies to RFMOs and many developed States adopted more structured approaches to assessments and management advice, built on the precautionary approach, legitimized by Agenda 21 and operationalized by FAO (1996).

The ecosystem focus of fisheries management increased with the 1993 National Oceanic and Atmospheric Administration Global Environment Facility (NOAA-GEF) initiative on Large Marine Ecosystems (LME; Sherman, 2008). The ICES WG on Multispecies Assessments concluded its work on trophodynamic interactions in assessments and management, noting that more traditional issues such as over-capacity and ineffective Monitoring Control and Surveillance (MCS) were more consequential than predator–prey relationships. The ICES WG on the ecosystem effects of fishing diversified the scope of investigations. The Ecosystem Approach to Fisheries (EAF) was internationally adopted at the 2001 Iceland-FAO Conference on Responsible Fisheries in the Marine Ecosystem (Reykjavik), a decision endorsed by the following Committee on Fisheries (COFI). In 2003, FAO EAF guidelines explicitly accounted for the 1998 CBD 'Malawi Principles' for the Ecosystem Approach adopted by the 4th Conference of the Parties (CoP4) of the CBD (in Bratislava, Slovakia). The 2002 WSSD Plan of Implementation (POI) called on implementation of existing instruments to reach agreed time-bound objectives for fisheries and conservation. Its arbitrary and optimistic milestones proved to be unrealistic, however.

From 2003 to 2012: The biodiversity decade

This period started with the FAO-New Zealand International Conference on Governance and Management of Deep-Sea Fisheries (the Deep Seabed Conference), held in Queenstown, New Zealand. The new UNGA focused on seamounts biodiversity and vulnerable marine ecosystems in areas beyond national jurisdiction (ABNJ), flagging the international focus of the decade. It ended with the Rio + 20 UN Conference on Sustainable Development (Rio de Janeiro, Brazil). The reported marine 'fish' landings oscillated during the decade between 77 and 80 million tonnes.

The Millennium Ecosystem Assessment (MEA, 2005) depicted a negative state of environment and biodiversity. The IUCN Sustainable Use concept based on the 1980 World Conservation Strategy principles and developed since 1994 was finally endorsed in 2000. Collaboration increased between FAO, CBD, United Nations Environment Program (UNEP), CITES and IUCN and some of the most important Environmental Non-Governmental Organizations (ENGOs). Results included: (1) the UN Regular Process for a global assessment and reporting on the state of the Ocean (still ongoing) and the establishment by UNEP and IOC of a database for Global and Regional Assessments of the Marine Environment (GRAMED); (2) signing of an FAO-CITES memorandum of understanding on the listing process (2006); (3) adoption of a legally binding Port States Agreement at FAO to fight against IUU (2009); (4) joint work of FAO, UNEP and CBD, with the participation of the IUCN Fisheries Expert Group, on destructive fishing practices (2009) and the consideration of biodiversity in fisheries assessments (2011), illustrating the will of international agencies to collaborate despite the sensitivity of some of issues; and (5) establishment of the Intergovernmental Platform on Biodiversity and Ecosystem Services (IPBES 2011). CITES attempted again to list Atlantic Bluefin tuna (in 2009) in a renewed effort to introduce international trade controls in fisheries management. The failure of the attempt created tensions between FAO and CITES (cf. Chapter 12). The leadership role of the FAO Code of Conduct for Responsible Fisheries in the incorporation of biodiversity concerns in fishery-related policies recognized by the CBD in 2004 (Decision VII/4) was also recognized by Ramsar in 2005.

Fisheries agencies themselves increased efforts towards biodiversity conservation in the high seas, deep-sea habitats (seamounts, hot vents, cold seep, mineral crusts) and their particular vulnerable species assemblages and ecology. This focus was progressively extended to Vulnerable Marine Ecosystems (VMEs, cf. Chapter 13) in response to increasingly strident calls for a moratorium on deep-sea bottom trawling, at least in ABNJ. UNGA had increasing difficulties in negotiating compromise language in its Sustainable Fisheries resolutions (e.g. resolutions 61/105 and 64/72) and fisheries issues appeared increasingly often in its broader Ocean and Law of the Sea Resolutions.

The Green Growth concept of economic development, adopted at the UN Economic and Social Commission for Asia and the Pacific (ESCAP), is based on the more extensive use of economic incentives and market-based instruments to shift economic activity in directions more compatible with sustainable use. Its relevance to reduce fisheries collateral damage on non-target species and ecosystems is being studied (see also 'From 2000 to 2012') with respect to biodiversity conservation. The increasing role of these economic and financial instruments in the governance toolbox will probably be a major trend over the next decade.

Historical developments in biodiversity conservation

The chapter focuses on intergovernmental processes since 1850; it does not cover conservation strategies of private foundations, national societies, large ENGOs and large aquaria, the influence of which should not be underestimated. Nature conservation obviously started on land where its paradigms were born, developed and tested before expanding on oceans. The following is based on accounts given by Biot et al. (1995), Blaikie and Jeanrenaud (1996), Brandon et al. (1998), Ciriacy-Wantrup (1968), Johannes (2003), Larson et al. (1996), Sand (2001) and Thomas and Middleton (2003).

Following Ciriacy-Wantrup (1968), Evans (1987) and de Klemm and Shine (1993), we will refer in this chapter to two types of conservation related to their rationale defined as follows.

- *A utilitarian or anthropocentric conservation*. This considers that biodiversity needs to be conserved as the source of essential resources, goods and services for present and future human livelihood. Its value is measured by its direct contribution to human wellbeing. This form of conservation view has existed for centuries (Evans, 1987) and rests on a community development ethic. It was initially aimed at wise use (Blaikie and Jeanrenaud, 1996) and, more recently, at sustainable development and use.
- *An aesthetic or ecocentric conservation*. This considers that biodiversity needs to be conserved for its charismatic species, beautiful landscapes and, more recently, its intrinsic value as an indispensable part of ecosystems. It rests on an environmental ethic. It calls for protection (or preservation) from human activities in the name of 'general interest'. Enjoying much public support and private funding, it has been promoted by numerous institutions through a variety of causes and campaigns for specific 'pet' targets (landscapes, habitats or emblematic species) or against specific threats (e.g. deforestation, whaling or destructive fishing).

The evolution of the tensions between these two philosophies has shaped the evolution of governance styles which have evolved differently in different countries and parks at different times. In the Kruger Park management history, Venter (2008) distinguishes five philosophies: establishment (1898–1926); optimization (1902–1980); command and control (1955–1985); transition (1985–1995); and strategic adaptive management (1995–2008). During that evolution, the role of science increased after the mid-1950s and people participation after 1995. At the global level, the average trajectory probably has a large variation. Nonetheless, starting from a neo-Malthusian explanation of environmental degradation, Biot et al. (1995) and Blaikie and Jeanrenaud (1996) distinguished three main conservation governance paradigms (classic, neo-populist and neo-liberal[11]) emerging in sequence but coexisting at times. Adding a traditional pre-industrial predecessor to the three paradigms above, an evolutionary trajectory may be drawn from the initial integration of preservation and use in traditional communities to the exclusion of humans and finally to modern 'sustainable use' with its pillars of equitable distribution of conservation benefits and people's participation.

The timeline (Annex 1) and the literature of the last 5 decades show a turbulent progress towards more humanist perceptions of our relationships with nature. They also show socio-politico-economic divisions between conservation and development, entrenched in their respective legal and institutional frameworks and poorly coordinated (if not conflictual) implementation by specialized agencies. The process is energized by the tensions existing between: (1) conservation and development; (2) international, national and local interests; (3) traditional and industrial development models; (4) developed and developing nations; (5) science and politics; (6) officials and folk; and (7) power relations at all levels (Blaikie and Jeanrenaud, 1996).

Before 1850: Traditional conservation

The period before 1850 could be called pre-industrial, pre-capitalist and colonial. The world had a large diversity of government systems from simple tribal organizations to complex feudal states. Pre-modern fishing communities (e.g. in the South Pacific) developed a traditional conservation ethic, an 'awareness that one can deplete or otherwise damage one's natural resources, coupled with a commitment to reduce or eliminate the problem' (Johannes, 2003; Ruddle, 2008). A utilitarian traditional form of self-conservation was in place to promote wise usage of resources for sustainable livelihoods. In animist societies, emblematic animals were protected for cultural/religious reasons. Both were integrated in community governance. Utilitarian conservation (e.g. in subsistence fisheries) was community-based, largely self-governed (Biot et al., 1995) and used closed areas and seasons, gear regulations and some forms of limited entry and exclusive use rights, enforced through social pressure, rituals and even violent conflicts under the leadership of traditional authorities. In these cultures, regulation of fishing and what today we call 'conservation' co-existed in a mutually beneficial concept of 'wise use'. The communities (and guilds of users) directly exerted the authority and resolved conflicts themselves.[12] The conservation performance during that phase, sometimes presented as a model to revive, has not been always good (cf. Bolster, 2012),[13] and worked better at sea than inland (Johannes, 2003).

Central governments had a non-systematic role, for example: (1) royal decrees protected special areas in India more than 2000 years ago; (2) protected areas for mammals, birds, fish and forests were established by the Emperor of India in 252 BC; (3) a nature reserve was established in the island of Sumatra by the King of Srivijaya in 684 AD; and (4) private hunting reserves were created in Europe for centuries by kings and lords, such as the New Forest created by William the Conqueror in 1079 (D. Laffoley, personal commication, 2012). These authorities closed and enforced reserves of charismatic or vulnerable species (e.g. sturgeons, swans and whales) for themselves and sometimes allocated some use rights to their people.

From 1850 to 1970–80: Increasing role of the State

This period corresponds to the classic governance paradigm mentioned in the introductory section, subdivided into time periods. The population of Western industrial countries grew from 117 million people in the early 19th century to 2.4 billion in 1960 (Ciriacy-Wantrup, 1968).

The conservation rift

By the middle of the 19th century, the fragile balance between use and conservation was profoundly changed in the Western economies by the consequences of demographic explosion, industrialization and emergence of the capitalistic market economy (Bolster, 2012; Ciriacy-Wantrup, 1968; Johannes, 2003; Ruddle, 2008). Colonization spread that change worldwide, affecting developing countries even after their decolonization. The massive industrial development led to burgeoning concerns for the environment and the progressive distinction between aesthetic/ecocentric and utilitarian/anthropocentric conservations (Ciriacy-Wantrup, 1968; de Kemm and Shine, 1993; Evans, 1987). In the first half of the 20th century, utilitarian conservation gave birth to the management of wildlife, fisheries, forestry and agriculture (Ciriacy-Wantrup, 1968), building on common concepts.[14]

Aesthetic conservation focused on endangered and emblematic species (primates, large mammals, birds and, later on, marine mammals, turtles, sharks, etc.) as well as remarkable sceneries and landscapes. The literature establishes the origin of this modern conservation movement in the USA in the 1870s with the establishment of the first national forestry parks. The term 'conservation' was apparently adopted in 1907 in the US (Ciriacy-Wantrup, 1968, p. 27) and applied to natural and non-renewable resources. The Yosemite Park (established in 1865) and the Yellowstone Park (first National Park established in 1872) were established as 'pleasuring ground for the benefit and enjoyment of the people' to maintain a link between increasingly industrialized and urbanized citizens and Nature, rather than for the conservation of Nature *per se*. The Yellowstone Park became an example of conservation by exclusion of people, forcibly if necessary,[15] and other examples exist in India[16] and South Africa. The US parks initiative was followed by Canada in 1886, Australia in 1886, New Zealand in 1894, Sweden in 1909, Switzerland in 1914, Italy in 1923 and Japan in 1934 (Raffin, 2005). In parallel, a movement for the protection of selected species developed between 1860 and 1900 for birds (1868), seals (1883) and salmon (1885) (Raffin, 2005). Since then, the control of areas (parks, wetlands and marine protected areas or MPAs) and the protection of individual species (e.g. in IUCN and CITES listings) have remained the two pillars of conservation until the introduction of market-based instruments (next main phase).

Conservation by exclusion

The concern for nature intensified after the Second World War and the conservation movement progressed towards its modern more cohesive shape, stimulated by the negative or perceived effects of large-scale agriculture, forestry and fisheries. People (as opposed to just 'industry') started being considered as the main threat to biodiversity. The movement developed first in rich countries and expanded to their colonies (Blaikie and Jeanrenaud, 1996). Conservation was seen as being in conflict with human uses of resources, disconnecting conservation from livelihoods and human welfare (Brown, 2002; Salafsky and Wollenberg,[17] 2000). Conservation was imposed or supported by the State,[18] which also massively supported development (with questionable coherence). More areas were strictly protected with top-down 'command and control' management (Venter et al., 2008) as in fisheries, and with different degrees of intervention in different countries (Uekoetter, 2006) including the eviction of people from traditional dwellings and livelihoods when necessary. The instruments of what was later referred to as 'fortress conservation' or conservation by 'fines and fences' (Biot et al., 1995) spread, turning traditional foresters, hunters and fishers into poachers by abolition of their traditional rights and provoking occasional bursts of violent reaction from affected populations. In the 1950s and 1960s, colonial powers applied these policies to their colonies where they persisted after independence, supported by international so-called development strategies.

The emergence of marine conservation

Marine conservation emerged slowly in the late 1960s and early 1970s. For example, Ratcliffe's (1977) selection of UK sites of national importance for nature conservation did not include any marine habitats on rocky shores or below the low water mark. Marine conservation considered protected areas as the instrument of choice for most needs. Strongly influenced by the forestry parks paradigm, marine parks and MPAs developed in coastal marine systems in the mid-1970s (cf. Chapter 8). The Great Barrier Reef, which extends 250 km offshore in its broadest part, was an exception.

Protection of marine species followed the terrestrial model, starting with iconic vulnerable species (e.g. whales, dolphins and turtles). The desire to identify threatened species and protect them from extinction led to many separate initiatives; the IUCN emerged as the global leader in species conservation. Its Red List of threatened and endangered species began with terrestrial species and only slowly added aquatic species, with an initial domination of freshwater fish threatened by habitat loss and only later harvested marine fish (see Chapter 12). The command and control approach of a listing based on advice by conservation experts followed by stringent and mandatory (often court-ordered) prohibitions on harm, implemented by central governments, followed a the 'fortress conservation' approach, the social and economic costs of which were also seen as prohibitive.

From 1970–80 to 2000: Increasing the role of the actors

By the 1970s, conservation failures were conspicuous and top-down policies were progressively complemented by more participative and consensual policies. These addressed both protection and livelihood concerns, with some conflicts between conservation institutions such as IUCN and WWF and even within WWF itself (Blaikie and Jeanrenaud, 1996, p. 25).

The creation of UNEP in 1972, the adoption of UNCLOS in 1982, the cross-sectoral summits (UNCHE, UNCED, WSSD and Rio + 20) and the establishment of the CBD in 1992 provided the legal, institutional and policy frames needed to boost marine biodiversity conservation. During this period, Venter et al. (2008) identify three successive and partially overlapping governance styles in the Kruger Park: a conventional command and control phase (1955–1985) with no participation at all; a transition phase (1985–1995) towards more ecosystem approaches still with no participation; and a strategic adaptive management phase (1995–2008) recognizing systems complexity and facilitating active participation.

In terrestrial systems, the shift from exclusion to participation was apparently prompted by the high failure rate of exclusion and an emerging vision of parks and surroundings as complex socio-ecological systems. The spectre of global warming strengthened the awareness of relatedness between natural and human subsystems. The relations between the utilitarian and aesthetic streams of conservation governance remained tumultuous (Tickell, 1995, cited by Blaikie and Jeanrenaud, 1996). By the 1990s, there was little support left for fortress conservation (Blaikie and Jeanrenaud, 1996). The neo-populist approach that emerged recognized the social and economic roots of biodiversity. It involved and empowered communities and indigenous populations, mobilizing their traditions including use rights (Biot et al., 1995; Blaikie and Jeanrenaud, 1996). It rested on local ownership and control, self-determination and rights of representation. It was facilitated by greater involvement of social sciences (since the 1980s), a better understanding of common pool resources management (Berkes, 1989; Ostrom, 1990) and recognition of the relation between conservation and poverty.

The approach was advocated in international policy documents such as Caring for the Earth (IUCN/UNEP/WWF in 1991), the Global Biodiversity Strategy (WRI/IUCN/UNEP 1992) and UNCED Agenda 21. Integrated Conservation and Development Projects (ICDPs) (cf. Chapter 2) were developed in the 1980s to generate alternative revenues, e.g. in WWF and in UNESCO Biosphere reserves and Man and Biosphere Programmes. By the 1990s, however, ICDPs were criticized for: (1) unacceptable tradeoffs; (2) a naïve view of the 'community'; (3) difficulties of effective participation; (4) lack of direct linkage between the substitution activities[19] and conservation; and (5) the excessively intrusive role of large NGOs (Blaikie and Jeanrenaud, 1996, p. 26; Brown, 2002; Salafsky and Wollenberg, 2000; Weigel et al., 2007; cf. Chapter 25).

The development of modern marine conservation

Space-based marine conservation governance (see Chapter 8) first developed its financial and public support strategies and instruments for terrestrial issues; it started covering coastal and marine areas in the 1970s, where it developed together with the emerging concept of sustainable development. Its development was boosted by the Stockholm Declaration in 1972, the World Conservation Strategy in 1980, UNCLOS in 1982, the Brundtland report in 1987, the IUCN Sustainable Use Initiative 1980–2000, the establishment of the CBD in 1992 and its 1995 Djakarta mandate, as well as the large UN cross-sectoral summits (UNCHE, UNCED, WSSD and Rio + 20 from 1972 to 2012). The total area covered by MPAs and the number of publications concerning them are probably good indicators of the development of modern marine conservation (Figure 1.2).

In Australia, the first offshore marine parks were established in 1937 and 1974 (R. Kenchington, personal communication). The MPA movement started in Europe in the late 1960s on the wake of the Torrey Canyon oil spill. The first Conference on Marine Parks and Reserves in Tokyo in 1975 confirmed its nascent dynamism. The emblematic Great Barrier Reef Marine Park established that same year in Australia served as a flagship. By the 1990s, virtually every coastal country had implemented some form of MPA, a high proportion of them unfortunately unmanaged 'paper parks' (Kelleher et al., 1995). Nonetheless, the movement gained momentum in science, funding and policy following UNCED and WSSD (Willis et al., 2003), and expanded from coastal areas to the high seas and deep seas thereafter (Toropova et al., 2010).

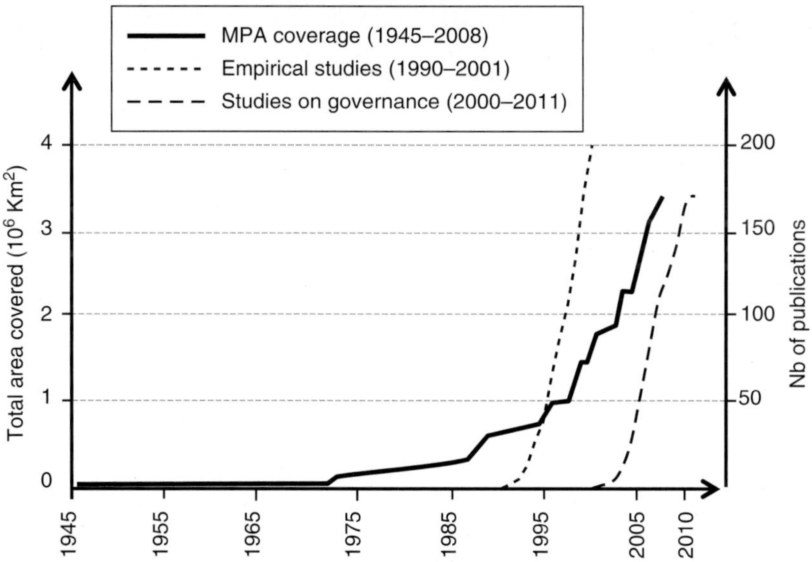

Figure 1.2 Evolution of the MPA coverage (1945–2008) in the ocean and in the literature. Area covered from IUCN and UNEP-WCMC (2009). Empirical studies from Willis et al. (2003). Studies on governance (Google Scholar search in French, English and Spanish, September 2011).

This growing MPA coverage was accompanied by the emergence of good governance principles (PNUD, 1997) and the growing requirement of the CBD and IUCN for management effectiveness in MPAs. The period 1990–2010 was clearly one of rapid expansion of marine biodiversity conservation and particularly of its science-based advocacy and governance (cf. Chapter 9).

From 2000 to 2012: Growing role of incentives and the market

In recent years, there has been growing focus on the creation and utilization of economic incentives and the tapping into market forces to shift human behaviour in conservation directions (see discussions by Blaikie and Jeanrenaud, 1996 and Brown, 2002). This has led to new governance approaches that seek to simultaneously address conservation failures (e.g. in MPAs and in ICDPs) and institutional, market and policy failures, and specifically to add value to biodiversity as an incentive to conserve it in the long term. Livelihoods drive conservation instead of being simply compatible with it. The approach has the potential to integrate fisheries and biodiversity and is an application of the 'green economy' paradigm (UNEP, 2011) and of the Economics of the Ecosystems and Biodiversity (Kumar, 2012; TEEB, 2011). The 2012 IUCN Resolution WCC-2012-Res-110 on biodiversity offsets and related compensatory approaches underlined the significantly increased interest of conservation institutions in the approach.

This approach calls for the creation of appropriate incentives while eliminating perverse incentives, as well as internalization of environmental costs. In principle, economic and market-based incentives offer policymakers new and cheaper ways to reach conservation objectives, using market forces to complement traditional measures, that is, generating revenue to support management (Bräuer et al., 2006). Options include taxes, fees, payments for ecosystem services, subsidies, grants, tradable use rights, eco-labelling, financial mechanisms and liability and compensation payments (offsets). These approaches may be applied to habitat, ecosystem and species conservation to complement impact reduction measures (e.g. closed areas or seasons, technological adaptation) and to increase compliance. Their cost must be passed to the consumer, internalizing the cost of biodiversity damage and making it explicit.

The approach creates direct linkages and interdependence between conservation and livelihoods (Salafsky and Wollenberg, 2000) while re-focusing the State's role on norm and target setting, together with oversight. This requires that difficult challenges be addressed, for example: the utilitarian appropriation of nature that is implied by these measures; the estimation of benefit/cost ratios and required compensation (Boncoeur et al., 2011); the effective generation and capture of benefits; and the major challenge of achieving equity. Implementation will certainly require capacity-building in innovative economic and banking domains (Bräuer et al., 2006).

Return to the barriers

In terrestrial conservation a recent trend called 'return to the barriers' (recalling the past 'fortress conservation' concept) argues that failing conservation should return to the values of aesthetic beauty and intrinsic value (Büscher and Diets, 2005) and 'that protected areas are the last bastions of biodiversity and wildlife conservation and hence should be protected at all costs and focus on preservation, the strictest form of conservation' (Terborgh and Van Shaik, 2002). It relates to the perceived failure of many ICDPs and participation processes (Brockington, 2003; Butz et al., 1991; Cernea and Schmidt-Scholtau, 2003). It is too soon to evaluate the trend, but the controversy is symptomatic of the century-old tension in the Humans–Nature relation.

It looms as a strong warning for the oceans in case of failure to successfully integrate fisheries and biodiversity conservation governance.

Conclusions

The history of the two streams is rich in international activity, reflecting growing concerns and efforts to improve performance through modification and diversification of the two governance paradigms. The comparison of the two timelines in Annex 1 and of the historical changes reminds us that the governance streams for fishery management and biodiversity conservation have important common roots in pre-capitalistic pre-colonial communities. Historically joined in traditional community governance (with a possible bias towards a utilitarian vision), the two streams diverged during the 19th century under growing impacts of industrialization, as distinct utilitarian and aesthetic branches of conservation with tumultuous relationships (cf. Chapter 2). Forms of management involving organized corporations and the State emerged much earlier on land (during the 13–15th centuries) than in the ocean (middle 20th century). From 1970 onwards however, the UN cross-sectoral summits established bridges between the two streams, accelerating their conceptual convergence. This increase of high-level cross-sectoral events (after 1970) has funnelled the two governance streams into closer and closer evolution through phases of indifference, conflict and apparent convergence in the last four decades. The recent focus on economic incentives and market-based approaches in both fisheries and biodiversity conservation, the efficiency of which as a win-win solution still needs to be tested, may significantly accelerate the present convergence and improve the scope for real integration.

Notes

1. Exterminations of large mammals in North America, for example, occurred as early as 9500 years BC (McGoodwin, 1984 in Troadec, 1989, p. 43). A North Sea resource of turbot (*Psetta maxima*) was collapsed in only 17 years (1823–1840) using simple technology (Alward, 1932 in Cushing, 1968, p. 91). Bolster (2012) describes early extermination of seabirds and walrus by the Vikings in the 9th century.
2. T.H. Huxley influenced the Royal Commissions on Sea Fisheries of the United Kingdom (1863–1865) and the subsequent Sea Fisheries Act of 1868 which repealed more than 50 fisheries acts. He maintained this position 20 years later at the 1882 Fisheries Exhibition in London, recommending 'that unrestricted freedom of fishing be permitted hereafter'. He stressed that 'with the present forms of fishing', the stocks such as 'cod, herring and mackerel were inexhaustible' (Roswadowski, 2002). Forgetting the underlined caveat, he has been repeatedly accused of having unleashed overfishing. Other scientists such as E.R. Lancaster disputed (in vain) his conclusions, proving by a decade-long experiment starting in 1885 in Scotland that harvesting had serious impacts and resources needed protection (Lankester 1884; http://en.wikipedia.org/wiki/Marine_Biological_Association_of_the_United_Kingdom).
3. http://www.britannica.com/EBchecked/topic/641450/whaling/65551/Regulation
4. Warming's work was eventually translated into English by Andersen (1983).
5. This concept is different from the 'optimum yield' proposed later by Roedel (1985) that combines ecological and socio-economic criteria.
6. Refer to the 1937 London Convention for the Regulation of Meshes for Fishing Nets and the Size Limits for Fish (the 1st London Overfishing Convention).
7. Limitations of total allowable catch (TACs) were introduced.
8. The FAO COFI subcommittee on fish trade was established in 1985, signalling the growing importance of fish trade.

9. The proposal was withdrawn but resurfaced in 2010.
10. Agreement to Promote Compliance with the International Conservation and Management measures by Fishing Vessels in the High Seas (the FAO Compliance Agreement).
11. Or 'new conservation' (Brown, 2002)
12. In case of failure, higher-level authorities were called in.
13. Bolster (2012) illustrates very well the way in which the conservative concern of traditional fishers for sustainability of changing practices and local/traditional resource depletion is constantly overwhelmed by technology development and geographical and economic fisheries expansion.
14. T.H. Huxley, for example, stressed the similarities between fisheries and agriculture to justify his views on management.
15. In the USA in 1863, the establishment of the Yellowstone Park (sometimes considered as a model of conservation by exclusion) led to forced evictions and the death of 300 Shoshoni Native Americans in a single day (Walsh, 2011).
16. In India, 7000 people were evicted over a 5 month period to create a lion protection area (Blaikie and Jeanrenaud, 1996, p. 58).
17. According to Salfsky and Wollenberg (2000), 60% of all IUCN MPAs do not have any development objective.
18. It can be assumed that, under Kings and Lords' ruling, something very similar already existed in many countries.
19. Activities of substitution replaced activities considered as producing unacceptable impacts with others not affecting the reserve.

References

Alward, G.L. (1932) *The Sea Fisheries of Great Britain and Ireland*. Albert Gait, London.

Andersen, P. (1983) On rent of fishing grounds: a translation of Jens Warming 1911 article with an introduction. *History of Political Economy* **15**(3), 381–396.

Anderson, E.D. (1998) The history of fisheries management and scientific advice – the ICNAF/NAFO history from the end of World War II to the present. *Journal of Northwest Atlantic Fishery Science* **23**, 75–94.

Atkinson, C.E. (1988) Fisheries management: An historical overview. In: *Marine Fisheries History* (ed. W.L. Hobart) pp. 111–123. *Marine Fisheries Review 50th Anniversary Issue* **50**(4), 244.

Berkes, F. (ed.) (1989) *Common-Property Resources: Ecology and Community-Based Sustainable Development*. Belhaven Press, London.

Biot, Y., Blaikie, P.M., Jackson, C. and Palmer-Jones, R. (1995) *Rethinking Research on Land Degradation in Developing Countries*. World Bank Discussion Paper No. 289. The World Bank, Washington, DC.

Blaikie, P., and Jeanrenaud, S. (1996) Biodiversity and human welfare. UN Research Institute for Social Development. UN RISD Discussion Paper 72, pp. 81.

Bolster, J.W. (2012) *The Mortal Sea. Fishing the Atlantic in the Age of Sail*. Harvard University Press, Harvard, 378 pp.

Boncoeur, J., Frédérique, A. and Thébaud, O. (2011) Bioeconomy–bioeconomic analysis of marine protected area fisheries effects. In: *Marine Protected Areas. A Multidisciplinary Approach* (eds J. Claudet and J. Cambridge), pp. 190–225. Cambridge University Press, Cambridge.

Brandon, K., Redford, K.H. and Sanderson, S.E. (1998) *Parks in Peril: People, Politics and Protected Areas*. Island Press, Washington, DC.

Bräuer, I., Müssner, R., Marsden, K., Oosterhuis, F., Rayment, M., Miller, C. and Dodoková, A. (2006) *The Use of Market Incentives to Preserve Biodiversity*. Final Report. Project ENV.G.1/FRA/2004/0081. Ecologic European Commission (DGE). 51 pp. http://ec.europa.eu/environment/enveco/biodiversity/pdf/mbi.pdf (accessed 12 February 2014)

Brockington, D. (2003) Injustice and conservation – is "local support" necessary for sustainable protected areas? *Policy Matters* **12**, 22–30.

Brown, K. (2002) Innovation for conservation and development. *The Geographical Journal* **168**(1), 6–17.

Butz, D., Lonergan, S. and Smit, B. (1991) Why international development neglects indigenous social reality. *Canadian Journal of Development Studies* **12**(1), 143–58.

Caddy, J.F. and Cochrane, K.L. (2001) A review of fisheries management past and present and some future perspectives for the third millennium. *Ocean & Coastal Management* **44**(9–10), 653–682.

Cernea M.M. and Schmidt-Soltau, K. (2003) The end of forcible displacements? Conservation must not impoverish people. *Policy Matters* **12**, 42–50.

Ciriacy-Wantrup, S.V. (1968) Resource Conservation: Economics and Politics, 3rd edn. University of California Press.

Cushing, D.H. (1968) *Fisheries Biology. A Study in Populations Dynamics*. University of Wisconsin Press, Madison, Milwaukee and London.

de Klemm, C. and Shine, C. (1993) Biological Diversity Conservation and the Law. Legal Mechanisms for Conserving Species and Ecosystems. IUCN, Gland, Switzerland and Cambridge, UK. IUCN Environmental Policy and Law Paper **29**, xix + 292 pp.

Evans, D. (1987) A History of Nature Conservation in Britain. Routledge, London.

FAO (1981) Marine Fisheries in the New Era of National Jurisdiction. In: *FAO: The State of Food and Agriculture 1980 (SOFA)*, pp. 83–129. Rome, FAO.

FAO (1984) Report of the FAO World Conference on fisheries management and development. Rome, 27 June–6 July 1984. FAO, Rome.

FAO (1996) The State of World Fisheries and Aquaculture. FAO, Rome.

Garcia, S.M. (1989) La recherche halieutique et l'aménagement: grandeur et servitude d'une symbiose. In: *L'homme et les Systèmes Halieutiques et Aquacoles Marins* (ed. J.P. Troadec). Masson et Cie, Paris.

Garcia, S.M. (1992) Ocean Fisheries Management. The FAO Programme. In: *Ocean Management in Global Change* (ed. P. Fabbri), pp. 381–418. Elsevier Applied Science.

Garcia, S.M. and Newton, C.H. (1994) Responsible fisheries: an overview of FAO policy developments (1945–1994). *Marine Pollution Bulletin* **29**(6–12), 528–536.

Garcia, S.M. and Newton, C. (1997) Current Situation, trends and prospects in world capture fisheries. In: *Global Trends: Fisheries Management* (eds H.K. Pikitch, D.D. Huppert and M.P. Sissenwine), pp. 3–27. *Proceedings of the American Fisheries Society Symposium* **20**.

Garcia, S.M. and Grainger, R. (2005) Gloom and doom? The future of marine capture fisheries. *Philosophical Transactions of Royal Society B* **360**, 21–46.

Garstang, W. (1900) The impoverishment of the sea. A critical summary of the experimental and statistical evidence bearing upon the alleged depletion of the trawling grounds. *Journal of the Marine Biological Association of the United Kingdom* **6**(1), 1–69.

Graham, G.M. (1935) Modern theory of exploiting a fishery and its application to the North Sea Trawling. *Journal du Conseil/Conseil Permanent International pour l'Exploration de la Mer* **10**, 264–274.

Gulland, J.A. (1971) *The Fish Resources of the Oceans*. Fishing News (books) Ltd, Sussex.

Gulland, J.A. and Carroz, J.E. (1968) Management of fishery resources. *Advances in Marine Biology* **6**, 1–71.

Holm, P. (2010) Fishing down the North Sea. In: *Beiträge zum Göttinger Umwelthistorischen Kolloquium 2007–2009* (ed. B. Herrmann). Göttingen, Germany, Universitätsverlag Göttingen.

Holm, P., Marboe, H.A., Poulson, B. and MacKenzie, B.R. (2010) Marine animal populations: a new look back in time. In: *Life in the World's Oceans. Diversity, Distribution and Abundance* (ed. A. McIntyre), pp. 3–23. Wiley-Blackwell, Oxford.

IUCN and UNEP-WCMC (2009) The world database on protected areas (WDPA). January 2009. UNEP-WCMC, UK. Available at http://www.wdpa.org/AnnualRelease.aspx (accessed March 2014).

Johannes, R.E. (2003) Fishers' knowledge and management: differing fundamentals in artisanal and industrial fisheries. In: *Putting Fishers' Knowledge to Work: Conference Proceedings* (eds N. Haggan, C. Brignall and L. Wood), pp. 15–19. Vancouver, Fisheries Centre, University of British Columbia.

Kelleher, G., Bleakley, C. and Wells, S. (1995) *A Global Representative System of Marine Protected Areas*. Volumes I to IV. The World Bank and the World Conservation Union, Washington, DC.

Kittinger, J.N., Pandolfi, J.M., Blodgett, J.H., Hunt, T.L., Jiang, H., Maly, K., McClenachan, L.E., Schultz, J.K. and Wilcox, B.A. (2011) Historical reconstruction reveals recovery in Hawaian coral reefs. *PlosOne* **6**(10), e25460.

Kumar, P. (ed.) (2012) *The Economics of Ecosystems and Biodiversity (TEEB): Ecological and Economic Foundations*. UNEP, New York and Earthscan from Routledge.

Lankester, E.R. (1884) The scientific results of the International Fisheries Exhibition, London 1883. *Literature* **4**, 443–446.

Lankester, E.R. (1890) The scientific results of the International Fisheries Exhibition, London 1883. In: *The Advancement of Science. Occasional Essays and Addresses* (ed. E.R. Lankester), pp. 193–223. Macmillan, London.

Larson, P., Freudenberger, M. and Wyckoff-Baird, B. (1996) Lessons from the field, a review of World Wildlife Fund's experience with integrated conservation and development projects 1985 ± 1996. Final Draft Report. World Wildlife Fund, Washington, DC.

McGoodwin, J.R. (1984) Some examples of self-regulatory mechanisms in unmanaged fisheries. Contribution to the Expert Consultation on the regulation of fishing effort (fishing mortality). FAO Fisheries Technical Paper, No. 289, suppl. 2: 46–61. FAO, Rome.

McLoughlin, R., Rayns, N., Smith, A.D.M., Johnson, D., Hobday, A. and Smith, T. (2008) Implementing the ecosystem approach in Australian Commonwealth (federally)-managed fisheries. In: *The Ecosystem Approach to Fisheries* (eds G. Bianchi and H.R. Skjoldal), pp. 293–300. CABI, Wallingford, UK and FAO, Rome, Italy.

MEA (2005) *Ecosystems and Human Well-Being: Synthesis*. Millennium Ecosystem Assessment. Island Press, Washington, DC.

Miller, M.L. (1989) L'organisation sociale des systèmes d'aménagement de la pêche. In: *L'homme et les Ressources Halieutiques. Essai sur l'usage d'une Ressource Commune Renouvelable* (ed. J.P. Troadec), pp. 557–583. IFREMER, Paris.

Ostrom, E. (1990) *Governing the Commons: The Evolution of Institutions for Collective Action*. Cambridge University Press, Cambridge.

PNUD (1997) La gouvernance en faveur du développement humain durable. Document de politique générale du PNUD. PNUD, New York.

Raffin, J.P. (2005) De la protection de la nature ą la gouvernance de la biodiversité. *Ecologie & Politique* **30**, 97–109.

Ratcliffe, D.A. (ed.) (1977) *A Nature Conservation Review. The Selection of Biological Sites of National Importance to Nature Conservation in Britain*. Cambridge, Cambridge University Press, for Nature Conservancy Council & Natural Environment Research Council.

Roberts, C. (2007) *The Unnatural History of the Sea*. Island Press, Washington DC.

Roedel, P.M. (1985) Optimum sustainable yield as a concept in fisheries management. *Special Publications of American Fisheries Society* **9**, 59.

Roswadowski, H.M. (2002) *The Sea Knows No Boundaries: A Century of Marine Science under ICES*. ICES, University of Washington Press, Copenhagen, Seattle, London.

Royce, W.F. (1988) Centennial Lecture IV: The historical development of fishery science and management. Marine Fisheries Review. 50th Anniversary Issue. *Marine Fisheries History* **50**(4), 30–39.

Ruddle, K. (2008) Introduction to the collected works of R.E. Johannes, publications on marine traditional knowledge and management. *SPC Traditional Marine Resource Management and Knowledge Information Bulletin* **23**, 13–24.

Saetersdal, G. (1992) Fishery resources and their environment, management and development. Paper presented at the International Conference on Responsible Fishing, Cancun, Mexico, 6–8 May 1992. FAO.

Salafsky, N. and Wollenberg, E. (2000) Linking livelihoods and conservation: a conceptual framework and scale for assessing the integration of human needs and biodiversity. *World Development* **28**(8), 1421–1438.

Sand, P.H. (2001) A century of green lessons: The contribution of nature conservation regimes to global environmental governance. *International Environmental Agreements: Politics, Law and Economics* **1**, 33–72. Revised version in *Managing Global Issues: Lessons Learned* (eds P.J. Simmons and C. de JongeOudraa), pp. 281–309. Carnegie Endowment for International Peace, Washington DC.

Sherman, K. (2008) The large marine ecosystem approach to marine resources assessment and management. In: *The Ecosystem Approach to Fisheries* (eds G. Bianchi and H.R. Skjoldal), pp. 47–75. CABI, Wallingford, UK and FAO, Rome, Italy.

TEEB (2011) *The Economics of Ecosystems and Biodiversity in National and International Policy-making* (ed. P. ten Brink). Earthscan, London and Washington.

Terborgh, J. and Van Shaik, C. (2002) Why the world needs parks. In: *Making Parks Work. Strategies for Preserving Tropical Nature* (eds J. Terborgh, C. Van Shaik, L. Davenport and M. Rao). Island Press, Washington DC.

Thomas, L. and Middleton, J. (2003) Guidelines for Management Planning of Protected Areas. No. 10. IUCN, Gland.

Tickell, O. (1995) Animal passions. *The Guardian*, 1 February, p. 5.

Toropova, C., Meliane, I., Laffoley, D., Matthews, E. and Spalding, M. (eds) (2010) *Global Ocean Protection: Present Status and Future Possibilities*. Agence des aires marines protégées, Brest, France; Gland, Switzerland; IUCN WCPA, Washington DC and New York; UNEP-WCMC, Cambridge, UK; TNC, Arlington, USA; UNU, Tokyo, Japan; WCS, New York.

Troadec, J.P. (1989) *L'homme et les Ressources Halieutiques. Essai sur l'Usage d'une Ressource Commune Renouvelable*. IFREMER, Paris.

Typhaigne de la Roche, G.F. (1760) *Essai sur l'Histoire Economique sur les Mers Occidentales de France*. C.J.B. Bauche, Paris.

Uekoetter, F. (2006) *The Green and Brown. A History of Conservation in Nazi Germany*. Cambridge University Press, Cambridge.

UNEP (2011) Restoring the natural foundation to sustain a Green Economy: a century-long journey for ecosystem management. UNEP Policy Series, Policy Brief 6.

Venter, F., Naiman, R., Biggs, H. and Plenaar, D. (2008) The evolution of conservation management philosophy: science, environmental change and social adjustments in Kruger National Park. *Ecosystems* **11**(1), 173–192.

Walsh, B. (2011) Going Green. Winning the Conservation War: How to Manage the World We're Stuck . *Time Science & Space*, December 20. Available at http://content.time.com/time/health/article/0,8599,2102832,00.html (accessed March 2014).

Warming, J. (1911) Vor grundrenteaf fiskgrunde. *Nazional Okonmiske Tidsskrift* **49**, 499–505 (translated by Andersen 1983).

Weigel, J.Y., Feral, F. and Cazalet, B. (2007) Les Aires Maritimes Protégées d'Afrique de L'ouest. Gouvernance et Politiques Publiques. Presses Universitaires de Perpignan (PUP).

Willis, T.J., Miller, R.B., Babcock, R.C. and Tolimieri, N. (2003) Burdens of evidence and the benefits of marine reserves: puting Decartes before des horse? *Environmental Conservation* **30**(2), 97–103.

WRI/IUCN/UNEP (1992) Global biodiversity strategy. Guidelines for action to save, study, and use Earth's biotic wealth sustainably and equitably. WRI, IUCN and UNEP in consultation with FAO and UNESCO. 244 pp. http://pdf.wri.org/globalbiodiversitystrategy_bw.pdf (accessed 12 February 2014).

Chapter 2

Governance of marine fisheries and biodiversity conservation: Convergence or coevolution?

S.M. Garcia[1], J. Rice[1,2] and A. Charles[1,3]

[1]IUCN Commission on Ecosystem Management (IUCN/CEM/FEG), Gland, Switzerland
[2]Fisheries and Oceans Canada, Ottawa, Canada
[3]Sobey School of Business and the Environmental Science Department, Saint Mary's University, Halifax, Canada

Abstract: The governance of fisheries and of biodiversity conservation are themselves complex socio-ecological systems that have evolved with minimal explicit intention to come closer to each other. An analysis of the two governance streams, their parallel evolution and that of their component strands shows incontrovertible signs of increasing overlap and similarities, however. It is argued that this might be the result of: (1) a common origin in utilitarian conservation; (2) convergence 'forced' by external factors of political and economic nature; and (3) coevolution through direct interactive adaptation to lessons learned. These factors and processes most probably operate at all scales and on all the themes addressed in this book, but their relative importance appears to be strongly case-specific. This suggests that no single model would probably correctly capture, explain and forecast the process of convergence and its outcomes.

Keywords: governance; adaptation; convergence; coevolution; fisheries; biodiversity; drivers

Introduction

Over time, fisheries and biodiversity governance streams have gradually adapted to changing societal demands with directions depending on contextual pressure for change and their intrinsic capacity to respond. In principle, independent adaptation processes could have led to increasing similarities or differences.

Chapter 1 identified a series of key political, legal and institutional international events that, over the course of centuries, marked the gradual changes of the two interacting streams of governance (Annex 1). Because the systems of use of natural resources are complex socio-ecological systems (*sensu* Folke et al., 2000), different properties may emerge at different times from the interaction between components. Being both a driver and an emergent property of these systems, it would not have been surprising for *governance* to evolve differently in the two streams. Notwithstanding, and despite remaining tensions, striking similarities are observed in these evolutions that can hardly be coincidences. We assume that they reflect either:

(1) *Convergence*, which occurs when two or more independent adaptation processes unintentionally produce similar responses to common 'forcing' contextual factors (e.g. overarching policies, demography, markets); or
(2) *Coevolution*, which occurs when two or more interdependent adaptation processes intentionally (albeit sometimes reluctantly) adapt to each other in response to their direct interaction (cooperation or competition).

Governance of Marine Fisheries and Biodiversity Conservation: Interaction and Coevolution, First Edition.
Edited by Serge M. Garcia, Jake Rice and Anthony Charles.

A closer analysis of Annex 1, the published literature and the legal and institutional instruments adopted by States (cf. Chapters 7 and 10–12) suggests that the overall trajectories of the two streams are complex aggregates of the trajectories of their key components (referred to as 'strands'). Some of these appear rather specific to one of the streams (e.g. the fisheries management paradigm shift), while others appear to be running in parallel in both streams (e.g. the shift to market-based instruments).

In this chapter, we first describe very succinctly a selection of stream-specific strands in fisheries and biodiversity conservation before identifying 'parallel' strands identifiable in both streams. In conclusion, we analyse the similarities observed, and examine their origin in terms of convergence or coevolution of the two streams towards sustainability.

While the focus of this chapter is on conservation in the ocean, some strands begin with a focus on terrestrial systems where most, if not all, of the concepts used in biodiversity conservation governance originated.

Selected strands in fishery governance

From industrial growth to responsible fisheries

The main shift in fisheries policies has been from those characterized by central planning, open access to natural resources, development subsidies and search for social peace in the 1950s to those characterized by more local planning, user rights, reduction/elimination of subsidies, increased financial efficiency and reduction of environmental impact in the 2000s (Rey-Valette and Cunningham, 2004). Generally, developed countries have moved further on such shifts than developing countries, where some proposed changes raise significant social and/or capacity issues. The first policies, under the post-Second World War development paradigm that accompanied the Marshal Plan implementation,[1] focused on mechanization and industrialization with little consideration for collateral environmental impact and artisanal fishing. Fishery policies started shifting towards sustainable development after the 1972 UN Conference on the Human Environment (UNCHE), and to more environmentally responsible fisheries after the 1992 UN Conference on Environment and Development (UNCED) and the adoption of the FAO Code of Conduct for Responsible Fisheries in 1995. In the process, the conventional 'technical' fishery management measures were progressively complemented by measures to limit harvest, control effort and reduce fishing capacity and ecological impact. The concerns regarding climate change and socio-economic impacts add complexity to a set of goals (including economic viability, food security, sustainable livelihoods and reduction of environmental impact) which are being operationalized at different speeds in different places.

From technical measures to rights, incentives and spatial measures areas

Overfishing has been haunting fishery management since its first mention in 1854 by de Wick (Roswadowski, 2002). After the First and Second World War in the first FAO Technical Meeting on Fisheries in 1946 and by the London Conference on Overfishing in 1946, as well as during the fisheries expansion of the 1960s and 1970s, overfishing was regularly emphasized. It was in the background of the third United Nations Conference on the Law of the Sea (UNCLOS III) process (1973–1982) and has been regularly highlighted in the media since. The central goal of fishery management has been to come as close as possible to Maximum Sustainable Yield (MSY) while avoiding overfishing. This strategic goal was pursued through a mix of technical measures (mesh sizes, minimum landing sizes), catch limitations (total allowable catches or TACs, quotas), effort limitations (limited entry, effort quotas), spatial measures (e.g. closed areas) and controls on fishing capacity, usually developed from the top down and applying only to target resources.

In recent years, these measures have been progressively accompanied by new approaches including: (1) adoption of good governance principles (PNUD, 1997), broadly agreed in the late 1990s and referred to in most agreements and guidelines, albeit with slow implementation; (2) trade-related controls (Convention on International Trade on Endangered Species or CITES, catch certification); (3) incentives (use rights, eco-labels and other market-based measures); and (4) measures from conservation biology (see 'Selected strands in conservation governance'). These developments have increasingly brought fisheries management directly into contact with biodiversity conservation, in particular for protected species and Marine Protected Areas (MPAs) (cf. Chapter 8). For example, the explicit analysis of the risk of extinction of some exploited marine populations is becoming a subject of active debate (cf. Chapter 13).

From small-scale to large-scale fisheries and back

Under the industrialized growth model, small-scale fisheries (SSFs) and their governance systems were neglected if not discouraged as an impediment to modernization. This perspective lasted for several decades, but interest in SSFs has been growing again since the 1980s due to:

(1) a new focus on food security/poverty alleviation in development policies (e.g. the sustainable livelihood approach to fisheries[2]), for which SSFs were seen to play a crucial role (FAO, 1984; ICSF, 1984);
(2) an increasing focus on environmental issues and expansion of MPA networks since the 1990s which, in coastal areas, required SSF buy-in; and
(3) an expanded prominence of decentralized management approaches and spatial management through the ecosystem approach, and interest in participatory, community-based management with a progressive realization that SSFs link closely to both approaches.

Scientific interest in SSFs and fishing communities grew commensurately since the mid-1970s. In particular, 1984 was marked by both the FAO Conference on Fisheries Development and Management – which held a session on 'The special role and needs of SSFs and rural fishing and fish farming communities' (FAO, 1984) – and the International Conference of Fishworkers and their Supporters (ICFWS) organized by the International Collective in Support of Fishworkers (ICSF). Further interest in SSFs developed after the 1992 Study on International Fisheries Research (SIFR) (World Bank, 1992). This complex story is expanded further in Chapters 22–25.

Selected strands in conservation governance

From protection to sustainable use

This central evolutionary strand started on land where conservation was born, and has gained momentum in the oceans only since the 1960s–1970s.

In line with the then-dominant conservation paradigm, the International Union for the Protection of Nature (IUPN), was established in 1948 to defend nature against human 'threats' (fortress conservation). The change of name in 1958 to the International Union for the Conservation of Nature (IUCN) reflected a rising awareness that the mission had to be more responsive to human concerns (Christophersen, 1997). Some of this original tension between protection/preservation[3] and more humanist conservation is still present in the conservation community.

During the following five decades, the clear negative consequences of the exclusion of people from forestry parks (Du Toit et al., 2003) changed the conservation paradigm in two major ways: (1) from a *fortress conservation* philosophy to a strategy of participative and adaptive management, and (2) from forced exclusion of human communities to their integration. This required a shift from predictive, deterministic and reductionist (Cartesian) science aiming to control Nature to a (post-modern) science of uncertainty and risk, more appropriate for the management of complex socio-economic systems (Venters et al., 2008). A strong focus also developed on identifying and quantifying the economic benefits that local people actually derive from biodiversity (Alban et al., 2011; Boncoeur et al., 2011; Dixon and Sherman, 1990; Grimes et al., 1994; McNeeley, 1988; Peters et al., 1989).

The cross-sectoral initiatives of the United Nations (e.g. UNCHE in 1972; the World Charter for Nature[4] in 1982; UNCED in 1992; WSSD in 2002 and Rio+20 in 2012) are important milestones in the conceptual convergence of the modern conservation and development concepts.

IUCN played a seminal conceptual role with the World Conservation Strategy (WCS) drafted in 1980 with UNEP and WWF with the collaboration of FAO and UNESCO (IUCN, UNEP and WWF, 1980). Its aim was to promote 'sustainable development through the conservation of living resources; maintain essential ecological processes; preserve genetic diversity; and ensure *sustainable utilization* of species and ecosystems'. It defined conservation very broadly as: 'the management of human use of the biosphere so that it may yield the greatest benefit to present generations while maintaining its potential to meet the needs and aspirations of future generations… embracing preservation, maintenance, sustainable utilization, restoration and enhancement of the natural environment'. These concepts were at the root of the Brundtland Report (WCED, 1987), which developed and popularized the *sustainable development* paradigm.[5]

The *sustainable use* concept was formally adopted by the Conservation International NGO in 1987. It was endorsed again by IUCN as a programme area in its 18th General Assembly (Resolution 18.24, Perth in 1990). It was then integrated and defined by the Convention on Biological Diversity (CBD) (Article 2) in 1992 and defined as: 'the use of the components of biological diversity in a way and at a rate that does not lead to the long-term decline of biological diversity, thereby maintaining its potential to meet the needs and aspirations of present and future generations'. The concept acknowledges formally that, under proper conditions, conservation can be achieved through wise use and it 'played a huge role in lending legitimacy to the role that people can play in conservation' (Steve Edwards, IUCN-CEM, personal communication 2013). It was formally endorsed by IUCN in 2000 as a fundamental policy principle, committing all programmes to its mainstreaming and is still explored through the rapidly developing 'resilience movement' (Ferwerda, 2012; Hughes et al., 2005).

For two decades, the concept had been resisted by part of the conservation community who argued that 'protection' had failed, not because the 'protection' concept was flawed but because its implementation had been ineffective due to: (1) lack of additional biological, social and economic measures to support it;[6] (2) exclusive focus on biology, neglecting political and socio-economic considerations; and (3) insufficient science to produce credible conservation (Brandon et al., 1998, pp. 2, 4).

The CBD aims explicitly at the sustainable use of biodiversity from species, populations, or 'stocks' to genes (Sand, 2001). Its 14 Addis Ababa Principles of Sustainable Use (adopted by CBD CoP 7 in 2004) call for: (1) supportive policy, legal and institutional frameworks; (2) users' empowerment and rights; (3) removal of perverse market and other incentives; (4) adaptive management, science and traditional knowledge; (5) minimum ecosystem impact; (6) inter-disciplinary research; (7) compatible time–space scales; (8) international cooperation and coordination; (9) participation; (10) economic and non-market values and market forces; (11) waste and impact minimization for optimum benefits; (12) equitable distribution of benefits to local communities; (13) internalizing management costs; and (14) education and awareness raising. The policy changes that accompanied the adoption of 'sustainable use' are described in the section on 'Types of governance: local, centralized, shared, market-based'.

Towards Integrated Conservation and Development (ICD)

This strand is tightly connected to the preceding strand and, indeed, reflects one of its practical manifestations in the field. The coexistence of development and conservation has always been a challenge. The ICD projects (ICDPs) emerged in Zambia in the 1960s, in FAO forestry park projects (Garnet et al., 2007) to better face that challenge, 'with the view to reconcile the biodiversity conservation and the socio-economic interests of multiple users at local, sub-national, national and international levels' (Franks and Blomley, 2004), recognizing that people's involvement was crucial for conservation. In 1985, the WWF program on the needs of wildlife areas and humans incorporated 20 ICD projects to improve the quality of life of rural populations (Hughes and Flintan, 2001).

Results have been mixed, leading to questions about the ability to effectively integrate and achieve conservation and development (Brandon and Wells, 1992; Brandon et al., 1998; Du Toit et al., 2003; Kramer et al., 1997; Robinson, 1993; Salafsky and Wollenberg, 2000; Wells and Brandon, 1992; Western and Wright, 1994).

A modern reincarnation of ICD is the *Sustainable Livelihood Approach* used to promote balanced improvement of the natural, human, social, physical and financial capitals of sustainable development (Allison and Horemans, 2006; Chambers and Conway, 1992). This helps to broaden the range of objectives in biodiversity conservation, particularly in areas where SSFs are important, recognizing the interdependence between coastal communities and ocean resources and habitats.

Parallel strands in conservation and fishery governance

Some of the 'parallel' developments described in the following are obviously energized by the same mandatory driver (e.g. UNCLOS; cf. 'From *mare liberum* to *mare clausum*?'). Others seem more related to ambient change in systems of values, societal goals, best practices and economic incentives.

From *mare liberum* to *mare clausum*?

Opposing Grotius' free ocean concept (*Mare liberum* in 1609), Selden (*Mare clausum* in 1635) thought that the sea could indeed be appropriated in conformity with the laws of States and nature (Miller, 1989, p. 560). However, it took 350 years of competition and a rising awareness of the resulting damage before the 1982 UNCLOS established a system of appropriation of the ocean space and the fundamental rules for its use. Modern fisheries emerged in the 50 years preceding UNCLOS and are still adapting to new governance developments, particularly in the high seas.

Fisheries governance evolved through: (1) a changing legal framework through the UNCLOS process, from total freedom beyond a few miles to a complex system of overlapping space-based constraints that limit and condition that freedom; (2) strengthening the role of coastal, flag and port States and their regional institutions; and (3) progressively reducing the high seas area and constraining use of its resources.[7] In this process, for the purpose of allocating rights and responsibilities, authority over marine fishing was progressively subdivided spatially as follows: (1) the ocean was split between the high seas and the area within the 200 mile limit; (2) the latter area was split into Exclusive Economic Zones (EEZs) by maritime boundaries; and finally (3) EEZs and their resources were split into smaller management units through use of quotas (subdivisions of the TAC) and territorial use rights.

Marine conservation started practically at the same time as the UNCLOS process but, with its initial concentration in coastal areas, it was little affected by the new jurisdictions of EEZs. However, contrasting with the process of subdivision mentioned above, the increasing focus on management of biodiversity and on the ecosystem approach produced progressively larger spatial frameworks: protected areas, networks of MPAs, Large Marine Ecosystems (LMEs), integrated ocean management areas, etc. (Garcia and Hayashi, 2000). As it extended to the high seas after the 1990s however, conservation met with the problem of establishment and management of large-scale MPAs (Ecologically or Biologically Significant Areas or EBSAs and Vulnerable Marine Ecosystems or VMEs; Chapter 14) in the high seas and deep seas, as well as with the incomplete global legal framework for the management of marine genetic resources (MGRs) (cf. Chapters 7 and 11).

UNCLOS has impacted on both streams, which face similar socio-economic difficulties in crowded coastal areas and legal difficulties in the high seas. Nonetheless, traditional freedom in the oceans has decreased and the scope for conservation of marine resources and biodiversity has increased.

From bio-ecology to socio-economy and the market

Despite the fact that economic theory and methods have been available for as long as their biological equivalents, formal direct application of economics in decision-making and management – within both streams of governance – has been limited for a long time. In fisheries (cf. 'From technical measures to…'), conventional governance used technical, biology-based (conservation) measures and reference points (MSY) for decades. Similarly in the conservation stream the focus was historically strong on biological and ecological aspects, notably in relation to individual species at risk. However, that quasi-exclusive focus on bio-ecology as the foundation of governance (typical in natural resource management) gave way to a progressive integration of economics, sociology and the market as causal factors as well as providers of instruments to improve governance.

Economic arguments started to explicitly influence decisions in the 1970s. For fisheries in particular, economic efficiency has progressively complemented if not replaced resource yield as a performance criterion. In both streams of governance, there has been growing emphasis on the use of economic tools to alter human behaviour, whether the motivation was in terms of 'conservationist' or 'sustainable use' goals. Such tools include a range of incentives from taxes (on undesired behaviours) and penalties/fines (for illegal behaviour) to market-based mechanisms that provide positive incentives for desired behaviour. The latter include eco-labels that are meant to provide financial benefits (e.g. higher product prices in the marketplace) to economic actors assessed to be environmental stewards, and market-based resource use rights which are intended to give the holders of the rights (e.g. fishers) an incentive to properly engage in sustainable use of the resource.

In the late 1990s a new thread emerged, attempting to bridge the conservation and sustainable use interests through economic vehicles of valuation, incentives and markets. In contrast to a focus on protection of ecosystems for their intrinsic value reflecting the aesthetic branch of nature conservation (Noss and Peters, 1995; Olson and Dinerstein, 1998), the new trend of ecosystem valuation is gaining momentum as a basis for justifying biodiversity conservation as *economically* beneficial. The rapid development of the latter perspective, that is as a more utilitarian concept of protecting ecosystems for the 'goods and services' they produce for humans, was embraced as the central framework for the Millennium Ecosystem Assessment and is viewed as being among the core approaches of the Ecosystem Approach to Fisheries (cf. Chapter 12; de Young et al., 2008).

All of these economic-oriented measures have combined in recent years as a foundation for initiatives promoting the so-called blue, green or blue–green economy. These initiatives might be seen as simply a repackaging of sustainable development thinking (i.e. an environment–economy balance) or as illustrating the neo-liberal and market-based approach (UNEP, 2010, 2011) but, in either case, they certainly reflect the growing popularity of viewing issues of conservation, sustainable use and governance through an economics lens. This is common to both the biodiversity conservation and the fisheries streams of governance. The introduction of economics has a much longer history for fisheries, but the shift to neo-liberal economic theories and methods is contemporaneous in fisheries and in biodiversity conservation (cf. Chapters 4–6 and 10–12).

Types of governance: local, centralized, shared, market-based

Governing structures and processes have always existed at all organizational levels. Until the 1850s, fisheries were mainly managed at the community level, often by professional guilds that were delegated degrees of power by central (feudal, royal or democratic) authorities (Chapter 1). The intervention of States in natural resources management increased in the late 18th century. Subsequently, the original community-based systems of management were ignored, weakened or eliminated as State-controlled science-based management systems developed in the North Atlantic, Northeast Pacific and more widely in colonial areas. Community-based systems have re-emerged since the mid-1980s, particularly in small-scale fisheries, in an attempt to correct the conspicuous deficiencies of top-down management. The shift reflects an increasing understanding of the socio-ecological system, a better integration of scientific disciplines (cf. Chapter 9) and a progressive integration of social concerns in management objectives (e.g. equity, participation, etc.) (cf. Chapter 6).

The shift in the nature of biodiversity conservation governance is illustrated in Table 2.1 and the implications for MPAs are shown in Table 2.2. Information on similar shifts in fisheries management policies can be seen on Table 2.3.

It is important to note that the tables refer to the 'paradigms', that is, the rhetoric and the principles underpinning

Table 2.1 The four conservation paradigms (Source: Modified from Biot et al., 1995).

Variable	Traditional	State-based	Participative	Market-based
People's image	Arcadian: virtuous, rational community-minded	Uninformed, irrational, requiring paternalism	Virtuous, rational community-minded; natural stewards	Rational but egocentric; drawn by profit
Causes of the environmental problems	Limited problems; local overuse (demography)	Growing problems; mismanagement by users; subsidies; technology	Mismanagement by state, capitalists, big business	Poor government policies, bureaucratic rules and regulations
Alleged causes of degradation	Limited degradation; over-population	Over-population; backwardness, lack of foresight, ignorance	Resource distribution; inappropriate technologies; lack of defendable tenures	Inappropriate property rights; institutions; prices; rapid population growth
Prescription	Laisser-faire; self-governance	Top-down decision-making; enforcement of State norms	Top-down State norms and bottom-up participation	Market policies; property rights; resource pricing; self- targeting safety nets
Science	Ethnography; geography; traditional knowledge	Fishery and conservation science; bio-ecology and (later) economy	As before, plus sociology; NGO advocates	Economics, political science, development expertise
Gender orientation	Diversified (local inequalities)	Gender blindness	Role of women and youth	Gender myopia
Research framework	None (empirical knowledge); individual observation	Systematic empiricism; ecological science; growing role of economics including green economics	Social constructivism; rapid rural appraisal; community as unit of analysis	Methodological individualism

(continued)

Table 2.1 *(Continued)*

Variable	Traditional	State-based	Participative	Market-based
Market	Subsistence; community integration	Priority on food security; trade controls, eco-labels; market distortions (subsidies)	Subsistence, recreational and commercial	Pareto optimality and externalities
Societal model	Social; conservative	Authoritarian; conservative	Paternalistic; egalitarian	Democratic; liberal
Collective action	Effective (not always!)	Bureaucratic; deficient	Essential, unproblematic	Conditional rationality; political entrepreneurs
Conservation technology	Indigenous conservation; utilitarian conservation; social pressure, taboos, traditions	Modern conservation; fortress conservation; fences and fines; exclusion	Post-modern conservation; community-based conservation; social learning	Market-based conservation
Leading actor	Community leaders (laic and religious)	The State through its agencies	The State and community leaders	The market and the State

Table 2.2 Shift in the MPA paradigm. (Source: Thomas and Middleton 2011. Reproduced with permission of IUCN.)

Topic	As it was: protected areas were...	As it is becoming: protected areas are...
Objectives	• Set aside for conservation • Established mainly for spectacular wildlife and scenic protection • Managed mainly for visitors and tourists • Valued as wilderness • About protection	• Run with social and economic objectives • Often set up for scientific, economic and cultural reasons • Managed with local people more in mind • Valued for the cultural importance of so-called ' wilderness' • About restoration and rehabilitation
Governance	• Run by central government	• Run by many partners and involve an array of stakeholders
Local people	• Planned and managed against people • Managed without regard to local opinions	• Run with, for and in some cases by local people • Managed to meet the needs of local people
Wider context	• Developed separately • Managed as 'islands'	• Planned as part of national, regional and international systems • Developed as 'networks' (strictly protected areas, buffered and linked by green corridors)
Perceptions	• Viewed primarily as a national asset • Viewed only as a national concern	• Viewed also as a community asset • Viewed also as an international concern
Management techniques	• Managed reactively within a short time scale • Managed in a technocratic way	• Managed adaptively in a long-term perspective • Managed with political considerations
Finance	• Paid for by taxpayer	• Paid for from many sources
Management skills	• Managed by scientists and natural resource experts • Expert led	• Managed by multiskilled individuals • Drawing on local knowledge

Table 2.3 Fishery and biodiversity conservation policies and their general policy and science backdrop. (Source: Adapted from Rey-Valette and Cunningham 2004.)

	1945	1955	1965	1975	1985	1995	2005
Dev. policy	Fordist dev. Model/Growth/ innovation/infrastructure dev.			Sustainable development/Integration			
	Investment		Lowering tariff barriers/Equity		Poverty reduction programmes/Entitlements		
Economic Policy	Support to growth/subsidies				Liberalization/Structural adjustment/Privatization		Green growth
Public policy	Centralized/bureaucratic			Decentralized/Contractual	Participative/shared/delegated		
Legal framework	Pre-UNCLOS process	UNCLOS process		UN LOSC is adopted		UN LOSC is in force CBD CCRF UNFSA	
Env. Policy	Little concern			Focus on pollution		Focus on biodiversity	
Fishery Science	Discovery/Stock focus			Multispecies focus	System/Ecosystem focus		
	Cartesian determinism/Equilibrium/prediction			Post-modern/uncertainty Adaptive flexibility/Dynamics			
Fishery policy	Conflict reduction/ Technical optimization		Catch control	Licensing/ Effort control	Capacity control/Fishing rights/		
	Subsidized expansion				Crisis	Reduction of environmental impact	
	Focus on industrial fisheries				New focus in SSFs		High seas
Conservation policy	Limited interest in marine conservation			Growing interest for MPAs			MPAs in fisheries
	Protection first			Sustainable use process			Sustainable use

them, and may not reflect the implementation reality everywhere. For example, increasing and achieving real 'participation' or democratic processes have not been always without problems.

The parallel evolution of management approaches is particularly striking and illustrated in Figure 2.1. In this figure, the timing and relative importance of the different governance approaches (drawn from the literature) are only approximations because of differences between countries and institutions, but there is no doubt that similar evolutions have taken place in the two streams despite little evidence of intentional co-design between the two.

Governance targets: From species to ecosystems and genetic resources

In both streams of governance, the targets to be protected have been broadened with time from single species (emblematic animals, hunting trophies, target fishery resources) to dependent and associated species (by UNCLOS), their habitats, genomes and whole ecosystems (by the CBD and the Ecosystem Approach).

In conservation governance

In the early days of conservation, the protection targets were animal species. General wildlife conservation measures were only enacted in respect of non-game species, while hunting legislation remained in force for conservation of game species. The lists of species to be protected featured large mammals and birds and gradually incorporated small mammals, reptiles, amphibians and, much later, invertebrates. Often 'useful' species were objects of positive attention, while dangerous or 'noxious' species (pests, competitors, predators) were eliminated or at least combatted.

Aquatic animals were considered for protection much later. In the meantime, in parallel to what had happened on land, fisheries legislation developed for sustainable use of exploited populations and gradually incorporated conservation of dependent and associated species and living habitats. For more than a century, marine species (except mammals and turtles) were almost universally excluded from the scope of nature conservation legislation but were part of 'fisheries conservation' (as fisheries management was called).

Historically, protection has often been accorded to 'areas' because of their special scientific interest and, exceptionally, to ensure survival of a particular species (de Klemm and Shine, 1993, pp. 66, 125–127). The establishment of the IUCN Red List in 1963 and of CITES in 1973 illustrated the long-standing conception of species-oriented conservation. With the Bonn Convention on Migratory Species (CMS) in 1979, the Berne Convention on the Conservation of European Wildlife and Natural Habitats and the EC Directive on the Conservation of Wild Birds (79/409), habitats were identified as entities needing protection in order to protect the target species. In conservation documents and law in recent years, the broad term 'species' has been gradually changed to 'genetic resources' and 'biological diversity', encompassing the diversity of ecosystems, species and genes (de Klemm and Shine, 1993). The establishment of Ramsar in 1971 for the protection and wise use of wetlands, their fauna and flora, represented a major shift towards protection of entire areas and life assemblages, including marine shorelines down to depths of 6 m, through an ecosystem approach in the context of sustainable development. The IOC-UNESCO Biosphere reserves also reflected this broadening of the natural targets to be protected while also taking human populations into account. MPAs, most of which were established after the 1970s, were originally often selected as single areas based on their perceived local importance (diversity, abundance, beauty, scientific interest) as *in situ* conservation sites of emblematic species (white dolphins, vaquita, manatees, turtles, etc.) and not as ecosystems per se. However, when successful, the exclusion of humans from these areas resulted *de facto* in the protection of many, if not all, the species they contained. The concepts of protecting whole ecosystems and of functional networks of protected areas emerged more recently, probably first in forests (Watkins, 1998), then in the oceans in the last two decades. The IUCN decision in 2010 to create a Red List of Endangered Ecosystems to complement their Red List of Endangered Species was a significant benchmark in that regard (Keith et al., 2013).

The same sequence (from species to areas and to ecosystems) developed more recently but much more rapidly for the deep seas, first with a call for protection based on a few vulnerable species (e.g. orange roughies) and living habitat structures (cold corals, sponge reefs) or particular geomorphological structures (hot vents) before generalizing the call to the protection of EBSAs (by the CBD) and VMEs (by the FAO). According to Pirot et al. (2000), the ocean conservation targets have not changed sequentially but emerged, co-existed and eventually merged in a complicated fashion, similarly to what took place on land. Spatial targets have also tended to be larger (from small reserves to larger multiple-use MPAs, MPA networks,

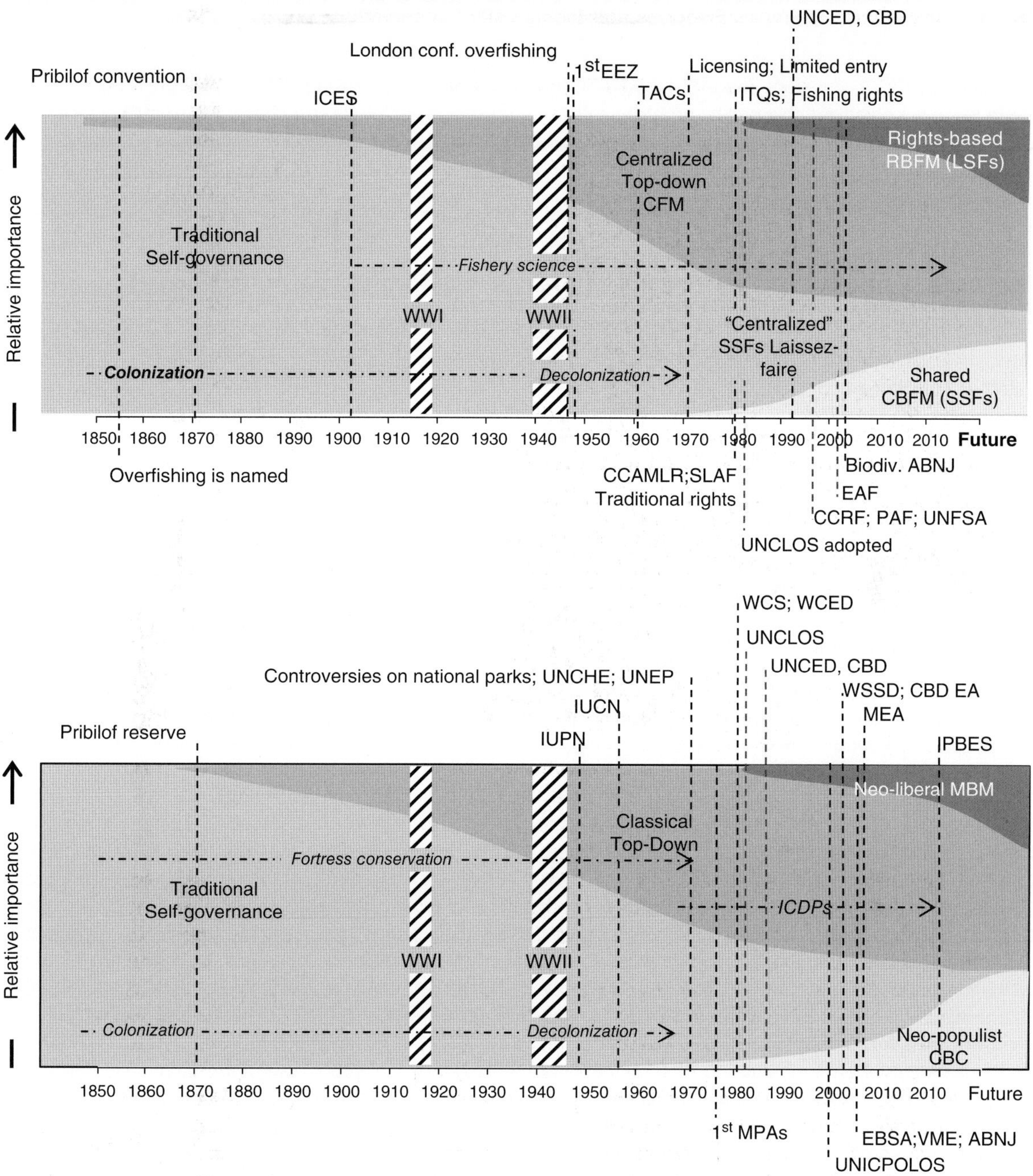

Figure 2.1 Evolution of governance paradigms and selected events for fisheries governance. The relative importance of the different management strands (on the y-axis) is only indicative (see sources below). Upper: fisheries, based on Caddy and Cochrane (2001), Garcia (1992, 2010), Garcia and Cochrane (2009) and the historical review Chapter 1. Lower: Biodiversity conservation based on Biot et al. (1995), Blaikie and Jeanrenaud (1996), Brown (2002), Salafsky and Wollenberg (2000) and the historical review given in Chapter 1. (CBC: Community-based conservation; CBD: Convention on Biological Diversity; CFM: Conventional fishery management; EA: Ecosystem approach; EAF: Ecosystem approach to fisheries; IPBES: Intergovernmental Platform on Biodiversity and Ecosystem Services; IUCN: International Union for Conservation of Nature; IUPN: International Union for Protection of Nature; MBM: Market-based management; RBM: Rights-based management; SLAF: Sustainable livelihoods approach to fisheries; UNCED: UN Conference on Environment and Development; UNCHE: UN Conference on Human Environment; UNCLOS: UN Convention on the Law of the Sea; UNEP: UN Environment Programme; WCED: World Commission on Environment and Development.)

biosphere reserves, eco-regions and LMEs), based on the yet-to-be-tested paradigm[8] that protecting large-scale targets will also protect the smaller targets and hence improve overall performance.

At the other end of the size continuum, marine genetic resources (MGRs) represent the smallest natural elements to be conserved and a non-quantified but possibly immense potential for the development of the biotechnology industry and wealth from the oceans (UNU, 2007). Their legal status in areas beyond national jurisdiction (ABNJ) (e.g. subject to high-seas freedom or considered as patrimony of humankind) is still unclear and the politics surrounding their appropriation (ownership, patenting) and the equitable sharing of benefits from their exploitation (particularly between developed and developing nations) dominate international discussions of MGR. It is however generally accepted that the issues are not threats of over-harvesting organisms with valuable genes, but rather of equity and intellectual property rights.

In fishery governance

The concern about direct and collateral environmental impacts has grown over time from a focus on the target species to the associated and dependent species, the habitats critical to the target species and finally all vulnerable habitats and species. All international fishery conventions established since the late 19th century had the conservation of target resources as an objective. The need to ensure their sustainability has been explicit in the MSY concept since the mid-1950s. The 1982 LOSC referred to MSY as well as associated and dependent species, connecting fisheries conservation to the food web in which the target species occurs. The 1992 UNCED Agenda 21 Chapter 17 on oceans and coastal areas called for even broader 'protection and sustainable development of the marine and coastal environment and its resources' and, more particularly: (1) sustainable use and conservation of marine living resources of the high seas; (2) maintenance and restoration of marine species at MSY levels; (3) protection of habitats; and (4) minimization of waste and bycatch. Nevertheless, associated actions in fisheries management only developed slowly through the 1990s. The 1995 UNFSA and the FAO Code of Conduct simultaneously introduced the precautionary approach, but its initial implementation frameworks rarely considered more than the target species (FAO, 1996). The FAO Code of Conduct also reflected the broader environmental concerns expressed at UNCED. The issues of uncertainty, risk and precaution remain important subjects of tension between fisheries and biodiversity governance (cf. Chapter 13).

The tensions are widespread with most actions to broaden the accountability of fisheries management agencies encountering opposition from established interests, and a reluctance to acknowledge non-standard tools for use in these contexts. For example, during the FAO Technical Consultation on the International Guidelines on Bycatch Management and Reduction of Discards (in Rome in December 2010), States with strong fisheries interests insisted on removal of all references to MPAs from the Guidelines and few RFMOs have broadened their VME identification beyond corals, sponges and hydrothermal vents (cf. Chapter 14).

Institutional development

Both streams have evolved towards a nested set of global-to-local institutions resting on an architecture of hard- and soft-law agreements (cf. Annex 1; Chapters 7 and 10–12).

In conservation governance

In conservation governance, the institutional development strand comprises two interacting components: (1) a quest by some parties for a global conservation authority; and (2) the development of a regional network of institutions.

Establishing one global institution for nature conservation instead of a kaleidoscope of specialized institutions has been a challenge (cf. also Chapter 10). The quest for such an institution started in 1909 with President Roosevelt's unsuccessful attempt to convene a Peace Conference on Global Nature Conservation in The Hague. Subsequently, the World Treaty on Conservation (in Bern in 1913) collapsed without implementation because of the First World War. Revived at IUCN in 1948, the idea of a global treaty was discussed at the first UNESCO-IUCN International Technical Conference on Nature Protection in 1949 but no agreement was reached and focus shifted to regional treaties. The momentum was maintained by IUCN through the creation of specialized global institutions (e.g. the International Commission on National Parks in 1960, the Ramsar Convention in 1971 and CITES in 1973) or regional institutions such as the African Convention on the Conservation of Nature and Natural Resources in 1968. The Creation of the UN Environment Programme (UNEP) in 1972 gave global environmental governance an important new dimension (see below) with, for example, adoption of the WCS in 1980. The UN sustained these efforts through international cross-sectoral summits (UNCHE in 1972,

UNCED in 1992, WSSD in 2002 and UNCSD/Rio + 20 in 2012. These summits lacked binding decision-making powers but they exerted significant influence on global and national agendas. The global cross-sectoral governance was also strengthened by the United Nations open-ended Informal Consultative Process on Oceans and the Law of the Sea (UNICPOLOS/ICP) created by the United Nations General Assembly (UNGA) in 1999. ICP has been important in analysing oceans issues for the UNGA consideration (cf. Chapters 10, 11).

The launch of the UNGA *Regular Process* for Global Marine Assessments and of the Intergovernmental Science–Policy Platform on Biodiversity and Ecosystem Services (IPBES) in 2011 created bodies that will play an important role in the future (cf. Chapters 1 and 10). Finally, further strengthening is sought through the ongoing but unresolved discussions in the CBD, Rio + 20 and UNGA about establishing a new implementing agreement for biodiversity in the high seas under UNCLOS.

The development of regional institutions connecting global with national governance and internationally agreed concepts with coordinated implementation started initially as a more feasible alternative to a global authority. The Antarctic Treaty in 1959 may be the first important regional environmental institution specific to a unique geopolitical and legal context. The network of regional environmental institutions developed effectively after the establishment of UNEP, a programme of the UN, through the signing of Regional Seas Conventions adopted between 1976 and 2002 (cf. Annex 1; Chapter 15). Many conventions were initially mandated to address general environmental issues (e.g. marine and land-based pollution), although their mandates quickly evolved to address the protection of living resources.

This web is not without its challenges, however. For example, the UNEP Governing Council's decision in 2009 that UNEP is to be the lead UN agency for all assessments of biodiversity and environmental issues has not yet been reconciled with the historical roles of FAO and RFMOs in conducting marine assessments. Similarly, the CBD/UNEP engagement in defining and striving to eliminate destructive fishing practices has not been fully reconciled with the role of the FAO in such issues. Among the as-yet-unresolved issues are those concerning whether environmental institutions (1) should deal only with environmental matters (as they do) or have a more cross-sectoral mandate and (2) should only have a role in awareness raising and coordination or be involved in sectoral implementation.

In fishery governance

In the case of fisheries, the international legal and institutional frameworks for management evolved with: (1) the first species-based hunting conventions for Pribilof seals in 1870 and whales (IWC) in 1946;[9] (2) the creation of Regional Fishery Bodies (RFBs) after the Second World War; (3) the comprehensive UNCLOS process (from 1958 to 1994); and (4) the 1995 UNFSA and the establishment of modern Regional Fishery Management Organizations (RFMOs). The global-level institutions competent in fisheries are the FAO, with an advisory mandate, and the UNGA, for cross-sectoral issues involving fisheries. More details are available in Chapters 7 and 10–12. The most recent binding instruments (e.g. the 1993 FAO compliance agreement, the 1995 UNFSA and the 2009 FAO Port States Agreement) aim at improving the order and application of law in the high seas, reducing the traditional freedom.

The 1991 UNGA moratorium on large-scale driftnet fishing[10] signalled the growing UN political intervention in fishery-related environmental matters, a role underlined by the 1995 UNFSA. The establishment of the United Nations open-ended Informal Consultative Process on Oceans and the Law of the Sea (UNICPOLOS/ICP) in 1999 facilitated and strengthened the UNGA debate on oceans-related issues, and prompted improved coordination and cooperation of fisheries and biodiversity governance. At a national level, the institutional development process has tracked the international changes, supported by international cooperation to improve statistical systems and fisheries research (particularly surveys and stock assessment), modernize legislation (in the wake of UNCLOS and UNCED[11]) and develop national management capacity.

Science–policy relations

In conservation governance

The biodiversity conservation stream has long associated itself with strong science, arguing that most of the calls for policy actions emerged from scientific studies documenting serious trends in biodiversity components and/or threats to biodiversity from unsustainable practices (cf. Chapter 9). Some of the first high-profile forays of the biodiversity conservation governance stream into the marine realm (and into direct conflicts with fisheries) came, in fact, from the application of general science-based criteria for assessing risk of extinction to exploited marine fish species (cf. Chapter 12). The conservation biology stream has also had a long-standing explicit acknowledgement of the value of traditional knowledge as an important part of the information relevant to policy developed, with Article 8(j) of the Convention on Biological Diversity promoting the culturally appropriate use of the knowledge, innovations and practices of indigenous and local communities relevant for the conservation and sustainable use of biological diversity.

Arguments about 'whose science' is the appropriate basis for policy development remain unresolved. At present, at least some major science–policy initiatives (e.g. the UNEP International Partnership for Sustainable Resource Management) maintain a principle that employees of government regulatory ministries, including Fisheries Ministries, are not eligible for membership in their assessment panels. The concern is that such experts might bias their science towards the policies of their governments, whereas academic experts and experts working for ENGOs are considered unbiased, even when closely affiliated with groups having strong conservation-oriented policy goals themselves. The opposite concern has also been expressed (Kearney, 2013).

A few cases of collaboration between science experts rooted in the two different streams have documented how quite different conclusions can be drawn from the same data, depending on how the questions are framed and the analyses conducted (Worm et al., 2006, 2009). Although both streams avow belief that sound policy benefits from a strong science-based foundation, such collaborations focus attention on the need for the streams to have a common view of the boundary between science and policy, and what practices comprise 'advocacy science' (Rice, 2011).

In fishery governance

The notion that decisions should be based on the *best scientific evidence available* is enshrined in UNCLOS but has been at the basis of fisheries governance for more than a century as illustrated by the creation of ICES in 1902.

Fisheries management emerged from the broader field of natural resource management and is probably the first and most fully documented example of systematic application of science in governance. Indeed, even prior to UNCLOS, scientific analysis provided a starting point for domestic and international fisheries policy and management decisions. National marine laboratories developed in the 1870s. Initially, biology and technology dominated the scientific input to fisheries management, with little collection or analysis of socio-economic data. The shift from fundamental marine biology to applied fishery science was boosted by the creation of ICES and its shift to providing fishery management advice in the 1930s (Roswadowski, 2002). The concept of science-based management spread more widely after THE Second World War. In the 1960s and 1970s, FAO (with UNDP funding) facilitated the development of most of the fisheries research laboratories that exist today in the developing world. FAO also established scientific groups and committees in all the Regional Fishery Bodies it created.

Discussion and conclusion

The governance streams for fishery management and biodiversity conservation have common roots (cf. Chapter 1). Historically merged in one governance process in traditional communities, the two streams diverged during the 19th century as utilitarian and aesthetic branches of conservation in the wake of a growing sectoral and public concern regarding the impact of industrialization (de Klemm and Shine, 1993; Evans, 1987). The tension between protection and sustainable use stiffened the relationship between the streams for many decades and even within them. It was still dividing WWF in the mid-1990s, for example (Tickell, 1995, cited by Blaikie and Jeanrenaud, 1996) and still divides supporters and sceptics of 'conservation through sustainable use'. Moving from terrestrial environments to ocean, biodiversity conservation started dealing with marine resources (and fisheries) after the 1960s. From 1970 onwards, the cross-sectoral summits (UNCHE, UNCED, WSSD and Rio+20) established conceptual bridges between the two streams, accelerating change in similar directions.

The strands identified above are only a sample illustrating how each of the streams evolved as its component strands emerged, evolved and were eventually mainstreamed or abandoned. Most of these changes were interconnected, energizing each other as success or failure in one strand fuelled the emergence and development of another as a replacement or complement.

Many developments taking place in one stream seem to happen together, slightly before or after a similar change in the other. These similarities may result from direct and conscious responses of one stream to changes in the other. They may also reflect a similarity in non-coordinated responses of both streams to a common external driver. Much more attention seems to have been paid by the media to conflicts between the two streams than to the similarities observed in, for example:

- *governance principles*: evolution of the types of governance adopted, converging towards very similar 'good governance' principles, considered international norms;
- *hard sustainability principles*, as embedded in the MSY concept and the Ecosystem Approach, prohibiting the replacement of the natural capital by other forms of capital;
- *integration* of development and sustainability concerns across multiple uses, social and biological sciences and of science and policy;
- *common approaches*: the adoption of the precautionary and ecosystem approaches, sustainable livelihoods, ICDs and sustainable use and adaptive management;
- *cross-scale institutions*: development of an effective institutional set-up with nested global, regional and national levels;
- *broadening management paradigms*: shift from species-based to ecosystem-based science and management paradigms;
- *corresponding broadening of policy objectives* from either growth or protection to sustainable and responsible use, with greater recognition of the legitimacy of the need to integrate current human wellbeing and socio-economic outcomes with ecological ones;
- *increasing participation*: the past trend from traditional community-based management to centralized state management is being reversed more recently with a return to shared governance;
- *progressive integration* of economic valuation, incentives and market-based instruments;
- *common 'hot' issues*: high seas, deep seas, bycatch reduction, overfishing, capacity-building and governance failures;
- *small-scale fisheries* generating similar recent interest as important components of both sustainable use and food security; and
- *science*: the need for a scientific foundation (including traditional knowledge) for decision-making.

The timeline given in Annex 1 highlights events of common interest in both streams. It shows that the commonalities between the two streams seem to have increased rapidly after the UNCHE in 1972 and even more after UNCED in 1992. The harsh conflicts of the 1960s and 1970s (e.g. on whaling) have progressively been replaced by more open debates on conflictual issues and parallel and simultaneous work towards solutions (cf. Chapter 3). The emergence of progressively more overlapping (if not explicitly coordinated) agendas illustrates the fact that inter-institutional cooperation has grown despite the natural resistance of resilient bureaucratic structures and cultures, even though the progress in translating this overlapping conceptual rhetoric into joint implementation is rather slow.

Convergence or coevolution?

How did this happen? As stressed in the title of this chapter, and using a biological evolution theory analogy, the reduced 'functional distance' between the two streams of governance could be the result of fortuitous convergence or purposeful coevolution (and most probably of both).

The similarities in the evolution of the two governance streams, illustrated in Chapter 1 and in this chapter, have apparently attracted much less attention than the (persisting) conflicts between them. In separate but interacting complex socio-ecological systems, different emergent properties might emerge at different times as a result of (1) interaction between their components, and (2) reaction to external stimuli (drivers). Strong and sustained similarities between such systems are therefore remarkable features, unlikely to be fortuitous. The two governance streams of fisheries and biodiversity conservation operate in overlapping natural and human domains, pursue partially overlapping objectives and share global policy, economic and environmental drivers. In such circumstances, *coevolution* or/and *convergence* (as defined in the introduction of this chapter) are very likely to emerge as a result of their direct interaction or/and as a common response to common external drivers, respectively. It is important for the future evolution of the relation between the two streams to determine the relative contribution of both mechanisms to the limited but noticeable progress made since UNCED.

In light of the additional material discussed in the various chapters listed below, it can be said that the similar responses observed in the two streams may have resulted from the following:

(1) *Their common origin*: Fisheries management and biodiversity conservation emerged from utilitarian resources conservation (cf. Chapter 1) and thus share concerns on vulnerable resources and human communities,

with differences on time frames, risk perceptions and priorities.

(2) *A process of convergence* prompted by external drivers, including the following.
- *The new Law of the Sea* (LOS) which started to emerge in the mid-1970s (during the UNCLOS III process) as customary law, conditioning choices in both streams including the nascent marine conservation. The LOS integrates the legal regimes for use and protection, within each of which concepts have evolved over time (cf. Chapters 7 and 12).
- *The cross-sectoral summits* (from UNCHE in 1972 to Rio + 20 in 2012) and their follow-up. With the strong moral back-up of the UNGA, they gave common policy directions to both streams (e.g. on sustainable development, precautionary approach, ecosystem approach, integrated management, participation, equity and economic efficiency) (cf. Chapters 10–12).
- *The global policy backdrop*. Sectoral ministries implement their sectoral policies but must also mainstream national overarching policies within their respective sectors; this provides powerful factors of alignment and coordination. Fisheries and biodiversity conservation policies developed against (and were influenced by) a common backdrop of cross-sectoral policies that conditioned their choices (e.g. on investment, subsidies, trade, human rights, indigenous people, international aid, environment and foreign relations; Table 2.3).
- *The moderate pressure exerted by soft-law instruments* including UNGA resolutions; Summits Declarations; Codes of Conduct; 'decisions' in FAO, CBD, UNEP, UNGA, etc.; (cf. Chapters 7 and 11). In the last 40 years many of these have focused efforts on the concept of Sustainable Development and Use and the ways and means of their implementation. Put forward in the 1980 World Conservation Strategy, enshrined in the 1982 UNCLOS, mainstreamed by the WCED process between 1983 and 1987 and the UN since then, the concept is still controversial in both streams but it has provided a policy framework to integrate sectoral development and conservation, enabling a clarification of points of divergence and consensus and joint or parallel work to resolve differences in the pursuit of solutions that serve both streams.

Together, all these drivers have delimited a sort of 'viability envelope' within which the two streams had to remain while unfolding. Oscillations have occurred with time while the focus shifted (in fisheries between growth and responsible fishing, and in conservation between protection and sustainable use).

(3) *A process of coevolution* fed by interaction between the two streams and learning of common lessons.
- The interaction between the two streams has occurred along the chain from assessment and decision to implementation at all scales in the UN, in regional organizations and at national and community level. It has been mediated by the UNGA and ICP at global level (cf. Chapters 10 and 11) and by courts, at international level[12] and national level (cf. Chapters 20 and 28). It has also occurred in a more or less formal way in national policy frameworks established for that purpose (cf. Chapter 21) that courts have helped to clarify and normalize.
- Lessons were learned through that interaction about effective alliances (Chapters 10 and 11), about the cost/benefit ratio of open conflicts, the importance and condition of effective compliance, the role of participation and, most importantly, perhaps the common realization that the poor performance of both streams originated in common threats (open access, poverty, inequity, inadequate systems of incentives, etc.) calling for similar solutions (tenure, participation, protection of vulnerable elements).
- The modern pressure towards space-based integrated governance frameworks (ICAM, MSP, multiple-use MPAs; cf. Chapter 8) using an ecosystem approach has increased the scope and probability of further coevolution as the two streams will start interacting even more closely within constraining policy frames and towards explicitly harmonized and ranked objectives.

Figure 2.2 illustrates how, from 1950 to 2010 (with an extrapolation to 2030), the two governance streams have come closer to each other by broadening their respective paradigms, shifting from originally opposed objectives (pure protection on the one hand and growth on the other) to more easily shareable paradigms (sustainable development, responsible fishing and sustainable use). That being said, the original opposite strategies (protection and growth) persist as outliers of the main trend, thus continuing to create tensions within and between streams.

Both streams have experimented with similar approaches (ICDP/SLA/Parks for People; EAF/EBFM;

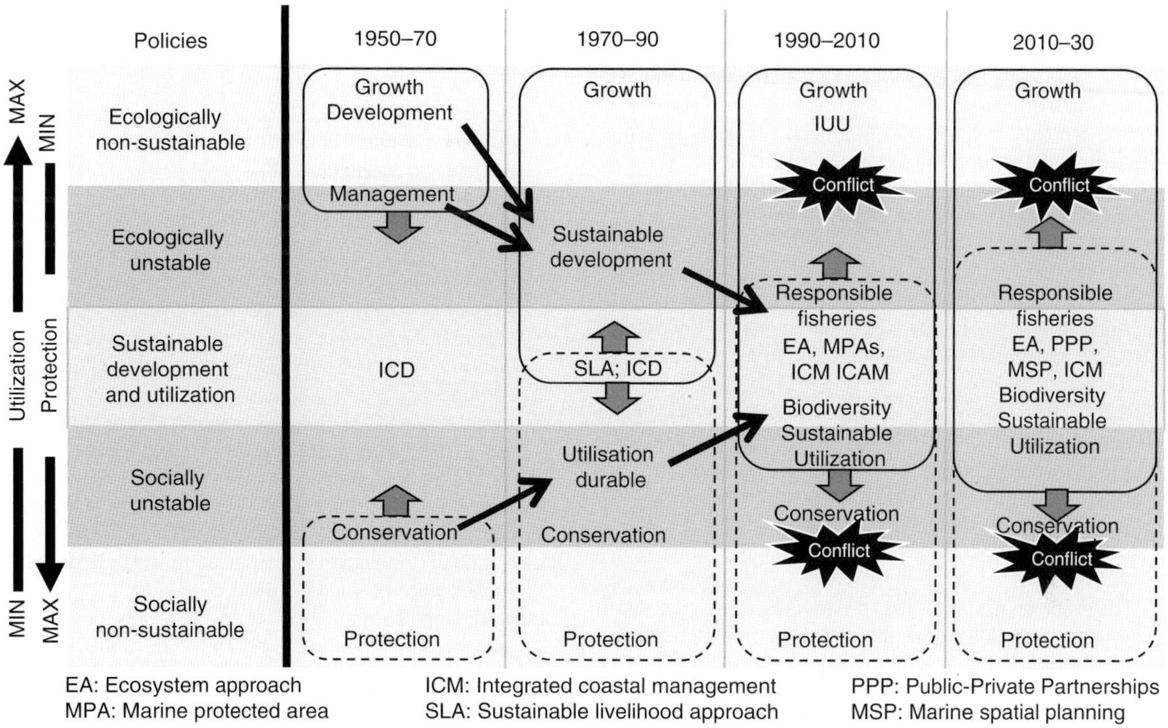

Figure 2.2 Schematic trend in fisheries management and biodiversity conservation concepts (Source: Garcia 2010). The grey horizontal bands reflect the degree of 'sustainability' resulting from the interaction between the amount of use and of protection (left axis and left column). The progressive extension of the governance of fisheries (upper rectangles) and biodiversity (lower dotted rectangles) from unsustainable to more sustainable practices illustrate their convergence and progressive overlap of the two streams (after 1990) against the sustainability backdrop. A selection of convergent or common approaches is shown in each decade. The conflicts represented are internal to each governance stream, between their moderate and extreme elements. Other tensions exist between streams mainly at an operational level.

precautionary approach; and traditional rights) and are experimenting now with trade controls/traceability, eco-labelling and market-based tools. Both governance systems are looking forward to improvement through the ecosystem approach, more integrated management and marine spatial planning and economic instruments (perhaps including private/public partnerships). Misunderstandings, suspicions and related frictions have not completely disappeared and tend to materialize at implementation level. Even within each stream, hard-rooted 'fundamentalist' views have continued to persist, sometimes clouding the emerging potential for integration (such as those questioning the benefits of people's participation in conservation or arguing that environmentalists unduly threaten fisheries). However, these views are increasingly marginalized in both streams.

The evolution has not been linear and smooth but chaotic with phases of progress, stagnation and even regression, and with combinations of conflict, cooperation and collaboration. The processes are ongoing as the overarching policy background keeps changing. We would not dare to predict the reactions of both streams to potentially important emerging factors not considered in this chapter such as: the global financial crisis of the late 2000s; the outcome of the public–private partnerships or blue growth initiatives; or the north–south tensions over capacity building, equitable access to and benefits from uses of biodiversity and marine genetic resources. These might affect both streams in different ways in different socio-economic country groupings, and in different jurisdictions (as in EEZs versus high seas).

Notes

1. The FAO was initially established for the implementation of the Marshall Plan.
2. The Sustainable Livelihood Approach considers simultaneously the natural, human, social, physical and financial forms of capital, providing the first really integrated approach to development of SSFs (Allison and Horemans, 2006).
3. The first terms means avoiding further impact. The second means keeping in its original 'untouched' state.
4. Adopted by the UNGA based on preparatory work of IUCN, and the 1993 Publication *Caring for the Earth* by IUCN and WWF.
5. In the conservation community, 'sustainable development' is perceived as relevant for top-down national development planning while 'sustainable use' focuses on fostering systems in which local people are empowered to sustainably use the natural resources that meet their needs (S. Edwards, IUCN-CEM, personal communication 2013).
6. For example: nesting MPAs in networks; seascapes, ecoregions and LMEs; and alternative income generating activities.
7. The modern milestones are the 1947 unilateral extension by Peru and Chile of their jurisdiction to 200 miles; the UNCLOs I, II and III processes (in 1958, 1960 and 1973); UNCLOS in 1982 and its entry into force in 1994; the 1993 FAO Compliance agreement; the 1995 UNFSA; the 1995 Code of Conduct and its International Plans of action; and the 2009 Port States Agreement, etc.
8. The broadening of the area to conserve implies a return to the preponderant role of States, increased distance between decision-makers and the people concerned and increased management costs.
9. Although these activities are closer to hunting than to fishing.
10. Resolution A/RES/46/215 of 20 December 1991.
11. The role of the FAO legal office in this process was significant.
12. Very occasionally in the UN Tribunal for the Law of the Sea (ITLOS).

References

Alban, F., Boncoeur, J. and Roncin, N. (2011) Assessing the impact of marine protected areas on society's wellbeing: an economic perspective. In: *Marine Protected Areas. A Multidisciplinary Approach* (ed. J. Claudet), pp. 226–246. Cambridge University Press, Cambridge.

Allison, E.H. and Horemans, B. (2006) Putting the principles of the sustainable livelihoods approach into fisheries development policy and practice. *Marine Policy* **30**(6), 757–766.

Biot, Y., Blaikie, P.M., Jackson, C. and Palmer-Jones, R. (1995) Rethinking Research on Land Degradation in Developing Countries. World Bank Discussion Paper No. 289. The World Bank, Washington, DC.

Blaikie, P. and Jeanrenaud, S. (1996) Biodiversity and human welfare. UN Research Institute for Social Development. UN RISD Discussion Paper, DP 72. 81 pp.

Boncoeur, J., Alban, F. and Thebaud, O. (2011) Bioeconomic analysis of marine protected areas fisheries effects. In: *Marine Protected Areas. A Multidisciplinary Approach* (ed. J. Claudet), pp. 190–225. Cambridge University Press, Cambridge.

Brandon, K. and Wells, M. (1992) Planning for people and parks: design dilemmas. *World Development* **20**(4), 557–570.

Brandon, K., Redford, K.H. and Sanderson, S.E. (1998) *Parks in Peril: People Politics and Protected Areas.* Island Press, Washington DC.

Brown, K. (2002) Innovation for conservation and development. *The Geographical Journal* **168**(1), 6–17.

Caddy, J.F. and Cochrane, K.L. (2001) A review of fisheries management past and present and some future perspectives for the third millennium. *Ocean & Coastal Management* **44**(9–10), 653–682.

Chambers, R. and Conway, G.R. (1992) Sustainable rural livelihoods: practical concepts for the 21st century. Institute of Development Studies. Brighton, Sussex, UK. Discussion paper, 296: 33 p.

Christophersen, L.E. (1997) IUCN, a bridge builder for Conservation of Nature. *Green Globe Yearbook* 59–69.

de Klemm, C. and Shine, C. (1993) Biological diversity conservation and the law. Legal mechanisms for conserving species and ecosystems. Gland (Switzerland). IUCN Environmental Policy and Law Paper, **29**: xix + 292 pp.

De Young, C., Charles, A. and Hjort, A. (2008) Human dimensions of the ecosystem approach to fisheries: An overview of context, concepts, tools and methods. *Fisheries Technical Paper*, **489**. Food and Agriculture Organization of the United Nations, Rome, Italy.

Dixon, J.A. and Sherman, P.B. (1990) *Economics of Protected Areas: A New Look at Benefits and Costs.* Island Press, Washington, DC.

Du Toit, J., Rogers, K.H. and Briggs, H.C. (2003) *The Kruger Experience.* Island Press, Washington DC.

Evans, D. (1987) *A History of Nature Conservation in Britain.* Routledge, London.

FAO (1984) Report of the FAO World Conference on fisheries management and development. Rome, 27 June–6 July 1984. FAO, Rome: 60 pp. + 6 annexes.

FAO (1996) *The State of World Fisheries and Aquaculture.* FAO, Rome.

Ferwerda, W. (2012) Nature resilience. Organizing ecological restoration by partners in business for next generations. Rotterdam School of Management. Erasmus University. IUCN-CEM. 98 pp.

Franks, P. and Blomley, T. (2004) Fitting ICD into a project framework: a CARE perspective: 77-97. In: *Getting Biodiversity Projects to Work: Towards Better Conservation and Development* (eds T.O. McShane and M.P. Wells). Columbia University Press, New York, USA.

Folke, C., Berkes, F. and Colding, J. (2000) Ecological practices and social mechanisms for building resilience and sustainability. In: *Linking Social and Ecological Systems. Management Practices and Social Mechanisms for Building Resilience* (eds F. Berkes, C. Folke and J. Colding), pp. 414–435. Cambridge University Press, Cambridge.

Garcia, S.M. (1992) Ocean Fisheries Management. The FAO Programme. In: *Ocean Management in Global Change* (ed. P. Fabbri), pp. 381–418. Elsevier Applied Science, London and New York.

Garcia, S.M. (2010) Fisheries Management, Nature Conservation and the Ecosystem Approach: Collision or Syncretism? From Armageddon to the New Atlantis? *Proceedings of the 15th IIFET Annual Conference*, Montpellier (France), July 13–16 2010. Economics of fish resources and aquatic ecosystems: balancing uses, balancing costs. Abstracts, 44 pp.

Garcia, S.M. and Hayashi, M. (2000) Division of the oceans and ecosystem management: a contrastive spatial evolution of marine fisheries governance. *Ocean & Coastal Management* **43**, 445–474.

Garcia, S.M. and Cochrane, K.L. (2009) From past management to future governance: a perspective view. In: *A Fishery Manager Guidebook*, 2nd edition (eds K.L. Cochrane and S.M. Garcia), pp. 447–472. Wiley-Blackwell and FAO, Oxford (UK) and Rome (Italy).

Garnett, S T., Sayer, J. and Du Toit, J. (2007) Improving the effectiveness of interventions to balance conservation and development: a conceptual framework. *Ecology and Society* **12**(1), 2.

Grimes, A., Loomis, S., Jahnige, P., Burnham, M., Onthank, M., Alarcon, K., Palacious Cuenca, W., Ceron Martinez, C., Neill, D., Balick, M., Bennett, B. and Mendelsohn, R. (1994) Valuing the rainforest: the economic value of nontimber forest products in Ecuador. *Ambio* **23**(7), 405–410.

Hughes, R. and Flintan, F. (2001) Integrating conservation and development experience. A review and bibliography of the ICDP literature. *Biodiversity and Livelihoods Issues* **3**. IIED, London.

Hughes, T.P., Bellwood, D.R., Folke, C., Steneck, R.S. and Wilson, J. (2005) New paradigms for supporting the resilience of marine ecosystems. *TREE* **20**(7), 380–386.

ICSF (1984) Report of the international Conference of Fishworkers and their supporters, Rome 4–8 July 1984. Available at http://www.icsf.net/en/proceedings/article/EN/3-report-of-the-i.html?start=10 (accessed February 2014).

IUCN, UNEP and WWF (1980) World Conservation Strategy. Living resources conservation for sustainable development. IUCN, UNEP, WWF. 77 pp.

Kearney, R. (2013) Australia's out-dated concern over fishing threatens wise marine conservation and ecologically sustainable seafood supply. Open Journal of Marine Science **3**, 55–61.

Keith D.A., Rodrıguez J.P., Rodrıguez-Clark K.M., Nicholson E., Aapala K., et al. (2013) Scientific Foundations for an IUCN Red List of Ecosystems. *PLoS ONE* **8**(5): doi:10.1371/journal.pone.0062111.

Kramer, R., van Schaik, C. and Johnson, J. (1997) *Last Stand, Protected Areas and the DefenseoOf Tropical Biodiversity*. Oxford University Press, Oxford.

McNeeley, J.A. (1988) Economics and biological diversity: developing and using economic incentives to conserve biological resources. IUCN. Gland, Switzerland. xiv + 223 pp.

Miller, M.L. (1989) L'organisation sociale des systèmes d'aménagement de la pêche. In: *L'homme et les Ressources Halieutiques. Essai Sur l'usage d'une Ressource Commune Renouvelable* (ed. J.P. Troadec), pp. 557–583. IFREMER, Paris.

Noss, R.F. and Peters, R.L. (1995) *Endangered Ecosystems. A Status Report on America's Vanishing Habitat*. Defenders of Wildlife, Washington, DC.

Olson, D. and Dinerstein, E. (1998) The Global 200: a representation approach to conserving the earth's most biologically valuable ecoregions. *Conservation Biology* **12**, 502–515.

Peters, C.M., Gentry, A.H. and Mendelsohn, R.O. (1989) Valuation of an Amazonian rain forest. *Nature* **339**, 655–656.

Pirot, J.-Y., Meynell, P.J. and Elder, D. (2000) *Ecosystem Management: Lessons from around the World*. IUCN, Gland, Switzerland.

PNUD (1997) La gouvernance en faveur du développement humain durable. Document de politique générale du PNUD. PNUD, New York. Available at http://mirror.undp.org/magnet/policy/ (accessed February 2014).

Rey-Valette, H. and Cunningham, S. (2004) Interactions between industrial and artisanal fisheries in the history of West Africa. In: *Pêches maritimes, écosystèmes et sociétés en Afrique de l' ouest: un demi siècle de changement* (eds P. Chavance, M. Ba, D. Gascuel, M. Vakily and D. Pauly), pp. 495–506. Collection des Rapports de Recherche Halieutique ACP/UE, 15(1). Paris, Institut de Recherche pour le Développement and Commission Européenne, Bruxelles.

Rice, J.C. (2011) Advocacy Science and Fisheries Decision-Making. *ICES Journal of Marine Science* **68**, 2007–2012.

Robinson, J. (1993) The limits to caring sustainable living and the loss of biodiversity. *Conservation Biology* **7**, 20–28.

Roswadowski, H.M. (2002) *The Sea Knows No Boundaries. A Century of Marine Science under ICES*. ICES, University of Washington Press (Copenhagen, Seattle, London). 410 pp.

Salafsky, N. and Wollenberg, E. (2000) Linking livelihoods and conservation: a conceptual framework and scale for assessing the integration of human needs and biodiversity. *World Development* **28**(8), 1421–1438.

Sand, P.H. (2001) A century of green lessons: the contribution of nature conservation regimes to global environmental governance. *International Environmental Agreements: Politics, Law and Economics* **1**, 33–72. (Revised version in *Managing Global Issues: Lessons Learned* (eds P.J. Simmons and C. de Jonge Oudraat), pp. 281–309. Carnegie Endowment for International Peace, Washington DC.)

Thomas, L. and Middleton, J. (2011) Lignes directrices pour la planification de la gestion des aires protégées. UICN, Gland Suisse. *Lignes directrices sur les meilleures pratiques pour les aires protégées* **10**, 67 p.

UNEP (2010) The role of ecosystems in developing a sustainable Green Economy. UNEP Policy Series, Policy Brief 2. 17 pp.

UNEP (2011) Restoring the natural foundation to sustain a Green Economy: a century-long journey for ecosystem management. UNEP Policy Series, Policy Brief 6. 25 pp.

UNU (2007) An update on Marine Genetic Resources: scientific research, commercial uses and a database on marine bioprospecting. A document presented at UNICPOLOS, 8th Meeting, New York, 25–29 June (2007) UN University, UNESCO-MAB. 71 pp.

Venters, F.J., Naiman, R.J., Biggs, H.C. and Pienaar, D.J. (2008) The evolution of conservation management philosophy: science, environmental change and social adjustments in Kruger national park. *Ecosystems* **11**, 173–192.

Watkins, C. (ed.) (1998) *European Woods and Forests: Studies in Cultural History*. CAB International, New York.

WCED (1987) *Our Common Future. World Conference on Environment and Development*. Oxford University Press, Oxford.

Wells, M. and Brandon, K. (1992) *People and Parks: Linking Protected Area Management With Local Communities*. The World Bank, Washington, DC.

Western, D. and Wright, R.M. (1994) *Natural Connections: Perspectives in Community-Based Conservation*. Island Press, Washington DC.

World Bank (1992) A study of international fisheries research. UNDP, CEC, FAO, The World Bank, Washington DC, USA Policy and Research Series, **19**. 103 pp.

Worm, B., Barbier, E.B., Beaumont, N. et al. (2006) Impacts of biodiversity loss on ocean ecosystem services. *Science* **314**, 787–790.

Worm, B., Hilborn, R., Baum, J.K., Branch, T.A., Collie, J.S., Costello, C., Fogarty, M.J., Fulton, E.A., Hutchings, J.A., Jennings, S., Jensen, O.P., Lotze, H.K., Mace, P.M., McClanahan, T.R., Minto, C., Palumbi, S.R., Parma, A.M., Ricard, D., Rosenberg, A.A., Watson, R. and Zeller, D. (2009) Rebuilding global fisheries. *Science* **325**, 578–585.

Chapter 3

Governance of marine fisheries and biodiversity conservation: The integration challenge

S.M. Garcia[1], J. Rice[1,2] and A. Charles[1,3]

[1]IUCN Commission on Ecosystem Management (IUCN/CEM/FEG), Gland, Switzerland
[2]Fisheries and Oceans Canada, Ottawa, Canada
[3]Sobey School of Business and the Environmental Science Department, Saint Mary's University, Halifax, Canada

Abstract: Fisheries and biodiversity governance have generated mixed results. The reasons for this, which are numerous, well known and, to a large extent, similar, are briefly reviewed in this chapter against a sustainable development backdrop. Insufficient integration between policies, institutions and processes of the two governance streams is one of the important factors and the focus of the chapter. The degree of integration and the processes and steps needed to increase it are explored. Factors of convergence and divergence as well as impediments are examined. The integration field on which the two governance processes interact is described in terms of gradients of risk for people and resources. Against this background, the initiatives taken in recent decades towards integration within and across the two governance streams are briefly described in terms of strategic tensions, competition and synergy. Proposals are made to increase integration.

Keywords: conservation; governance; integration; interaction; cooperation; competition; synergy; tension

Introduction

Fisheries management has always had as its central objective, mandate and responsibility (enshrined in UNCLOS) the conservation of target fishery resources with little explicit concern and few operational measures for broader biodiversity conservation. In the last two decades, following the United Nations Conference on Environment and Development (UNCED), there has been an increasing extension of this narrow utilitarian sectoral concern to address the conservation of marine biodiversity, with objectives going beyond the strict long-term wellbeing of fisheries. The two types of objectives are closely interconnected however, as the conservation of marine biodiversity requires sustainable fisheries and *vice versa*. Consequently, there must be positive interaction between the two respective streams of governance resulting in coherence and, in many cases, integration of policies and practices.

Coherence and integration are facilitated by a number of factors including obvious common interests, lessons learned from past failures, societal pressure and a common legal framework (UNCLOS). They are slowed down by a number of tensions between: (1) conservation and development; (2) different administrations (Chapter 2); (3) short-term sectoral economic interests and long-term societal interests; (4) the rural producing sector and the urban consuming sector; and (5) developing

Governance of Marine Fisheries and Biodiversity Conservation: Interaction and Coevolution, First Edition.
Edited by Serge M. Garcia, Jake Rice and Anthony Charles.

countries (major exporters) and developed countries (major importers). There is also a lack of agreement on transition pathways, even to agreed objectives and outcomes.

Resolving these tensions to reach the globally agreed objective of sustainable use should be a central task of governance. This chapter focuses on that proposition, putting forward the argument – to be explored throughout this book – that better integration of the two administrative processes and possibly structures in charge of fisheries and biodiversity conservation is essential to both fields of governance.

There is a general agreement that the two separate streams of governance have not delivered a sufficient level of harmonization and alignment of visions, conceptual goals and operational objectives. The need to better connect environmental and sectoral policies has been repeatedly expressed for decades in fundamental texts on sustainability such as the World Conservation Strategy (IUCN/UNEP/WWF, 1980), the Brundtland Report (WCED, 1987) and the 2000 Millennium Summit Declaration. The Brundtland report states:

> 'The integrated and interdependent nature of the new challenges and issues contrasts sharply with the nature of the institutions that exist today. These institutions tend to be independent, fragmented, and working to relatively narrow mandates with closed decision processes. Those responsible for managing natural resources and protecting the environment are institutionally separated from those responsible for managing the economy. The real world of interlocked economic and ecological systems will not change; the policies and institutions must.' (WCED, 1987, p. 310)

A quarter of a century later the tension between development and conservation – that is, between natural resources available and *de facto* expectations for ever-increasing benefits from their use – has not been resolved; this provides a challenge in fisheries and far beyond. Some progress has been made on the conceptual, legal and institutional planes but the above quotation remains largely valid today. The symptoms of the multidimensional misfit within and between the two systems are apparent in the frequent inadequate performance of management, illustrated by risky decreases in the abundance of fishery resources and key elements of biodiversity, poor economic performance, social distress, low compliance and conflicts. In each sector, experts from various disciplines (including biologists, statisticians, ecologists, environmental scientists, political scientists, economists, sociologists and lawyers) explained the failures from their disciplinary angle, proposing related solutions. The total result is a poorly harmonized plethora of 'solutions' reflecting the complexity of the socio-ecological systems (*sensu* Berkes et al., 2000). None of the solutions – alone – are sufficient. Finding the proper mix and balance among them, achieving an adequate level of coordination and integration to allow implementation and adapting the mix dynamically as the contexts evolve is the governance challenge (Charles, 2001; Garcia and Charles, 2007; Grafton et al., 2010). Failure leads to fragmented and incomplete implementation, and the risk is that both streams of governance will fail to reach their objectives (Brown, 2003; Persson, 2004; Rice and Garcia, 2011).

Better integration is a *sine qua non* part of the solution. The challenge for fisheries and biodiversity conservation is to build or reinforce connections among the respective policies, strategies and plans. This calls for simultaneous integration of conservation concerns and objectives into fisheries development and management processes, and of livelihood and food security concerns of fishing communities into biodiversity conservation.

Chapters 1 and 2 have shown that convergence between the two fields of governance has already started, despite a range of difficulties and conflicts. In agreement with Rice and Ridgeway (2010), this chapter argues that further progress will be limited in the absence of clear empowerment of existing institutions for more integrated governance, keeping in mind that the challenge this represents is only a subset of the more complex and challenging ocean governance integration.

This chapter examines the sustainable development backdrop to integration ('Sustainable development backdrop', next), the aim, process and desired degree of integration ('Integration process'), the factors of convergence and divergence, impediments to overcome and lessons learned by the two governance streams through history ('Integration factors'). The ongoing interaction and conceptual bridges and frictions are described in 'Integration through interaction' before providing some concluding thoughts.

Sustainable development backdrop

The interaction between and potential integration of the governance systems of marine fisheries and of biodiversity conservation is a subset of the more general problem of oceans sustainable development. This section places the analysis within that context.

Definitions

Sustainable development (SD) has been loosely defined by the World Commission on Environment and Development as: 'development that meets the needs of present generations without compromising the ability of future generations to meet their own needs' (WCED, 1987). Using this as their foundation, more specific definitions have been elaborated in relation to more specific mandates. As an environmental institution, the United Nations Environment Programme (UNEP) stresses the objective of 'improving the quality of life for all of the Earth's citizens without increasing the use of natural resources beyond the capacity of the environment to supply them indefinitely and the need for taking action, changing policy and practice at all levels, from the individual to the international' (UNEP, 2009). The Food and Agriculture Organization of the United Nations (FAO), as a sectoral institution, stresses 'the conservation of the natural resource base' and the requirement for development to be 'technologically appropriate, economically viable and socially acceptable' (FAO, 1989, in Garcia and Grainger, 1997).

In the environment arena, SD has been viewed sceptically as implying sustainable 'growth' and disregarding the finite nature of natural resources. This arena embraced instead (and with some hesitation; cf. Chapter 1) the closely related concept of 'sustainable use' (SU), coined in 1980 in the World Conservation Strategy (IUCN/UNEP/WWF, 1980) and formally adopted in the Convention on Biological Diversity (CBD) in 1992 and International Union for the Conservation of Nature and Natural Resources (IUCN) in 2000. The CBD (Article 2) defines 'sustainable use' as:

> 'The use of components of biological diversity in a way and at a rate that does not lead to the long-term decline of biological diversity, thereby maintaining its potential to meet the needs and aspirations of present and future generations.' (CBD, accessible at: http://www.cbd.int/convention/articles/?a=cbd-02)

The IUCN 2000 Policy Statement on sustainable use[1] recognizes that the use of wild resources, if sustainable, is an important conservation tool because the ongoing social and economic benefits derived from such use provide incentives for people to conserve them.

In this chapter, we will consider the concepts of sustainable development, sustainable use, ecologically sustainable development (ESD) and responsible fisheries as equivalent with similar long-term implications (albeit possibly different priorities in the short term).

Sustainability in fisheries

Based on the definitions above, few of the world fisheries could be considered "sustainable" judging from their performance. Problems and solutions have been abundantly identified, for example in Charles (1994), Garcia and Grainger (1997, 2005), Garcia and Newton (1997), Garcia and Rosenberg (2010), Grafton et al. (2010), Mace (1997), World Bank (2009) and Ye et al. (2012). During the last 60 years, fisheries governance has gone through a dynamic process of testing solutions to conserve target resources, with mixed outcomes. Significant institutional developments have progressively taken place, elaborating States' rights and responsibilities and reinforcing the legal means available to them in a changing Law of the Sea context (cf. Chapters 7, 10 and 11). The 1995 UN Fish Stock Agreement (UNFSA) and FAO Code of Conduct for Responsible Fisheries (CCRF) complemented the 1982 United Nations Convention on the Law of the Sea (UNCLOS), highlighting the States' duty of care, promoting stricter accountability of States and transparency in decision-making as well as more effective people participation and community empowerment, calling for more careful consideration of environmental impact and stressing the need for compatibility of management measures across resources distribution range. The general agreement of conservation institutions with these measures is signalled, for example in Greenpeace (1996).

Sustainability in biodiversity conservation

The weak sustainability of present economic developments in the ocean is also reflected in the unsatisfactory state of the ocean environment and biodiversity (MEA, 2005). Some problems are related to fisheries impacts which, in some regions, predominate. Others have to do with land-based and marine pollution and degradations, including global climate change. The biodiversity conservation framework itself has evolved (cf. Chapters 1 and 2) from an early focus on protection to the more dynamic and socially conscious sustainable use.

The IUCN Policy Statement (referred to above) specifies that sustainable use should minimize biodiversity loss, adapt management to risk and uncertainty (intrinsic to species and ecosystems) and make use of incentives and penalties. The statement: (1) calls for applying the principles of good governance, institutionalizing participation; (2) stresses the importance of systems of rights and of ethical values; (3) recommends removal of incentives and internalization of costs and benefits; and (4) highlights the need to reduce risks of resources degradation, conversion or depletion and to pay constant attention to changing global drivers. Most of these requirements are also part of the principles of responsible fisheries governance as reflected in the CCRF.

Sustainable development and integration

The need for integration of sectoral policies, objectives and programmes has been expressed in major international processes, for example: the 1992 Rio Declaration (and Agenda 21; Chapter 17); the 2000 Millennium Summit; the 2002 World Summit on Sustainable Development (WSSD); the 2005 Millennium Ecosystem Assessment (MA); and *The World We Want* Declaration of the 2012 Rio+20 Summit. It is a central concern of the United Nations General Assembly (UNGA) and environment programmes (UNEP, 2009, 2012) but the practical implementation of national commitments in that regard is still insufficient.

The term 'integration' is often used without clarifying its meaning and implications. In dictionaries, the meaning closer to our concern is: 'to unite something with something else; to blend, bring together, parts into a larger functioning whole; to combine two or more things to increase effectiveness'.[2] Integration is both a process and its outcome. In this chapter, 'integration' is viewed as the process of reduction of the 'functional distance' between the two streams of governance, filling gaps, increasing coherence and promoting synergy with the view to improve stakeholders' efficiency in reaching agreed objectives. Integration increases information exchange, consultation and cooperation in critical steps of the decision and implementation cycles of the two streams, to reduce the cost-benefit ratio of the processes. Integration follows 'good governance' principles, looking for efficiency, coherence, effective participation and systematic performance assessment.

The integration process might be imposed or endogenous and requires shared priorities and objectives. Its outcome may range from improved harmless harmonization to more demanding full integration (or even an aggressive takeover) and may be achieved with ease or difficulty (see next section).

Chapters 1 and 2 have shown that the two streams of governance have co-evolved in the right direction within a common legal framework (UNCLOS), but in parallel implementation frameworks and processes not sufficiently conducive to the level of coherence needed. Factors of friction include old entrenched habits and paradigms (e.g. fortress protection, freedom of the seas), administrative cultures and differences in perceptions of the state of the system and appropriate reference values. Friction also results from disagreement on the parameters of action such as the risk tolerance (aversion) in relation to ecological, economic and social wellbeing and the ensuing tension between the risks of not protecting biodiversity 'enough' and of constraining economic activities 'more than necessary'. Last but not least, it emerges from inequitable distribution of costs and benefits of change in any direction.

Integration process

The aim of integration

The main objective of integration is the improvement of the overall performance of policy-making and implementation in relation to the long-term achievement of bio-ecological and socio-economic goals. However, there is a wide range of not fully coinciding objectives and perceptions of what constitutes 'best performance' as well as different opinions on what type of 'integration' will work best. These differences can be at the root of some of the greatest challenges to progress on policy coherence and efficiency.

Some conservation champions may see integration as a way to ensure that environmental objectives become principal or overarching societal objectives. For example, for Lafferty (2002, p. 13) integration implies: (1) 'the incorporation of environmental objectives into all stages of policy making in non-environmental policy sectors'; (2) 'the aggregation of presumed environmental consequences into an overall evaluation of policy'; and (3) 'a commitment to minimize contradictions between environmental and sectoral policies by *giving priority to the former over the latter*' (emphasis added). Environmental Policy Integration (EPI) takes a similar perspective, as an approach to mainstream environmental objectives into development policies and environmental policies into the policies of economic sectors.

In both of the above, the proposed integration is *de facto* one-way in that the idea is to integrate environmental considerations into 'non-environmental policy sectors' but not the other way around. There is also an unbalanced distribution of costs and assessment of consequences (working with 'presumed' rather than demonstrated impacts). The latter might arise in circumstances when environmental risks are very large and/or alternative cost-effective ways to achieve social and economic objectives are available, but structuring such an outcome in as a precondition is unlikely to be widely embraced as a starting point for integration. In particular, this contrasts with a joint process of finding the most effective solution at the lowest possible aggregate cost for both sets of stakeholders. It is therefore not surprising that EPI has attracted great scholarly interest and some political backing, particularly in the developed world, but that 'its practical fulfillment appears to lag well behind aspirations, with deep disagreement regarding its actual application… few examples of joint best practices and scarce knowledge of policy outcomes' (Jordan and Lenschow, 2010).

The spirit of SD and SU is that integration, through a *balanced* governance approach, facilitates a compromise *acceptable to all concerned*,[3] balancing human and natural wellbeing.[4] In general, the intergovernmental debates on policy integration tend to aim more at harmonization between the two fields than a merger of the two or takeover by one of the two. Indeed, key policy actions to bring biodiversity considerations into fisheries policy have improved internationally agreed principles (e.g. good governance), approaches (e.g. precautionary and ecosystem approach to fisheries), processes (participation) and instruments (1995 UNFSA, CCRF, etc.) and brought key issues to the table. However, while broadening and better specifying States' responsibilities, these debates left fisheries policy-making and management practices largely unaltered in most countries (e.g. CBD, 2011; Gilman et al., 2012).

Integrating decision processes

As ongoing governance is a necessity, steps to better integrate the two streams of governance must obviously be undertaken dynamically while the processes are functioning. One might think of two trains moving on not-quite parallel railways, trying to synchronize timing and speed so that their cargo can be transferred, identifying moments and places where this can take place at least risk to both trains. The idea of achieving the benefits of integration while keeping the overall financial, institutional and political costs to the minimum is a challenging one.

Chapter 8 of Agenda 21, adopted by UNCED in 1992, addressed the issue of integrating environment and development in policy and decision-making and recommended action to: (1) integrate environment and development at all levels of the decision chain (policy, planning and management); (2) provide an effective legal and regulatory framework; (3) make effective use of economic instruments and incentives; and (4) develop systems to integrate environmental and economic accounting. The subsequent summits (WSSD in 2002 and Rio+20 in 2012) focused on implementation, particularly at the national level. In marine fisheries, implementation at the regional level is also considered a very useful step to promote the necessary cooperation.

Effective integration requires a reduction of the *functional distance* between the two streams at the various steps of their decision/implementation cycle (see below). It also requires a better integration between the various scales of governance at local, national, regional and global levels, the interaction of which determines the degree of legitimacy and performance.

Figure 3.1 illustrates the decision cycle followed in rationally managed fisheries and biodiversity conservation institutions (left and right panels) from the high-level policy principles and instruments developed in interaction between the global and national levels, to implementation, monitoring and iterative adaptation. The parallel and loosely coordinated processes use an overlapping set of instruments and it is argued that the institutional distance (degree of independence) between these processes could be further reduced for better joint outcomes by building more or stronger bridges between the different steps in the process.

The types of integrative instruments that are or could be used in the process are mentioned and explained in more detail and context within the different chapters of this book. Following and adapting Eggenberger and Partidario (2000, p. 204) and Persson (2004, p. 27, 36), the integration toolbox includes the following.

- *Legal instruments*: consisting of 'hard' and 'soft' law (cf. Chapters 7, 15).

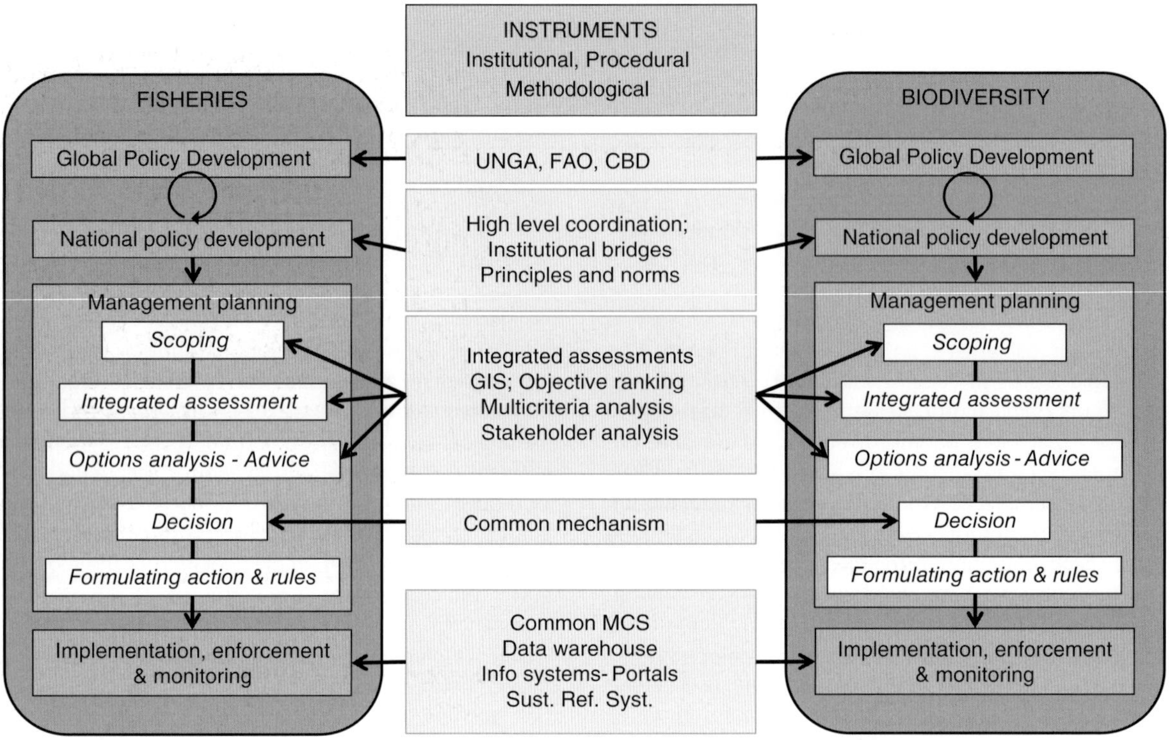

Figure 3.1 Types of integrative instruments that can be used to bridge key steps in the fisheries and biodiversity governance decision cycles.

- *Overarching policy frameworks* such as the Australian ESD (cf. Chapters 10, 22) to incorporate key principles in both governance systems (e.g. cross-sectoral integration, spatial basis, user rights and good governance) as well as to clarify mandates and roles, establish common definitions, objectives, constraints, norms[5] and approaches[6] and harmonize national positions in international organizations.
- *Institutional instruments*: development of the capacity of institutions to interact; establishment of 'bridging' hybrid institutions (common committees, task forces, memorandums of understanding or MOUs, audits, joint assessment processes) for integrated and strategic environmental assessment; institutional information exchange (e.g. on best practices); mandatory performance evaluation; and ultimately – if cost-effective – the merging of ministries (cf. Chapters 10, 12).
- *Common procedures* for integrated assessment of status and trends (cf. Chapter 9), risk assessments, performance evaluation and reporting and conflict resolution.
- *Common decision-support tools*: GIS, scenario analysis, stakeholder analysis, ecological accounting, multicriteria analysis, agent-based modelling, databases and shared data collection systems; and integrated sets of indicators and sustainability representation systems (cf. Chapters 13 and 14).
- *Communication*: joint formal communication with stakeholders.

The degree of integration

On a scale of integration ranging from little to total integration, we can identify the following degrees of integration (Figure 3.2).

- *Independent decision-making*. In this situation of 'infinite functional distance', the two streams operate separately with total independence, sometimes mutual ignorance and usually competing for power and budgets. In reality, total 'independence' is very rare as in any reasonably

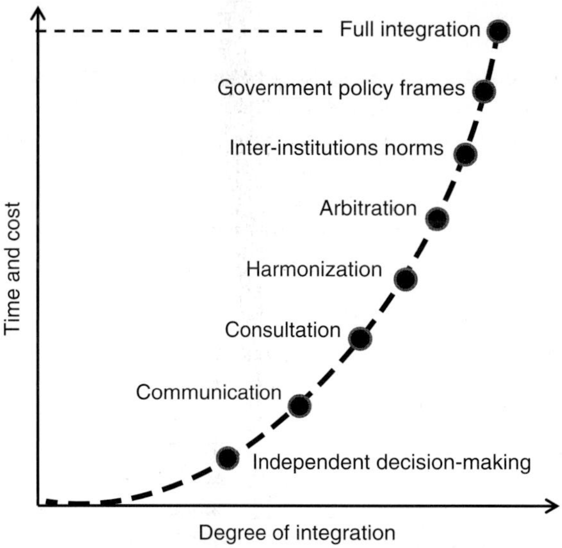

Figure 3.2 Schematic representation of the evolutions of cost and time needed for integration processes as a function of the degree of integration. (Source: Adapted from Metcalfe 1994.)

organized government the two streams would be forced to apply similar overriding national policies of a social, economic or environmental nature.

- *Communication.* In this 'dual monologue', the two streams keep each other informed, voluntarily exchanging information on objectives, constraints, activities and programmes and their outcomes. They may develop a common set of standards for joint databases as well as common toolboxes.
- *Consultation.* Institutions in each stream consult each other before or during the decision-making, sharing blueprints and white papers and organizing technical consultation meetings, but eventually each chooses its own path and pursues it independently.
- *Harmonization.* Both streams seek to build consensus, reduce/eliminate conflicts and look for synergies. They follow similar general lines, aiming to adopt compatible objectives and pursuing them in similar or complementary ways. Planning and implementation remain autonomous in each however, possibly enhanced by 'dialogue meetings', joint declarations and inter-ministerial mechanisms for collaboration.
- *Arbitration.* When horizontal collaboration between fisheries and conservation ministers is stalled, arbitration could be provided by higher authorities. Specifically, at a national level the office of the President or Prime Minister might intervene and decide with the procedure either institutionalized (e.g. in an inter-ministerial body) or *ad hoc.* In areas beyond national jurisdiction (ABNJ) only the UN General assembly has a mandate to play that role, with high moral albeit less binding authority.
- *Inter-institution norms.* These are established by higher authorities and represent a permanent mechanism of integration between the two streams. They provide a continuous integration frame and their correct application is evaluated regularly, with appropriate feedback and mandatory adaptive response.
- *Government policy framework.* As briefly addressed above, this option can be enhanced if a common strategy joint or coordinated implementation plans are adopted.
- *Full integration.* With zero functional distance, this combines all previous elements of the two streams in a (hopefully) better functioning whole, eliminating redundancies and filling gaps and reorganizing the distribution of powers. Generally, this means that both streams are put under a single ministry. There will still be cultural, substantial and operational differences to overcome between ministerial departments. For example, marine fisheries have already been in and out of other ministries such as agriculture, commerce or environment.

Policy integration requires comprehensive inputs (with the right scope), a specific and effective process and a systematic check on performance. The transaction costs of the integration process increase rapidly with the degree of integration (even though there may be overall economies of scale in the end and better outcomes). The problem is to find the optimal point offering a type of Pareto equilibrium[7] between institutions/stakeholders concerned (Underdal, 1980, p. 165).

The integration road map

The processes and conditions leading to better integration of the governance of marine fisheries and biodiversity conservation are similar to those advocated for achieving sustainable development (IUCN/UNEP/WWF, 1980; OECD, 2002, p. 813, box 1; WCED, 1987). The road map includes the following.

- *Review of policy and legal frameworks.* Review of national fishery and marine conservation policies, strategies and plans, identifying gaps, conflicts and potential synergies; review of relevant new and existing laws and regulation to check compatibility between objectives; mechanisms for effective feedback between the relevant government levels; and review of the two agendas to identify and enhance joint efforts on common issues, mainstreaming the principles of sustainable development/use in both agendas.

- *Development of a common understanding.* Progressive agreement on common principles, facts and tradeoffs in a pragmatic manner that may not take place in that order. Harmonization of goals and clear operational objectives that are clearly understood by the government, leaders of both streams and other stakeholders. This may include establishment of a catalogue of best practices with clear examples of benefits, as well as information devices for the sector, the consumers and the public.
- *Demonstration of commitment and leadership.* Clear commitment towards integration at the highest level, demonstrated over time by effective communication; explicit efforts to bridge gaps between the respective administrative and political agendas; determination of unambiguous joint priorities; maintenance of a sense of urgency; active promotion of joint pilot activities to demonstrate feasibility and benefits.
- *Provision of adequate means* to implement the integration plans, including plans for incentives and rewards for performance and provision for transitional costs.
- *Institutional development.* Strengthening of both sets of institutional structures, developing hybrid institutions and other institutional bridges at all levels of decision-making from policy-making to ground-level implementation.
- *Conflict resolution mechanisms.* Institutions and processes to foster healthy debates between scientists, managers and users on conflicts related to chronic or emerging problems; opening new arenas for negotiation in a dynamic integration process; adopting solutions that minimize conflict and balance sectoral and societal interests.
- *Stakeholder involvement.* Use approaches that increase credibility, legitimacy and compliance by the various stakeholders; adopt legal provisions and elaborate guidelines for active (possibly joint) consultation and participation of interested parties concerned at all levels; systematic evaluation of participative performance in consultations and implementation; mechanisms for ensuring transparency of decisions.
- *Procedural enhancements.* Elaboration of specific procedures (e.g. in systems of indicators, performance assessments, auditing, etc.) that effectively integrate environment and development concerns; adopt and apply the principles of good governance (UNDP, 1997).
- *Knowledge enhancement.* Development and management of knowledge bases; establishment of information flows among scientists and between them and the community and decision-makers; mechanisms for managing conflictual knowledge; development of scenarios and options; efforts to promote multidisciplinary research and integrated assessments (as proposed, for example, for small-scale fisheries by Garcia et al., 2008 and in Chapters 9 and 23–26).
- *Performance assessment.* Institutionalization of a framework for assessing performance of organizations in the integration process relative to sustainable development; evaluation mechanisms to support sustainability appraisal in the public sector and use of these evaluations in an adaptive manner, checking and optimizing performance.

The approach to integration might be *direct* or *indirect* (Persson, 2004). In a direct (or linear directive) approach, integration is 'imposed' from the top down with objectives, means, milestones and expected outcomes as in an engineering project. The indirect approach recognizes the difficulties of this strategy and puts in place the conditions needed for the integration to materialize from the interactions themselves. As the governance of complex systems is itself usually a complex system of systems (Garcia, 2009), Persson's indirect approach to integration amounts to promoting it as an emergent property of the complex governance system more than as an explicit design factor of social engineering.

Given the reality of system complexity, the outcomes of different integration pathways will likely not be completely predictable. For the same reason, no single pathway should be expected to emerge as universally 'best'. The common set of external drivers provided by economic globalization and international cooperation will, however, constrain the *a priori* numerous potential pathways to a limited set, characteristic of places, States, ecosystems or policy issues with similar histories and sets of constraints.

Integration factors

The streams of governance have been functional for about a century for fisheries resources conservation and about half a century for marine biodiversity conservation (cf. Chapter 1). Since the Brundtland Report (WCED, 1987), continued efforts have been made to improve fisheries and biodiversity conservation separately. As these efforts progressed, opportunities for integration have arisen and, at least in some cases, have been taken (Chapter 2). Present efforts towards better integration do not, therefore, develop in a vacuum. Legal frameworks, policies, strategies and plans have been developed and are being implemented and lessons have been learned. The potential integration field is encumbered by few successes, many disappointments and strong prejudices. It is complicated by jurisdictional boundaries (EEZs, high seas and designated areas of all sorts), by administrative inertia and turf-building, as well as by the complexity of the marine ecosystem and the interaction of human and natural subsystems. In addition, substantial parts of the concepts, strategies and plans for marine biodiversity conservation were developed for terrestrial ecosystems and populations, and jurisdictions where spatially based ownership was better established than on land. Not all parts of that heritage function with comparable efficiency in marine systems (Rice, 2011; Chapter 8).

The result of that interaction, against a backdrop of parallel histories, common policy backgrounds and global drivers, is that the 'integration' field is presently loaded with factors of convergence and divergence that may foster or stall integration. These factors are examined below.

Factors of convergence

As stated in Chapter 1 (developed in Chapters 12, 15 and 21), converging policy backgrounds and common global drivers have facilitated convergence of the two streams of governance notwithstanding the tensions between them.

From a bird's-eye view, there are strong similarities in: (1) the conservation targets (resources, biodiversity); (2) the jurisdictional framework (UNCLOS); (3) the law-making processes; (4) most long-term goals and some short-term goals; (5) major management approaches such as the precautionary, ecosystem, participative, good governance and rights-based approaches; (6) the historical shift from species-based to ecosystem-based strategies, from centralized to decentralized governance; and (7) a common desire for higher levels of spatial and sectoral integration of policies and management in highly conflictual areas.

There are also similarities in the challenges to governance effectiveness in both streams: (1) implementation difficulties at the national/local level; (2) capacity and costs of assessment, monitoring and reporting; (3) non-compliance and free riders; (4) decision loopholes, whether involuntary or planned (such as opting-out procedures); (5) mixed performance and high levels of failure;[8] (6) the complexity of socio-ecological systems (Folke et al., 2000); and (7) insufficient attention to socio-economics, social dynamics and equity. Some of these points are examined in the following sections.

There is complete agreement in the marine biodiversity and fishery resources conservation arenas on the fact that ocean fauna and flora are living renewable resources (or are essential to the maintenance of such resources) and part of the Earth's complex system supporting life and human livelihoods. The two streams share the vision of reaching *simultaneously* long-term ecosystem and human wellbeing, recognizing that conserving biodiversity and maintaining productive ecosystems for future generations requires consideration of the social and economic conditions of present generations (IUCN/UNEP/WWF, 1980; WCED, 1987).

The long-term goals of marine biodiversity and fishery resources conservation (responsible fisheries and sustainable use of biodiversity) are already aligned at the global level, thanks to efforts made by States since the 1972 Stockholm conference within international institutions (FAO, UNEP, IUCN, CBD, etc.) under the UNGA umbrella.

A quick comparison of the conditions for 'responsible fisheries' (e.g. as provided in the 1995 FAO Code of Conduct) and for 'sustainable use' of aquatic resources (e.g. as provided in the 2000 IUCN Policy Statement) indicate that there is also: (1) a high level of concordance in States' rhetoric about what constitutes an adequate policy framework both for sustainable fisheries and for marine biodiversity conservation; as well as (2) a high level of commitment to implement such policies. This concordance has led to some progress but much remains to be done (CBD, 2011; Gilman et al., 2012).

The process of reflection over international agreements and commitments in regional institutions and national systems is ongoing, boosted by the increasing accountability of States to report at the international level (e.g. the annual UNGA or the 2012 Rio + 20 summit) on progress (e.g. made in relation to the 2000 Millennium Development Goals, the 2002 WSSD plan of implementation targets, the 1992 CBD targets and their 2010 Aichi targets or the various UNGA resolutions on sustainable fisheries). However, a number of old and some emerging problems remain and the high-level conceptual coherence of the two streams is not yet (or too slowly) leading to coherence in ground-level implementation.

Factors of divergence

Conspicuous tensions exist between, as well as within, the two streams of governance that interfere with the integration process at various levels. They relate *inter alia* to differences in: (1) the balance in the respective sets of objectives and their relative weightings; (2) perceptions and tolerances of risks faced by ecosystems, resources and people; (3) preference for the allocation of costs and benefits; and (4) weighting of present versus future positive or negative outcomes. Some of these points are expanded below.

First, while the long-term goals underpinning the governance of both marine fisheries resources and biodiversity conservation are largely overlapping, they are not completely so (Salomon et al., 2011) and probably never will be; this naturally generates tensions.

The governance of the fishery sector aims at maintaining a responsible sectoral contribution to national SD, sustaining both ecosystem and human wellbeing. This implies constraining fishing operations with the view to maintaining as far as possible: (1) the resources (target and non-target) at or above their level of maximum biological productivity (MSY);[9] (2) the ecosystem structures and functions in altered but unimpaired states; and (3) viable enterprises and sustained community livelihoods.

In contrast, the overriding goal of marine biodiversity conservation is to maintain marine life, its supporting environment and ecosystem processes far above the minimal critical level below which their decline (natural or human-induced) may become difficult to reverse, and often in a state where the human footprint does not dominate over natural variation. This implies constraining human development activities to a level compatible with 'typical' ecosystem functioning.

The core objective of one governance stream therefore faces the core constraint of the other and finding common ground (positive overlap) is challenging. For example, limiting the reduction of a (target or non-target) fish population to say 20–40% of its un-fished biomass with concomitant changes in age structure may be considered necessary, acceptable and sustainable by those in charge of fisheries. However, this may be viewed by those in charge of conserving biodiversity as unacceptably adverse impacts, excessive perturbation of ecosystem processes and increased risks of extinction in the face of other stressors (e.g. harsh climatic conditions). This last concern is reinforced by the fact that many stocks appear to be still exploited beyond their MSY (Garcia and Grainger, 2005; Piet and Rice, 2004; Ye et al., 2012) and many recovery plans have failed to achieve their objectives in the timeframes initially set (Mora et al., 2009). The difference between the two streams appears to be, as usual, the different weights given to short-term costs and long-term benefits of reducing the sector's activity as well as the different perception of what 'broad ecosystem health' looks like.

While UNCLOS offers a legal reference (MSY) for yield, mortality and biomass, it does not offer any similar binding guidance in terms of proportion of the habitat that can be impacted; this vacuum is illustrated in the debate about the relative coverage of MPAs which the CBD tries to fill in its strategic plan objectives.

Common lessons: achievements and impediments

The last half-century of numerous failures and insufficient successes should have taught both streams similar lessons. For example: (1) the 'common pool' nature of marine biodiversity and fishery resources increases the risk of excessive rates of removal and degradation of productive habitats; (2) there is a maximum level of tolerable impact, beyond which the risk of irreversible damage becomes unacceptable; (3) scientific knowledge is limited and needs to be supplemented by traditional and informal knowledge; (4) scientific advice is necessary and important but the decision belongs to representative policy-makers; (5) there are inconsistencies between development and management strategies; (6) unresolved conflicts between users are a source of failure; (7) free riders weaken management strategies and increase non-compliance; (8) the rules of 'good governance' and adaptive management are essential and similar for both streams; (9) economic and social incentives have an important role to play in changing attitudes and behaviour; (10) poverty is a major obstacle to the conservation of resources and the ecosystem; (11) corruption is another; (12) systems of sound indicators may facilitate management and communication; and (13) the proper valuation of non-market resources must be addressed by environmental economists and social scientists.

These lessons should lead both streams to adopt similar devices in their strategies (e.g. on use rights, economic incentives or spatial integration) and to better explore the potential for synergies (e.g. on joint assessments, monitoring and surveillance systems). Many chapters of the book explore these points in more depth, often showing that despite sharing many challenges and opportunities, the respective institutions may not always take away the same lesson. Where the lessons drawn happen to be the same, the ultimate results generally turn out to be real improvements in governance (cf. Chapters 2, 10, 11 and 16). Where differing expectations and risk tolerances lead to the two streams drawing different lessons from a common experience, governance challenges may even have increased (cf. Chapters 8 and 9).

The real differences between the two streams of governance have generated conflicts, even though in some instances these conflicts created arenas for more intensive interaction towards solutions acceptable to both streams.[10] Examples of this include: conservation versus traditional use rights, MPAs (1975-ongoing); aboriginal fishing rights (1982-ongoing); bycatch for example of dolphins in tuna purse seine fisheries (1980–2000), of seabirds in longline fisheries (since the late 1980s), of turtles in shrimp trawl fisheries (1994–2001) and of many of the same species in large-scale pelagic driftnets (1998–2001); ghost-fishing by abandoned or lost gear (since the mid-1980s); exploitation of forage species; destructive fishing practices; and the interaction of fisheries, MPAs and biodiversity in ABNJ since the early 2000s. It is encouraging to see that many of the recent issues have been approached less as partisan 'conflicts' to be won than as 'issues' to be resolved through cooperation and even compromise, perhaps because of the increasing convergence and growing awareness that neither stream can achieve its goals without cooperation with the other.

Some divergences/conflicts affecting integration are also occurring within the two streams of governance. Such conflicts (albeit sometimes marginal in appearance) relate for example to the cost/benefit assessment of integrating social and economic arguments into conservation, and the pros and cons of soft law versus hard law at the fisheries/biodiversity interface (cf. Chapter 2). In fisheries governance, the adoption of eco-labelling in FAO and its broad up-take by markets and industry sectors for example involved good collaboration between NGOs and the fishery sector. However, until recent accommodations to certification processes were made, the high cost of the eco-labelling process and the required data acquisition, as well as the challenge of meeting stringent sustainability criteria (with the potential to be subject to market discrimination if unable to pay or to meet the criteria) led to opposition from groups of fish-exporting countries, developing countries and small-scale fisheries. Similarly the FAO process to reinforce port States responsibilities (to deter IUU) was originally resisted by a number of countries worried about their sovereign rights and duties (S. Beslier, personal communication, 2013). In the governance of biodiversity conservation, the most emblematic internal conflict is between protection and sustainable use, reflecting the original divergence between utilitarian (anthropocentric) and aesthetic (ecocentric) stands of conservation (cf. Chapter 1; de Klemm and Shine, 1993).

Divergences between the two governance streams and failures at integration can be traced to internal heterogeneity (and 'politics') within each stream. Each stream – and perhaps any governance system at all – has a range of participants from the pragmatic 'compromisers' to the radical 'ideologues'. Within-stream tensions may be amplified in the context of a negotiation process with the other stream if the latter fear that their stream may lose sight of its fundamental objectives by compromising through the acceptance of some of the others' arguments. The dynamics of consensus-building is analysed in some detail in Chapter 10.

At a higher level, ongoing change management faces a number of well-known problems and challenges including: (1) lengthy processes for parliamentary endorsement of UN agreements, modification of national legal frameworks and change of strongly entrenched social behaviours, particularly in areas where incentives and alternative livelihoods are limited; (2) insufficient capacity in science and administration at local and central levels to address the complex problems posed by fishery sustainability and marine conservation and their complex interaction; and (3) insufficient attention to social dynamics, equity and the need to design viable transitional pathways.[11]

Integration through interaction

The preceding sections have provided a brief and partial description of the issues related to integration and the opportunities offered to advance it. Clearly, the 'integration field' is complex and there is potential for further cooperation as well as conflict. Full integration may result from authoritative decisions from a government to merge two ministries into one or potentially, at an international level, for an RFMO and a Regional Seas organization to merge. More targeted integration may happen when a decision is made to formally establish institutional bridges, for example for management in large MPAs. However, the functional distance between the two streams of governance may also be reduced organically as common understandings, visions, objectives and approaches emerge from daily interaction as an 'evolutionary adaptation'.

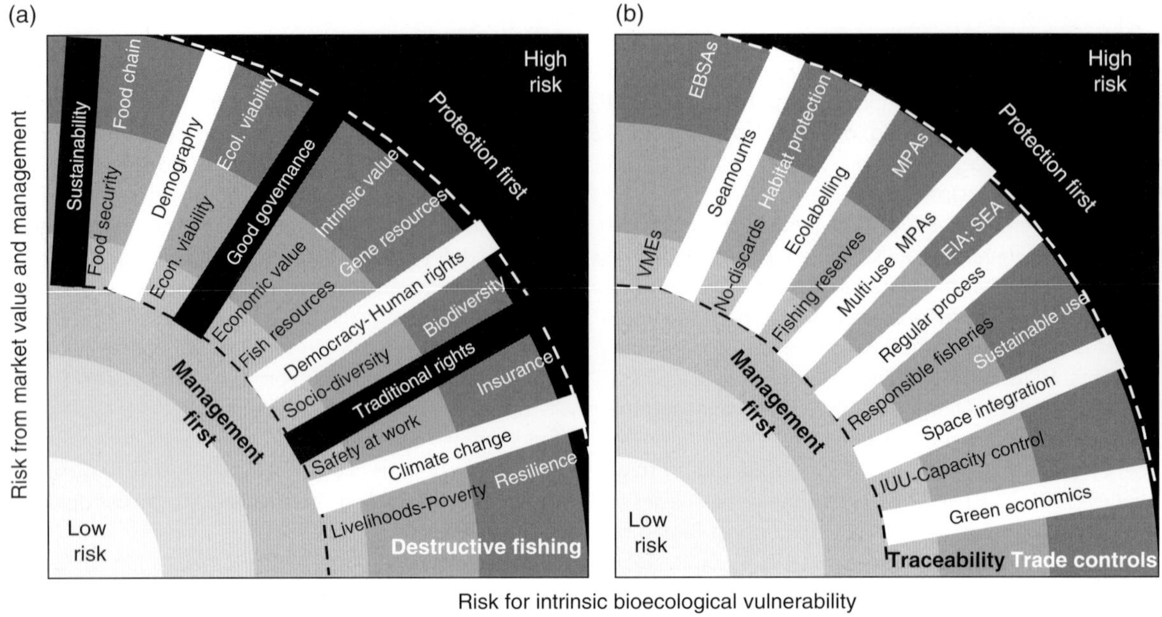

Figure 3.3 Schematic representation of the interaction between fisheries and biodiversity conservation on a risk background: (a) common drivers (white) and principles (black) and (b) issues (white) that act as bridges for improved integration. (Source: Adapted from Garcia 2010.)

To illustrate this point, the interaction field on which governance of biodiversity conservation and of fisheries meet as they exert their mandates is depicted in Figure 3.3 in relation to levels of risks. The shades of grey (from white to black) reflect the overall level of risk for the resources (or biodiversity) resulting from the interaction between: (1) their intrinsic bio-ecological vulnerability (for example, to fishing pressure or climate change), and (2) their vulnerability to market, economic and management aspects, such as attractiveness to industry (e.g. market value, abundance) or management inefficiencies (e.g. distance from shore, assessment and/or management errors).[12]

Where risk is low, fisheries management (and human socio-economic objectives) tend to have the priority and we might expect that FAO, RFMOs and national fishery management agencies would claim primacy. Where risk of lasting harm and slow recovery is high, the concern for protection takes priority; in this case CITES, CBD, regional and national environmental agencies may claim primacy. This is obviously a caricature as conservation attention is also given in low-risk situations and fisheries governance must also intervene to rebuild high-risk fishery resources.

Two particular challenges to governance should be noted here. First, the areas of low and high conservation risk[13] may be sufficiently contrasted to legitimate the respective primacy of fisheries or conservation governance. However, there is an area of intermediate levels of risk in which both streams may claim primacy, leading to either competition or cooperation. Second, the various institutional actors may well differ in their assessment of the risks, and thus may differ in where they consider the governance issue to lie in relation to their respective 'domains'; this could potentially lead to significant conflict.

In Chapter 2, we have argued that a convergence has been ongoing for at least 4 decades, albeit with difficulties. Figure 3.3 (left panel) illustrates the fact that incentives for better integration are provided by common drivers such as demography, democracy, human rights or climate change that impact the two streams in a similar manner and should call for similar types of strategies and actions. Convergence is also facilitated by adoption of common governance principles such as long-term sustainability, good governance or consideration of traditional rights. On the other hand, friction may arise from opposition, misalignment or tensions between contrasting ideas, over a diverse spectrum such as: economic vs. ecological viability; food security vs. forage species; and human vs. animal community resilience. Many of these tensions are higher on the short term than on the longer term.

In many ways, bridging is also facilitated by issues faced together, even from different angles (Figure 3.3, right panel). For example, vulnerable seamounts, eco-labelling, MPAs, integrated spatial management and the potential role of economics all offer platforms for encounter, differences testing and agreement reaching. Traceability in fisheries and trade controls in CITES have elements in common, as do responsible fisheries and sustainable use strategies, VMEs and EBSAs and fishery reserves and MPAs. There may be mutual suspicion and grounds for conflict, but there is also room for improved mutual understanding and, potentially, compromise and synergy.

Whether a result satisfactory to all parties can be achieved depends, of course, on the context and the power relationships among the actors involved. Furthermore, we recognize that the behaviour of both streams with respect to integration is not necessarily driven by societal optimization; instead, it is affected by sectoral ('within-stream') motivations which can include a desire to maintain or expand jurisdiction. Overall however, we hypothesize that more progress will be made in both streams through 'integration' of efforts across the two streams than through competition, and that this integration implies a lower aggregate cost to society. These hypotheses will be explored throughout this book.

Concluding thoughts

In order to fully tackle the integration process, the institutions and processes involved must have suitable maturity, internal coherence and capacity. Unfortunately, that is often not yet the case. In many cases, integration may therefore need to be applied to individual components of each stream of governance, to accelerate positive evolution as well as the converging coevolution (through joint learning and common moves). For example, the science in both streams needs to become integrated itself: a multidisciplinary science of complex systems with both social and natural sciences and including complex modelling, traditional knowledge and participatory assessment and advisory mechanisms This is occurring slowly, but the coevolution could be accelerated by the development of joint assessment processes with the above attributes. Availability of a common science foundation then allows institutional design of integrated planning as well as a strategy for conflict resolution, without requiring more difficult merging of policy and management authorities across streams (cf. Chapter 9; Garcia et al., 2008).

Similarly, there is a need to foster the integrated use of existing legal instruments through development or better use of existing integrative policy frameworks for sustainability, at national (cf. Chapter 22), regional (cf. Chapters 12, 15–20) and global levels (cf. Chapters 7, 10 and 11). There are particular issues in the high seas where the rights and duties of States are still being specified and the instruments are still being developed to help fulfil them. The ongoing debate about the conservation and sustainable use of biodiversity in ABNJ is testing the maturity of the systems intervening in ocean governance and the real willingness of States to accommodate both conservation and development concerns under sustainable use. Agreement on what a sustainably used ocean in ABNJ would look like seems a precondition for constructive dialogue between the streams on the degree of structural integration of governance that is needed in ABNJ.

A better integration of conservation with fisheries requires taking into account the resilience of the already heavily stressed coastal communities concerned in relation to the change required for impact mitigation or ecosystem rebuilding, or when facing loss of livelihoods and/or development opportunities. Such integration should give significant attention to transition costs and not just outcomes, and the development of pathways that minimize the stress on populations while pursuing the intended ecosystem rehabilitation or protection. Community engagement in decision-making is essential (cf. Chapters 6, 22–25 and 27).

With the globally rising awareness of ecosystem-level and ocean-wide interconnections, cumulative impacts and the need for inter-sectoral management, important discussions are also ongoing at global and national levels about the potential need and options for high-level integration both among sectoral management authorities and between them and their biodiversity conservation counterparts. Integration is advocated at all levels from the local, community-relevant level to the EEZ (Integrated Coastal Area Management or ICAM; Marine Spatial Planning or MSP), the large marine ecosystem (LME) or whole oceans (Integrated Ocean Management or IOM). The implications of that perspective are complex; experience is still limited and the debate is still open (cf. Chapters 8 and 10). Additional problems emerge in scaling up, notably an ignorance about and mismatch of the most relevant boundaries and scales of the ecological, social and economic subsystems. As the geographical scale of integration grows to higher-level issues, the risk increases that the needed contact and support from local populations will be lost and discontinuous institutions will be overstretched, leading to efficiency loss and failure.

Integration appears therefore as a utopia: a real reachable opportunity or a source of undue burden depending on the context. Accordingly, its promotion and implementation should be strongly based on local realities and capacity.

Notes

1. http://intranet.iucn.org/webfiles/doc/SSC/SSCwebsite/Policy_statements/The_IUCN_Policy_Statement_on_Sustainable_Use_of_Wild_Living_Resources.pdf
2. Merriam-Webster's Collegiate Dictionary (http://www.m-w.com); Dictionary.com (http://dictionary.reference.com/browse/Integrate)
3. However, those who study power dynamics argue that there is no reason to expect that outcomes be '*satisfactory to all concerned*' since the outcome is entirely dependent on who has the power when these 'compromises' are being worked out.
4. *Human wellbeing* is a condition in which all members of society are able to determine and meet their needs and have a range of choices to meet their potential. *Ecosystem wellbeing* is a condition in which the ecosystem maintains its diversity and quality (and thus its capacity to support people and the rest of life) and its potential to adapt to change and provide a wide range of choices and opportunities for the future (Prescott-Allen, 2001, p. 5).
5. For example, for Environmental Impact Assessment or Strategic Environmental Assessment.
6. For example, the precautionary and ecosystem approaches or integrated spatial planning.
7. Using Pareto optimality as a criterion when making tradeoffs may meet with objective difficulties relative to insufficient knowledge, for example (Persson, 2004).
8. With notable exceptions, the global performance of both streams of governance has been dismal as shown by the state of the environment and fishery resources (Cunningham and Bostock, 2005; FAO-SOFIA, 2010; Millennium Ecosystem Assessment, 2005; OECD, 1994; Sand 2001; Worm et al., 2009). Failures outnumber successes but, as stressed by Sand for conservation, one cannot tell how much worse the current situation would be without past efforts.
9. It is recognized that this stock-related reference level cannot be achieved simultaneously for all resources and that an acceptable compromise, at the ecosystem level, must be found. The single population MSY level of abundance, initially considered to be at around 50% of the unexploited level is now considered to be around 20–40% of the unexploited level (Hilborn and Walters, 1992).
10. The elements of history are described in Chapter 1 and Annex 1.
11. While the objectives are often agreed, the contrast between immediate conservation costs and delayed conservation benefits create unsustainable tensions in the absence of social safety nets. The Brundtland report (WCED, 1987, p. 313) stated that '... policy paths that are sustainable requires that the ecological dimensions of policy be considered at the same time as the economic, trade, energy...and other dimensions on the same agendas and in the same national and international institutions. That is the chief institutional challenge of the 1990s.' This remained the main challenge more than 20 years later at the Rio20 summit.
12. This reflection on sources of risk is borrowed from FAO work on the CITES listing criteria (FAO, 2000).
13. Corresponding respectively to biomass levels at MSY and at recruitment collapse.

References

Berkes, F., Folke, C. and Colding, J. (2000) *Linking Social and Ecological Systems. Management Practices and Social Mechanisms for Building Resilience*. Cambridge University Press, Cambridge.

Brown, K. (2003) Integrating conservation and development: a case of institutional misfit. *Frontiers in Ecology and the Environment* **1**, 479–487.

CBD (2011) Report of the Joint Expert Meeting on addressing biodiversity concerns in sustainable fisheries. Bergen, Norway, 7–9 December (2011) CBD, Nairobi. UNEP/CBD/JEM.BC-SF/1/4.

Charles, A. (1994) Towards sustainability: the fishery experience. *Ecological Economics* **11**, 201–211.

Charles, A. (2001) *Sustainable Fishery Systems*. Blackwell Science, Oxford. Fish and Aquatic Resources Series.

Cunningham, S. and Bostock, T. (eds) (2005) *Successful Fisheries Management*. Eburon Academic Publishers Delft, The Netherlands.

de Klemm, C. and Shine, C. (1993) Biological diversity conservation and the law. Legal mechanisms for conserving species and ecosystems. Gland (Switzerland). IUCN Environmental Policy and Law Paper, 29: xix + 292 pp.

Eggenberger, M. and Partidario, M. (2000) Development of a framework to assist the integration of environmental, social and economic issues in spatial planning. *Impact Assessment and Project Appraisal* **18**(3), 201–207.

FAO (1989) Sustainable development and natural resources management. Food and Agriculture Organization of the United Nations, Rome. C 89/2, Suppl. 2, August.

FAO (2010) The state of fisheries and aquaculture 2010. FAO, Rome.

FAO (2000) An appraisal of the suitability of the CITES criteria for listing commercially exploited aquatic species. *FAO Fisheries Circular* **954**, 66 pp.

Folke, C., Berkes, F. and Colding, J. (2000) Ecological practices and social mechanisms for building resilience and sustainability. In: *Linking Social and Ecological Systems: Management Practices and Social Mechanisms for Building Resilience* (eds F. Berkes, C. Folke, J. Colding), pp. 414–435. Cambridge University Press, Cambridge.

Garcia, S.M. (2009) Governance, science and society. The ecosystem approach to fisheries. In: *Handbook of Marine Fisheries Conservation and Management* (eds R. Quentin Grafton, R. Hilborn, D. Squires, M. Tait and M. Williams), pp. 87–98. Oxford, Oxford University Press.

Garcia, S.M. (2010) Fisheries management, nature conservation and the ecosystem approach: collision or syncretism? From Armageddon to the New Atlantis. Presentation at the *5th Biennial Conference of the international Institute of Fisheries Economics (IIFET)*. Economics of fish resources and aquatic ecosystems: balancing uses, balancing costs. Montpellier (France) 13–16 July (2010) Abstracts, 43 pp.

Garcia, S.M. and Grainger, R. (1997) Fisheries management and sustainability: A new perspective of an old problem? In: *Developing and Sustaining World Fisheries Resources* (eds D.A. Hancock, D.C. Smith, A. Grant and J.P. Beumer), pp. 631–654. The State of Science and Management. 2nd World Fisheries Congress, CSIRO.

Garcia, S.M. and Newton, C. (1997) Current Situation, trends and prospects in world capture fisheries. In: *Global Trends: Fisheries Management* (eds H.K. Pikitch, D.D. Huppert and M.P. Sissenwine), pp. 3–27. American Fisheries Society Symposium **20**.

Garcia, S.M. and Grainger, R.J. (2005) Gloom and doom? The future of marine capture fisheries. In: *Fisheries: A Future?* (eds J.R. Beddington and G.P. Kirkwood). *Philosophical Transactions of the Royal Society, B* **360**(1453), 21–46.

Garcia, S.M. and Charles, A.T. (2007) Fishery systems and linkages: from clockwork to soft watches. In: *Fishery Management Strategies*. Oxford University Press. (*ICES Journal of Marine Science* **64**(4), 580–587.)

Garcia, S.M. and Rosenberg, A.A. (2010) Food security and marine capture fisheries: characteristics, trends, drivers and future perspectives. *Philosophical Transactions of the Royal Society B* **365**, 2869–2880.

Garcia, S.M., Allison, E.H., Andrew, N.J., Bené, C., Bianchi, G., de Graaf, G.J., Kalikoski, D., Mahon, R. and Orensanz, J. M. (2008) Towards integrated assessment and advice in small scale fisheries: principles and processes. FAO Fisheries Technical Paper 515, 84 pp.

Gilman, E., Passfield, K. and Nakamura, K. (2012) Performance assessment of bycatch and discards governance by regional fisheries management organizations. IUCN Commission on Ecosystem Management, Regional Office for Oceania, Lenfest Ocean Program.

Grafton, Q.R., Hilborn, R., Squires, D., Tait, M. and Williams, M. (eds). (2010) *Handbook of Marine Fisheries Conservation and Management*. Oxford, Oxford University Press.

Greenpeace (1996) Greenpeace principles for ecologically responsible fisheries. Preliminary document. 6 pp. Available at http://www.greenpeace.org.uk/files/pdfs/migrated/MultimediaFiles/Live/FullReport/6640.pdf (accessed March 2014).

Hilborn, R. and Walters, C. (1992) *Quantitative Fisheries Stock Assessment*. Chapman & Hall, New York.

IUCN/UNEP/WWF (1980) *World Conservation Strategy: Living Resource Conservation for Sustainable Development*. IUCN, Gland, Switzerland.

Jordan, A. and Lenschow, A. (2010) Environmental policy integration: a state of the art review. *Environmental Policy and Governance* **20**, 147–158.

Lafferty, W. (2002) Adapting government practice to the goals of sustainable development. Improving Governance for Sustainable Development. OECD Seminar 22–23 November (2001) OECD, Paris.

Mace, P. (1997) Developing and sustaining world fisheries resources: The state of science and management. A keynote presentation. In: *Developing and Sustaining World Fisheries Resources: The State of Science and Management* (eds D.A. Hancock, D.C. Smith, A. Grant and J.P. Beumer), pp. 1–22. 2nd World Fisheries Congress, CSIRO.

Millennium Ecosystem Assessment (2005) *Ecosystems and Human Well-Being: Synthesis*. Island Press, Washington, DC.

Metcalfe, L. (1994) International policy coordination and public management reform. *International Review of Administrative Sciences* **60**, 271–290.

Mora, C., Myers, R.A., Coll, M., Libralato, S., Pitcher, T.J., Sumaila, R.U., Zeller, D., Watson, R., Gaston, K.G. and Worm, B. (2009) Management effectiveness of the world's marine fisheries. *PLos Biology* **7**(6), 1–11, e1000131.

OECD (1994) *Review of Fisheries in OECD Member Countries*. OECD, Paris.

OECD (2002) *Improving Policy Coherence and Integration for Sustainable Development: A Checklist*. OECD, Paris.

Persson, A. (2004) *Environmental Policy Integration: An Introduction*. Stockholm Environmental Institute. Available at http://www.sei-international.org/mediamanager/documents/Publications/Policy-institutions/pints_intro.pdf (accessed February 2014).

Piet, G.J. and Rice, J.C. (2004) Of precautionary reference points in providing management advice on North Sea fish stocks. *ICES Journal of Marine Science* **61**, 1305–1312.

Prescott-Allen, R. (2001) *The Well-Being of Nations: A Country-By-Country Index of Quality of Life and Environment.* Island Press, Washington.

Rice, J. (2011) Achieving coherent policies for conservation and sustainable use of marine ecosystems. *Conservation Biology* **25**(6), 1065–1068.

Rice, J. and Ridgeway, L. (2010) Conservation of biodiversity and fisheries management. In: *Handbook of Marine Fisheries Conservation and Management* (eds Q.R. Grafton, R. Hilborn, D. Squires, M. Tait and M. Williams), pp. 87–98. Oxford, Oxford University Press.

Rice, J. and Garcia, S.M. (2011) Fisheries, food security, climate change and biodiversity: characteristics of the sector and perspectives on emerging issues. Paper presented at the *International Symposium on Climate Change Effects on Fish and Fisheries*, Sendai, Japan, 24–26 April (2010) (also published in *ICES Journal on Marine Science* **68**(6), 1343–1353).

Salomon, A.K., Gaichas, S.K., Jensen, O.P., Agostini, V.N., Sloan, N.A., Rice, J., McClanahan, T.R., Ruckelshaus, M.H., Levin, P.S., Dulvy, N.V. and Babcock, E.A. (2011) Bridging the divide between fisheries and marine conservation science. *Bulletin of Marine Science* **87**(2), 251–274.

Sand, P.H.A. (2001) Century of green lessons: the contribution of nature conservation regimes to global environmental governance. *International Environmental Agreements: Politics, Law and Economics* **1**, 33–72.

Underdal, A. (1980) Integrated marine policy: What? Why? How? *Marine Policy* **4**(3), 159–169.

UNDP (1997) Governance for sustainable human development. A UNDP policy document. United Nations Development Programme. Available at http://mirror.undp.org/magnet/policy/ (accessed February 2014).

UNEP (2009) Integrated policymaking for sustainable development. A reference manual. UNEP, Nairobi.

UNEP (2012) *21 Issues for the 21st Century: Result of the UNEP Foresight Process on Emerging Environmental Issues.* United Nations Environment Programme (UNEP), Nairobi, Kenya.

WCED (1987) *Our common future. World Conference on Environment and Development.* Oxford University Press, Oxford.

World Bank (2009) *The Sunken Billions. The Economic Justification for Fisheries Reform.* World Bank, Washington and FAO, Rome.

Worm, B., Hilborn, R., Baum, J.K. et al. (2009) Rebuilding global fisheries. *Science* **325**(5940), 578–585.

Ye, Y., Cochrane, K., Bianchi, G., Willmann, R., Majkowski, J., Tandstad, M. and Carocci, F. (2012) Rebuilding global fisheries: the World Summit Goal, costs and benefits. *Fish and Fisheries* **14**(2), 174–185.

Part II
Governance dimensions

Chapter 4

Bio-ecological dimensions of fisheries management, biodiversity and governance

J. Rice[1,2] and P.M. Mace[3]

[1]Fisheries and Oceans Canada, Ottawa, Canada
[2]IUCN Commission on Ecosystem Management (IUCN/CEM/FEG), Gland, Switzerland
[3]New Zealand Ministry for Primary Industries, Wellington, New Zealand

Abstract: The biological basis of fisheries management is reviewed in the context of the sustainable use of marine resources and conservation of biodiversity. A roughly chronological approach is taken in discussing developments with regard to managing impacts of fishing on target species, bycatch, marine habitats and food web relationships. The roles of fisheries management agencies and biodiversity conservation interests are discussed as they evolved initially with little overlap and more recently with increasing interactions. Attention is paid to the similarities and differences between the perspectives in the priorities afforded different objectives of management, and the tools preferred for pursuing the objectives. Although progress towards addressing the broadening set of bio-ecological objectives has been erratic with numerous conflicts, a number of recent developments show promise for faster progress and more commonality of efforts.

Keywords: fisheries management; bycatch; ecosystem effects of fishing; fisheries productivity; (fisheries) management tools; sustainable fisheries

Introduction and background

Fisheries affect aquatic ecosystems in many ways which, for policy and management purposes, are often grouped into four categories: target species mortality; bycatch mortality; habitat alterations; and trophodynamic changes. The first two refer to direct effects on species or stocks. Many but not all fisheries use gears and fishing practices intended to capture a subset of all the species in the ecosystem, and such subsets are called the target species in the fishery. However, most fisheries commonly capture some unintended species and/or sizes or 'bycatch', a term that may or may not include species that are landed if taken, retained but used for fishmeal or fishoil or discarded, depending on jurisdictions (FA0, 2010a). The characterization or fate of catch as target or bycatch can vary depending on the demands of local or global fish markets. Because policies and regulations often differentiate 'bycatch' from 'targeted catch', in this chapter they will be considered two separate categories of fishery effects. In many artisanal fisheries all catch is retained and used (FAO, 2011a).

The other two categories of impacts cover broader effects on ecosystems. Many fishing gears impact the seafloor, with impacts ranging from negligible to destructive depending on the nature of the habitat, the gear and how it is deployed (CBD, 2012; FAO, 2008; Lokkeberg, 2005). The possible impacts of bottom-contacting gears include not only the destruction of structure-forming species such as corals and sponges, but such gears may also re-suspend the upper layers of the sediments and hence re-mobilize

Governance of Marine Fisheries and Biodiversity Conservation: Interaction and Coevolution, First Edition.
Edited by Serge M. Garcia, Jake Rice and Anthony Charles.

contaminants and fine particulate matter into the water column (Kaiser et al., 2000). In some circumstances seabed impacts may cause modest benefits to fast-growing high-turnover species or increase nutrient cycling in systems (de Madron et al., 2005). Collectively, these fishing gear impacts on habitats comprise the third category of fishery impacts.

Finally, fishing alters the abundance and size composition of the harvested species. As the abundances and size compositions are altered, predator–prey and competitive relationships among species may also be changed and these changes comprise the fourth category of fisheries effects. The nature of these food-web responses to harvesting are complex and not fully understood (Yodzis, 1994). Notwithstanding their complexity, several key conservation issues related to fisheries and marine food-web impacts have emerged.

(1) When systems have key forage species, their depletion by fisheries can have serious short- and longer-term consequences for many dependent predators (Baum and Worm, 2009).
(2) Large predators are important as a mechanism for top-down control and density dependent feedback (Heithaus et al., 2008; Myers and Worm, 2005).
(3) Equilibrium species-based models provide only limited insight to system dynamics for informing management decision-making, due to multiple pathways connecting many predators and prey and changing size-dependent prey selection as predators grow (Yodzis, 1994).
(4) The multiple linkages of predators and prey lead to 'functional redundancy', making predictions of future trajectories of food webs under different assumptions about fisheries highly uncertain (Schindler et al., 2010).
(5) Non-linearities and tipping points in food-web responses to exploitation can be large when they occur. Populations and communities may show few indications of impacts from fishing until all pathways for compensation are stressed, when further pressure may result in abrupt and substantial impacts (Bakun et al., 2009).

Although not a separate class of impacts, many jurisdictions have identified some species as protected, endangered, threatened or of high cultural value (PET species). PET species are generally top predators of low productivity including marine mammals, seabirds, turtles and sharks. They frequently pose special concerns for fishery impacts on biodiversity because of their low productivity, their key ecological roles and their vulnerability to detrimental impacts from fishing that reduce the availability of their prey species (Furness, 2002; NRC, 2003, 2006). When PET species are impacted, management of these fisheries may be of increased priority to both fisheries and conservation of biodiversity governance.

In addition, there is growing evidence that fisheries may alter the genetic composition of harvested species, particularly in cases of severe overfishing (Conover et al., 2005); such effects are not universally seen, however (Larsson et al., 2010). These genetic impacts have not yet been addressed in policies and management, and are not treated further in this chapter. They may become an emergent governance issue, although measures used to ensure the other impacts of fisheries are sustainable will usually result in sustainable levels of genetic impact as well.

As a generalization, the greatest impacts of fisheries should be on targeted species because the fisheries are actively trying to capture them. Other impacts are incidental and should be smaller. However, this generalization can be challenged on several grounds. For species with low biological productivity, even a low rate of bycatch mortality may impact population status (Patrick et al., 2010; Zhou et al., 2012). Habitat impacts can be extensive in fragile habitats fished with bottom-contacting gears, with potential for widespread consequences for species using those habitats. In addition, some types of habitat impacts are often difficult or impossible to reverse. Similarly, food-web impacts can be spread widely through interacting species and be long lasting (NRC, 2006) and the spatial pattern of removals can be as important as the magnitude, particularly for sedentary or semi-sedentary species (Frid et al., 1999). These potential complexities mean that although fisheries that are sustainable relative to their target species should have smaller overall impacts than unsustainable fisheries, merely keeping fisheries sustainable for the target species may not ensure that all biodiversity impacts will also be sustainable.

In the rest of this chapter we first review the historical context of fisheries agencies' accountabilities for the effects of fishing, then describe how ecosystem considerations became incorporated in policy and management. This evolution does not partition neatly into actions by separate streams of fisheries management governance and actions by conservation of biodiversity governance, but the role of each is highlighted as trends are discussed. We then examine each category of impact more thoroughly, to clarify the fisheries and the biodiversity features that are relevant to conservation and the sustainable use of fish stocks and ecosystems. In each case we also consider which tools in fisheries management and biodiversity conservation can be used effectively in managing these aspects of fisheries impacts. In the concluding sections we outline how national and international government agencies and non-government initiatives are adopting these tools. These new developments highlight the potential for conflicts and synergies, as well as opportunities for increasing the consistency and coherence among these initiatives in the context of achieving full bio-ecological sustainability of fisheries in marine ecosystems.

Fisheries management up to the 1990s

For many decades the agencies managing fisheries have been responsible for keeping the impacts of fisheries on the stocks sustainable while providing large catches. Although the contemporary language of sustainability (see Chapter 2) was not being used in the first half of the 20th century, the early fisheries population dynamics models were all trying to estimate 'optimum' harvest levels that were as large as possible without impairing the productivity of the stock (Baranov, 1918; Ricker, 1940; Thompson and Bell, 1934).

The nature of fisheries management changed fundamentally when operational biological reference points started to be developed and used. Amongst the first of these was the concept of maximum sustainable yield (MSY), originally developed during the 1930s to 1950s (Graham, 1936; Russell, 1931; Schaefer, 1954). With the development of the United Nations Convention on the Law of the Sea (UNCLOS) during 1973–82, national fisheries legislation and RFMO conventions have increasingly adopted MSY-based approaches, often using wording similar to that in Article 61(3) in UNCLOS:

> '3. Such measures shall also be designed to maintain or restore populations of harvested species at levels which can produce the maximum sustainable yield, as qualified by relevant environmental and economic factors, including the economic needs of coastal fishing communities and the special requirements of developing States, and taking into account fishing patterns, the interdependence of stocks and any generally recommended international minimum standards, whether subregional, regional or global.'

Early fisheries science acknowledged that environmental conditions could be important to stock production and fishery yields (Hjort, 1914; Silliman, 1943), but fisheries management was at most expected to react to how the stock changed due to environmental conditions ('as qualified by relevant environmental... factors'), not to take accountability for the environmental effects of fisheries themselves. Major advances in the analytical tools available for modelling fish population dynamics starting in the 1950s (Beverton and Holt, 1957; Ricker, 1955) led fisheries managers to expect more accurate assessments. Efforts to provide greater accuracy and precision in the advice to fisheries managers caused fisheries assessment scientists to try to include effects of the environment, predators and food supply in assessments of stock productivity (Anderson and Ursin, 1977; Kerr and Dickie, 2001). This made the ecosystem increasingly a consideration in fisheries management, but primarily as a complicating factor in keeping yields high and target stocks productive. Fisheries policy and management remained the domain of fisheries agencies as the factors considered in assessing stock productivity broadened.

As fisheries management organizations developed through the 20th century, they initially relied primarily on input controls such as restrictions on vessel numbers, vessel size or horsepower, gear type and size, gear characteristics such as mesh size, and area closures. Beginning as early as the 1970s, output controls such as catch quotas or Total Allowable Catches (TACs) and minimum or maximum sizes of fish that can be landed have increasingly been employed. The focus remained on managing the harvest of the target species, however; with some exceptions (see Chapter 18), output controls were rarely used directly to manage ecosystem impacts of fisheries.

Over this period, the policy environment for the conservation of biodiversity was developing independently (see Chapter 2), primarily focused on terrestrial, freshwater and near-shore systems. This included approaches for evaluating the risk of extinction of species and mitigating associated threats (see Chapter 13), and for protecting special ecosystems and habitats (see Chapter 14). Organizations developing these tools had few links to agencies assessing and managing fisheries, aside from concerns about bycatch of seabirds and marine mammals. When these organizations began to turn their attention to marine ecosystems they brought with them concepts and tools that they considered well tested (although they had been primarily applied in terrestrial systems). However, they lacked the jurisdiction to implement these concepts and tools in fisheries management and fisheries management agencies questioned many aspects of these tools. Merging these two independently developed frameworks into coherent governance has gone smoothly in some cases (see Chapter 14) and not so well in others.

The ecological categories of impacts of fishing and their management

The following sections primarily cover developments since the 1990s.

Impacts on target species

The main target species in large-scale fisheries tend to be those that are abundant, productive and/or have high economic value, particularly if they aggregate in predictable locations and times. Some fisheries (e.g. those on some small pelagics and shellfish) focus on individual species, while in others (e.g. many demersal finfish fisheries) there may be multiple target species with varying desirability and value. Small-scale fisheries may harvest broadly, with different parts of the catch used for local trade, personal consumption or distant markets (see Chapters 22 and 25). A wide variety of fishing gears are used to target species, usually taking advantage of their patterns of aggregation. These include purse seines for small and large pelagic species, hooks and lines, pots, traps, gillnets, driftnets, mid-water trawls, bottom trawls and dredges. These fishing gears have variable selectivity and impacts on associated species and habitats, with the effects depending on when, where and how the gears are deployed.

The impacts on target species focus on whether the amount harvested is sustainable and if the harvest is taken from appropriate ages, sizes and places. Despite continuing debate about MSY as a suitable management reference point (e.g. Larkin, 1977, but see Mace, 2001; Punt and Smith, 2001), it remains in wide use (e.g. WSSD, 2002). However, the calculation and implementation of MSY-based targets has been highly variable, and many fisheries agencies have chosen to substitute alternative approaches. In particular, from the 1970s onward, many fisheries management organizations have preferred fishing mortality-based reference points such as $F_{0.1}$ and F_{max} from yield per recruit analysis, various levels of $F_{\%SPR}$ from spawning per recruit analysis, F_{MSY} itself from a number of vastly different types of models and other levels of fishing mortality variously denoted F_{med}, F_{high}, F_{low}, F_{PA} and F_{crash} (ICES, 2002). Most of these have been treated as fisheries management targets, but some have been treated as limits. Although many of these reference points are conceptually capable of taking trophodynamic considerations into account, such practices are not routine and are usually incorporated in *ad hoc* ways when done at all (CBD, 2012).

Important recent developments in single-species fisheries management and governance, and the science needed to support it, have arisen from recognition of major structural problems in fisheries, including:

- the role of fishing overcapacity as an underlying cause of overfishing (FAO, 1999; Garcia and Newton, 1997; Mace, 1997) and measures taken to control fishing capacity;
- fisheries subsidies which contribute to perpetuating exploitation of already severely depleted fish stocks and are incongruous with longer-term roles of fisheries in food security and livelihoods (Sumaila et al., 2008);
- IUU fishing is a major contributor to the overfishing of national and international marine resources (FAO, 2001); and
- formulating fisheries management plans and regulations is insufficient if effective management, control and surveillance systems are not put in place, or if fishers are not involved in their formulation and co-management.

Important actions to address these concerns include the following:

- the FAO-led Code of Conduct for Responsible Fisheries (with its guidelines and supplements) and International Plans of Action (IPOA) on selected topics (FAO, 1999, 2001);
- involvement by developing countries in international fisheries management initiatives;
- the formulation and implementation of catch share systems such as Individual Transferable Quotas (ITQs), territorial use rights (TURFs) and community-based fisheries management programs and related initiatives (Christy, 1982; Grafton, 1996);
- the implementation of limit reference points for either fishing mortality or fish biomass that are to be avoided with high probability;
- formal definitions of terms such as 'overfishing' and 'overfished' with a legal or policy requirement to develop

formal, time-constrained plans to rebuild fish stocks (DAFF, 2007; New Zealand Ministry of Fisheries, 2008; US Department of Commerce, 2007);
- harvest control rules that specify management rules as a function of fish biomass or the fishing mortality rate;
- improved stock assessment models that can integrate a range of different types of fishery-dependent and fishery-independent information, improved analytical techniques and management approaches for dealing with uncertainty; and
- Management Strategy Evaluations (MSEs), also called Operational Management Procedures (OPMs), that incorporate harvest strategies and analytical and monitoring uncertainties (Butterworth and Punt, 2003; Kell et al., 2006).

The above advances have all contributed to much more sophisticated fisheries assessment and modelling methods, and much more realistic interpretations of effective fisheries management and the precautionary approach. Implementation of the policies and measures has been almost exclusively by fisheries management bodies, with very limited roles for the conservation of biodiversity governance stream. However, biodiversity interests have been active in the discussions of appropriate levels for the reference points and control rules in those bodies, as well as in the development of provisions of the Code of Conduct and its IPOAs. Moreover, because these developments give greater consideration to the human dimension of fisheries, particularly with respect to economic and social considerations and the necessity of participatory governance (de Young et al., 2008; Garcia and Charles, 2008; Hilborn, 2007) they have provided avenues for biodiversity and fisheries interests to find common ground on which to move forward on shared long-term goals. Opening this dialogue between fisheries management agencies and conservation of biodiversity interests also created the conditions needed to increase the prominence in policy of the other effects of fishing on ecosystems.

Impacts on bycatch species

Early concerns about the effects of fisheries on bycatch species, particularly species with low productivity or high charisma, were focused on marine mammals and seabirds. Particularly hot issues included dolphin–tuna interactions in purse seine fisheries in the eastern Pacific Ocean since the 1960s and bycatch of marine mammals and seabirds in drift nets reaching lengths of up to 60 km, leading to the 1991 ban on drift nets longer than 2.5 km on the high seas.

More recently, an increasing number of other bycatch species has received public and legal attention, including oceanic sharks taken both as bycatch and targeted (Gallagher et al., 2012). In some cases fisheries management agencies have initiated measures to manage bycatch rates and composition, particularly for fisheries with exceptionally high bycatch such as shrimp fisheries (Alverson et al., 1994). Bycatch issues involving PET species increased in prominence after IUCN adopted quantitative criteria for assessing risk of extinction in formulating its Red List, and highlighted several marine species (see Chapter 13). By the 2008 IUCN Congress no fewer than seven different resolutions on fisheries impacts on vulnerable species, particularly sharks, had been proposed (IUCN, 2008). Target 6 of the CBD Aichi Biodiversity Targets also explicitly calls for Parties to ensure that fishery bycatch is within safe biological limits by 2020, posing additional demands on fisheries management agencies.

Responding to the increasing profile of fisheries bycatch, FAO organized an expert consultation on bycatch management and discard reduction in 2009 (FAO, 2010a) followed by a Technical Consultation, both with participation by biodiversity experts from IGOs and NGOs, leading to the International Guidelines on Bycatch Management and Discard Reduction (FAO, 2011a). These guidelines provide a full framework for the sustainable management of bycatch. A review at the end of 2011 (CBD, 2012) found that all RFMOs have adopted policies intended to manage bycatch, but that assessments and evidence of active bycatch monitoring and bycatch reduction are infrequent.

For this ecosystem effect of fishing, the fisheries and conservation of biodiversity governance streams have worked together, and there are examples of significant progress in reducing unsustainable bycatch. However, only a few RFMOs (e.g. CCAMLR) and States have comprehensive plans and measures for bycatch management in place.

Impacts on habitats

Impacts of fishing gears on seafloor habitats had emerged as a policy issue by the early 1990s (ICES, 1990), and was the subject of a major review in the EU in the mid 1990s (Lindeboom and deGroot, 1998). For the next decade or more, fisheries management jurisdictions generally treated the issue as relatively isolated with particular hotspots of focal interest (Rijnsdorf et al., 1998). However, conservation of biodiversity interests have increasingly called for widespread and large-scale MPA networks (see Chapter 8) and even moratoria on the use of bottom trawls in ABNJ (Black, 2005) and, in some cases, also in national EEZs.

Both fisheries agencies and biodiversity conservation interests have long histories of advocating spatial closures or gear restrictions to protect habitats, but their emphasis has differed. Biodiversity interests promote the use of protected areas as a primary tool to protect important and sensitive habitats from the impacts of human uses, including fishing, with implementation and management by environmental agencies (Rice et al., 2012; Chapter 8). This contrasts with the long-standing use of permanent and seasonal area as a fisheries management tool, primarily for protecting spawning grounds or nursery areas for key commercial fish species (FAO, 2011b). These separate avenues converged in the 2000s, with WSSD (2002) calling for networks of MPAs globally and regionally by 2012 and the 2006 UNGA Sustainable Fisheries Resolution 61/105 requiring States and RFMOs to seriously address the impacts of fishing practices on vulnerable marine ecosystems (paragraph 80–89), and encouraged fisheries management agencies to further explore the contribution that MPAs can make to fisheries management (para. 92).

The more recent evolution of those initiatives is covered in Chapters 8 and 14. In general, the efforts of fisheries management agencies and environmental agencies have remained largely on independent but parallel tracks. Fisheries agencies are increasingly according long-term protection to specific habitats; for example, the Norwegian Ministry of Fisheries and Coastal Affairs began protecting cold-water coral reefs in 1999; the US National Marine Fisheries Service banned trawling in a large proportion of areas between 3 and 200 NM (nautical miles) off its west coast in 2006; and the New Zealand fishing industry worked with the then Ministry of Fisheries to implement a series of Benthic Protected Areas covering about 30% of that country's EEZ in 2007 (Helson et al., 2010).

FAO has recently produced guidance on how MPAs can contribute to fisheries management objectives (FAO, 2011b). An FAO-sponsored expert workshop concluded that many of the ecological objectives for spatial approaches to fisheries management and conservation of biodiversity are the same, although incompatibilities remain in the social and economic objectives of fisheries and conservation biologists (Rice et al., 2012) and their respective risk tolerances. Concomitantly, conservation of biodiversity interests are making quite specific recommendations for networks of MPAs to address the impacts of fishing on seabed communities and habitats, with little or no engagement by agencies with competence for the management of fisheries (Rogers and Gianni, 2010). Although both fisheries management and environmental agencies acknowledge the importance of managing the impacts of fisheries on marine habitats, there is far from full convergence on the best strategies including the ways, means, costs and benefits of doing so.

Impacts on food webs

Forage species

There are several examples of fisheries management applying greater risk aversion in the management of fisheries on forage species, including a lower exploitation rate than implied by the stock's life history parameters, effectively setting aside an 'ecosystem allocation' of forage species biomass for predator needs, and spatial or temporal measures to prevent fishing at times and places when the forage species or its predators are particularly vulnerable (Anonymous 1997; NRC, 2003; Shelton et al., 1992). Such measures have been far from universally applied, however (Montecino and Lange, 2009). Conservation advocates have called for even more stringent restrictions on fisheries on forage species (Everson, 2000; Furness, 2002; http://www.oceanconservationscience.org/foragefish/), or even outright bans (http://blueplanetsociety.blogspot.com/). In some cases however, environmental agencies and interest groups have worked with commercial fisheries to develop and implement management strategies for forage species that link spatial and/or temporal restrictions on harvest to the status of dependent predators (CBD, 2012; Crawford, 2004; Daunt et al., 2008; Okes et al., 2009; Smith et al., 2011).

Maintaining large predators

Simply achieving the goals of target species management should allow a portion of populations with large maximum lengths to reach large sizes. Beyond that, fisheries managers have occasionally placed a cap on aggregate removals of large predators, such that fisheries are closed when the sum of catches of the suite of predators reaches the cap, regardless of whatever quota may remain unharvested for some of the species (e.g. ICNAF for the Northwest Atlantic in the 1970s, Anderson, 1998; North Pacific Fisheries Management Council for the Bering Sea in the late 2000s, Witherell et al., 2000). Conservation of biodiversity interests tend to promote approaches focused on overall reductions in fishing effort, large no-take protected areas and, in some cases, bans on fishing species groups considered particularly vulnerable (Ferretti et al., 2010; Roberts, 2007). Again the goals and preferred tools of the two governance schemes appear to differ little; the major differences are in perceived urgency and desired severity of actions.

Areas of overlap and potential for inconsistencies between fisheries and conservation of biodiversity approaches

Populations: Target and bycatch species

Since the late 1990s there have been increasing calls to move from single-species fisheries management to ecosystem approaches to fisheries. To support this change a large number of existing or new organizations have been modified or set up, numerous meetings held and many new, mostly voluntary, agreements signed. As recent UNGA Sustainable Fisheries Resolutions affirm, the widespread endorsement of ecosystem approaches to fisheries management should not divert attention from tractable issues and tactical measures such as reducing excessive fishing mortality, eliminating fishing fleet overcapacity and harmful fishing subsidies and instituting more effective controls and sanctions on IUU fishing. It is also important that ecosystem approaches do not dilute efforts at improving the databases and scientific underpinnings of managing fisheries, by spreading limited resources more broadly over a wider range of ecosystem and fishery properties, each monitored and studied with less depth.

Eliminating overfishing – and fishing fleet overcapacity – on individual target species is a crucial first step towards ensuring the persistence and integrity of marine ecosystems. Conservative single-species fishing mortality rates are more likely to satisfy common ecosystem objectives, and achieving an operational objective such as 'reduce single-species fishing mortalities to or below F_{MSY}' is highly likely to achieve positive results on many ecosystem components, not just the target species. Moreover, the ways to achieve such an objective are known and the jurisdictions of agencies who can take the needed actions are clear. We know what to do if the goal is to rebuild individual fish stocks to a target (e.g. B_{MSY}) level: reduce fishing mortality on that species. If rebuilding several species in an ecosystem is the aim, reduce fishing mortality on all of them which will also result in a reduced impact on bycatch.

On the other hand, it is far from clear what actions are needed to achieve a vague objective such as 'maintain ecosystem integrity' or 'consider ecosystem impacts', or which governance mechanisms can be held accountable for taking the actions. Although there has been an increasing number of success stories of rebuilding of individual fish stocks, particularly in national jurisdictions (Mace, 2004; Worm et al., 2009), even the basic concepts for 'ecosystem rebuilding' are still being debated (Rodriguez et al., 2011). At the same time, there is little reason for complacency; fisheries management agencies are not universally successful in addressing even the species-level effects of fishing on marine ecosystems (Hilborn and Hilborn, 2012). Widespread overfishing continues despite the clear mandates to reduce it, and many fish stocks that were depleted years to decades ago remain depleted. Moreover, the implementation gap between the growing number of international fisheries and marine conservation instruments and agreements and any effective associated action is cause for concern, particularly given that most international agreements are non-binding. A more collaborative approach that attempts to identify common ground and work towards common goals has a far greater chance of success ('The economic context of governance' in Chapter 5 provides examples).

Habitats

There is universal agreement that representative marine habitats should be protected from damage by fishing (FAO, 2003, 2011b; UNGA Resolution 61/105). Sensitive habitats are also a top priority for conservation of biodiversity interests (cf. Chapter 14). However, there are strongly divergent views on whether designation of large areas as MPAs or marine reserves is essential to such protection. Depending on where they are placed and how they are managed, it is becoming increasingly acknowledged that MPAs can make positive contributions to fisheries management objectives. It is also being recognized that MPAs do not deal with the underlying problems of excessive fishing capacity and fishing mortality. In some cases MPAs may even have overall negative consequences, particularly if the consequences of displaced fishing effort and the needs of coastal communities are not considered carefully in their design (see Chapters 8 and 24).

Food webs and communities

In terms of priorities for policy and management action, the food-web impacts of fisheries are higher on the agenda of environmental agencies than fisheries agencies. The FAO Guidelines on the Ecosystem Approach to Fisheries (FAO, 2003) call for attention to such considerations but neither the sections on use of management tools nor those on developing management plans directly address the effects of fishing on food webs and communities. In contrast, fisheries on forage species and general benefits to marine food webs from increased restrictions on or exclusion of fisheries are featured in several UNEP and CBD documents (UNEP, 2011). In addition to the differences in priority given to addressing food-web impacts of fisheries, the actual approaches proposed for doing so tend to reflect the different preferences of the various agencies. Fisheries agencies consider reduction in overfishing and rebuilding of depleted stocks to be a primary strategy for addressing food-web issues, whereas biodiversity interests view closures of important feeding grounds and protection of key forage species from fisheries as key.

Venues for change

Over the last two decades, but particularly since about 2005, a substantial number of relatively new initiatives provide optimism for the future convergence of fisheries governance and conservation of biodiversity governance streams. A few examples follow.

Interagency collaborations

An increasing number of collaborative arrangements are being developed between fisheries agencies and marine conservation agencies. Many of these are in the form of Memoranda of Understanding (MoUs) or similar instruments. The ability to implement these MoUs can be hampered by: (1) the disparate goals of the signing parties; and (2) the resources and logistical difficulties of actually being able to work together to implement programmes designed to reach common goals, when staff tend to be already over-burdened with large workloads geared towards the core objectives of their respective agencies.

Examples include MoUs between FAO and CITES and between FAO and CBD. The former of these was signed in 2006, while the latter has involved a number of individual agreements in various areas of mutual interest. The FAO–CITES MoU formalized an evolving relationship between the two parties to improve the monitoring and management of fisheries resources in international trade (cf. Chapter 13). FAO–CBD agreements and initiatives relevant to marine fisheries include a memorandum of cooperation to collaborate on the UN Atlas of the Oceans in 2001 and two more general memorandums of cooperation in 2005 and 2011. These agencies are also increasingly co-sponsoring joint expert meetings, such as an FAO–CBD workshop on Destructive Fishing Practices (FAO, 2010b) and the FAO–UNEP workshop on MPAs and Fisheries (Rice et al., 2012).

UN Working Groups

Chapter 11 reviews several specific initiatives within the United Nations to increase collaboration and coherence of policy and management efforts in ABNJ. Three of these are summarized here.

The Informal Consultative Program (ICP) was established in UNGA Resolution 54/33 with a broad mandate to guide the UN General Assembly efforts regarding coordination and cooperation on the full range of ocean issues at the intergovernmental and inter-agency levels. Topics that have been addressed include:

- the strengthening of the information base for the management of ocean activities and conservation of biodiversity;
- capacity-building and technology transfer;
- cooperation and coordination for integrated ocean management and ecosystem approaches;
- conduct of environmental impact assessments; and
- use of area-based management tools, in particular marine protected areas.

The second example is the annual 'Ad Hoc open-ended Informal Working Group to study issues relating to the conservation and sustainable use of marine biological diversity beyond areas of national jurisdiction (ABNJ)'. This meeting has become the central forum for discussions of global aspects of high-seas MPA networks, environmental impact assessments and marine genetic resources. These first two examples have specific implications for developing more common ground between the fisheries and conservation of biodiversity governance streams.

The third example is the 'Regular process for global reporting and assessment of the state of the marine environment, including socio-economic aspects'. The overall mandate of the Regular Process is to improve understanding of the oceans and to develop a global system for delivering science-based information to decision makers and the public, by serving as the mechanism to keep the world's oceans and seas under continuing review and providing regular assessments at global and supra-regional levels. It is premature to speculate on how the Regular Process will contribute to improved ocean governance and greater coherence of policy and management between fisheries and conservation of biodiversity, but the explicit mandate to produce fully integrated policy-relevant assessments and the comprehensive structure of the first report create a promising context for progress.

Non-government initiatives

Eco-certification

Eco-certification of fishery products is rapidly gaining momentum, particularly in developed countries, largely as a result of environmental groups placing pressure on retailers and restaurants to avoid fish caught 'unsustainably' and to set a date for moving towards sourcing and selling certified products only. In 2005, FAO published a set of Guidelines for the Eco-labelling of Fish and Fishery Products from Marine Wild Capture Fisheries. The best-known and most widely used fisheries certification program is that of the Marine Stewardship Council (MSC). Fish that have the MSC eco-label indicate that the fishery from which it comes operates in an environmentally responsible way, that overfishing is not occurring and the stock is not overfished, taking into account maintenance of the target fish stock (Principle 1) and of the ecosystem (P2), and effectiveness of the fishery management system (P3). Almost 10 million tonnes of seafood (more than 11% of the annual global harvest) come from MSC-certified fisheries.

The MSC certification program is not without its detractors, particularly those who believe that trawl fisheries are by definition unsustainable. As a result, several rival certification schemes have appeared.

Environmental NGOs

The US Environmental Defense Fund (EDF) was among the first of the environmental NGOs to recognize the potential benefits of forming collaborative partnerships with industry and other resource users. Over the last 20 years, it has developed a large number of partnerships with companies and explored market incentives to make industries more environmentally conscious and responsible. Recently, EDF has promoted the propagation of catch share systems as ways to rationalize fisheries, which usually results in an increase of the abundance of fish stocks and increased profitability of fisheries.

The World Wildlife Fund (WWF) also recognized that industries have a crucial role to play in supporting conservation efforts and being part of the solution to conservation challenges. The organization works with major companies and their supply chains to promote market change in the way key global commodities are produced, processed, consumed and financed worldwide. The creation of the Marine Stewardship Council jointly with Unilever in 1996 is one of many examples relevant to fisheries and marine conservation.

The International Union for the Conservation of Nature (IUCN) works on marine issues through a number of subgroups. The Species Survival Commission's Marine Conservation Sub-Committee addresses marine species at risk issues, as described in Chapter 13. It also has a Fisheries Expert Group in the Commission on Ecosystem Management which has undertaken specific activities, including preparation of this book. The Protected Areas Working Group and the Marine and Polar Program are also active at the interface of fisheries management and conservation of biodiversity.

Some other environmental NGOs have yet to embrace the collaborative conservation–industry model. For example, the Pew Environment Group sees their Global Oceans Legacy project as a priority for marine conservation. This project aims to establish a worldwide system of very large highly protected marine reserves where fishing and other extractive activities are prohibited. Other marine initiatives of the Group include global shark conservation, global tuna conservation, an Antarctic krill conservation project and campaigns for ending fisheries on forage species.

NGOs with industry affiliations

An increasing number of NGOs that include industry partners or board members are materializing, primarily in response to perceived needs to: work more collaboratively with other organizations; promote the effective use of science; and, in some cases, to counter anti-fishing lobbies by disseminating objective information about the failures and successes of fisheries management. Many of them are members of the International Coalition of Fisheries Associations (ICFA), a coalition of national fish and seafood industry trade associations or fisheries councils that represents countries harvesting more than 85% of the world's fish. The group was formed in 1988 and aims to 'advocate policies for the long-term sustainable use of living marine resources for the benefit of global food security and prosperity'. A recent example of a coalition between an environmental NGO and industry is the International Seafood Sustainability Foundation (ISSF), a global coalition of scientists, the tuna industry and the World Wildlife Fund (WWF) formed to 'promote science-based initiatives for the long-term conservation and sustainable use of tuna stocks, reduce bycatch and promote ecosystem health'.

Conclusions

The bio-ecological basis for fisheries management has been rooted for over a century in simply keeping the impacts of fishing on various ecosystem components within sustainable bounds. Initially the target species in fisheries was the ecosystem component of concern, and 'sustainable' meant that yields were high and able to be kept high. The main management tools used were gear regulations, catch and effort limits, and spatial and temporal measures. Management advice was based on models of the dynamics of the target species. Ultimately, other parts of ecosystems became relevant to fisheries management to the extent that they affected the size or consistency of yields from target species. However, the population models and management tools were adapted only as much as needed to accommodate the increasing number of ecosystem considerations affecting yields.

In the decades following UNCLOS in 1982, UNCED in 1992 and particularly the FSA in 1995, there were a number of important changes to this evolving framework. Management Reference Points, now supported by Harvest Control Rules, and Management Procedures placed decision-making on more rigorous foundations, while economic instruments such as allocation of rights and certification became more widespread in the developed world and a few developing countries. At the same time, conservation biology interests focused attention on fishery impacts on seabed habitats, bycatch species and food webs as issues in their own right, rather than in the context of their impacts on fishery yields. These interests also promoted a different hierarchy of management tools with spatial measures (particularly no-take reserves), protection of specific species and prohibitions of specific gears as preferred tools.

A major expansion of the policy basis for fisheries management has occurred since the mid-2000s, with UNGA Resolutions, CBD Decisions and FAO Guidelines all contributing. Fisheries management agencies are only part-way through the transition to accommodating this larger framework for management, and various partnering arrangements with governance agencies responsible for the conservation of marine biodiversity, NGOs and the private sector are still being established and evolving. However, continued progress towards the classical goals for fisheries management should also continue to reduce unsustainable impacts of fishing on non-target species, habitats and food webs. These efforts will not be sufficient to address all impacts of fisheries on marine biodiversity, but eliminating overfishing is a necessary condition for progress on the conservation of biodiversity. Regardless of what other progress is being made in harmonizing the governance of fisheries and the conservation of marine biodiversity, building governance capacity and improving the effectiveness of fisheries management should remain a priority.

References

Alverson, D.L., Freeberg, M.H., Murawaski, S.A. and Pope, J.G. (1994) A global assessment of fisheries bycatch and discards. FAO Fisheries Technical Paper No. 339. FAO, Rome.

Anderson, E.D. (1998) The history of fisheries management and scientific advice – the ICNAF/NAFO history from the end of World War II to the present. *Journal of Northwest. Atlantic Fisheries Science* **23**, 75–94.

Anderson, K.P. and Ursin, E. (1977) A multispecies extension to the Beverton and Holt theory of fishing, with accounts of phosphorous circulation and primary production. *Meddelelser fra Danmarks Fiskeri-og Havundersøgelser* **7**, 219–435.

Anonymous (1997) Forage Fishes in Marine Ecosystems. *Lowell Wakefield Fisheries Symposium Series* 14, Alaska Sea Grant AK-SG-97-01. Fairbanks. 828 pp.

Bakun, A., Babcock, E.A. and Santora, C. (2009) Regulating a complex adaptive system via its wasp-waist: grappling with ecosystem-based management of the New England herring fishery. *ICES Journal of Marine Science* **66**, 1768–1775.

Baranov, F.I. (1918) On the question of the biological basis of fisheries. *Izvestiya otdela rybovodstva i nauchno-promyslovykh issledovanii* **I**, 81–128 (in Russian).

Baum, J.K. and Worm, B. (2009) Cascading top-down effects of changing oceanic predator abundances. *Journal of Animal Ecology* **78**, 699–714.

Beverton, R.J.H. and Holt, S.J. (1957) On the dynamics of exploited fish populations. *Fishery Investigations Series II* **19**, 533.

Black, R. (2005) Call for a ban on destructive fishing practices. BBC World News. Available at http://news.bbc.co.uk/2/hi/science/nature/4308156.stm (accessed February 2014).

Butterworth, D.S. and Punt, A.E. (2003) The role of harvest control laws, risk and uncertainty and the precautionary approach in ecosystem-based management. In: *Responsible Fisheries in the Marine Ecosystem* (eds M. Sinclair and G. Valdimarsson), pp. 311–319. FAO, Rome.

CBD (2012) Report of Joint Expert Meeting on Addressing Biodiversity Concerns in Sustainable Fishery. UNEP/CBD/SBSTTA/16/INF/13.

Christy, F.T. (1982) Territorial use rights in fisheries: definitions and conditions. FAO Fisheries Technical Paper 227. 10 pp.

Conover, D.O., Arnott, S.A., Walsh, M.R. and Munch, S.B. (2005) Darwinian fishery science: lessons from the Atlantic silverside (*Menidia menidia*). *Canadian Journal of Fisheries and Aquatic Sciences* **62**, 730–737.

Crawford, R.J.M. (2004) Accounting for food requirements of seabirds in fisheries management – the case of the South African purse-seine fishery. *African Journal of Marine Science* **26**, 197–203.

DAFF (2007) *Commonwealth Fisheries Harvest Strategy: Policy and Guidelines*. Australian Department of Agriculture, Fisheries and Forestry, Australia.

Daunt, F., Wanless, S., Greenstreet, S.P.R., Jensen, H., Hamer, K.C. and Harris, M.P. (2008) The impact of the sand eel closure on seabird food consumption, distribution and productivity in the northwestern North Sea. *Canadian Journal of Fisheries and Aquatic Sciences* **65**, 362–381.

de Madron X.D., Ferre, B., Le Corre, G., Grenz, C., Conan, P., Pujo-Pay, M., Bodiot, O. and Buscai, R. (2005) Trawling-induced resuspension and dispersal of muddy sediments and dissolved elements in the Gulf of Lion (NW Mediterranean). *Continental Shelf Research* **25**, 2387–2409.

de Young, C., Charles, A. and Hjort, A. (2008) Human dimensions of the ecosystem approach to fisheries: an overview of context, concepts, tools and methods. *FAO Fisheries Technical Paper* No. 489. FAO, Rome. 152 pp.

Everson, I. (ed.) (2000) *Krill: Biology, Ecology and Fisheries*. Fish and Aquatic Resources Series **6**. Blackwell Science, Oxford.

FAO (1999) International Plan of Action for reducing incidental catch of seabirds in longline fisheries; International Plan of Action for the conservation and management of sharks; International Plan of Action for the management of fishing capacity. FAO, Rome. 26 pp.

FAO (2001) International Plan of Action to prevent, deter and eliminate illegal, unreported and unregulated fishing. FAO, Rome. 24 pp.

FAO (2003) The ecosystem approach to fisheries. FAO Technical Guidelines for Responsible Fisheries. N04, Suppl. 2. FAO, Rome. 112 pp.

FAO (2008) Report of the Expert Consultation on International Guidelines for the Management of Deepsea Fisheries in the High Seas. Bangkok, 11–14 September 2007. FAO Fisheries Report No. 855. FAO, Rome. 39 pp.

FAO (2010a) Report of the Expert Consultation on International Guidelines for Bycatch Management and Reduction of Discards Rome, 30 November–3 December 2009. FAO Fisheries and Aquaculture Report No. 934.

FAO (2010b) Report of the FAO/UNEP Expert Meeting on Impacts of Destructive Fishing Practices, Unsustainable Fishing, and Illegal, Unreported and Unregulated (IUU) Fishing on Marine Biodiversity and Habitats. FAO Fisheries and Aquaculture Report No. 932. Rome. 33 pp.

FAO (2011a) Report of the technical consultation to develop international guidelines on bycatch management and reduction of discards. Rome, 6–10 December 2010. FAO, Rome.

FAO (2011b) Fisheries Management 4. MPAs and Fisheries Management. FAO Technical Guidelines for Responsible Fisheries No. 4. Supp. 4. FAO, Rome. 198 pp.

Ferretti, F., Worm, B., Britten, G.L., Heithaus, M.R. and Lotze, H.K. (2010) Patterns and ecosystem consequences of shark declines in the ocean. *Ecology Letters* **13**(8), 1055–1071.

Frid, C.J.L., Clark, R.A. and Hall, J.A. (1999) Long term changes in the benthos of a heavily fished ground off the NE coast of England. *Marine Ecology Progress Series* **188**, 13–20.

Furness, R.W. (2002) Management implications of interactions between fisheries and sandeel-dependent seabirds and seals in the North Sea. *ICES Journal of Marine Science* **59**, 261–269.

Gallagher, A.J., Kyne, P.M. and Hammerschlag, N. (2012) Ecological risk assessment and its application to elasmobranch conservation and management. *Journal of Fish Biology* **80**, 1727–1748.

Garcia, S.M. and Newton, C. (1997) Current situation, trends and prospects in world fisheries. In: *Global Trends: Fisheries Management* (eds E.K. Pikitch, D.D. Huppert and M.P. Sissenwine). American Fisheries Society, Bethesda, Maryland.

Garcia, S.M. and Charles, A.T. (2008) Fishery systems and linkages: implications for science and management. *Oceans and coastal Management* **51**, 505–427.

Grafton, R.Q. (1996) Individual transferable quotas: theory and practice. *Reviews in Fish Biology and Fisheries* **5**, 5–20.

Graham, M. (1936) Modern theory of exploiting a fishery, and application to North Sea trawling. *Journal du Conseil International pur l'Exploration de la Mer* **10**, 264–274.

Heithaus, M.R., Frid, A., Wirsing, A.J. and Worm, B. (2008) Predicting ecological consequences of marine top predator declines. *Trends in Ecology and Evolution* **23**, 202–210.

Helson, J., Leslie, S., Clement, G., Wells, R. and Wood, R. (2010) Private rights, public benefits: industry-driven seabed protection. *Marine Policy* **34**, 557–566.

Hilborn, R. (2007) Managing fisheries is managing people: what has been learned? *Fish and Fisheries* **8**, 285–296.

Hilborn, R. and Hilborn, U. (2012) *Overfishing: What Everyone Needs to Know*. Oxford University Press, Oxford.

Hjort, J. (1914) Fluctuations in the great fisheries of northern Europe viewed in the light of biological research. *Rapport et Proces-Verbaux de Conseil Permanent International pour l'Exploration de la Mer* **20**, 1–228.

ICES (1990) Report of the Working Group on Ecosystem Effects of Fishing. ICES CM 1999/ACFM-12. ICES Copenhagen.

ICES (2002) Report of the Study Group on Further Development of the Precautionary Approach to Fisheries Management. ICES CM 2002/ ACFM: 10. 67 pp.

IUCN (2008) The 2008 World Conservation Congress Report. Available at http://www.iucn.org/congress_08/ (accessed March 2014).

Kaiser, M.J., Ramsay, K., Richardson, C.A., Spence, F.E. and Brand, A.R. (2000) Chronic fishing disturbance has changed shelf sea benthic community structure. *Journal of Animal Ecology* **69**, 494–503.

Kell, L.T., De Oliveira, J.A.A., Punt, A.E., McAllister, M.K. and Kuikka, S. (2006) Operational management procedures: an introduction to an evaluation framework for management systems. In: *The Knowledge Base for Fisheries Management* (eds L. Motos and D. Wilson), pp. 379–407. Elsevier, Amsterdam.

Kerr, S.R. and Dickie, L.M. (2001) *The Biomass Spectrum: A Predator–Prey Theory of Aquatic Production*. Columbia University Press, New York.

Larkin, P.A. (1977) An epitaph for the concept of maximum sustainable yield. *Transactions of the American Fisheries Society* **106**, 1–11.

Larsson, L.C., Laikre, L., Andre, C., Dahlgren, T.G. and Ryman, N. (2010) Temporally stable genetic structure of heavily exploited Atlantic herring (*Clupea harengus*) in Swedish waters. *Heredity* **104**, 40–51.

Lindeboom, H.J. and de Groot, S.J. (1998) IMPACT-II: The effects of different types of fisheries on the North Sea and Irish Sea benthic ecosystems. NIOZ- Report 1998-1/RIVO-DLO Report C003/98. Den Burg, Texel, Netherlands, Netherlands Institute for Sea Research. 404 pp.

Lokkeberg, S. (2005) Impacts of trawling and scallop dredging on benthic communities. FAO Fisheries Technical Paper No. 472. FAO, Rome.

Mace, P.M. (1997) Developing and sustaining world fisheries resources: the state of science and management. In: *Developing and Sustaining World Fisheries Resources: the State of Science and Management* (eds D.A. Hancock, D.C. Smith, A. Grant and J.P. Beumer), pp. 1–20. Proceedings of the Second World Fisheries Congress, Brisbane, Australia, 1996. CSIRO Publishing, Australia.

Mace, P.M. (2001) A new role for MSY in single-species and ecosystem approaches to fisheries stock assessment and management. *Fish and Fisheries* **2**, 2–32.

Mace, P.M. (2004) In defence of fisheries scientists, single-species models and other scapegoats: confronting the real problems. In: *Perspectives on Ecosystem-Based Approaches to the Management of Marine Resources* (eds H. I. Browman and K. I. Stergiou), pp. 285–291. Marine Ecology Progress Series, **274**.

Montecino, C. and Lange, C.B. (2009) The Humboldt Current System components and processes, fisheries, and sediment studies. *Progress in Oceanography* **83**, 65–79.

Myers, R.A. and Worm, B. (2005) Extinction, survival or recovery of large predatory fishes. *Philosophical Transactions of the Royal Society B-Biological Sciences* **360**(1453), 13–20.

New Zealand Ministry of Fisheries (2008) Harvest Strategy Standard for New Zealand Fisheries. Ministry of Fisheries, Wellington, New Zealand.

NRC (2003) *Decline of the Steller Sea Lion in Alaskan Waters: Untangling Food Webs and Fishing Nets*. National Academy Press, Washington, DC.

NRC (2006) *Dynamic Changes in Marine Ecosystems. Fishing, Food Webs, and Future Options*. National Academy Press, Washington, DC.

Okes, N.C., Hockey, P.A.R., Pichegru, L., van der Lingen, C.D., Crawford, J.M. and Grémillet, D. (2009) Competition for shifting resources in the southern Benguela upwelling: Seabirds versus purse-seine fisheries. *Biological Conservation* **142**, 2361–2368.

Patrick, W.S., Spencer, P., Link, J., Cope, J., Field, J., Kobayashi, D., Lawson, P., Gedamke, T., Cortes, E., Ormseth, O., Bigelow, K. and Overholtz, W. (2010) Using productivity and susceptibility indices to assess the vulnerability of United States fish stocks to overfishing. *Fishery Bulletin* **108**, 305–322.

Punt, A.E. and Smith, D.M. (2001) The gospel of maximum sustainable yield in fisheries management: birth, crucifixion and reincarnation. In: *Conservation of Exploited Species* (eds J.D. Reynolds, G.M. Mace, K.H. Redford and J.G. Robinson). Cambridge University Press, Cambridge, UK.

Rice, J., Moksness, E., Attwood, C., Brown, S.K., Dahlee, G., Gjerde, K.M., Grefsrude, E.S., Kenchington, R., Kleiven, A.R., McConney, P., Ngoilei, M.A.K., Næsje, T.F., Olsen, E., Olsen, E.M., Sanders, J., Sharma, C., Vestergaard, O. and Westlund, L. (2012) The role of MPAs in reconciling fisheries management with conservation of biological diversity. *Ocean and Coastal Management* **69**, 217–230.

Ricker, W.E. (1940) Relation of "catch per unit effort" to abundance and rate of exploitation. *Journal of the Fisheries Research Board of Canada* **5**, 43–70.

Ricker, W.E. (1955) Computation and Interpretation of Biological Statistics of Fish Populations. *Bulletin of the Fisheries Research Board of Canada* **191**, 382 pp.

Rijnsdorp, A.D., Buys, A.M., Storbeck, F. and Visser, E. G. (1998) Micro-scale distribution of beam trawl effort in the southern North Sea between 1993 and 1996 in relation to the trawling frequency of the sea bed and the impact on benthic organisms. *ICES Journal of Marine Science* **55**, 403–419.

Roberts, C. (2007) *The Unnatural History of the Sea*. Island Press, Washington, DC. 465 pp.

Rodríguez, J.P., Rodríguez-Clark, K. M., Baillie, J. E., Ash, N., Benson, J., Boucher, T., Brown, C., Burgess, N. D., Collen, B., Jennings, M., Keith, D.A., Nicholson, E., Revenga, C., Reyers, B., Rouget, M., Smith, T., Spalding, M., Taber, A., Walpole, M., Zager, I. and Zamin, T. (2011) Establishing IUCN Red List criteria for threatened ecosystems. *Conservation Biology* **25**(1), 21–29.

Rogers, A.D. and Gianni, M. (2010) The Implementation of UNGA Resolutions 61/105 and 64/72 in the Management of Deep-Sea Fisheries on the High Seas. Report prepared for the Deep-Sea Conservation Coalition. International Programme on the State of the Ocean, London, United Kingdom. 97 pp.

Russell, S.F. (1931) Some theoretical considerations on the "overfishing" problem. *Journal du Conseil de l'exploration de la Mer* **6**, 3–27.

Schaefer, M.B. (1954) Some aspects of the dynamics of populations important to the management of the commercial marine fisheries. *Bulletin of the Inter-American Tropical Tuna Commission* **1**, 27–56.

Schindler, D.E., Hilborn, R., Chasco, B., Boatright, C.P., Quinn, T.P., Rogers, L.A. and Webster, M.S. (2010) Population diversity and the portfolio effect in an exploited species. *Nature* **465**, 609–613.

Shelton, P.A., Carscadden, J. and Hoenig, J. M. (1992) Risk evaluation of the 10% harvest rate procedure for capelin in NAFO Division 3L. In: *Risk Evaluation and Biological Reference Points for Fisheries Management* (eds S.J. Smith, J.J. Hunt and D. Rivard), pp. 193–201. Canadian Special Publication in Fisheries and Aquatic Sciences, **154**.

Silliman, R.P. (1943) Studies on the Pacific pilchard or sardine (*Sardinops caeruela*). 5. A method of computing mortalities and replacements. *US Fish and Wildlife Service Special Scientific Report* **24**, 1–10.

Smith, A.D.M., Brown, C.J., Bulman, C.M., Fulton, E.A., Johnson, P., Kaplan, I.C., Lozano-Montes, H., Mackinson, S., Marzloff, M., Shannon, L.J., Shin, Y.J. and Tam, J. (2011) Impacts of fishing low-trophic level species on marine ecosystems. *Science* **333**, 1147–1150.

Sumaila, U.R., Watson, L., Teh, R., Tyedmers, P. and Pauly, D. (2008) Fuel price increase, subsidies, overcapitalization and resource sustainability. *ICES Journal of Marine Science* **65**, 832–840.

Thompson, W.F. and Bell, F.H. (1934) Biological statistics of the Pacific halibut fishery. 2: effect of changes in intensity on total yield and yield per unit of gear. Report of the International Fisheries (Pacific Halibut) Commission, 8: 49 pp.

UNEP (2011) Taking Steps toward Marine and Coastal Ecosystem-Based Management: An Introductory Guide. UNEP Nairobi 68. Available at http://www.unep.org/ /Publications/Publication/tabid/439/language/en-US/ Default.aspx?BookID=6200 (accessed February 2014).

US Department of Commerce (2007) Magnuson-Stevens Fishery Conservation and Management Act. 170 pp.

Witherell, D., Pautzke, C. and Fluharty, D. (2000) An ecosystem based approach for Alaska groundfish fisheries. *ICES Journal of Marine Science* **57**, 771–777.

Worm, B., Hilborn, R., Baum, J.K., Branch, T.A., Collie, J. S., Costello, C., Fogarty, M.J., Fulton, E.A., Hutchings, A., Jennings, S., Jensen, O.P., Lotze, H.K., Mace, P.M., McClanahan, T.R., Minto, C., Palumbi, S.R., Parma, A.M., Ricard, D., Rosenberg, A.A., Watson, R. and Zeller, D. (2009) Rebuilding global fisheries. *Science* **325**, 578–585.

WSSD (2002) Report of the World Summit on Sustainable Development. Johannesburg, South Africa, 26 August–4 September 2002. United Nations. Available at http://www.un.org/jsummit/html/documents/summit_docs/131302_wssd_report_reissued.pdf (accessed March 2014).

Yodzis, P. (1994) Predator-prey theory and management of multispecies fisheries. *Ecological Applications* **4**, 51–58.

Zhou, S., Milton, D.A. and Fry, G.C. (2012) Integrated risk analysis for rare marine species impacted by fishing: sustainability assessment and population trend modelling. *ICES Journal of Marine Science* **69**, 271–280.

Chapter 5

The economic dimension: Addressing behaviour, incentives and context for effective governance

S. Hanna

Oregon State University, Corvallis, USA

Abstract: Economics is critical to the success of marine governance. Fisheries and marine conservation are each influenced by economic motivations, context and processes. Utility and incentives shape behaviour; expectations and time horizons influence decisions, transaction costs and externalities form processes; and equity effectiveness and efficiency determine distributional outcomes. The economic context of governance has biophysical, social and institutional components. The biophysical component is influenced by variability and scarcity; the social component by values, complexity and uncertainty; the institutional component by fragmentation, scale and paths. Fisheries and marine conservation have placed different emphasis on economics and economic instruments. Economic instruments encompass a range of tools that encourage desired fishery and conservation outcomes through incentives to reduce costs and increase revenues and have the potential to promote effective integration of the two governance areas.

Keywords: fisheries; conservation; governance; incentives; externalities; transaction costs; values; instruments; distribution

Introduction

Marine governance is complex, dynamic and contested. Economics permeates governance at every level; it drives motivations, provides context, shapes structures and influences outcomes. The significance of economic dimensions is further enhanced by the expanding connections between fisheries, marine conservation and other policy areas such as climate change, food security, seafood trade and coastal community resilience. The failure to understand and accommodate these connections leads to major governance problems. The Convention on Biological Diversity (CBD) recognizes the importance of the economic context as one of its fundamental principles of the ecosystem approach to management (UNEP, 2003).

This chapter reviews key economic aspects of marine governance. It discusses the governance foundations of objectives, motivations, decisions, processes and outcomes. It addresses the economic context as shaped by biophysical, social and institutional components. The scope of governance and its expression in economic frameworks and instruments are discussed. The chapter concludes with an assessment of the potential of economics to contribute to the integration of fisheries and marine conservation.

Economic foundations of governance

The economic foundations of fishery and marine conservation governance are reflected in the objectives articulated for marine ecosystems and in the economic

Governance of Marine Fisheries and Biodiversity Conservation: Interaction and Coevolution, First Edition.
Edited by Serge M. Garcia, Jake Rice and Anthony Charles.

principles that influence behaviour, decisions, processes and outcomes.

Economic objectives in governance

Fisheries and marine conservation are managed for different outcomes. Fisheries are managed for the production and use of ecosystem goods through multiple objectives related to food, recreation, employment, communities and tradition. Managers seek a balance among these objectives to achieve optimal yield (DiCosimo et al., 2010; Hilborn, 2007; Salomon et al., 2011).

In contrast, marine conservation is focused on the preservation of ecosystem services such as the reproductive capacity of fish populations. Conservation actions are taken to protect, conserve and recover biodiversity and habitats and to minimize risk from fishing and other activities.

At first glance the different objectives of fisheries and marine conservation would appear to be contradictory and to require separate governance tracks. This has indeed been the case until relatively recently, but in some contexts the two tracks are moving towards integration. Some fisheries are broadening the scope of management to include multiple species, ecosystem effects and the protection of biodiversity with tools such as protected areas and bycatch controls (At-Sea Processors Association, 2012; Pacific Whiting Conservation Cooperative, 2012; Reuter et al., 2010). These fisheries represent an operational convergence with marine conservation objectives. Other fisheries continue to focus on single-stock management and have not adopted broader marine conservation approaches.

Fisheries and marine conservation each require sustainable marine ecosystems. Although it is generally recognized that sustainability has ecological, economic and social dimensions, stakeholders in each area may give different weights to each of these three components. Vaguely stated objectives permit multiple interpretations that foster conflict over the 'correct' weight to assign the different dimensions and make it difficult to assess tradeoffs (Hilborn, 2007; Millennium Ecosystem Assessment, 2005; Salomon et al., 2011). Conflict may also arise over the appropriate time horizon for action. Marine conservation tends to focus on long-term outcomes, while fishery management operates over a mix of time horizons.

Economic principles

The economic principles underlying fisheries and conservation performance are relatively few. They concern the behavioural motivations of utility and incentives, the decision frameworks of assurance and time horizons, the process elements of transaction costs and the governance outcomes of distribution and efficiency.

Behavioural motivations: Utility and incentives

Utility is the satisfaction received from consumption, reflecting a preference for some set of goods or services. Utility embodies self-interest, a concept that is impossible to measure but useful to understand. It affects behaviour through willingness to pay for particular goods or services or to be compensated for their loss. In the marine setting these payments may be provided directly through investment or disinvestment in fishing capacity, or indirectly through government funding of fish stock assessments or marine protected areas.

Incentives work with utility to motivate behaviour. A classic illustration of incentives at work is Hardin's 'tragedy of the commons' (Hardin, 1968), in which open access resources lead individuals to take actions that collectively lead to resource overuse. A painful lesson of fishery management is that the incentives of open-access fisheries, although often implemented with good intentions, will in the long term benefit neither resource users nor ecosystems.

Incentives are central to decisions made by individuals and organizations and to the costs and benefits of those decisions. Incentive design is trickier than it appears and perverse incentives abound, as the long history of unintended policy consequences in fisheries attests.

Fishery and conservation stakeholders may be motivated by different incentives. Financial incentives are dominant in some fisheries and are often tied to social rewards, for example when a 'highliner' captain is not only a top earner but is also held in high esteem by other fishers. Non-financial incentives, such as coercion or moral suasion (persuasive pressure to 'do the right thing') may also play a strong role. The degree of compliance with regulations also rests on incentives (Anderson and Holliday, 2007).

Financial incentives may also be important to marine conservation, albeit as a means to fund a particular action rather than as a financial reward. Marine conservation is also motivated by intangible incentives such as association with a particular environmental cause perceived to be in the social good.

Decision frameworks: Expectations and time horizons

The reliability of decisions rests on the degree to which people can be assured that actions will have intended outcomes (Runge, 1984). For fishery stakeholders, assurance is the expectation that fish conserved in the present will be available to be caught in the future. This is a fundamental incentive on which conservation actions such as stock rebuilding rest. For marine conservation stakeholders, assurance may be the knowledge that bycatch is monitored or protected areas enforced.

Decisions are also affected by time horizons. Marine ecosystems are natural capital, producing value in both stocks and flows. Future expected benefits of these stocks have less value – because they are uncertain and because there is an opportunity cost to waiting – than do known current benefits, and so the future is discounted relative to the present. The rate at which the future is discounted relative to the present determines the rate of flow. The task of sustainable governance is to maintain the stock and flow of benefits into the indefinite future.

Marine conservation and fishery interests often bring different perspectives on time horizons to the governance arena (Hanna, 1998). Fisheries participants may resist capacity reduction and stock rebuilding because the expected future benefits seem small compared to known present costs. Marine conservation interests are motivated by the expectation that future benefits will exceed present costs. As a result, conservation receives a greater emphasis in countries that apply low discount rates and place a high value on ecosystem services than in countries that place a higher value on consuming ecosystem goods (Hanna, 2009; Sumaila, 2005).

Process fundamentals: Transaction costs and externalities

Governing creates costs of coordination, design, decision-making, implementation, monitoring and enforcement. These are transaction costs; some vary with the way decisions are made, while others remain fixed regardless of the decision process (Eggertsson, 1990; North, 1990). Context also influences transaction costs (Williamson, 1985, 1994); as a marine ecosystem becomes depleted, governance must account for more tradeoffs among direct and indirect, as well as present and future, uses.

The structure of governance influences the magnitude and distribution of transaction costs. Command and control structures in fisheries are associated with low costs of fishery description and regulation design (*ex ante* to implementation) as negotiations with stakeholders are absent, but high levels of costs for implementation, monitoring and enforcement (*ex post*) as the absence of stakeholders results in a lack of regulatory legitimacy (Jentoft, 2000). In contrast, decentralized cooperative management processes involving more interest groups in decision making tend to have higher *ex ante* costs of coordination and information gathering, but lower *ex post* costs of monitoring and enforcement (Hanna, 1995; Raakjær Nielsen and Vedsman, 1999).

Externalities exist when people do not take full account of the costs or benefits of their actions, and so do not pay for harm they cause or receive compensation for benefits they create. In fisheries, a stock externality is created when one person's catch decreases the amount of fish available to others (Agnello and Donnelley, 1976; Smith, 1969). Similarly, a crowding externality results when interference among fishermen increases the cost of fishing and decreases catch. A management externality is the consequence of a failure to account for the costs or benefits of biodiversity protections. Externalities generate transaction costs through the actions people take in the attempt to resolve them; the greater the number and diversity of fishery interests, the higher the transaction costs will tend to be.

The incentive to address externalities depends on the existence of property rights to benefit or be protected from. The form of property rights and the protections they provide is a social choice that will differ among regions and countries. Some may opt to protect subsistence or small-scale commercial fishing while others may focus on industrial development, sacrificing the protection of biodiversity to other commercial goals (Charles, 2009). If property rights are absent or unenforceable marine ecosystems remain vulnerable to external costs, leading to a loss of both biodiversity and economic value.

Distributional outcomes: Equity, effectiveness and efficiency

The distribution of benefits and costs among stakeholders is one of the most frequent sources of conflict in marine governance. All decisions have allocative effects and the equitable distribution of these effects becomes increasingly difficult as the number of ecosystem uses expands. The complex issue of biodiversity expands equity questions about the distribution of benefits and costs among rich and poor nations and across generations. Those receiving the benefits of biodiversity protection may be geographically and temporally separated from those incurring the costs.

The separation of benefits from costs also matters for governance effectiveness. Effective governance depends on legitimacy, that is, on the acceptance of rules and procedures by participants. When people cannot accept the distributional outcomes of regulations, they have an incentive to undermine them (Jentoft, 1989).

Distributional questions affect the efficiency of governance. Whether recognized or not, economic efficiency is of critical importance to governance because resources are limited and have competing uses; being economically efficient means that the greatest value is obtained from their use (Hanna, 2006).

The economic context of governance

The economic context of governance comprises biophysical, social and institutional components.

The biophysical component: Variability and scarcity

Marine ecosystems vary in biological and physical dimensions and may be characterized by scarcity in the production of desired ecosystem goods. Variability and scarcity form an economic context.

Variability

Variability is a key feature of marine ecosystems, where changes in the distribution and productivity of populations are common throughout the entire food web. The frequency and amplitude of these changes vary widely and are driven by physical mechanisms such as temperature, currents, winds, precipitation and pH level. Ecosystems can respond to physical or biological forcing in linear or nonlinear ways, and uncertainty about the type of response limits the ability to forecast change.

Climate-induced changes exaggerate the normal range of variability (Barange and Perry, 2009; Rice and Garcia, 2011) and are already affecting the seasonality of some biological processes (e.g. the timing of life history events) and altering marine food webs (Barange and Perry, 2009; Brander, 2007; Cochrane et al., 2009; Daw et al., 2009). These effects not only generate their own change but also alter rates of change and erode the ability to predict change. The resulting scientific uncertainty combined with social and institutional factors will create uncertainties for governance (McIlgorm et al. 2010).

Scarcity

The productive capacity of the ocean is finite, as is the capacity of a fish population to sustain mortality. Finite may or may not mean scarce; scarcity is created when demand exceeds supply. Scarcity becomes the impetus for discussing integrated fisheries and conservation governance when it becomes apparent that a dual-track approach cannot meet the demand for ecosystem goods and services. However, scarcity can at the same time be a deterrent to governance integration if it erodes legitimacy and enhances rent-seeking.

The social component: Competing values, complexity and uncertainty

People are embedded within a social context. Motivations, utility, expectations and incentives are all shaped by social networks of family, community and culture, requiring that governance involve the social context as a first-order consideration (Hanna and Jentoft, 1996; Jentoft, 2007). Three key elements of the social component of governance are the competing values assigned to the marine ecosystem, the complexity of connections and the influence of uncertainty on behavioural incentives.

Competing values

Economic value is measured by the amount people are willing to pay for a good or service, or the amount they are willing to accept as compensation for not using the good or service. Marine governance involves tradeoffs among a range of values as represented by different mixes of goods and services.

Total economic value is the sum of all 'use' and 'non-use' values that people hold for ecosystem goods and services. Use values typically involve some human interaction with the resource and include both direct and indirect use; for example the direct use value from catching fish or the indirect use value from fish habitat (Dziegielewska, 2007). Non-use values, sometimes called 'passive use', are potential values that rely on the continued existence of a resource. They may result from protecting endangered species against extinction, from ensuring that certain ecosystem services are preserved as a bequest for future generations and from maintaining options for future actions (Dziegielewska, 2007; FAO, 2003).

Value can also be categorized according to how it is expressed, whether through market transactions or independent of markets. The market value of a fish stock is expressed directly in the price of commercially caught fish and indirectly in the cost of a recreational fishing trip. The non-market-values assigned to that same fish stock include the reproductive capacity necessary to sustain the stock or as a bequest to future generations.

The categories 'use' and 'non-use' and 'market' and 'non-market' illustrate the potential for values to lead to conflicts and tradeoffs between fisheries and marine conservation, where proponents of the respective actions may assign very different types of values to the same ecosystem components.

Complexity

Fisheries and marine conservation encompass a wide range of people with different demographics, geography and culture. In striving to reflect the diverse social context, governance moves inexorably towards greater complexity (Raakjær, 2009), requiring as much variety in its available actions as there exists in the systems it regulates (Jentoft, 2007; Ostrom, 1995).

Governance complexity can challenge effectiveness. In fisheries management a typical progression of management changes can result in internal complexity that adds inconsistency, diminishes effectiveness and limits flexibility (Anderson and Holliday, 2007; Hanna, 2006). Complexity can also diminish transparency and introduce moral hazard, where the opportunity for unobservable actions increases and it becomes easier to avoid complying with regulations (Hanna, 2006; Healey and Hennessey, 1998).

The scale of governance is another source of complexity. Transaction costs are magnified over large areas, or in a social context of heterogeneous interests requiring multiple layers of coordination (Hanna, 2011). Garcia and Charles (2008) raise the relevant economic question: at which point does the cost of taking on more complexity overtake the costs of not doing so? However, social and economic dimensions of complexity have not yet been seriously considered as an essential element within governance arenas (Rice, 2007).

Uncertainty

Uncertainty is a powerful piece of the social context. It creates incentives to shorten time horizons, intensify rates of use and avoid investing in the future, all in direct contradiction to the precautionary approach (Garcia, 1994). Uncertainty is a major reason why long-term actions such as biodiversity protection or fish stock rebuilding are difficult to implement (Hanna, 2009).

Three forms of uncertainty are particularly important: fact, tenure and process. Fact uncertainty results from incomplete knowledge about management actions and leaves them vulnerable to opportunistic attempts to undermine them (Williamson, 1985). Tenure uncertainty results from insecure access to marine resources, which can be created by poorly specified property rights or by abrupt changes in management. In fisheries, tenure uncertainty underpins the destructive race for fish, overcapitalization and conflicts between fishing and conservation interests. Process uncertainty can exist when managers and stakeholders have incomplete information about requirements or procedures, or when power is ambiguous. It can prevent the formation of patterns of cooperation and preclude the making of credible commitments across fishery and marine conservation interests, a situation exemplified by the ongoing conflict over marine protected areas (Runge, 1984; Williamson, 1985).

Governance uncertainty generates a psychology that runs counter to needs for precautionary actions. It encourages the delusion that decisions can be postponed until perfect information is achieved. If the scale of uncertainty is large enough, problems become diffused and the benefits of localized action become unclear. In these ways uncertainty can exacerbate existing problems rather than lead to a more conservative stance (McIlgorm et al., 2010).

The institutional component: Fragmentation, scales and paths

Institutions comprise the rights, rules and responsibilities of organizations and individuals (Eggertsson, 1990; North, 1990). They are both formal and informal, including the laws, organizations and procedures that shape the way people interact with each other and with their environment. They determine who makes decisions and how those decisions are made, regulations are implemented and adaptation takes place.

People make decisions through the structure of municipal, state, federal and international organizations.

Ideally, institutions should coordinate these decisions over different scales, reconciling actions taken at one level with those in other 'nested' levels. They should also accommodate diverse objectives, develop processes for assessing tradeoffs and address the distribution of the costs and benefits of actions. Finally, institutions should be flexible enough to respond to changing conditions and to evolve in ways that allow their benefits to outweigh their costs (Hanna, 2008). This institutional ideal is challenged by problems of fragmentation, incompatible scales and historical paths.

Fragmentation

Agencies and organizations make decisions across fragmented jurisdictions. In many countries decisions about fisheries management, stock rebuilding, endangered, threatened and protected (ETP) species, marine protected areas and critical habitat are the responsibility of entities with overlapping boundaries, competing objectives and incomplete authorities to accommodate the full scale of causes or effects. The decision landscape may be partitioned across regions, national governments and international bodies. Further, interest groups and communities may be socially or economically fragmented. Rosenberg and McLeod (2005) describe US marine governance as fairly typical in having different uses managed under different legal authorities and under different organizational structures, with no single mandate to coordinate interactions across agencies. Fish stocks, water quality and coastal zone use are managed along parallel tracks.

Scale

Scale – spatial as well as temporal – is a fundamental institutional consideration. Too large a spatial scale can mean the loss of feedback, representation and accountability, creating ineffective governance. Too small a scale can mean incomplete coverage and the creation of spillover effects from one activity to another. Interactions vary with scale, for example where fishing effort is a driver of stock abundance in the short term; in contrast, on a larger scale and in the longer term abundance may be the driver of effort (Garcia and Charles, 2008).

Scale can contribute to externalities. For example, the costs and benefits of actions of biodiversity protection are usually unevenly distributed over space, with costs incurred by those in proximity to the site of protective action and benefits realized over much larger areas. Similarly, the costs of restricting fish harvests to protect other elements of the ecosystem are borne by fishery managers, commercial fishermen and anglers, but the benefits are widely dispersed among the population at large. Marine biodiversity protection is often an international issue; as such there are global costs and benefits that are unaccounted for in national systems of property rights.

Paths

Actions taken for fishery management or marine conservation follow a path over time where each step is affected by the cumulative effect of previous steps, influencing the shape and direction of new actions.

A fishery management path follows a number of representative steps. Sequential overexploitation of stocks leads to attempts to remedy the subsequent losses through development of stricter harvest regulations or through marine conservation set-asides. New expectations – forms of implicit property rights – develop and are layered over old. Expectations eventually build to zones of conflict within fisheries or across fishery and marine conservation interests.

It is possible to be stuck on a given path because the strength of expectations and the benefits of inertia are great enough that people are willing to absorb large levels of transaction costs, or they are able to shift those costs externally (Dixit, 1998). The influence of the path often creates resistance to the full-scale evaluation of performance effectiveness.

Evolving economic scope of governance

Just as the biological scientific traditions in fisheries and marine conservation have different histories, cultures and priorities (Salomon et al., 2011), so too do the traditions in the economic sciences. The emphasis on economics has been different in fisheries and marine governance, reflecting the different sets of objectives and evolution of thinking about the application of economic instruments.

Economics in fishery management

The role of economics in fishery management has evolved in response to changing incentives, costs and benefits. Fisheries were managed under systems of open access throughout much of the 20th century (OECD, 1997). Few regulations were in place, and these primarily focused on limiting the amount and timing of catch or the size of fish caught. The incentives of open access meant that in profitable fisheries ever-increasing numbers of participants raced for fish, placing pressure on the fishery resource.

Although the loss of biological productivity appears as a concern in 19th century documents, the damage to economic productivity was not the subject of general attention until much later. In 1911 Jens Warming, writing in Danish, published an article on the loss of economic rents in fisheries (Andersen, 1983). It took until 1954 for the topic to again be addressed, with Gordon's (1954) paper about the economic costs of open access followed by Scott's (1955) publication contrasting rent dissipation under open access with rent generation under a theoretical sole ownership. Both pieces brought economics directly into the arena of fishery governance and became cornerstones of fishery economics (Gordon, 1954; Scott, 1955; Wilen, 2000).

An early FAO conference on the economics of fishery management raised the concept of economic objectives, after which an international community of fishery economists helped expand awareness of economic objectives and the power of economic incentives. Despite these efforts, economics remained confined to the assessment of management impacts and had little immediate effect on management approaches (Wilen, 2000).

Confronted with the problems of governing open access fisheries, managers implemented regulations to limit how people fished, first as input controls over fishing gear, areas and seasons, followed by output controls limiting total allowable catch or bycatch. Neither type was sufficient to address the incentive problem created by the absence of property rights. Because fish were owned only upon capture, output controls created incentives for fishermen to develop new ways to fish faster and catch more fish, before total output limits were reached. While it might make economic sense to forego harvest of fish until they were larger and more valuable, without assurance that the fish would be available to be caught at a later date there was no incentive to conserve. This was the race for fish.

New and increasingly restrictive regulations were developed to try to control the problems created by the race for fish (Anderson and Holliday, 2007; USCOP, 2004). Limits on the number of people fishing were attempts to control fishing pressure on the stocks. But although seemingly radical at the time, access limitation programs only controlled another dimension of inputs and did not address the fundamental problems of the race for fish (Anderson and Holliday, 2007; USCOP, 2004).

The role of economics changed as managers, observing the costly futility of managing a race for fish, began to look more favourably on instruments designed to end it. Guaranteeing an individual right to catch a share of the total quota would remove the incentive to race for fish and would provide an incentive to maximize the value, instead of quantity, of the catch. In 1973 Christy introduced the idea of individual quotas (Christy, 1973; Wilen, 2000). Such share-based instruments, while socially and politically controversial, elevated the position of economics in management to one of serious consideration.

Notable exceptions to the typical fishery progression from open access to eventual rights-based management were isolated small-scale fisheries whose management evolved in the opposite direction. Some Pacific Island cultures had traditional forms of management based on rights to territory and species that kept the fishery in balance with the ecosystem (Carrier, 1987; Goodenough, 1951). However, these rights-based approaches were exceptions and were vulnerable to large-scale social change and to interactions with larger markets (Cordell, 1989; Johannes, 1978) that, by destroying the foundations of the more restrictive management, recreated conditions of open access.

Economics in marine conservation

The application of economics in marine conservation, and specifically to biodiversity conservation, focuses on valuation and impacts. The valuation problem is that marine ecosystems produce many services unvalued by markets and whose value often remains unaccounted and uncompensated. The impact question is that conservation actions have economic impacts on individuals, businesses and communities. Both non-market valuation and impact assessment can be addressed by economic techniques.

Ecosystem valuation is challenged by questions of scale. Early valuation work was limited by its tendency to use a 'benefit transfer' approach to extend site-specific values to other unrelated areas or to extrapolate broad-scale policy implications. More recent valuation work has focused on spatially explicit economic and ecological models that allow estimation of values for site-specific ecosystem services. Working at scales that are more policy-relevant has also allowed broader incorporation of stakeholder values (Groot et al., 2010).

A precursor for ecosystem services valuation is their identification and measurement. Some services are measured directly; others are measured indirectly through indicators. The function of ecosystem services can be categorized according to whether they are provisioning, regulating, habitat or cultural and amenity (Groot et al., 2010). Measurement of an ecosystem service distinguishes between the ecosystem's capacity to produce it (e.g. reproduction within a fish stock) and its actual use (fishery catch based on reproductive capacity). Importance is assessed in terms of a service's contribution to human wellbeing, as a source of food, income, or way of life (Groot et al., 2010).

The valuation of conservation impacts has also benefited from more spatially explicit assessments. Analyses focused on a specific policy action area allow consideration of context and scale in assessing the distribution of benefits and costs (Balmford et al., 2008; Groot et al., 2010). Distributional effects are critical in influencing the response, and therefore likelihood of success, to marine conservation actions.

Current economic frameworks and approaches

Economists use a wide variety of frameworks and approaches to analyse issues related to fisheries and marine conservation. These include the following.

- *Bioeconomics*: Models the interaction among biological and economic processes, typically combining biological models of population dynamics with economic models of fleet operations to estimate optimal management strategies (Clark, 1987), ecological and technological interdependencies and spatial management within ecosystems (Anderson and Seijo, 2010).
- *Ecological economics*: Treats the economy as a subsystem of natural systems and considers a range of issues related to ecosystem sustainability including natural capital, intergenerational equity, market and non-market values, sustainable development, resource management and conservation and green accounting (Costanza and Wainger, 1991).
- *Green economics*: Integrates the economy with ecosystems with an emphasis on applied policy, considering benefits to be achieved through investing in natural capital in forests, water and fisheries, resource and energy efficiency in cities, transportation and tourism (UNEP, 2011).
- *Blue–green economics*: Emphasizes the potential to increase economic benefits from ocean resources through better management practices, based on four tenets: (1) productivity limits; (2) rebuilding fish populations; (3) protection of essential marine habitats; and (4) energy-efficient fishing and other ocean activities (UNEP, 2011).
- *Institutional economics*: Broadens the approach of conventional economics in explicitly addressing the influence of organizations, rights, rules and incentives on behaviour, applied in fisheries and marine conservation to questions of regulatory design and response (Williamson, 1985).
- *Socio-economics*: Addresses economic and social dimensions of sustainability at individual, family and community levels to understand impacts of regulatory change, degree of cultural or economic dependence, decision-making, economic and social traditions, employment patterns and social relationships (Charles, 2005).
- *Livelihoods approach*: Places fisheries and marine conservation in the context of households and communities having the skills, resources and access to make and sustain a living through a portfolio of livelihood sources related to other economic sectors (Allison and Ellis, 2001; Cochrane and Garcia, 2009; De Young et al., 2008; Glavovic, 2006).
- *Development economics*: Considers the economic dimensions of development in low-income countries, including economic growth, distributional equity and human wellbeing (Bell, 1987). Worldwide, the difference between the potential and actual net economic benefits generated by fisheries over 30 years is an estimated economic loss of two trillion dollars (World Bank, 2009).

Economic instruments in fisheries and marine conservation

Economic instruments

Economic instruments, also called market-based or incentive-based, encourage desired fishery and conservation outcomes through incentives to reduce costs and increase revenues. In contrast to command and control approaches, economic instruments operate on a decentralized level and rest on the ability to assign rights as well as responsibility for performance outcomes. The idea is that the increased flexibility afforded by a less prescribed approach will lead to greater efficiency in outcomes (Wolf, 1997).

Economic instruments encompass a range of tools that address the economic issues of incentives, expectations, externalities and assurance. They are typically directed towards improving efficiency, but can also address issues of distribution through intentional design (Charles, 2009).

Economic instruments in application

Economic instruments are applied to a range of fishery components: fishery output (individual fishing quotas or IFQs, cooperatives and community development quotas), fishing space (TURFs), consumer demand (eco-certification) and governmental action (trade agreements and subsidies).

Fishery output

Individual fishing quotas (IFQ)

Also called quota shares and catch shares, IFQs allocate the rights to harvest a specified proportion of the total allowable catch to individual fishers. If the shares can be sold or leased to others they are called individual transferable quotas (ITQs). IFQ holders are responsible for and accountable to staying within their assigned catch limit.

Example: Denmark manages its diverse pelagic and demersal fisheries with ITQs. Program objectives are primarily economic, but also include community and conservation objectives (Bonzon et al., 2010; MRAG et al., 2009; Schou, 2010).

Performance outcomes: The program has reduced fishing capacity, increased fleet profitability, increased investment in value-added techniques and increased revenue to coastal communities. Conservation outcomes have also been achieved; a Code of Conduct includes collaboration with scientists, bycatch reduction and avoidance of ETP species. Several of the pelagic and demersal fisheries have achieved MSC certification (MSC, 2012).

Cooperatives (Co-ops)

Co-ops make internal allocations and conduct joint management of pooled individual shares using contractual arrangements. Cooperatives are able to develop operational objectives such as maximizing per-unit value of fish by group decisions on how the pooled quota shares are fished.

Example: The US Pacific Whiting Conservation Cooperative (PWCC) was formed to improve profitability and reduce bycatch by eliminating the race for fish among quota holders. It comprises three fishing companies, each owning multiple vessels, licensed to operate in the at-sea catcher-processor sector of the Pacific whiting fishery. Quotas shares are transferable within the cooperative.

Performance outcomes: By allowing contracts allocating quota among vessels and setting the timing and rate of fishing, the cooperative has improved fishery performance and fish stock conservation. Product recovery rates have increased and fishing effort and bycatch of protected species has declined. The fishery achieved MSC certification in 2009 (Anderson and Holliday, 2007; MSC, 2009; PWCC, 2012; Sylvia et al., 2008).

Community development quotas (CDQs)

CDQs grant a specified portion of the total allowable catch to a community, which then controls its allocation.

Example: The Western Alaska CDQ Program allocates a percentage of all Bering Sea and Aleutian Islands quotas for groundfish, halibut and crab to 65 western Alaska communities organized into six regional CDQ corporations. The program was created to achieve social and economic objectives of economic development and poverty alleviation (NPFMC, 2012).

Performance outcomes: The program has resulted in increased social and economic benefits to community residents in the form of asset value, wages and educational and medical facilities. Conservation benefits are achieved through the larger fishery management system (NPFMC, 2012; WACDA, 2010).

Fishing space

Territorial use rights (TURFS)

TURFs define rights of access to a specific area. They are used for sedentary species such as crustaceans and shellfish, where boundaries are definable (Charles, 2009).

Example: The Chilean National Benthic Resources TURF Program controls access to abalone and other nearshore benthic resources. The voluntary program includes over 17,000 artisanal small-boat fishermen and over 550 distinct coastal areas in cooperative management with government. Qualifying groups of fishermen are granted exclusive access through area concessions.

Performance outcomes: The program has maintained access for artisanal fishermen and improved economic performance. Conservation and biodiversity outcomes are mixed; in some TURF areas no-take areas have been established, while in others overharvesting and illegal fishing exists (Bonzon et al. 2010; Castilla and Gelgich, 2006, 2008; Castilla and Fernandez, 1998; Gallardo Fernández, 2008).

Combined community shares and TURFS

Hybrid applications of instruments address multiple fishery objectives.

Example: The Baja California Regional Federation of Fishing Cooperative Societies (FEDECOOP) is a hybrid species and area-based cooperative catch share program that targets benthic species along the central Baja California

Pacific coast of Mexico. The cooperative is a confederation of nine community cooperatives representing the majority of the lobster and abalone harvest. Access to fishing areas is limited to members of community coop members.

Performance outcomes: Members conduct fishery management, monitoring, enforcement and scientific research with little government assistance. The co-op has developed profitable international markets for its products and received MSC certification in 2004 (Bonzon et al. 2010; Bourillon and Ramade, 2006; Leal et al., 2008).

Consumer demand

Eco-certification

Certification is based on the assumption that consumers will be willing to pay for sustainably managed fisheries through higher prices for certified products. Higher prices provide producers with incentives to become certified by meeting sustainability standards (FAO, 2001, 2003, 2005; MSC, 2012).

Example: In 2009 the United Nations Environment Program (UNEP) reviewed the environmental and economic impact of certification on fisheries profitability, poverty alleviation, food security and sustainability. The review found that: the perceived and actual benefits of certification differed among consumers, producers and retailers; cost can be a burden in some fisheries; and demand for certification is uneven between countries.

Performance outcomes: Certification has the potential to improve the performance of fishery management but it also faces significant challenges, especially in developing countries where data and funding limitations combine with weak governance. It is likely that long-term benefits to producers will be derived from continued market access rather than from significant price premiums (UNEP, 2009).

Governmental action

Trade Agreements and Multilateral Environmental Agreements (MEAs)

Economic instruments play an important role in the implementation of trade and, to a lesser extent, other international agreements. The World Trade Organization's (WTO) Doha mandate includes negotiations on the reductions or elimination of tariff and non-tariff barriers to environmental goods and services as well as the issue of harmful fish subsidies (UNEP, 2004).

Example: Tariff exemptions for environmental goods and services affect biodiversity conservation (Wolf, 1997). The 1992 Convention on Biological Diversity (CBD) includes the use of economic incentives as 'essential elements of conservation and sustainable use'. The 1975 Convention on International Trade in Endangered Species of Wild Fauna and Flora (CITES) regulates international trade in animals and plants, including marine species, according to their conservation status (UNEP, 2004).

Performance outcomes: The CBD contains language endorsing the use of economic instruments and COP meeting decisions support their utility, but this support appears to remain theoretical rather than operational. In contrast, CITES is in itself an economic instrument; in granting or denying access to valuable international markets it provides a powerful incentive for conservation (UNEP, 2004).

Subsidies

Fishery subsidies are government payments for the purpose of producing economic value. They have had harmful effects of expanding fishing capacity and effort (Milazzo, 1998; World Bank, 2009), but they have also had beneficial effects through buying back excess fishing capacity or protecting ecosystem services through financial incentives to develop more selective gear.

Example: Subsidies express public willingness to pay for ecosystem services for which there are publicly held values but for which markets are absent, such as marine biodiversity protection through marine protected areas (MPAs). MPAs achieve non-use values, non-consumptive-use values (recreation or ecotourism) or indirect use values (fish population productivity) (Anderson, 2002). The number of government-sponsored MPAs is expanding worldwide (Anonymous, 2012).

Performance outcomes: The expansion in the number of MPAs has not been matched by a corresponding expansion in funding. Estimates of the funding gap required to achieve the three global objectives of the CBD fall within the range US$ 10–50 billion per year. The IUCN notes that reducing subsidies for activities that harm biodiversity is therefore one of the most cost-effective ways to slow biodiversity loss (IUCN, 2010).

Discussion: Economic instruments and prospects for governance integration

The thread of economics runs throughout fisheries and marine conservation and their governance. It serves as a powerful motivating force, behavioural influence and performance determinant. There are common economic foundations to fisheries and marine governance and it will be through recognizing this commonality that the gap between the approaches may be bridged.

Economic instruments are a way to link the two governance areas. The most frequent application of economic instruments is in fisheries. Some economic instruments, such as subsidies and international agreements, fit both fisheries and marine conservation contexts.

An advantage of economic instruments is their flexibility in application to different scales and varying objectives as well as their accommodation of social objectives and individual contexts. They are based on some form of rights that, by ending the race for fish, reorient incentives toward long-term stewardship and provide incentives for conservation. The wider application of economic instruments in fisheries suggests that the primary potential for integrating fisheries and marine conservation governance is through broadening the scope of fishery management to accommodate marine conservation concerns.

Greater opportunities for integrating governance of fisheries and marine conservation exist if:

- fisheries are stabilized through the application of instruments that provide tenure assurance and build incentives for marine conservation;
- fisheries are managed to eliminate the race for fish;
- fisheries make wider application of economic instruments;
- fisheries instrument choice reflects the social and economic context;
- marine conservation actions are designed with recognition of principles that motivate economic behaviour;
- marine conservation actions are taken with recognition of social and economic context and likely impacts of actions;
- more research is conducted into the valuation of marine ecosystem services;
- research on ecosystem service valuation includes costs as well as benefits;
- economic productivity of marine ecosystems is routinely monitored;
- greater use is made of economic tools to assess tradeoffs;
- economic instruments are embedded in international agreements;
- fishery and marine conservation stakeholders are integrated through planning, design and advisory processes; and
- economic and other social sciences comprise a standard element of governance scientific advice.

In conjunction with social and natural sciences, economics can contribute to these opportunities through joint research efforts in valuation, monitoring and tradeoffs and through the integrated provision of scientific advice.

References

Agnello, R.J. and Donnelley, L.P. (1976) Externalities and property rights in fishing. *Land Economics* **52**, 518–529.

Allison, E.H. and Ellis, F. (2001) The livelihoods approach and management of small-scale fisheries. *Marine Policy* **25**, 377–388.

Andersen, P. (1983) 'On Rent of Fishing Grounds': a translation of Jens Warming's 1911 article, with an introduction. *History of Political Economy* Fall 1983, **15**(3), 391–396.

Anderson, L.G. (2002) A bioeconomic analysis of marine reserves. *Natural Resource Modeling* **15**(3), 311–334.

Anderson, L.G. and Holliday, M.C. (eds) (2007) The design and use of limited access privilege programs. NOAA Technical Memorandum NMFS-F/SPO-86, National Marine Fisheries Service, National Oceanic and Atmospheric Administration, US Department of Commerce. http://spo.nwr.noaa.gov/tm/tm86.pdf (accessed February 2014).

Anderson, L.G. and Seijo, J.C. (2010) *Bioeconomics of Fisheries Management*. John Wiley & Sons.

Anonymous (2012) MPA News 13 (5) March–April. Available at http://depts.washington.edu/mpanews/MPA125.htm (accessed February 2014).

At-Sea Processors Association (2012) The Pollock Conservation Cooperative. Available at http://www.opc.ca.gov/webmaster/ftp/project_pages/dctf/AK_Pollock_Coop.pdf (accessed February 2014).

Balmford, A., Rodrigues, A., Walpole, M. J., ten Brink, P., Kettunen, M., Braat, L. and de Groot, R. (2008) *Review of the Economics of Biodiversity Loss: Scoping the Science*. European Commission, Brussels.

Barange, M. and Perry, R.I. (2009) Physical and ecological impacts of climate change relevant to marine and inland capture fisheries and aquaculture. In: *Climate Change Implications for Fisheries and Aquaculture: Overview of Current Scientific Knowledge* (eds K. Cochrane, C. De Young, D. Soto and T. Bahri), pp. 7–106. FAO Fisheries and Aquaculture Technical Paper No. 530. FAO, Rome.

Bell, C. (1987) Development economics. In: *The New Palgrave: A Dictionary of Economics* (eds J. Eatwell, M. Milgate and P. Newman), pp. 818, 825. Palgrave Macmillan, London.

Bonzon, K., McIlwain, K., Strauss, C.K. and Van Leuvan, T. (2010) *Catch Share Design Manual: A Guide for Managers and Fishermen*. Environmental Defense Fund, San Francisco.

Bourillón, L. and Ramade, M. (2006) Community-based fisheries management and eco-labeling in Baja California, Mexico. Available at http://www.icsf.net/icsf2006/jspFiles/cedepesca/

presentaciones/sharer_brasil/california_lobster_ certification.pdf (accessed February 2014).

Brander, K.M. (2007) Global fish production and climate change. *Proceedings of the National Academy of Sciences* **104**(50), 19709–14.

Castilla, J.C. and Fernandez, M. (1998) Small scale benthic fisheries in Chile: on co-management and sustainable use of benthic invertebrates. *Ecological Applications* **8**(1), S124–S132.

Castilla, J.C. and Gelcich, S. (2006) Chile: Experience with management and exploitation areas for coastal fisheries as building blocks for large-scale marine management. In: *Scaling Up Marine Management: The Role of Marine Protected Areas*. World Bank Report No. 36635 – GLB. Environment Department, Sustainable Development Network, Washington, DC. World Bank.

Castilla, J.C. and Gelcich, S. (2008). Management of the loco (Concholepas concholepas) as a driver for self-governance of small-scale benthic fisheries in Chile. In: *Case Studies on Fisheries Self-governance* (eds R. Townsend and R. Shotton). FAO Fisheries Technical Paper 504. Food and Agriculture Organization of the United Nations.

Carrier, J.G. (1987) Marine tenure and conservation in Papua New Guinea: problems in interpretation. In: *The Question of the Commons* (eds B. McCay and J. Acheson), pp. 142–167. University of Arizona Press, Tuscon.

Charles, A.T. (2005) The big picture: a fishery system approach links fishery management and biodiversity. In: *Proceedings of the Workshop on Biodiversity Challenges for Fishery Management*. Conference on Biodiversity: Science and Governance. IFREMER, Paris.

Charles, A.T. (2009) Rights-based management: the role of use rights in managing access and harvesting. In: *A Fishery Manager's Guidebook*, 2nd edn (eds K.L. Cochrane and S.M. Garcia), pp. 253–282. FAO and Wiley-Blackwell Publishing.

Christy, F.T. Jr. (1973) Fishermen's quotas. Occasional Paper 19, The Law of the Sea Institute, University of Rhode Island, Kingston.

Clark, C.W. (1987) Bioeconomics. In: *The New Palgrave: A Dictionary of Economics* (eds J. Eatwell, M. Milgate and P. Newman). Palgrave Macmillan, London.

Cochrane, K.L. and Garcia, S.M. (eds) (2009) *A Fishery Manager's Guidebook*. Second Edition. FAO, Rome and Wiley-Blackwell Publishing.

Cochrane, K., De Young, C., Soto, D. and Bahri, T. (eds) (2009) Climate change implications for fisheries and aquaculture: overview of current scientific knowledge. FAO Fisheries and Aquaculture Technical Paper No. 530. Rome, FAO.

Cordell, J. (ed.) (1989) A sea of small boats. Cultural Survival Report 26. Cultural Survival Inc, Cambridge, MA.

Costanza, R. and Wainger, L. (1991) *Ecological Economics: The Science and Management of Sustainability*. Columbia University Press, New York.

Daw, T., Adger, W.N., Brown, K. and Badjeck, M.C. (2009) Climate change and capture fisheries: potential impacts, adaptation and mitigation. In: *Climate Change Implications for Fisheries and Aquaculture: Overview of Current Scientific Knowledge* (eds K. Cochrane, C. De Young, D. Soto and T. Bahri), pp. 107–150. FAO Fisheries and Aquaculture Technical Paper No. 530. FAO, Rome.

De Young, C., Charles, A. and Hjort, A. (2008) Human dimensions of the ecosystem approach to fisheries: an overview of context, concepts, tools and methods. FAO Fisheries Technical Paper No. 489. FAO, Rome.

DiCosimo J., Methot, R.D. and Ormseth, O.A. (2010) Use of annual catch limits to avoid stock depletion in the Bering Sea and Aleutian Islands management area (Northeast Pacific). *ICES Journal of Marine Science* **67**, 1861–1865.

Dixit, A. (1998) Transaction cost politics and economic policy: a framework and a case study. In: *Institutions and Economic Organization in the Advanced Economies: The Governance Perspective* (eds M. Baldassarri, L. Paganetto and E.S. Phelps), pp. 139–176. St Martin's Press, Inc., New York.

Dziegielewska, D., Tietenberg, T. and Niggol Seo, S. (2007) Total economic value. In: *Encyclopedia of Earth* (ed. C. J. Cleveland). Environmental Information Coalition, National Council for Science and the Environment, Washington, DC. (First published in 2006).

Eggertsson, T. (1990) *Economic Behavior and Institutions*. Cambridge University Press, Cambridge.

FAO (2001) Product certification and eco-labeling for fisheries sustainability. FAO, Rome.

FAO (2003) Fisheries management 2. The ecosystem approach to fisheries. FAO Technical Guidelines for Responsible Fisheries, No. 4 Suppl. 2.

FAO (2005) Report of the technical consultation on international guidelines for the eco-labeling of fish and fishery products from marine capture fisheries, 19–22 October 2004. FAO, Rome.

Gallardo Fernández, G.L. (2008) *From Seascapes of Extinction to Seascapes of Confidence: Territorial Use Rights in Fisheries in Chile: El Quisco and Puerto Oscuro*. Co-Action Publishing, Stockholm.

Garcia, S.M. (1994) The precautionary principle: Its implications in capture fisheries management. *Ocean and Coastal Management* **22**, 99–125.

Garcia, S. and Charles, A.T. (2008) Fishery systems and linkages: Implications for science and governance. *Ocean and Coastal Management* **51**, 505–527.

Glavovic, B. (2006) Coastal sustainability – an elusive pursuit? Reflections on South Africa's coastal policy experience. *Coastal Management* **34**, 111–132.

Goodenough, W.H. (1951) *Property, Kin and Community on Truk*. Yale University Publications in Anthropology, New Haven, CT.

Gordon, H.S. (1954) The economic theory of a common property resource: the fishery. *Journal of Political Economy* **62**, 124–142.

Groot, R.S. de, Fisher, B., Christie, M., Aronson, J., Braat, L., Haines-Young, R., Gowdy, J., Maltby, E., Neuville, A., Polasky, S., Portela, R. and Ring, I. (2010) Integrating the ecological and economic dimensions in biodiversity and ecosystem service valuation. In: *The Economics of Ecosystems and Biodiversity (TEEB): Ecological and Economic Foundations* (ed. P. Kumar). Earthscan Publications Ltd., London.

Hanna, S. (1995) Efficiencies of user participation in natural resource management. In: *Property Rights and the Environment: Social and Ecological Issues* (eds S. Hanna and M. Munasinghe), pp. 59–67. World Bank, Washington, DC.

Hanna, S. (1998) Institutions for marine ecosystems: economic incentives and fishery management. *Ecological Applications* **8**(1), Supplement: Ecosystem Management for Sustainable Marine Fisheries: S170–S174

Hanna, S. (2006) Implementing effective regional ocean governance: perspectives from economics. *Duke Environmental Law and Policy Forum* **16**, 205–216.

Hanna, S. (2008) Institutions for managing resilient salmon (Oncorhynchus Spp.) ecosystems: the role of incentives and transactions costs. *Ecology and Society* **13**(2), 35.

Hanna, S. (2009) Managing the transition: distributional issues of stock rebuilding. In: *The Economics of Rebuilding Fisheries* (ed. A. Cox), pp. 141–162. OECD, Paris.

Hanna, S. (2011) Economics in the service of fisheries policy and practice. *Marine Resource Economics* **26**, 87–94.

Hanna, S. and Jentoft, S. (1996) Human use of the natural environment: an overview of social and economic dimensions. In: *Rights to Nature: Ecological, Economic, Cultural and Political Principles of Institutions for the Environment* (eds S. Hanna, C. Folke and K-G. Maler), pp. 35–56. Island Press, Washington, DC.

Hardin, G. (1968) The tragedy of the commons. *Science* **162**, 1243–1248.

Healey, M.C. and Hennessey, T. (1998) The paradox of fairness: the impact of escalating complexity on fisheries management. *Marine Policy* **22**(2), 109–118.

Hilborn, R. (2007) Defining success in fisheries and conflicts in objectives. *Marine Policy* **31**, 153–158.

IUCN (2010) Saving biodiversity: an economic approach. *World Conservation* **40**(1), 6–10.

Jentoft, S. (1989) Fisheries co-management: delegating responsibility to fishermen's organizations. *Marine Policy* **13**(2), 137–54.

Jentoft, S. (2000) Legitimacy and disappointment in fisheries management. *Marine Policy* **24**, 141–148.

Jentoft, S. (2007) Limits of governability: institutional implications for fisheries and coastal governance. *Marine Policy* **31**, 360–370.

Johannes, R.E. (1978) Traditional marine conservation methods and their demise. *Annual Review of Ecology and Systematics* **9**, 349–364.

Leal, D., De Alessi, M. and Baker, P. (2008) *Beyond IFQs in Marine Fisheries*. Property and Environment Research Center, Bozeman, MT.

Marine Stewardship Council (MSC) (2009) Pacific hake mid-water trawl fishery is MSC-certified 21 October 2009. Available at http://www.msc.org/track-a-fishery/fisheries-in-the-program/certified/pacific/pacific-hake-mid-water-trawl (accessed March 2014).

Marine Stewardship Council (MSC) (2012) Certified fisheries in the north-east Atlantic. http://www.msc.org/track-a-fishery/certified/north-east-atlantic (accessed February 2014).

McIlgorm, A., Hanna, S., Knapp, G., Le Floc'H, P., Millerd, F. and Pan, M. (2010) How will climate change alter fishery governance? Insights from seven international case studies. *Marine Policy* **34**, 170–177.

Milazzo, M. (1998) Subsidies in world fisheries: a reexamination. World Bank Technical Paper No. 406, Fisheries Series. The World Bank, Washington, DC.

Millennium Ecosystem Assessment (2005) *Ecosystems and Human Well-being: Biodiversity Synthesis*. World Resources Institute, Washington, DC.

MRAG, IFM, CEFAS, AZTI Tecnalia and PolEM (2009) An analysis of existing Rights Based Management (RBM) instruments in Member States and on setting up best practices in the EU. Final Report. MRAG Ltd, London.

North, D.C. (1990) *Institutions, Institutional Change, and Economic Performance*. Cambridge University Press, Cambridge.

North Pacific Fishery Management Council (NPFMC) (2012) Community Development Quota Programs. Available at http://www.fakr.noaa.gov/npfmc/catch-shares-allocation/CDQ.html (accessed February 2014).

OECD (1997) *Towards Sustainable Fisheries: Economic Aspects of the Management of Living Marine Resources*. OECD, Paris.

Ostrom, E. (1995) Designing complexity to govern complexity. In: *Property Rights and the Environment: Social and Ecological Issues* (eds S. Hanna and M. Munasinghe), pp. 33–46. World Bank, Washington, DC.

Pacific Whiting Conservation Cooperative (PWCC) (2012) http://www.pacificwhiting.org/ (accessed February 2014).

Raakjær, J. (2009) *A Fisheries Management System in Crisis: The EU Common Fisheries Policy*. Aalborg University Press, Aalborg, Denmark.

Raakjær Nielsen, J. and Vedsman, T. (1999) User participation and institutional change in fisheries management: a viable alternative to the failure of "top-down" driven control? *Ocean and Coastal Management* **42**(1), 19–37.

Reuter R.F., Conners, E., DiCosimo, J., Gaichas, S., Ormseth, O. and Tenbrink, T. (2010) Managing non-target, data-poor species using catch limits: lessons from the Alaskan groundfish fishery. *Fishery Management and Ecology* **17**, 323–335.

Rice, J. (2007) An ecologist's view of economic instruments and incentives. *International Journal of Global Environmental Issues* **7**(2/3), 191–204.

Rice, J.C. and Garcia, S.M. (2011) Fisheries, food security, climate change, and biodiversity: characteristics of the sector and perspectives on emerging issues. *ICES Journal of Marine Science* **68**(6), 1343–1353.

Rosenberg, A.A. and McLeod, K.L. (2005) Implementing ecosystem-based approaches to management for the conservation of ecosystem services. *Marine Ecology Progress Series* **300**, 270–274.

Runge, C.F. (1984) Institutions and the free rider: the assurance problem in collective action. *Journal of Politics* **46**(1), 154–181.

Salomon, A.K., Gaichas, S.K., Jensen, O.P., Agostini, V.N., Sloan, N.A., Rice, J., McClanahan, T.R., Ruckelshaus, M.H., Levin, P.S., Dulvy, N.K. and Babcock, E.A. (2011) Bridging the divide between fisheries and marine conservation science. *Bulletin of Marine Science* **87**(2), 251–274.

Schou, M. (2010) Sharing the Wealth. *Samudra Report* **55**, 18–23.

Scott, A.D. (1955) The fishery: The objectives of sole ownership. *Journal of Political Economy* **63**, 116–124.

Smith, V.L. (1969) On models of commercial fishing. *Journal of Political Economy* **77**, 181–198.

Sumaila, U.R. (2005) Differences in economic perspectives and implementation of ecosystem-based management of marine resources. *Marine Ecological Progress Series* **300**, 279–282.

Sylvia, G., Munro Mann, H. and Pugmire, C. (2008) Achievements of the Pacific whiting conservation cooperative: rational collaboration in a sea of irrational competition. In: *Case Studies in Fisheries Self-Governance* (eds R. Townsend, R. Shotton and H. Uchida), pp. 425–441. FAO Fisheries Technical Paper No. 504. Rome, FAO.

UNEP (2003) Review of the principles of the ecosystem approach and suggestions for refinement: a framework for discussion. CBD Expert Meeting on the Ecosystem Approach, Montreal 7–11 July 2003. Available at http://www.cbd.int/doc/meetings/esa/ecosys-01/official/ecosys-01-03-en.pdf (accessed February 2014).

UNEP (2004) The use of economic instruments in environmental policy: opportunities and challenges. United Nations Environment Programme Division of Technology, Industry and Economics.

UNEP (2009) Certification and sustainable fisheries. United Nations Environment Programme Division of Technology, Industry and Economics. Available at http://www.unep.ch/etb/publications/FS%20certification%20study%202009/UNEP%20Certification.pdf (accessed February 2014).

UNEP (2011) Fisheries: investing in natural capital. Towards a Green Economy: Pathways to Sustainable Development and Poverty Eradication. Available at http://www.unep.org/greeneconomy (accessed February 2014).

US Commission on Ocean Policy (USCOP) (2004) An ocean blueprint for the 21st century. Final report. US Commission on Ocean Policy, Washington, DC.

Western Alaska Community Development Quota Association (2010) Western Alaska Community Development Quota Program. Available at http://www.wacda.org/media/pdf/SMR_2010.pdf (accessed February 2014).

Wilen, J.E. (2000) Renewable resource economists and policy: what difference have we made? *Journal of Environmental Economics and Management* **39**, 306–327.

Williamson, O. (1985) *The Economic Institutions of Capitalism*. The Free Press, New York.

Williamson, O.E. (1994) Transactions cost economics and organization theory. In: *The Handbook of Economic Sociology* (eds N.J. Smelser and R. Swedberg), pp. 77–107. Princeton University Press, Princeton, NJ.

Wolf, A. (1997) *Quotas in International Environmental Agreements*. Earthscan Publications, Ltd., London.

World Bank (2009) The Sunken Billions: The Economic Justification for Fisheries Reform, Agriculture and Rural Development Department. The World Bank, Washington, DC.

Chapter 6

The social dimension: The challenge of dealing with equity

B. Hersoug

Norwegian College of Fishery Science, University of Tromsø, Norway

Abstract: The social dimension of fisheries management and conservation is complex, ranging from cultural aspects to the socio-economic status of the fishers and their families. This chapter deals with one aspect of the social dimension, equity issues and, in particular, the allocation of benefits and costs. The claim is that fisheries management has largely been concerned with creating wealth (neglecting the distribution of that wealth) while conservation has focused on creating (natural) values but with an unequal distribution of costs. Using the establishment of marine protected areas (MPAs) as a case, it is shown that reconciling fisheries management and conservation requires honesty from both parties. Fishers may support MPAs in order to obtain larger catches (in the long run), while No-Take-MPAs advocated by conservationists are implemented in order to conserve marine biodiversity. A good start could be made if all management issues are discussed in the light of their distributional effects, keeping an eye especially on those people with the fewest alternative options.

Keywords: social dimension; equity; MPAs; unequal distribution; co-management; EAF; ICZM

Introduction: The two cultures

In 1959 the British scientist and novelist C.P. Snow delivered his famous speech on *The Two Cultures*. He claimed that intellectual life of the whole western society was split into the sciences and the humanities, and that this was a major hindrance to solving important problems in the world (Snow, 1959). Twenty years later, from 1980 onwards, much the same can be said about the two main protagonist groups involved with marine and inland waters: fishers and fisheries administrators on the one hand and conservationists on the other. The first group will, in spite of their inherent conflicts, talk about resource *utilization*, while the second is concerned with resource *conservation*. From the outset it would seem that they are worlds apart but, on closer inspection, we find that their goals may be more similar over time although their strategies still differ considerably.

If we return to the ultimate goals, fisheries management has over the last 30 years increasingly been involved and constrained by the demand for *sustainable utilization*. Over the last ten years the demand for an ecosystem-based approach to fisheries management (EAF) has come to the fore, although much remains to be done in terms of implementation (Cochrane and Garcia, 2009). In responsible fishery management, the goal is to achieve the highest human benefits compatible with the necessary long-term conservation of the productive ecosystem. In conservation, the goal is to achieve the maximum protection of the ecosystem structure and

Governance of Marine Fisheries and Biodiversity Conservation: Interaction and Coevolution, First Edition.
Edited by Serge M. Garcia, Jake Rice and Anthony Charles.

functions compatible with the necessary satisfaction of human needs. While the ultimate goals may not be completely overlapping, one would believe that there should be sufficient common ground to start a discussion between the two camps. However, co-operation must be based on an honest assessment of the different goals, of proper scientific evaluation of benefits and costs and, not least, of communication and consultation with the stakeholders most affected by the various management measures.

This chapter briefly describes how fisheries management and conservation deals with one aspect of the social dimension. The social dimension is a complicated concept and has a variety of meanings ranging from the cultural aspects of fishing and coastal communities to the more precise socio-economic status of fishers and their families. If we accept that the ecosystem also includes human beings, the challenge of EAF is how to regulate the interaction between the natural and the social system, that is, the institutions involved. And if institutions are central parts of the social dimension, so are participation and democracy. The social dimension can further be studied on the level of the individual, family or community. In order to make the task manageable this chapter concentrates on *equity issues*, in particular on the allocation of benefits and costs. After a brief review of fisheries management and conservation practices in the following two sections, making a distinction between models and reality, the chapter proceeds by looking for common ground ('Reconciling fisheries management and conservation'). The fifth section deals with the need for consultation and co-management, both in fisheries management and conservation. The chapter then attempts to place both fisheries management and conservation within larger planning and policy frameworks, such as integrated coastal and ocean management (ICOM), integrated coastal zone management (ICZM), ecosystem-based management (EBM) and marine spatial planning (MSP); a short summary of lessons learnt is provided at the close of the chapter.

Fisheries management: Creating wealth, forgetting about distribution

The most common fisheries management measures are described more in detail in Chapter 12. They can be classified into four broad categories: input control, output control, technical regulations and financial measures (taxes, fees and subsidies). What is striking about them is that they hardly encompass any measures aiming directly at the social dimension of fisheries, in this case, equity and distribution. With the exception of certain types of subsidies nearly all known management measures are aimed at biological or economic goals. Only by implication do these measures offer solutions to the overriding social issue: the distribution of catch and income. While fisheries biologists have been concerned with the wellbeing of the fish and increasingly with the quality of the ecological system, fisheries economists have concentrated on wealth creation and on how to maximize the resource rent. While most fisheries biologists have disregarded the distributional issues entirely fisheries economists have concentrated on wealth creation, leaving distributional aspects to the market or to politics. By implication the argument has been that fish stocks in good condition will benefit all fishers, albeit not to the same degree. In a similar manner, fisheries economists have argued that creating the largest value of the fisheries in question will, by way of redistribution, benefit all members of society including the poor.[1]

There are at least two hitches to this argument. In most developing countries the State apparatus is not geared towards collecting resource rent or any taxes from the fisheries at all. If rent is collected, as in the case with leasing fishing rights to other States, the money is often squandered by the groups or families controlling the State; this is also the case with resource rent from petroleum (Hodges, 2001). In any case the small-scale fishers, making up the bulk of fishers in nearly all developing countries, will hardly see any money redistributed through the State.[2] In developed countries, and specifically in the cases where individual transferable quota (ITQ) or individual vessel quota (IVQ) systems have been introduced, it is the *vessel owners* that have been compensated while the crew members, making up the majority of the fishers, have received nothing. With few exceptions this also applies to scrapping schemes.

Most fisheries regulations have *social implications*, that is, they influence the distribution of catches between different fleet groups and fishers. Some technical fisheries regulations may also contain distributional goals by design, as in the case of the famous Lofoten fishery (the largest cod fishery in the world). Here the central fishing area is regulated into different zones according to gear groups, a system that dates back more than 100 years (Jentoft and Kristoffersen, 1989). This is done in order to secure all fishers, regardless of vessel size and gear type, a share of the total catch (Holm et al., 2000). The system is still in use and similar zoning arrangements are frequently used in other countries, for example the 5 nautical miles zone used in Chile to protect the artisanal fishers and the 3 nautical miles limit in Mozambique (Castilla, 2010; Menezes et al., 2011). These regulations secure the small-scale fishers access to the resources, although the efficiency of the measures depends entirely on the administrative will and capacity to ensure that the industrial fleets are actually kept outside these limits.

While many countries have introduced minimum wages, minimum prices (of fish) or support schemes for fishers experiencing hard times, the most important measure in terms of influencing equity is the allocation of fish resources, that is, the distribution of global quotas or the allocation of effort. The resource allocation keys that are used in the various fisheries determine to a large degree *who is getting what*, which in turn largely influences the social equity dimension of the fishery in question. In the case of Norway, the main issue in the important cod fisheries has been the conflict between the trawlers and the coastal fleet, starting back in the 1930s. After years of conflict a long-term allocation key was established in 1989 (the 'trawl ladder'), while a similar allocation key was established within the coastal fleet from 1992 onwards (Hersoug, 2005). When fishing rights were later made transferable, restrictions were made in terms of vessel size and separate geographical markets in order to maintain a diversified fleet, protecting also the fishers in the small-scale fleet. Similar arrangements are also in place in fully fledged ITQ systems, as in Iceland and New Zealand where the effective ITQ-markets are divided according to size or gear groups and geographical location of the resources (Hersoug, 2002b; Karlsdóttir, 2008).

It is not being claimed here that fisheries administrators do not have social goals. Most fisheries plans contain goals relating to food, income (State and private) and employment, necessitating hard choices (Bailey and Jentoft, 1990). Clark (1985) offers a menu of 22 objectives, of which 8 can with some reasonable justification be termed equity-based. The claim here is that most management measures are derived from theoretical models aiming at biological or economic goals. When these models and management measures meet the social reality, they have to be adjusted. The ideal TAC or effort cannot be achieved without Draconian interventions. Adjustment of the scientific advice is therefore commonplace, occasionally ending up with unsustainable quotas or effort regulations, always with a reference to 'social considerations' and in order not to impose sudden changes to the industry. Holden (1994) offers a vivid illustration of how quota recommendations in the EU were adjusted according to what was politically acceptable in the respective political constituencies. Whether these adjustments really serve the proclaimed social concerns is an empirical question. Part of the problem is that socio-economic research about the actual consequences of various policy interventions is still largely missing, with only rudimentary knowledge of income levels and living standards in the various fishing communities. Sverdrup-Jensen and Eliasen (2009) demonstrate with case examples from the UK and Denmark that data on the social dimension may be reasonably well obtained on community level (village or municipality), but that the link to various management strategies may be much harder to establish. In Australia however, socio-economic impacts are regularly evaluated when establishing MPAs and an extensive toolkit has been developed (Australian Government, 2005). In a similar manner, social considerations are written into the US legislation (through the Magnuson-Stevens Fishery Conservation and Management Act) where regulatory impacts on fisheries communities have to be considered (Clay and Olson, 2008).

As countries such as Norway, New Zealand, Iceland, the US and the EU nations have modernized their fleets, the demand for labour has been reduced. In these countries the redundant fishers have, by and large, found occupations in other sectors, primarily in industry and the service sector, although the transitional periods may have been more or less painful. Norway can serve as poignant illustration: in 1946 the number of fishers was about 120,000 while in 2012 it was less than 12,000 (Figure 6.1). This reduction has been implemented without any decrease in total catches and without creating any mass unemployment in the coastal areas, largely due to the opportunities offered by development in other sectors including aquaculture (Hersoug, 2007). Unfortunately it has been difficult for most developing countries to replicate this transition, which has given rise to an extensive debate about strategies to reduce poverty in the fishing areas (Allison and Ellis, 2001; Béné et al., 2010; FAO 2000; Jentoft and Eide, 2011; Jul-Larsen et al., 2003; Panayotou, 1982; Pearse, 1982; World Bank, 1992).

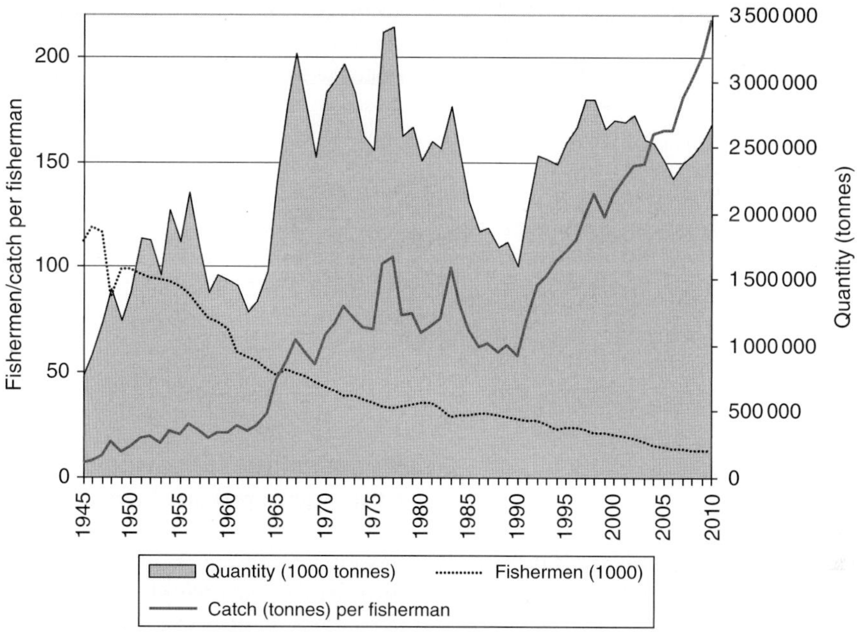

Figure 6.1 Norwegian total catch, catch per fisher and number of fishers (1945–2010).

Conservation: Creating values with unequal distribution of costs

As in fisheries management, nature conservation contains a number of different goals and management measures. For the sake of simplification and illustration, we shall concentrate here on the use of marine protected areas (MPAs), considered the most important tool by most conservationists. While terrestrial nature conservation has a long and distinguished history, the preservation or protection of marine areas is a relatively new invention. The original system of classification of conservation areas made in the 1970s by the International Union for Conservation of Nature (IUCN) excluded marine ecosystems, but was subsequently modified in 1988 and 1994 to include marine protected areas (Al-Abdulrazzak and Trombulak, 2011). This was not least spurred by the 1992 Convention for Biological Diversity, setting a target for 10% of the global marine area to be designated as MPAs by 2010. Relative to this ambitious goal, progress has been slow and, according to McCay and Jones (2011), MPAs are still covering only 1.3% of the total marine area and 3.2% of the marine areas under national jurisdiction; the deadline was therefore extended to 2020. Even more ambitious were the goals of IUCN, claiming that that 20–30% of the area of each marine habitat should be designated not only as MPAs but as *no-take* MPAs (NTMPAs), that is, with fishing strictly prohibited. Similar initiatives have been made by a large number of scientists both in 1998 and 2001 (Jones, 2007).

Nevertheless, in the past two decades there has been a rapid increase in the number of MPAs established around the world and in the research connected to the effects of MPAs both as a tool for fisheries management and as a conservation measure. According to Marinesque et al. (2011), they were able to identify altogether 6834 MPAs in 182 countries (out of a total of 210 coastal countries). The distribution is rather skewed however, with two-thirds of the MPAs (isolated or as part of networks) established in the developed countries (first and foremost in the United States, Canada, Australia and New Zealand, with the European countries following next). The distribution is rather erratic, with huge variations within the different groups of countries ranging from advanced economies to the least-developed countries (LDCs) (Marinesque et al., 2011).

Marine protected areas have a variety of synonyms and definitions: national parks, wildlife refuges, monuments and marine sanctuaries, fisheries closures, critical habitat, habitat areas of particular concern, State parks, conservation areas, estuarine reserves and preserves and numerous others. None of these really tell the user anything specific about the nature of the MPA. Further, terms such as 'park' or 'reserve' or 'preserve' may have different meanings and provide varied levels of resource protection in different situations. The US National Marine Protected Areas Center has therefore provided a classification system to describe MPAs in purely functional terms using five objective characteristics common to most MPAs[3]:

- conservation focus;
- level of protection;
- permanence of protection;
- constancy of protection; and
- ecological scale of protection.

Other scientists still claim 'the label of marine protected area has been haphazard, providing little rigor in terms of what is being protected at what level and why' (Al-Abdulrazzak and Trombulak, 2011, p. 576). They suggest a system of classification based on the level of protection, operating with seven different indicators which combined can be used to place MPAs in one of five different classes. Whatever the classification used, most of the interest and debate is connected to the no-take MPAs which is equivalent to 'closed areas' in conventional fisheries management terminology or to Category Ia (strict nature reserve) under the IUCN's protected area management categories (IUCN, 1994). According to PFMC (2004) the objectives and rationales of reserves can largely be divided into five main groups:

- reserves as an insurance policy;
- reserves as a source of fishery benefits;
- reserves as a source of ecosystem benefits;
- reserves as a means of achieving social objectives; and
- reserves as an opportunity to advance scientific knowledge.

There is little doubt that NTMPAs fulfil their conservation goals, although the empirical knowledge of what is being achieved in the large number of MPAs is still unknown. NTMPAs act as an insurance policy, they provide ecosystem benefits (such as biological diversity) and they definitely contribute to the advancement of scientific knowledge. The more debatable claims refer to the NTMPAs as a source of fishery benefits and to what extent they contribute to fulfilling social objectives. A closer inspection of the philosophies and goals of the environmental NGOs (eNGOs) offers limited references to *people* and, in particular, to the people most affected: the fishers. This is not to say that all conservationist schemes are top-down and little concerned with the people involved; there is however a certain bias in many projects where the main preoccupation is with the natural marine ecosystem aspects, to the detriment of the human components. This is probably also due to the staffing of the eNGOs and the particular conservation projects, with a majority coming from a biology background (marine biology, fisheries biology, ecology, etc.) and relatively few from the social sciences (sociology, social anthropology, etc.).

Reconciling fisheries management and conservation

Area-based measures are definitely not new to fisheries management, but normally area regulations are temporary (for example protecting spawning fish) or, if permanent or semi-permanent, they aim at keeping certain fleet or gear groups out of designated areas in order to protect the fish, the fishers or the habitat (Hall, 2009). Permanent no-take zones have seldom been used in traditional fisheries management. However, with fisheries management gradually moving towards an ecosystem approach, MPAs may be more common, a fact that is clearly reflected in FAO's technical guidelines to responsible fisheries (FAO, 2011).

As pointed out by McCay and Jones (2011) there are various reasons why it has been difficult to integrate MPAs with traditional fisheries management. First of all, it is technically difficult to sort out the site, the size (of the MPAs) and the disturbance level. Marine ecosystems are closely interconnected and with a high degree of variability. While MPAs in coral areas may be small and directed at protecting fish in specific areas, MPAs in temperate waters, aiming at protecting groundfish species, have to be extremely large. Hall (2009) mentions the modelling done by ICES on the prospect of closing one-quarter of the North Sea in order to protect the cod stocks. It was found that such a measure would 'do little or nothing' to protect the widely dispersed and mobile cod.

Second, it is politically difficult to impose area restrictions in a setting where most fishers are already hard pressed by declining catches, often strong competition (from the more industrialized fleets) and by intensified controls.

Third, there seems to be different views regarding the ecosystem, where one group of stakeholders considers humans as intruders in the marine ecosystem while another consider the marine seascapes as inhabited by humans, having a long trajectory in utilizing various marine resources.

A fourth obstacle may be the promotion of MPAs as successful management devices. Many scientific studies of MPAs claim that the closed areas serve to enhance sustainable fisheries by way of two mechanisms: spillover and larval transport (Mascia et al., 2010). The biological argument is fairly simple: MPAs will maintain productive fisheries by protecting brood stocks within their borders. These stocks can enhance catches through adults that grow up inside the reserve and then migrate to fishing areas (spillover), or through enhanced recruitment to fishing areas due to increased population fecundity from the reserve (larval transport). While a number of eNGOs have enthusiastically promoted both the MPAs and the arguments cited above, claiming a win–win situation, the scientific community has been much more divided (Grimes and Ralston, 2003; Hilborn, 2003; Hilborn et al., 2004, 2006; Kaiser, 2004, 2005).

According to Kolding (2011), we find marine and conservation biologists on one side and resource biologists on the other where the divergences are based on two key factors: different objectives and different science (Jones, 2006a, b). We are not going to follow this debate further here, although the scientific underpinning of such a grand scheme is extremely important in its own right. My contention is that the MPAs have been 'oversold' as if they also contribute to increased wellbeing of the fishers, in terms of higher income, due to larger catches. Here the scientific backing is shaky, since in most instances we do not have reliable studies covering the situation before and after the introduction of an MPA. As pointed out by Voyer et al. (2011) most studies have been implemented afterwards, asking the larger community about their attitudes towards the MPAs. While these may generally be positive, they seldom cover the important minority that has been directly involved, that is, the fishers losing their fishing grounds or those being evicted due to the establishment of MPAs. It may be difficult to adequately map reactions in certain developing countries where fishers are hardly literate, except by noticing poaching as a silent response (Isaacs, 2011).

While fishers in most developed countries have their organizations and are perfectly able to voice their protests against encroachment in their fishing areas, poor fishers in developing countries may find it difficult to oppose NTMPA proposals, due to a strong alliance of government, conservation organizations and tourist interests. Menezes et al. (2011) offer a strong warning against this tendency:

> 'NGOs aiming at protecting sharks, whales and dolphins have promoted the introduction of marine protected areas (MPAs) in several places along the Mozambican coast. The aim is often to protect resources not even being exploited, only disturbed by fishers operating in the same waters. Some of these areas have developed tourist industries within the context of the MPA, combining significant economic interests and environmentalism. This development is increasingly challenging the fishing rights of the poor artisanal fishers, as the poor subsistence fishers are usually the first to be excluded from MPAs.'
>
> *(Menezes et al., 2011, p. 422).*

Similar stories are also heard from other countries (Isaacs, 2011). As pointed out by Sharma (2008) MPAs have increasingly become tools that limit, forbid and control use-patterns and human activity through a structure of rights and rules. While numerous studies have examined the ecological and biological impacts of MPAs, few have focused on their social implications for communities and other stakeholders in the area who depend on fisheries resources for a livelihood. 'A particular MPA may be both a "biological success" and a "social failure", devoid of broad participation in management, sharing of economic benefits, and conflict-resolution mechanisms', according to Christie et al. (2003. p. 22). In order to investigate the social effects of MPAs the International Cooperative in Support of Fishworkers (ICSF) commissioned studies in six different countries, all with a number of different MPAs (Brazil, India, Mexico, South Africa, Tanzania and Thailand). Experiences in Brazil were the most positive, where communities were at the forefront of setting up marine extractive reserves (MERs); results from other countries were less convincing however, clearly demonstrating that the communities do not consider themselves equal partners in the MPA process. They were expected to participate in implementation, but were not part of the process of designing the projects. Furthermore, the alternative livelihood options provided limited (if any) support (Christie et al., 2003). Other studies have presented a more diversified picture, showing that fishers are also leading initiatives towards establishing MPAs to protect special habitats (spawning grounds), limit illegal fishing and the use of damaging fishing methods or simply protect their own fishing rights.

Nevertheless, several authors have argued that the primary aim of NTMPAs is to restore marine biodiversity and whatever occurs in terms of improved fishing should be considered a bonus (Ballantine, 2002). Halpern et al.

(2004) move a step further, claiming that NTMPAs 'need not, and perhaps should not, be designed with fisheries management as a primary goal'. Jones (2007) maintains 'it is increasingly accepted that the primary goal of NTMPAs is to conserve marine biodiversity, with fisheries management being a secondary objective'. If this view were accepted, it would certainly make the debate between fishers/managers and conservationists less acrimonious. While fishers and fisheries managers are dealing with fish stocks and increasingly with the larger ecosystem, conservationists are forwarding various types of MPAs in order to restore ecosystems. In the end, compromises have to be found, and in democratic States these have to be worked out in the political sphere. It goes without saying that as soon as the requirement of no-take is eased it is considerably easier to find possible compromises, thus zoning an area into sub-areas where different activities are allowed (although this clearly demands more in terms of supervision and control). In any case, both successful fisheries management and conservation require consultations with the stakeholders and preferably some type of co-management, to which we now turn.

Consultation and co-management

Contrary to popular perception, the idea of consultation and co-management is not new in fisheries management. This chapter started by mentioning the Lofoten cod fishery in Norway, where co-management institutions were established more than 100 years ago. Management committees for each fishing area were elected by all participating fishers (more than 30,000 in the 1930s), being part of a nested system working in tandem with the State-nominated fisheries manager of the fishery. Decisions regarding area demarcation for different fishing gears, time regulations and surveillance functions were largely made by these committees, thus instituting a long tradition of co-management in Norway. However, the true co-management debate was initiated in the 1980s (Jentoft, 1989; Pinkerton, 1989; Pomeroy, 1991). Based on the perception that top-down State-dominated fisheries management had failed, co-management was seen as a blanket recommendation to improve not only the resource situation but also in terms of providing equity. While it is clearly utopian to expect each fisher to receive the same quota or catch, different outcomes may more easily be accepted if the process has been considered legitimate and fair. Co-management was then defined as 'the sharing of responsibility and authority between the government and the community of local fishers to manage a fishery' (Pomeroy and Berkes, 1997, p. 466). In spite of being based on a rather weak analytical understanding, co-management soon turned out to be a near-mandatory requirement in most fisheries projects (Béné and Neiland, 2006). Banks, development organizations and NGOs all required co-management structures to be set up in order to obtain support, whether for fishery management or conservation projects.

Unfortunately, just like other management interventions, co-management is not one model that can easily be fitted into any type of environment. As pointed out by Béné and Neiland (2006), much of the co-management literature has a rather simplified model of co-management meaning and refers to direct participation by all involved fishers, in a form of direct democracy similar to the ancient Greek republics. However, in a modern setting with a large number of fishers fishing for different species with different boats and gears, in different environments along the entire coast, direct democracy could not possibly function.

A system of corporative governance developed in the Norwegian setting, whereby fishermen's organizations were consulted by the Ministry of Fisheries in nearly all issues of fisheries policy. As demonstrated by Hersoug and Rånes (1997) this system gave tangible benefits for both parties. First, decisions taken after stakeholders have been heard are generally considered more legitimate. Secondly, by involving group members and organizations, they then have to share responsibility for decisions, including the unpopular and difficult ones. There are also advantages for the participating groups. The corporatist system offers a separate channel of political influence, whereby groups that are numerically small or have less voting power are still able to exert considerable influence with regard to important decisions (Rokkan, 1966).

Wildawsky (1973) once claimed 'if planning was everything, maybe it is nothing'. The same applies to co-management. Whereas some authors seem to presume a legal framework that institutionalizes both autonomous and shared decision-making between the government and industry (McGoodwin, 1991), others expand the concept to less formal agreements that delegate some power to user groups or joint industry–government bodies (Hersoug, 2004; Pinkerton, 1989). Consequently, both ideas presuppose some form of power sharing between the State apparatus and the stakeholders. To describe the co-management system more precisely, it is useful to address the classical questions: what, how, when, where and who?

Starting with 'what', co-management has mainly been connected to issues related to resource management in the more restricted sense. We shall however connect co-management to a wider set of activities under the umbrella of 'fisheries policy'. The logic for doing so is simple: to maximize, optimize or 'satisfy' their options, most fishers will try to influence a variety of issues and not only issues connected to resource management proper. This means that the idea of co-management is not exclusively linked to management issues, but also plays an important role in development activities (Hersoug and Paulsen, 1996).

Referring to the question of 'how', the analytic framework that is commonly used by the revisionist common property rights tradition is the imaginary 'ladder of participation', where the two extremes are State power and fishermen power. In the first instance, fishermen are at the receiver's end as the decision-making process is entirely 'top-down'; government decides and acts unilaterally, user groups adhere. At the other extreme, user groups have full control and organize and run their own management system. These positions are probably only interesting from a heuristic point of view. More common in democratic policies however is the vast 'in between' arena by which the user groups and government interact through instructions, consultations, cooperation, advice and information. The imaginary ladder therefore has several rungs, according to the degree of fisher influence in the policy-making process. Even if co-management is limited to consultation, cooperation and advice, the model is becoming rather complex because different issues can require different modes of operation.

In general, this 'ladder of participation' gives a much too simplified picture of policy making and co-management. If we consider a simple model of the project cycle consisting of four stages – planning, decision, implementation and feedback – it is obvious that influence exercised at an early stage is not equal to being consulted in the feedback phase. Furthermore, it may be that fishers are excluded from the planning phase, but exercise considerable influence during the implementation stage of a plan or measure. Timing (the question of 'when?') is therefore considered essential, and should be examined as a separate, third dimension.

A fourth important dimension is the level of co-management ('where?'). Does it take place primarily at the national level or at regional and local level as well? Again, there is no right answer. Some issues, such as the assessment of total quotas, are normally decided at a national or even international level whereas others, such as the distribution of quotas or subsidies, may be determined at a lower level. Other issues such as local regulations are decided upon through complicated interactions at several different levels.

The fifth and final dimension is related to user group or stakeholder representation. In other words, *who* should participate in co-management? Should it include fishers only, representatives from the processing industry or, more generally, people living in the coastal communities who are dependent upon fish? If the answer is fishers only, the next question is: which fishers? Should representation be based on functional gear groups (fishers using the same gear) or on location, that is, where the fishers live? The answers to these questions are essential to the establishment of any co-management system. The five dimensions suggested here show that co-management can be a very complex idea, covering a whole range of different options.

Despite more than 20 years of experience in co-management in the fisheries, it seems as if the lessons have taken time to filter down to many NGOs and managers involved in conservation management. In a comprehensive report on governing the MPAs, Jones et al. (2011) discuss how to get the balance right: the balance between top-down, bottom-up and market-driven approaches to MPA management. They analyse 20 established MPAs around the world and try to 'deconstruct' how the governance systems are working, and to what extent the conservation goals have been achieved. Not surprising, they find that there is no 'one size fits all' and that the challenge is to find the right mix of steering systems using a variety of incentives, depending on the political setting and the local challenges.

In the past, the set-up of co-management regimes within fisheries management tended to concentrate on the immediate stakeholders and, in particular, on the fishers. With ecosystem-based fisheries management the number of stakeholders and interests increases, and the set-up will not be very different from what is required for conservation purposes. In the end, the rationale for entering into co-management arrangements is essentially the same. It is a democratic right to participate in decision-making activities affecting local livelihoods; the quality of management is improved when local resource users are involved, providing better information and, finally, the costs of implementation are reduced if local users are involved, as a function of increased legitimacy (McCay and Jentoft, 1996). However, as strongly emphasized by Béné and Neiland (2006) the actual participation may not be the most important aspect 'but the nature of the institutional constraints that determine the degree of (downward) accountability of the agencies'. Hence, the formal set-up of a co-management regime is no guarantee of a successful outcome in terms of equity. At least we have to clarify whether the equity refers to *endowment equity* (access to the resource), *economic equity* (rent distribution), *institutional equity* (the distribution of costs) or *political equity* (participation in the decision-making process). Unfortunately, most co-management studies – whether dealing with fisheries management or marine conservation – cannot deliver precise answers, largely because baseline data are lacking (Béné and Neiland, 2006, p. 54). The central message is that co-management, whether for fisheries or for conservation, is more than a technical fix (Degnbol et al., 2006).

Fisheries management and conservation within larger frameworks

So far we have discussed distributional aspects of the social dimension of fisheries management and conservation as if the main conflict in most marine areas is between fishing and conservation. This is not the case, even if part of the current scientific debate over MPAs and their role in fisheries management could contribute to such an impression. While the use dimension was mainly dominated by fishing and sea transport until the early 1970s, more recent developments have introduced a number of new stakeholders and interests. They include (among others): the oil and gas industry, aquaculture, the tourist industry, the recreational sector, the energy industry (wind and tidal turbines), the mining industry (using the sea for waste deposits) and the navy (using certain areas for arms' testing), to mention the most important. They all lay claims to the use of marine space, and quite often they compete for the use of the same areas. Even in Norway, with a coastline of c. 2500 km (101,000 km if all fjords and islands are included) and only 5 million people, the competition for space is becoming harder by the day. This is even more so, since not all sea space is equally valuable. In aquaculture all operators are out for the 'super localities', that is, the sheltered areas with reasonably strong currents, sufficient depth and close to established infrastructure on land. Furthermore, most of the Norwegian population is concentrated along the coast; with increased incomes and more leisure time the general public are making increased use of the marine areas for recreational purposes.

Within the nearshore areas efforts are made to resolve these conflicting demands through integrated coastal zone management (ICZM), a planning tradition which in the Norwegian case can be traced back to the early 1980s. However, the planning system is complicated and very time-consuming; area planning takes place through the coastal municipalities (276 for the time being), while county plans and national sector plans frequently intercept local priorities. In 1989 the municipalities obtained the right to also plan their sea areas, that is, out to the base lines (the straight lines connecting the outlying land areas and islands), an area of approximately 90,000 km^2 or the size of Portugal. Planning of sea areas was only optional and not mandatory, however. Later in 2009, the municipalities were allowed to extend their sea areas even further (to the baselines plus one nautical mile). This was clearly done in order to fulfil the EU Water Directive and facilitate better planning for the aquaculture industry, now moving from the interior waters to further out from the coast. In most EU countries the situation is more or less similar, although a fourth supranational level is also added to the already complex set-up.

In the Norwegian case, fishing is firmly established in a separate Ministry of Fisheries and Coastal Affairs, while conservation and planning is the responsibility of the Ministry of Environmental Affairs. Oil, gas and energy belong to a third ministry, while tourism is the responsibility of a fourth. Hence, making tradeoffs and compromises is a complicated affair, and the planning system often demonstrates a sluggish reaction regarding new development needs and initiatives. The main device being used is zoning: trying to allocate specific activities to specific zones or combining other less-demanding activities in multipurpose zones.

On top of this complicated system of sector planning and physical area planning, Norway also obtained a complete new system of integrated management plans for large marine ecosystems in 2005. At present, plans for the Barents Sea and the Norwegian Sea have been finalized while the plan for the North Sea is in preparation. These are plans integrating the effects of petroleum development, fisheries and sea transport in a context of global warming, including the effects of long-distance contamination of air and sea. In these plans zoning plays an important role by clearly demarcating particularly vulnerable and/or valuable areas, trying to direct petroleum and sea transport outside these areas (Knol, 2010).

In the Norwegian context these plans are equivalent to similar attempts of large-scale ocean planning which were attempted in New Zealand, Australia and in Canada.

Within the EU such initiatives have long been in the offing but national plans have been hampered by conflicting responsibilities, not only within different national ministries but also between different General Directorates at the EU level. Nevertheless, according to Douvere (2008) marine spatial planning (MSP) may be developed further towards an eco-system-based planning, under certain conditions. This is very much in line with the recommendations of Agardy et al. (2011) of using MSP to encompass strategic plans for representative networks of MPAs. In Scotland, the zoning exercise has already resulted in a complete marine map, with zones for different activities (Smith, 2011). Douvere and Ehler (2009) also report a number of European marine plans, but find that their goals are rather unspecific; consequently, it is difficult to measure goal achievements. They also point out that so far most planning efforts have been dominated by the natural sciences, leaving social and economic aspects on the sideline. This observation was echoed by St Martin and Hall-Arber (2008) in the case of the US.

The point here is not to review the comprehensive literature on MSP, ICZM, integrated coastal ocean management (ICOM) or the broader ecosystem-based management (EBM), only to remind the factions battling over the merits of MPAs that both fishers and conservationists will have to deal with more interests and stakeholders in the future. It is by no means clear whether this will increase the chances of integrating social concerns in the planning and implementation of the new policies. Experiences from Norway seem to indicate that economic priorities take the front seat, with the oil and gas sector emerging as the strongest stakeholder. Facing the threats of much stronger stakeholders, fishers as well as conservationists may find that they have more in common than what appears to divide them at present.

Lessons learnt

From establishing the fact that explicit social concerns play a very modest role in fisheries management and conservation, we could have argued in favour of a massive research effort to establish baseline data and distributional patterns in order to find out how the various fisheries and conservation management measures affect the fishers, their families and communities, whether in developed or in developing countries, with special emphasis on how these measures affect poverty. This would have been in line with the 'maximum approach' preferred by many natural scientists regarding the need for collecting data regarding the entire ecosystem (Knol, 2010). The human part of the ecosystem is no less intricate than the natural part and the possible connections, the spill-over effects and the limits of the system are equally difficult to establish. Hence, it is rather unrealistic that we would be able to establish a dataset which could guide the allocation of quotas, the socio-economic effects of all types of regulations and the displacement effects of establishing MPAs in different forms. In the same manner as a truly ecosystem-based fisheries management system would imply massive costs, changes in the institutional set-up and the involvement of many new stakeholders, the introduction of socio-economic effects as a central parameter would imply considerable changes and costs. In addition, there would be endless discussions over parameters and critical points of interventions. Socio-economic research connected to the fisheries, used as an allocation parameter, would run into precisely the same problems which the 'neutral' biologists are facing today trying to give unbiased advice to the fisheries administrators. Hence, in this chapter we follow Degnbol's (2002) prescription of *keeping it simple*.

The starting point would be that allocation of rights and quotas largely determine the equity issues in fisheries. (The same applies to credit, support, infrastructure, training, etc., but with a more complicated relationship between input and output.) Greater allocation to the smaller poorer fishers would certainly improve many of the socio-economic inequalities we see today. This sounds simple, but here we have firmly established interests that will resist any changes in resource allocation, whether in developed or in developing countries. Even in countries with a strong political commitment to equality, such as in South Africa, redistribution has turned out to be extremely difficult (Hersoug, 2002a; Isaacs et al., 2007). While the effects of unequal distribution in most developed countries to some extent are solved by jobs in other sectors or by a social security net of variable quality, similar effects may in many developing countries be solved by keeping access open for the poorest segments of the fisher population. While this may seem easy in fisheries largely determined by natural variations where fishers operate with very simple technology, the challenge is much greater in marine fisheries where technologies are continuously improved and where market access is extended (Hersoug, 2004; Jul-Larsen et al., 2003). A partial solution may be to reserve the inshore area for the small-scale fishers (the artisanal fleet) in combination with limiting the establishment and/or access of an industrial fleet, thus securing an element of fair competition between very different fleet groups. In the end, the most difficult challenge may be to limit access (reduce effort) without harming the poor or those with few or no other alternatives to fishing (Béné et al., 2010).

The use of time and area restrictions have a long tradition in fisheries management and, as pointed out by Degnbol (2002), simple restrictions on certain types of gear or size of vessels could do much to achieve conservation goals.[4] The same applies to multi-use MPAs, where restrictions are imposed on certain types of fishing depending on the goal of conservation. The difficult conflict applies to the use and claim for conservation measures such as *no-take MPAs*. While such claims may be perfectly legitimate from a conservationist point of view, they are bound to create conflict and resistance from fishers and fisheries administrators. While such conflicts may, in the end, be determined by democratically elected politicians, they will be somewhat easier to handle if we are able to sort out what is best for the fishers and what is best for conservation, not *a priori* 'selling' conservation measures as (successful) fisheries management. If for example a NTMPA turns out to give fishers increased catches and increased income, let that be a bonus. If it does not, it would be fair to pay compensation to fishing vessel owners as well as crew. The present alliance of environmental NGOs, central or local governments and tourist interests has to be critically examined. The whole conservation movement stands to lose legitimacy if conservation (and monetary) values are created for the few selected, while costs are borne by the already poor fishers. On the other hand, 'what is best for the fishers' also has to be critically looked into. Quite often this claim is made by the more powerful vessel owners who do not necessarily represent the poorer small-scale fishers.

Whether MPAs are created for fisheries or for conservation purposes or both, there is an urgent need for consultation and preferably some type of co-management. However, it is important that consultation starts *before* the MPA proposal is finished, so that stakeholders are not invited to participate only in the implementation phase. According to Chuenpagdee et al. (2012), the failure of many MPA projects is mostly attributed to the process leading up to their establishment, that is, the initial stage when the idea was conceived, communicated and discussed among stakeholders. Co-management is difficult and demanding, and more than 20 years of experiences in the fishing sector demonstrate the many pitfalls and also the many possible arrangements that can be made, depending on the type of resources, the natural and political setting and the knowledge and engagement of the stakeholders. Without proper consultation and some form of co-management, the legitimacy of any type of conservation measure will soon deteriorate, resulting in poaching and increasing costs of control and surveillance: a vicious circle. FAO (2011) has given extensive recipes for how to establish MPAs that are considered beneficial also from a fisheries management point of view. A key issue seems to be the political commitment to the establishment, thus securing funding and follow-up. Establishment of MPAs just to fulfil the goals of international agreements runs the risk of creating non-viable areas, where both fisheries and conservation goals may suffer. In the end, conservation measures such as MPAs are not just a technical management measure but also a socio-political enterprise (Chuenpagdee et al., 2012).

With fisheries management approaching EAF, although at very different speeds and in different fashions, the need for greater stakeholder consultation resembles what is required when establishing conservation measures such as MPAs. Many interests are normally involved, and many stakeholders would like to participate in the process. So far fisheries co-management has primarily aimed at fishers, processors, managers and scientists. With EAF, a greater number of other stakeholders would be eligible for participation. Whether their goals and concerns can be negotiated with the traditional stakeholders remain to be seen. Both in fisheries and conservation management there is a difficult tradeoff between efficiency in the decision-making system versus participation and democracy. The more stakeholders and diverging interests, the harder it is to reach joint decisions, the larger the areas to be managed and the greater the need to create representative structures. Decentralized co-management is not necessarily the best recipe for every type of fishery or conservation purpose.

For this reason alone, it can be worthwhile to move the debate for and against MPAs into a larger framework. MPAs are not goals in themselves, but means to achieve favourable outcomes, whether in fisheries or in conservation. Fishing is definitely not the only activity disturbing the ecosystem and, in many countries, there are a number of competing interests not only in the coastal zone, but in offshore areas as well. It therefore makes sense to regulate all activities influencing the ecosystem and not only the fisheries. Integrated coastal zone management, ecosystem-based management and marine spatial planning may be such systems, although their major tools (e.g. zoning) are still on a relatively primitive level. In principle, social concerns have to be observed in precisely the same manner as with fisheries and conservation management; the only difference is that the stakeholders and their concerns are more numerous.

Regarding the equity issues in fisheries management and conservation, a good start could be made if all management measures were discussed according to their distributional effects, trying to keep an eye on those people with the least alternative options. The bottom line is always that regulations that are not (to some degree) considered legitimate will be undermined by poaching and non-compliance. While equal distribution of costs and benefits is clearly utopian, disregarding distributional concerns will normally backfire. This is especially true in countries where it is explicitly stated that the marine resources belong to the nation, that is, all inhabitants, with the State acting as their trustee.

Notes

1. In the case of promoting ITQs, the argument has been that the more efficient operators (fishers or vessel owners) will be able to buy out the less effective ones, thus compensating the losers (Anderson, 1991; Gibbs, 2009).
2. Namibia is a remarkable exception; see Sumaila et al. (2004).
3. See http://marineprotectedareas.noaa.gov/nationalsystem/.
4. Degnbol (2002) claimed that if overcapacity was reduced, this would resolve 80% of the ecosystem concerns. Supplementary technical measures to reduce the impact of gear damage on habitats and some closed areas to protect sensitive habitats would cater for the remaining 18%, leaving 2% for the more complicated fine-tuning of EAF.

References

Agardy, T., Notarbartolo di Sciarab, G. and Christie, P. (2011) Mind the gap: Addressing the shortcomings of marine protected areas through large scale marine spatial planning. *Marine Policy* **35**(2), 226–232.

Al-Abdulrazzak, D. and Trombulak, S.C. (2011) Classifying levels of protection in Marine Protected Areas. *Marine Policy* **36**(3), 576–582.

Allison, E.H. and Ellis, F. (2001) The livelihood approach and management of small-scale fisheries. *Marine Policy* **25**, 377–388.

Anderson, L.G. (1991) A note on market power in ITQ fisheries. *Journal of Environmental Economics and Management* **21**, 229–296.

Australian Government (2005) Socio-economic Impact Assessment Toolkit; A guide to assessing the socio-economic impacts of Marine Protected Areas in Australia. Department of the Environment and Heritage, Canberra.

Bailey, C. and Jentoft, S. (1990) Hard choices in fisheries development. *Marine Policy* **14**, 333–344.

Ballantine, W.J. (2002) MPA perspective: MPAs improve general management, while marine reserves ensure conservation. *MPA News* **4**(1), 5.

Béné, C. and Neiland, A.E. (2006) From Participation to Governance. A critical review of governance, co-management and participation in natural resources management, with particular reference to inland fisheries in developing countries. CGIAR and WorldFish Center, Penang, Malaysia.

Béné, C., Hersoug, B. and Allison, E.H. (2010) Not by rent alone: analysing the pro-poor functions of small-scale fisheries in developing countries. *Development Policy Review* **28**(3), 325–358.

Castilla, J.C. (2010) Fisheries in Chile: small pelagics, management, rights, and sea zoning. *Bulletin of Marine Science* **86**(2), 221–234.

Christie, P., McCay, B., Miller, M.L., Lowe, C., White, A.T. and Stoffle, R. (2003) Towards developing a complete understanding: A social science research agenda for marine protected areas. *Fisheries* **28**(12), 22–26.

Chuenpagdee, R., Pascual-Fernández, J.J., Szeliánszky, E., Alegret, J.L., Fraga, J. and Jentoft, S. (2011) Marine protected areas: Re-thinking their inception. *Marine Policy* **39**, 234–240.

Clark, C.W. (1985) *Bioeconomic Modelling and Fisheries Management*. Wiley, New York.

Clay, P. and Olson, J. (2008) Defining 'fishing communities': vulnerability and the Magnuson–Stevens Fishery Conservation and Management Act. *Human Ecology Review* **15**(2), 143–160.

Cochrane, K.L. and Garcia, S.M. (eds) (2009) *A Fishery Manager's Guidebook*. FAO and Wiley-Blackwell. FAO, Rome.

Degnbol, P. (2002) The ecosystem approach and fisheries management institutions: The noble art of addressing complexity and uncertainty with all onboard and on a budget. IIFET paper no. 171. Available at http://citeseerx.ist.psu.edu/viewdoc/download?doi=10.1.1.161.8674&rep=rep1&type=pdf (accessed March 2014).

Degnbol, P., Gislason, H., Hanna, S., Jentoft, S., Raakjær Nielsen, J., Sverdrup-Jensen, S. and Wilson, D.C. (2006) Painting the floor with a hammer: Technical fixes in fisheries management. *Marine Policy* **30**(5), 534–543.

Douvere, F. (2008) The importance of marine spatial planning in advancing ecosystem-based sea use management. *Marine Policy* **32**(5), 762–771.

Douvere, F. and Ehler, C.N. (2009) Ecosystem-Based Marine Spatial Management: An Evolving Paradigm for the Management of Coastal and Marine Places. Intergovernmental Oceanographic Commission and Man and the Biosphere Programme, UNESCO, Paris, France.

FAO (2000) Poverty in coastal fishing communities. In *Advisory Committee on Fishery Research*. Third session, December 5–8 2000. FAO, Rome.

FAO (2011) Technical Guidelines for Responsible Fisheries. No. 5, Suppl. 6. FAO, Rome.

Gibbs, T. (2009) Individual transferable quotas and ecosystem-based fisheries management: it's all in the T. *Fish and Fisheries* **10**(4), 470–474.

Grimes, C.B. and Ralston, S. (2003) Marine reserves: the best option for our oceans? *Frontiers in Ecology and the Environment* **1**(9), 496–497.

Hall, S.J. (2009) Time and area regulations. *In A Fishery Manager's Handbook* (eds K.L. Kochrane and S.M. Garcia). Wiley-Blackwell, Oxford.

Halpern, B.S., Gaines, S.D. and Warner, R.R. (2004) Confounding effects of the export of production and the displacement of fishing effort from marine reserves. *Ecological Applications* **14**(4), 1248–1256.

Hersoug, B. (2002a) *Fishing in a Sea of Sharks. 2002: Reconstruction and Development in the South African Fishing Industry.* Eburon, Delft (Netherlands).

Hersoug, B. (2002b) *Unfinished Business. New Zealand's Experience with Rights-Based Fisheries Management.* Eburon, Delft (Netherlands).

Hersoug, B. (2004) Exporting fish, importing institutions – fisheries development in the Third World. In: *Fisheries Development: The Institutional Challenge* (eds B. Hersoug, S. Jentoft and P. Degnbol). Eburon, Delft (Netherlands).

Hersoug, B. (2005) *Closing the Commons. Norwegian Fisheries from Open Access to Private Property.* Eburon, Delft (Netherlands).

Hersoug, B. (2007) Always too many? The Human Side of Fishery Capacity Adjustment in Norway. Structural Change in Fisheries. OECD, Paris.

Hersoug, B. and Paulsen, O.A. (1996) *Monitoring, Control and Surveillance in Fisheries Management.* University of Namibia, Windhoek.

Hersoug, B. and Rånes, S.A. (1997) What Is good for the fishermen, is good for the nation: co-management in the Norwegian fishing industry in the 1990s. *Ocean & Coastal Management* **35**, 157–172.

Hilborn, R. (2003) Marine reserves: the best option for our oceans? *Frontiers in Ecology and the Environment* **1**(9), 495–502.

Hilborn, R., Stokes, K., Maguire, J.J., Smith, T., Botsford, L.W., Mangel, M., Orensanz, J., Parma, A., Rice, J., Bell, J., Cochrane, K.L., Garcia, S., Hall, S.J., Kirkwood, G.P., Sainsbury, K., Stefansson, G. and Walters, C. (2004) When can marine reserves areas improve fisheries management? *Ocean and Coastal Management* **47**, 197–205.

Hilborn, R., Micheli, F. and De Leo, G. A. (2006) Integrating marine protected areas with catch regulation. *Canadian Journal of Fisheries and Aquatic Science* **63**, 642–649.

Hodges, T. (2001) *Angola: From Afro-Stalinism to Petro-Diamond Capitalism.* James Currey, London and Indiana University Press, Bloomington.

Holden, M. (1994) *The Common Fishery Policy.* Blackwell, Oxford.

Holm, P., Hersoug, B. and Rånes, S.A. (2000) The missing T. Path dependency within an individual vessel quota system — the case of Norwegian cod fisheries. *Marine Policy* **24**(4), 319–330.

Isaacs, M. (2011) Creating action space: small-scale fisheries policy reform in South Africa. In: *Poverty Mosaics: Realities and Prospects in Small-Scale Fisheries* (eds S. Jentoft and A. Eide). Springer, Dordrecht.

Isaacs, M., Hara, M. and Raakjær, J. (2007) Has reforming South African fisheries contributed to wealth redistribution and poverty alleviation? *Ocean & Coastal Management* **50**, 301–313.

IUCN (1994) Guidelines for protected area management categories. IUCN, Gland, Switzerland and Cambridge, UK.

Jentoft, S. (1989) Fisheries co-management: delegating government responsibility to fishermen's organizations. *Marine Policy* **13**, 137–154.

Jentoft, S. and Kristoffersen, T. (1989) Fishermen's co-management: the case of the Lofoten fishery. *Human Organization* **48**(4), 355–365.

Jentoft, S. and Eide, A. (eds) (2011) *Poverty Mosaics: Realities and Prospects in Small-Scale Fisheries.* Springer, Dordrecht.

Jones, P.J.S. (2006a) Collective action problems posed by no-take zones. *Marine Policy* **30**(2), 143–156.

Jones, P.J.S. (2006b) Arguments for conventional fisheries management and against no-take marine protected areas: only half of the story? *Reviews in Fish Biology and Fisheries* **17**, 31–43.

Jones, P.J.S. (2007) Point-of-View: Arguments for conventional fisheries management and against no-take marine protected areas: only half of the story? *Reviews in Fish Biology and Fisheries* **17**(1), 31–43.

Jones, P.J.S., Qiu, W. and De Santo, E.M. (2011) Governing marine protected areas: getting the balance right. Technical Report, United Nations Environment Programme, Nairobi. Available at http://www.mpag.info (accessed February 2014).

Jul-Larsen, E., Kolding, J., Overå, R., Raakjær Nielsen, J. and van Zwieten, P.A.M. (2003) Management, co-management or no management? Major dilemmas in southern African freshwater fisheries. Synthesis report. FAO Fisheries Technical Paper 426/1. FAO, Rome.

Kaiser, M.J. (2004) Marine protected areas: the importance of being earnest. *Aquatic Conservation* **14**, 635–638.

Kaiser, M.J. (2005) Are marine protected areas a red herring or fisheries panacea? *Canadian Journal of Fisheries and Aquatic Science* **62**(5), 1194–1199.

Karlsdóttir, A. (2008) Not sure about the shore! Transformation effects of Individual Transferable Quotas on Iceland's fishing economy and communities. In: *Enclosing the Fisheries: People, Places and Power* (eds M. Lowe and C. Carothers). Bethesda, MD.

Knol, M. (2010) Marine Ecosystem Governance in the Making: Planning for petroleum activity in the Barents Sea-Lofoten area. Doctoral thesis, University of Tromsø, Tromsø.

Kolding, J. (2011) MPAs in relation to fisheries: what are the biological and fish stock implications? In: *Marine Protected Areas (MPAs) in Relation to Fisheries Management: Challenges and Experiences from Developing Countries* (eds I. Bryceson and F. Francis). WIOMSA Book Series No. 9, Zanzibar, Tanzania.

Marinesque, S., Kaplana, D.M. and Rodwellc, L.D. (2011) Global implementation of marine protected areas: Is the developing world being left behind? *Marine Policy* **36**(3), 727–737.

Mascia, M.B., Claus, C.A. and Naidoo, R. (2010) Impacts of Marine Protected Areas on Fishing Communities. *Conservation Biology* **24**(5), 1424–1429.

McCay, B. and Jentoft, S. (1996) From the bottom up: Participatory issues in fisheries management. *Society & Natural Resources* **9**(3), 237–250.

McCay, B.J. and Jones, P.J.S. (2011) Marine protected areas and the governance of marine ecosystems and fisheries. *Conservation Biology* **25**, 1130–1133.

McGoodwin, J.R. (1991) *Crisis in the World's Fisheries: People, Problems and Policies*. Stanford University Press, CA.

Menezes, A., Eide, A. and Raakjær, J. (2011) Moving out of poverty: Conditions for wealth creation in small-scale fisheries in Mozambique. In: *Poverty Mosaics: Realities and Prospects in Small-Scale Fisheries* (eds S. Jentoft and A. Eide). Springer, Dordrecht.

Panayotou, T. (1982) Management concepts for small-scale fisheries: Economic and social aspects. FAO Fisheries Technical Paper 228, FAO, Rome.

Pearse, P.H. (1982) Turning the tide: a new policy for Canada's Pacific fisheries. Final report of the Commission on Pacific Fisheries Policy. Vancouver, BC.

PFMC (2004) Marine Reserves: Objectives, Rationales, Fishery management implications and Regulatory requirements. White Paper September 2004. Marine Reserves Subcommittee, Scientific and Statistical Committee, Pacific Fishery Management Council.

Pinkerton, E. (1989) *Co-operative Management of Local Fisheries: New Directions for Improved Management and Community Development*. University of British Columbia Press, Vancouver.

Pomeroy, R.S. (1991) Small-scale fisheries management and development: Towards a community-based approach. *Marine Policy* **15**(1), 39–48.

Pomeroy , R.S. and Berkes, F. (1997) Two to Tango: the role of government in fisheries co-management. *Marine Policy* **21**(5), 465–480.

Rokkan, S. (1966) Norway: Numerical Democracy and Corporate Pluralism. In: *Political Oppositions in Western Democracies* (ed. R. Dahl). Yale University Press, New Haven.

Sharma, C. (2008) A study of marine protected areas and their impact on traditional small-scale fishing communities in South Africa. In: *Marine Conservation and Coastal Communities: Who Carries the Costs?* (eds J. Sunde and M. Isaacs). SAMUDRA Monograph, International Collective in Support of Fishworkers, Chennai.

Smith, G. (2011) Maps, networks and a sea that won't conform: thinking critically about marine spatial planning in Scotland. MSc thesis in Environment and Development, School of GeoSciences, University of Edinburgh.

Snow, C.P. (1959) *The Two Cultures. The Rede Lecture*. Cambridge University Press, Cambridge.

St Martin, K. and Hall-Arber, M. (2008) Creating a Place for "Community" in New England Fisheries. *Human Ecology Review* **15**(2), 161–170.

Sumaila, U.R., Boyer, D., Skog, M., and Steinshamm, S.I. (2004) *Namibia's Fisheries: Ecological, Economic and Social Aspects*. Eburon, Delft (Netherlands).

Sverdrup-Jensen, S. and Eliasen, S. (2009) *Socio-Economic Indicators and their Application*. IFM, Aalborg University.

Voyer, M., Gladstone, W. and Goodall, H. (2011) Methods of social assessment in Marine Protected Area planning: Is public participation enough? *Marine Policy* **36**(2), 432–439.

Wildawsky, A. (1973) If planning is everything, maybe it's nothing. *Policy Sciences* **4**(2), 127–153.

World Bank (1992) A strategy for fisheries development. Fisheries Series Discussion Papers No. 135, World Bank, Washington, DC.

Chapter 7

The global legal dimension: Navigating the legal currents of rights and responsibilities

A.H. Hoel[1] and D.L. VanderZwaag[2]

[1]Institute of Marine Research, Bergen, Norway
[2]Marine & Environmental Law Institute, Dalhousie University, Halifax, Canada

Abstract: This chapter describes the global legal framework for the management and conservation of living marine resources, addressing legally binding as well as non-binding instruments. The 1982 United Nations (UN) Law of the Sea Convention and its main provisions in relation to living marine resources are first described, followed by an overview of further global fisheries agreements such as the 1995 UN Fish Stocks Agreement. FAO documents, particularly the FAO Code of Conduct for Responsible Fisheries, are discussed further. The importance of multilateral environmental agreements, such as the 1992 Biodiversity Convention, is then highlighted. Before a concluding discussion of possible future developments, the contributions of global environment and development conferences/summits and the UN General Assembly resolutions and processes are reviewed.

Keywords: framework; binding/non-binding instrument; convention; agreement; UNCLOS; IUU; responsibilities

Introduction

The law of the sea has evolved over centuries. Until the end of the Second World War, the generally accepted view was that the oceans were subject to the freedom of the seas for navigation and other uses. Coastal State jurisdiction prevailed only in a narrow belt of sea along the coasts (Churchill and Lowe, 1999). In 1958, the First United Nations Conference on the Law of the Sea (UNCLOS I) was convened, resulting in four conventions: the Convention on the Territorial Sea and the Contiguous Zone; the Convention on the High Seas; the Convention on Fishing and Conservation of the Living Resources of the High Seas; and the Convention on the Continental Shelf. The conventions formalized the core of the law of the sea[1] but did not solve many issues including the breadth of the territorial sea (Rothwell and Stevens, 2010).

The global framework for the conservation and management of living marine resources has evolved substantially over the past decades (Birnie et al., 2009). The adoption of the 1982 United Nations (UN) Law of the Sea Convention (LOSC) as well as the emergence of a number of other instruments has forged a legal framework addressing the sustainable use and conservation of living marine resources (Edeson, 2005; Orrego Vicuna, 2001). These instruments establish global rules for ownership of resources, for their management, for how international cooperation is to take place and for how rules are to be enforced and disputes resolved (Ebbin et al., 2005). They constitute a dynamic framework evolving over time in response to new challenges, the latest addition to which is the 2009 Port State Measures Agreement in response to an international growth in landings from illegal, unreported and unregulated (IUU) fisheries.

Governance of Marine Fisheries and Biodiversity Conservation: Interaction and Coevolution, First Edition.
Edited by Serge M. Garcia, Jake Rice and Anthony Charles.

The evolution in the global legal framework for the management of living marine resources coincides in time with an increasing emphasis on environmental responsibilities in the use of natural resources (Hoel, 1998). This is reflected in the living marine resources related instruments in a number of ways, for example by the introduction of the precautionary approach in fisheries management (Garcia, 1994; VanderZwaag, 2002) and an increasing emphasis on the impacts of fishing activities on marine ecosystems (Pikitch et al., 2004). A growing number of environmental agreements relevant to management of living marine resources, such as the 1992 Convention on Biological Diversity (CBD), have also emerged.

Two overall currents, sometimes raising major tensions, run through the global governance regime for living marine resources. One current is the bestowal and claiming of *rights* by States to exploit marine living resources and to exert enforcement measures and compel conservation practices on third parties for the protection of those rights. The other current is the setting out of numerous *responsibilities* on States, and in some cases on their resource users, to effectively conserve and manage the taking of living marine resources and broader marine biodiversity (Hutchings et al., 2012; Rayfuse, 2007).

This chapter provides a six-part navigational tour of the array of international instruments and institutional mechanisms established by the global community to govern fisheries and marine biodiversity conservation (see also Annexes 1 and 2 in this volume). The following section reviews the role of the 1982 Law of the Sea Convention in establishing rights and responsibilities. We then highlight further legally binding global fisheries agreements, including the 1995 UN Fish Stocks Agreement, in the third section. The contributions of the FAO Code of Conduct for Responsible Fisheries and its subsidiary documents (four international plans of action and various guidelines) towards sustainable fisheries are then examined in the fourth section. Other key multilateral environmental agreements of relevance, such as the Convention on Biological Diversity, are summarized in the fifth section. The sixth section briefly describes the importance of global environment and development conferences/summits. The penultimate section emphasizes the contributions of UN resolutions and processes to the global quest for healthy marine ecosystems and sustainable coastal communities. The paper concludes with some observations on possible future developments.

It is important to note that the global framework discussed here is only a part of the governance framework for living marine resources. In addition to the global rules, regional bodies have become increasingly important (Lodge et al., 2007) and a large number of bilateral agreements between neighbouring States exists to manage transboundary resources (FAO, 2002; Munro et al., 2004).

The Law of the Sea Convention

UNCLOS III and the status of the Law of the Sea Convention

The Third United Nations Conference on the Law of the Sea (UNCLOS III) took place between 1973 and 1982 and resulted in the 1982 United Nations Convention on the Law of the Sea (LOSC) (1833 UNTS 397). The Convention represents a comprehensive, legally binding instrument that defines a global order of the oceans, including the formal recognition of the 200 nautical mile (NM) exclusive economic zone (EEZ) and a more precise definition of the seaward extent of coastal state rights and jurisdiction over the continental shelf (Hannesson, 2004; Miles, 1998).

As of 21 January 2013, 164 countries had ratified the Convention, which entered into force in November 1994. The Convention has an important position in international law. It is widely considered to be the fundamental instrument on the law of the sea, and therefore other agreements must be compatible with it (LOSC Article 311). It establishes general principles and mandates, directly and through international organizations and diplomatic conferences, to develop further rules, measures and regulations. Many of its provisions are also widely considered to reflect customary international law (Churchill and Lowe, 1999).

Fisheries rights under LOSC

The Law of the Sea Convention recognizes extensive rights by coastal States in their various offshore zones to marine fisheries. Coastal States enjoy full sovereignty over internal waters, such as harbours and some bays,[2] and over an adjacent territorial sea out to 12 NM from the baselines, subject to the right of innocent passage of foreign vessels. Coastal States have the total right of access to fisheries and broad discretion in how to regulate fisheries, subject to various conservation responsibilities under the Convention and possible other rules of international law (Article 2).

The Law of the Sea Convention also defined the new category of an exclusive economic zone, the seaward limit of which may extend to 200 NM from the baselines. In the EEZ the coastal State has sovereign rights:

> '...for the purpose of exploring and exploiting, conserving and managing the natural resources, whether living or non-living, of the sea-bed and sub-soil and the superjacent waters, and with regard to other activities for the economic exploitation of the zone such as the production of energy from the water, currents and winds' (Article 56(1)(a)).

The coastal State also has jurisdiction over matters such as marine scientific research and environmental protection (Article 56(1)(b)).

The Convention also contains provisions regarding enforcement of laws and regulations of the coastal State relating to management of living resources in the EEZ. Subject to certain limitations, the coastal State may take the measures it deems necessary to ensure compliance with its laws and regulations (Article 73).[3]

Other States have to comply with the coastal State's regulations regarding exploring and exploiting the resources in the EEZ, but they also continue to enjoy specified customary rights. The Convention reaffirms the right of all States to exercise various freedoms of the high seas under conditions laid down in the Convention. These include scientific research, navigation and overflight, and the freedom to lay submarine cables and pipelines (Article 58).

The coastal State also exercises sovereign rights to explore and exploit the natural resources of the continental shelf. The continental shelf extends either to 200 NM or further out, depending on legal and geological criteria described in the Convention.[4] While granting coastal States all rights to mineral and other non-living resources of the seabed and subsoil of the continental shelf, Article 77(3) restricts coastal State rights over living marine resources on an extended continental shelf beyond 200 NM to sedentary species. This provision raises uncertainties as to precisely what species are covered beyond sessile species such as mussels and corals, but crabs and lobsters are also considered sedentary (Rothwell and Stevens, 2010). The exercise of coastal State rights over the continental shelf must not infringe or result in an unjustifiable interference with navigation and other rights and freedoms of other States (Article 78(2)).

Furthermore, the Convention provides that the seabed and subsoil thereof beyond the limits of national jurisdiction (the Area) and its mineral resources is the common heritage of mankind. The International Sea-Bed Authority (ISA) is the body through which State Parties to the Convention organize and control activities in the Area (Warner, 2009). The 1994 Agreement relating to the implementation of Part XI of the United Nations Convention on the Law of the Sea of 10 December 1982, an integral part of the Convention, addresses difficulties with the seabed mining regime (Part XI) of the Convention that prevented many industrialized countries from becoming a party to it.[5] A major field of controversy has been the alleged *lacunae* in LOSC regarding the management of seabed living resources (Freestone, 2012) and the regime applicable to the exploration/exploitation of marine genetic resources in areas beyond national jurisdiction (Leary, 2007, 2012).

Pursuant to Article 87 of LOSC, all States enjoy the freedom of fisheries on the high seas beyond 200 NM EEZs. However, that freedom must be exercised with due regard for the interest of other States in their exercise of high seas freedoms and also with due regard for the rights under the Convention with respect to activities in the Area.

Responsibilities under LOSC

Responsibilities under LOSC may be summarized under two categories: responsibilities relating specifically to living marine resource management and those relating to broader marine environmental protection.

Living marine resources

The rights of coastal States to exploit and manage fisheries in the EEZ are subject to numerous conservation responsibilities. Coastal States are required to determine the allowable catch of living resources within their EEZs (Article 61(1)). They are to take into account the best scientific evidence available to ensure the living resources are not endangered by over-exploitation. The coastal State and competent international organizations, whether subregional, regional or global, shall co-operate to this end (Article 61(2)). Measures must be designed to maintain or restore populations of harvested species at levels which can produce maximum sustainable yield as qualified by relevant environmental and economic factors (Article 61(3)). In establishing conservation measures, the coastal State shall take into consideration the interdependence of stocks and the effects on species associated with or dependent upon harvested species, with a view to maintaining or restoring populations of such associated or dependent species above levels at which their reproduction may become seriously threatened (Article 61(4)).

A wide range of international cooperation responsibilities in marine living resource conservation is imposed by LOSC. Cooperation is required, either directly or through regional or subregional organizations, to manage shared fish stocks in EEZs (Article 63(1)) and fish stocks straddling EEZs and high seas (Article 63(2)). Cooperation duties are also spelled out in general terms for conserving highly migratory species (Article 64), marine mammals (Article 65), anadromous stocks (Article 66) and catadromous species (Article 67).

On the high seas, the flag States of the fishing vessels are to respect certain conditions, such as the duty to take such measures for their respective nationals as may be necessary for the conservation of the living resources and the duty to cooperate with other States in the conservation and management of such resources (Articles 117 and 118). In determining such conservation measures, States are to take into account the same criteria noted above for coastal State fisheries (interdependence of stocks, associated or dependent species, generally recommended international minimum standards; Article 119). These provisions establish a foundation for the 1995 UN Fish Stocks Agreement (UNFSA).

Responsibilities for the utilization of living resources in the EEZ are also spelled out under LOSC. Coastal States are to promote the objective of optimum utilization (Article 62(1)). Coastal States must determine their capacity to harvest EEZ living resources and, where they do not have the capacity to harvest the entire allowable catch, they are required to give other States, particularly developing States, the surplus allowable catch (Article 62(2)). In practice this provision does not mean much, as it is up to the coastal State to decide both its own harvesting capacity and the total allowable catch (Burke, 1994).

The LOSC and marine environmental responsibilities

Part XII of LOSC sets out numerous responsibilities for States to protect and preserve the marine environment (Birnie et al., 2009). A coastal State's sovereign right to exploit its natural resources is to be done in accordance with the duty to protect and preserve the marine environment (Article 193). Steps taken by coastal States must include measures necessary to protect and preserve rare or fragile ecosystems and habitats of depleted, threatened or endangered species (Article 194).

States are generally urged to cooperate on a global and regional basis to further elaborate international rules, standards and best practices for the protection of the marine environment (Article 197). They are encouraged to specifically develop global and regional rules and standards for land-based marine pollution (Article 207), seabed activities (Article 208), ocean dumping (Article 210) and pollution through the atmosphere (Article 212). States are also required to further establish international rules and standards to control vessel-source pollution (Article 211).

LOSC also provides a fundamental environmental assessment obligation. Where States have reasonable grounds for believing that planned activities under their jurisdiction or control may cause substantial marine pollution or significant and harmful changes to the marine environment, they must assess the potential effects of such activities and must publish the results and notify competent international organizations (Article 206).

Dispute resolution

The Law of the Sea Convention requires its Parties to settle disputes concerning its interpretation and application by peaceful means (Part XV). Where no settlement can be reached, a Party may refer the dispute to a court or tribunal having jurisdiction in regard to the matter at hand. The Convention points to four such courts or tribunals: the International Tribunal for the Law of the Sea (ITLOS), the International Court of Justice, an arbitral tribunal established pursuant to Annex VII of the Convention or a special arbitral tribunal (Article 287).

The dispute resolution system in the Law of the Sea Convention is mandatory for its Parties, which is unusual in international law (Klein, 2005), and therefore an important development for international law in general. Limitations and exceptions are provided however, for example a coastal State is not subject to binding resolution regarding its discretionary powers over fisheries management in the EEZ, such as determining the allowable catch and allocating surplus to other States (Article 297(3)).[6]

Legally binding global fisheries agreements

The 1982 Convention needed to be strengthened in several respects in relation to living marine resources (Burke, 1994). Over time, a system of additional instruments has evolved addressing compliance, high seas fisheries and the role of port States in combating IUU fisheries.

The 1946 International Convention for the Regulation of Whaling

The International Convention for the Regulation of Whaling[7] entered into force in 1948. It establishes an International Whaling Commission (IWC), which currently has 89 member governments. Against the background of the overfishing of a number of stocks of large whales before the Second World War, the Convention was concluded to provide for the proper conservation and management of whale stocks, enabling the orderly development of the whaling industry.

The International Whaling Commission must adopt regulations with respect to the conservation and utilization of whale resources. The Commission's scientific committee provides scientific advice and recommendations to the Commission. The IWC established a commercial moratorium on whaling for the 1986–1990 period, to which several countries filed an objection and are thereby not bound by it. The moratorium has not been lifted, as the status of whales as natural resources remains controversial. The IWC therefore remains a divided body (Iliff, 2008).

The 1993 Compliance Agreement

The Agreement to Promote Compliance with International Conservation and Management Measures by Fishing Vessels on the High Seas[8] was adopted in 1993 and entered into force in 2003. There are 39 parties to the Agreement.[9] The scope of the agreement is global, as it applies to all fishing vessels that fish on the high seas.

The objective of the agreement is to promote compliance with international conservation measures on the high seas, and it applies to fisheries of all fish stocks located there. The agreement requires flag States to take necessary measures to ensure that fishing vessels flying their flag do not engage in activities that undermine the effectiveness of international conservation and management measures. A key requirement is that no Party allow fishing vessels entitled to fly its flag to be used for fishing on the high seas unless they have been authorized to do so by the Party. Also, States are not to authorize vessels entitled to fly their flag to fish on the high seas unless they can exercise effective control over these vessels.

The flag State must take measures against fishing vessels that act in contravention of the provisions of the agreement. Sanctions shall be of sufficient gravity as to be effective in securing compliance.

The Agreement also requires the State Parties to cooperate on enforcement, by exchanging information on activities of fishing vessels in order to assist the flag State in identifying fishing vessels flying its flag reported to have engaged in activities undermining international conservation and management measures.

The 1995 UN Fish Stocks Agreement

The United Nations Agreement for the Implementation of the Provisions of the United Nations Convention on the Law of the Sea of 10 December 1982 relating to the Conservation and Management of Straddling Fish Stocks and Highly Migratory Fish Stocks (UNFSA)[10] was adopted in 1995 and entered into force in 2001.[11]

The UNFSA is an implementing agreement of the Law of the Sea Convention regarding the conservation and management of straddling and highly migratory fish stocks (Balton, 1996). The objective of the Agreement is '...to ensure the long-term conservation and sustainable use of straddling fish stocks and highly migratory fish stocks through effective implementation of the relevant provisions of the Convention' (UNFSA, Article 2). The Agreement applies to the conservation and management of straddling fish stocks and highly migratory fish species beyond areas under national jurisdiction and, to some extent, to the conservation and management of such stocks within areas under national jurisdiction.

The Agreement sets out provisions for the conservation and management of such fish stocks as well as principles for management, notably the precautionary approach and an ecosystem approach. Since straddling fish stocks and highly migratory fish stocks migrate across maritime zones, the Agreement requires that their management in the coastal States' zones and in the adjacent high seas are compatible and do not undermine the conservation efforts of coastal States. The Agreement also establishes a framework and benchmarks for the development of regional and subregional fisheries agreements (Stokke, 2001) as well enforcement arrangements. It also contains provisions on dispute resolution.

The conservation and management principles in the agreement include the precautionary approach (Articles 6 and 7) as well as an ecosystem approach (Article 5(e)). The Agreement also affirms the duty set out in the Law of the Sea Convention to cooperate concerning straddling and highly migratory fish stocks. Where regional fisheries management organizations or arrangements (RFMO/As) already exist, these are to be used. In areas where a fishery occurs and no organization or arrangement exists, States fishing on the high seas shall establish one.

Building on the Compliance Agreement, the UN Fish Stocks Agreement requires the flag State to ensure compliance by its vessels with subregional and regional conservation and management measures for straddling and highly migratory fish stocks. States cooperating through RFMO/As are to establish cooperative mechanisms for effective monitoring, control, surveillance and enforcement (Hayashi, 1996).

An annual State Party meeting and an ongoing review conference are mandated by the agreement. The review conference has met twice and considered issues relating to the implementation of the Agreement. Informal Consultations and the Review Conference are called for on an *ad hoc* basis by the UNGA Resolution on Sustainable Fisheries.

The 2009 FAO Port State Agreement

The Agreement on Port State Measures to Prevent, Deter and Eliminate Illegal, Unreported and Unregulated Fishing (FAO Port State Agreement) was adopted in 2009 and is not yet in force (25 ratifications required).[12] The objective of the Agreement is to prevent, deter and eliminate illegal, unreported and unregulated fishing (IUU fishing) through the implementation of effective port State measures, and thereby to ensure the long-term conservation and sustainable use of living marine resources and marine ecosystems.

The Agreement recognizes that port State measures are considered a cost-effective means to fight IUU fishing. It includes measures such as requesting permission for entry into designated ports by foreign fishing vessels ahead of arrival and transmission of information on their activities and catch to the port State. Port States must conduct inspections of ships according to standards set out in the Agreement. The flag State is responsible for follow-up action.

The 1995 FAO Code of Conduct for Responsible Fisheries and its subsidiary documents

In parallel with the development of the legally binding UN Fish Stocks Agreement, a broader framework addressing all aspects of fisheries and aquaculture was negotiated under the auspices of the FAO. The Code of Conduct is a non-binding voluntary instrument which aims to provide principles for responsible fisheries activities, guidance for the implementation of international agreements and sets the standard for good conduct in all matters relating to fisheries (FAO, 1995, Article 2). The Code is based on relevant rules of international law, including those reflected in the Law of the Sea Convention. It contains provisions that may be or have already been given binding effect by other legally binding instruments such as the UNFSA, agreements creating RFMOs and the FAO Compliance Agreement, which forms an integral part of the Code. The Code also reflects important UNCED provisions such as the precautionary approach of the Rio Declaration.[13] The FAO secretariat intended the Code to express the position of the FAO regarding ongoing environmental conflicts without engaging directly in them.

The Code contains a number of concepts and principles States should make use of, starting with General Principles, and lists principles for fisheries management (including the precautionary approach), fishing operations, aquaculture development, integration of fisheries into coastal area management, post-harvest practices and trade and fisheries research.

The Code has given rise to and is implemented through a number of international plans of action:

- International Plan of Action for reducing incidental catch of seabirds in longline fisheries (IPOA-SEABIRDS) in 1999 (http://www.fao.org/fishery/ipoa-seabirds/);
- International Plan of Action for the conservation and management of sharks (IPOA-SHARKS) in 1999 (http://www.fao.org/fishery/ipoa-sharks/en);
- International Plan of Action for the management of fishing capacity (IPOA-CAPACITY) in 1999 (http://www.fao.org/fishery/ipoa-sharks/en); and
- International Plan of Action to prevent, deter and eliminate illegal, unreported and unregulated fishing (IPOA-IUU) in 2001 (http://www.fao.org/DOCREP/003/y1224e/y1224e00.htm).

The FAO has also developed and adopted a series of technical and international guidelines in support of the implementation of the Code of Conduct for Responsible Fisheries.[14] These include guidelines on fishing operations, the implementation of the precautionary approach and fisheries management. Recent examples are the International FAO Guidelines for the Management of Deep-sea Fisheries in the High Seas in 2008 and the International Guidelines on Bycatch Management and Reduction of Discards, adopted in 2011 (http://www.fao.org/docrep/011/i0816t/i0816t00.htm). Following discussions in the FAO Committee on Fisheries (COFI), a document aiming at establishing guidelines for flag State performance is under discussion in an ongoing technical consultation.[15]

There is an ongoing debate concerning the effectiveness of the Code, and the global reviews of the state of the world's fisheries resources provided biannually by the FAO indicate room for improvement (Agnew et al., 2009; Pitcher et al., 2009). Progress reports on Code implementation are provided every two years by COFI and the 2011 report showed 'mixed' results.[16] This may explain the discrepancy between a high proportion of countries (90%) claiming to have management plans and the high proportion of stocks which are still exploited at the limit of natural productivity and beyond (FAO, 2012, p. 56–57).

Multilateral environmental agreements relevant to the conservation of living marine resources

Introduction

In parallel with the development of the global instruments pertaining to the conservation and management of living marine resources previously discussed, there has been a substantial growth in the number of multilateral environmental agreements.[17] Some of these establish standards and mechanisms which also apply to living marine resources. Further, fisheries agreements have been inspired by elements found in environmental instruments, for example in the case of the precautionary principle adopted at UNCED (Principle 15) which found its way into the 1995 UN Fish Stocks Agreement and in the 1995 Code of Conduct for Responsible Fisheries (CCRF) (General Principle 6.5) in the form of a 'precautionary approach' (VanderZwaag, 2002) and ecosystem-based management (Garcia et al., 2003).

The 1992 Convention on Biological Diversity (CBD)

The Convention on Biological Diversity (CBD)[18] was adopted in 1992, and entered into force in 1993. As of 28 January 2013 the Convention has 193 Parties (http://www.cbd.int). The objectives of the CBD are (1) the conservation of biological diversity; (2) the sustainable use of its components; and (3) the fair and equitable sharing of the benefits arising from the utilization of genetic resources (Article 1). Biological diversity is the diversity within species, between species and of habitats and ecosystems (Article 2). The concept of biological diversity covers the entire variability of life. The provisions of CBD apply to areas under the Parties' national jurisdiction and to 'processes and activities' regardless of where their effects occur, carried out under a party's national jurisdiction or control.

The Convention includes a number of obligations to achieve its three objectives. It calls for general measures for conservation and sustainable use, and requires its Parties to identify and monitor components of biological diversity and the effects of activities that are likely to have significant adverse impacts on biodiversity (Articles 6, 7 and 10). Parties are also to take steps to assess environmental impacts of proposed projects that are likely to have significant adverse effects on biodiversity, and to cooperate with other States to notify and address such effects (Article 14).

The COP has adopted two legally binding Protocols on Biosafety (http://bch.cbd.int/protocol/) and on Access and Benefit-sharing (http://www.cbd.int/abs/). Of importance in relation to the marine realm is the Programme of Work on Marine and Coastal Biodiversity (http://www.cbd.int/marine/), which includes the development of criteria for identifying ecologically and biologically significant areas and for the selection, establishment and management of Marine Protected Areas.[19] This work also includes cooperation in providing information relevant to the establishment of marine protected areas beyond the limits of the national jurisdiction.[20]

The Strategic Plan for Biodiversity 2011–2020,[21] adopted by the Conference of the Parties in October 2010, contains 20 key targets (the Aichi Biodiversity Targets) for 2015 or 2020 with three particularly important targets in the marine context. Target 6 urges:

> 'By 2020 all fish and invertebrate stocks and aquatic plants are managed and harvested sustainably, legally and applying ecosystem based approaches, so that overfishing is avoided, recovery plans and measures are in place for all depleted species, fisheries have no adverse impacts on threatened species and vulnerable ecosystems and the impacts of fisheries on stocks, species and ecosystems are within safe ecological limits.'

Target 10 calls for the minimization of anthropogenic pressures on coral reefs and other vulnerable ecosystems impacted by climate change or ocean acidification by 2015. Target 11 recommends the establishment of representative and well-connected systems of protected areas covering at least 10% of coastal and marine areas by 2020.

With regard to the actual management of living marine resources, it can be argued that the Law of the Sea

Convention is a far more important instrument than the CBD as it mandates the actual regulation of the use of living marine resources and the enforcement of those regulations (Hoel, 2003). The relevance of the CBD in this regard is more at the normative level, with the emphasis on the need to conserve biological diversity as an end in itself. Also, the competence of CBD in areas beyond national jurisdiction is subject to discussion (Molenaar, 2007).

Other global environmental agreements relevant to the management and conservation of living marine resources

The 1971 Convention on Wetlands of International Importance (the Ramsar Convention)

The Ramsar Convention (2 February 1971, 996 UNTS 245) has 163 Contracting Parties[22] and is generally applicable to wet lands. The Convention's definition of 'wetlands' includes brackish or salt water and areas of marine water the depth of which at low tide does not exceed 6 m (Article 1). The objective of the Ramsar Convention is to protect and conserve wetlands of international significance. The Convention requires the Parties to designate suitable wetlands within their territory for inclusion in a List of Wetlands of International Importance, and promote their conservation (Articles 2 and 3). Each Party is to further promote the conservation of wetlands by establishing nature reserves on wetlands, whether included on the List or not, and to provide adequately for their wardening (Article 4).

The 1973 Convention on International Trade in Endangered Species of Wild Fauna and Flora (CITES)

CITES (3 March 1973, 993 UNTS 243) has 177 Parties[23] and regulates international trade in plant and animal species that are or may become threatened with extinction due to trade. The Convention has three appendices with species listed according to the degree of protection they are perceived to require (cf. Chapter 13 for more details). Trade in species that are threatened with extinction (appendix I) is banned, except for non-commercial purposes (Article III). Appendix II pertains to trade in species that may be threatened with extinction unless trade is subject to strict regulation (Article IV). Appendix III includes species which are subject to regulation in some States, needing international cooperation to control trade. States can list species here unilaterally (Article XVI). States are required to report on all trade in species of flora and fauna that are listed in the appendices (Article VIII).

To assist CITES in considering proposals for the listing of commercial fish species, a cooperation arrangement has been established between CITES and the FAO. Such proposals are assessed by experts appointed by the FAO.

Listing of marine species having commercial interest has become politicized and difficult. Some countries believe such species are more appropriately managed through existing regional fisheries management organizations or arrangements (Hutchings et al., 2012). At the Fifteenth Conference of the Parties to CITES in 2010, various marine listing proposals for marine species were rejected.

The 1979 Convention on the Conservation of Migratory Species of Wild Animals (CMS)

The Convention (23 June 1979, 1651 UNTS 333), having 118 parties,[24] addresses the conservation of species of wild animals that migrate across national boundaries. The central mechanism of the Convention is listing species in two appendices and the establishment of obligations for their protection.[25] Endangered migratory species are listed in appendix I. Takings of such species are prohibited with limited exceptions, such as for scientific purposes (Article III). Appendix II deals with migratory species that would significantly benefit from international co-operation and those having an unfavourable conservation status which require international agreements for their conservation (Article IV). A number of regional agreements and memoranda of understanding (MOUs) have been concluded for marine species.

Global environment and development conferences/summits

The first global environmental summit was held in Stockholm in 1972. Often forgotten today, it initiated the establishment of the United Nations Environment Program (UNEP). In relation to living marine resources it is remembered for initiating the discussions on a moratorium on whaling.[26] Subsequently, three global environment and development conferences and summits have been held in 1992, 2002 and 2012.

The 1992 UN Conference on Environment and Development (UNCED)

UNCED resulted in a global action plan for the sustainable development, Agenda 21[27] and the Rio Declaration on Environment and Development.[28] In Agenda 21, chapter 17 sets forth a program for the international community in pursuing the objective of sustainable development of the oceans and coasts. A number of program areas are listed, among them integrated management of coastal areas and implementation of obligations for international cooperation to conserve marine living resources found on the high seas. The latter called for the convening of an intergovernmental conference on straddling and highly migratory fish stocks (chapter 17, para. 17.50) which led to negotiation and adoption of the UN Fish Stocks Agreement; in that respect, the significance of Agenda 21 for the management of living marine resources has been substantial.

The Rio Declaration sets out 27 principles supportive of achieving sustainable development. Among key principles relevant to fisheries management are the principles of public participation (Principle 10), the precautionary approach (Principle 15) and environmental impact assessment (Principle 17).

The 2002 World Summit on Sustainable Development

The 2002 World Summit of Sustainable Development resulted in the Johannesburg Plan of Implementation (JPOI).[29] The JPOI also deals with oceans issues and calls for a number of institutional improvements with deadlines set (Garcia and Doulman, 2005). To provide for sustainable development of the oceans, it calls for: the application of the ecosystem approach by 2010; the promotion of integrated oceans management at the national level; the establishment of marine protected areas consistent with international law and based on scientific information, including representative networks by 2012; and the establishment of the Regular Process to study the status of the global marine environment by 2004 (paras 30–36). The JPOI sets certain targets to achieve sustainable fisheries including the goal of maintaining or restoring stocks to levels that can produce the maximum sustainable yield, with the aim of achievement for depleted stocks on an urgent basis and where possible no later than 2015 (para. 31(a)).

The JPOI has also had substantial impact in terms of initiating new processes and targets for the management of the marine environment and living marine resources. The Regular Process has started and will deliver its first report in 2014 (see below). While unrealistic, the 2015 MSY target is nevertheless being addressed in practice by many countries.

The 2012 UN Conference on Sustainable Development (Rio+20)

While oceans texts dealing with living marine resources in the 1992 Agenda 21 and in the 2002 Plan of Implementation are somewhat concrete and, in the case of the latter, establish specific objectives with deadlines, the Rio + 20 'The Future We Want' oceans text[30] is short (4 pages) and general in nature. Of the 20 paragraphs dealing with oceans issues, about half deal directly with living marine resources.

The 2002 JPOI commitment to maintain or restore stocks at MSY levels is reiterated, pointing to science-based management plans, reduction of discards and reduction of adverse impacts of fishing on ecosystems as relevant measures. Also, the implementation of the 1995 UN Fish Stocks Agreement and the FAO Code of Conduct for Responsible Fisheries is urged, and measures in accordance with the FAO IPOA to combat IUU fishing are referred to. The document furthermore calls on States to ratify the FAO Agreement on Port State Measures in order to bring it into force. RFMO transparency, the commitment to eliminate fisheries subsidies and the needs of developing States are also addressed.

The text notes the ongoing discussions under the UN General Assembly (see below) on the conservation and sustainable use of marine biodiversity beyond areas of natural jurisdiction and sets a target date for deciding whether a new global instrument addressing marine biological diversity in areas beyond national jurisdiction is needed.

The Rio + 20 text reiterates previous commitments and refers to ongoing processes in the UN system. This is probably just as well: the JPOI commitments are quite ambitious and time is needed to achieve them. A new issue raised by the 2012 Rio text is that of food security and nutrition in the context of living marine resources.

UN General Assembly resolutions and processes

UN General Assembly resolutions

The General Assembly adopts annual resolutions on oceans, the law of the sea and sustainable fisheries that address a wide range of issues. The resolutions are not binding on States, but carry a substantial normative weight.

The annual oceans and law of the sea resolutions are very comprehensive and have run to some 40 pages and over 250 paragraphs.[31] They address a broad range of issues including: implementation of the Law of the Sea Convention and related instruments; capacity building; the work under various bodies and processes established by the Convention; and substantive issues such as maritime safety, marine biodiversity and marine science.

As regards to living marine resources, many paragraphs in the oceans and law of the sea resolutions are relevant. For example, Resolution 61/222 of 20 December 2006 urged application of ecosystem approaches in ocean management and noted key elements of an ecosystem approach (para. 119).

The annual fisheries resolutions[32] are somewhat briefer and address issues that States need to tend to in order to achieve sustainable fisheries. The 2011 Resolution 66/68 addressed factors for achieving the objective of sustainable fisheries, implementation of the 1995 UN Fish Stocks Agreement, IUU fishing, enforcement, fishing overcapacity and shark finning,[33] among other things.

Sustainable fisheries resolutions have been helpful in setting the international wheels in motion for strengthening the regulation of bottom fisheries (Russell and VanderZwaag, 2010). For example, Resolution 61/105 of December 2006 called upon RFMO/As to assess whether individual bottom fishing activities would have significant impacts on vulnerable marine ecosystems (VMEs, cf. Chapter 14) not later than 31 December 2008; to close VME areas to bottom fishing unless preventive conservation and management measures have been established; and to require members to mandate vessels flying their flag to cease bottom fishing activities where VMEs are encountered and to report the encounter (para. 83).

The most important provisions with regard to living marine resources vary from year to year and depend on the current political agenda. In 2011 for example, the establishment of the Regular Process for Global Reporting and Assessment of the State of the Marine Environment (see below) was subject to in-depth treatment in the Ocean and Law of the Sea resolution.[34]

Ongoing processes under the UN General Assembly

The General Assembly has also initiated several processes and meetings that consider specific ocean issues in more detail and may make recommendations for consideration by the UN General Assembly.

Following the entry into force of the Law of the Sea Convention in 1994, the UN General Assembly started annual oceans debates and resolutions. In 1999, it established the United Nations Open-ended Informal Consultative Process on Oceans and the Law of the Sea (the Informal Consultative Process or UNICPOLOS, resolution 54/33).[35] UNICPOLOS facilitates the annual review by the General Assembly of developments in ocean affairs and the law of the sea by considering the Secretary-General's report on oceans and the law of the sea and by suggesting issues and possible elements for addressing them. The emphasis of UNICPOLOS has been on identifying areas where coordination and cooperation at the intergovernmental and inter-agency levels needs to be enhanced. The consultation process has reviewed the status of, for example, marine science, vulnerable marine environments and ecosystem approaches to management (UNGA, 2006).

The UN General Assembly has also initiated processes for specific issues. The Ad Hoc Open-ended Informal Working Group to study issues relating to the conservation and sustainable use of marine biological diversity beyond areas of national jurisdiction (referred to as the BBNJ meeting) was launched in 2004.[36] The Working Group held its sixth meeting in August 2013 and has discussed the scientific, socio-economic and environmental dimensions of marine biodiversity in areas beyond national jurisdiction. Substantial disagreement exists among States regarding the applicable principles governing access to marine genetic resources and whether a further implementing agreement under LOSC should be negotiated to strengthen the governance framework for marine biodiversity in areas beyond national jurisdiction. Led by the Working Group, the UN General Assembly initiated a process through resolution 66/231 to ensure that the legal framework for the conservation and sustainable use of marine biodiversity in areas beyond national jurisdiction effectively addresses those issues by identifying gaps and ways forward, including through the implementation of existing instruments and the possible development of a multilateral agreement under the United Nations Convention on the Law of the Sea (paras 166 and 167).[37]

A recommendation from the 2002 WSSD was to strengthen the monitoring of the status of the world's oceans. The UNGA Assembly in 2005 initiated the start-up phase of a Regular Process for Global Reporting and Assessment of the State of the Marine Environment.[38] A framework was endorsed in 2009, and the formal establishment of the Regular Process and its institutional mechanisms was agreed in 2010. Following considerable discussions on format, content and process, the actual work to produce a first global marine assessment by 2014 started in 2012. The intention is that this will be an ongoing process, producing reports at regular intervals.

Conclusion: Future developments

The global legal regime for living marine resources might be described as a partial success. More than 160 countries have ratified the Law of the Sea Convention which, together with a number of other global instruments, provides a broad framework for the management of living marine resources. Without this regime, which establishes rights and obligations for States, the status of the world's living marine resources would undoubtedly be worse than it is now.

Moreover, the regime has secured a stable framework for the further development of oceans governance at international as well as domestic levels of governance. One major power, the US (along with Venezuela and Iran, among others), has not acceded to the Convention. The United States of America (USA) does however consider most of the provisions of the Convention customary international law and abides with them. Its non-ratification is more of a problem to the US itself than to others, as it is prevented from participating in the work in the institutions established by the Conventions such as the Commission on the Limits of Continental Shelf.

The implementation and enforcement of the principles and provisions embodied in the evolving global framework accounted for here (see also Annex 1 and 2) is a key issue. Regional Fisheries Management Organizations and national governments implement the provisions of the global regime with varying degrees of success. It appears that, over time, results in terms of reduced fishing pressure are manifesting themselves (Hilborn and Hilborn, 2012, p. 123). While a lot is still to be desired with respect to the status of many fish stocks, the evidence is that more than two-thirds of the world's fish stocks are exploited at MSY levels or below, that is, at 'sustainable levels'. About one-third of fish stocks are overexploited according to the FAO (2012), and the trend is one of an increasing number of depleted stocks.

The global regime is gradually evolving in response to new and emerging problems. When fisheries on the high seas were growing during the 1980s (Burke, 1994), the eventual response was the 1993 Compliance Agreement and the 1995 UN Fish Stocks Agreement. When IUU fishing was increasing, the global response came in the form of FAO IPOAs and the 2009 FAO Port State Measures Agreement. A set of performance criteria targeting flag States are progressing under the auspices of the FAO. A major challenge which has not yet been sufficiently addressed is allocation of fish resources between states (Henriksen and Hoel, 2011), an issue that may well be accelerated with climate change (Chen et al., 2011).

Over time, environmental considerations have become more important in the management of living marine resources. One important driver in this regard is the advances in fisheries economics, demonstrating that living marine resources are common pool resources susceptible to overexploitation and needing regulation (Hannesson, 2004). Another driver is learning in the fishing industry itself, a realization in the industry that it stands to gain from government intervention. Ideas and experiences from the environmental sector have been important, for example in the introduction of ecosystem-based management. While such ideas catch on early at a theoretical level in the management of living marine resources – the first international ecosystem-based management agreement for living marine resources was adopted as early as 1982 – it takes considerable time to translate the concepts into practice (Fluharty, 2005; Garcia et al., 2003).

The most important challenge for the continued development of the global legal regime for living marine resources is therefore the actual implementation of new concepts and ideas at the regional and domestic levels of governance. Considerable attention has been devoted to studying how Regional Fisheries Management Organizations face up to such challenges (Molenaar, 2003; Russell and VanderZwaag, 2010) and how their performance in this regard can be measured (Kvalvik, 2010). Up to now, less attention has been paid to implementation at the State level, although studies exist to indicate that much can be improved even in countries considered to have strong management regimes (Pitcher et al., 2009) and that strong performance over time is possible (Hønneland, 2012).

Tracking all the future evolutions in international law relating to fisheries and marine biodiversity will continue to be a challenging task. Future decisions and initiatives can be expected to multiply under a fragmented array of legally binding and soft law instruments. International legal development remains a voyage in progress.

Notes

1. The UN held the Second Conference on the Law of the Sea (UNCLOS II) in 1960, which did not result in any new agreements.
2. LOSC, Article 10 allows for internal waters status to be exerted over historic bays and geographical bays where a maximum closing line of 24 NM may be drawn to delineate internal waters.
3. Arrested vessels and their crews must be promptly released upon the posting of reasonable bond or other security and penalties for fisheries law violations may not include imprisonment.
4. The Convention provides that the coastal State is to delineate the outer limits of its continental shelf where the shelf extends beyond 200 NM and submit information on the limits to the Commission on the Limits of the Continental Shelf (CLCS). The CLCS then makes recommendations to the coastal State on the outer limits of its continental shelf. The limits of the shelf established by a coastal State on the basis of these recommendations shall be final and binding (LOSC, Article 76 and Annex II regarding the Commission on the Limits of the Continental Shelf).
5. 28 July 1994 33 ILM 1309. See http://www.un.org/Depts/los/convention_agreements/convention_overview_part_xi.htm.
6. Only a non-binding conciliation procedure may be invoked for such disputes.
7. 2 December 1946, 161 UNTS 72.
8. 24 November 1993 (1994) 33 ILM 969.
9. As per 30 January 2013. See also http://www.fao.org/fileadmin/user_upload/legal/docs/1_012s-e.pdf.
10. 4 August 1995, 34 ILM 1542.
11. See http://www.un.org/Depts/los/convention_agreements/convention_overview_fish_stocks.htm. As of 21 January 2013 there were 80 Parties.
12. 22 November 2009, available at http://www.fao.org/docrep/013/i1644t/i1644t00.pdf.
13. Rio Declaration, Prinicple 15. See http://www.unep.org/Documents.Multilingual/Default.asp?documentid=78&articleid=1163.
14. Over 20 technical guidelines may be found at http://www.fao.org/fishery/ccrf/publications/guidelines/en.
15. See ftp://ftp.fao.org/FI/DOCUMENT/tc-fsp/2011/prospectus.pdf. The Committee of Fisheries (COFI) of the UN Food and Agriculture Organization (FAO) meets every second year to consider issues relating to fisheries and aquaculture.
16. Progress in the Implementation of the Code of Conduct for Responsible Fisheries and Related Instruments, Including International Plans of Action and Strategies, and other Matters. COFI/2011/2.
17. The University of Oregon International Environmental Agreements database project estimates that there are >1100 multilateral environmental agreements; http://iea.uoregon.edu/page.php?file=home.htm&query=static
18. 5 June 1992 Convention on Biological Diversity 1760 UNTS 79. The text of the convention is also available at www.cbd.int.
19. An elaborated programme of work was adopted in Decision VII/5, available at http://www.cbd.int/decision/cop/?id=7742.
20. Decision IX/20 available at http://www.cbd.int/decision/cop/?id=11663.
21. Decision X/2 available at http://www.cbd.int/decision/cop/?id=12268.
22. As of 5 December 2012; see http://www.ramsar.org/cda/en/ramsar-home/main/ramsar/1_4000_0__.
23. As of 21 January 2013; see http://www.cites.org/eng/disc/parties/alphabet.php.
24. As of 1 January 2013, http://www.cms.int/documents/convtxt/cms_convtxt.htm.
25. A species may also be listed under both of the appendices.
26. 1972 Stockholm United Nations Conference on the Human Environment, recommendation 33, http://www.unep.org/Documents.Multilingual/default.asp?DocumentID=97&ArticleID=1506&l=en.
27. Available at http://www.unep.org/Documents.Multilingual/Default.asp?documentid=52
28. Available at http://www.un.org/documents/ga/conf151/aconf15126-1annex1.htm.
29. Available at http://www.un.org/esa/sustdev/documents/WSSD_POI_PD/English/WSSD_PlanImpl.pdf.
30. http://www.uncsd2012.org/content/documents/727The%20Future%20We%20Want%2019%20June%201230pm.pdf, paras 158–177.
31. For example, Resolution 66/231, Oceans and the law of the sea.
32. The fisheries resolution is termed 'Sustainable fisheries, including through the 1995 Agreement for the Implementation of the Provisions of the United Nations Convention on the Law of the Sea of 10 December 1982 relating to the Conservation and Management of Straddling Fish Stocks and Highly Migratory Fish Stocks, and related instruments'.
33. Para. 15 called upon States to consider taking further measures to prohibit or restrict fisheries conducted solely for harvesting shark fins, such as requiring that all sharks be landed with each fin naturally attached. See

http://www.un.org/Depts/los/general_assembly/general_assembly_resolutions.htm.

34. Resolution 66/231, para. 198–217.
35. See http://www.un.org/Depts/los/consultative_process/consultative_process_background.htm.
36. See A/RES/59/24, para. 73; and http://www.un.org/Depts/los/biodiversityworkinggroup/biodiversityworkinggroup.htm.
37. Inter-sessional workshops under the auspices of the Working Group were endorsed.
38. See http://www.un.org/Depts/los/global_reporting/global_reporting.htm.

References

Agnew, D., Pearce, J., Pramod, G., Peatman, T., Watson, R., Beddington, J. and Pitcher, T. (2009) Estimating the worldwide extent of illegal fishing. *PlosOne* **4**(2), 1–8.

Balton, D. (1996) Strengthening the law of the sea: the new agreement on straddling fish stocks and highly migratory fish stocks, *Ocean Development & International Law* **27** (1–2), 125–151.

Birnie, P., Boyle, A. and Redgwell, C. (2009) *International Law and the Environment*. Oxford University Press, Oxford.

Burke W. (1994) *The New International Law of Fisheries*. Oxford University Press, Oxford.

Chen, I.C., Hill, J.K., Ohlemuller, R., Roy, D.B. and Thomas, C.D. (2011) Rapid range shifts of species associated with high levels of climate warming. *Science* **333**(6045), 1024–1026.

Churchill, R.R. and Lowe, A.W. (1999) *The Law of the Sea*. Manchester University Press, Manchester.

Ebbin, S.A., Hoel, A.H. and Sydnes, A.K. (eds) (2005) *A Sea Change. The Exclusive Economic Zone and Governance Institutions for Living Marine Resources*. Springer, Netherlands.

Edeson, W. (2005) A brief introduction to the principal provisions of the international legal regime governing fisheries in the EEZ. In: *A Sea Change. The Exclusive Economic Zone and Governance Institutions for Living Marine Resources* (eds S.A. Ebbin, A.H. Hoel and A.K. Sydnes), pp. 17–32. Springer, Netherlands.

FAO (1995) The Code of Conduct for Responsible Fisheries. FAO, Rome.

FAO (2002) Norway–FAO Expert Consultation on the Management of Shared Fish Stocks, Bergen, Norway 7–10 October 2002. FAO Fisheries Report No 695, FIPP/R695, Rome.

FAO (2012) The State of World Fisheries and Aquaculture. FAO, Rome.

Fluharty, D. (2005) Evolving ecosystem approaches to management of fisheries in the USA. *Marine Ecology Progress Series* **300**, 241–296.

Freestone, D. (2012) International governance, responsibility and management of areas beyond national jurisdiction. *International Journal of Marine and Coastal Law* **27**, 191–204.

Garcia, S.M. (1994) The precautionary principle: Its implications for fisheries management. *Ocean and Coastal Management* **22**, 99–125.

Garcia, S.M. and Doulman, D.J. (2005) FAO's fisheries programme and the plan of implementation of the world summit on sustainable development. In: *A Sea Change. The Exclusive Economic Zone and Governance Institutions for Living Marine Resources* (eds S.A. Ebbin, A.H. Hoel and A.K. Sydnes), pp. 169–194. Springer, Netherlands.

Garcia, S.M., Zerbi, A., Aliaume, C., Dho Chi, T. and Lasserre, G. (2003) The Ecosystem Approach to Fisheries. FAO Fisheries Technical Paper No. 443. FAO, Roma.

Hannesson, R. (2004) *The Privatization of the Oceans*. The MIT Press, Cambridge.

Hayashi, M. (1996) Enforcement by non-flag states on the high seas under the 1995 agreement on straddling and highly migratory fish stocks. *The Georgetown International Environmental Law Review* **9**(1), 1–36.

Henriksen. T. and Hoel, A.H. (2011) Determining allocation: from paper to practice in the distribution of fishing rights between countries. *Ocean Development & International Law* **42**(1–2), 66–93.

Hilborn, R. and Hilborn, U. (2012) *Overfishing: What Everyone Needs to Know*. Oxford University Press, Oxford.

Hoel, A.H. (1998) Political uncertainty in international fisheries management. *Fisheries Research* **763**, 1–12.

Hoel, A.H. (2003) Marine biodiversity and institutional interplay. *Coastal Management* **31**, 25–36.

Hønneland, G. (2012) *Making Fishery Agreements Work. Post-Agreement Bargaining in the Barents Sea*. Edward Elgar, London.

Hutchings, J.A., Coté, I.M., Dodson, J.J., Fleming, I.A., Jennings, S., Mantua, N.J., Peterman, R.M., Riddell, B.E, Weaver, J. and VanderZwaag, D.L. (2012) Sustaining Canadian marine biodiversity: responding to the challenges posed by climate change, fisheries, and aquaculture. Expert panel report prepared for the Royal Society of Canada, Ottawa.

Iliff, M. (2008) Compromise in the IWC: Is it possible or desirable? *Marine Policy* **32**(6), 997–1003.

Klein, N. (2005) *Dispute Settlement in the UN Convention on the Law of the Sea*. Cambridge University Press, Cambridge.

Kvalvik, I. (2010) The Northeast Atlantic Fisheries Commission and the implementation of sustainability principles: lessons to be learned? In: *Recasting Transboundary Fisheries Management Arrangements* (eds D.A. Russell and D. L. VanderZwaag), pp. 387–418. Martinus Nijhoff Publishers, Leiden and Boston.

Leary, D.K. (2007) *International Law and the Genetic Resources of the Deep Sea*. Martinus Nijhoff, Leiden/Boston.

Leary, D.K. (2012) Moving the maritime genetic resources debate forward: some reflections. *International Journal of Marine and Coastal Law* **27**, 435–448.

Lodge, M.W., Anderson, D., Løbach, T., Munro, G., Sainsbury, K. and Willock, A. (2007) Chapter Four: Recommended best practices for regional fisheries management organizations.

In: *Report of an Independent Panel to Develop a Model for Improved Governance by Regional Fisheries Management Organizations*. Chatham House, London.

Miles, E. (1998) *Global Ocean Politics*. Martinus Nijhof, Dordrecht.

Molenaar, E.J. (2003) Participation, allocation and unregulated fishing: The practice of Regional Fisheries Management Organisations. *International Journal of Marine and Coastal Law* **18**, 468.

Molenaar, E.J. (2007) Managing biodiversity in areas beyond national jurisdiction. *International Journal of Marine and Coastal Law* **8**(1), 928.

Munro, G., Houtte, A.V. and Willmann, R. (2004) The conservation and management of shared fish stocks. FAO Fisheries Technical Paper 465. FAO, Rome.

Orrego Vicuna, F. (2001) The International law of high seas fisheries: from freedom of fishing to sustainable use. In: *Governing High Seas Fisheries: The Interplay of Global and Regional Regimes* (ed. O.S. Stokke), pp. 23–52. Oxford University Press, Oxford.

Pikitch, E.K., Santora, C., Babcock, E.A., Bakun, A., Bonfil, R., Conover, D.O., Dayton, P., Doukakis, P., Fluharty, D., Heneman, B., Houde, E.D., Link, J., Livingston, P.A., Mangel, M., McAllister, M.K., Pope, J. and Sainsbury, K. J. (2004) Ecosystem-based fishery management. *Science* **305**, 346–347.

Pitcher, T.D., Kalikoski, G., Pramod, G. and Short, K. (2009) Not honoring the code. *Nature* **457**, 658–659.

Rayfuse, R. (2007) Biological resources. In: *Handbook of International Environmental Law* (eds D. Bodensky, J. Brunée and E. Hey), pp. 362–393. Oxford University Press, Oxford.

Rothwell, D.R. and Stevens, T. (2010) *The International Law of the Sea*. Hart Publishing, Oxford and Portland, Oregon.

Russell, D.A. and VanderZwaag, D.L. (eds) (2010) *Recasting Transboundary Fisheries Management Arrangements in Light of Sustainability Principles*. Martinus Nijhoff Publishers, Leiden and Boston.

Stokke, O. (ed.) (2001) *Governing High Seas Fisheries: The Interplay of Global and Regional Regimes*. Oxford University Press, Oxford.

United Nations General Assembly (2006) Report on the work of the United Nations open-ended consultative process on oceans and the law of the sea at its seventh meeting. A/61/156. UN Department of Oceans and Law of the Sea, New York.

VanderZwaag, D.L. (2002) The precautionary principle and marine environmental protection: slippery shores, rough seas, and rising normative tides. *Ocean Development & International Law* **33**(2), 165–188.

Warner, R. (2009) *Protecting the Oceans Beyond National Jurisdiction: Strengthening the International Legal Framework*. Martinus Nijhoff, Leiden and Boston.

Chapter 8

Spatial dimensions of fisheries and biodiversity governance

R. Kenchington[1,3], O. Vestergaard[2] and S.M. Garcia[3]

[1] Australian National Centre for Ocean Resources and Security, University of Wollongong, Australia
[2] United Nations Environment Programme, Nairobi, Kenya
[3] IUCN Commission on Ecosystem Management (IUCN/CEM/FEG), Gland, Switzerland

Abstract: Marine biodiversity and fisheries operate in 3D dynamic space while UNCLOS jurisdictions are mainly based on 2D boundaries, challenging space-based governance. Spatial dimensions of fisheries governance become explicit in space–time operational restrictions in a system aiming at conserving fishery resources by maintaining stocks maximum sustainable yield level. Environmental governance uses Marine Protected Areas as a main measure for conserving biological diversity and aims at covering 10% of the world oceans by 2020. This target and the issue of no-take MPAs have been a source of tension between the two streams of governance. The chapter considers spatial governance in fisheries and biodiversity conservation, the impacts of the terrestrial heritage of conservation, the spatio-temporal issues in both governance fields and the potential offered by spatially structured multiuse integrated management frameworks. It concludes that the spatial dimensions of both governance streams and the similarity of issues arising from an increasing range of human uses and impacts would require an effective integrated approach to spatial and temporal management.

Keywords: space-based management; spatio-temporal restrictions; MPAs; integrated management

Introduction

Despite early expressions of concern in the middle of the 19th century, the capacity of human activities to have significant impacts on marine environments was not widely regarded as a serious issue until the second half of the 20th century. Since then, increasing technological capacities have made new uses of marine space possible and extended the range and impacts of existing uses (Halpern et al., 2008, 2012).

The dialogue between marine fisheries management (aiming at ecologically and socio-economically sustainable use of living resources) and biodiversity conservation (aiming to protect biological diversity and ecosystem processes from cumulative human uses and impacts) has highlighted some differences in the meanings and/or scope of conservation and of approaches to it. This has led to conflicts characterized, *inter alia*, by conflictual views on the roles of Marine Protected Areas (MPAs)[1] in fisheries management with two main considerations:

- the capacity of fisheries management to meet socio-economic expectations and objectives while halting and reversing increasing fishery impacts on biodiversity and ecosystem processes, e.g. using an ecosystem approach to fisheries (EAF); and
- the capacity of MPAs (as no-take zones or areas where an ecosystem-based approach to multiple-use management could be applied) to enhance or complement existing fishery management measures,

Governance of Marine Fisheries and Biodiversity Conservation: Interaction and Coevolution, First Edition.
Edited by Serge M. Garcia, Jake Rice and Anthony Charles.

with a view to maintaining biodiversity and ecosystem processes without unacceptable impacts on human populations.

We reckon that despite conflicts, there is substantial common ground between fisheries management and biodiversity conservation. Both are confronted by limited predictability of nature and an increasing range of human uses and impacts that require effective integrated and adaptive approaches to spatial and temporal management.

In marine systems area-based management, the water column represents an important third dimension of the ecosystem. It is an overwhelmingly important habitat and system of material mass transport linking seabed, sea-surface, coastal and open-ocean habitats to each other and to inputs from the land. Governance of fisheries and marine biodiversity conservation therefore needs to be three-dimensional while the jurisdictions established by UNCLOS are essentially two-dimensional with boundaries defined by geodesic reference points on the sea surface. Moreover, the boundaries of the natural and socio-economic processes in ocean ecosystems rarely coincide, complicating the quest for coherence within and between governance systems.

Faced with the evidence of growing human impacts, marine biodiversity conservation interests have proceeded on the basis that more and larger areas should be protected as closed reserves or through overarching area-based management regimes for ecologically sustainable multiple-use to address the negative impacts of economic activities including fisheries, coastal development, mining, oil drilling and shipping.

Faced with expanding fishing capacity and declining catches, fisheries resources conservation interests have proceeded with regulation of access (e.g. fishing rights) and operational constraints (catch and gear regulations). Fisheries have evolved and expanded on the basis that virtually all open marine space and the resources therein are potentially accessible and usable by fishers, unless formally and exclusively allocated to other activities or closed for fishery management purposes.

Recognition of the need to better integrate the governance of biodiversity conservation, fishery resources conservation and of other marine activities presents three challenges: (1) the terrestrial governance heritage of sectoral primacy does not transfer easily to the biophysical processes of marine ecosystems; (2) in most jurisdictional systems, marine natural resources are common property; and (3) there has been substantial tension between biodiversity conservation and fishery interests concerning primacy of objectives and roles.

The following sections will consider the development of marine spatial governance, impacts and issues arising from the terrestrial governance heritage, space–time issues in fisheries management, spatial issues in biodiversity conservation and ecosystem management and the governance shift towards spatially structured multi-use management frameworks, before offering a few conclusions.

Development of marine spatial governance

Legal framework

Until the 20th century, the need to manage marine activities was largely addressed by legislation and management of the various economic sectors with little 'communication' between them. With an increasing range of human uses and activities affecting marine biodiversity and ecosystem function and productivity, it has become important to understand how the joint and cumulative impacts of different sectors may be kept within acceptable limits.

Historically, management of for example shipping addressed use of the sea surface and shallow waters with core focus on safety of human life at sea, security of trade routes and navigational rights. Management of fisheries focused on access rights and long-term use of biological resources of the water column and seabed. Apart from conflict or forced interaction in specific cases, for example accidental spilling of toxic substances, the two management systems were complementary with limited need for formal sectoral interaction beyond the rules of navigation.

New technologies expanded the capacity, reach and potential impacts of both sectors and rapid emergence of new uses of marine space. Fishery conservation concerns emerged towards the end of the 19th century because of the spreading of local overfishing. The International Council for the Exploration of the Sea (ICES) was established in 1902 and the first international conference on overfishing was organized in 1946. The 1967 Torrey Canyon oil spill raised biodiversity and fishery concerns regarding the increasing size and number of vessels engaged in shipping hazardous cargoes and the need for management to address their operational and accidental impacts on marine ecosystems. This coincided with the emergence of a broader environmental ethic and calls for multiple-objective management of human activities.

By the mid-20th century, the rapidly growing capacity and potential for marine industries had changed the context of marine governance. The discovery of hydrocarbon and mineral development potential added momentum to a global process. A prolonged series of United Nations Conferences on the Law of the Sea (UNCLOS I–III) were held, involving technical debates and political negotiations about space-based jurisdictions. In 1958, UNCLOS I developed conventions on the territorial seas and the contiguous zone, the continental shelf, the high seas and on fishing and the conservation of living resources of the high seas. Two subsequent conferences (UNCLOS II and III) refined these conventions and developed the 1982 United Nations Convention on the Law of the Sea (hereafter referred to as UNCLOS[2]) (United Nations, 1982) which came into force in 1994. UNCLOS provides the first integrated space-based legal framework for conservation and utilization of living resources and protection and preservation of the marine environment (cf. Chapters 10 and 11).

Before UNCLOS I, the reach of national sovereignty in most coastal states was the 3 nautical mile (NM) territorial sea. The rest (the high seas) was legally accessible and usable by everyone. Nations could make and enforce their laws with respect to activities within territorial waters. Beyond that, they had jurisdiction over the activities of their citizens, on vessels flying their flag, and their navies would protect their trade routes and the interests of their citizens. Put simply, all vessels could pass and operate freely everywhere, subject only to constraints of safety, prudent navigation and security with respect to piracy or warfare.

UNCLOS defines the rights and duties of all States and imposes a duty of cooperation building from a coastal state geographic baseline that takes account of bays, indentations and islands. It establishes that *internal waters* (landward of the baseline) and *territorial waters* (seaward, to 12 NM beyond the baselines) are part of sovereign States. In the further 12 NM of *Contiguous Waters*, the State has the authority to enforce laws in four specific areas: customs, taxation, immigration and pollution impacts on living resources and the marine environment. The rights of coastal nations with respect to exploitation of all natural resources from the sub-soil to the surface are provided by an Exclusive Economic Zone (EEZ) extending up to 200 NM from the baseline. States have also sometimes claimed the area from the 12 NM territorial baseline and up to 200 NM as an Exclusive Fishing Zone. Where the continental shelf extends beyond 200 NM (and to a maximum of 350 NM from the baseline), the living resources closely linked to the seabed are under national jurisdiction while those in the water column are in international waters. Beyond the EEZ and extended shelf are the high seas, and the parts of the seabed placed under international jurisdiction are referred to as areas beyond national jurisdiction (ABNJ).

Spatial approaches to managing marine biodiversity and sustainability

The first MPA was apparently established in 1879, primarily for nature-based recreation purposes in the subtidal areas of the Royal National Park in New South Wales, Australia (Brown, 2002). Marine biodiversity conservation in the sense of seeking to ensure survival and health of marine biodiversity and ecosystem processes is a more recent concept. In 1962 the World Conference on National Parks was one of the first international meetings to address marine biodiversity conservation, calling upon 'governments ... to examine as a matter of urgency the possibility of creating marine parks or reserves to defend underwater interests' (Adams, 1962).

During the processes leading from UNCLOS II (in 1960) to the coming into force of the Law of the Sea (in 1994), the United Nations conducted a series of activities addressing broader concerns about environment and resource conservation and sustainability, including: (1) the 1972 UN Conference on the Human Environment (UNCHE; United Nations, 1972) and the establishment the United Nations Environment Programme (UNEP); (2) the elaboration of the 1980 World Conservation Strategy (WCS) by IUCN, UNEP and WWF in collaboration with UNESCO and FAO, which signalled growing agreement that conservation and development needed to be considered together; (3) the World Commission on Environment and Development (WCED), also called the Brundtland Commission, which further developed the concepts of 'sustainable development' identified in the 1980 WCS (WCED, 1987); and (4) the 1992 United Nations Conference on Environment and Development (UNCED) which adopted the Convention on Biological Diversity (CBD) and Agenda 21 setting out actions needed for change in the 21st century in order to achieve sustainability of environment and development. Chapter 17 of Agenda 21 specifically addresses 'the protection of the oceans, all kinds of seas, including enclosed and semi-enclosed seas, and coastal areas and the protection, rational use and development of their living resources'.

The term 'biodiversity' is not mentioned in UNCLOS but the convention explicitly addresses both conservation and sustainable utilization of living resources (Articles

61–68, 77 and 116–120). It also addresses what is today referred to as 'biodiversity' in Part XII, calling for 'the protection and conservation of the natural resources… the prevention of damage to the flora and fauna of the marine environment' (Article 145b) and for protection and preservation of 'rare or fragile ecosystems… habitat of depleted, threatened or endangered species and other forms of marine life' (Article 194.5).

The 2002 World Summit on Sustainable Development (WSSD; United Nations, 2002) urged nations, *inter alia*, to establish networks of MPAs in their marine jurisdictions by 2012 (Plan of Implementation Para. 32 c). The 7th Conference of the Parties of the CBD (2004) adopted a minimum target of 10% of marine ecosystems to be included in representative networks of MPAs by 2012 (Decision VII/30, Annex II). In 2010, the 10th CBD COP held at Nagoya in the Aichi Prefecture of Japan reasserted this target for 2020 as Aichi Target 11 (CBD, 2011; United Nations, 2012). WSSD also established a specific fishery target: 'to maintain or restore stocks to levels that can produce the maximum sustainable yield with the aim of achieving these goals for depleted stocks on an urgent basis and where possible not later than 2015'. Doubts have been expressed as to the likelihood of the fisheries or the MPA targets being achieved in the foreseeable future without substantial and difficult action and expensive economic adjustment (Ye et al., 2012), although they have been accepted as priority areas for the United Nations agencies working with Member States (UN, 2012). Recent analyses of the World Database on Protected Areas (WDPA) indicate that in many areas and countries the 10% MPA coverage might be reached in 2012 but with an important proportion of MPAs of uncertain validity and effectiveness, therefore missing the Aichi target criteria of ecological and socio-economic effectiveness (Spalding et al., 2013).

Recognizing common interests

The core objectives of fisheries and biodiversity conservation are linked. Both depend on maintaining marine ecosystem function and on the capacity to manage the impacts of the growing number and footprint of other uses of marine space and the downstream consequences of all land and freshwater activities and management. In crowded or sensitive areas, the complexity of integrating multiple objectives has led to the introduction of zoning (e.g. Agardy 2010; Kenchington, 1990, 2010; Kenchington and Day, 2011; Patterson and Pernetta, 2006) through marine spatial planning (Crowder and Norse, 2008; Ehler and Douvere, 2009; Gilliland and Laffoley, 2008). Developing space-based approaches for overlapping human uses in a mobile and changing three-dimensional environment, with unallocated or common property resources, presents significant inter-sectoral and jurisdictional challenges. Zoning or marine spatial planning can be seen in the broader context of an ecosystem-based perspective for management and governance of both biological diversity (CBD, 1992) and fisheries (FAO, 1995, 2002). These provide focus for a clearer environmentally specified concept of sustainable development, integrating the management of multiple uses on a spatial basis as provided for in Agenda 21, Chapter 17 (United Nations, 1992).

However, biodiversity and fishery resource conservation proponents have taken apparently contrasting approaches to space management (Garcia and Hayashi, 2000). Those concerned mainly with biodiversity conservation have called for increasingly large protected areas and integrated governance across large marine ecosystems, requiring centralized governance. Those concerned mainly with fisheries have promoted the subdivision of marine space by allocation of defined 'rights' to smaller and smaller lots (e.g. high seas; EEZs; coastal zones; fisheries and species, aiming at decentralized governance, community-size TURFs and individual transferrable quotas or ITQs). The two trends may be reconciled only under some sort of Marine Spatial Planning (MSP) and within an Ecosystem Approach.

Within a spatially defined jurisdiction, voluntary or regulatory operational controls of economic activities may be applied to any purposes of use or entry, to manage what people may do with respect to marine resources, biological diversity or ecosystem function. Such controls may prohibit certain activities and/or regulate their access (licenses, rights), equipment, practices, removals and discharges. These controls may apply within a range of space and time scales, from permanent measures applying throughout a spatial jurisdiction to temporary measures applying only to subcomponents.

Management of human behaviour and expectations in relation to marine biodiversity, ecosystems and resources cannot be viewed in isolation from the overall balance of societal expectations. The governance challenge is to develop an integrated governmental (and often intergovernmental) system that can address the constraints, opportunities, objectives, responsibilities and aspirations of the various sectors and interest groups that use or impact marine space. This requires some form of space-based multi-objective management framework that addresses and keeps under review a reasonable and sustainable balance of objectives and outcomes and an equitable distribution of costs and benefits.

The challenge is great and solutions typically reflect very specific situations. This has led to various forms of integrated planning and management approaches of coastal and marine areas under various names and acronyms. Results have generally been rather disappointing, challenging any implicit assumption that addressing conservation of biological diversity and fishery resources through development planning and management coordination can quickly resolve unsustainable behaviours. Integration of coastal and ocean management. This presents complex social, cultural, environmental, economic and legal issues whose joint solution requires context-specific spatial frameworks.

The terrestrial heritage of marine governance

Concepts of title and right to use land reflect a presumption of separation ('good fences make good neighbours') whereby substantially different suites of uses or activities may take place on either side of a boundary with limited need for cross-boundary management (except on rivers and large lakes). At any location, the characteristics of soils, local rainfall regime, unidirectional drainage flows (down and within catchments) and exposure regimes are primary determinants of site-associated ecological communities and potential economic uses of land. In terms of terrestrial biodiversity, the third dimension of the atmosphere provides linkages of fungal spores, viruses, bacteria, pollen, airborne seeds, insects, birds and bats, but essentially plant and animal life in the atmosphere is directly linked to and dependent upon primary productivity by plants on the surface of the Earth.

In contrast with the atmosphere, the active mobile third dimension of the coastal and oceanic water column is an important functional habitat. Seawater is a chemically buffered solvent 800 times as dense as air, providing a benign and stable medium for biological cells. The water column sustains complex assemblages of species for all or part of their lifespan. It is a highly active and variable transport medium, providing horizontal and vertical connection between habitats and biodiversity populations through currents, tides and surface wind/wave interactions that have little relationship to jurisdictional boundaries.

The relationships of species in the ecological communities of the water column to those of the seabed are extremely diverse. At one end of the range there is close connection, for example feeding on the seabed by water column species and vice versa and for reproduction of seabed species and dispersion of their larval forms in the water column. At the other end, there is no relation between the column and the bottom except for decay after death. Few species have life cycles that are confined to a single domain.

Four broad but overlapping strategic groups of scales of life-cycle processes relevant to area management for biodiversity conservation were proposed by Kenchington (1990):

(1) species attached to the bottom or with very limited (1 km^2) adult territory and no planktonic or nektonic dispersal: scale of 1 km^2–10 km^2;
(2) species attached or with limited (10 km^2) adult territory and restricted planktonic or nektonic dispersal: scale of 10 km^2–10^2 km^2;
(3) species with moderate (10 km^2–10^2 km^2) adult territory and extensive planktonic or nektonic dispersal: scale of 10^3 km^2; and
(4) species living in the water column (planktonic, nektonic) and with long-range dispersal or migration (scale of 10^2 km^2 –10^4 km^2).

Typically, in at least the course of a year, an area of seabed and the water column above it will sustain interacting young or adults of species from all four groups. There can be considerable variety in the extent to which their activities interact, compete or affect each other. A number of economic activities may also use an area of seabed and the overlying water column at the same or slightly different times, with little mutual impact or even awareness (e.g. navigation; tourist activities such as whale watching or diving; fisheries). Some other activities requiring more permanent facilities (e.g. alternative energy production, oil and gas extraction, cables and pipelines, military target areas) may require more exclusive site allocation for operational or security purposes.

Space–time issues in fisheries management

Overview

After the mid-1980s fishery policies started to shift from aggressive industrial growth to ensuring responsible fisheries and sustainable use to rebuild overfished stocks, reverse falling catches and reduce environmental impact and economic losses (Garcia and Newton, 1997; Hilborn et al., 2003; World Bank, 2009). In this endeavour, they face heavy socio-economic and political costs (Ye et al., 2012) with an increasingly complex management framework requiring an ecosystem approach with a combination of technical, structural and spatio-temporal measures.

Fish stocks and fisheries are not uniformly distributed. While not always explicit, space is a fundamental element of fishing operations and fisheries management, which often use some area-based measures to protect juveniles and spawners or the habitat of target species. Operational controls can exclude some or all categories of fishing from all or part of an EEZ or fishing ground:

- *permanently*: for example, dynamite fishing everywhere; mammals in sanctuaries; pelagic driftnets in the high sea; closed nursery areas; and zones reserved for small-scale fisheries;
- *seasonally*: for example, during periods of recruitment or high vulnerability; or
- *temporarily*: for example, on short-time concentration of juveniles.

The geographical coordinates as well as the opening and closing dates can be fixed or variable (context-sensitive). As overcapacity tended to defeat their purpose, closures have lost ground in the past in favour of regulations (e.g. of gear characteristics, catch, effort and capacity) and forms of fishing rights (e.g. TURFs, ITQs). Improved technology for navigation, fish-finding, seabed scanning, gear positioning, etc. has led to a significant reduction of areas formerly avoided or inaccessible (isolated pinnacles, reefs, rocks, shipwrecks, canyons, seamounts and deeper seabed areas, greatly reducing the extent of natural 'sanctuaries').

Objectives of fishery closures

Conventional fisheries management provides an operational framework for measures to optimize or rebuild a resource, a fishery or a subsector by modifying the behaviour of fleets and the spatial distribution of fishing. With the adoption of EAF (FAO, 2002) the environmental scope of these measures is increasing, accelerating the convergence between fisheries and biodiversity conservation and between MPAs and fishery reserves. The objectives of closures, ranked from 'classical' to 'modern' might include (inspired from Hall, 2009):

(1) protection of vulnerable life stages of target species: concentrations of juveniles, spawners or their prey species;
(2) protection of habitats from destructive impacts of fishing gears and practices (e.g. trawls, dredges);
(3) reduction of overlap and conflict between subsectors and preferential allocation of resources: e.g. between small-scale and large scale fisheries; national and foreign fleets; fishing and aquaculture;
(4) attempts to reduce impact of overcapacity (not very successful if capacity is not effectively reduced);
(5) protection and restoration of stocks with a high risk of collapse; and
(6) protection of genetic pools from the effect of selective fishing.

In an ecosystem approach to fisheries, objectives focus on reduction of bycatch of vulnerable or protected species or their critical life stages, limitation of discards and protection of habitats of high ecological value. This is the case for most spatio-temporal restrictions (STRs) established to protect deep-sea fisheries stocks or habitats such as seamounts. In 2005, the General Fisheries Commission for the Mediterranean (GFCM) adopted a total ban on trawling in all areas deeper than 1000 m. MPAs in ABNJ have also been established in the North Atlantic in cooperation between NEAFC, OSPAR and Portugal (more examples in Garcia et al., 2013).

Types of space–time restrictions (STRs)

Fishery reserves

Permanent or multi-annual closures may be called closed areas, fishery no-take zones, fishery protected zones or fishery MPAs. These areas are legally established, typically through a governance instrument under the responsibility and control of a ministry of fisheries. They may be closed to all or only some fishing activities (considered compatible with the central objective of the reserve). They may be scientifically monitored but lack of or poorly defined management objectives, and the confounding effect of other measures in complex management arrangements make it hard to establish their effectiveness (STECF, 2007). Where objectives are specified they include: optimizing exploitation, protecting critical habitat or vulnerable species assemblages (e.g. in reef fisheries) and reserve of food (in traditional communities).

Fishery refugia

The term is often used as synonym for fishery reserve but may cover broader objectives. In Asia, they are 'spatially defined areas within which specific management measures are applied to fish habitats during critical phases of their life cycle, to ensure their sustainable utilisation' (Pernetta et al., 2010; http://refugia.unepscs.org). Their primary objective is healthy habitats to sustain life-stages of important target fish species. They are integrated in fishery management plans (SEAFDEC, 2006; UNEP, 2007, 2008). In Asia (e.g. in Thailand, Vietnam and Philippines), fishery refugia and sanctuaries are much more willingly accepted by fishing communities than conventional MPAs (FAO, 2011, pp. 93; Pernetta et al., 2010; UNEP, 2007, 2008). Their development is promoted in Southeast Asia in an approach very similar to that used for MPA networks (Patterson and Pernetta, 2006), and they may represent a good instrument to reconcile fisheries and conservation objectives.

Territorial use rights

TURFs have been used in traditional small-scale fisheries (SSFs) for centuries, for example in Japan, Côte d'Ivoire, South Korea, Sri Lanka, Canada, Chile and in the Pacific (Castilla and Gelcich, 2008; Charles, 2009; Johannes, 1981, 1998). The major objectives of TURFs are better economic organization of the fishery and allocation of local target fishery resource access to coastal communities with limited operational range. A TURF is legally defensible and geographically formally designated. In practice, TURFs are owned by coastal communities or municipalities; in theory however they might belong to an individual, an extended family or tribe, a company, enterprise or association, a local or provincial government, a trust, an international or non-governmental agency or a combination of them (Christy, 1982).

Traditionally used, TURFs have often been neglected or combated by governments or overwhelmed by demography, urban development or industrialization. Their renewed use was suggested by Christy (1982) who noted their vulnerability to market economy. This concern was echoed in the case of MPAs by Weigel et al. (2011). The Chilean 'benthic resources exploitation and management areas' have been shown to improve resources and economic performance, reducing capacity (Castilla and Gelcich, 2008). The tension between the needs and expectations of TURF owners and the broader communities as well as between residents and migrants or new arrivals is a typical dynamic of coastal area management, irrespective of the area-based instrument used.

Spatial issues in biodiversity and ecosystem management

The Protected Areas concept originated in the terrestrial environment where areas of wilderness or unchanged ecological communities are the exception because most have been substantially cleared, modified or completely transformed by human activities. Remaining areas have biodiversity, scarcity, cultural and recreational values that may be addressed by assigning a land-use category and legal title to establish restrictive conditions of use and management for preservation, restoration and maintenance of these values. Such areas may be impacted by cross-boundary effects from the atmosphere or from upstream sources, but their governance can largely be addressed on a basis of primacy of the conservation objective within boundaries on the Earth's surface. Conservation in the sense of ecosystem preservation thus becomes a form of land use where the title-holder – the State or private owner of the area – determines and controls purposes of use and entry within the property boundary.

Extending terrestrial management approaches to the sea, particularly in densely populated coastal areas, presents substantial conceptual and practical challenges because of the interactions of complex spatial and temporal linkages of marine life-cycle strategies discussed above, as well as the socio-economic implications.

Protected area management of significant representative marine habitat areas can play substantial roles in biodiversity conservation of species that are attached or have limited adult territories (cf. 'Development of marine spatial governance'). For wide-ranging or migratory species, such management may be important for protecting sites critical for specific life-cycle events such as spawning, nursery, birthing or migratory feeding, but additional approaches through fishery or other management measures may be needed elsewhere in their range (e.g. Game et al., 2009a, b; Kaplan et al., 2010).

The need for a systematic approach to establishing protected areas in marine environments was first clearly articulated at an IUCN International Conference on Marine Parks and Protected Areas in Tokyo in 1975, which adopted resolutions calling for a comprehensive, effective and well-monitored global system of marine parks and reserves representing all coastal and oceanic ecosystems (IUCN, 1976).

In 1988, the need for a global representative system of MPAs was revisited at the 17th General Assembly of the IUCN, which adopted a resolution recognizing the urgency of the need for a spectrum of measures 'to provide for the protection, restoration, wise use, understanding and enjoyment of the marine heritage of the world in perpetuity through the creation of a global representative system of marine protected areas and through the management, in accordance with the principles of the World Conservation Strategy, of human activities that use or affect the marine environment'. This resolution clearly addresses both biodiversity and fishery resource conservation and their sustainable use.

The development of IUCN guidelines for establishing MPAs (Kelleher and Kenchington, 1992) started in 1987 but proved to be a complex task. Through an exhaustive and iterative process of drafting and widespread review, the guidelines came to address a broad range of situations relating to scale, linkage, buffering, objectives, practicality of management, legislative and other controls.

Recognition and documentation of continuing decline of marine ecosystem conditions (as in Halpern et al., 2008, 2012) has occurred at a time of growth of excessive fishery capacity and consequent declining percentage of stocks fished within safe limits (Garcia and Newton, 1997; World Bank, 2009; Ye et al., 2012), and a growing range and intensity of existing and new uses of marine space. As with terrestrial ecosystems, these pressures and concern about diminishing areas in a natural state have led to the global call for establishment of MPAs.

Broad ecosystem-scale spatial management has now been in place for the Great Barrier Reef for more than three decades (Kenchington and Day, 2011; Lawrence et al., 2002) and, more recently, for the Papahanaumokokea Marine National Monument (NOAA, 2006) and Phoenix Islands Protected Area (PIPA, 2006). These remote areas are or were lightly used at the time their planning was initiated. In most situations, however, the need for conservation is greatest in areas that are accessible, increasingly used and impacted, and where the establishment of protected areas presents complex social, environmental and economic challenges.

Objectives of MPAs

The creation of MPAs may be approached as a separate initiative of the environment management sector or as part of broader integrated planning of many sectoral activities. As a key instrument of marine biodiversity conservation, MPAs can encompass a range of operational management objectives:

- preservation of some areas of each marine ecological region in their natural state undisturbed by man except for the purposes of scientific research, in order to provide reference, control or baseline areas for evaluation of management effectiveness and understanding human components of environmental change;
- protection of representative examples of all marine ecoregions with particular consideration for areas important for vulnerable life stages of rare, at risk, endangered or natural resource target species;
- maintenance of genetic pools of species and communities through replication in networks of protected areas in order to provide security in the event of substantial or catastrophic impacts;
- reservation of some areas of the marine environment in their natural state for their appreciation and enjoyment by the public;
- provision for reasonable and multiple uses of natural areas and resources consistent with an overall objective of conservation of biodiversity and ecosystem processes;
- protection of habitats and of fishery non-target species from destructive impacts of human uses, fishing gears and practices, construction and seabed installations.

The core biodiversity conservation strategy is typically the creation of reserves (or no-take) protected areas with a view to creating a representative system protecting viable examples of marine biological diversity for their intrinsic values. These may also serve as baseline or reference areas against which the impacts of human use and environmental change can be robustly assessed and from which it may be possible to restore areas and populations of species that have been impacted. The broader multiple-use strategy provides a setting for consistent ecologically sustainable purposes of use, entry and impact at the ecosystem scale.

There is a clear overlap with several objectives and methodologies for fisheries resource conservation, but biodiversity conservation extends them to address species other than those of recognized or potential commercial value (for food or livelihoods) and to important and vulnerable habitats.

Types of MPAs

While there are clear obligations in UNCLOS to address protection of biodiversity and sustainability of natural resources use, there is typically confusion over the roles and implications of MPAs. The IUCN guidelines (Day et al., 2012; Dudley, 2008) recognize 6 types of MPAs covering a broad range of operational management systems. The primary objective for all is conservation of marine biodiversity but three main approaches or groups of MPAs can be identified.

- *Reserve (or no-take) MPAs*: In line with the terrestrial National Parks equivalent they correspond to intact areas serving, to the greatest extent possible, as strict nature reserves, refuges or reference sites and as components of national and global systems of viable representative examples of marine biological diversity and ecosystem processes. Human activities are totally prohibited apart from management activities, including monitoring and scientific research that cannot be undertaken elsewhere.

- *Limited-use no-take MPAs*: With operational controls reflecting the primary function of protection of biodiversity, ecosystem process and habitat while allowing demonstrably compatible uses including research, education, nature-based recreation and tourism activities.
- *Multiple-use MPAs*: Incorporating reserve and limited-use MPAs in an overall zoning system providing overarching regimes with operational controls addressing both senses of conservation: (1) protection of biological diversity, ecosystem processes and habitat through operational controls or prohibitions on access; and (2) resource conservation (sustainable use) through operational management of multiple human uses and impacts.

Protection of biodiversity and sustainable use of natural resources are intimately connected; they both therefore need to be efficient, transparent and accountable. Both have value for society and are costly in terms of operational and of current and longer-term opportunity costs.

As with TURFs, MPAs can be distinguished by their governance approach which may be: (1) directly led by central government structures; (2) decentralized to commissions or authorities created by government; (3) devolved to communities, private institutions, companies or non-government organizations; (4) absent (paper parks), or a combination of (1) to (3) (Garcia et al., 2013; Jones et al., 2011).

Spatial structure of MPAs

MPAs have been created at scales ranging from 10^{-1} to 10^5 km^2. Smaller MPAs are typically created by site allocation to address immediate local conservation and nature-based recreational issues within an area currently or expected to be subject to competing pressures, and where it is not possible to address the social and economic costs of developing a different regime for longer-term sustainability.

Large area designations should generally involve cross-sectoral spatial planning or zoning to create a hierarchy of protection of biodiversity from human uses and impacts. With respect to fishing, a core area in which no fishing access is allowed may be buffered by an outer area in which, for example, benthic habitat is protected but mid-water line trolling is allowed for pelagic fish, and which in turn borders a zone with general fishing activity. In deeper waters (Environment Australia, 2001) this can include vertical zoning whereby, for example, a seamount is strictly protected from trawling and dredging but the overlying waters are open to pelagic fishing.

The Great Barrier Reef Marine Park (GBRMP) provides an example of a process in which the regulation of purposes of use and entry is considered at an ecosystem scale and managed through a multiple-use zoning system. A summary table of activities allowed, prohibited and allowed subject to permit is provided (Table 8.1) as an example of use of categories in such an overarching regime, designed to provide for conservation and reasonable use of a large marine ecosystem.

Governance shift to a multiple-use framework

The provisions of the LOSC place clear obligations on participating Nations to address equitable and efficient use, conservation of marine living resources and the protection and preservation of the ocean and seas environment. These obligations address conservation in the senses of living resource target species, ecosystem function and protecting 'the habitat of depleted, threatened or endangered species and other forms of marine life, vulnerable species, habitats, ecosystems' (Article 194). It is therefore usually agreed that UNCLOS covers what is referred to today as 'biodiversity' (cf. 'Spatial approaches to managing marine biodiversity and sustainability'). This raises fundamental questions related to the clarity of objectives, demonstrable performance criteria and processes for establishing priority and balance between 'equitable and efficient use and conservation' of living marine resources and the 'protection and preservation' of biological diversity. This dilemma is central to the question of the use and management of fishery resources, other marine activities and MPAs.

Fisheries and biodiversity planners and managers have to address inherent challenges of scale and equity in planning and management. As the management area increases, the role of the State tends to increase while that of the local communities decreases. Generally however, the greater the extent to which communities are effectively engaged, the greater the likelihood of compliance. The scale at which costs and benefits are perceived and received and the degree of equity in their distribution is a central issue. At one extreme, disenfranchised poor rural populations with few livelihood alternatives bear the cost of losing subsistence resources and livelihood from natural

Table 8.1 Zoning of the GBRMP. Detailed information contained in Great Barrier Reef Zoning plan and regulations (Source: GBRMPA 2003). (Y: Allowed; conditions may apply; P: Permit required; X: Prohibited.)

	Type of zone						
Activity	**General use**	**Habitat protection**	**Conservation park**	**Buffer**	**Scientific research**	**Marine nat. park**	**Preservation**
Aquaculture	P	P	P	X	X	X	X
Bait netting	Y	Y	Y	X	X	X	X
Boating, diving, photography	Y	Y	Y	Y	Y	Y	X
Harvest: aquarium fish, coral, beach worms	P	P	P	X	X	X	X
Harvest: trochus, sea cucumber, lobster,	P	P	X	X	X	X	X
Limited collecting	Y	Y	Y	X	X	X	X
Limited, snorkel only spearfishing	Y	Y	Y	X	X	X	X
Line fishing	Y	Y	Y	X	X	X	X
Netting, other than bait	Y	Y	X	X	X	X	X
Research other than limited impact	P	P	P	P	P	P	P
Shipping outside designated area	Y	P	P	P	P	P	X
Tourism program	P	P	P	P	P	P	X
Traditional use of marine resources	Y	Y	Y	Y	Y	Y	Y
Trawling	P	X	X	X	X	X	X
Trolling					X	X	X

resource use, while most benefits accrue to urban-dwelling or foreign tourists or to the global society. At the other end, a combination of inadequately addressed current, growing and unsustainable pressures and expectations is degrading marine environments and ecosystem services and reducing the quality of life and opportunities for the local, national and global community.

There is no standard method for establishing effective strategies and programs to address the spatial allocation and integration of management of marine uses and impacts. Each situation has to be addressed within a specific governance context of social, economic and biophysical objectives and considerations, and their ongoing dynamics. Guidance and case studies for integrated planning and tradeoff processes are provided by an increasing number of publications (e.g. Katsanevakis et al., 2011; Kidd et al., 2011; McLeod and Leslie, 2009; UNEP, 2011; White et al., 2012).

A key challenge to creating an overarching strategic governance regime addressing both domains is to clearly understand and establish the precedence or primacy of objectives and their interdependency. Fisheries and biodiversity conservation managers share the objective of maintenance of healthy productive marine ecosystems through protection of vulnerable habitats, areas important for species at vulnerable points in their life cycle and genetic pools. Both use closed areas as a means to achieve conservation, but in fisheries all such areas are modifiable directly by the manager while permanent closure to fishing is the core strategic biodiversity conservation objective of no-take MPAs. However, subject to an overarching objective of biodiversity conservation, in large limited-use and multiple-use MPAs conditions of use and entry can include provisions for the manager to declare seasonal or post-traumatic impact closures and openings.

Following the release of the FAO Guidelines on MPAs and fisheries (FAO, 2011) and a UNEP Technical Report on MPA governance (Jones et al., 2011), the role of MPAs in reconciling fisheries and biological diversity conservation and management was addressed at an international workshop in Bergen, Norway comprising equal numbers of fishery and biodiversity conservation experts (Rice et al., 2012). The workshop used an expert opinion process to provide insight into the extent to which objectives might be shared and where there could be potential conflicts between fisheries management and biodiversity conservation interests in MPA management. Half of all fisheries objectives and 40% of biodiversity objectives were considered likely to receive support from both perspectives. Only 25% of fisheries objectives and 30% of biodiversity objectives were considered to be potential sources of conflict. This outcome may not be surprising given the overlap of common fisheries and biodiversity objectives and the constraints of growing activities and impacts from other sectors that affect ecosystems and living resources. Such systematic exploration of objectives is however useful for developing an informed basis for integrating fisheries and biodiversity conservation in a multiple-use framework. No-take zones may best be seen as a complement to other approaches in a comprehensive approach to marine resource management, as well as reference areas against which the effects of human uses and their management can be monitored.

Conclusions

Global targets and efforts for marine biodiversity conservation and fisheries management highlight complex social, political and economic challenges, operational difficulties of enforcement and financing and inter-sectoral mistrust. Improved integration requires a spatial framework.

The MPA coverage and fishery stock restoration targets adopted by the CBD are not necessarily in conflict but, with few exceptions, and despite some successes for both, the challenges of re-dimensioning fishing capacity to match natural resource productivity while establishing and enforcing strictly protected areas have generated substantial inter-sectoral resistance to the latter and some conflict.

Correcting overfishing and establishing MPAs for biodiversity conservation imply limitations to fisheries growth and reduction or loss of access to resources in some fisheries and of livelihoods for those excluded from fishing or from the areas.

No-take MPAs (sensu IUCN Category I/II) have a role in reducing immediate and longer-term collateral impact of fisheries that is hard to dispute, but addressing immediate fishery overcapacity or food security problems is not their primary objective. As a consequence, many coastal and fishing communities are sceptical at the onset of MPA planning. They may be more positive when planning is inclusive and participative and when it reinforces their tenure systems and reduces or eliminates competition, for example from industrial or non-resident, migrating fishers (Garcia et al., 2013). All types of MPAs therefore require a consequent effort to understand and explicitly address issues of expectations, equity, sustainability and transitional hardship for the fishing industry and coastal communities, using stepwise and participatory approaches as well as sustainable pathways for change.

The comprehensive rights and obligations provided by UNCLOS are unlikely to be addressed in any jurisdiction without some form of overarching spatial regime to address multiple objectives and impacts of human activities. For any sector, the key issues in such a regime are authority or precedence of the agency in charge of the strategic planning and the equity of outcomes. These may be addressed by an explicit overarching authority of government designated by the head of government or cabinet. A sectoral agency (in charge of fisheries or conservation) usually has too limited a mandate and legitimacy to plausibly address and resolve the conflicts of interest with other competing sectors in an overarching strategic planning approach alone.

The need for policies to address issues of biodiversity, habitat and natural resource conservation has added significant and increasing pressure on fisheries. Such policies are unlikely to be effective without mature recognition of common interests and engagement by both sectors. This will involve compromises between those who seek restoration to more pristine ecosystems and those who seek a higher sustainable level of capture fisheries and related livelihoods. Future fisheries management should increase efforts to develop and assess the effects of more spatially explicit approaches, including permanent, temporary and rotating closures, and be consistent with ecosystem-based management principles. Similarly, biodiversity conservation should explicitly address socio-economic considerations in conservation initiatives to establish and manage protected areas.

Notes

1. There is also a debate on the role of species listings in fisheries management, but this issue is beyond the scope of this chapter and is dealt with in Chapter 12.
2. To distinguish between them, UNCLOS is sometimes used to refer to the conferences (i.e. UNCLOS I, II and III) while LOSC is used to refer to the convention. In this book, we will use UNCLOS in all cases.

References

Adams, A.B. (ed.) (1962) First world conference on national parks: Seattle, Washington, June 30–July 7, 1962. US National Park Service and IUCN, Washington, DC. xxxiv + 471 pp.

Agardy, T. (2010) *Ocean Zoning: Making Marine Management More Effective*. Earthscan, London.

Brown, I. (2002) Oldest MPA in the world: Royal National Park, New South Wales, Australia. Designated 1879. *MPA News* **3**(6), 5.

Castillla, J.C. and Gelcich, S. (2008) Management of the loco (*Concholepas concholepas*) as a driver for self-governance of small-scale benthic fisheries in Chile. In: *Case Studies in Fisheries Self-Governance* (eds R. Townsend, R. Shotton and H. Uchida), pp. 441–450. FAO, Rome.

CBD (1992) Text of the Convention. Available at http://www.cbd.int/convention/text/ (accessed February 2014).

CBD (2004) COP 7 Decision VII/30. Available at http://www.cbd.int/decision/cop/?id=7767 (accessed February 2014).

CBD (2011) COP 10 Decision X/2. Available at http://www.cbd.int/decision/cop/?id=12268Re (accessed February 2014).

Charles, A. (2009) Chapter 10. Rights-based fisheries management: the role of rights in managing access and harvesting. In: *A Fishery Managers' Handbook,* 2nd edition (eds K.L. Cochrane and S.M. Garcia), pp. 253–282. Wiley Blackwell and FAO, Chichester (UK) and Rome (Italy).

Christy, F. Jr. (1982) Territorial use rights in marine fisheries: definitions and conditions. FAO Fishery Technical Paper 227, pp. 9.

Crowder, L. and Norse, E. (2008) Essential ecological insights for marine ecosystem-based management and marine spatial planning. *Marine Policy* **32**, 772–778.

Day, J., Dudley, N., Hockings, M., Holmes,G., Laffoley, D., Stolton, S. and Wells, S. (2012) *Guidelines for Applying the IUCN Protected Area Management Categories to Marine Protected Areas*. IUCN, Gland, Switzerland.

Dudley, N. (ed.) (2008) *Guidelines for Applying Protected Area Management Categories*. IUCN, Gland, Switzerland. x + 86 pp.

Ehler, C. and Douvere, F. (2009) *Marine Spatial Planning: A Step-by-step Approach Toward Ecosystem-based Management*. UNESCO, Paris. Available at http://www.unesco-ioc-marinesp.be/uploads/documentenbank/d87c0c421da4593fd93b-bee1898e1d51.pdf (accessed February 2014).

Environment Australia (2001) Tasmanian Seamounts Marine Reserve Management Plan. Department of Environment and Heritage. Australian Government. Available at http://www.comlaw.gov.au/Details/F2007B00980 (accessed February 2014).

FAO (1995) Code of conduct for responsible fisheries. FAO, Rome. 41 pp.

FAO (2002) Report of the Reykjavik Conference on Responsible Fisheries in the Marine Ecosystem, Reykjavik, Iceland, 1–4 October 2001. FAO Fisheries Report 658, 128 pp.

FAO (2011) Fisheries management. 4. Marine protected areas and fisheries. In: *FAO Technical Guidelines for Responsible Fisheries*, No. 4, Suppl. 4. Food and Agriculture Organization of the United Nations, Rome.

Game, E.T., Grantham, H.S., Hobday, H.J., Pressey, R.L., Lombard, A.T., Beckley, L.E., Gjerde, K., Bustamante, R., Possingham, H.P. and Richardson, A.J. (2009a) Pelagic protected areas: the missing dimension in ocean conservation. *Trends in Ecology and Evolution* **24**, 360–369.

Game, E.T., Grantham, H.S., Hobday, H.J., Pressey, R.L., Lombard, A.T., Beckley, L.E., Gjerde, K., Bustamante, R., Possingham, H.P. and Richardson, A.J. (2009b) Pelagic protected areas: the devil you know. *Trends in Ecology and Evolution* **24**(7), 63–64.

Garcia, S.M. and Newton, C. (1997) Current situation, trends and prospects in world capture fisheries. In: *Global Trends: Fisheries Management* (eds E.L. Pikitch, D.D. Huppert and M.P. Sissenwine), pp. 2–27. Proceedings of the American Fisheries Society Symposium 20, Seattle, Washington, 14–16 June 1994. American Fisheries Society, Bethesda, Maryland.

Garcia, S.M. and Hayashi, M. (2000) Division of the oceans and ecosystem management: a contrastive spatial evolution of marine fisheries governance. *Ocean & Coastal Management* **43**, 445–474.

Garcia, S.M., Cazalet, B., Féral, F. and Beslier, S. (2013) 4ème partie: Gouvernance de la pêche et des AMP. In: *La Pêche et les Aires Marines Protégées: Bio-écologie, Socio-économie et Gouvernance* (eds J. Boncoeur, S.M. Garcia and D. Gascuel). Presses Universitaires de Perpignan.

GBRMPA (2003) Great Barrier Reef Marine Park Zoning Plan (2003) Great Barrier Reef Marine Park Authority. Available at http://www.gbrmpa.gov.au/__data/assets/pdf_file/0015/3390/GBRMPA-zoning-plan-2003.pdf (accessed March 2013).

Gilliland, P.M. and Laffoley, D. (2008) Key elements and steps in the process of developing ecosystem-based marine spatial planning. *Marine Policy* **32**, 787–796.

Hall, S.J. (2009) Chapter 8. Area and time restrictions. In: *A Fishery Manager's Guidebook*, 2nd edn (eds K.L. Cochrane and S. M. Garcia), pp. 196–219. Wiley-Blackwell, Chichester (UK) and FAO, Rome (Italy).

Halpern, B.S., Walbridge, S., Kimberly A. et al. (2008) A global map of human impact on marine ecosystems. *Science* **319**(5865), 948–952.

Halpern, B.S., Longo, C., Hardy, D. et al. (2012) An index to assess the health and benefits of the global ocean. *Nature* **488**, 615–620.

Hilborn, R., Branch, T.A., Ernst, B. et al. (2003) State of world fisheries. *Annual Review of Environment and Resources* **28**, 15.1–15.40.

IUCN (1976) *Proceedings of International Conference on Marine Parks and Reserves*. Tokyo, Japan, 12–14 May 1975. IUCN, Gland.

Johannes, R.E. (1981) *Words of the Lagoon: Fishing and Marine Lore in the Palau District of Micronesia*. University of California Press, Berkeley, CA.

Johannes, R.E. (1998) The case for dataless marine resources management: example from nearshore tropical fin fisheries. *TREE* **13**(6), 243–246.

Jones, P.J., Qiu, W. and De Santo, E. (2011) *Governing Marine Protected Areas: Getting the Balance Right*. UNEP, Nairobi.

Kaplan, D.M., Chassot, E., Gruss, A. et al. (2010) Pelagic MPAs: the devil is in the details. *Trends in Ecology and Evolution* **25**(2), 62–63.

Katsanevakis, S., Stelzenmüller, V., South, A. et al. (2011) Ecosystem-based marine spatial management: Review of concepts, policies, tools, and critical issues. *Ocean & Coastal Management* **54**(11), 807–820.

Kelleher, G. and Kenchington, R. (1992) *Guidelines for Establishing Marine Protected Areas*. A Marine Conservation and Development Report. IUCN, Gland, Switzerland.

Kenchington, R.A. (1990) *Managing Marine Environments*. Taylor and Francis, New York and London.

Kenchington, R.A. (2010) Strategic roles of marine protected areas in ecosystem scale conservation. *Bulletin of Marine Science* **86**(2), 303–313.

Kenchington, R.A. and Day, J.C. (2011) Zoning, a fundamental cornerstone of effective Marine Spatial Planning: lessons learnt from the Great Barrier Reef, Australia. *Journal of Coastal Conservation* **15**(2), 271–278.

Kidd, S.A., Plater, A. and Frid, C. (2011) *The Ecosystem Approach to Marine Planning and Management*. Routledge, London, UK.

Lawrence, D.R., Kenchington, R.A. and Woodley, S.J. (2002) *The Great Barrier Reef: Finding the Right Balance*. Melbourne University Press, Melbourne.

McLeod, K. and Leslie, H. (eds) (2009) *Ecosystem-Based Management for the Oceans*. Island Press, Washington, DC.

NOAA (2006) Establishment of the Northwestern Hawaiian Islands National Marine Monument: A Proclamation by the President of the United States of America. Available at http://www.gc.noaa.gov/gcil_papahanaumokuakea.html (accessed February 2014).

Patterson, C. and Pernetta, J. (2006) Marine protected areas and the concept of fisheries *refugia* developed by the regional working group on fisheries. Prepared as a discussion document for the FAO Workshop on Marine Protected Areas and Fisheries Management: Review of Issues and Considerations Rome, 12–14 June 2006. UNEP, GEF, South China Sea Project.

Pernetta, J.C., Paterson, C.J. and Siriraksophon, S. (2010) Fisheries refugia and marine protected areas: can they help sustain the contribution of fisheries towards food security in Southeast Asia? SEAFDEC. *Fisheries for the People* **8**(2), 1–9.

PIPA (2006) Phoenix Islands Protected Area. Available at http://www.phoenixislands.org/index.php (accessed February 2014).

Rice, J., Moksnsess, E., Attwood, C. et al. (2012) The role of MPAs in reconciling fisheries management with conservation of biological diversity. *Ocean and Coastal Management* **69**, 217–230.

SEAFDEC (2006) Section D. Regional guidelines on the use of fisheries refugia for capture fisheries management in Southeast Asia. In: *Supplementary Guidelines on Co-management using Group User Rights, Fishery Statistics, Indicators and Fisheries Refugia*. Southeast Asian. Fisheries Development Center, Bangkok, Thailand, pp. 84. Available at http://www.unepscs.org/Refugia_Training/Fisheries%20Refugia%20Concept/11-ASEAN-SEAFDEC-Regional-Guidelines-on-the-Use-of-Fisheries-Refugia.pdf (accessed March 2014).

Spalding, M.D., Meliane, I., Milam, A., Fitzgerald, C. and Hale, L. Z. (2013) Protecting marine spaces: global targets and changing approaches. *Ocean Yearbook* **27**, 213–248.

STECF (2007) Evaluation of closed areas schemes (SGMOS-07-03). European Commission staff working document. Subgroup on management of stocks (SGMOS-07-03) of the Scientific Technical and Economic Committee for Fisheries (STECF). STECF opinion expressed during the Plenary Meeting of 5–9 November 2007 in ISPRA.

UNEP (2007) Procedure for Establishing a Regional System of Fisheries *Refugia* in the South China Sea and Gulf of Thailand in the context of the UNEP/GEF project entitled: "Reversing Environmental Degradation Trends in the South China Sea and Gulf of Thailand". South China Sea Knowledge Document No. 4. UNEP/GEF/SCS/Inf.4

UNEP (2008) Fisheries Refugia as a Tool for Integrated Fisheries and Habitat Management at Phu Quoc Archipelago, Viet Nam. South China Sea Lesson Learned Note No. 3.

UNEP (2011) *Taking Steps toward Marine and Coastal Ecosystem Based Management: An Introductory Guide*. UNEP Regional Seas Reports and Studies No 189. UNEP, Nairobi.

United Nations (1972) Proceedings of *UN Conference on the Human Environment*, Stockholm 5–16 June 1972. Available at http://www.unep.org/Documents.Multilingual/Default.asp?documentid=97&articleid=1503 (accessed February 2014).

United Nations (1982) United Nations Convention on the Law of the Sea. Available at http://www.un.org/Depts/los/convention_agreements/convention_overview_convention.htm (accessed February 2014).

United Nations (1992) UN Conference on Environment and Development. Agenda 21, Chapter 17. Available at http://www.un.org/Depts/los/consultative_process/documents/A21-Ch17.htm (accessed February 2014).

United Nations (2002) Report of the World Summit on Sustainable Development. Johannesburg, South Africa 24 August–4 September 2002. Available at http://www.johannesburgsummit.org/html/documents/summit_docs.html (accessed March 2014).

United Nations (2012) The Oceans for Prosperity: An initiative of the United Nations Secretary- General. Available at http://www.un.org/Depts/los/ocean_compact/oceans_compact.htm (accessed February 2014).

WCED (1987) *Our Common Future. World Commission on Environment and Development.* Oxford Universiy Press, Oxford and New York.

Weigel, J.Y., Féral, F. and Cazalet, B. (2011) Governance of marine protected areas in the least developed countries. Evidence from West Africa. FAO Fisheries Technical Paper no. 548.

White, C., Halpern, B.S. and Kappel, C.V. (2012) Ecosystem service tradeoff analysis reveals the value of marine spatial planning for multiple ocean uses. *PNAS* **109**(12), 4696–4701.

World Bank (2009) *The Sunken Billions. The Economic Justification for Fisheries Reform.* World Bank, Washington and FAO, Rome.

Ye, Y., Cochrane, K., Bianchi, G., Willman, R., Majkowski, J., Tandstad, M. and Carocci, F. (2012) Rebuilding global fisheries: the World Summit Goal, costs and benefits. *Fish and Fisheries* **14**(2), 12.

Chapter 9

Scientific foundation: Towards integration

J. Rice[1,4], S.J. Jennings[2] and A. Charles[3,4]

[1]Fisheries and Oceans Canada, Ottawa, Canada
[2]Centre for Environment, Fisheries and Aquaculture Science, Lowestoft, UK
[3]Sobey School of Business and the Environmental Science Department, Saint Mary's University, Halifax, Canada
[4]IUCN Commission on Ecosystem Management (IUCN/CEM/FEG), Gland, Switzerland

Abstract: Science has a central role in informing governance, with 'science' defined broadly to include traditional knowledge and concepts from post-normal science. Fisheries management and biodiversity conservation occurs in an integrated socio-ecological system, requiring support from ecological, social and economic sciences. Each component of the system is individually complex with many linkages within and among subcomponents. The uncertainties arising from the structural complexity of the socio-ecological system are amplified by incomplete monitoring of all components. Despite these uncertainties, governance of fisheries and biodiversity conservation have found ways to use the science that is available. Although the science-based decision-support systems are designed to take risks and uncertainties into account, the fisheries and biodiversity streams may evaluate and manage the risks with different tolerances. Even the best integrated social and ecological sciences may therefore be insufficient to result in full coherent decisions among the governance streams.

Keywords: socio-ecological system; science–policy interface; complexity; uncertainty; monitoring; integrated advice

Introduction

Regardless of governance jurisdiction and spatial scale, policy makers and managers present their approaches and individual decisions as 'science based'. There is a lengthy history of this practice, including the mandate in Article 61 of the United Nations Convention on the Law of the Sea (UNCLOS) for decisions to take into account 'the best scientific information available'. Science is invoked to convey messages that decisions made have taken into account all the relevant information, processed the information systematically and soundly and dealt with the information in objective, verifiable and balanced ways leading to decisions that are rational given the information available (Pielke, 2007). 'Science based' does not mean all the sources and types of information were given equal weight in the decisions, nor that 'scientific information' comprised the only factor considered in the decision-making (Mitchell et al., 2006). Moreover, even within the parts of the governance processes that are grounded in science, it is often a science of complex systems with implications for governance, the subject of this chapter.

Governance for sustainable fisheries and conservation of biodiversity must deliver social, economic and environmental objectives in socio-ecological systems where biological, physical, chemical, economic and social processes are individually complex and interact with each other. This complexity results in high uncertainty in much

Governance of Marine Fisheries and Biodiversity Conservation: Interaction and Coevolution, First Edition.
Edited by Serge M. Garcia, Jake Rice and Anthony Charles.

of the science advice provided to decision-makers in both governance streams.

The following section ('Ecological complexity') reviews the multiple causes of physical–biological complexity: the numerous bathymetric, physical and chemical properties of the water masses and the diversity and life histories of constituent species that have individually patchy spatial distributions and respond to numerous external drivers on timescales from seconds to centuries (Mann and Lazier, 1996). The interactions among the parts and their responses to the drivers are rarely linear and may vary over time due to the presence of multiple pathways and to the responses of some parts being conditional on the status of other parts.

We then consider the multiple causes of socio-economic complexity ('Human system complexity') arising from the demography of coastal or user communities, the diversity of harvesting strategies and markets for fishery products, the availability of alternative livelihoods, cultures and traditions of fishers and many externalities affecting the choices available (e.g. changes in fuel prices or social values of markets).

These individually complex systems with uncertain dynamics may be linked loosely or tightly, depending on properties of both the ecological and human subsystems. They are impossible to fully decouple however, making these integrated socio-ecological systems the units being used or managed (e.g. Charles 2001, 2004; MASR 2005; Polasky et al., 2005). Many linkages between the human and bio-physical components of the systems are conditional on the status of other features of the socio-ecological system and/or have feedback connections that may amplify or buffer their responses to changes in the interacting components or to external forces (Link, 2010; Ohl et al., 2007). This complexity helps to scope the true challenges faced by governance systems striving to use the best-available science as a basis for decision-making, and to highlight efforts of science advisers to improve science inputs to policy and management.

We then discuss the additional uncertainty that results from incomplete monitoring of marine socio-ecological systems in space and time and coverage of key system components ('Confronting the uncertainty'). Moreover uncertain data are used to parameterize models and frameworks, which represent processes that are incompletely understood and often non-linear and non-deterministic. This also makes prediction of future trajectories and responses to management interventions uncertain. Consequently, the best science advice will often have large uncertainties. The use of other forms of knowledge can often address some of the sources of scientific uncertainty, but bring with them their own complexities and uncertainties.

Socio-ecological systems are complex and the best science advice is uncertain. Nonetheless, natural sciences – and to some extent social sciences – are utilized substantially in decision-making. The section entitled 'Complexity, uncertainty, risk and science-based decision-making' reviews how science advisors and decision-makers communicate about the uncertainties and associated risks, how they approach the tradeoffs inherent in decision-making and how science-based decision-making is affected by the differing perceptions and tolerances of various risks by different actors in the two governance streams. Aware of the challenges of trying to manage activities in complex systems, and the limitations on the best science available, policy-makers and managers increasingly seek options robust to uncertainties, as well as policies and measures that reduce uncertainties through collection of better information or reducing non-compliance.

Experiences with complex decision-making, focusing on tipping points and science advice in economic, ecological and social aspects of decisions, are considered in 'Illustrations of complex systems, science and governance'. The penultimate section ('Summary of the challenges to governance') identifies governance options that may contribute to more coherent and sustainable policies and practices, thereby improving the performance of both fisheries management and conservation of biodiversity. Finally, we look beyond traditional science frameworks to consider how post-modern or post-normal science may help move past some of the difficulties encountered by decision-making within 'classical' interpretations of science as a foundation for policy and management.

Ecological complexity

A dynamic and complex physical environment influences ecological processes and interactions on scales from seconds to millennia and millimetres to global. The perceived significance of variability in a human context depends on the time and space scales over which people interact with the oceans and over which policies are implemented, reviewed and reformed. For instance, managers may establish multiyear plans to recover species or habitats to specified targets, even if multidecadal cycles or a changing climate means that recovered species or habitats are unlikely to persist at historically observed abundances on longer timescales (e.g. Pope and Macer, 1996; Southward et al., 1988).

Marine ecosystems may also show patterns and periods of stability across large space- or timescales and/or many species. These patterns may form the basis for policy and management, often focused on specific species and habitats that are recognized by both fishermen and conservationists, have distinctive characteristics and distributions, can be individually assessed, managed and conserved and can be targeted for fishing or conservation given their distinctive distribution or behaviour. Even so, humans remain concerned about the variability they encounter over short time periods, at local scales and for species or habitats of special interest (Spencer and Collie, 1997).

The abundance and distribution of marine species and the state of their environment is inherently uncertain due to factors such as underlying random variability, incomplete and imprecise observation and models based on imperfect observation and incomplete knowledge. All these sources bring uncertainty to estimates of status and trends of ecosystem components and attributes. This uncertainty can be reduced but not eliminated by statistically rigorous and more intensive data collection, validation of model assumptions and parameters and appropriate monitoring. Many analytical assessment methods take the residual uncertainty into account explicitly (e.g. Cochrane, 1999; Francis and Shotton, 1996; Quinn and Deriso, 1999).

For key commercially important species in developed countries, in contrast with many such species in less-developed countries (Chapter 22), the main processes contributing to variation in abundance and distribution are thoroughly studied and reasonably well known. Even for such species, a well-known process does not ensure low uncertainty in the population parameters. For example the causes of natural mortality are well known: predation, starvation, disease and transport into unfavourable oceanographic conditions (Gislason et al., 2010). Nonetheless variation in oceanographic and biotic community conditions is sufficiently large and difficult to quantify that few predictions of short-term changes in natural mortality are accurate enough to include in stock assessments (Griffith and Harrod, 2007; Lee et al., 2011). In less-developed countries, the ecological processes influencing populations are even more poorly known and other approaches are necessary (cf. Chapter 22).

Despite uncertainty about population status and trajectory for many larger species on timescales of years and decades, fairly reliable predictions can be made of the consequences of management actions that change mortality rates and schedules or the presence and intensity of pressures, including actions that build resilience to environmental variation (e.g. McGowan et al., 2011; White et al., 2010). Uncertainty is often addressed explicitly in those predictions, necessarily increasing as trends in abundance and effects of management are extrapolated further into the future. These approaches form the basis of adaptive, species or habitat focused management and conservation, and can be reasonably effective (cf. Chapter 4).

Even though complexity increases at the scale of ecological communities and habitats, science still has a significant role in informing policy development and contributing to management. For example, models of species interrelationships can explore alternate scenarios for systems managed in different ways. Although the results of these scenarios provide limited insight into local opportunities for catching or conserving individual species, they may guide choices among harvesting strategies, recovery strategies or protected area measures that are more or less compatible with healthy, resilient ecological communities. For habitats, science-based approaches can link human activities to increasing risks of specific habitat impacts, even if the relationship of scale of activity to scale of impact is poorly quantified (Rice, 2009). This knowledge is sufficient to guide compensation and mitigation decisions for major undertakings affecting habitat, especially if mechanisms for feedback and adaptive actions are included in the management plans, to address the uncertainty in both impacts of the undertakings and responses of the habitat to mitigation (Stankey et al., 2005; Walters, 2007). However, uncertainties about the effects of different rates and intensities of impact on 'special' habitats and the ecosystem services they provide, coupled with strong views about their 'intrinsic value', have tended to focus habitat-oriented policy and management on protecting specified (often arbitrary) proportions of the total habitat (e.g. Hourigan, 2009).

The differences among species, communities and habitats regarding their inherent complexities, uncertainties and responsiveness to management summarized above mean that fisheries and marine environmental management have evolved in different ways. Both strategic ('what') and tactical ('how' and 'how much') species-based advice, predicated on the thorough analysis of assumed cause–effect relationships, has been the dominant paradigm in fisheries management and considerable resources have been made available to support it. Scientific advice on the management of marine communities and habitats has often been strategic, less quantitatively nuanced and more precautionary: for example, generic harvest strategies for 'forage species' (see Chapter 4) or simply delineation of specific areas of special habitats for more risk averse management (see Chapters 8 and 11).

Human system complexity

There are remarkable parallels between the social and ecological subcomponents of marine ecosystems, in terms of complexity and how that complexity is understood. For example, 'old' fisheries management thinking placed the biological focus on a single target species, with relatively little management attention to the broader ecosystem, and on the human side, focused on a simple 'fishing fleet' as the entity exploiting the single fish stock. This led to considerable scientific effort studying and modelling 'fish and fleet' dynamics, reflecting a perception of the relevant socio-ecological system at that time. Furthermore, corresponding management effort focused largely on controlling the fleet and its activity to optimize performance of the fishery (Charles, 2001). Even in this simple fish-and-fleet world, high uncertainty was commonplace with uncertainties arising, from ignoring variability in fish prices, fishing costs, effectiveness of regulatory measures, fish population dynamics and other aspects of the ecosystem.

Accompanying a recognition of the biophysical and ecological complexities of marine ecosystems, and the importance of overall ecosystem health to fishery outcomes, there has been a parallel emergence of appreciation for complexity in the human system (e.g. Armitage et al., 2009; Berkes and Folke, 1998; McCay et al., 2011). This human system is also manifest in many timescales (with human dynamics occurring from the very short term to the very long) and spatial scales (e.g. with some fishers harvesting in small local bays and others over much of the world's ocean).

Complexity is also reflected in the structure of human society, community and resource use activities, from the diversity of coastal communities and households within them to the many types of fishers, fishing gears and non-fishery resource extraction methods (De Young et al., 2008; Perry and Ommer, 2003). Similar to uncertainty about linkages among ecosystem components, the manner in which people interact within their community and how that impacts on marine decision-making are not well understood. For example, the 'middlemen' in a fishing community have diverse potential roles from useful sources of investment funds and providers of access to markets to merciless exploiters of fishers. Similar uncertainties exist in terms of the complex interactions between multiple ocean uses, a reality to be faced given the rapidly expanding interest in integrated management and marine spatial planning (e.g. Cicin-Sain and Knecht, 1998; Ehler and Douvere, 2009; Norse, 2010).

All of this is driven by the complexity of underlying values, goals and objectives pursued by humans and human society and the intricacies of human behaviour. Differences in values between a government and a coastal community (e.g. in terms of development priorities) or between a fishery manager and a fishery organization (e.g. over the nature of co-management) or between components of a coastal community, can produce conflict and behavioural responses with enormous effects on the marine socio-ecological system (Charles et al., 2010; Kooiman et al., 2005). This highlights the importance of having a strong understanding of human considerations (De Young et al., 2008), although there has been little support for studies and practical implementation in this area to date.

A key aspect of the human system that differentiates it from the ecological system is having a 'governance system' embedded within it (Garcia and Charles, 2007; Kooiman et al., 2005). We not only 'use' the ocean, we 'manage' our use of it. Governance is a human affair and thus part of the human system, even though there are benefits to examining governance as a discrete subsystem in its own right (Charles, 2001). Uncertainties are also prominent in that governance system, partly structural and partly relating to values, goals and human behaviour. Furthermore, as for the marine socio-ecological system as a whole, governance faces challenges across scales, for example in dealing with issues of resource use and conservation under climate change (Charles, 2012).

Interestingly, as we increasingly adopt principles of 'good governance' and shift to participatory decision-making, these complexities become more apparent (Armitage et al., 2007). The extent of research on governance itself perhaps now exceeds that for the rest of the human system. Whether this leads to improved outcomes in terms of sustainable ocean use and marine conservation remains to be seen, but certainly a positive aspect of this trend is to emphasize the importance of decision-making processes. For example, a major present-day priority lies in implementing suitable processes for assessing and managing the risks arising from human use of marine space and resources.

Confronting the uncertainty

Uncertainty has profound effects on the form and interpretation of the evidence base that supports governance. As complexity increases and knowledge about socio-ecological interactions becomes generic rather than specific, objectives and management actions must be less concerned with achieving precise specified outcomes, and more with delivering general outcomes and improved status. For example, the objectives of single-species harvest management often focus on specific abundance targets for the fished populations (Smith et al., 1993) and restoration targets for specific sites may specify a particular state of the habitat (Andersen et al., 2004). In contrast, the objectives for protected area networks may specify targets for the extent of area protected rather than for the state that is to be achieved (UN, 2006).

Addressing uncertainty from monitoring

Within marine socio-ecological systems, monitoring of the human components (social, economic, cultural, political, legal, etc.) has been typically irregular and sporadic, both spatially and temporally. Many jurisdictions have little human-side data and it is rare to find time series of human-oriented indicators; data are typically obtained from one-time cross-sectional studies. The fishery sector does have some basic economic indicators (e.g. landed value, exports and employment) sometimes monitored over time, with reasonable spatial coverage. However, the growing popularity of integrated coastal and ocean management and marine spatial planning is leading to the recognition of the need for databases and monitoring to accompany new governance initiatives (De Young et al., 2008; Ehler and Douvere, 2009; Norse, 2010). Governmental efforts to collect data are expanding the spatial coverage, but it is very difficult and potentially misleading to extrapolate data or insights on human aspects from one scale to another (Perry and Ommer, 2003). For example, a solution from a socio-economic study of households in a given coastal community may be inapplicable to a regional integrated management initiative, while lessons learned on a broad scale may be insufficient for local governance of a specific coastal area. This situation could be improved through more systematic collection of time series data, covering multiple human dimensions and applicable at multiple scales, but this would be very expensive.

Monitoring ecological components of the system is also challenging, needing expensive platforms that will often have to sample large areas coarsely. Some ecosystem features, such as primary productivity and sea-surface temperature, can be monitored remotely at high resolution (Santos, 2000), but many of the features of greatest interest to both fisheries management and conservation of biodiversity require sampling the water column using capture gears or visual or acoustic methods with much more limited spatial coverage and higher costs. Consequently, samples are widely spaced relative to the spatial scales at which management is usually applied (Bednarek et al., 2011; Fraschetti et al., 2005; Jimenez-Alfaro et al., 2012; Piet and Quirijns, 2009), and data may not capture the spatial heterogeneity of seabed and water column features (ICES-JRC, 2010a; Smale et al., 2012). Substantial scientific guidance exists on how to extrapolate among samples separated often widely in space and/or time, and to estimate the resultant uncertainties in the properties being monitored (ICES-JRC 2010b; Quintino et al., 2006). As with the human system, such uncertainty can be reduced by more intensive sampling (but costs increase, often more than proportionately).

Addressing uncertainty about processes

The processes linking physical and biological components of these systems are rarely linear and often not even monotonic. Tipping points linked to trophic cascades (Anderson and Pedersen, 2010; Baum and Worm, 2009), regime shifts (Jiao, 2009; King and McFarlane, 2006) and multiple equilibria (Matsuda and Abrams, 2006; Petraitis and Hoffman, 2010) have been documented in an increasing number of systems. In addition there are often multiple pathways linking ecosystem components, such that systems can respond in different ways to repeated exposures to the same pressures (McCann and Rooney, 2009; Yodzis, 2000). Detailed research can improve knowledge of ecosystem processes, but sometimes the outcome is simply a better description of the sources of uncertainty. For example, despite decades of research on the functional feeding relationship between a predator and its prey, crucial to predictions of predator consumption or prey mortality (e.g. Hunsicker et al., 2011), there usually needs to be case-specific study to identify the relationship. Even when case-specific studies are conducted, there is uncertainty about the persistence of feeding relationships through time.

Social and economic processes are comparably complex. Non-linearities have been documented and analysed in economic systems, and notably in fishery 'bio-economic' systems, for decades (Clark, 1985). There is growing interest in social tipping points as well (Grimm and Schneider, 2011). Resilience is a dominant consideration, frequently defined as the ability of a system to withstand perturbations (e.g. Armitage et al., 2009; Berkes and Folke, 1998; Gunderson, 1999; Holling, 1973). Recent studies recognize the need to better integrate resilience with the various social dimensions of marine systems; for example, how resilience in fisheries interacts with social wellbeing (Armitage et al., 2012). The multiple pathways and multiple equilibria noted above for ecological processes also exist for social and economic processes; similarly, there is recognition of the need for case-specific analysis. Accordingly, there is a significant focus on building *adaptive learning* into human systems as a mechanism for people to adjust their behaviour to reflect change within an uncertain environment (Armitage et al., 2007, 2009).

Faced with this uncertainty about processes, science advice commonly encourages management to take the multiple pathways into account and be risk averse relative to tipping points. However, with uncertainty about which pathways will be preferred in any particular situation and with the locations of tipping points hard to document until they have been passed (and risk aversion is too late), this is easier to advise than to implement. Moreover, these multiple pathways, non-linearities and concomitant uncertainties make it challenging to develop management strategies that are robust to uncertainty (and to faulty information) about ecosystem relationships and about human dynamics (cf. Charles, 2008). Even for well-studied systems, these features limit the ability to predict trajectories of system properties into the future and the manner in which the system will respond to management actions. Policy and management measures still run greater risk of failure if they do not take the science advice into account (Hilborn and Hilborn, 2012; Rice and Cooper, 2004), but even the best science advice will not guarantee success.

Use of other sources of knowledge

So far, this chapter has focused on 'science' as the use of quantitative (and sometimes qualitative) scientific data, along with mathematical analysis and modelling methods. This is the notion of science at the core of science support for policy and management in most of the developed world. This in turn implies a significant institutional and human resource capacity to collect and manage data and to conduct and interpret analyses.

In many parts of the world this capacity is limited however, and other sources of knowledge play important roles in development of policy and management measures. In particular, what is referred to as 'traditional knowledge' has been built up through long-term observations and experiences of the state of ecosystems, natural resources and their spatial and temporal changes. In some cases, such knowledge is the key input to management decision-making, for example leading to effective management boundaries and rules in some marine resource-use tenure systems. Furthermore, there is a growing body of literature documenting that such traditional knowledge is not just a second-best option when the science provided by the developed world is not available (Gray and Hatcher, 2008), but rather can be a broadly rich source of insight and understanding into both ecosystems and human systems (Berkes, 1999; Gray, 2006). In some cases the cultural continuity of these sources of knowledge give the information a time perspective covering several decades to centuries, whereas in many areas 'western datasets' may only cover at most a few decades (Jentoft, 1999).

There are clear standards and guidelines for collection and quality assurance of traditional knowledge, often codified and scrutinized, because of the moral and ethical issues associated with the collection and use of such knowledge. Even when significant amounts of 'western science' have been available, there are many documented cases where augmenting that science with the experiential knowledge of resource users and serious amateur naturalists has led to improved policy and management. The processes best suited to bring such knowledge into policy-making and management have some differences from the typical processes for provision of science advice to government institutions, but those processes are also well studied and clearly codified (e.g. many examples in Gray, 2006). For the remainder of this chapter, references to 'science' should be read as including other appropriately obtained sources of knowledge.

Complexity, uncertainty, risk and science-based decision-making

Good physical, biological, economic and social science, often augmented by traditional knowledge, will help to clarify and quantify many types of uncertainties. Expectations that science could eliminate these uncertainties are proving unrealistic, however. Rather, the best science will provide an understanding of the major sources of uncertainties, their general magnitudes and their implications for decision-making. This leaves the governance processes with the dual tasks of: (1) finding optimal (or at least robust) tradeoffs of desired states of the ecological, economic and social systems; and (2) evaluating these tradeoffs with uncertainty about the costs, benefits and risks associated with the alternative outcomes of different possible tradeoffs.

The feasibility of these tasks will vary across jurisdictions, and may be particularly limited for developing countries. Nevertheless, all jurisdictions must pursue the tasks, to the extent possible, simultaneously in fisheries management and in conservation of biodiversity. The reality however is that fisheries and biodiversity conservation governance streams are likely to consider different tradeoffs to be optimal, even if they are working with the same science information and uncertainties (see Chapters 2, 3, 10). For example, where fishing grounds are productive and comparatively easy to both fish and regulate and coincide with areas important for seabird foraging, a fisheries manager may tend towards maintaining fishery access whereas a biodiversity conservation agency may tend more towards minimizing harvest. Moreover, the actions of one of the governance jurisdictions can be one of the important sources of uncertainty for the other. For example, the potential effectiveness of options available to a biodiversity conservation jurisdiction to recover an endangered species may depend crucially on decisions made by a fisheries management jurisdiction about fisheries where that species may be taken as bycatch.

Some of these differences in perceptions of optimal outcomes of decision-making under uncertainty stem directly from the different legal mandates of the jurisdictions (cf. Chapters 4, 10). Although all agencies may consider achievement of ecological objectives to be important, to meet their respective mandates resource management agencies are likely to give greater weight to achievement of economic and social objectives for use of an ecosystem than biodiversity conservation agencies. These different weightings on the objectives contribute to the different perceptions of the optimal outcome sought by the respective decision-makers.

Complexity and uncertainty do not just produce contrasting perceptions of optimal tradeoffs between the fisheries and biodiversity conservation governance streams. They also contribute to differences in the tolerances of the streams with regards to the different types of possible decision errors. This more subtle difference is particularly well illustrated in the case of protection and recovery of endangered species (e.g. Crawford et al., 2011).

In many jurisdictions, legislation has empowered managers to apply severe restrictions on fisheries that may kill threatened or endangered species. There is high uncertainty in the decision to protect or not protect a species, because the true population sizes and ranges usually cannot be quantified with accuracy and precision and the future trajectory of the population under various management scenarios is even more uncertain. Consequently, every listing decision has a risk of errors: either a 'miss' (failing to require prescriptive management when retrospectively these actions were necessary for a species' survival) or a 'false alarm' (imposing severe restrictions on fisheries when retrospectively they were not required for survival of the species).

Faced with inescapable uncertainty and risks of decision errors, the conservation of biodiversity stream gives priority to the irreversibility of extinction and the difficulty of reversing local extirpations. Because the consequences of failure to take conservation measures when they may be necessary are so serious, the preference is to have a very low miss rate, even if the cost is a larger number of false alarms. In contrast, the fisheries management stream and their main fisheries clients have a lower tolerance to the risk of unnecessary restrictions on fisheries than the biodiversity conservation stream. Furthermore, with harvested stocks assessed regularly, misses are considered less serious on the assumption that, if a stock continues to decline, there will be additional chances to implement appropriate conservation measures in the future and under conditions of less uncertainty because there will be more years of data (Davies and Baum, 2012; Rice and Legacè, 2007).

The governance challenges in making tradeoffs between misses and false alarms are rooted in the complexity of the ecological and social systems determining the stock trajectory, the role of fisheries as a threat and a factor in recovery and the ways that costs and benefits of conservation measures will be distributed. Science may reduce the uncertainties associated with the ecological and social systems, but cannot resolve the different risk tolerances of the fisheries and the biodiversity conservation governance streams for the errors in decision-making inevitable with incomplete and uncertain information on complex systems.

Illustrations of complex systems, science and governance

Given the significant challenges to governance summarized above, how do these challenges play out in particular contexts? Here we explore the interaction of complex systems, science and governance in relation to decision-making and potential tipping points.

Decision-making, tipping points and science advice

Although not labelled at the time as ecological tipping points, conceptually much of the work on incorporating the Precautionary Approach in fisheries in developed countries was based on concerns relating to non-linearities in stock–recruitment relationships (Rice, 2009; Chapter 4). It is argued that when stock biomasses fall below biologically based Limit Reference Points, there is a rapid increase in the risk that recruitment would be impaired and recovery would not be 'rapid and secure' if fishing mortality were reduced (ICES, 2002a, b). As research on ecological regime shifts and cascades (Andersen and Pedersen, 2009; Jiao, 2009) progressed, the ecological science community paid increasing attention to the importance of avoiding ecological tipping points (Rice et al., 2009). This reasoning leads to calls for highly precautionary management to protect biodiversity, due to the highly uncertain scientific knowledge on the vulnerability and resilience of rich biodiversity systems. This rationale is now prominent in justifications of large marine protected areas and the ecosystem approach in general (Agardy, 2012; Weaver and Johnson, 2012).

This dialogue is becoming more complex however because economics and social sciences have also recognized the existence of tipping points and the importance to management of not exceeding them. For example, the ability of the fishing industry to continue to absorb reductions in revenues drops sharply beyond economic tipping points (involving costs and earnings). More recently, social scientists are making the case that it is along the community and social dimension of sustainability that non-linearities and tipping points may be most serious because, when human communities and cultures have been stressed beyond their *de facto* Limit Reference Points, recovery may be even less likely than recovery of ecological communities (Arthur et al., 2011; Ommer, 2007).

The dialogue about the interplay of non-linearities in decision-making by both streams of governance is in its early stages. The motivation of governments to give greater or lesser priority to avoidance of economic, social and marine ecological tipping points has not been systematically quantified by political scientists, but is clearly important. We possibly face long periods of impasse, as the risk tolerances of the conservation of biodiversity stream for approaching the various tipping points may again be different from the risk tolerances of the fisheries governance stream. Nevertheless, the identification of where tipping points lie on all the dimensions of sustainability is a crucial science theme, and useful progress will require unprecedented collaboration between natural and social scientists.

This need for enhanced collaboration among the various science communities highlights one other emerging area of scientific efforts: decision-making that better integrates the ecological, social and economic outcomes of the choices available requires better integrated science advice to inform the decision-makers (Miller et al., 2010). Many different types of assessments have been called 'integrated ecosystem assessments' but, in the context of informing governance and policy-making, assessments need to integrate across three axes: the biophysical and ecological (from 'physics to fish'); the multiple industry sectors that are using and potentially altering a marine ecosystem; and the social, economic, institutional and ecological dimensions of the system being assessed, used and managed (UNEP and IOC-UNESCO, 2009). These fully integrated socio-ecological system assessments are forming a cornerstone of science in support of governance, and provide another pathway where science can contribute to greater coherence of governance across streams by providing all governance components with a common factual basis for dialogue on tradeoffs, costs and benefits and ensuring that the ecological, social and economic aspects of decisions are clear and available for consideration in the dialogue leading to decisions. This alone is a substantial improvement over each perspective, packaging the science and uncertainties to steer the dialogue towards its desired outcome (Rice, 2011).

Summary of the challenges to governance

(1) Decisions must be made under uncertainty. In complex systems even the best science, while potentially explaining and sometimes reducing uncertainty, cannot eliminate it. Governance therefore has to proceed without certainty of outcomes, even when rational decisions are made and implemented effectively. The science advice for management has to explain the risks as well as the potential benefits associated with each policy option available.

(2) Decision-makers have to strive to achieve multiple objectives with few win-win choices and sanguine tradeoffs. This requires information from natural, economic and social sciences, augmented by

traditional and experiential knowledge. There will often be different amounts of information available concerning the likelihoods of various ecological, social and economic aspects of a given policy option. Consequently, although outcomes of decision-making will typically be judged on many factors, the governance processes will rarely be equally well informed about their likelihood of suitable performance on each of the factors.

(3) Given 1 and 2, decision-makers may apply risk management practices to minimize likelihood of outcomes that are unacceptable to individual constituencies. It is more challenging to apply such risk-based or probabilistic scientific decision-support processes to determine optimal tradeoffs among potentially conflicting objectives if the risk tolerance profiles of different constituencies are different, since there may not be agreement on the size of any cost or benefit being traded off.

(4) Formal tradeoff analyses require tools and currencies that are in the early stages of development (e.g. avoidance of tipping points; goods and services valuation) and whose ultimate value in decision-making is unresolved. The currencies for measurement of ecological, social and economic costs and benefits will not often be available for a single decision and the knowledge about tipping points on the various dimensions of the tradeoffs is unlikely to be equally complete. Governance processes therefore will take a subjective approach in trying to balance ecological risks, costs and benefits with those measured in social and economic terms, possibly supported by accelerated research to locate tipping points and to quantify the value of ecosystem goods and services.

(5) Management for effective outcomes often requires coherence among multiple agencies and/or authorities. In these complex systems, often the likelihood of a particular outcome may be influenced by decisions made in several sectors. Governance processes therefore have to deal with not just the complexities and uncertainties of their immediate decisions, but the decisions made by other authorities dealing with similarly complex and uncertain decisions. Fully integrated assessments better inform the governance process in such multi-use multi-stakeholder contexts because the same information goes to all groups of decision makers, but groups with different perspectives may still interpret the same information in different ways.

Concluding thoughts

This chapter is written as if science always lives up to the ideals of empiricism, impartiality and objectivity that are used to legitimize science as a superior form of knowledge. In the past decades, this 'science is best' paradigm has been challenged on both sound and sometimes spurious grounds. Some of the challenges are the broad questions that social constructionists have asked about science in general for nearly half a century (Burr, 1995; Funtowicz, 2006) and about fisheries for at least 20 years (Finlayson, 1994). The discourse about science, fisheries management and conservation of biodiversity had gone deeply into the limitations of exclusively science frameworks for informing governance of fisheries and marine biodiversity (Garcia and Charles, 2007, 2008). This chapter cannot carry out a full review of the perspectives and messages emerging from the dialogue sparked by these challenges, but it is necessary to acknowledge the paradigm of 'post-normal science' and its implications.

Building coherence and effective governance of fisheries and conservation of marine biodiversity involves making a coherent picture out of pieces not designed to fit together. The reliance of scientific explanations on parsimony does not fit well in a coherent governance framework (Garcia, 2005). The biodiversity and the fisheries communities have different conceptual boundaries for decisions about what is 'minimally realistic' (Punt and Butterworth, 1995), as will the social, economic and ecological experts from each perspective. With one group's 'practical simplification' perceived as leaving out some crucial considerations by another group, singular science answers are unlikely to be useful as inputs to governance systems that have to address many needs.

The inability to set useful boundaries extends well beyond the boundaries of parsimonious science. The scales and boundaries appropriate for the dynamics of the social and the ecological systems are not only likely to differ within each stream (e.g. for fisheries, managing harvest at meta-population scales to supply markets responding to globalization), but the meaningful boundaries in each dimension may differ between a fishery (harvested stock) and a biodiversity (impacted ecosystem or a highly migratory bycatch species) perspective. Moreover, the governance frameworks themselves extend from global agreements (Chapter 10) to local customary practices (Chapters 22–26) in each stream. As boundaries for analyses and modelling become soft and inconsistent, the practice of 'hard' science rapidly becomes intractable.

The role of science as the foundation for governance must also help in dealing with the mismatch of the complexity of the ecological and social systems with the constrained nature of law and jurisdiction (Garcia and Charles, 2007, 2008) and with the continuing presence of a 'fallacy of controllability' (Charles, 2007). Coherence among governance streams requires tradeoffs that, as explained above, are hard to find. Furthermore, when acceptable tradeoffs are found, they may not be within the power of existing jurisdictions to deliver because of geographic or sectoral restrictions or limitations on human resources or governance (e.g. in many developing countries). Moreover, there is ample evidence that agencies cannot fully control many parts of the system over which they have authority (physical, biological or human) so there is never certainty that compromise outcomes among the governance streams can actually be delivered, at least not in the ways expected when the compromises were reached.

This change in how science, governance and the socio-ecological systems are viewed, from 'clockworks to soft watches' (Garcia and Charles, 2007), means science must find ways to support management that are adaptive, evolving and both scientifically defensible and socially robust. Conversely, the development, structure and operation of management systems will influence the evolution, perception and value of science. For example, some systems may reduce scientific demands by prioritizing issues with risk assessment or placing the burden of proof on users, thus reducing the scope and frequency of requests for evidence. Improving the two streams of marine governance, and supporting both existing and emerging management systems, will require an improved understanding of tradeoffs between human and ecological objectives, the development of better tools for embracing uncertainty and the determination of approaches that find a feasible and cost-effective balance of knowledge sources and requirements. This may help manage the escalating demands for ever more complex integrative science that even the funding of science in the developed world is unable to sustain.

References

Agardy, T. (2012) In trade-offs and choices, there is one simple rule. *Marine Ecosystems and Management* **5**, 4–5.

Andersen, J.H., Conley, D.Y. and Hedal, S. (2004) Palaeoecology, reference conditions and classification of ecological status: the EU Water Framework Directive in practice. *Marine Pollution Bulletin* **49**, 283–290.

Andersen, K.H. and Pedersen M. (2010) Damped trophic cascades driven by fishing model marine ecosystems. *Proceedings of The Royal Society B: Biological Sciences* **227**, 795–802.

Armitage, D., Berkes, F. and Doubleday, N. (eds) (2007) *Adaptive Co-Management: Collaboration, Learning, and Multi-Level Governance*. UBC Press, Vancouver, Canada.

Armitage, D., Plummer, R., Berkes, F., Arthur, R., Charles, A., Davidson-Hunt, I., Diduck, A., Doubleday, N., Johnson, D., Marschke, M. et al. (2009) Adaptive co-management for social–ecological complexity. *Frontiers of Ecology and Environment* **7**, 95–102.

Armitage, D., Bene, C., Charles, A., Johnson, D. and Allison, E. (2012) The interplay of wellbeing and resilience concepts: Towards a transdisciplinary social-ecological perspective. *Ecology and Society* **17**(4).

Arthur, R., Macfadyen G., Cappell, R., Delaney, A., Triantaphyllidis, G., Caillart, B. and Agnew, D. (2011) Regional social and economic impacts of change in fisheries-dependent communities Final Report European Commission Fish/2006/06. Available at http://ec.europa.eu/fisheries/documentation/studies/regional_social_economic_impacts/final_report_en.pdf (accessed February 2014).

Baum, J.K. and Worm, B. (2009) Cascading top-down effects of changing oceanic predator abundances. *Journal of Animal Ecology* **78**, 699–714.

Bednarek, A.T., Cooper, A.B., Cresswell K.A., Mangel, M., Satterthwaite, W.H., Simpfendorfer, C.A. and Wiedenmann, J.R. (2011) The certainty of uncertainty in marine conservation and what to do about it. *Bulletin of Marine Science* **87**, 177–195.

Berkes, F. (1999) *Traditional Ecological Knowledge and Resources Management*. Taylor and Francis, Philadelphia and London.

Berkes, F. and Folke, C. (eds) (1998) *Linking Social and Ecological Systems: Management Practices and Social Mechanisms for Building Resilience*.Cambridge University Press, Cambridge.

Burr, V. (1995) *An Introduction to Social Constructionism*. Routledge, London, UK.

Charles, A. (2001) *Sustainable Fishery Systems*. Wiley-Blackwell (formerly Blackwell Science), Oxford, UK.

Charles, A. (2004) Sustainability and resilience in natural resource systems: Policy directions and management institutions. In: *Encyclopaedia of Life Support Systems (EOLSS)*. Developed under the auspices of the UNESCO. Eolss Publishers, Oxford, UK.

Charles, A. (2007) Adaptive co-management for resilient resource systems: Some ingredients and the implications of their absence. In: *Adaptive Co-Management* (eds D. Armitage, F. Berkes and N. Doubleday), pp. 83–102. UBC Press, Vancouver, Canada.

Charles, A. (2008) Robust Management, Risk and the Ecosystem Approach to Fisheries. ICES CM Documents 2008/O:05. International Council for the Exploration of the Sea.

Charles, A. (2012) People, oceans and scale: Governance, livelihoods and climate change adaptation in marine social–ecological systems. *Current Opinion in Environmental Sustainability* **4**, 351–357.

Charles, A., Wiber, M., Bigney, K., Curtis, D., Wilson, L., Angus, R., Kearney, J., Landry, M., Recchia, M., Saulnier, H. and White, C. (2010) Integrated management: a coastal community perspective. *Horizons* **10**, 26–34.

Cicin-Sain, B. and Knecht, R.W. (1998) *Integrated Coastal and Ocean Management: Concepts and Practices*. Island Press, Washington, DC.

Clark, C.W. (1985) *Bioeconomic Modelling and Fisheries Management*. Wiley-Interscience.

Cochrane K.L. (1999) Complexity in fisheries and limitations in the increasing complexity of fisheries management. *ICES Journal of Marine Science* **56**, 917–926.

Crawford, R.J.M., Altwegg, R., Barham, B.J. et al. (2011) Collapse of South Africa's penguins in the early 21st century. *African Journal of Marine Science* **33**, 139–156.

Davies, T. and Baum, J. (2012) Extinction Risk and Overfishing: Reconciling Conservation and Fisheries Perspectives on the Status of Marine Fishes. *Scientific Reports* **2**, 561.

De Young, C., Charles, A. and Hjort, A. (2008) Human dimensions of the ecosystem approach to fisheries: An overview of context, concepts, tools and methods. Fisheries Technical Paper. FAO, Rome.

Ehler, C. and Douvere, F. (2009) Marine Spatial Planning: a step-by-step approach toward ecosystem-based management. Intergovernmental Oceanographic Commission and Man and the Biosphere Programme. UNESCO, Paris. IOC Manual and Guides, 53. ICAM Dossier, 6.

Finlayson, C. (1994) *Fishing for Truth: a Sociological Analysis of Northern Cod Stock Assessments 1977–1990*. Institute of Social and Economic Research, St John's, Newfoundland.

Francis, R.I.C.C. and Shotton, R. (1996) "Risk" in fisheries management: a review. *Canadian Journal of Fisheries and Aquatic Sciences* **54**, 1699–1715.

Fraschetti, S., Terlizzi, A., Bussotti, S. et al. (2005) Conservation of Mediterranean seascapes: analyses of existing protection schemes. *Marine Environmental Research* **59**, 309–332.

Funtowicz, S. (2006) Post-normal science. Environmental policy under conditions of complexity. EC-JRC/ISIS, Ispra, Italy.

Garcia, S.M. (2005) Fishery science and decision-making: Dire straights to sustainability. *Bulletin of Marine Science* **76**, 171–196.

Garcia, S.M. and Charles, A. (2007) Fishery systems and linkages: from clockworks to soft watches. *ICES Journal of Marine Science* **64**, 580–587.

Garcia, S.M. and Charles, A. (2008) Fishery systems and linkages: Implications for science and governance. *Ocean and Coastal Management* **51**, 512–529.

Gislason, H., Daan, N., Rice, J. and Pope, J.G. (2010) Size, growth, temperature and the natural mortality of marine fish. *Fish and Fisheries* **11**, 149–158.

Gray, T.S. (ed.) (2006) *Participation in Fisheries Governance*. Kluvier–Springer, Dordretch, The Netherlands, pp. 249–268.

Gray, T. and Hatcher, J. (2008) A complicated relationship: stakeholder participation in the ecosystem-based approach to fisheries management. *Marine Policy* **32**, 158–168.

Griffiths, D. and Harrod, C. (2007) Natural mortality, growth parameters, and environmental temperature in fishes revisited. *Canadian Journal of Fisheries and Aquatic Sciences* **64**, 249–255.

Grimm, S. and Schneider, G. (2011) Predicting Social Tipping Points: Current Research and the Way Forward. Hamburg. Deutsches Institut für Entwicklungspolitik gGmbH. Available at http://www.die-gdi.de/CMS-Homepage/openwebcms3.nsf/(ynDK_contentByKey)/ANES-8G2GNY/$FILE/DP%208.(2011)pdf (accessed February 2014).

Gunderson, L. (1999) Resilience, flexibility and adaptive management – antidotes for spurious certitude? *Conservation Ecology* **3**(1), 7–12.

Hilborn, R. and Hilborn, U. (2012) *Overfishing: What Everyone Needs to Know*. Oxford University Press, Oxford.

Holling, C.S. (1973) Resilience and stability of ecological systems. *Annual Review of Ecology and Systematics* **4**, 1–23.

Hourigan, T.F. (2009) Managing fishery impacts on deep-water coral ecosystems of the USA: emerging best practices. *Marine Ecology Progress Series* **397**, 333–340.

Hunsicker, M.E., Ciannelli, L., Bailey, K.M., Buckel, J.A., White, J.W., Link, J.S. et al. (2011) Functional responses and scaling in predator–prey interactions of marine fishes: contemporary issues and emerging concepts. *Ecology Letters* **14**, 1288–1299.

ICES (2002a) Report of the Advisory Committee on Fisheries Management (2001) ICES Cooperative Research Report 255 (3 vol).

ICES (2002b) Report of the Study Group on Further Development of the Precautionary Approach to Fisheries Management. ICES CM 2002/ ACFM. 10 pp.

ICES-JRC (2010a) Task Group 6 Report Seafloor Integrity. Available at http://ec.europa.eu/environment/marine/pdf/6-Task-Group-6.pdf (accessed March 2014).

ICES-JRC (2010b) Management Group Report: MSFD Descriptor Implementation. Available at http://publications.jrc.ec.europa.eu/repository/bitstream/111111111/13626/1/management%20group%20report_final_vii.pdf (accessed March 2014).

Jentoft, S. (1999) Social science in fisheries management: a risk assessment. In: *Reinventing Fisheries Management* (eds T.J. Pitcher, P.J.B. Hart and D. Pauly), pp. 177–188. Kluwer Academic Publishers, London, UK.

Jiao, Y. (2009) Regime shift in marine ecosystems and implications for fisheries management, a review. *Reviews in Fish Biology and Fisheries* **19**, 177–191.

Jimenez-Alfaro, B., Draper, D. and Nogues-Bravo, D. (2012) Modeling the potential area of occupancy at fine resolution may reduce uncertainty in species range estimates. *Biological Conservation* **147**, 190–196.

King, J.R. and McFarlane, G.A. (2006) A framework for incorporating climate regime shifts into the management of marine resources. *Fisheries Management and Ecology* **13**, 93–012.

Kooiman J., Bavinck J.M., Jentoft S. and Pullin, R. (2005) *Fish for Life: Interactive Governance for Fisheries*. Amsterdam University Press, Amsterdam.

Lee, H.H., Maunder, M.N., Piner, K.R. and Methot, R.D. (2011) Estimating natural mortality within a fisheries stock assessment model: An evaluation using simulation analysis based on twelve stock assessments. *Fisheries Research* **109**, 89–94.

Link, J.M. (2010) *Ecosystem-Based Fisheries Management: Confronting Trade-offs*. Cambridge University Press, Cambridge.

Mann, K.H. and Lazier, J.R.N. (1996) *Dynamics of Marine Ecosystems*. Blackwell Science, London.

MASR (2005) Millennium Ecosystem Assessment Scientific Report. Ecosystems and Human Well-being. Biodiversity Synthesis World Resources Institute, Washington, DC.

Matsuda, H. and Abrams, P.A. (2006) Maximal yields from multi-species fisheries systems: Rules for systems with multiple trophic levels. *Ecological Applications* **16**, 225–237.

McCann, K.S. and Rooney, N. (2009) The more food webs change, the more they stay the same. *Philosophical Transactions of the Royal Society B: Biological Sciences* **364**, 1789–1801.

McCay, B.J., Brandt, S. and Creed, C.F. (2011) Human dimensions of climate change and fisheries in a coupled system: The Atlantic surfclam case. *ICES Journal of Marine Science* **68**, 1354–1367.

McGowan, C.P., Runge, M.C. and Larson, M. A. (2011) Incorporating parametric uncertainty into population viability analysis models. *Biological Conservation* **144**, 1400–1408.

Miller, K., Charles, A., Barange, M., Brander, K., Gallucci, V.F., Gasalla, M.A., Khan, A., Munro, G., Murtugudde, R., Ommer, R.E. and Perry, R.I. (2010) Climate change, uncertainty and resilient fisheries: institutional responses through integrative science. *Progress in Oceanography* **87**, 338–346.

Mitchell, R.B., Clark, W.C., Cash, D.W. and Dickson, N.M. (eds) (2006) *Global Environmental Assessments: Information and Influence*. The MIT Press, Cambridge, MA.

Norse, E.A. (2010) Ecosystem-based spatial planning and management of marine fisheries: Why and how? *Bulletin of Marine Science* **86**(2), 179–195.

Ohl, C., Krauze, K. and Grünbühel, C. (2007) Merging socio-economic and environmental research toward understanding of long-term ecosystem dynamics. *Ecology and Economics* **63**, 383–391.

Ommer, R.E. (ed.) (2007) *Coasts Under Stress: Restructuring and Social-Ecological Health*. McGill University Press, Montreal.

Perry, R.I. and Ommer, R.E. (2003) Scale issues in marine ecosystems and human interactions. *Fisheries Oceanography* **12**, 513–522.

Petraitis, P.S. and Hoffman, C. (2010) Multiple stable states and relationship between thresholds in processes and states. *Marine Ecology-Progress Series* **413**, 189–200.

Pielke, R. Jr. (2007) *The Honest Broker: Making Sense of Science in Policy and Politics*. Cambridge University Press, Cambridge.

Piet, G.J. and Quirijns, F.J. (2009) The importance of scale for fishing impact estimation. *Canadian Journal of Fisheries and Aquatic Sciences* **66**, 829–835.

Polasky, S., Costello, C. and Solow, A. (2005) *The Economics of Biodiversity. Handbook of Environmental Economics*. Elsevier, Amsterdam.

Pope, J.G. and Macer, C.T. (1996) An evaluation of the stock structure of North Sea cod, haddock, and whiting since 1920, together with a consideration of the impacts of fisheries and predation effects on their biomass and recruitment. *ICES Journal of Marine Science* **53**, 1157–1169.

Punt, A.E. and Butterworth, D.S. (1995) The effects of future consumption by the Cape fur seal on catches and catch rates of the Cape hakes. 4. Modelling the biological interaction between Cape fur seals (Arctocephalus pusillus pusillus) and Cape hakes (Merluccius capensis and M. paradoxus). *South African Journal of Marine Science* **16**, 255–285.

Quinn, T.J. and Deriso, R. (1999) *Quantiative Fish Dynamics*. Oxford University Press, Oxford. 560 pp.

Quintino, V., Elliot, M. and Rodrigues, A.M. (2006) The derivation, performance and role of univariate and multivariate indicators of benthic change: case studies at different spatial scales. *Journal of Experimental Marine Biology and Ecology* **330**, 368–382.

Rice, J.C. (2009) A generalisation of the three-stage model for advice using a Precautionary Approach in fisheries, to apply to a broadly to ecosystem properties and pressures. *ICES Journal of Marine Science* **67**, 433–444.

Rice, J.C. (2011) Advocacy science and fisheries decision-making. *ICES Journal of Marine Science* **68**, 2007–2012.

Rice, J.C. and Cooper J.A. (2004) Management of flatfish fisheries: what factors matter. *Journal of Sea Research* **50**, 227–243.

Rice, J.C. and Legacè, È. (2007) When control rules collide – a comparison of fisheries management reference points and IUCN criteria for assessing risk of extinction. *ICES Journal of Marine Science* **64**, 718–722.

Rice, J.C., Borges, M.F., Grehan, A., Kenny, A., Loeng, H., Maynou, F., Santos, R.S., Skjoldal, H.R., Thébaud, O., Vassilopoulou, V. and Volckaert, F. (2009) Science Dimensions of an Ecosystem Approach to Management of Biotic Ocean Resources. European Science Foundation Marine Board Position Paper, 14. 117 pp.

Santos, A.M.P. (2000) Fisheries oceanography using satellite and airborne remote sensing methods: a review. *Fisheries Research* **49**, 1–20.

Smale, D.A., Kendrick, G.A., Harvey, E.S. Langlois, T.J., Hovey, R.K., Van Niel, K.P., Waddington, K.I., Bellchambers, L.M., Pember, M.B., Babcock, R.C., Vanderklift, M.A., Thomson, M.P., Jakuba, M.V., Pizarro, O. and Williams, S.B. (2012) Regional-scale benthic monitoring for ecosystem-based fisheries management (EBFM) using an autonomous underwater vehicle (AUV). *ICES Journal of Marine Science* **69**, 1108–1118.

Smith, S.J., Hunt, J.J. and Rivard, D. (1993) Risk evaluation and biological reference points for fisheries management. *Canadian Special Publication of Fisheries and Aquatic Sciences* **120**, 1–422.

Southward, A.J., Boalch, G.T. et al. (1988) Fluctuations in the herring and pilchard fisheries of Devon and Cornwall linked to changes in climate since the 16th century. *Journal of the Marine Biological Association of the United Kingdom* **68**, 423–445.

Spencer, P.D. and Collie, J.S. (1997) Patterns of population variability in marine fish stocks. *Fisheries Oceanography* **6**, 188–204.

Stankey, G.H., Clark, R.N. and Bormann, B.T. (2005) *Adaptive Management of Natural Resources: Theory, Concepts and Management Institutions.* US Department of Agriculture, Washington, DC.

UN (2006) Framework for monitoring implementation of the achievement of the 2010 target and integration of targets into the thematic programmes of work. Decisions adopted by the Conference of the Parties to the Convention on Biological Diversity at its 8th meeting (Decision VIII/15). UN Department of Economic and Social Affairs, Division of Sustainable Development, New York.

UNEP and IOC-UNESCO (2009) An Assessment of Assessments: Findings of the Group of Experts. Start-up Phase of a Regular Process for Global Reporting and Assessment of the State of the Marine Environment including Socio-economic Aspects. IOC, Paris.

Walters, C.J. (2007) Is adaptive management helping to solve fisheries problems? *Ambio* **36**, 304–307.

Weaver, P. and Johnson, D. (2012) Think big for marine conservation. *Nature* **483**, 399.

White, J.W., Botsford, L.W., Moffitt, E.A., et al. (2010) Decision analysis for designing marine protected areas for multiple species with uncertain fishery status. *Ecological Applications* **20**, 1523–1541.

Yodzis, P. (2000) Diffuse effects in food webs. *Ecology* **81**, 261–266.

Part III

Global governance

Chapter 10

Global level institutions and processes: Frameworks for understanding critical roles and foundations of cooperation and integration

L. Ridgeway

Formerly with Fisheries and Oceans Canada, Ottawa, Canada

Abstract: This chapter provides a context and describes frameworks for understanding the global governance challenge, integrated institutional governance, foundations of institutional legitimacy and the nature of consensus decision-making in large groups. It provides a framework for understanding the potential roles and contributions of key global institutions and processes described in Chapter 11, and indeed may have relevance for other subsequent institutional chapters in Part III. In a unified and interdependent marine governance system, all players and their primordial organizations can and must play robust roles as part of the solution to better governance.

Keywords: frameworks; institutions; processes; cooperation; integration; governance; legitimacy; consensus

Introduction

Preceding chapters of the book have focused on the knowledge, spatial and legal foundations of fisheries and marine biodiversity and ecosystem conservation. The focus now turns to the scales and institutions across which these perspectives develop and integrate, while this chapter and the next focus on the global level. While they describe key institutions, organizations and processes, their focus is not on the complexity of the global system but rather on roles: how institutions and organizations are linked to different stakeholders and interests; how this affects debates on ocean governance; where the global community may best focus its efforts; and how fisheries experts need to navigate in the global arena. In this chapter some analytical prisms for understanding institutional tensions and successes at the global level are provided; in this light, the following chapter (Chapter 11) describes and assesses key global institutions and processes and their roles in integrating fisheries and conservation approaches.

Global oceans governance is based on collective State action to address issues that transcend national boundaries. Global oceans discussions bring together a formidable geographical mix of States with diverse socio-economic drivers and vulnerabilities, cultures, politics and capacities and environmental opportunities and constraints.[1] States have interdependent, but different, stakes in ocean

Governance of Marine Fisheries and Biodiversity Conservation: Interaction and Coevolution, First Edition.
Edited by Serge M. Garcia, Jake Rice and Anthony Charles.

resources and the marine environment, as well as global rules, processes and approaches. States also have distinct views on how environmental issues mainstream into agendas that matter to them. These diverse interests can be difficult to resolve, and this affects what can reasonably be achieved in global decision-making.

Establishing some frameworks

This chapter describes five prisms or frameworks through which a global institutional assessment can be understood, namely: choosing a 'governance' perspective that best illuminates issues; integrated institutional governance; the role of 'legitimacy' in understanding institutional dynamics and global decision-making; consensus-based decision-making in large groups; and inherent strengths and weaknesses of global institutions. These prisms offer a frame for understanding global oceans governance issues for (and even beyond) the specific institutions or processes described in Chapter 11.

Framing the global 'governance' challenge

Descriptions of the 'global governance challenge' often emphasize the complexity, fragmentation and overlap of relevant institutions, processes, norms and tools and the gap between this elaborate structure and ongoing deterioration of the marine environment (e.g. Najam et al., 2006; Weiss, 2009). Such descriptions emphasize the potential inefficiency and 'transaction costs' of global governance, and motivate reform proposals of varying credibility and feasibility. The debate is particularly polarized between those favouring top-down centralized models of governance that limit State sovereignty or restrict sectoral management mechanisms and those valuing distributed roles and sectoral approaches to achieve agreed ends.

Description of the 'complexity' of the system as a coherence challenge is only partly useful in understanding divergent tensions and implementation gaps in marine policy and governance. More helpful is a focus on why some institutions, roles and decisions are supported or implemented by relevant players (or not), and what global institutions and organizations can realistically contribute to improved governance.

Using a lens of institutional economics, Paavola (2005) examines concepts that are critical to 'decoding' global debates on governance and marine sustainability. He describes environmental resources as collectively owned[2] among interdependent[3] users. Governance institutions work to resolve conflicting interests in the use or preservation of environmental resources, with conflicts over governance approaches involving relative roles in certain governance functions (e.g. excluding unauthorized resource users; affirming entitlements to resources; sharing benefits and costs; monitoring and enforcement; resolving use-conflicts; and modifying governance). The role and distribution of power in resolving these conflicts introduces notions of distributive and procedural justice into environmental decision-making.

Paavola's framework helpfully emphasizes that the multiple services provided by environmental resources (e.g. oceans, ecosystems and biodiversity) are interdependent but distinct assets, each requiring a separate governance institution. A kaleidoscope of resource rights and governance arrangements is therefore entirely rational, and implies a distributed system of interdependent and linked institutions (Scott, 2011; Figure 10.1). As described below, linking governance to explicit ecosystem services also helps decode nuances of how nature conservation and fisheries approaches and institutions differ (see Chapter 11).

Integrated institutional governance

In light of the wide range of institutions and bodies involved in marine governance, Ridgeway and Rice (2010) emphasized the *mutual dependence* among *robust components* of an *integrated system*. This dependence arises from different roles that organizations play in encouraging horizontal and vertical policy and program coherence in fisheries globally, regionally and sub-regionally.

The work stressed that institutions and organizations create (or inhibit) coherence through the debates they sponsor; diversity of stakeholders involved; knowledge created and diffused; conceptual, legal, regulatory and operational frameworks developed; legal and technical tools created; partnerships encouraged; monitoring and review undertaken; and accountability mechanisms built. The 'personalities' of institutions derive from the disciplines and expertise involved in their activities: scientific and other technical; legal and policy; and industry, diplomats and the like. These factors together determine whose interests are realized and who enjoys the benefits and bears the consequences of institutional decisions.

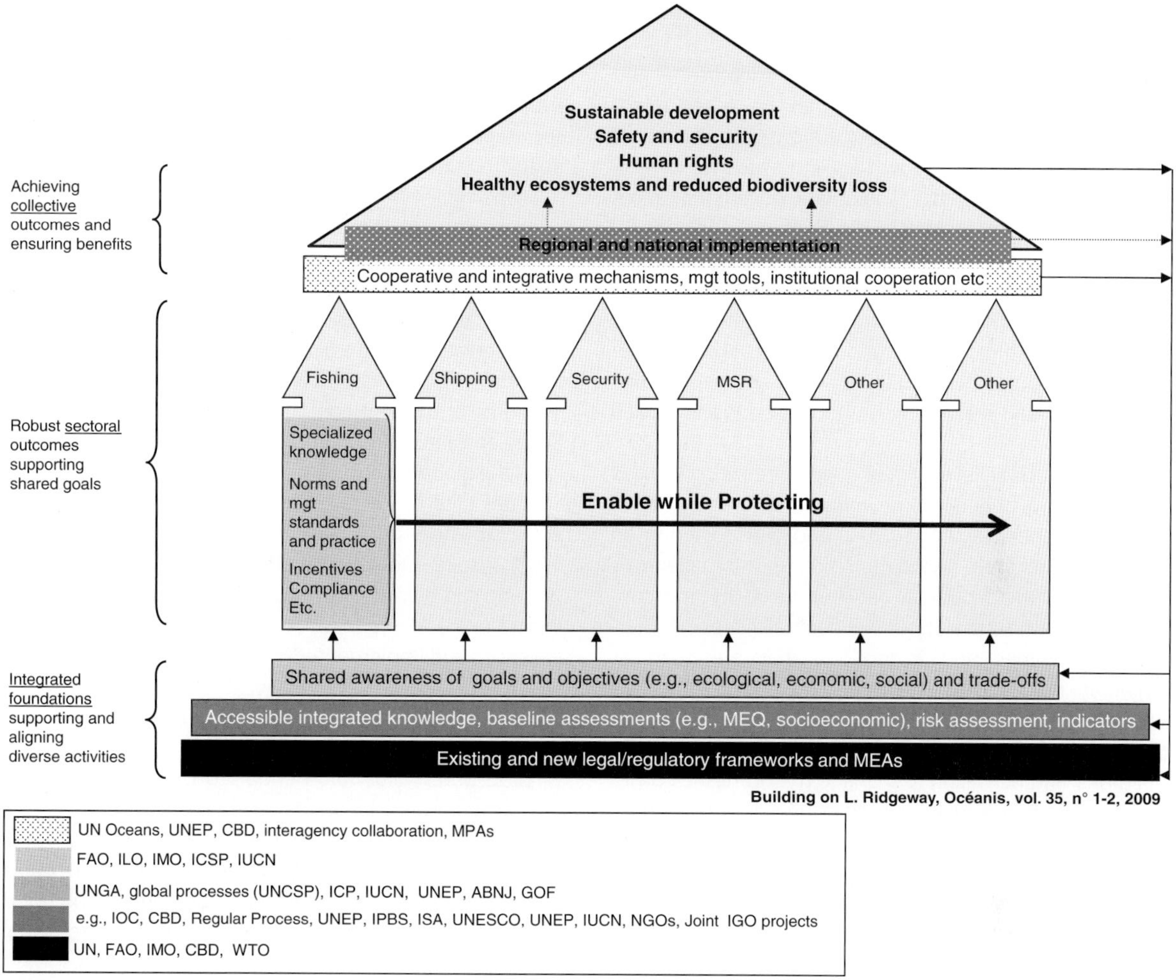

Figure 10.1 A model for integrated oceans governance.

Ridgeway (2009) presents a model emphasizing the diverse components or *roles* in a coherent and integrated system of ocean governance, which is adapted in Figure 10.1. The model rebuts arguments that sector-based legal frameworks and governance amount, *de facto,* to an 'absence' of governance in an integrated world. Similar to Paavola's sector-based network of governance institutions across oceans uses, the model illustrates the contributions of different institutional players. In the spirit of integrated management however, it adds common foundations for decision-making and places sectoral and integrated management in mutually dependent roles. The model therefore emphasizes the inherent contributions and accountabilities of an integrated interdependent system. Each component is a necessary but not sufficient part of the system.

This interdependent system needs *common and integrated foundations* (knowledge, legal, regulatory and goal-setting), *robust sectoral management* (involving specialized knowledge, legal foundations, standards, management tools[4] and aligned incentives) and *integrative mechanisms and tools* bringing sectoral interests together to deliver on collective goals. Defined roles for the activities of different global institutions and organizations are described in Annex 2 of the book.

Legitimacy

Tensions in global oceans governance debates – and how issues and communities find their voice and role – often relate to the underlying perceptions of legitimacy of institutions and processes and their outputs for diverse and relevant communities. 'Legitimate' governance relies on implicit or explicit agreement built through rational discussion and decision-making among interested States, in collaboration with relevant stakeholders with deep interests. If legitimacy is disputed or absent, then tensions can be expected and global institutions are less likely to generate strong outcomes. Legitimacy also helps to predict which issues can reasonably be addressed within which institutions, if at all.

At its most basic level, global governance is concerned with laws and general frameworks for managing transboundary issues (Fabienne, 2010) and with fostering an efficient and effective negotiation climate. The challenge is to find the correct balance between flexibility and constraint on State action. States will generally support international regimes and institutions that are seen as legitimate mechanisms to achieving their goals. Only then will States endorse outputs and processes and allow their powers or interests to be constrained by collective decisions, and be willing to manage the domestic consequences of change that such collective decisions imply. Implementation and compliance are more likely when governance solutions and decision-making are broadly considered to be legitimate. Legitimacy can therefore replace the need for centralized top-down global *compliance* authorities, as some would otherwise propose (see also the following section on 'Consensus decision-making'). As for top-down approaches to force governance *coherence*, it would seem, on the same basis, that legitimacy is a necessary precondition for consensus on a multi-faceted toolkit as much as a single approach. Top-down approaches not seen as legitimate will bring neither consensus, coherence of accepted approaches or willing implementation. Without willing implementation, all governance approaches are hollow.

There is also an important relationship between *institutional performance* and both the successful resolution of issues and the willingness of States to allow their prerogatives and activities to be constrained (Breitmeier, 2008). This includes indirect constraints on States such as rule-making outside their direct control,[5] agreements and instruments that enable institutions or organizations to make and enforce rules beyond their direct membership,[6] or institutional collaboration involving institutions or frameworks that States may not endorse (Scott, 2011). Weak institutional performance can prevent approval or support of activities, including autonomous activities of secretariats that may be perceived as lacking accountability to States, individuals and others.

Responsible oceans use relies on the cumulative impact of constraints and incentives on State and citizen behaviour in favour of sustainable use and conservation. However, the factors raised above compound the strains of bringing together, at the global level, different streams and approaches to conservation and sustainable use that are able to adapt to new knowledge, pressures and threats, while respecting the strengths of different contributors as well as the needs of States and interested communities. This complexity limits what the global system can realistically achieve.

Issues raised by researchers in this field (e.g. Breitmeier, 2008; Fabienne, 2010; Paavola, 2005; Steffek, 2000; Voerten, 2003) inspire the following general criteria that, when present, can help determine the regimes and activities of institutions and organizations that are most likely to be endorsed and implemented:

- respect for the primacy and sovereignty of States;
- institutional de-politicization through norms and procedures valued and endorsed by the international community;
- capacity to reach non-arbitrary decisions;[7]
- decisions based on deliberative and mature discourse with procedural fairness[8] and transparency, allowing affected stakeholders to express views and see how interests and concerns are considered in resolving differences;
- debates and decisions based on objective knowledge that has been subject to review and criticism and are consistent with legal and institutional underpinnings (as opposed to ideologically driven diagnoses or options);
- democratic processes that are not perceived as, or are otherwise *de facto*, dominated by special interests and which include adequate time and opportunities for consultation (e.g. thereby avoiding excuses for non-collaboration);
- participants act in good faith, are open to persuasion and are consistent in their engagement and undertakings;
- decisions take account of the ability of potential parties to implement collective commitments and recognize knowledge/technology transfer and capacity-building needs;

- norms and commitments are based on endorsed policy (e.g. not arbitrary targets); and
- outcomes otherwise meet tests of moral acceptability and distributional justice, and are consistent with other high-level global values, principles, objectives and priorities.

Breitmeier's work (2008) goes further into institutional activities, and notes that the nature of organizational programming helps stakeholders develop their priorities and preferences in solving collective problems or developing new regimes. Participation enhances learning – especially for developing States – and ensures that priorities and preferences are taken into account in planning and decision-making. Breitmeier's criteria for *programming legitimacy* are, foremost, those that:

- reduce uncertainty regarding the knowledge base (e.g. for framing of issues and solutions);
- have causal impact on problem-solving and attaining goals;
- help achieve compliance with regime norms;
- ensure that distributional impacts meet tests of equity or sustainability; and
- are transparent and include participation by civil society.

The first point is of particular importance, both regarding *specialized knowledge* regarding specific ocean uses and *integrated knowledge* for developing common approaches (see Figure 10.1). Options to protect the marine environment and its resources are ultimately about who will bear the costs of adaptation and implementation. States will usually make the development of *consensus* on *knowledge* a precondition for framing and testing the validity of options and supporting resulting policies and regimes.

In this light, Haas (2004) observes that the design of scientific assessment bodies is critical to decision-making processes, including the need for resulting information to be understandable to decision-makers.[9] His criteria for science legitimacy for institutional decision-making are: accuracy (widely perceived to be true); legitimacy (achieved through impartial processes insulated from direct political influence) and saliency (policy relevant and politically tractable). Science is less successful in anchoring solutions when there is: public suspicion over its development or methodology; institutional design questions; lack of clarity on priorities for scientific assessment; and governmental misgiving or unwillingness to cede 'authority' to science if the resulting policy decisions have unfavourable consequences.

Consensus decision-making

Closely linked to the issue of legitimacy is that global decisions are most often taken by consensus, and that States are generally not bound by decisions they explicitly oppose (by breaking consensus, registering reservations in instruments, etc). It is not common for decisions to be taken by vote and, in any case, consensus is sought in advance of decisions in order to bind the largest group possible. However, governance solutions involving complex natural or human ecosystems take time to be developed and accepted. This is especially true at the global level, given the diverse interests involved. Negotiations of new international instruments, for example, typically involve a 'packaging' of rights and obligations, which require agreement on all elements before consensus can be reached.

Original and innovative ideas are usually, and most efficiently, formulated into initial (and often hard-fought) consensus in smaller groups of experts or stakeholders – often under the auspices of global or regional institutions – or groupings of States (e.g. OECD, G-77, EU or other arrangements). These ideas then need to find early allies or advocates in larger groupings (e.g. like-minded States, geo-political alliances or those linked by obligations or agreements such as RFMOs) before moving to broader agreement.

As seen in Figure 10.2, tipping-points to broad global acceptance are very closely linked to legitimacy issues. Support for innovation involves considerable – and often repeated – debate and advocacy. Conservative or dissenting players can also exert pressure against consensus on innovations, reflecting various concerns including political feasibility, legitimacy, conflicting national priorities and geo-political interests or lack of capacity.

Many institutions and stakeholders underestimate the gestation period required for new issues and ideas to develop and diffuse at the global level. This is true of very specific issues within institutions as well as across them, especially for complex issues involving unfamiliar or new approaches; this is especially so if ideas are taking place in institutions unfamiliar to players (or otherwise suffering a perceived lack of legitimacy on some grounds). Slow change is often ascribed, perhaps unfairly, to lack of political will, which may induce conflict or Draconian proposals and external pressure for strong-arm advocacy to force change. In some cases, it may be impractical to expect certain innovations or best practices to be

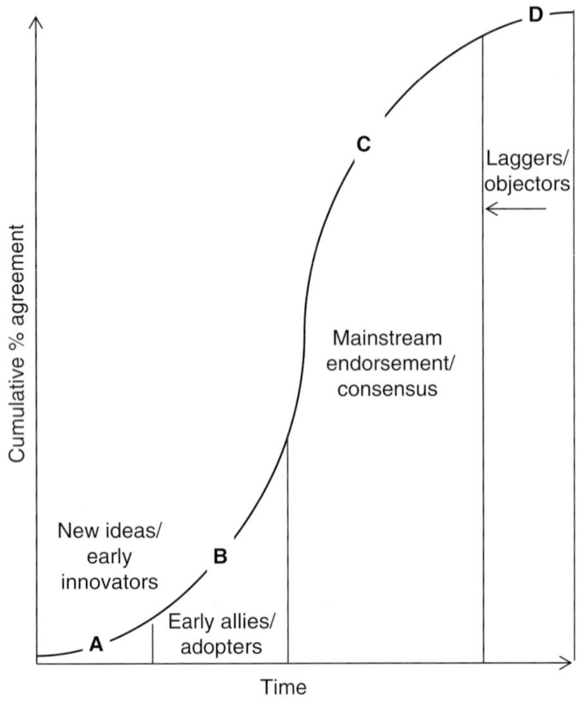

Characteristics:

A) Smaller expert and stakeholder groups (e.g. expert groups, think-tanks, task forces, regional arrangements, NGOs, Secretariats): issues-oriented; knowledge and capacities may trump inclusiveness; objective, outcomes-oriented, evidence-based. **Tipping point is To engage**: crisis, risk analysis, assessment.

B). Larger groups and alliances or interested communities (e.g., regional groupings, OECD or other alliances (G-77), user groups/ stakeholders, IGO or specialized agency commitments): may have obligations in arrangements with innovators or other allying characteristics; capacities and expertise to adopt and implement; some rationale or pressure for joining. **Tipping point is To agree**: advocacy, deliberation, dissemination of analysis, strategic alliances

C) Broad political feasibility as well as agreed conceptual desirability: Initiatives align with or do not threaten broad goals, including human dimension of mainstream groupings; meets due-process tests; equitable outcomes and perceived fair burden-sharing; commitments to capacity-building and expertise; applies across diverse contexts; or part of effective 'packaging' with other initiatives. **Tipping point is To agree:** advocacy, deliberation, political momentum, Issues alignment, packaged negotiations

D) Dissenting, conservative or isolated players; disagreement with fundamental diagnoses, domestic political pressures priorities or perceptions; geopolitical concerns or a strategic vulnerability; threats to perceived sovereignty; perceived or legitimate lack of capacities; vetos in alliances prevent adoption by others. **Tipping point is To agree or abstain**: advocacy, cohesiveness with alliances, resolution of equity or geo-political concerns, capacity building.

Figure 10.2 Building to consensus.

reflected in consensus declarations or resolutions,[10] which may be drafted in higher-level general language to garner agreement.

Potential strengths and weaknesses of global institutions, organizations and bodies

Potential strengths

Irrespective of specific institutional features, tremendous potential exists for institutions, organizations and bodies operating at the global level to improve coherence and alignment of fisheries and biodiversity agendas and, thus, ocean outcomes. This potential results from the following characteristics shared by many global bodies:

- *state-centric organizations* that respect the primary role of the State in governance decisions;
- membership is *universal* and representative, so bodies can articulate global goals;
- discussions *unite diverse perspectives* and agreed solutions are resilient in diverse contexts;
- solutions can *internalize global spillovers* and encourage universal application of rules, policy, standards, tools and evaluation (and constituencies can help unite divergent interests);
- possibility of aligning and integrating issues with *other global priorities;*
- *high levels of political influence*, commitment and visibility (e.g. UNGA) and traditional tools (e.g. resolutions, declarations) to establish priorities, set agendas or course corrections, mobilize commitment and funding, set expectations and accountabilities;
- links across institutions and processes allow for *coordination and multiple entry points* for common issues;
- access to *financial resources* to implement decisions and mobilize reforms (e.g. operating budgets of large institutions, large-scale special finance mechanisms such as Global Environmental Facility or GEF, specialized trust funds, international aid flows and leveraging and pooling of voluntary resources from governments and civil society);

- global conventions and their principles, obligations and norms provide coherence to regional, sub-regional and national implementing legislation, customized standards and guidelines, management and compliance tools and incentives mechanisms, monitoring, review and capacity building;
- *compliance* provided by enforceable international rules (e.g. WTO) and dispute resolution processes (e.g. ITLOS) helps ensure a *global 'level playing field'*;
- *permanent secretariats* with access to high-level multi-disciplinary experts and links to other organizations can establish large-scale deliberative processes and networks and enhance cooperation and coordination;
- capacity to launch large-scale *review processes* and *coordination mechanisms* to improve implementation, harmonization and compliance; and
- pathways to *publish and disseminate* information that will attract global attention, including that of opinion-makers.

Potential weaknesses

Institutions and bodies operating at the global level also face corresponding weaknesses (the exceedingly cogent analysis by Najam et al., 2006 addresses many of these issues):

- state-centric institutions can *marginalize* the contributions of important non-State actors;
- decision-making processes may be predisposed to *negotiating outcomes*, resulting in ever-increasing decisions and agreements;
- universal membership slows consensus on innovations and risks *lowest-common-denominator outcomes*, principles or processes rather than action;[11, 12]
- solutions that need to meet the diversity of needs, preoccupations, capacities and priorities are usually *less progressive or innovative* than what is originally proposed or needed;
- decision-making is increasingly subject to *bloc negotiation* (e.g. G-77, regional groupings, EU)[13] which reduces open policy debate and risks negotiation rigidities;
- effectiveness of decisions depends on implementation, which may require *degrees of freedom* for willing regional, sub-regional or national implementation; however, the lower the implementation flexibility, the more elusive may be consensus (e.g. WTO);
- high-level political commitments or targets may not be grounded in *technical feasibility*, capacities and enablers or buy-in, leading to implementation gaps;
- *proliferation* of processes, meetings and associated activities (e.g. reporting, coordinating mechanisms) can overwhelm States (especially developing States), diverting significant resources and capacity from implementation;
- *jurisdictional overlap* among bodies and institutions can lead to fragmentation and inconsistency in outcomes addressing similar issues;
- *global priorities*, such as sustainable development, poverty alleviation, human rights, security etc., can overwhelm or censor sector-specific or technical issues and solutions;
- breadth of institutions with different disciplinary mixes, constituents and experts can lead to *'forum-hopping'* of issues seeking sympathetic players, reducing the contesting of ideas;[14]
- global legal frameworks and principles are contextual and difficult to update, which can create *resistance to integrated solutions*[15] or leave innovations without solid legal foundation (e.g. high-seas MPAs); and
- *assumptions of resourcing and capacity of large organizations are often unrealistic*: global institutions face non-discretionary programming and constrained budgets, often with little room for new normative work, and voluntary funding is variable; tight budgets and strict conditions on voluntary funding capacities of developed States are often overestimated.

Implications

Efforts to address global governance challenges to improving ocean sustainability need to focus on the right aspects of the issue. The issue is not simply (or even mainly) a question of the complexity of oceans governance, but also the role, personality and behaviour of the institutions and organizations involved in governance improvements and how they bring about a willingness to change. The observations noted above can help decode the dynamics and processes of many global (or, in their own scale, regional or sub-regional) debates on ocean governance, far beyond the parameters of Chapter 11.

There are important differentiated institutional roles to be played in a unified and interdependent marine governance system, some contributing to specialized functions and others to integrating functions. These interdependent roles need to be transparent and robust for the system to perform successfully. In a *unified* and *interdependent* marine governance *system*, all players and their primordial organizations can, and must, be part of the solution to better governance. This can be achieved if the system 'plays to strengths', that is, draws upon the comparative advantages of relevant bodies, organizations and stakeholders. In the final analysis, institutions that encourage integrated approaches and willing engagement of diverse States and stakeholders will be more important to sustainable outcomes than the structure or complexity of the global marine governance system itself. This contrasts with top-down global governance proposals, which raise complex issues regarding State sovereignty and would not likely, without strongly coercive approaches, secure agreement or make a real difference to marine outcomes.

Generally speaking, recommendations and decisions will only be adopted and implemented by States and their key stakeholders when there is sufficient buy-in to the processes that generated them. This is true also of broad communities of interest, such as fishing and nature-conservation communities. In addition, there is a need for a common basis of understanding and the framing of issues and solutions, one of the most important foundations to decision-making, within and across all institutions and organizations. The design of the knowledge-creating processes and access to their outputs is important to the utility and credibility of governance.

Notes

1. There are 166 parties to the United Nations Convention on the Law of the Sea (UNCLOS), 80 parties to the United Nations Fish Stocks Agreement (UNFSA) and 193 parties to the Convention on Biological Diversity (CBD). There are also 192 members of the Food and Agriculture Organization (FAO).
2. Collective ownership means: (1) interdependence among users; and (2) the combined effect of national policies and international conventions.
3. Interdependence creates conflicts over who gets to consume or use particular goods or resources (Paavola, 2005). Pluralism among users implies diverse views on governance solutions.
4. Both traditional and emerging (market-based) tools (e.g. eco-labelling and other certifications).
5. For example, the World Trade Organization (WTO) or the International Monetary Fund (IMF) rules can have significant economic impacts on States.
6. For example, limited-membership RFMOs that enforce rules against non-members.
7. Deliberative decision-making helps provide reasons for and against options and generates alternatives. Notwithstanding procedural fairness, deep-seated arguments for and against options will always remain. The extent of shared values among players will help determine the feasibility of emerging solutions.
8. Paavola's test is: (1) whose *interests* are recognized and how, in processes of planning and decision-making; (2) who can *participate* in planning and decision-making, and how; and (3) the effective *distribution of power* in planning and decision-making.
9. Najam et al. (2006) noted that the scientific and technical bodies of many multilateral environment agreements (MEAs) have become politicized over time.
10. For example, UNGA resolutions which may seem conservative with respect to innovations not already endorsed by other global-level institutions or carried out in UN processes.
11. Consensus wording in one process may be imported into others to help broker agreement, sometimes out of context, which can impede innovation and dynamism.
12. International rule-making aiming to harmonize *domestic* standards among States with links to sovereignty and domestic processes can make agreement very difficult to achieve.
13. The RIO+20 (20–22 June 2012) outcome formalized the importance of constituencies, alongside States.
14. Different global organizations favour delegations and experts from different ministries and civil society organizations (Annex 2).
15. Some States resist discussing IUU fishing in relation to trans-national organized crime, such as smuggling or terrorism, to avoid mixing solutions and approaches in the security framework with the fishing sector (Ridgeway and Badji, 2008).

References

Breitmeier, H. (2008) *The Legitimacy of International Regimes*. Fernuniversität, Hagen, Germany.

Fabienne, P. (2010) Political legitimacy. In: *Stanford Encyclopaedia of Philosophy* (ed. E. Zalta), Summer 2010 edn. Available at http://plato.stanford.edu/archives/sum2010/entries/legitimacy (accessed February 2014).

Haas, P. (2004) Scientific expertise, a question of legitimacy. In *Proceedings of IDDRI, International Environmental Governance Conference*, March 15–16 2004, Paris.

Najam, A., Papa, M. and Taiyab, N. (2006) *Global Environmental Governance, A Reform Agenda*. International Institute for Sustainable Development, Winnipeg, Manitoba, Canada.

Paavola, J. (2005) *Environmental Conflicts and Institutions as Conceptual Cornerstones of Environmental Governance*

Research. Centre for Social and Economic Research on the Global Environment, University of East Anglia, UK.

Ridgeway, L. (2009) Governance Beyond Areas of National Jurisdiction: Linkages to Sectoral Management. *Océanis* **35**(1–2), 245–260.

Ridgeway, L. and Baji, P. (2008) Report on the Work of the United Nations Open-ended Informal Consultative Process on Oceans and the Law of the Sea at its Ninth Meeting, Summary of Discussion on Maritime Security and Safety. Report to the General Assembly, 63rd Session of the General Assembly, New York.

Ridgeway, L. and Rice, J. (2010) International organisations and fisheries governance. In: *Handbook of Marine Fisheries Conservation and Management* (eds Q. Grafton, R. Hilborn, D. Squires, M. Tait and M. Williams), pp. 485–504. Oxford University Press, New York.

Scott, K. (2011) International environmental governance: managing fragmentation through institutional connection. 2010 Four Societies Conference: International Law in the New Era of Globalization. *Melbourne Journal of International Law* **12**(1), 177–216.

Steffeck, J. (2000) *The Power of Rational Discourse and the Legitimacy of International Governance*. European University Institute papers, Robert Schuman Centre for Advance Studies. RSC 2000/46.

Voerten, E. (2003) The Political Origin of the UN Security Council's Ability to Legitimize the Use of Force. Working Paper. Annual Meetings of the American Political Science Association, Portland, Oregon.

Weiss, T. (2009) What Happened to the Idea of World Government. Presidential address, 50th Anniversary of the International Studies Association, Feb 2009. New York.

Chapter 11

Global level institutions and processes: Assessment of critical roles, foundations of cooperation and integration and their contribution to integrated marine governance

L. Ridgeway

Formerly with Fisheries and Oceans Canada, Ottawa, Canada

Abstract: In light of frameworks provided in Chapter 10, this chapter[1] provides the overall global policy context in which fisheries and nature conservation issues and debates nest, describes changing dynamics of global debates and decision-making, describes and assesses key global-level institutions and processes favoured by both fisheries and conservation communities and describes new initiatives developing in the area of integrated knowledge foundations and how they contribute to an integrated system of marine governance. The increasing importance of linking conservation and fishing priorities to higher-level global priorities is emphasized, particularly with regard to the emerging importance of the global sustainable development agenda. In a unified and interdependent marine governance system, all players and their primordial organizations can and must play robust roles as part of the solution to better governance.

Keywords: global governance; global institutions; global processes; global roles contributions; summits; United Nations agencies; G77; voting blocs; institutional coherence

Introduction

Chapter 10 outlined several prisms through which one can assess the role and effectiveness of global institutions involved in fisheries and oceans conservation. Indeed, the prisms can apply to institutions at the regional or sub-regional level, especially where consensus must be forged from disparate views. This chapter describes and assesses key relevant global institutions and processes and, in doing so, makes explicit or implicit reference to these frameworks.

The current fabric of key global institutions in marine governance

Annex 2 of this book describes key global institutions, organizations, bodies and processes that address oceans governance (and, more specifically, fisheries and biodiversity issues) and paints a high-level picture of the nature and diversity of global marine governance issues being addressed. It covers: key UN-based institutions, programs and processes; MEA-established bodies; key non-UN or

Governance of Marine Fisheries and Biodiversity Conservation: Interaction and Coevolution, First Edition.
Edited by Serge M. Garcia, Jake Rice and Anthony Charles.

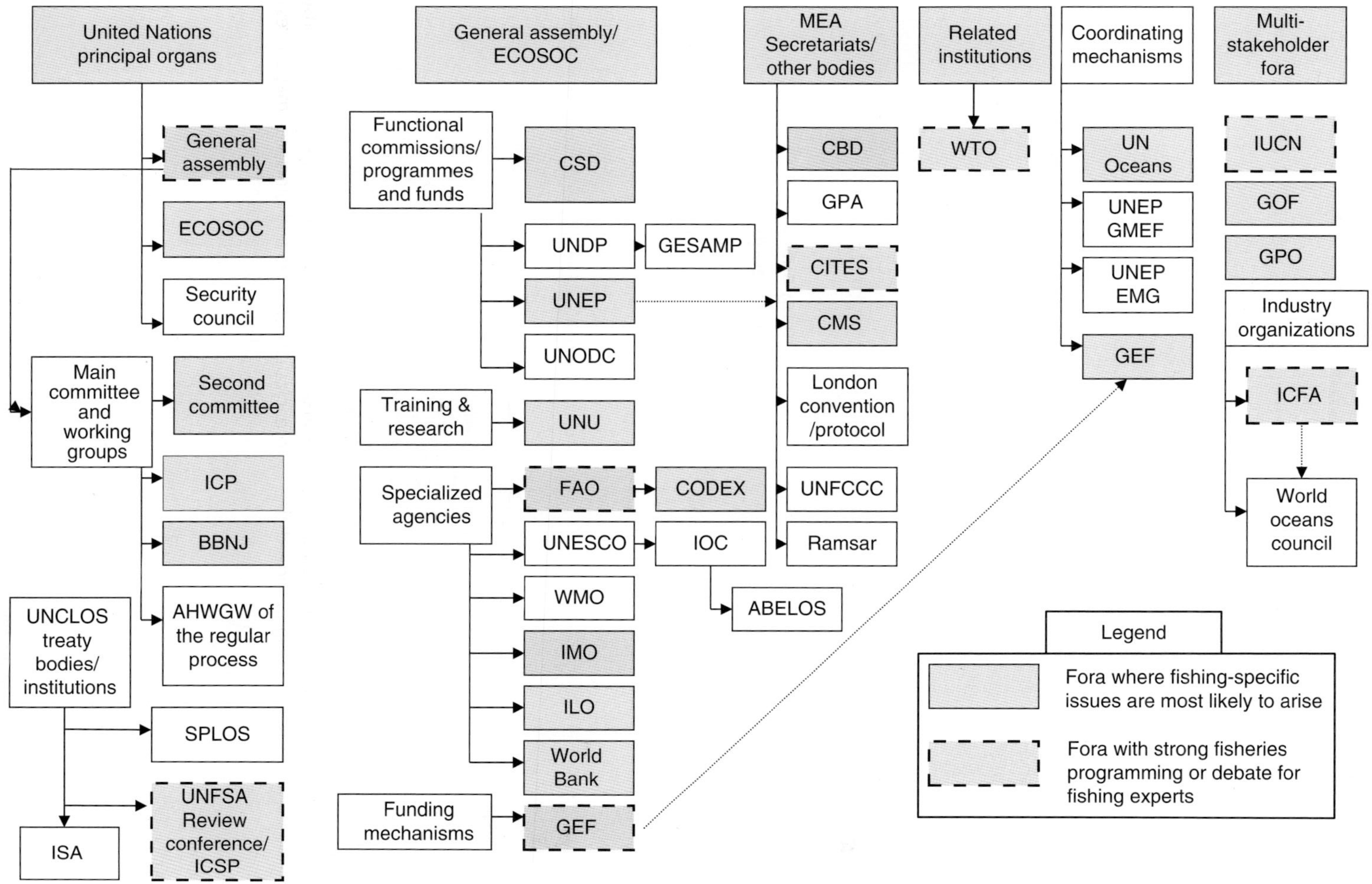

Figure 11.1 Architecture of global organizations.

non-environmental institutions dealing with fisheries or biodiversity; a key global financial mechanism; some coordination mechanisms; and select NGO organizations that are visible and active in global processes. It also describes typical delegates, how IGOs and NGOs are engaged, and primordial areas of focus, that is, prisms through which issues tend to be addressed. The Annex can be read horizontally (within organizations) or vertically (within issues).

An overall architecture for Annex 2 is shown in Figure 11.1. Fishing issues most often arise in organizations with shaded boxes. The few fora where fisheries interests can or do engage directly are shown in dashed boxes. A general UN system organizational chart can be viewed at www.UN.org/en/aboutun/structure/org_chart.shtml.

Assessment

Understanding links between fisheries and biodiversity issues and global priorities

In practice, what may matter most at the global level is how governance approaches in the biodiversity/conservation and fishing communities 'mainstream' into other global agendas. Participants in debates in any global forum, whether on biodiversity, oceans or fisheries, need to be aware of – and be able to manage and integrate – linkages across a wider range of agendas, some of which may be unfamiliar to them.

The 'nesting' and intersection of these issues is roughly illustrated in Figure 11.2. Fishing issues nest in marine biodiversity issues, which nest within oceans issues and, more broadly, within sustainable development. Sustainable development issues intersect with other identifiable agendas such as security, health and human rights. These overlapping agendas often affect the ability to generate consensus on fisheries and biodiversity issues. Accordingly, attempts to 'compartmentalize' debate on conservation or biodiversity issues may find little traction, especially at the UN level, unless they can link to broader agendas.

Two specific and related issues or trends affecting global debates on oceans, biodiversity and fisheries are: (1) the emergence of the sustainable development agenda; and (2) the growing role of blocs used to organize a common voice and force in negotiations.

The emerging sustainable development agenda

Najam (2005) describes how 'global environmental politics has been transformed into the global politics of sustainable development'. This is demonstrated in the evolution of the environmental agenda since 1972, starting with the UN Conference on the Human Environment in Stockholm, which launched environmental issues as a global priority.

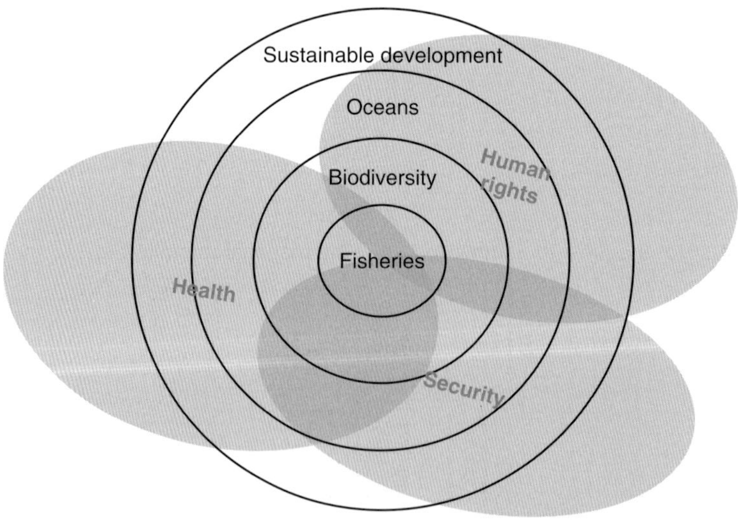

Figure 11.2 Overlapping agendas at the global level.

Najam argues that developing States saw the 1972 environmental 'problematique' as illegitimate, as it threatened to sabotage the development aspirations of the South. These strains persist.

By 1992 the global dialogue had shifted to 'sustainable development', the focus of commitments made at the UN Conference on Environment and Development (UNCED). The goal of developed States at UNCED was to design effective environmental institutions and instruments, while developing States were preoccupied with the legitimacy of these institutions and instruments. These States even contested capacity-building mechanisms (such as the Global Environment Facility or GEF) on the basis of their role in supporting environmental over distributional outcomes (Najam, 2005).

The UN Commission on Sustainable Development (CSD) (Annex 2) emerged from this process as the main body to advise the UN on progress on Agenda 21, the Rio Declaration on Environment and Development and the sustainable development process. The United Nations Environment Programme (UNEP) also emerged from UNCED, strongly supported by developing States. In addition, the Convention on Biological Diversity (CBD) opened for signature, emphasizing the conservation and sustainable use of biological diversity and, notably, the fair and equitable sharing of benefits. The continuing strain over the *legitimacy* of environmental governance versus its *effectiveness* has continued to affect discourse through the years (Najam, 2005) and is still apparent in global-level debates on marine governance.

By the 2002 World Summit on Sustainable Development (WSSD) in Johannesburg, the interests of both developed and developing States were focused on 'implementation'. However, a new polarity emerged between States focused on implementation for ecological outcomes and other States focused on implementing a sustainable development agenda and the role of developing States in environmental governance.

More recently in June 2012, the Rio + 20 Summit focused on the green economy in the context of sustainable development and poverty eradication and institutional frameworks for sustainable development (IFSD) and international environmental governance, including for oceans. Conservation outcomes are firmly embedded in focal socio-economic development aspirations, culminating in commitment to collective development of new Sustainable Development Goals under the auspices of the UN.

The mainstreaming of environmental governance issues into a sustainable development framework brings potential for more unifying policy frameworks and potential 'buy-in' to issues and solutions. Conversely, it can lead to the 'editing' or censoring of sector-specific or technical issues in global governance debates when they are perceived to be inconsistent with sustainable development aspirations of some participants.

The growing role of blocs, such as the Group of 77 and the European Union

An associated development is the growing role of 'blocs' in global debates, including the Group of 77 (G77) and the European Union (EU). These groups have *de facto* served to 'level out the playing field' in global negotiations, while making it more difficult for individual States to influence agendas in institutions, organizations, bodies and processes.[2]

The G77 originated as a loose caucus of 77 States in 1964 at the launch of the UN Commission on Trade and Development (UNCTAD). Its genesis was a perception of developing States being disempowered and disenfranchised by the global economic system (Najam, 2005) and of a multilateral system that lacked commitment to 'southern' issues. The creation of UNCTAD as a permanent UN body for trade issues gave the G77 a forum and institutional framework with which to influence international governance generally. Now at approximately 130 members, the G77 strongly influences the dynamics of global debates, and the evolution of multilateralism in the UN (Behnam, 2005). The G77, bolstered by China's alignment with its positions, has emerged as a key negotiating machinery of developing States, many of which otherwise lack the capacity to participate effectively in negotiations in a number of global fora (e.g. FAO, UNEP, UNESCO, IMF, World Bank and GATT/WTO).

The role of the G77 in contesting the positions of developed States, leveraging north–south cooperation or enhancing south–south cooperation has evolved alongside global issues. The modern manifestation of past tensions over sustainable development involves: implementation of international commitments and the harmonization of environmental, trade and financial standards and approaches; inequities resulting from the international economic system; concerns over inequities being institutionalized in proposed governance systems or approaches; and the need for capacity-building and the transfer of technology.

The G77 has affected global debates on oceans, biodiversity and fisheries, particularly with regard to the following:

- sensitivity to lack of access to global oceans resources, including as coastal States with weak capacity or other

impediments to exploiting even adjacent marine resources (including in some cases a perceived lack of realized benefits from bilateral access agreements) and including other benefits of biodiversity (e.g. marine genetic resources) which are believed to have been 'expropriated' through high-seas freedoms and intellectual property rules;
- related concerns over whether developing States are bearing the costs of conservation and research for the benefit of developed States;
- sensitivity to approaches perceived to erode State sovereignty and development choices;
- concerns that developed States' interests determine the issues and solutions dominating global debate;
- resistance of developing States to market-based mechanisms and technical or certification standards that may affect economic opportunities and reduce access to markets;
- sensitivity of developing States to procedural fairness and resistance to legitimate knowledge, dialogue, innovations or best-practices from non-inclusive fora or processes;
- favouring of fora and instruments grounded in principles of resources as a common heritage of mankind (e.g. ISA) and strong preference that this principle apply in law or in spirit to other institutions and conventions;
- favouring of fora focused on sustainable development (e.g. UN fora and UNEP) in which to articulate their needs, establish agendas and seek solutions and capacity-building;
- acceptance of governance and management solutions conditional on their application to developing States (e.g. small-scale fisheries);
- need for support on capacity building and technical assistance as a precondition to taking on new commitments and obligations; and
- preference for soft-law tools enabling State flexibility in implementation regarding conservation and management, but for hard-law instruments to address issues of benefit sharing and capacity building.

A number of geographically based sub-blocs have developed within the G77. Many of these blocs are supported by framework or constitutive instruments, including in the context of the UNEP Regional Seas Programmes.

Another key 'bloc' is the European Union (EU), which now has enhanced observer status at the UN. The EU has a different framework and structure than other blocs, including a complex division of competencies with member States, which can confuse non-EU States. For example, the European Union has exclusive competence over the conservation of marine biological resources contained in the EU Common Fisheries Policy. The EU may be a member of an international body or a party to an international treaty in its own right, with the positions led by the European Commission (EC). However, the EU shares competency with member States on the environment as well as some aspects of fisheries, in which case the EU voice is still a collective one (represented by the State that has the current – rotating – EU Presidency), but individual States may also speak on the matter. Because of the complicated procedures underlying the development and political validation of collective European positions, flexibility in positions during debates and negotiations may be affected.

The impact of bloc voting and influence

The growing role of bloc negotiation and voting has affected the global agenda, processes and outcomes. Constructively, blocs unite the divergent views of individual States, thus simplifying global debates and adding efficiencies to global decision-making. Moreover, blocs can also add collective strength to the views of individual weaker, potentially marginalized, States. There is however increased risk of protracted debates and negotiations due to inflexibility in positions, as the carefully forged consensus and negotiating mandates of blocs need to be preserved. Negotiation schedules must also accommodate time for bloc coordination and adjustment of positions, and innovative ideas of individual States or delegates that could break negotiation deadlocks may be hampered by bloc interventions, which may limit individual delegations to supportive roles.

Intra-bloc transparency is also often weak, affecting open deliberative dialogue in global discussions. Real policy debates important to non-bloc members may disappear behind closed doors in bloc coordination meetings. Reversal of positions and compromises from informal discussions, after bloc coordination, may also lead to accusations of lack of good faith. Flexibility in negotiations has thus diminished at the same time as the complexity of issues being debated has increased. These issues link directly back to legitimacy considerations in Chapter 10.

The importance of tactics as well as dialogue

Debates focused solely on ecological aspects of conservation and sustainable use – whether for fisheries or biodiversity – will no longer find traction in global processes without meeting multiple agendas. At the same time, formal global-level debate on these issues has become more rigid, polarized and complex. This gives rise to the tactical importance of 'issues packaging'.

For example, some developed States are attempting to promote a new implementing agreement under UNCLOS for *biodiversity conservation*, on the basis of perceived gaps in existing instruments. Failing to build consensus for such a diagnosis or instrument, these States have now aligned with developing States who have encountered resistance in clarifying principles on marine genetic resources beyond areas of national jurisdiction, including the *sharing of benefits*. A 'packaging', as opposed to merging, of these interests has increased support for the launching of a process by UNGA on the legal framework for the conservation and sustainable use of marine biodiversity in areas beyond national jurisdiction. The interests of States remain fundamentally divergent, however. On one hand, many developed States continue to oppose any reinterpretation of high-seas freedoms in UNCLOS. For developing States however, the omission or lack of clarity of common-heritage principles in UNCLOS needs to be rectified (Ridgeway, 2009).

International organizations and fishing stakeholders

Chapter 10 and the section above enable a better understanding of roles and issues at play in various global institutions.

The primordial importance of the FAO

The FAO – a specialized agency of the UN system – is the most important global organization for fishing interests (e.g. policy, management and industry). This is clear from both debates and in the pre-eminent position FAO is given in global declarations and resolutions regarding fishing. States participate in FAO bodies and processes to: review programs of work in fisheries and aquaculture; enhance understanding; frame issues and priorities for normative reform and capacity building for responsible fishing and aquaculture; and adopt instruments, guidelines and spur collective action. The FAO provides a more homogenous community of interests, approach and literacy on fisheries developments and issues than in broader fora. The FAO Secretariat itself engages in considerable expert work, and sponsors important projects and programmes.

While the FAO is promoted by its members as the most legitimate global fisheries forum, there is resistance to granting it any formal oversight role as a 'top-down' management forum. Views even differ on FAO's role in reviewing implementation by reporting on the quality of fisheries management programs or policies of individual RFMOs or States. (For example, the G77 considers that inadequate capacity-building has left them exposed to unfair criticism for gaps in implementation.) In response, the FAO has undertaken overall strategic reform to increase its effectiveness in both capacity-building and normative work across its entire mandate.

The Code of Conduct for Responsible Fisheries[3] is the FAO's key framework instrument. It outlines principles and requirements for sustainable fisheries, which are further developed in guidelines and four International Plans of Action.[4] Most States see the CCRF, a soft-law tool, as complementary to obligations in binding instruments (e.g. UNCLOS and the United Nation Fish Stocks Agreement or UNFSA). However, non-parties to these instruments often promote the CCRF as an alternative.

Initially, the FAO Secretariat outlined best-practice 'technical' guidelines for aspects of the CCRF. Implementation issues have become more complex over time, involving interactions with biodiversity concerns and/or economic and trade issues sufficiently important to States that such guidelines increasingly command sophisticated State-to-State negotiations.[5] New guidelines and agreements typically require four formal FAO stages: a small consultation with *experts* nominated by the FAO in their personal capacity; an ensuing 'technical consultation' with State-to-State negotiation by *specialists*; *diplomatic* adoption at the FAO's Committee on Fisheries (COFI); and political endorsement at the level of the FAO Council. State-to-State negotiations may intermingle fisheries issues with broader global goals (e.g. food security, sustainable development and human rights), which can complicate the development or adoption of even technical guidance.

Industry representatives participate frequently in FAO discussions as members of State delegations or as industry-association observers. Conservation experts often participate also, as members of State delegations or ENGOs observers. Both perspectives have contributed, formally and informally, to understanding issues and developing solutions. However, these groups cannot actively participate in decision-making, except indirectly by influencing the positions of States. ENGOs have not historically raised broader environmental or biodiversity issues in FAO meetings, preferring to reserve them for other fora where the issues may have more traction and the proponents more leverage.

There has been strong collective progress at FAO on the CCRF since its adoption, especially regarding illegal, unreported and unregulated (IUU) fishing (cf. Chapter 4).[6] Significant new efforts have been made to combat IUU fishing through innovations affecting economic incentives, such as collective port State measures against vessels on IUU lists. However, reaching agreement among States on such new tools is often difficult due to concerns over distributional impacts, implications for market access and free trade aspirations, as harmonization of domestic approaches may restrict State economic and trade prerogatives. Moreover, developing States are increasingly reluctant to take on new obligations without reciprocal commitments on capacity-building to help ensure compliance and cover incremental costs. Part of the bargain for new and possibly contentious CCRF guidelines is the need to achieve balance with guidance and new policy regimes for small-scale fisheries (cf. Chapters 22–25).

In the context of capacity development, some of the 'battle ground' between fisheries and conservationists is now in the EEZs of developing States. Interests supporting the greater use of MPAs may target States with lower capacity but high needs to promote closures as the 'tool of choice' for managing fisheries, especially for States with significant vulnerable ecosystems (e.g. coral reefs). Some area-based management tools (e.g. no-take zones) are easier to put in place than the full suite of capacities needed to manage sensitive species in an ecosystem approach to fisheries (cf. Chapter 4). There is some risk that over-reliance on such tools may impede development of capacities needed to manage fisheries, which risks creating a dichotomy of closed areas and overexploited areas, rather than a unified system using a mix of appropriate management tools.

Notwithstanding progress on responsible fishing, the FAO still suffers the challenges of any 'global institution', as described in Chapter 10. Development of, and compliance with, new global fishing norms is neither immediate nor without conflict, leading some civil society organizations to dismiss the sincerity of fisheries reforms and argue for new oceans governance to force change.

Only recently have there been discussions among fishing stakeholders at the FAO on the role of fisheries in broader marine governance and other fora.[7] The FAO Secretariat could play a major role by educating fishing experts on wider agendas, helping to reduce distrust in fishing communities about the nature of discussions in other fora and helping to ensure that fisheries stakeholders views are well represented and influential in broader debates, internationally and nationally. FAO could benefit from a stronger mandate to represent fisheries issues in international debates in non-fisheries fora. In this context, the approach of fishing States in broader ocean governance discussions can be 'defensive' rather than supportive, especially when those discussions criticize fishing or downplay or ignore real developments in fisheries management. In addition, CCRF implementing guidelines could be updated to take account of broader contextual developments (FAO, 2012; Ridgeway and Rice, 2010).

As an example of institutional mistrust, States have not agreed on cooperative efforts between the FAO and CBD to build a common archiving system for information related to VMEs and EBSAs (cf. Chapter 14) that would bring together data and analysis to build a common basis for understanding oceans issues. Resistance in FAO may be due to underlying suspicions that the CBD favours specific conservation tools, such as no-take zones, which would severely limit fishing opportunities. Such cooperation was also initially slow to be approved by the Parties to the CBD, although this has changed more recently.[8] Some States have also expected 'State oversight' in FAO of cooperative and integrative efforts with other types of institutions (FAO, 2012), reflecting such concerns as links of such activities to underlying instruments not having universal membership, such as UNFSA or CBD (cf. Chapter 10).

In light of the above, the relevance and role of fisheries issues in broader oceans and biodiversity governance tends to be left to other fora. However, fisheries stakeholders may have a smaller voice in such fora with which to educate other stakeholders on issues with planned tools, regimes or approaches to fisheries, especially because lead delegates in such fora often represent environment or foreign affairs departments (and the fisheries sector may be myopic about the need to be at these debates). However, improvement in oceans governance needs the best elements from both fishing and conservation approaches, shared in common fora.

Fisheries interests and the United Nations

After the FAO, the second institution of choice for fishing interests is the United Nations. The UNGA addresses fisheries issues in the context of its annual resolutions on oceans and the law of the sea and on sustainable fisheries. The general legal framework for discussions on all activities in the oceans and seas is UNCLOS. One of its Implementing Agreements, UNFSA, provides the legal regime for the conservation and sustainable use of straddling and highly migratory fish stocks. States Parties to UNFSA address implementation issues as relevant preparatory steps for the Review Conference on the Agreement and, recently, philosophical and legal issues preventing wider participation in the Agreement in the Informal Consultations of States Parties to the United Nations Fish Stocks Agreement (ICSP).

ICSP meetings deal with fishing issues directly, and are open to participation by IGOs, NGOs and ENGOs as observers (see Annex 2). However, delegates to these meetings are usually representatives from Permanent Missions and Foreign Affairs departments. Some delegations are supplemented by fisheries policy experts from State capitals, who can link discussions to other agendas including at the FAO and other regional and global fisheries processes relevant to implementation of UNFSA obligations.

For some States Parties, ICSP has fallen short of its potential as a forum for substantive regular discussions on fisheries governance. The UNFSA Review Conference has been a major agenda-setting opportunity to better address continuing problems in the conservation and management of fish stocks. However, legal and policy issues have impeded endorsement of the outcomes of these meetings by some State non-parties. For example, RFMOs are the primary vehicle under UNFSA for cooperation among States in the conservation and management of fish stocks. However, some State non-parties argue that RFMOs have closed membership, yet make decisions that affect non-members (e.g. IUU lists, port State measures). These philosophical or legal impediments in relation to UNFSA have led some States to actively oppose its endorsement in global fora. Significant resistance was also encountered in efforts to incorporate the collective recommendations from the Review Conference into the UNGA resolution on sustainable fisheries (see 'Consensus decision-making' in Chapter 10), even recommendations that had been tempered to accommodate the interests of some State non-parties.

The debates in ICSP and the Review Conference have been complicated by issues relating to: procedural fairness and the participation of States non-Parties; State sovereignty and broader geopolitical dynamics; and the need for increased capacity-building and links to sustainable development. These complicated dynamics have slowed progress on collective means to address critical gaps in responsible fishing. This is sometimes misinterpreted as a lack of broad political will of the international fishing community, resulting in accusations and defensiveness between fishing and nature conservation communities.

Fisheries issues are also addressed in informal UNGA processes, namely the Open-ended Informal Consultative Processes on Oceans and the Law of the Sea (ICP) and the *Ad Hoc* Working Group on Biodiversity Beyond Areas of National Jurisdiction (BBNJ). These processes, open to broad participation, rely on approval from the UNGA to meet on a specified agenda so that processes and agendas do not get too far ahead of global consensus. Other standing UN bodies, some little known by fisheries interests, are also increasingly dealing with oceans and fisheries issues (e.g. UNGA Second Committee, ECOSOC; see Annex 2).

The ICP was established to facilitate the annual review by the UNGA of developments in ocean affairs and the law of the sea. Its deliberations are anchored on the annual report of the Secretary General[9] on oceans and the law of the sea, with special topics of in-depth focus each year.[10] The emphasis of the ICP is on identifying areas where coordination and cooperation at the intergovernmental and inter-agency levels should be enhanced.

The ICP uniquely exposes delegates to a range of experts in the selected focal area through state-to-state plenary sessions and special panels of experts drawn from diverse IGOs, academics experts, NGOs, ENGOs and industry groups. All participants, including observers, fully participate in panel-discussion portions of the meeting on an equal footing to States. Until recently, collective recommendations were negotiated by States in these meetings and later incorporated into UNGA resolutions, but difficulties in reaching agreement rendered this increasingly impractical for contentious or new issues. Fisheries issues have arisen in every meeting in this forum since its inception. Issues regarding oceans governance are also prominent, irrespective of the topic. Recently it has not even been possible to secure States' consensus on the ICP agenda without the oceans issues being examined within the context of sustainable development.

BBNJ was established to study issues relating to the conservation and sustainable use of marine biological diversity, including fishing issues, beyond areas of national jurisdiction. It was not possible to create consensus on establishing BBNJ without ensuring a package of issues in areas beyond national jurisdiction would be addressed. It is not a single-issue process. Although originally envisioned as a single meeting in 2006, it has since met four times. Among a range of issues, various rationales for a new implementing agreement under UNCLOS have been featured in these discussions. Delegates to BBNJ have been representatives from Permanent Missions, Foreign Affairs, legal advisors and international policy representatives, depending on the capacities and topic of the meeting, with relatively fewer fisheries experts.

Bottom trawling and the impact of fishing on ecosystems and fish stocks have figured prominently in these UN fora, and in moving towards more unified governance debates. Starting around 2002, ENGOS used the ICP as a UN forum to raise concerns about impacts of bottom fishing on vulnerable marine ecosystems (VMEs) and sustainability of fish stocks. After two years of difficult ICP debate, in 2004 the UNGA decided to conduct a review in 2006 of the actions taken by States and RFMOs to address the impacts of destructive fishing practices, including bottom trawling having adverse impacts on VMEs. Following lengthy debate on fisheries management and VMEs, the 2006 UNGA resolution on sustainable fisheries (61/105) was adopted by consensus, and represented a major regime-shift in linking fishing to environmental issues.[11]

The commitments in resolution 61/105 focused on protecting areas of vulnerability and addressing threats from bottom fishing, rather than banning specific fishing gears, and anchored a new set of FAO guidelines on deep-sea fishing.[12] The resolution encouraged other important developments, through the work of jointly recognized experts and collaborative efforts of secretariats, in bringing the biodiversity and other communities together in efforts to objectively define vulnerable areas (cf. Chapter 14). The framework in resolution 61/105 has also been adapted to other activities in other fora, such as the development of regulations on prospecting and exploration for polymetallic sulphides by the International Seabed Authority.

These developments would not likely have occurred but for: initial endorsement by the UNGA of the new framework for management regimes for bottom fishing; subsequent efforts of the policy, legal, fishing and biodiversity communities to address the technical aspects of bottom fishing and render the debate less value-laden; increased focus by States and RFMOs on ecosystem approaches and areas needing protection from specific threats; and efforts to ensure that political commitments were technically and legally feasible, could be implemented and did not unreasonably override established management methods. A related – and equally important – influence has been the development of global commitments on MPAs including representative networks, in particular in the context of WSSD and CBD and highlighted in the oceans resolutions in the UN. These varied and interrelated elements demonstrate the contributions that can be made by all sectors and communities to issues relating to oceans governance.

The bottom trawling issue woke up the collective fishing community (and fishing Ministries) to the need to monitor and participate in broad global processes where fisheries issues were being considered and addressed. In many States, the fishing sector was unprepared for the strength of support for a moratorium on bottom trawling, in large part due to being disconnected from the scientific and environmental fora where these issues had been percolating. This issue also emboldened ENGOs in these debates, but also demonstrated the need for more sophisticated approaches in their efforts to secure agreement across States and stakeholders. Prior to the negotiation and adoption of resolution 61/105 ENGOs (operating through a coalition) had taken media-based strategies aimed at polarizing opinion rather than moving consensus.

As this debate moves forward, there is room for optimism. In September 2011, a 2-day workshop was held at the UN to discuss implementation of the resolutions relating to bottom fishing. The meeting was attended by States, representatives of IGOs and other bodies including RFMOs, and NGOs. The discussions were particularly interactive and were subsequently taken into account by the UNGA in deciding on additional urgent actions for bottom fishing in areas beyond national jurisdiction, as reflected in UNGA resolution 66/68.

In the context of the CBD, processes and criteria to identify EBSAs are not inconsistent with VMEs by design. Initiatives to apply the criteria are taking place at the regional level, in collaboration with FAO and the involvement of RFMOs. As shown in Annex 2, there is also a promising new capacity-building program in this area under the Global Environment Facility (GEF), jointly sponsored by FAO, UNEP and the World Bank with input from others, including the CBD Secretariat (described in Chapter 14).

Despite various emerging collaborative initiatives to describe areas that meet VME/EBSA criteria, a sense of mistrust continues to pervade dealings between the nature conservation and fishing communities. The compromise in the UNGA resolutions on a new bottom fishing regime did not resolve differences in their fundamental approaches. As these issues continue to be debated, some conservation stakeholders have promoted and pursued approaches (some of which are blunt, such as moratoria and no-take zones) due to frustration with the slow rate of progress on the part of the VME regime. Fisheries stakeholders, on the other hand, remain frustrated by the degree to which their efforts are often discounted, arguing that time is needed to design and implement measures in RFMOs (some aspects of Chapter 10 apply) and assess effectiveness. As a result, they have tended to prefer the continued consideration of these issues by the FAO and RFMOs, rather than the General Assembly. Currently, it is the work of diligent experts trying to align these approaches through the development of objective and common tools that is making the biggest difference to bridging these differences.

Fisheries and other fora

The fora described above are particularly oriented to building a robust 'fisheries pillar' in the integrated system shown in Chapter 10 (Figure 10.1). Other UN fora having an influence on fisheries-specific regimes include the specialized agencies of the International Maritime Organization (IMO) (e.g. safety for fishing vessels) and the International Labour Organization (ILO) (e.g. safety standards for fishers). In the context of the Convention on the Conservation of Migratory Species of Wild Animals (CMS), corroborating activities for fishing reforms may also occur. Outside the UN, the International Union for Conservation of Nature (IUCN) also sponsors expert work aimed at furthering improvements in fishing-specific standards (cf. Annex 2).

One of the more divisive UN bodies for many fisheries stakeholders and Ministries is UNEP, which addresses both fisheries-specific issues as well environmentally based issues. UNEP was originally designed to play a catalytic and coordinating role in the environmental area in the UN, which includes fisheries. UNEP is also involved in specific implementation projects, which can compete with its integration role and create strains regarding the target beneficiaries of joint work. In this regard, UNEP's focus (and organizational culture) on approaches, which is often best aligned with developing-country small-scale fisheries, does not always transfer well to large industrial fisheries. In meeting the needs of its key members including members of FAO, UNEP has undertaken considerable expert work – often excluding government expertise by design and favouring 'independent' and academic experts – which may raise concerns of legitimacy with States and risk duplication with FAO guidance.

A strengthening and clarification of the role of UNEP in environmental governance has been the subject of many reviews, including from within the UN (see Aziz et al., 2006; UN, 2006). At Rio + 20, delegations committed to strengthening the role of UNEP by establishing universal membership in its Governing Council as well as other measures to strengthen its governance and its responsiveness and accountability to States.[13]

The fisheries community has a direct and commercial interest, and plays a principal role, in two international fora dealing with trade issues: the Convention on International Trade in Endangered Species (CITES) and the World Trade Organization (WTO) (cf. Annex 2). CITES aims to ensure that species at risk (terrestrial and aquatic) are not rendered unsustainable as a result of international trade. Delegates to CITES are generally from environment departments, with differing degrees of involvement of fishing experts across States.

In recent years, CITES has shown a desire to play a larger, more general role in the protection of marine species, including in reviewing some aspects of the efficacy of fisheries management of stocks with commercial importance. The relationship between CITES and the fishing community, including FAO, has been difficult for a number of reasons, and is examined in detail in Chapter 13.

The WTO is devoted to ensuring that trade is not hampered by unnecessary border restrictions or domestic practices in a range of areas. These issues are generally the purview of trade lawyers and trade policy experts, whose primordial focus tends to be on liberalizing trade. WTO deliberations have become strained however, as they move beyond border measures to harmonizing domestic practices to place trade on a 'level playing field' as relates domestic advantages. Fishing interests rely on the WTO to solve disputes about legitimate restrictions in fish trade, including technical and other barriers, and whether trade barriers have a legitimate basis in either global standards (e.g. see CODEX in Annex 2) or in provisions of multilateral environmental agreements (MEAs). A committee on trade and environment mainly addresses the latter issue.

Controversially, the recent Doha Round, widely recognized as trying to make a contribution to sustainable development, launched a special negotiation on fisheries subsidies; this was not necessarily for trade reasons, but for environmental sustainability through an intended impact on global fishing capacity. The fishing community is directly affected by these deliberations and large commercial interests are at stake. One issue has been the extent to which the regime would apply to developing States including large industrial-fishing nations such as China, another driver for clarity on management norms and standards for small-scale fisheries. The debate has not been resolved and DOHA negotiations lag.

Lessons from the WTO are germane to understanding impediments to global debates on governance, especially when those debates address issues affecting State freedom, to whom decisions on fisheries management will be delegated or who will bear the costs of sustainability adjustment. The more those global discussions try to harmonize domestic standards (in legally binding tools that affect economic interest), the more elusive will be agreement.

As noted, the institutions above relate to institutions hosting specialized fisheries discussions. Many of the activities in fora that affect fisheries are integrating processes (cf. Chapter 10, Figure 10.1 and Figure 11.1). In these processes, fisheries interests should have a major voice and role in order to provide assistance and guidance on implications of integrated regimes for sectoral management. However, it is unclear whether the fisheries community is aware of the range of global activities that could affect their interests. Aside from a few very connected technical experts with varying motivations, government fisheries experts are mainly not involved in these fora. It is clearly difficult to develop an integrated approach to oceans management, even nationally, without a common understanding of the extent and breadth of activity in fisheries and marine biodiversity needed to anchor a coherent approach. Additionally, fisheries stakeholders may not be in a position to influence the shape and feasibility of policy options and tools, which they may ultimately need to implement.

The biodiversity/conservation community and global governance

Not unlike the fishing community, those promoting nature conservation-oriented tools or approaches also seek out organizations that support their issues and knowledge and provide them access to relevant processes. However, there are important differences. This community does not generally 'manage' ocean activities; some of these preferred organizations are therefore less likely to be exposed to practical implementation issues associated with proposed new frameworks. Also these stakeholders tend to favour iconic species and area-based management tools (e.g. MPAs), often drawing on experience in fora that also address terrestrial issues. This community (especially ENGOs) also shares a predominant focus on cumulative threats, with an interest in tools protecting marine ecosystems from a range of activities at once.

Two critical differences of perspective related to this divergence in disciplinary or institutional experience affect both (1) the choice of preferred fora within which players frame governance approaches, and (2) global governance debates.

First is the nature of the design process for conservation and the derivation of socio-ecological benefits. In the case of fisheries, the focus or driver is the provision and sustainability of a single *tangible ecosystem service* (e.g. commercial or food fisheries). In designing optimal or pragmatic tools within ecosystem planning constraints, the question is how best to optimize that service (e.g. management regimes with area-based management tools *designed for fisheries outcomes*; FAO, 2006, 2011b). However, environmental approaches argue that management needs to consider the full suite of ecosystem services discussed in the Millennium Ecosystem Assessment (MEA, http://www.millenniumassessment.org). This perspective gives a central position to consideration of conservation protection, existence values and cumulative threats which more rarely enter fisheries debates. This perspective often concludes that preservation and representativeness (e.g. no-take zones and 'parks' approaches) are necessary to ensure all the ecosystem services are sustained. Those favouring more extreme tools – including arguing for arbitrary protection targets – sometimes argue that they are not incompatible with socio-economic outcomes because they can promote enhanced ecosystem services such as enhanced fisheries as a long-term by-product of protection.

While both approaches support the ultimate provision of the ecosystem service (i.e. sustainable fisheries), the motivation/drivers and design elements are different. Global dialogue taking place in the ambience of these approaches, but without awareness of these differences, is often ineffective, particularly in terms of resolving debates. Issues are often left unresolved, with the common language or consensus defaulting to reiterating ideal normative high-level outcomes.

A second critical difference is the closely related issue of ecological tolerances in the design, scale and scope of protection or application of conservation measures. Fisheries managers are experienced in applying the concept of sustainable use and planning against measurable proxies where the mortality is greater than zero (e.g. MSY, TAC, quotas). Even in the case of coral or species bycatch the design question is one of threshold tolerances, partly related to ecological role. In fact, some advanced integrated fisheries management systems address ecosystem impacts through market-based measures such as tradable quotas for bycatch species. Risk assessment thus focuses on the resilience of components of the ecosystem, and precautionary approaches focus on 'limits' and triggers for changed activity levels.

For classic conservation protection approaches, where optimizing a single ecosystem service in an ecosystem context such as fisheries is not a goal, the debate centres more on the scale and scope of protection that would yield balanced benefits to the greatest number of ecological services (e.g. IUCN, 2010; Laffoley, 2006). Discussions on activities associated with resource exploitation are more centred on precautionary tools justifying their legitimacy (e.g. environmental impact assessments).

It is therefore not surprising that in non-fishing global fora there is a strong emphasis on area-based management tools to address threats in aggregate, and prior environmental impact assessments of each use. This includes bodies or fora with a strong development focus, such as UNEP. As noted above, area-based measures have some appeal to developing States, even for fishing activities, as such measures may take less knowledge and technical capacity to implement, can be fairly distribution-neutral and can demonstrate incidental benefits. As illustrated in the UNGA debates on bottom fishing, these measures can also meet other agendas such as excluding 'outsiders' who would otherwise have access to fishing grounds. Only since recently has there been a systematic application of sophisticated management tools to information-poor and small-scale fisheries; area-based measures in developing States are often the key, and default, ecosystem-based fishing approach.

For those attending international debates on fisheries and broader oceans governance where multiple stakeholders are in attendance, it becomes immediately clear that both fishing and conservation communities invoke scientific foundations to justify preferred design elements: one stream focusing on the *function* of ecosystem components and their link to fishing sustainability, which would suggest certain design elements (stock assessment, productivity, critical habitats, role and status of large predators or forage species, ecosystem resilience); and the other focusing on the *integrity* of the entire ecosystem, with an emphasis on risk analysis, identification of special areas, design of spatial tools, measurement and prediction of benefits. The situation is made only more confusing by 'agenda-driven' science advocacy, which can creep into deliberations from many perspectives and sources. The latter often enters charged debates, especially from the most extreme stakeholders, a distraction that can hinder appreciation of the fundamental differences in the two perspectives.

Philosophical or normative debates between these groups do not easily resolve at the global level, particularly when these differences in perspective are not understood. Differently framed debates in different global fora will persist as long as they are not inclusive of 'cross-disciplinary teaching' and a collective framing of issues anchored in objective integrated knowledge, combined with a careful targeting of tools against needs in an integrated tool kit. Critically, the key stakeholders in global processes need the tools of the other, but mistrust and lack of coherence can interfere with synergies that might otherwise exist.

At the regional or national level, it is likely that these issues will be more easily resolved in the implementation stage where common or more homogeneous interests may be more dominant. Examples include jointly collaborated research projects for specific protection, or specific integrated management and/or capacity-building projects in a multi-stakeholder context (e.g. Tallis et al., 2009).

Most biodiversity/conservation-oriented debates in non-fisheries fora are attended by government (environment) representatives and debates in UN fora are dominated by international lawyers and/or international diplomats (Annex 2). Few participants may have familiarity with the nuances of fishing issues or the practical implications for management. Apart from some informal processes involving technical experts, 'cross-disciplinary teaching' related to fishing involves 'mainstreaming' of issues into broader agendas and legal frameworks and on socio-political tolerances, rather than serious science–policy–management debate.

On the basis of the above, there is a strong reticence (especially among developed States) to engage fisheries issues at the CBD. This is especially true for high-seas issues. If not in FAO, issues relating to governance of the high seas are instead raised by fishing States in the UNGA and its related processes and by experts familiar with UNCLOS (and its treatment of sectoral activities) and its related instruments. Some CBD technical documents in the past were perceived to disparage fishing as a legitimate activity in the oceans (e.g. in promoting a ban on bottom trawling). However, CBD processes now carefully negotiate language on fishing issues and activities in a range of areas. The CBD is also limited in the nature of its mandate in regards to areas beyond national jurisdiction,[14] and has neither mandate nor capacity for actual management implementation. The effectiveness of CBD activities in relation to fisheries will therefore depend, in large part, on the extent to which these activities are integrated in the relevant processes of other institutions, organizations and bodies, as well as by States.

The mainstreaming of resource conservation and environmental protection into the sustainable development agenda will likely have a mixed impact on the bridging of these two approaches and bridging across institutions. On the positive side:

- it may enable some high-level unifying global strategic frameworks that are lacking (Inomata, 2008); for example, clarifying links between ecosystem services and socio-economic and development outcomes (Tallis et al., 2009) and facilitate reframing of issues to better support both sectoral tools and integrated management methods;
- collective solutions addressing impacts on sustainable development may help develop legitimacy, political buy-in and willingness to adopt and implement innovations;
- serious capacity-building efforts may increase, such as the new multi-agency GEF project (cf. Annex 2) that will raise capacities for ecosystem approaches; and
- it may encourage full management measures in fisheries of all kinds (e.g. including small-scale fisheries) so that area-based measures are not the sole default option for some States.

On the negative side, however:

- global debate may edge even further from the resolution of foundational issues around sustainability perspectives and management options; and
- it may limit or censor deliberative debate on issues or options for better sustainability that are politically infeasible at a point in time, as has already occurred in some UN processes.

Integrated knowledge foundations

As noted in Chapter 10, a key contribution to an efficient, unified approach to governance is integrated, shared and authoritative knowledge. It helps to build common understanding, frame issues among diverse players in a coherent fashion and identify practical and feasible solutions, tools, monitoring methods and indictors of success and efficacy of approaches. Alongside the legal framework, it is one of the most important foundational contributions to an effective and unified system of marine governance (cf. Figure 10.1).

However, to be useful for framing issues or decision making, knowledge foundations must be credible and 'legitimate'. Equally important is the need for such knowledge to be both understood by policy makers and capable of translation into practical terms; this is not an easy task, given the diversity and scale of the ocean community.

International organizations providing research and useable information of potential importance to fisheries management and biodiversity conservation include CBD, FAO, IOC-UNESCO, IUCN, UNEP, UNU and WMO, as well as marine science organizations such as ICES and PICES. Some organizations produce new knowledge on emerging issues (e.g. impacts of climate change on ecosystems/fisheries, use of space observational tools for fisheries and integrated planning). It is however unclear the extent to which these tools are known by fisheries interests and actually integrated or mainstreamed into management, even where sophisticated capacity exists.

One key and promising development is the work being undertaken globally on integrated assessments. The UNGA has long recognized the need for better scientific understanding to inform and improve decisions on managing human pressures on the oceans. This featured in the 2008 meeting of the BBNJ which, irrespective of views on management solutions for improving protection of marine biodiversity, emphasized the need for integrated and shared knowledge about biodiversity and threats. The model of the Intergovernmental Panel on Climate Change (IPCC), which provides integrated information on the impacts of changing climate to governments, was cited.

In 2010, the General Assembly launched the first cycle of the Regular Process, which is tasked with producing the first global integrated marine assessment by 2014.[15] The assessment will have an emphasis on integrating natural science aspects with socio-economic information, to be scaled up from available regional and national assessments.

The Regular Process fulfils the commitment made in 2002 at WSSD to improve the scientific understanding and assessment of marine and coastal ecosystems as a fundamental basis for sound decision-making. Although a 'bottom-up process' integrating expert advice and existing information and assessments, it has a broad and inclusive governance structure (the UNGA *Ad Hoc* Working Group of the Whole or AHWGW) intended to ensure that its processes and results are endorsed broadly. Important to its legitimacy is the division between scientific experts who conduct the assessments, and States who provide oversight through the AHWGW and the General Assembly.

UNEP has also had a historical primordial role in the UN relating to environmental assessments, including ecosystem health. The most well-known project is the MEA which includes a chapter on oceans. It used procedures and processes also modelled on IPCC but innovated in several ways, especially in its promotion of multi-stakeholder governance, engagement and outreach. States expressed doubts about its top-down approach and the utility of the assessment for practical outcomes however, possibly due to low levels of government engagement in the assessment (Miller and Dublin, 2004; Wells et al., 2006).

UNEP has recently established a new process – the Intergovernmental Science–Policy Platform on Biodiversity and Ecosystem Services (IPBES, http://www.ipbes.net) – which is also intended to be a State and expert-led process to provide objective information and assessment of the health of both terrestrial and marine biodiversity. Its gestation took a fundamentally different approach to the Regular Process, with much more grassroots expert and ENGO-driven planning and design and loose and high-level State coordination under UNEP. Only recently has attention turned to actual operating modalities, including specific planning on links to, and relationship with, other assessment processes including the Regular Process.

Other key organizations providing integrated information include IOC-UNESCO, which plays an especially important role in integrating ocean observation information (including space-based) into useful tools for both sectoral and integrated planning. Tools are also being developed by the CBD on environmental assessments and by IUCN on criteria for listing 'ecosystems at risk'. As noted, progress is also being made on EBSAs and VMEs by CBD and FAO in collaboration with States, regional seas organizations, RFMOs and other relevant regional initiatives, and with the assistance of GEF to encourage regional processes where capacity is weak. In the case of the CBD, it will remain for States or other processes (such as assessment processes) to pick up and use the information.

These developments have the potential to unite communities of interest around basic assessments if they integrate well but divide them if they do not, especially given the different project governance and stakeholders involved. It remains to be seen whether these efforts will provide an integrated knowledge base for decision-making, or a fragmented and competing system.

Conclusions

This chapter applies frameworks elaborated in Chapter 10 for understanding institutional legitimacy, global consensus decision-making, the general architecture and texture of some key global institutions addressing oceans issues and potential roles and contributions within global oceans governance.

In a *unified* and *interdependent* marine governance system, all players and their primordial organizations can, and must, be part of the solution to better governance. As explained in the preceding chapter, this can be achieved if the system 'plays to strengths' of all agents and, in a world of mobile activities and threats, ensures that irresponsible but rational behaviour does not seek out points of global weakness in governance with adverse consequences to conservation and sustainable development. Often these points of weakness are in areas that are already facing the greatest vulnerabilities in terms of sustainable development: the emerging unifying paradigm for global discussions, as discussed above. To prevent this, both sufficient global capacity and alignment of incentives is needed. As seen in Chapter 10 however, there is unlikely to be agreement on approaches and regimes not accounting for the goals and aspirations of the majority of States, or that suffer from a lack of legitimacy on other grounds. Flexibility in approach is needed, particularly when rigidities are emerging that risk making the global system vulnerable to gridlock. The current increasing difficulty of reaching global agreement on even basic declarations and procedures is testimony to these difficulties.

It is worth reflecting on what can be practically achieved at the global level. This chapter has reviewed how current governance streams use global institutions to try to prevent unnecessary stress from human activities on already burdened oceans ecosystems. Global institutional focus follows differences in the underlying perspectives for fishing and conservationist approaches. Global agreement is not easily achieved, however. The driver for one community is to ensure protection from all threats in order to manage risk to future ecosystem services, including through prior assessment of the sustainability of activities. For the other the goal is optimization, that is, to design the best tool kit to maximize the benefits of specific ecosystem services currently and over time, while not reducing the ability of a given ecosystems to provide other services. It is unlikely that the two basic motivations are easily reconcilable, just as a car designed to last forever is not likely to meet performance standards desired by drivers even though the ultimate intent is to provide sustainable transportation.

At the global horizontal level, it can easily be agreed that all ecosystem services – including fishing – are valuable and must be sustained in order to support sustainable

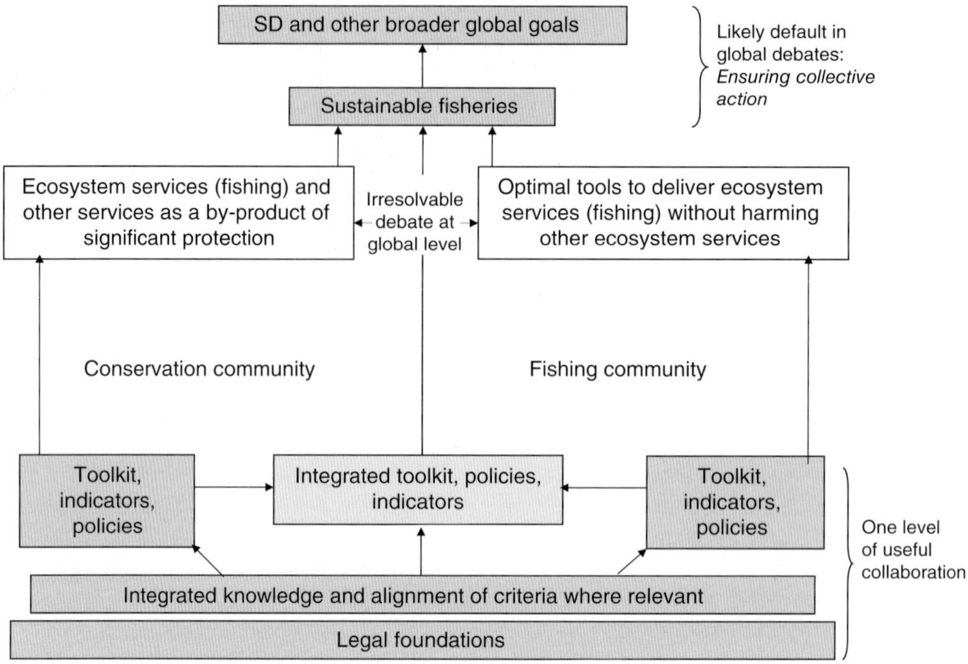

Figure 11.3 Practical areas of global collaboration.

development aspirations (see Figure 11.3). Further, irresponsible behaviour can not only threaten the aspirations of specific States, but also create global spillovers that affect others. These messages are already being reflected in global declarations.

Members of the global international community have an especially strong platform for political advocacy to: (1) promote collective responsibility; and (2) ensure that the best tools and standards applicable to different contexts are available, their efficacy in different contexts understood and global incentives for adoption established (or disincentives and impediments minimized). At the global level, what is needed is commitment to adoption of required tools, compliance with selected governance models, a shared expectation of goals and outcomes and avoidance of monitoring approaches and indicators that focus on only one set of tools.

In contrast, normative debates or declarations trying to resolve or assert the 'best' tools are unlikely to be achievable or productive, as much as efforts to do this take place (Figure 11.3). There is no single correct approach to solving or even measuring progress on sustainability issues; a multifaceted approach is needed. The actual choice of tools is an implementation issue, and likely best resolved at regional and national levels according to the needs and capacities of the particular context.

Knowledge, tools, standards and guidance for adoption by regions and States should be suitable, practical and coherent, and developed in institutions best equipped to do the work using processes credible and legitimate to all perspectives. The global governance system can encourage more cooperation and transparency on the tools that apply in different contexts, on efforts being undertaken, on the impacts being achieved and as necessary capacity-building.

The chapter reinforces the fundamental elements of a well-functioning global governance system, which are as follows:

- *global institutions* that engage relevant players effectively, ensure informed deliberation and have other procedures in place to maximize, within their mandates, the potential for practical contributions to coherent and integrated marine governance;
- *integrated, verifiable and useable knowledge, assessment and monitoring*, including information on risks, vulnerabilities, threats, options and impacts, which can provide a common basis of understanding and maximize the

possibilities of a successful framing of needs and issues relevant to different organizations;

- *international legal frameworks* that adequately support tools and approaches to address and monitor key risks and allow objective monitoring of implementation and impacts;
- *technical cooperation and collaboration among global institutions* to minimize unnecessary duplication and competition and ensure tools and conservation methods can be integrated among sectors and across different contexts;
- *marshalling global resources for capacity-building* and transfer of knowledge/technology; and
- *political advocacy to avoid weak links* in the global fabric of protection and sustainable use.

While there are many challenges for marine sustainability, there are grounds for optimism on governance improvements based on recent developments on marine knowledge, tools and capacities in global institutions. Increasingly, collaborative efforts are aimed at a common basis of knowledge and understanding of the health of marine ecosystems and oceans, as well as steps to enable the design of complementary and mutually reinforcing governance tools.

Annex 2 highlights the extensive collaboration which is already in place. The issue is one of awareness and accessibility, including results. Within the fishing community, governance is becoming more robust and capacity development is finally linking to global mechanisms. Implementation challenges at all levels still threaten the credibility of the system, however. Processes (especially at the global level) are difficult to rush, notwithstanding the urgency of issues as implementation ultimately depends on acceptance by States. To this end, there is a need for trust building and dialogue among stakeholders at all levels, including the development of information and analysis to help frame issues collectively and pragmatically.

One risk is the newness and fragility of truly collaborative efforts in global marine processes. Current progress is partly based on the efforts of experienced personalities bridging differences among communities whether as delegates, experts, international civil servants or NGO members. Unless these cooperative efforts are better supported, formalized or fully institutionalized by States in governing bodies and conferences of global institutions and their mechanisms, they will be transient.

Institutionalized collaboration is needed at three levels: formal cooperation on the issues being addressed; integrated institutional management of projects and programs; and relevant compliance and monitoring (Scott, 2010). However, ultimate responsibility for coherence in global debates and actions across organizations and bodies (including encouraging inter-agency collaboration) lies in State capitals and with their representatives in governing bodies. 'Integration and trust-building starts at home.'

For specialized interests navigating the global arena, the key messages of this chapter are as follows.

(1) Fisheries and broader biodiversity-related issues are addressed in a much wider range of institutions and organizations than specialized communities may know, or be able to specifically influence.
(2) Leveraging specialized issues (e.g. fisheries) into global processes requires literacy of the various processes, cultures, rules, practices and norms of key organizations and bodies. These processes can be difficult for specialists unaware of linkages with other primordial goals and objectives (including geopolitical issues, sustainable development, human rights, international justice, etc.) and the influence of non-marine specialist interlocutors (e.g. international lawyers and diplomats) in debates and outcomes. Realistically, the importance given to any perspective or issue – whether responsible fisheries or biodiversity-oriented – may depend on whether it is leveraged into, or packaged with, these broader global goals and objectives. In addition, well-meaning initiatives can fail to find traction and support once they reach the global level unless done through established UN processes or in accordance with UN rules and standards.
(3) Fisheries may find themselves at the 'implementing end' of commitments, standards, targets and other decisions that have been developed in non-fishing institutions and organizations. This may be difficult if the resonance of proposed measures with scientific, technical, management or sectoral feasibilities and capacities is lacking. Potential implementation gaps can reinforce lack of trust and engagement and feed into a vicious cycle of 'competition' of ideas among diverse communities of interest. Those with fisheries interests must be aware of and influence global processes *before* decisions on frameworks, etc. are reached, so that outcomes are practicable and can be incorporated into regional or domestic law and regulation.
(4) These challenges are further complicated by institutional culture or protocols at the global level and the

influence of blocs or coalitions in global processes that may be unfamiliar to experts used to technical settings. Participants may need to adapt aims, objectives and initiatives to align with other issues, such as sustainable development or capacity-building, in order to ensure support.

(5) Most global declarations and decisions are adopted by consensus as political commitments and are not, in and of themselves, legally binding on States unless they are incorporated into binding international or regional instruments, enabled into domestic law and regulation and/or become imbedded as international norms. However, most States take global decisions and outcomes seriously, and are politically and morally obligated to support and implement these commitments, particularly if they have the means to do so. Although global commitments often lack sanction and enforceability, implementation is monitored in many contexts; this could single out certain States as non-compliant, unless they have explicitly disagreed with the decision or outcomes.

In the end, well-functioning global governance is necessary but not sufficient to enhance the overall sustainability of oceans and fisheries resources. Its coherence with and effective integration into regional and domestic fisheries and oceans management is integral to the delivery of both strong fisheries management and healthy oceans ecosystems.

Notes

1. This chapter benefited from the guidance of an expert who wishes to remain anonymous.
2. Importantly, the Rio+20 outcome formalized recognition of constituencies, alongside States.
3. For further information see FAO (2011a).
4. IPOA for Reducing Incidental Catch of Seabirds in Longline Fisheries; IPOA for Conservation and Management of Sharks; IPOA for the Management of Fishing Capacity; and IPOA to Prevent, Deter and Eliminate Illegal, Unreported and Unregulated Fishing.
5. For example, International Guidelines for the Management of Deep-sea Fisheries in the High Seas in 2009; FAO Guidelines on By-catch Management and Reduction of Discards in 2010; and FAO Guidelines on Flag State Control in 2013.
6. IUU fishing is defined in terms of compliance with rules of domestic and regional fishing bodies.
7. Provided as a background document to the FAO Committee on Fisheries for discussion, July 9–13 2012. See 'Information on the Recommendations relating to Fisheries Management and Conservation and Sustainable Use of Aquatic Living Resources: recently adopted by the Subsidiary Body on Scientific, Technical and Technological Advice (SBSTTA) of the CBD FAO 2012/COFI/2012/SBT.10.
8. Readers can refer to Decisions adopted by the Conference of the Parties to the CBD at the 10th and 11th meetings, specifically decision X/29 (in 2010) and X/17, 18 (in 2012).
9. Reports of the Secretary-General on oceans and the law of the sea are available at: http://www.un.org/Depts/los/general_assembly/general_assembly_reports.htm
10. A list of topics addressed since 2000 is available on the Law of the Sea website (http://www.un.org/Depts/los/).
11. Available at: http://www.un.org/Depts/los/general_assembly/general_assembly_resolutions.htm
12. FAO International Guidelines for the Management of Deep-seas Fisheries (2009).
13. A/CONF.216/L.1, The Future We Want (UN, 2012).
14. The CBD has a key role in supporting the work of the General Assembly with regard to marine protected areas beyond national jurisdiction by focusing on provision of scientific and, as appropriate, technical information and advice relating to marine biological diversity, the application of the ecosystem approach and the precautionary approach.
15. Regular Process for Global Reporting and Assessment of the State of the Marine Environment including Socio-economic Aspects.

References

Aziz, S., Diogo, L. and Stoltenberg, J. (2006) Delivering as One. Report of the Secretary General's High Level Panel. United Nations, New York.

Behnam, A. (2005) Developing countries in the group of G-77: a journey through multilateral diplomacy 1964-2004. In: *Towards World Constitutionalism* (eds R.J. Macdonald and D.M. Johnston). Koninklijke Brill BV, Netherlands.

FAO (2006) Fisheries Report 85, Report and Documentation of the Expert Workshop on Marine Protected Areas and Fisheries Management, Review of Issues and Considerations, 12–14 June 2006, FIEP/R825.

FAO (2009) *International Guidelines for the Management of Deep Seas Fisheries in the High Seas*. Rome, FAO.

FAO (2011a) *Code of Conduct for Responsible Fisheries*. Rome, FAO

FAO (2011b) Fisheries management, Marine protected areas and fisheries, FAO Technical Guidelines for Responsible Fisheries, No. 4, Suppl. 4.

FAO (2012) Committee on Fisheries, Final Report 30th Session (FIPI/R1012), Rome. 9–13 July 2012.

Inomata, T. (2008) Management Review of Environmental Governance within the United Nations System. United Nations Joint Inspection Unit, Geneva.

IUCN (2010) New Vision for Biodiversity Conservation Post-2010 Strategic Plan of the Convention on Biological Diversity (CBD). Position Paper for the *Fourteenth Meeting of the Subsidiary Body on Scientific, Technical and Technological Advice (SBSTTA14)*, Nairobi, Kenya, 10–21 May 2010. (Also Third Meeting of the Ad-Hoc Open-Ended Working Group on Review of Implementation of the Convention (WGRI3), Nairobi, Kenya, 24–28 May 2010.)

Laffoley, D.d'A. (ed.) (2006) The WCPA Marine Plan of Action. Working together to secure a global representative system of lasting networks of Marine Protected Areas (consultation version). IUCN WCPA, Gland, Switzerland. 26 pp.

Miller, C. and Dublin, H. (2004) Millennium Ecosystem Assessment Mid-term Evaluation. UNEP/GEF/UNFIP PROJECT - MP/FP/1010-04. Evaluation and Oversight Unit, UNEP, Nairobi, Kenya.

Najam, A. (2005) Developing countries and global environmental governance, from contestation to participation to engagement. *International Environmental Agreements* **5**, 303–321.

Ridgeway, L. (2009) Marine genetic resources: outcomes of the United Nations informal consultation process (ICP). *International Journal of Coastal and Marine Law* **24**(2), 309–331.

Ridgeway L. and Rice, J. (2010) International organisations and fisheries governance. In: *Handbook of Marine Fisheries Conservation and Management* (eds Q. Grafton, R. Hilborn, D. Squires, M. Tait and M. Williams), pp. 485–504. Oxford University Press Inc., Oxford.

Scott, K. (2010) International environmental governance: managing Fragmentation through institutional connection. *Melbourne Journal of International Law* **12**(1), 177–216.

Tallis, H., Goldman, R., Uhl, M. and Brost, B. (2009) Integrating conservation and development in the field: implementing ecosystem service projects. *Frontiers in Ecology and the Environment* **7**(1), 12–20.

UN (2006) Delivering as One. New York. Report of the Secretary-General's High-level Panel on UN System-wide Coherence in the Areas of Development, Humanitarian Assistance, and the Environment.

UN (2012) The Future We Want. United Nations Conference on Sustainable Development. Outcome of the Conference, document A/CONF.216/L.1.

Wells, M.P., Grossman, D. and Navajas, H. (2006) UNEP, Terminal Evaluation of the UNEP/GEF Project 'Millennium Ecosystem Assessment'. Project Number MT/FP/CP/1010-01-04.UNEP Evaluation and Oversight Unit, Nairobi, Kenya.

Chapter 12

Integrative policy and legal instruments, approaches and tools: Fisheries and biodiversity conservation

B. Kuemlangan, J. Sanders, P. Deupmann and C. De Young
Food and Agriculture Organization of the United Nations (FAO), Rome, Italy

Abstract: This chapter discusses the relationships between international legal and policy frameworks and their relevant features as well as the management approaches and tools that facilitate integrated action in the biodiversity conservation and fisheries governance streams. It demonstrates how these frameworks provide the basis of and influence integrative approaches and tools and provides some examples. The chapter shows the integrative features of the fisheries and biodiversity conservation legal and policy frameworks and the approaches and tools which are increasingly being implemented. It concludes that despite significant progress made by the fisheries and conservation governance streams in pursuing integrated action, much remains to be done and that cross-sectoral communication and cooperation is key to success.

Keywords: fisheries; biodiversity; integrative policy and legal instruments; integrative approaches; integrative tools; convergence

Introduction

For decades before the advent of the United Nations Convention on the Law of the Sea of 10 December 1982 (UNCLOS) and thereafter, fisheries management and governance practices largely followed a sectoral approach.[1] This has led to commonly held views in the biodiversity conservation community that fisheries management is solely about development and production, without regard for broader environmental concerns. On the other hand, the broader biodiversity or environment conservation agenda has been commonly perceived by a large part of the fisheries community as being mainly concerned with the preservation of the environment in almost its pristine state at any cost, while disregarding user values in an aquatic system. These misconceptions are partly due to the lack of communication between the two communities. Insufficient understanding and consideration of the other community's interests and concerns have contributed to insufficient convergence in approach, tools and interventions between the biodiversity conservation and fishery governance streams.

In this chapter, frequent references are made to the legal and policy framework, instrument, approach, process and tool. For the purpose of this chapter, a *framework* is the legal or policy *instrument* that provides the overarching architecture for governance. Its scope could be sectoral (e.g. the fishery governance framework) or multi-sectoral (e.g. UNCED Agenda 21). The term *instrument* is used generically as encompassing legal and policy frameworks

Governance of Marine Fisheries and Biodiversity Conservation: Interaction and Coevolution, First Edition.
Edited by Serge M. Garcia, Jake Rice and Anthony Charles.

including agreements, conventions, resolutions, etc. A *process* is an intergovernmental or other systematic series of actions directed to some end, and affects or takes policy and management level decisions. At the intergovernmental level, the *process* (e.g. UNGA) can result in an overarching governance framework that could be multi-sectoral or sectoral. A process could also be important and referred to at the implementation or management level. An *approach* (e.g. the precautionary approach) is understood as the detailed way to implement a principle (e.g. the Precautionary Principle) – which itself may not contain much implementation guidance – or to achieve policy goals that could be reached in different ways. A *tool* is the means used to give effect to management or the implementation of an approach (e.g. EIA, multi-criteria analysis, eco-labels). Together, instruments, approaches, processes and tools, among others, translate and implement international policy.

A review of fisheries and biodiversity conservation legal and policy instruments and governance processes shows that sectoral governance is fundamental in order to enable tailored responses. It is also clear that the drafters of these instruments envisaged the integration of biodiversity conservation into sectoral or cross-sectoral policies, plans and programmes. In this chapter we review the fabric of policy and legal frameworks, approaches, processes and tools that are *integrated*, that is, they demonstrate elements related to both biodiversity conservation and fisheries, or are *integrative*, that is, they aim to increase coherence between the two governance streams. In practice, most of the integrative and integrated approaches and tools are complex, relatively new and often not yet fully implemented or adopted in management[2] (United Nations, 2006). Furthermore, coordination at the national level is often lacking or inadequate. This has delayed actual integration between the two governance streams and is one of the reasons for the persistence of the assumption that governance 'prisms' of the fishery management and biodiversity conservation governance streams disregard each other's concerns.

This chapter first of all examines the use and integration of two of the integrative approaches adopted by both the biodiversity conservation and fisheries governance streams: (1) application of precaution (the precautionary principle or approach); and (2) the holistic management approach (the ecosystem approach). We describe the policy and legal instruments in which these approaches are incorporated, as well as the processes that have facilitated their adoption. We then discuss a number of management level processes, approaches and tools that facilitate integration of the precautionary principle or approach and the ecosystem approach. The chapter concludes with examples at the regional level of enhanced integration in management.

The chapter attempts to demystify the issues of compatibility and integration of the approaches and tools preferred in the biodiversity conservation and fisheries governance streams and affirms the need for common and integrated foundations *sensu* Chapters 3 and 10.

Integrative concepts and approaches

At the management level, the application of the precautionary and the ecosystem approach are examples of mainstreaming concerns related to both biodiversity conservation and fisheries governance. These approaches are interwoven into policy and legal instruments of past decades, as set out below. They were codified into the international regime at different times (cf. Annex 1) and are intricately linked.

Precautionary approach

The concept of precaution emerged in 1854 as an environmental and social issue in the area of securing public health and safety (Harremoës *et al.*, 2001). In the wake of the UNCHE (in Stockholm in 1972), the concept was progressively extended to cover the general state of the environment and all species. As a result, many countries began to reflect the concept of precaution in their national legislation and policy.

The precautionary principle (PP) first appeared at the international level, in the context of the marine environment, in the North Sea Ministerial Conferences of 1984 to 1990. The event which stressed the need for precaution as a key principle for development was UNCED, as reflected specifically in Principle 15 of UNCED's landmark Rio Declaration of 1992. In the same year, the precautionary approach was incorporated into the Convention on Biological Diversity (CBD). Subsequently, in the marine biodiversity conservation stream, the concept of precaution has been largely championed by the CBD through decisions by its Conference of the Parties (CoP) such as the Jakarta Mandate on Marine and Coastal Biodiversity and its programme of work for implementation. IUCN has also contributed to the development of the concept through its Guidelines for Applying the Precautionary Principle to Biodiversity Conservation and Natural Resource Management (in 2007). At the species level, CITES Resolution 24 of the 9th Conference of the Parties in 1994 (Conf. 9.24, as revised) explicitly referred to UNCED Principle 15 and endorsed the application of the precautionary approach in consideration of proposals for amendments of appendices I and II.

In fisheries, the need for 'precautionary, adaptive and anticipatory management' was identified as early as 1973 at the FAO Technical Conference on Fisheries Management and Development (Vancouver) while the consequences of UNCED Principle 15 for fisheries were first discussed by COFI in 1993. In 1995, the Precautionary Approach to Fisheries was adopted in the FAO CCRF (General Principles and Article 6.5) and, in a binding form, in the UNFSA.[3] In 1999, the ICES Advisory Committee on Fisheries Management (ACFM) established an operational definition for the precautionary approach to fisheries that has influenced stock assessment and scientific advice in the North Atlantic region since then.

The concept of precaution has been embedded into a number of instruments pertinent to both fisheries and biodiversity conservation. Today the precautionary approach is considered as an integral part of the ecosystem approach (cf. FAO, 2003, p. 22, 85) and is embedded in modern legal frameworks that govern fisheries at the global, regional and national[4] levels.

Whether one refers to the precautionary approach (PA), which is preferred in the fisheries community, or the precautionary principle (PP), which is preferred in the environmental community, the real struggle associated with the concept of precaution is in its practical application. Indeed, the choice of term indicates how precaution is interpreted and applied. On one hand, the use of the PA in the fisheries community has been largely based on interpreting the 'approach' as an 'effective and pragmatic operationalization of the PP' (Garcia, 1996), implying flexibility in the application of a coherent set of explicit measures. As intended by UNCED Principle 15 this allows other considerations, including socio-economic implications, to be taken into account. The need to take into account the social and economic risks in the PA is clearly specified in the FAO Ecosystem Approach. *In extremis,* the inclusion of these factors could undermine the PA in terms of its protection aspects. On the other hand, *precautionary measures* are often explicitly interpreted and applied in 'the best interest of the conservation of the species concerned' as in CITES Resolution 25 of the 9th Conference of the Parties, 1994 (Conf. 9.25, as revised) which *in extremis* could unduly hinder the sustainable use of marine resources. This has often been referred to as the PP in environmental literature, perhaps to increase its moral strength.

Interpreting the PA and the PP in these dividing terms is not helpful and has caused entrenched exclusive positions rather than promoting integration between the fisheries and biodiversity conservation communities. Instead of dwelling on definitions, since adopting its Resolution 61/105 the UNGA as a high level policy-making body has consistently called on all States to apply precaution in managing fish stocks and protecting vulnerable marine ecosystems (VMEs), recognizing the importance and value of deep-sea ecosystems and the biodiversity they contain. It must be recognized, however, that most of the applications of the PP and the PA (except under the FAO EAF projects) deal with ecological risk and hardly consider social and economic risks.

Ecosystem approach

As the focus in both the fisheries and biodiversity communities turned to broader issues of biodiversity and ecosystem conservation, more holistic governance approaches were developed. The ecosystem approach (EA) can be traced to environment and biodiversity-related conservation instruments. Among these are the 1971 RAMSAR Convention on Wetlands, CITES and the 1979 Bonn Convention on Migratory Species of Wild Animals. In 1992, the EA was explicitly incorporated and brought into prominence by chapter 17 of Agenda 21[5] and, in 1995, by the CBD 2nd Conference of the Parties, Decision 8, 1995 (COP Decision II/8). These instruments also provided the basis for sectoral variants of the EA such as the Ecosystem Approach to Fisheries (EAF) adopted in FAO in 2001 or the Ecosystem-Based Fishery Management (EBFM) adopted, for example, in the USA.

The EA as defined in the CBD includes 12 principles and operational details (CBD 5th CoP, Decision 6, 2002V (CoP Decision V/6); Vierros, 2008, p. 39). The IUCN has developed Guidelines on the EA (Shepherd, 2004) following the definitions and principles established under the CBD, and established a Commission on Ecosystem Management. The EA was enshrined in the WSSD with a specific target to implement before 2012. The principles of the EA as defined by the CBD are cross-sectoral, are not specific to any sector or biome and include: considerations of effects of activities on adjacent ecosystems; involvement of all relevant sectors of society and disciplines; and seeking the appropriate balance between and integration of conservation and use of biological diversity (Vierros, 2008). The EA also explicitly promotes the incorporation of uncertainty and application of precaution within an adaptive management framework that is based on the best available knowledge.

The fishery-specific variant of the EA (the EAF) was developed to take into account the interactions between fisheries and ecosystems (as well as impacts on fisheries from non-fishery activities; cf. Chapter 4). An EAF framework was adopted in the 2001 Reykjavik Declaration (Bianchi and Skjoldal, 2008; FAO, 2003). This approach was developed for practical application of the concepts and the principles that form the basis of sustainable (responsible) fisheries as defined in the FAO Code of Conduct for Responsible Fisheries. The details and norms associated with this approach have been developed in a number of countries (e.g. Australia, USA) and regional institutions (e.g. CCAMLR, ICES), and were elaborated and promoted at the global level by FAO.

EAF is designed as an expansion of conventional fisheries management and therefore adds a systemic consideration of ecosystems to fisheries management (FAO, 2003). EAF encompasses a number of fundamental issues, including reduction of overcapacity and illegal unreported and unregulated (IUU) fishing, and the promotion of the use of alternative management tools such as fishing rights. It integrates sustainable development concepts by aiming at both human and ecological wellbeing. The EAF explicitly incorporates the PA. It has the potential to be integrative within the sector, thus providing a common framework for small-scale and large-scale fisheries and aquaculture.[6] It could facilitate fisheries integration in broader planning and management frameworks (Bianchi *et al.*, 2008) and could integrate fisheries concerns into a broader EA. However, the degree and timeliness in realizing this potential for integration will depend greatly on how much progress the two streams make towards common data sources, methodologies, scientific assessments and joint meetings.

Sectoral variants of EA are obviously needed for sectoral participative implementation and their formal adoption should be encouraged. However, some argue that an overarching governance mechanism is needed to coordinate sectoral management, for example under integrated governance frameworks such as ecological sustainable development (ESD), integrated ocean management (IOM) or MSP (cf. Chapters 8, 21).

The choice of terminology used for holistic or ecosystem conservation or management can vary (e.g. EAF, EBM, EBFM, etc.). The expression 'ecosystem approach' is perhaps used more by those significantly concerned with humans, socio-economic impact, equity and ultimately sustainable use in both the fisheries and biodiversity streams. The expression 'ecosystem-based', implying the primacy of nature and ecological consideration, is perhaps used more frequently by those concerned primarily with nature protection. These preferences become blurred however when looking at the definitions and the different guidelines for implementation, which are all based on the same fundamental concepts.

Relationship between framework and implementation agreements and related instruments

In order to appreciate the inherently integrative nature of policy and legal instruments, it is important to understand the fabric of *framework and implementation instruments*, and other *instruments* that govern the sustainable use of oceans and the wellbeing of marine ecosystems and their users.

Key global *framework instruments* such as UNCLOS have codified principles, policy and legal approaches at the time of their negotiations into provisions such as those related to the conservation and management of highly migratory and straddling fish stocks (Articles 63 and 64). These provisions are further outlined in more detailed *implementation instruments* (such as the UNFSA or the constituting agreements of regional fisheries management organizations and arrangements (RFMO/As)) that either fix perceived gaps as new issues emerge, or further delineate concepts and principles codified in higher-level framework instruments to assist in implementation. These concepts and principles are further outlined in regional and national instruments, which form the basis for the application of these concepts and principles in the field.

In some cases, concepts, principles, standards and guidelines emerge from soft-law[7] instruments that guide implementation of hard law by providing regulatory details. These instruments also help States to adopt uniform minimum standards. They may also contribute to the codification of current norms or be the first step in developing hard-law instruments.

Market-based tools can also support the application of principles, goals and commitments set out in soft-law instruments, for example through incentives provided by eco-labeling, to increase coherence in the whole value chain from natural resource extraction to final consumption (cf. 'Incentives' in 'Fishery management tools').

Over time, a whole fabric of instruments, approaches and tools (conventional and modern) has emerged as discussed in 'Operational plans and processes' and 'Fisheries management tools', implemented through command and control, the market, co-management and community-based approaches. This fabric affects societal choices and encourages behaviour that is conducive to conservation and sustainable use of the oceans and applies to sectoral issues as well as broader cross-sectoral conservation interests. Ideally, the whole fabric should be coherent (even integrated where feasible) across uses of marine ecosystems as well as within each sector.

Basis and application of integrated approaches and principles in framework instruments and global policy instruments

Overarching policy and legal framework instruments

The linkages between resource utilization, through economic activities such as fisheries and conservation, can be discerned in a number of global policy and legal instruments. Of greatest importance are UNCLOS and the CBD.

UNCLOS is an overarching and multi-sectoral governance framework based on strong sectoral pillars covering key ocean activities including fisheries, mining and shipping. UNCLOS also addresses a number of cross-sectoral issues: its part XII provides for cross-sectoral considerations of conservation of the marine environment and sets out the basis for integrated management at the highest level (cf. Chapter 7 for an elaboration of UNCLOS obligations). Biodiversity or ecosystems as such are not mentioned but traces of a holistic approach to utilization of living resources are evident (e.g. UNCLOS Articles 61, 119, 145, 192 and 193).

The CBD deals specifically with biodiversity and links conservation to sustainable use and requires that States share benefits derived from the use of genetic resources across sectors. The integrative nature of the CBD and its relevance for marine areas is evident in a number of its provisions. Notably, the CBD applies to any process and activity carried out under the jurisdiction or control of Contracting Parties, *regardless of where its effects occur* (Article 4b). Furthermore, Contracting parties are required to integrate conservation and sustainable use in cross-sectoral and sectoral plans, programmes and policies (Article 6b) and within overall management, including by considering ecosystems (Article 8).

The CBD clarifies its relationship with other international legal frameworks, and underlines the equally important and mutually supporting nature between the CBD and UNCLOS by:

- ensuring that the rights and obligations deriving from existing international agreements are not affected, except when exercising those rights and obligations would cause serious damage or threaten biological diversity (Article 22(1)); and
- specifying that, with respect to the marine environment, CBD shall be implemented consistent with the rights and obligations of States under 'the law of the sea' (Article 22(2)).

UNCED (in 1992) led to the adoption of Agenda 21, an important non-binding policy instrument for both environment conservation and fisheries management. Agenda 21 in its Preamble calls for greater attention to the integration of environment and development concerns in order to fulfil basic needs, improve living standards, better protect and manage ecosystems and prepare a safer and more prosperous future. The principles and approaches stipulated in its forty chapters link environmental concerns with development (and therefore management) concerns as a core recurring theme (cf. Chapter 10).

The United Nations General Assembly (UNGA) addresses fisheries and marine conservation issues in the context of its two annual resolutions on: (1) Oceans and the Law of the Sea; and (2) Sustainable Fisheries. It has initiated a number of processes designed to facilitate multilateral discussions on ocean matters including: the UN Open-ended Informal Consultative Processes on Oceans and the Law of the Sea (ICP), which looks at emerging ocean issues; the Regular Process for Global reporting and Assessment of the State of the Marine Environment, which aims at developing institutional arrangements and strategies to conduct the first integrated global marine assessment; and the Ad Hoc Open-Ended Informal Working Group to Study Issues Relating to the Conservation and Sustainable Use of Biodiversity Beyond Areas of National Jurisdiction (cf. Chapter 11 for more details).

The UNGA Resolutions are not binding. However, they have significant moral weight and can initiate dynamic processes through which they continue to be referred to, affirmed and enhanced, and they may instigate the development of other high level international frameworks.[8] In several resolutions, the UNGA reaffirmed 'the importance it attaches to the long-term conservation, management and sustainable use of the living marine resources…and the obligations of States to cooperate to this end…as reflected in the relevant provisions of [UNCLOS].'[9]

WSSD[10] (in 2002) and the United Nations Conference on Sustainable Development (UNCSD or Rio + 20 in 2012) are multi-sectoral processes, which followed UNCED and maintained the impetus on implementation by States of their non-binding commitments and triggered the development of global policy instruments. In their sections on oceans and seas, Agenda 21, the Johannesburg Plan of Implementation (POI) and 'The Future We Want' call for strengthening integration and refer to the importance of the ecosystem approach (cf. Chapter 10).[11]

Implementation instruments

The fabric of policy and legal instruments described above establish principles and approaches including EA, the duty to cooperate, use of best scientific evidence, balance of rights and obligations and fair and equitable sharing of benefits. These are further elaborated in implementing instruments. The implementing instruments are subsequently translated into management and conservation approaches and related tools. The following section is an analysis of the UNFSA and the CCRF to demonstrate how, in fisheries, provisions from high-level instruments trickle down into management approaches and tools.

The United Nations Fish Stock Agreement (UNFSA)

The UNFSA (cf. Chapter 7) aims at ensuring conservation and sustainable use of straddling and highly migratory fish stocks. It builds on and adds to UNCLOS and implements the general duty of States to cooperate[12] in relation to the management of straddling and highly migratory fish stocks. Among others, it requires States to participate in and establish RFMO/As (Article 8(3), (5)).

The UNFSA[13] translates the framework obligations in UNCLOS at a more detailed level. In elaborating the duty of States to cooperate, it requires States to ensure compatibility of agreed conservation and management measures (Article 7(2)) and, in the absence of such agreed measures, to agree on provisional arrangements or measures (Article 7(6)), both of which imply the need for coordination. The UNFSA also combines sustainable management of target fish stocks with the protection of biodiversity in the marine environment by requiring States to cooperate to this end (Article 5 g). It emphasizes cooperation by designating RFMO/As as the main means for cooperation, and calls for States to join RFMO/As and adopt and comply with their conservation and management measures. In regions where RFMO/As do not exist, the UNFSA requires States to establish them (Article 8(5)). For the first time in any binding fisheries instrument, the UNFSA boldly enunciates the need to apply the precautionary approach, which requires the adoption of conservation and management measures even in the absence of scientific evidence (Article 6(2)).

In addition, the UNFSA explicitly requires consideration for other species within the same ecosystem, including dependent and associated species (Article 5e) reflecting, in a more detailed way, UNCLOS (Article 61(4)).

In addition to their obligation to apply the general principles listed in article 5 of UNFSA to fisheries and areas within national jurisdiction (Article 3), States should ensure that conservation and management measures that apply in areas beyond national jurisdiction are compatible with and do not undermine measures that apply within areas under national jurisdiction, ensuring that such measures also apply across the population range of straddling and highly migratory stocks.

Membership to RFMO/As is open to States with a real interest in fisheries (Article 8.3). The UNFSA attempts to be more integrative by requiring that representatives from intergovernmental organizations and non-governmental organizations concerned with straddling and highly migratory fish stocks have the opportunity to participate in RFMO meetings and have timely access to reports and records (Article 12(2)). While the participation of outside representatives could open up RFMO/As to external involvement, participation at meetings and access to reports and records are subject to the RFMO rules and procedures. In this way, RFMO/A member States ultimately determine the degree to which non-member representatives, IOs and NGOs are afforded the opportunity to participate effectively.

The Code of Conduct for Responsible Fisheries (CCRF)

The 1995 CCRF (cf. Chapter 7) is an excellent example of a global soft-law implementation instrument that further enhances the integration of environmental concerns (as specifically expressed at UNCED) into fisheries

Box 12.1 Areas Beyond National Jurisdiction and the Deep Seas: An example of integrative implementation instruments

The International Guidelines for the Management of Deep-sea Fisheries in the High Seas (FAO Deep Seas Guidelines) explicitly refer to the global framework that they intend to implement (UNGA Resolution 61/105, paragraphs 76–95). They provide guidance on how to achieve the objectives set forth in UNGA Resolution 61/105 in relation to managing vulnerable marine ecosystems, addressing concerns about destructive fishing practices already expressed in previous UNGA Resolutions (A/RES/59/25, paragraph 66–69). The FAO Deep Seas Guidelines aim at the long-term conservation and sustainable development of marine living resources in deep-sea areas and at preventing significant adverse impacts on VMEs. The Guidelines require States to adopt measures in accordance with the precautionary approach and the ecosystem approach to fisheries, and in accordance with international law. They also call upon States to identify areas where VMEs are known or likely to occur and to take action using the best information available (paragraph 12). The Guidelines are currently being implemented by a number of RFMO/As and have to a certain extent provided the bridge between the global soft-law standards set by the UNGA and the management practices of States through RFMO/As.

management. It consists of a collection of principles, goals and action items seeking to place fisheries management within broader environmental, social and economic concerns. Its specific aspects are further elaborated in associated multilaterally negotiated soft-law instruments: International Guidelines and International Plans of Actions. The CCRF requires fishers and States to consider the environment. It refers to instruments adopted for other sectors, in relation to safety at sea and pollution (Articles 8.4.1 and 8.7)[14] and incorporates many principles of the EAF, such as moving away from a target-species approach and considering environmental, social and economic factors (Articles 6.2 and 6.4). The CCRF embraces the precautionary approach (Articles 6.5 and 7.5). It refers in general to aquatic ecosystem conservation and not only to 'resources' (Article 6.1). It recognizes interaction between fisheries and the ecosystem (Article 6.4) and promotes the use of environmentally safe gear (Article 6.6). The CCRF urges States to protect and rehabilitate critical fisheries habitats in marine and freshwater ecosystems (Article 6.8) and to ensure that fisheries interests, including the need for conservation of resources, are taken into account and integrated into coastal area management, planning and development (Article 6.9).

The call for taking care of fisheries interests is balanced by the need to ensure that decision-making processes are transparent and timely, that consultations with environmental and other interested organizations are undertaken in accordance with appropriate procedures (Article 6.13) and that relevant fishery institutions are informed and consulted before decisions are made that may affect conservation and management measures taken by a competent RFMO/A (Article 7.2.5). Despite its non-binding nature, the CCRF does not deviate from the UNFSA in relation to participation of IOs and NGOs in meetings of RFMO/As, which is left to the discretion of respective RFMO/As (Article 7.1.6).

The CCRF reflects the UNFSA by requiring States to cooperate through fisheries management organizations and other international agreements or arrangements to promote conservation and management. To this end, States should consider the distribution range of the fisheries resources and take into account the need for compatibility of measures that apply in areas beyond national jurisdiction with those that apply within national jurisdiction (Article 6.12).

Management processes, approaches and tools for integrated implementation

The effective implementation of globally agreed principles and approaches that promote integration between sectoral development and biodiversity concerns requires their translation into operational frameworks and specific plans and the use of specific tools. Some elements of that 'implementation chain' are described in the following sections.

Management plans, processes and approaches

Integrated Coastal Area Management (ICAM)

ICAM can be described as a 'dynamic process in which a coordinated strategy is developed and implemented for the allocation of environmental, socio-cultural and institutional resources to achieve the conservation and sustainable multiple use of the coastal zone' (CAMPNET, 1991). ICAM has been promoted for decades and figured prominently in Agenda 21. It features in the 1995 CBD Jakarta mandate on marine and coastal biodiversity. It is applied primarily in the coastal area and focuses on the development of integrated cross-sectoral management plans and strategies that reduce conflict. It is a variant of Integrated Coastal and Ocean Planning (ICOM) and contains many of the principles that will be integrated later in marine spatial planning (MSP) as described in the following section.

These approaches offer a means of balancing the competing demands of different users of the same resources and of managing the resources to optimize the benefits to be derived on a sustainable basis that is consistent with a country's goals (Scialabba, 1998). This concept is embedded in the CBD and is recognized as an integrative management process by FAO and within many countries and regions. ICAM aims at bringing together sustainable use of coastal resources and environmental conservation. Integrating fisheries into ICAM is recommended by the CCRF which calls on States to engage in cooperation at regional levels and establish suitable institutional frameworks, policy measures and mechanisms for cooperation and coordination among national authorities involved in planning, development, conservation and management of coastal areas (Article 10). However, the momentum from the international community appears to be focused on MSP to address the management of multiple uses of marine spaces as evidenced by the more than 20 countries that have already developed plans.[15]

Marine Spatial Planning (MSP)

MSP provides a practical way of organizing the use of marine space and resources in a transparent, integrated and strategic manner. MSP is a method of planning for the application of an EA and builds off other broader approaches to management such as ICAM. Spatial management allows for management of potentially conflicting uses and is a mechanism through which space can be allocated for the variety of human activities taking place in the marine environment. While usually considered to be a relatively new public process, MSP has the potential to produce outcomes that are positive both from a conservation and sustainable use viewpoint. Key features of successful MSP programmes include consideration of multiple scales; a long-term perspective; recognition that humans are an integral part of ecosystems; an adaptive management perspective; and concern for sustaining ecosystem goods and services. In many countries, the need for MSP has been driven by the intensifying and increasing uses and competition in the marine environment.

Although MSP and ICAM are potentially excellent practical management processes and approaches for integrating designated policy goals, policy makers and managers struggle with the complexity and diversity of challenges they pose. In addition, in practice some sectors (e.g. fisheries) may not be included in MSP or ICAM processes or may decide not to participate due to concerns about the process or outcomes. Other concepts and their related processes exist (Chapter 21) to promote integration such as the Australian Ecologically Sustainable Development (ESD).

Operational plans and processes

To implement management approaches, implementation will require sectoral plans (and related processes) aligned with applicable norms, whether they derive from sectoral or cross-sectoral instruments. Examples from the fishery sector that illustrate efforts to integrate biodiversity concerns into sectoral management are given in the following sections.

Fishery Management Plan (FMP)

An FMP is defined as 'a formal or informal arrangement between a fishery management authority and interested parties, which identifies the partners in the fishery and their respective roles, details the agreed objectives for the fishery and specifies the management rules and regulations which apply to it and provides other details about the fishery which are relevant to the task of the management authority' (FAO, 1997, p. 58). It integrates management action across all sustainability dimensions and issues of the sector (CCRF Article 7.3.3; FAO, 1997). FMPs are essential for the management process to achieve goals and objectives, and the fisheries legislation of many countries requires their formulation. Within the progressively broadening mandates of fisheries authorities, environmental concerns and regulations can be addressed and integrated with user needs in FMPs. The CCRF and many countries require that the development process of FMPs engage all stakeholders, including the conservation community and local fisher communities, and that they consider a broad suite of management approaches and tools including MPAs (FAO, 2012). When effectively applied, this inclusiveness promotes consideration and even mainstreaming of environmental concerns related to fisheries. In addition, FMPs can and should provide a link from the fisheries sector to broader management processes and plans such as ICAM and MSP when they exist. MSP and ICAM can also relate explicitly to biodiversity management tools such as conservation plans for MPAs related to fishery grounds.

When appropriately designed and implemented, FMPs result in more extensive and inclusive integration of sector development planning and management processes. However, the complexity of MSP and ICAM and insufficient capacity in many countries must be kept in mind when considering linking FMPs to integrative frameworks. FMPs increasingly require the application of precaution and the ecosystem approach to fisheries mentioned in 'Integrative concepts and approaches'. A review of the required content of FMPs as set out in the Code and national fisheries legislation indicates a trend towards better consideration of environmental constraints and specific requirements for broader conservation measures (Die, 2009; Kuemlangan, 2009).

It should be stressed that the mandatory character of modern FMPs gives them a fundamental role in promoting or blocking progress towards integration of concerns, depending on how effectively they are formulated and implemented.

The degree of integration of fisheries and biodiversity governance goals and constraints in an FMP (or an MPA management plan) depends primarily on the efforts made at higher governmental levels to develop the proper cross-sectoral policy and necessary institutional bridges (cf. Chapter 3), clearly allocating responsibilities including for oversight and performance assessment. However, there is still a long way to go in most countries in the establishment of the integrative management tools such as MSP, ICAM or ESD (cf. Chapter 21).

Impact and risk assessments

Well-integrated FMPs will provide for a number of processes facilitating the integration of environmental concerns. Participative research and decision-making are among these. Below, the processes needed to assess impacts (on biodiversity) and the related risks are dealt with.

The use of impact assessments (IAs) in sectoral, cross-sectoral or transboundary settings to determine if anthropogenic activities have potential adverse impacts on the environment is now standard practice (Druel, 2013). UNCLOS describes a general obligation to carry out such assessments if a state has reasonable grounds to believe that planned activities will cause significant harm to the marine environment (Article 206). IAs are also required in many of the instruments for biodiversity conservation described above including the CBD, CITES and the Convention on Migratory Species (CMS). In order to assist states in a more comprehensive inclusion of biodiversity-related concerns of IAs, the CBD developed voluntary guidelines on biodiversity-inclusive environmental impact assessment (CBD 11th CoP, Decision XI/18).

At the sectoral level, the CCRF sets the framework for the fisheries sector which is expected to reduce its ecosystem impacts to a limit that is both acceptable to society and compatible with its own sustainable existence. Chapters 4 and 9 review the rationale and basis for assessing the risks that fishing poses to various ecosystem components, typically through environmental impact assessments (EIA) and ecological risk assessments (ERA), including cumulative impacts. EIA and ERA are both proposed as an integral part of implementing an ecosystem approach across fisheries sectors.

An EIA or ERA for fisheries should assess the potential impacts of fishing on target species, bycatch species, associated habitat, the broader ecosystem as well as the impact on the resource from other fishing activities and, as much as possible, other non-fishing activities. These assessments should also include social and economic impacts and risks, helping to bridge conservation with fisheries objectives. However, their systematic application in real management situations is still evolving. Except in marine aquaculture where EIAs or environmental impact studies and reports are now required in many fisheries and aquaculture legislation, environmental impact studies or reports are only beginning to emerge as a mandatory requirement for specific fisheries activities, for example in South Africa[16] in relation to the granting of a fishing right and in Namibia[17] in relation to the granting of a right to undertake exploratory harvesting of marine living resources.

Fishery management tools

To attain fisheries management objectives, a broad range of tools are available within the sector to shape or define limits to fishing activities and to create incentives to comply with these limits. Fisheries management tools may be classed into *input controls* (how and when to fish), *output controls* (how many fish to catch), *spatial and temporal controls*

Box 12.2 Environmental Impact Assessment in Areas Beyond National Jurisdiction (ABNJ)

Noting the difficulties of undertaking EIAs in ABNJ, the Conference of the Parties of the CBD (CoP) began to elaborate the scientific and technical aspects specific to EIAs and Strategic Environmental Assessments (SEAs) for ABNJ. Voluntary guidelines for the consideration of biodiversity in EIAs and strategic environmental assessments for biodiversity in marine and coastal areas, including in ABNJ, were discussed and noted by parties at the CoP in 2012.

The FAO Deep-seas Fisheries Guidelines addresses impact assessments and recommend that they be conducted to determine whether deep-sea fishing activities are likely to produce significant adverse impacts on vulnerable marine ecosystems. The impact assessment should include a description of the fishing activity and the related ecosystem, any known or likely vulnerable marine ecosystem, an evaluation of uncertainties in information presented, description of the likely impacts and risk of those impacts occurring and the proposed mitigation and management measures. If applied fully, such an impact assessment would be in line with other similar cross-sectoral recommendations such as the 1987 UNEP Goals and Principles of EIA. In some regions, flag States have completed initial EIAs for fishing activities. In other regions RFMOs are debating which areas and activities require EIAs. The South Pacific RFMO has been active on this topic and has developed an Interim Benthic Assessment Framework to use as a guide in assessing EIAs developed by its member states.

(where and when to fish) and *technical measures* (what technology to use for fishing). The ecological basis for these tools is presented in Chapter 4. In addition, there are also several approaches to ensuring compliance or uptake of these tools, such as the creation of legal, market and social incentives (DeYoung and Charles, 2008). Any given management context will require an analysis of the available suite of tools and how each tool or combination of tools may assist in attaining the objectives in that particular context.

This section provides a brief overview of fisheries management tools and how they are linked to better use and conservation of fisheries resources, and considers their potential or actual role in relation to the broader objective of biodiversity conservation.

Input and output controls

These types of measures either restrict the intensity of use of fishing gear (i.e. input controls, such as restrictions on the number or size of vessels or the amount of effort permitted) or the amount of fish taken by a specific fishery (i.e. output controls, such as quotas or total allowable catch limits and constraints on authorized length or age). These tools are often used to control the amount taken of target species, but can also be used for management of bycatch species (retained or discarded). Both sets of tools aim to maintain overall fishing within sustainable limits, and the control/reduction of fishing pressure they aim at has a direct effect on the pressure exerted on the ecosystem and thus on the conservation of marine biodiversity (Pope, 2009; Chapter 4). Reducing fishing pressure will also reduce collateral impact on biodiversity.

Technical measures

As explained in Chapter 4, fishing gear can also have impacts on the habitat or ecosystem. Technical measures may include gear regulations designed to reduce or modify these impacts (for example regulating gear type, structure, size and mesh size). They may also aim to reduce unintentional fishing such as reducing bycatch and avoiding 'ghost fishing' by lost or abandoned gear, for example by using biodegradable materials. The potential impact on biodiversity management is obvious.

Trade-related measures for endangered or threatened species

Actual measures to avoid or mitigate impact on endangered or threatened species (ET species) usually include those listed above such as technical measures (changes in fishing gear used to reduce capture or modifications such as turtle excluder devices to reduce bycatch) or spatial or seasonal

Box 12.3 The FAO Agreement on Port State Measures to Prevent, Deter and Eliminate Illegal, Unreported and Unregulated (IUU) Fishing

The FAO Agreement on Port State Measures to Prevent, Deter and Eliminate Illegal, Unreported and Unregulated (IUU) Fishing (the Port State Measures Agreement) was adopted by the FAO Conference in 2009, but has not yet entered into force. The main purpose of the Agreement is to prevent, deter and eliminate IUU fishing through the implementation of a suite of port State measures. These measures include the denial of landing of fish caught in contravention of international conservation and management measures and the denial of port entry of vessels known to have engaged in IUU fishing. The Port State Measures Agreement thus provides a powerful tool against vessels carrying fish derived from IUU fishing activities by physically blocking entry into markets, to promote sustainable fisheries.

closures to avoid impact on certain species groups (e.g. during spawning aggregations). Economic incentives and disincentives (including taxes or compensatory payments for residual mortality) are also being tested.

CITES was developed to minimize the effect of international trade on the survival of flora and fauna. Several species of interest to the fishing industry have been listed in one of the appendices of CITES. International trade in species listed in CITES Appendix I is allowed only in exceptional circumstances and subject to import and export permits; for those listed in CITES Appendices II and III, trade is controlled through export permits. The listing of several species of interest to fisheries in one of the appendices of CITES has led to the establishment (through an MOU between CITES and FAO) of an ad hoc expert advisory panel, comprising experts identified by FAO. The panel assesses proposals for listings of commercially exploited aquatic species taking into account the panel membership's collective knowledge and experience and the information received by the CITES Secretariat and from other organizations, such as IUCN on the status of the species concerned. Chapter 13 looks at cooperation and concerns between FAO and CITES on the listing issue.

Spatial management tools

Spatial management tools (SMTs) are used by both the fisheries and biodiversity community to control fishing and protect the ecosystem respectively, and provide opportunities for integrating management. Many of the instruments described above encourage the use of SMTs including marine protected areas for both fisheries and conservation purposes, for example Agenda 21 (chapter 17.7), CBD (Article 8) and the WSSD-POI (para. 32). In fisheries, the CCRF (Article 6.8 and 7.6.9) and the Technical Guidelines on EAF (FAO, 2003) and on MPAs and fisheries (FAO, 2012) elaborate the importance of MPAs as a fisheries management tool. Spatio-temporal management tools have always been used in fisheries to reduce bycatch and protect critical habitats. The requirements for deep-sea fisheries to avoid significant adverse impacts on vulnerable marine ecosystems in UNGA Resolution 61/105 have augmented the role of spatial tools in the high seas, as precautionary closures of areas that host VMEs are required until appropriate management measures for the area have been developed (cf. Chapter 14). In the conservation community, spatial tools including MPAs also are used for a variety of objectives ranging from total protection to multiple-use management frameworks. The interaction of the fisheries and biodiversity conservation streams with regard to views on appropriate uses of spatial management tools is complex and has been addressed in Chapter 8 and Chapters 22–24.

Incentives

The above-mentioned tools are often applied through regulatory incentive approaches. That is, a rule is defined and those who are not in compliance tend to receive some predefined penalty. The high costs associated with monitoring and enforcing regulatory incentives has led to other complementary measures, including through developing social, institutional, economic and market-based incentives. These incentives attempt to alter fishing practices by making responsible behaviour more rewarding than irresponsible behaviour. Incentives can contribute to compliance with regulation of fisheries or biodiversity conservation, and therefore have a huge potential for integrating the concerns of both fields (if they can be translated in the same currency).

Social or institutional incentives, such as the establishment of community-based management systems (e.g. in fisheries or spatial management zones), use more participatory approaches and peer relationships for defining and enforcing sustainable fisheries, sustainable use and conservation objectives and tools for a particular ecosystem. Such collaborative management arrangements may provide stakeholders with a sense of 'ownership' and stewardship of the resources and their ecosystem (cf. Chapter 23).

Economic and market-based incentives attempt to adjust behaviour through providing financial benefits (or costs) to actions that do (or do not) support the fishery management and conservation objectives. These incentives include: fees and fines; tradable or sellable quotas; other use rights (e.g. Dolphin Mortality Limits, DLM); taxes; credit schemes; insurance; compensatory payments to offset residual conservation costs; subsidies to promote the use of best-available technologies (e.g. turtle exclusion devices, vessel monitoring systems); and conservation payments to compensate for restrictions on fishing patterns (e.g. no-take zones or seasons).

Eco-labels and guides, fair-trade labels, good fish guides and *fish fairs* may also affect the markets for fish products. They are based on standards and are geared towards the provision of information to consumers regarding the circumstances leading to the availability of products (e.g. fishing practices, stock status and management regimes) with the view to generate a market advantage as well as a more ecologically friendly consumer behaviour with consequences across the value chain.

The fisheries and biodiversity conservation governance streams are both experimenting with this range of tools and the various approaches to implement them. The acknowledgement that no single tool will allow either fisheries or conservation objectives to be met and that a suite of tools tailored to each fishery system, combined with incentives mechanisms, make sense for the system, may prove the fastest way to achieving both sets of objectives.

Reinforcing integrated implementation at regional level

Integration can be achieved through the application of tools and approaches that are by definition integrative, or through institutional linkages with bodies typically concerned with conservation. Due to limited space, this section is restricted to just a few of the available examples of regional integration efforts.

Performance review of RFMOs

The growing concern for the general state of targeted fish stocks and fishing-related degradation of the marine environment, the emergence of the PA and EAF and the renewed focus on strong and accountable RFMO/As as key mechanisms for international cooperation (see above) triggered a call to review and strengthen RFMO performance. That call gained momentum through the United Nations Review Conference on the UNFSA, the recommendations of the ICP[18] and a number of annual UNGA Resolutions.[19]

Some RFMO/As have strengthened management and integration by amending their basic texts, while more recently established RFMOs took an integrated approach in the design of their founding legal instruments. This section provides some examples of RFMO/As and regional environmental organizations that integrate biodiversity concerns through formal arrangements such as amendments of their basic texts or development of *ad hoc* arrangements.

Performance reviews of the Commission for the Conservation of Southern Bluefin Tuna (CCSBT), the Indian Ocean Tuna Commission (IOTC), the International Commission for the Conservation of Atlantic Tunas (ICCAT), the Northwest Atlantic Fisheries Organization (NAFO) and others indicate that the conventions of these organizations need to be updated to include modern fisheries management principles including the EAF and the precautionary approach. In most cases, it is still too early to see whether these recommendations have led to adopted amendments in the constituting agreements of these RFMOs. Steps taken by RFMOs after their performance reviews indicate the good direction in which many are heading.

Although RFMO/As vary in their application of precaution and EAF, some have adopted them. The Performance Review Process has facilitated this trend. For example, in its new Convention, NAFO has moved away from the single-species approach and adopted the ecosystem approach to fisheries.[20] When setting catch levels, NAFO has used the PA to ensure long-term sustainability of resources (article III(a) and (c)). The term 'living marine resources' is now used to ensure that all living components of the marine ecosystems are considered when taking management decisions.

Regional cross-sectoral cooperation

The Secretariat of the Pacific Regional Environment Programme (SPREP) is the primary regional organization concerned with environmental management and sustainable development in the Pacific. SPREP has adopted a number of action plans and programmes relevant to fisheries, including the Pacific Island Marine Species Programme (MSP) framework 2003–2007/2008–2012, which includes the Marine Turtle Action Plan, the Whale and Dolphin Action Plan and the Dugong Action Plan. In the Marine Turtle Action Plan, SPREP lists collaboration and partnerships with relevant organizations as an objective and the establishment of formal communication with a number of organizations including Pacific Islands Forum Fisheries Agency (FFA), Western and Central Pacific Fisheries Commission (WCPFC) and the Secretariat of the Pacific Community (SPC) as a related activity.

As a follow-up, MoUs with WCPFC and FFA have been concluded (Pacific Island Marine Species Programme or MSP framework 2003–2012, Marine Turtle Action Plan, Theme 1, Collaboration and Partnerships 1.3). The Marine Turtle Action Plan also calls for the development and implementation of management and mitigation actions related to fisheries bycatch, including cooperation with relevant organizations, improved documentation, identification, reporting and an observer program (Theme 2, Threats 2.3, 2.7 and 2.8). The Whale and Dolphin Action plan also calls for the development of linkages with relevant regional organizations, including RFMOs and regional processes (Theme 1, Collaboration and Cooperation 1.6). This includes sharing information and developing mitigation techniques to combat IUU fishing (Theme 2, Threat Reduction). In October 2009, a Regional Plan of Action for Sharks (RPOA-Sharks) was developed jointly by SPREP, FFA and SPC. The RPOA-Sharks is a response to the FAO International Plan of Action for the Conservation and Management of Sharks and to the conservation and management measures adopted by the WCPFC (Lack and Meere, 2009). The intensive collaboration on conservation and fisheries management is an excellent example of how regional environmental and fisheries bodies are shaping and applying integrated approaches.

Conclusions

This chapter has shown that framework and implementation instruments related to fisheries, biodiversity conservation and those with a cross-sectoral scope all call, explicitly or implicitly, for integration of sustainable use and conservation. Since the early 1980s, global processes have consistently called for such integration in relation to the marine environment, particularly by urging concerned stakeholders to use precaution and ecosystem considerations as the basis for management approaches. Specific tools have been developed that ensure application of such approaches and higher-level objectives on the ground. These integrative tools in turn find their basis in implementing instruments and their application is supported by global governance processes.

The fabric of instruments, processes, approaches and tools discussed in this chapter provides the basis for fisheries management to ensure that there is appropriate concern for biodiversity considerations and conservation, to ensure that there is appropriate consideration for user needs. Achieving the appropriate balance is essential for true integration but this will remain a challenge, particularly at the implementation level. In the fisheries sector, for example, the right to participate in meetings of RFMO/As and to have access to documents is and remains subject to the rules and procedures of the individual RFMO/A which promotes exclusivity. Nevertheless, the positive practical examples of integration at the regional level given in this chapter or provided in other chapters (cf. Chapters 15–19) illustrate that integration is possible. The examples show that States that are committed to integrated management are able, with the instruments that are currently available to them, to govern oceans successfully. While specificities of the region can have a bearing on the outcome, much depends on political will and a certain level of coordination at the national level, which is crucial.

Fostering these integrative initiatives for better implementation, political will and more effective coordination should contribute to improved ocean governance.

Acknowledgements

The authors wish to acknowledge the valuable inputs from S.M. Garcia, L. Ridgeway, J. Rice (Department of Fisheries and Oceans, Canada), A. Charles (St. Mary's University) and Z. Shah (FAO). The views expressed in this article are those of the authors and do not necessarily reflect in any way those of the Food and Agriculture Organization of the United Nations.

Notes

1. In this chapter, 'sectoral' refers to distinct types of activities carried out in the marine environment such as fishing, shipping and mining. 'Multi-sectoral' refers to a combination of two or more sectoral activities. 'Cross-sectoral' refers to those issues that are not sector-specific and apply throughout sectors, such as conservation, climate change and biodiversity.
2. For example, the ecosystem, precautionary, sustainable livelihood or sustainable use approaches are still being tested in both streams and not yet fully mainstreamed and generalized at national level.
3. United Nations Agreement for the Implementation of the Provisions of the United Nations Convention on the Law of the Sea of 10 December 1982 relating to the Conservation and Management of Straddling Fish Stocks and Highly Migratory Fish Stocks (UNSFA).
4. National-level examples of the integration of precaution are not dealt with in this chapter.
5. Agenda 21 is the main output of the 1992 UN Conference on Environment and Development (UNCED) held in Rio de Janeiro, Brazil in 1992. It is divided into 40 chapters. Chapter 17 is entitled 'Protection of the oceans, all kinds of seas, including enclosed and semi-enclosed seas, and coastal areas and the protection, rational use and development of their living resources'. See Report of the United Nations Conference on Environment and Development, A/CONF.151/26, Annex II.
6. An EA variant has also been developed for aquaculture (Ecosystem Approach to Aquaculture; FAO, 2010).
7. The term 'soft law' refers to instruments that are not binding, contrary to 'hard law'.
8. For example, the development of the FAO International Guidelines for the Management of Deep-sea Fisheries in the High Seas was a response to UNGA Resolution 61/105, see example in Box 12.112.1.
9. For example, in A/RES/62/177, paragraph 1, A/RES/63/112, paragraph 1, A/RES/64/72, paragraph 1, A/RES/65/38, paragraph 1, A/RES/66/68, paragraph 1.
10. The World Summit on Sustainable Development, Johannesburg in 2000, recognized that the ocean-related objectives of Agenda 21 (chapter 17) were not being widely achieved and adopted a political declaration and a plan of implementation in relation to capture fisheries, ecosystem health and the conservation of biodiversity. In the Declaration, the Heads of States agreed to 'develop and facilitate the use of diverse approaches and tools, including the ecosystem approach, the elimination of destructive practices, the establishment of marine protected areas consistent with international law and based on scientific information, including representative networks by 2012 and time/area closures for the protection of nursery grounds and periods [...] and the integration of marine and coastal areas into key sectors'.
11. WSSD Plan of Implementation (WSSD-POI) paragraph 32c; The Future We Want, paragraph 158.
12. The general duty to cooperate is rooted in customary international law but its implementation requires the elaboration of specific directives: framework agreements, implementation agreements, guidelines, etc. at international, regional and national levels.
13. The call for States, under chapter 17 of Agenda 21 Programme Area C, 'to take effective action to ensure that high seas fisheries are managed in accordance with UNCLOS and to convene, as soon as possible, an intergovernmental conference under United Nations auspices' resulted in the UNFSA.
14. This includes ship-based pollution provisions of MARPOL 73/78, the safety-of-life-at-sea provisions of the SOLAS Convention and several IMO rules and regulations.
15. See IOC-UNESCO website: http://www.unesco-ioc-marinesp.be/msp_around_the_world
16. South Africa Marine Living Resources Act 18 of 1998, section 18.
17. Namibia, Marine Resources Act of 2000, Section 34.
18. The ICP addressed the role of RFMO/As at its 6th meeting in 2005 (UNGA Resolution A/60/99, para. 7) and proposed that the UNGA call on all relevant States to (1) become members of RFMO/As; (2) welcome and urge efforts by RFMO/As to fill gaps in their mandates in order to apply the EAF; and (3) include biodiversity considerations and the precautionary approach. It also mentioned the need for strengthening integration, coordination and cooperation among RFMO/As, regional seas arrangements and other relevant organizations.
19. In particular the 2005 UNGA Resolution A/60/31 (para. 31) on Sustainable Fisheries which encourages States to initiate performance reviews of RFMO/As and recommends broadening the mandate of RFMOs to include an EA.
20. Amendment to the Convention on Future Multilateral Cooperation in the Northwest Atlantic Fisheries, 2007, articles II and III(d).

References

Bianchi, G., Sandberg, P., Skjoldal, H.R. and Thorarinsson, K. (2008) The Bergen Conference on Implementing the Ecosystem Approach to Fisheries (Bergen, Norway, 26–28 September 2006): Summary and main conclusions. In: *The Ecosystem Approach to Fisheries* (eds G. Bianchi and H.R. Skjoldal), pp. 1–19. CABI, Wallingford, UK and FAO, Rome, Italy.

Bianchi, G. and Skjoldal, H.R. (eds) (2008) *The Ecosystem Approach to Fisheries*. CABI, Wallingford, UK and FAO, Rome, Italy.

CAMPNET (1991) *The Status of Integrated Coastal Zone Management: A Global Assessment* ed. J. Clark). Coastal Area Planning and Management Network. University of Miami/RSMAS, Miami, Florida.

Die, D.J. (2009) Fisheries management plans. In: *A Fishery Managers' Guidebook*, 2nd edn (eds K.L. Cochrane and S.M. Garcia), pp. 425–446. Wiley-Blackwell, Chichester, UK and FAO, Rome, Italy.

DeYoung, C. and Charles, A. (2008) Creating incentives for the ecosystem approach to fisheries management; a portfolio of approaches. In: *The Ecosystem Approach to Fisheries* (eds G. Bianchi, and H.R. Skjoldal), pp. 138–157. CABI, Wallingford, UK and FAO, Rome, Italy.

Druel, E. (2013) Environmental impact assessments in areas beyond national jurisdiction: identification of gaps and possible ways forward. Studies No.01/13. IDDRI, Paris, France. 42 pp.

FAO (1997) Fisheries management. FAO Technical Guidelines for Responsible Fisheries, No. 4. FAO, Rome.

FAO (2003) Fisheries management. 2. The ecosystem approach to fisheries. FAO Technical Guidelines for Responsible Fisheries, No. 4 (suppl. 2). FAO, Rome.

FAO (2010) Aquaculture development. 4. Ecosystem approach to aquaculture. FAO Technical Guidelines for Responsible Fisheries, No. 4 (Supp. 4). FAO, Rome.

FAO (2012) Fisheries management. 4. Marine protected areas and fisheries. FAO Technical Guidelines for Responsible Fisheries, No. 4 (Suppl. 4). FAO, Rome.

Garcia, S.M. (1996) The precautionary approach to fisheries and its implications for fishery research, Technology and Management: An updated review. FAO Fisheries Technical Papers 350/2 (FAO 1996).

Harremoës, P., Gee, D., MacGarvin, M., Stirling, A., Keys, J., Wynne, B. and Guedes Vaz, S. (eds) (2001) Late lessons from early warnings: the precautionary principle 1896–2000. Environmental issue report no. 22. European Environment Agency, Copenhagen.

Kuemlangan, B. (2009) Legal aspects. In: *A Fishery Managers' Guidebook*, 2nd edn (eds K.L. Cochrane and S.M. Garcia), pp. 105–134. Wiley-Blackwell and FAO, Chichester UK and Rome, Italy.

Lack, M. and Meere, F. (2009) Pacific Islands Regional Plan of Action for Sharks: Guidance for Pacific Island Countries and Territories on the Conservation and Management of Sharks. October 2009. Available at www.sprep.org/att/publication/000853_RPOA_Sharks.pdf (accessed February 2014).

Pope, J. (2009) Input and output controls: the practice of fishing effort and catch management in responsible fisheries. In: *A Fishery Managers' Guidebook*, 2nd edn (eds K.L. Cochrane and S.M. Garcia), pp. 220–250. Wiley-Blackwell and FAO, Chichester UK and Rome, Italy.

Scialabba, N. (ed.) (1998) Integrated coastal area management and agriculture, forestry and fisheries. FAO Guidelines. Environment and Natural Resources Service, FAO, Rome.

Shepherd, G. (2004). The ecosystem approach. Five steps to implementation. IUCN, Ecosystem Management Series No. 3. Available at http://data.iucn.org/dbtw-wpd/edocs/CEM-003.pdf (accessed February 2014).

United Nations (2006) Report on the work of the United Nations Open-ended Informal Consultative Process on Oceans and the Law of the Sea at its seventh meeting. A/61/156, Sixty-first session. 17 July 2006.

Vierros, M. (2008) The ecosystem approach of the Convention on Biological Diversity. In: *The Ecosystem Approach to Fisheries* (eds G. Bianchi and H.R. Skjoldal), pp. 39–46. CABI, Wallingford, UK and FAO, Rome, Italy.

Chapter 13

Conservation and risk of extinction of marine species

P.M. Mace[1], C. O'Criodain[2], J. Rice[3,4] and G.J. Sant[5]

[1]New Zealand Ministry for Primary Industries, Wellington, New Zealand
[2]WWF International, Gland, Switzerland
[3]Fisheries and Oceans Canada, Ottawa, Canada
[4]IUCN Commission on Ecosystem Management (IUCN/CEM/FEG), Gland, Switzerland
[5]Australian National Centre for Ocean Resources and Security, University of Wollongong, Australia

Abstract: We review the evolution of the activities of both the conservation of biodiversity and the fisheries management governance streams to identify and promote the protection of marine species at risk of extinction, particularly marine fish and invertebrates exploited in commercial fisheries. Attention is focused on the roles of IUCN, CITES, FAO, RFMOs and, to a lesser extent, national entities with the main emphasis being on developments in the late 20th and early 21st centuries. Conflicts in the determination of marine species at risk illustrate policy and management issues where the two governance streams have been interacting directly for nearly 20 years, yet differences of viewpoint and practice remain not only in determining risk, but in agreeing what is needed to manage that risk and who should be responsible. There are consequences for both the protection offered to species considered to be threatened and for opportunities to harvest fish to provide food and livelihoods.

Keywords: fisheries; biodiversity conservation; extinction risk; IUCN; CITES; FAO

Introduction

This chapter focuses on the evolution and interaction of the fisheries and conservation of biodiversity governance streams in addressing commercially exploited marine species at risk, covering only a few examples of the parallel evolution of the rich science that has greatly enhanced the information base informing both sides of the debate (e.g. FAO, 2001b; Mace and Lande, 1991; Mace et al., 2002; Mahon et al., 2000).

The key biodiversity conservation organization in the thematic area of species at risk has been the International Union for Conservation of Nature (IUCN), which was founded in 1948. IUCN has a complex internal structure with both non-governmental organizations and individual States having voting membership. Although its resolutions and recommendations do not have binding status on Parties, they provide a strong policy impetus for conservation action. The IUCN internal structure of Commissions and subsidiary expert groups also has strong science capacity so their products, including the various Red Lists of Threatened Species, have high credibility in the expert community.

Another major institution of relevance to species conservation is the Convention on International Trade in Endangered Species of Wild Fauna and Flora (CITES), which entered into force in 1975 to monitor and regulate trade in species at risk from international trade. It now has 180 member countries. Its primary tools are three Appendices where international trade in the species listed in each CITES

Governance of Marine Fisheries and Biodiversity Conservation: Interaction and Coevolution, First Edition.
Edited by Serge M. Garcia, Jake Rice and Anthony Charles.

Appendix is placed under progressively more stringent controls. Appendix I prohibits international trade, with limited exceptions, while Appendix II monitors international trade and regulates it by requiring non-detriment findings[1] for exported species. Species are listed in Appendices I or II by a minimum of a two-thirds majority of voting Parties, normally at meetings of the Conference of the Parties (CoP) which take place roughly every three years, whereas Parties are able to unilaterally place species on Appendix III where the controls are less stringent.

On the fisheries side, the major international players have been the United Nations Food and Agricultural Organization (FAO), founded in 1945, and the Regional Fisheries Management Organizations (RFMOs) founded (or significantly adapted) after the United Nations Convention on the Law of the Sea (UNCLOS) was negotiated during the period 1973–1982. FAO and the RFMOs have been involved in fisheries management issues since their individual foundations, but did not systematically differentiate work on species at risk in their core business until the 1990s. Within national jurisdictions, Fisheries Ministries address fisheries sustainability and management issues including FAO and RFMO mandates, and national Environment Ministries typically have the lead on species at risk activities in a broader biodiversity conservation context.

For the purposes of this chapter, the phrase 'species at risk' is not restricted to species in imminent danger of extinction but rather is defined as species that require explicit safeguards, beyond those based on common practice, to ensure their survival in the wild for the foreseeable future.

Early history of the development of criteria for classifying species in terms of extinction risk

IUCN

The early IUCN Red Lists were based on expert processes within the specialist groups of the Species Survival Commission (SSC), which attempted to provide standardized assessments without quantitative criteria for assigning species to the categories in use. The categories of Vulnerable, Endangered and Critically Endangered were considered the greatest challenge, while Extinct and Extirpated were considered more straightforward classifications to apply and the lower-risk categories were always acknowledged as including a number of considerations (Mace et al., 1993; Scott et al., 1987).

As the profile of the Red List grew through the 1980s, there were increasing concerns that the classifications were to some degree subjective and inconsistent (Collar et al., 1992). In response, in 1989 the SSC Steering Committee initiated work to establish standardized criteria to use in creating Red Lists and appointed an expert group to develop quantitative definitions of the boundaries between the categories of Red List species and quantitative criteria to assign species to the appropriate category.

CITES

Although the species listed on the CITES appendices when it first entered into force in 1975 were determined without the benefit of explicit criteria, CITES quickly passed resolutions to provide further guidance on which species to list on the appendices at its first CoP in Berne, Switzerland in 1976 (Wijnstekers, 2003). The resulting 'Bern criteria' were only semi-quantitative and were intended to provide very broad directions to be taken into account in the evaluation of proposals for listing any species of animal or plant. Their limitations were almost immediately realized. Even so, it was not until the mid-1990s that CITES adopted criteria with more quantitative elements, albeit with repeated disclaimers that any numeric guidelines are 'presented only as examples, since it is impossible to give numeric values that are applicable to all taxa because of the differences in their biology'.

FAO and RFMOs

Prior to the 1990s, the FAO Fisheries Department was heavily involved in developing and implementing UNCLOS and related initiatives. RFMOs were focused on developing and applying their own conventions, all of which incorporated elements of UNCLOS once this came into effect. The main impetus for fisheries organizations was to facilitate sustainable utilization of fisheries, which increasingly involved the need to reduce fishing pressure on commercially exploited fish stocks. The focus was on target species and major bycatch species, and concerns relating to species at risk of extinction were virtually absent. Although it was widely acknowledged that several fish stocks were overexploited and that there was overcapacity in many fishing fleets, fish stocks were generally not believed to be depleted to the extent that extinction was even remotely a risk.

Developments in the 1990s

The events summarized in the previous section set the stage for a rapid evolution of activities in the conservation of biodiversity stream to define criteria for risk of extinction, along with a complementary set of initiatives to compel agencies with appropriate competencies to review the IUCN assessments and to provide species that met such criteria with appropriate protection measures. Species harvested in commercial fisheries received no special treatment in the IUCN activities, but triggered strong reactions by fisheries agencies nonetheless.

IUCN

The working group assigned by IUCN to develop quantitative criteria designed a general framework in 1991 (Mace and Lande, 1991) and submitted its recommended category definitions and classification criteria to the 1994 meeting of the IUCN Council where they were adopted, with a review of performance scheduled for 2000 (Baillie and Groombridge, 1996; Mace et al., 1993). The quantitative criteria for the threatened categories of Vulnerable, Endangered and Critically Endangered provided category-specific metrics for five sets of attributes: (A) population reduction (Table 13.1); (B) geographic range; (C) numbers of mature individuals in combination with decline or fragmentation metrics; (D) numbers of mature individuals alone; and (E) the probability of extinction over a specified time frame obtained from a population viability analysis (IUCN, 1994). From the time the quantitative criteria were adopted, IUCN noted that the primary functions of the Categories and Criteria were intended to be 'a warning flag, signalling [the need for] global attention to the perilous state of many species' and 'calls for attention to the causal factors' (Baillie and Groombridge, 1996). The Guidelines interpreted the Categories and Criteria, and explained how they might be used in various situations.

The 1996 IUCN Congress was synchronous with the release of the first Red List based on application of the new quantitative criteria. A total of 734 bony fish species were listed in the risk categories, including several that were commercially harvested (Baillie and Groombridge, 1996). Several fisheries agencies raised strong objections about the inclusions of commercially exploited fish species in the Red List. Although there was little debate about the inclusion of several species of sturgeon (Acipenseridae), there was heated debate about listing Atlantic cod (*Gadus moruha*) and Atlantic haddock (*Melanogrammus aeglefinus*). The argument centred on the fact that although declines in the abundance of some populations of these species did meet the IUCN quantitative benchmarks for Endangered (Table 13.1), the remaining adult populations of these species still numbered in the tens of millions or more and their distributions were circum-North Atlantic which should make the risk of extinction negligible.

Although the Red List was already published and the Congress did not have to endorse it, no consensus was reached on the conservation measures needed for these and other commercial fish species. Rather, in SSC Resolution 1.4, provisions were inserted in the text on the scheduled review of the performance of the criteria stating that special attention be given to marine species in general and to species under active management programs, and directing the Red List to specify 'that the listings for some species of marine fish are based on criteria that may not be appropriate for assessing the risk of extinction for these species'. SSC Resolution 1.25 on implementation of the Red List was also amended to say that the SSC should 'ensure that these guidelines reflect that there is a caveat to the existing criteria indicating that they may not be appropriate for assessing the risk of extinction for some species of marine fish'.

The division between the fisheries perspective on these issues and view of the SSC was deep.

In 1999, Japan hosted the expert meeting called for in SSC Resolution 1.4.5. The meeting built on the results of a scoping workshop held the previous year (Isaac and Mace, 1998) as part of the planned IUCN review of the criteria. At the prior scoping workshop many of the points of dispute about the Red List had been developed into terms of reference for the focused expert meeting, which attracted at least ten fisheries experts among the 26 participants. From the outset, the workshop displayed a lack of common views on the fundamental issues, with fisheries experts arguing the focus should be on the risk of actual biological extinction and the IUCN SSC representatives arguing that a valuable function of the Red List was to flag species showing 'symptoms of risk of extinction'. There was also disagreement about the appropriate role of expert groups of fisheries agencies in the listing process. Issues discussed included the productivity of populations, timescales of decline and communication of the implications of applying the criteria. The same issues have continued to be hotly debated ever since.

After extensive debate, the workshop adopted nine recommendations addressing the themes of quantitative criteria for the area of distribution, units of assessment and definitions of mature individuals and generation time. Eight different options for recommendations on the decline criteria were considered (Mace, 1999), reflecting the

Table 13.1 Population decline criteria adopted by IUCN in 1994 (IUCN, 1994) and subsequently revised in 2001 (IUCN, 2001), and those adopted by CITES in 1994 (CITES, 1994) and subsequently revised in 2004 (CITES, 2004b). The IUCN (2001) and CITES (2004b) criteria are essentially identical to those that exist today. Note however that they are all in a highly summarized form; for full descriptions, refer respectively to http://www.iucnredlist.org/technical-documents/categories-and-criteria/1994-categories-criteria; http://www.iucnredlist.org/technical-documents/categories-and-criteria/2001-categories-criteria#critical; http://www.cites.org/eng/res/all/09/E09-24.pdf; and http://www.cites.org/eng/res/all/09/E09-24R13.pdf.

IUCN (1994)	**Critically Endangered**	**Endangered**	**Vulnerable**
Population reduction			
A population reduction of ≥ X over the last 10 years or 3 generations, whichever is the longer (or a future projection over this period)	X = 80%	X = 50%	X = 20%
IUCN (2001)			
Reduction in population size			
A reduction in population size of ≥ X over the last 10 years or 3 generations, whichever is the longer, where the causes of the reduction are clearly reversible and understood and ceased,	X = 90%	X = 70%	X = 50%
or			
A reduction in population size of ≥ X over the last 10 years or 3 generations, whichever is the longer, in other circumstances (including future projections)	X = 80%	X = 50%	X = 30%
CITES		**CoP9 (1994)**	**CoP13(2004)**
Population decline			
Marked[a] historical extent of decline			
≥X% of (pre-exploitation) baseline			
'Fisheries footnote' (Appendix I)		–	X = 70–95%[b]
Low productivity species		–	X = 80–85%[b]
Medium productivity species		–	X = 85–90%[b]
High productivity species		–	X = 90–95%[b]
'Fisheries footnote' (Appendix II)		–	= X – (5–10%)[b]
Marked[a] recent rate of decline			
≥X% in the past Y years or Z generations, whichever is the longer		X = 50% Y = 5 Z = 2	X = 50% Y = 10 Z = 3
Marked[a] recent rate of decline for a small population			
≥X% in the past Y years or Z generations, whichever is the longer		X = 20% Y = 10 Z = 3	X = 20% Y = 5 Z = 2

[a]The term 'marked' was not used in the CITES 1994 guidelines.
[b]In the CITES 2004 guidelines, the historical extent of decline is expressed as declines 'to' a percentage of the pre-exploitation level; here they are expressed as a percentage 'of' the pre-exploitation level, in order to make them more compatible with the IUCN guidelines.

inability to reach consensus among the experts on this divisive issue. Extracts from key recommendations include:

- 'The formulation of the values for decline rates in A [Table 13.1] should be reviewed, and these values should be tested against real data sets. Appropriate levels for revised, more exclusive thresholds should be investigated.' (Recommendation 5)
- 'The current formulation of criterion A [Table 13.1] is generally agreed to be too inclusive and has resulted in

the inappropriate listing of many species. The group ... generally agreed: (1) the threshold decline rates are probably too low (and therefore include too many species); more work needs to be done to set new levels; (2) the most precautionary new formulation of criterion A would be to develop an opt-out clause.' (Recommendation 6)

These recommendations clearly demonstrate both the inability of the two perspectives to agree on if and how the decline criteria should apply to marine species, and the shared belief that some future population research and modelling could resolve the differences of viewpoint. This set the stage for most activities on this topic in the 2000s.

CITES

At the 9th CITES Conference of the Parties (CoP9) in Ft Lauderdale, USA in 1994, CITES Parties adopted a variant of the criteria that were in the process of being finalized by IUCN. This had the effect of linking the classification of species directly to binding management actions that result when species are listed on the CITES appendices. This linkage has been a central component of the tension between the fisheries and conservation of biodiversity governance streams ever since.

The decision to develop a more structured set of criteria for the 'Criteria for Amendment of CITES Appendices I and II' was taken at CITES CoP8 in Kyoto, Japan in 1992. IUCN prepared the initial draft Resolution,[2] which was then considered by the Animals and Plants Committees (the scientific committees of CITES) and the Standing Committee (which governs CITES between CoPs) (CITES, 1994a). CoP9 subsequently approved what became known as Resolution Conf. 9.24 (CITES, 1994b), the Resolution enshrining the new listing criteria with their associated definitions and other guidance material. The involvement of IUCN in the initial draft resulted in similarities in much of the language and concepts used by the two organizations. However, the CITES CoP9 criteria were far less structured in terms of quantitative criteria or guidelines, particularly regarding the population decline metrics (Table 13.1).

With regard to the criteria for listing species in Appendix II, Resolution Conf. 9.24 in its original form retained criteria that were based on application of judgement rather than quantitative metrics. This approach was justified on the basis that, in Article II of the CITES Convention, it is stated that Appendix II should contain 'all species which although not necessarily now threatened with extinction may become so unless trade in specimens of such species is subject to strict regulation in order to avoid utilization incompatible with their survival'.

FAO

At the same time, the development of the UN Fish Stocks Agreement (FSA) in the early 1990s (finalized in 1995) triggered a round of new activities in the fisheries management stream. Although the FSA focused on the fisheries management responsibilities needed to ensure sustainable use of straddling (transboundary) and highly migratory species, it also provided policy endorsements to the precautionary approach and ecosystem approaches in fisheries, and called for quantitative target and limit reference points. Many of these elements have since been applied to fish stocks under national jurisdiction as well because of their universal applicability and their simultaneous incorporation in the 1995 Code of Conduct for Responsible Fisheries (FAO, 1995) and the ensuing FAO Technical Guidelines for Responsible Fisheries. These and other initiatives such as FAO's International Plans of Action (IPOAs) for seabird bycatch, the conservation and management of sharks, management of fishing capacity, and illegal, unreported or unregulated (IUU) fishing were generally well received by both governance streams.

Implementation of these provisions by RFMOs and national fisheries management agencies, particularly those related to the definition and adoption of limit reference points (LRPs) and precautionary and ecosystem approaches, greatly increased the overlap between the fisheries and conservation of biodiversity governance streams – and rapidly brought them into greater conflict with one another.

The key conservation benchmark in all of these documents was about the sustainability of harvest levels. In particular, the concept of limit reference points (LRPs) was fundamentally different to the key concepts of the IUCN criteria. If the LRPs promulgated by international and national fisheries organizations were estimated appropriately, they would be triggered at stock sizes much higher than the stock conditions that might trigger concern for risk of biological extinction, either in the short or longer term. This is an important distinction that has sometimes been lost in the nearly two decades since these quantitative criteria, developed in parallel, were put into use. Such misinterpretations have been an ongoing source of conflict with some in the conservation of biodiversity governance stream treating LRPs as if they were indicators of risk of extinction, whereas the fisheries management stream uses them simply as thresholds to avoid recruitment overfishing and potential loss of future fisheries yields.

Developments in the 21st century

Revision of the quantitative IUCN criteria

The current IUCN Red List Categories and Criteria were formally adopted in 2001 (IUCN, 2001). The population decline criteria were appreciably modified for the threatened categories of Vulnerable, Endangered and Critically Endangered (Table 13.1), but fewer changes were made to the other four classes of criteria. Key changes relevant to fisheries were made to the Guidelines for the use of the criteria however, noting that the decline criteria might give false alarms when applied to populations supporting managed harvests and that natural variations of species with strongly fluctuating populations should not be considered 'declines', but that the burden of proof lay with demonstrating that the variation was indeed 'natural'. These were considered adequate responses to the input from the expert meetings (cf. 'Developments in the 1990s') because, as the Guidelines emphasized, the recommended action if a criterion was triggered was a more thorough analysis of the cause(s) of the changes in population status.

Revision of the quantitative CITES criteria

When CITES adopted a set of criteria for species listings at CoP9 in 1994, the Resolution stated that the Parties to CITES would review the criteria and guidelines and evaluate their applicability to various taxonomic groupings before CoP12. CoP11 (in Gigiri, Kenya in 2000) set up a Criteria Working Group that met twice (in Canberra, Australia in August 2000 and Sigüenza, Spain in May 2001). At the same time (September 2000), the US National Marine Fisheries Service (NMFS) set up an interagency taskforce to provide input into the potential revision of the listing criteria used by CITES (Mace et al., 2002) and FAO commissioned an appraisal of the suitability of the CITES criteria for listing commercially exploited aquatic species (Mahon et al., 2000).

FAO

Concerned by the increased involvement of CITES with species exploited by fisheries following the CITES CoP10 in 1997, FAO Member countries discussed the issue at the Committee on Fisheries (COFI) Sub-Committee on Fish Trade meeting in 1998 and subsequently formed a working group of three fisheries experts to review the CITES criteria with regard to their applicability to commercially exploited aquatic species.

Mahon et al. (2000) reviewed the available evidence for extinctions of marine and freshwater fish and larger marine invertebrates. They concluded that documented true extinctions of species were very rare and habitat destruction played a more dominant role than exploitation; indeed, habitat destruction was frequently the sole cause. They questioned the validity of the 1994 CITES criteria for commercially exploited aquatic species. Based on their literature review and the application of simple fisheries population dynamics models, they concluded that the metrics for absolute numbers (5000 individuals for a population and 500 individuals for a subpopulation) and area of distribution (10,000 km^2 for a population or 500 km^2 for a subpopulation) were probably too low and may therefore lead to false misses, but that the decline criteria metrics (discussed in more detail below) would probably produce too many false alarms. They also noted the distinction between LRPs and the much lower fish population biomass levels at which risk of extinction might be a concern. They argued that in marine fisheries risk of extinction was usefully viewed as a surface in a three-dimensional space incorporating vulnerability (based on life history), value (commercial) and 'violability' (based on expected effectiveness of management to control exploitation). Thus, in terms of the relevance of CITES, they concluded that

> '... the species most at risk of extinction and for which criteria for listing under CITES might therefore be more closely examined would be those: (i) that allow particularly profitable operations (high prices and/or low costs); (ii) that are highly vulnerable to exploitation because of their life-history; and (iii) for which normal management rules are non-existent or systematically violated'.

An attempt has been made to develop the 'three Vs' approach into a usable risk assessment method for aquatic species in trade (Sant et al., 2012). Initial trials and peer review (Fleming et al., 2012) have supported the further development of a two-stage approach with an initial assessment against Vulnerability (biological characteristics) and then an assessment of Exposure (scale of fishery, value, management and compliance). The development of this method may assist in determining risk in relation to CITES deliberations, but also has wider application to FAO Guidelines such as those for responsible fish trade.

Mahon et al. (2000) was submitted to an FAO Technical Consultation in 2000 (FAO, 2001a). Countries did not achieve consensus at that meeting on the role of CITES as an instrument complementary to traditional fisheries management. However, they agreed on the need for greater input from national fisheries agencies in the elaboration of listing proposals dealing with aquatic species and from FAO in the evaluation of these proposals, and on the need to refine the criteria and guidelines for listing species in CITES appendices to reflect the specific characteristics of aquatic resources. At this time FAO came to use the term 'commercially exploited aquatic species', which was loosely defined as commercially exploited fish and invertebrates in marine and large freshwater bodies (excluding mammals, birds and reptiles).

Following up on the recommendation to refine the criteria and guidelines, FAO commissioned a second small expert group which again considered the then-existing CITES criteria and produced a background document (FAO, 2001b) for a second Technical Consultation held in Namibia in October 2001. Of the documents produced by these two expert groups and two Technical Consultations (FAO, 2001a, b, 2002; Mahon et al., 2000), the background document for the Second Technical Consultation (FAO, 2001b) was the most enduring and influential on subsequent developments.

NMFS

Independently, the NMFS/Interagency Working Group (Mace et al., 2002) reached similar conclusions to several aspects of the Mahon et al. (2000) paper, but focused mostly on the decline criteria. Although the membership of this group was primarily made up of fisheries scientists, it also included experts in forestry, terrestrial vertebrates and terrestrial plants, and their conclusions were intended to broadly apply across all taxa, not just marine species.

As for IUCN a decade earlier, the fisheries scientists were particularly keen to develop quantitative criteria that could be applied across all taxa in order to make the listing process more objective and repeatable. However, as was the case in previous exercises, it quickly became apparent that it would be difficult if not impossible to identify absolute numbers for metrics such as 'small population size', 'small geographic range' or 'small area of occupancy' that would be universally applicable.

Early in the deliberations, the NMFS/Interagency Working Group decided to investigate the possibility of defining generic percentage decline metrics. The goal was to identify thresholds below which there might be increased risk that a population would experience depensation.[3] Heated debates ensued about the level of decline that might trigger increased risk. Fisheries scientists, who can sometimes trace the history of fish populations back to pre-exploitation levels (or at least to the onset of heavy exploitation) and who consider declines of the order of 50–60% from pristine levels as necessary to achieve maximum productivity (cf. Chapter 4), also believed that declines of the order of 80% or more did not necessarily constitute a cause for alarm in terms of a population or species being at risk. This contrasts with the current IUCN criteria where, for example, an 80% reduction could trigger a designation of Critically Endangered (Table 13.1). The NMFS/Interagency Working Group experts from the other branches of science, who generally worked on species that have been heavily exploited for several centuries, felt that even a 10–30% decline might be alarming (Mace et al., 2002).

The divergence of views was primarily based on the length of time over which the decline criteria should be considered and the relevance of the full history of the abundance of the species. Fisheries experts argued that a 20% decline could have quite different implications for a species that has already experienced an 80% decline than for a previously unexploited species, and that time periods of the order of only 5–10 years or 2–3 generations for evaluating declines (Table 13.1) were likely to result in both false alarms and false misses if they were not placed in a historical context. For example, only considering a period of '10 years or 3 generations, whichever is the longer' for a highly variable species such as Pacific sardine, which has a mean generation time of about 3 years, would mean that if biomass sometime in the past decade had been historically high but had dropped back to near the long-term average, a false alarm could be generated (Mace et al., 2002).

The NMFS/Interagency Working Group ultimately reached a consensus that declines down to 5–30% of historical or potential levels – with high productivity species being nearer the 5% end, low productivity species being nearer the 30% end and average productivity being somewhere in the middle – constituted reasonable thresholds below which a species might be considered to be at risk, and should therefore be subject to further analysis. However, use of such thresholds also needed to take into account a number of vulnerability factors that might increase concern for a particular species (e.g. low absolute numbers or biomass, or specialized niche requirements), or mitigating factors that might decrease concern (e.g. high absolute numbers or biomass, or selectivity of removals restricted to a particular life history stage).

The range of 5–30% of historical levels was then fed into, and further developed by, various FAO Working Groups and

Expert and Technical Consultations, as well as CITES own Criteria Working Group (CWG). Ultimately however, the NMFS/Interagency Working Group developed decline criteria that it considered were broadly applicable across all taxa, while FAO considered only 'commercially exploited aquatic species'.

Developments at CITES CoP12 and CoP13

The CITES CWG made substantial revisions to the 'Criteria for Amendment of Appendices I and II' (then Resolution Conf. 9.24; CITES, 1994b) with considerable input from FAO and NMFS representatives, but when these were distributed to Parties and other organizations for comments prior to CoP12, they received extremely mixed reviews (CITES, 2002). The primary concerns were that the CWG had not fulfilled the original mandate to review the applicability of the criteria and definitions to different groups of organisms (by, for example, applying them to individual taxa as most respondents had envisioned), potential weakening of the precautionary principle, increased burden on Parties preparing and writing listing proposals and tension about the proposed new quantitative guidelines for the extent-of-decline metrics and their relationship to guidelines and metrics related to recent rates of decline.

As a result, CoP12 (in Santiago, Chile in 2002) was unable to agree to a new version of Resolution Conf. 9.24, although a working group formed during the conference adapted the CWG's text into a working draft for further refinement (the 'Chairman's text'). To undertake this, an inter-sessional process was established, led jointly by the chairs of the Animals and Plants Committees. The aim of this process was to test the new draft criteria by applying them to a representative range of taxa (as had been previously specified) including fish species, in order to assess the relevance of each sub-criterion and definition to the particular biological characteristics of these taxa. Case studies included 25 animal taxa and 20 plant and fungi taxa. Marine species included white sturgeon, California sardine, Georges Bank haddock, Norwegian herring and red coral. With the help of the material thus generated (CITES, 2004a), a new version of Resolution Conf. 9.24 was prepared by the Committees. Ultimately this resulted in relatively few amendments to the CWG and 'Chairman's text', particularly in terms of the quantitative decline metrics (Table 13.1). It was adopted with minor adjustments at CoP13 (in Bangkok, Thailand in 2004; CITES, 2004b). Aside from a small number of editorial changes made subsequently for the sake of consistency with other CITES Resolutions, this is the version of the criteria that is currently in force (CITES, 2013).

However, the transition to the new CITES criteria and their associated definitions and guidelines has been far from smooth. Controversy over the criteria and guidelines and their interpretation re-erupted prior to CoP14 when the CITES Secretariat released recommendations on the listing proposals for marine species that differed from those of FAO (cf. 'Outcomes for CITES listings of aquatic species'). This led to subsequent discussions in the Standing Committee and at CoP15, and the creation of an inter-sessional criteria working group after CoP15.

Relationship between the current IUCN and CITES listing criteria

There is now a greater divergence between the IUCN and CITES listing criteria than was the case in the 1990s, although common themes remain. The biggest difference is in the quantitative metrics used. CITES has now dropped any mention of quantitative metrics for the area of distribution (previously, a guideline of <10,000 km^2 was provided), whereas IUCN has retained and expanded its guidelines on the geographic range criterion. CITES retained its previous metrics for a small wild population (5000 individuals) and a very small wild subpopulation (500 individuals) but stresses that these numbers are only guidelines that may not apply to all taxa, whereas IUCN has further elaborated its Guidelines for use of the population size criterion. However, the most contentious difference between the two sets of criteria is in the decline criteria, particularly when it comes to their application to commercially exploited marine species. The CITES criteria now emphasize the need to consider the historical extent-of-decline along with the recent rate of decline (Table 13.1), whereas IUCN generally considers only the rate of decline over the recent time period of 10 years or 3 generations, whichever is the longer.

CITES' general guidance for a marked historical extent-of-decline is a percentage decline down to 5–30% of the baseline, depending on the biology and productivity of species. CITES was however reluctant to over-specify the decline criteria for all taxa, preferring instead to relegate most of the detail of the extent-of-decline criteria to a footnote in Annex 5 of its Criteria for Amendment of Appendices I and II (currently Resolution Conf. 9.24 Rev CoP16; CITES, 2013) that has subsequently often been referred to as the 'fisheries footnote'. For commercially exploited aquatic species, the threshold historical extent-of-decline range is reduced down to 5–20% in the 'fisheries footnote'

(Table 13.1). This contrasts with the IUCN criteria of declines down to 10–70% (i.e. declines of 30–90% with 80–90% signifying Critically Endangered, 50–70% signifying Endangered and 30–50% signifying Vulnerable; Table 13.1).The recent IUCN Guidelines (revised in 2006 and 2011) continue to highlight the need for special consideration of highly fluctuating populations and strengthen the language on managed populations, endorsing not listing populations where the cause of decline is known and under management control. However, explicit mention of false alarms for commercially exploited fish species has been dropped.

As a result, the approaches of the two organizations to listing marine species have diverged and have in fact become largely independent of each other.

Global marine species assessment project

Throughout the 2000s, the involvement of IUCN in influencing CITES criteria and metrics and listing decisions appears to have diminished, particularly with respect to marine species. However, IUCN has remained active in assessing the risk of extinction of marine species, with more marine species included in each Red List update. In 2005, the IUCN Species Programme Marine Biodiversity Unit launched the Global Marine Species Assessment project, an ambitious programme to review the threat of extinction for every marine vertebrate species as well as selected plants and invertebrates. The project aims to compile and analyse all existing data on approximately 20,000 marine species, and to determine the risk of extinction according to the IUCN Red List Categories and Criteria (IUCN, 2001). The initiative began with evaluations of 150 species of sharks and rays in the Atlantic Ocean in 2006. Species groups subsequently evaluated cover a wide diversity of taxa including sharks and rays in other areas, various groups of tunas and billfishes, groupers, wrasses, eels, sea turtles, sea cucumbers, reef-forming shellfish, mangroves, sea-grasses and algae. Some of the workshops in which the evaluations have been conducted have been very controversial, particularly those involving commercially exploited marine species. An example is a workshop convened in April 2011 to synthesize and evaluate the available information on tuna and billfish species that resulted in highly divergent views among the participating experts on the validity of the IUCN criteria and the resulting status assessments for these species, leading to several participants refusing to allow their names to be included on the resulting publication (Collette et al., 2011).

Collaboration between CITES and FAO

On a more positive note, in parallel with FAO's involvement in reviewing and revising the CITES listing criteria, the two organizations negotiated an MoU which was formally signed in 2007. Notwithstanding differences of opinion on the interpretation of the criteria, they have also collaborated on a number of projects to assist CITES Parties and FAO Members to evaluate the status of species of concern or to implement CITES listings.

In the lead-up to CoP13 (in Bangkok, Thailand in 2004), before either the revisions to the CITES criteria were finalized or the CITES-FAO MoU was signed, FAO began the practice of formally convening expert panels to evaluate all proposed listings for 'commercially exploited aquatic species'. This practice continued with a formal mandate prior to CoP14 (in The Hague, The Netherlands in 2007), CoP15 (in Doha, Qatar in 2010) and CoP16 (in Bangkok, Thailand in 2013). Panel participants are selected for their expertise in the species of concern or implementation of the CITES convention and are required to act in their individual capacities, not representing any governments or agencies, therefore providing objective unbiased advice and recommendations, consistent with the listing criteria and using the best available information. The resulting FAO reports on the CITES proposals (see FAO website) are the product of preparatory work by panel members and consultants as well as analyses conducted during the 4–5 day expert meetings.

The FAO expert panels are only one part of the substantive workplan for FAO on CITES in relation to commercially exploited aquatic species that was approved by COFI in 2003. FAO has also addressed legal and implementation issues associated with the application of CITES to commercially exploited aquatic species, and has played an active role in promoting capacity building of Member Countries on issues related to commercially exploited aquatic species listed on CITES appendices, including sharks, queen conch, sea cucumbers, sea horses and humphead wrasse.

Current status

Outcomes for CITES listings of aquatic species

In addition to the FAO expert panels, two other independent reviews encompassing all of the proposals tabled (not just those relating to aquatic species) are also undertaken for every CoP. The CITES Secretariat, which includes three scientific staff, undertakes an initial review followed by a period of consultation with Parties, after which the review is finalized with a recommendation for each proposal. IUCN and TRAFFIC also undertake a joint analysis of all proposals using expert input from individuals associated with the IUCN Species Programme, the SSC, external experts and TRAFFIC. This analysis only considers whether there is sufficient information to conclude whether a listing proposal meets the CITES listing criteria and does not make recommendations to CITES Parties to support or reject the listing proposals. On the basis of these analyses, TRAFFIC does however issue specific recommendations.

In general, the CITES Secretariat and TRAFFIC have tended to favour far more proposals for commercially exploited aquatic species than has the FAO panel. For example, of the 16 proposals for commercially exploited aquatic species submitted for consideration to CoPs 13–15, TRAFFIC only recommended rejecting one (primarily on the basis of implementation issues) while FAO recommended rejecting eight and was ambivalent about two others.

In addition, individual States have adopted widely differing approaches in formulating their positions on the aquatic species listings. Some have tended to follow the FAO Expert Panel recommendations; some have tended to support nearly all listings proposals, using the FAO Expert Panel recommendations where they support the case and not otherwise; and some have continued to advocate in favour of a belief that has frequently been stated or implied for a long time at CITES CoPs, namely that commercially exploited marine species fall within the mandates of national and regional fisheries management organizations and should therefore be exempt from CITES listings. Some Parties also argue that there are insurmountable implementation issues associated with marine species listings; however, others believe that these are qualitatively no greater than the implementation problems for terrestrial species.

The ultimate result is that, while the CITES appendices currently include almost 100 commercially exploited aquatic species of fish, molluscs and echinoderms, relatively few are of key commercial significance. Some of the more important commercial species or species-groups adopted since the formation of the FAO expert panels in 2004 and up to CoP15 in 2010 include:

- European eel, *Anguilla anguilla* (2007, Appendix II); adopted after a favourable recommendation from FAO;
- sawfishes, Pristidae (2007, Appendix I apart from one species on Appendix II); adopted after a favourable recommendation from FAO;
- humphead (Napoleon) wrasse, *Cheilinus undulatus* (2004, Appendix II); adopted after a favourable recommendation from FAO;
- Mediterranean date mussel, *Lithophaga lithophaga* (2004, Appendix II); adopted despite an unfavourable recommendation from FAO; and
- white shark, *Carcharodon carcharias* (2004, Appendix II); adopted despite an ambivalent recommendation from FAO who concluded that some populations met the criteria for listing and some may not have.

The following species were among those adopted before the formal commencement of FAO's role:

- basking shark, *Cetorhinus maximus* (2002, Appendix II);
- pipefishes and seahorses, *Hippocampus spp.* (2002, Appendix II);
- whale shark, *Rhincodon typus* (2002, Appendix II);
- sturgeons and paddlefish, Acipenseriformes (1997, Appendix II apart from 2 species on Appendix I);
- Caribbean queen conch, *Strombus gigas* (1994, Appendix II) ; and
- giant clams, Tridacnidae (1983, Appendix II).

From the first FAO expert panel up until CoP15, the following commercially exploited aquatic species proposals were rejected

- spiny dogfish, *Squalus acanthias* (2007 and 2010, Appendix II); negative recommendation from FAO on both occasions;
- porbeagle, *Lamna nasus* (2007 and 2010, Appendix II); negative recommendation from FAO in 2007 and positive recommendation in 2010;
- larger hammerhead sharks, *Sphyrna lewini, S. mokarran, S. zygaena* (2010, Appendix II); positive recommendation from FAO;
- oceanic whitetip shark, *Carcharhinus longimanus* (2010, Appendix II); positive recommendation from FAO;
- Banggai cardinalfish, *Pterapogon kauderni* (2007, Appendix II); negative recommendation from FAO;
- Atlantic bluefin tuna, *Thunnus thynnus* (2010, Appendix I); favourable recommendation from FAO but with one dissenting voice on the panel; also received favourable assessment from ICCAT's scientific committee; and
- red corals, *Corallium* spp; red and pink corals, Coralliidae (2007 and 2010, Appendix II); negative recommendation from FAO on both occasions.

Sharks and rays have been an ongoing issue of concern in CITES since 1994 when CoP9 adopted a Resolution on sharks, and 2004 when the Animals Committee was requested to convene a working group for these species. FAO has regularly participated in the deliberations of this group. However, successive CoPs have tended to dilute the recommendations of the shark working group.

One of the reasons why some proposals that received favourable recommendations from FAO in 2010 still fell short of the two-thirds majority required for CITES listings to be adopted at CoP15 was the view expressed by some Parties that trade in commercially exploited aquatic species should not be regulated by CITES. There is no text in the Convention to this effect; indeed, Articles III, IV and XIV of the Convention all make explicit references to its application to marine species, as do a number of Resolutions and Decisions adopted by successive CoPs. Those who support the view argue that such species should be the responsibility of national and regional fisheries management bodies. On the other hand, opponents argue that CITES is charged with the regulation of international wildlife trade and that commercially exploited fish are not intrinsically different from other commercially exploited species. Moreover, most of the commercially exploited aquatic taxa thus far considered have received little if any attention by relevant fisheries bodies, or else are not coherently managed across their entire range.

The other key reason that the conclusions and recommendations of the first three FAO expert panels did not generate the consensus that was hoped for is the controversy that has arisen over the decline criteria and the associated guidelines and interpretations for Appendix II listings post-CoP14 (most subsequent proposals have been for Appendix II). After inconclusive discussions in the Standing Committee and at CoP15, an inter-sessional criteria working group was set up under the auspices of the Animals Committee. It reported to the Standing Committee in July 2012 that:

- 'the variability in Parties' and those reviewing proposals' risk tolerance may be more pronounced when considering commercially exploited aquatic species'; and
- 'it is not possible to provide guidance preferring or favouring one approach over another'.

The Standing Committee accepted these conclusions and determined that there was little merit in continuing the discussion.

Following CoP15 the gap in perspectives seemed to be growing. We had gone from a situation where in 2004 CITES adopted one listing proposal despite a negative recommendation from FAO, to that in 2010 when four proposals that were found by the FAO expert panel to meet the listing criteria were nevertheless rejected. This seemed to reflect increasingly polarized views regarding the role of CITES in fisheries management and conservation.

However, at CoP16 in 2013 there was a marked change in the situation for sharks, to the surprise of most CITES participants and observers. The Parties present at CoP16 voted with the required two-thirds majority to list oceanic whitetip shark, three species of hammerhead shark, porbeagle shark and manta rays on CITES Appendix II, and to move the one sawfish species remaining in Appendix II to Appendix I. An FAO expert advisory panel that met prior to CoP16 provided favourable recommendations for sawfish and all of the shark species, but concluded that there was a paucity of reliable information for manta rays. It was the second attempt to list oceanic whitetip and the hammerheads and the third attempt to list porbeagle. In one sense, this did not represent a step change as all five shark species had only narrowly missed out on the two-thirds majority vote at CoP15 and received 68.7–70.5% of the vote at CoP16. However, it is a potentially significant milestone that the listing proposals were successful, particularly given that manta rays and all five shark species are heavily traded with poor reporting by most countries.

Debates on the proposals at CoP16 were highly charged and, even after acceptance, some countries expressed their doubts that the listings could be effectively implemented. Subsequently, Canada,[4] Guyana and Japan have lodged reservations for all five shark species; Yemen has lodged a reservation for the hammerheads; Denmark (on behalf of Greenland) and Iceland have lodged reservations for porbeagle sharks; and Canada and Guyana have lodged reservations for manta rays. This means that these countries will not be bound by these particular listings although the listings will still have an impact if they trade with other Parties who have not submitted reservations.

However, these countries represent a minority and a substantial amount of investment, time and money has gone into resolving implementation issues and addressing capacity concerns in developing countries prior to the regulations entering into force on 14 September 2014. This includes work undertaken by the CITES Secretariat, TRAFFIC, FAO, the European Commission, RFMOs and several individual countries and other environmental groups.

The current situation

Despite efforts on the part of many Parties to CITES, FAO Member Countries and IUCN – along with a formal MoU between CITES and FAO – to facilitate convergence between the recommendations of marine conservation and fisheries groups, progress has been limited. There remains a broad spectrum of views among experts and States, from those who still believe that CITES is not an appropriate instrument for application to commercially exploited marine species to those who believe that most if not all marine fisheries are in a state of crisis, fisheries management bodies have failed and additional means and agencies are therefore essential to reverse the crisis.

In between these extremes there are many and varied positions. For example, many experts and some Parties may agree that certain marine species are in immediate need of reduced exploitation, but that CITES listings need to be considered in the context of other mechanisms for achieving conservation. Even within that group, some feel that fisheries management should be the business of fisheries management bodies as much as possible, and a CITES listing should only be a tool of last resort.

Others argue that the RFMOs who have this responsibility in theory have at best a patchy record of keeping fisheries under their jurisdictions sustainable, and fall even shorter when it comes to regulating the catches of most bycatch species. In this regard, proposals for CITES listings can and have had a positive impact on RFMO actions. For example, following the failure of CITES Parties to adopt the proposals for listing oceanic whitetip and hammerhead sharks at CoP15 in 2010, there was a substantial international backlash that influenced the International Commission for the Conservation of Atlantic Tuna's (ICCAT's) recommendation later that year to prohibit the retention, trans-shipping, landing, storing or selling of oceanic whitetip and hammerhead sharks caught in association with ICCAT fisheries. Successful Appendix II proposals may also complement RFMO measures through their associated monitoring and documentation requirements. Some of these causes of conflict and opportunities for complementarity are analysed in more detail by Vincent et al. (2013).

The biggest reason for discrepancies in the recommendations made by the FAO Expert Panel and the CITES Secretariat and IUCN/TRAFFIC stems from differing interpretations of the Appendix II criteria, and particularly the robustness of the percent decline criteria to reflect the actual risk to the species. Even then there are further differences in viewpoints. Some have argued that the CITES Secretariat should follow the recommendations of the FAO Panel for aquatic listing proposals in developing its recommendations to Parties, acknowledging its greater disciplinary expertise in fisheries, while others argue that different experts can reach different conclusions.

Conclusions

The core areas of disagreement between the fisheries management and conservation of biodiversity streams were apparent at the first IUCN Congress in 1996 where assessments using the IUCN quantitative criteria were applied. All expert activities thereafter have amounted to increasingly detailed work on the same core concerns. Despite nearly two decades of efforts by science experts and policy makers, most core points of disagreement remain.

Science experts working with the same information and similar processes continue to reach different conclusions on some central issues, particularly regarding the biological implications of population declines. Governance processes continue to differ in their views on the appropriate roles for fisheries management organizations and environmental organizations with regard to the protection and sustainable use of commercially exploited aquatic species.

The debates within CITES have initiated cooperation between CITES and regional fisheries management authorities, notably ICCAT and CCAMLR, but this has been limited. To date there is little evidence of a real meeting of minds between fisheries and CITES practitioners. Regardless of jurisdictional challenges and impediments to taking effective conservation action, the substantial amount of resources required to implement fisheries management or conservation decisions remains an obstacle.

In terms of actual impacts on decision-making and management actions, it appears that the fisheries governance stream is much stronger than conservation of biodiversity governance stream. This stems partly from the way that governance meetings are conducted in the biodiversity stream, where diverse participants often have status, giving the fisheries stream the ability to influence wording or prevent consensus. In contrast, in many fisheries bodies such as RFMO Councils only Contracting Parties have full status and are usually represented by their fisheries ministries (although national delegations have increasingly tended to include representatives of environmental and conservation of biodiversity interests in recent years). More significantly, the fisheries governance stream generally has greater national legislative support and more empowering international agreements as foundations for action.

Of all the international organizations dealing with the conservation of biodiversity, the CITES Convention is arguably the most empowering because it is linked to strong national legislative requirements. However, these requirements only take effect if proposals are actually accepted by the Parties, and this has made it a focal point for international conflicts between the two governance streams. For countries with binding national species at risk legislation, that legislation often results in a similar focus of conflict between the streams at the national scale.

Looking forward, in order to facilitate a more holistic and balanced approach, it is imperative that the fisheries and conservation of biodiversity governance streams work more collaboratively together. The best way to further enhance such a process is for each governance stream to strive to find common ground with the other. This is already underway in a number of instances. For example, CITES now routinely invites FAO to participate in or lead fisheries-related initiatives, while a member of the CITES Secretariat participates in all FAO expert panel meetings. Working together on obvious common ground often leads to subsequent discoveries of further common ground. For both governance streams an increased focus on implementation and accountability, rather than making endless resolutions and recommendations that are often not fully enacted, would be beneficial. Fisheries and environmental organizations need to be increasingly held to account through in-depth reviews of their performance and effectiveness. Recognition of the need for shared responsibilities is also imperative if we are to see effective action to manage the risks associated with marine species in trade.

The recent shark listings will provide a test case to determine how well the two governance streams can work together and the role that CITES can play in complementing fisheries management actions by international and national fisheries organizations.

Notes

1. CITES Scientific Authorities of exporting (and sometimes importing) countries need to determine whether a particular export will be detrimental to the survival of a species. This is called a non-detriment finding (NDF) and is required along with a legal finding by the CITES Management Authority of the exporting country before an export permit can be issued.
2. Resolutions are the documents that set out the 'soft law' associated with the text of the CITES treaty itself.
3. A situation where the per capita population growth rate decreases with further reductions in population size.
4. Canada tends to lodge reservations until such time as they can change their domestic legislation to reflect the Appendices; Canada's reservations may therefore be removed in the future.

References

Baillie, J. and Groombridge, B. (compilers and editors) (1996) *1996 IUCN Red List of Threatened Animals*. IUCN, Gland, Switzerland and Cambridge, UK.

CITES (1994a) Interpretation and Implementation of the Convention: New criteria for amendment of Appendices I and II. CITES Document 9.41. Available at http://www.cites.org/eng/cop/09/doc/E9-Doc-41.pdf (accessed March 2014).

CITES (1994b) Criteria for Amendment of Appendices I and II. CITES Resolution Conf. 9.24. http://www.cites.org/eng/res/all/09/E09-24.pdf (accessed March 2014).

CITES (2002) Interpretation and implementation of the Convention. Amendment of the Appendices. Criteria for Amendment of Appendices I and II. CITES CoP12 Document 58. Available at http://www.cites.org/eng/cop/12/doc/E12-58.pdf (accessed March 2014).

CITES (2004a) Interpretation and implementation of the Convention. Amendment of the Appendices. Criteria for Amendment of Appendices I and II. CITES CoP13 Document 57. Available at http://www.cites.org/eng/cop/13/doc/E13-57.pdf (accessed March 2014).

CITES (2004b) Criteria for Amendment of Appendices I and II. CITES Resolution Conf. 9.24, Revised CoP13. Available at http://www.cites.org/eng/res/all/09/E09-24R13.pdf (accessed March 2014).

CITES (2013) Criteria for Amendment of Appendices I and II. CITES Resolution Conf. 9.24, Revised CoP16. Available at http://www.cites.org/eng/res/09/09-24R16.php (accessed March 2014).

Collar, N.J., Gonzaga, L.P., Krabbe, N., Madrono-Nieto, A., Naranjo, L.G., Parker, T.A. and Wege, D.C. (1992) *Threatened Birds of the Americas: The ICBP/IUCN Red Data Book*. International Council for Bird Preservation, Cambridge.

Collette, B.B., Carpenter, K.E., Polidoro, B.A., Juan-Jordá, M.J., Boustany, A., Die, D.J., Elfes, C. et al. (2011) High value and long-lived: double jeopardy for Tuna and Billfishes. *Science* **333**(6040), 291–292.

FAO (1995) Code of Conduct for Responsible Fisheries. FAO, Rome.

FAO (2001a) Report of the Technical Consultation on the Suitability of the CITES Criteria for Listing Commercially-Exploited Aquatic Species. FAO Fisheries Report No. 629.

FAO (2001b) A Background Analysis and Framework for Evaluating the Status of Commercially-Exploited Aquatic

Species in a CITES context. Second Technical Consultation on the Suitability of CITES Criteria for Listing Commercially-Exploited Aquatic Species. FI:SLC2/2001/2.

FAO (2002) Report of the Second Technical Consultation on the Suitability of the CITES Criteria for Listing Commercially-Exploited Aquatic Species. FAO Fisheries Report No. 667.

Fleming, L.V., Goodman, G., Williams, J., Crook, V., Littlewood, A., Oldfield, T. and Sant, G. (2012) Fish and Multilateral Environmental Agreements: developing a method to identify high risk commercially-exploited aquatic organisms in trade and an analysis of the potential application of MEAs. Joint Nature and Conservation Committee (JNCC) Report No. 453. Aberdeen, Scotland.

Isaac, N.L. and Mace G.M. (eds) (1998) *The IUCN Criterion Review: Report of the Scoping Workshop*. International Union for Conservation of Nature, Gland, Switzerland.

IUCN (1994) IUCN Red List Categories and Criteria. International Union for the Conservation of Nature, Gland, Switzerland. Available at http://www.iucnredlist.org/technical-documents/categories-and-criteria/1994-categories-criteria (accessed February 2014).

IUCN (2001) IUCN Red List Categories and Criteria. International Union for the Conservation of Nature, Gland, Switzerland. Available at http://www.iucnredlist.org/technical-documents/categories-and-criteria/2001-categories-criteria#critical (accessed February 2014).

Mace, G.M. (ed.) (1999) *IUCN Criteria Review: Report of the Marine Workshop*. Gland Switzerland International Union for Conservation of Nature.

Mace, G.M. and Lande, R. (1991) Assessing extinction threats: Toward a re-evaluation of IUCN threatened species categories. *Conservation Biology* **5**, 148–157.

Mace, G., Collar, N., Cooke, J., Gaston, K., Ginsberg, J., Leader-Williams, N., Maunder, M., and Milner-Gulland, E.J. (1993) The development of new criteria for listing species on the IUCN Red List. *Species* **19**, 16–22.

Mace, P.M., Bruckner, A.W., Daves, N.K. Field, J.D., Hunter, J.R., Kohler, N.E., Kope, R. et al. (2002) NMFS/Interagency Working Group Evaluation of CITES Criteria and Guidelines. NOAA Technical Memorandum NMFS-F/SPO-58.

Mahon, R., Pope, J. and Rice, J. (2000) An appraisal of the suitability of the CITES criteria for listing commercially-exploited aquatic species. FAO Fisheries Circular No. 954. FAO, Rome.

Sant, G., Goodman, G., Crook, V., Lack, M. and Oldfield T.E.E. (2012) Fish and Multilateral Environmental Agreements (MEAs): developing a method to identify high risk commercially-exploited aquatic organisms in trade and an analysis of the potential application of MEAs. JNCC Report No. 453.

Scott, P., Burton, I.A. and Fitter, R. (1987) Red Data Books: The historical background. In: *The Road to Extinction* (eds R. Fitter and M. Fitter), pp. 1–5. International Union for Conservation of Nature, Gland, Switzerland.

Vincent, A.C.J., Sadovy, Y.J., Fowler, S.L. and Lieberman, S. (2013) The role of CITES in the conservation of marine fishes subject to international trade. *Fish and Fisheries*, doi: 10.1111/faf.12035.

Wijnstekers, W. (2003) *The Evolution of CITES*, 7th Edition. CITES Secretariat, Geneva, Switzerland.

Chapter 14

Parallel initiatives: CBD's Ecologically or Biologically Significant Areas (EBSAs) and FAO's Vulnerable Marine Ecosystems (VMEs) criteria and processes

J. Rice[1,2], J. Lee[3] and M. Tandstad[4]

[1] Fisheries and Oceans Canada, Ottawa, Canada
[2] IUCN Commission on Ecosystem Management (IUCN/CEM/FEG), Gland, Switzerland
[3] Secretariat of the Convention on Biological Diversity, Montreal, Canada
[4] Food and Agriculture Organization of the United Nations (FAO), Rome, Italy

Abstract: This chapter first compares the criteria adopted by the CBD for identifying areas that are ecologically or biologically significant and by FAO for areas that are vulnerable marine ecosystems relative to deep-sea fishing. It traces their parallel histories of development and describes experiences with initial application of the criteria, sometimes in the same marine areas. The many similarities and some important differences in the pathways to development of the criteria and their application are reviewed. Finally, some key lessons are extracted from this review, highlighting opportunities for greater coherence in planning and actions between fisheries management agencies and institutions engaged with the conservation of marine biodiversity.

Keywords: ecologically or biologically significant area; vulnerable marine ecosystem; bottom fishing; Convention on Biological Diversity; FAO Deep-Sea Fishery Guidelines

Introduction

Both the Convention on Biological Diversity (CBD) Conference of Parties (CoP) and members of the UN Food and Agriculture Organization (FAO) (through the International Guidelines for the Management of Deep-sea Fisheries in the High Seas) have adopted criteria for identifying areas of the oceans that have special ecological properties and may require enhanced risk aversion in the management of human activities in those areas. This chapter first presents the criteria adopted by each agency, and contrasts their properties and their development. It then traces the separate pathways taken by the respective intergovernmental organizations (IGOs) that led to the creation of separate but similar criteria for supporting different governance processes. The current situation is summarized, as States and IGOs begin to learn from the use of the respective criteria. Particular attention is given to improvements in coherence of efforts and of outcomes when the two sets of criteria are applied to information from the same area.

Governance of Marine Fisheries and Biodiversity Conservation: Interaction and Coevolution, First Edition.
Edited by Serge M. Garcia, Jake Rice and Anthony Charles.

The criteria

The CBD criteria for Ecologically or Biologically Significant Areas

The CBD criteria for Ecologically or Biologically Significant Areas (EBSAs) are defined in Annex 1 to CBD CoP Decision IX/20 (CBD, 2008a). The detailed language in the definition of the criteria was reviewed extensively by scientific experts and negotiated by Parties, so subtleties of wording are important. The criteria include the following.

- *Uniqueness or rarity*: 'Area contains either (i) unique ('the only one of its kind'), rare (occurs only in few locations) or endemic species, populations or communities, and/or (ii) unique, rare or distinct, habitats or ecosystems; and/or (iii) unique or unusual geomorphological or oceanographic feature.' This criterion can be applied on many spatial scales, so 'unique' may be a global property but 'endemic' is a more regional or local property. Uniqueness or rarity may be applied to many scales of ecosystem properties from individual populations to whole communities and ecosystems and to physical features and habitats of the sea and seabed, even if their ecological roles are poorly known.
- *Special importance for life-history stages of species*: 'Areas that are required for a population to survive and thrive.' The apparent generality of this criterion is constrained by the word 'required', which is intended to restrict application to cases where loss or serious alteration of the area would have marked negative impact on the ability of the population to 'thrive'.
- *Importance for threatened, endangered or declining species and/or habitats*: 'Areas containing habitat for the survival and recovery of endangered, threatened, declining species or areas with significant assemblages of such species.' The important nuance in this criterion is that it can be used to identify areas where the habitat would be necessary for a threatened, endangered or declining (TED) population to recover to a secure target; the population does not have to be shown to be using the area in its depleted state. In addition the presence of several TED species in an area is sufficient to meet this criterion, without having to demonstrate that the area is necessary for the survival or recovery of any of them.
- *Vulnerability, fragility, sensitivity, or slow recovery*: 'Areas that contain a relatively high proportion of sensitive habitats, biotopes or species that are functionally fragile (highly susceptible to degradation or depletion by human activity or by natural events) or with slow recovery.' The intent of this criterion is straightforward, and the inclusion of species as well as habitats and biotopes was intentional, allowing the criterion to capture areas based on the species that use them and not solely on the physical features of the areas.
- *Biological productivity*: 'Areas containing species, populations or communities with comparatively higher natural biological productivity.'
- *Biological diversity*: 'Areas containing a comparatively higher diversity of ecosystems, habitats, communities or species, or which have higher genetic diversity.'
- *Naturalness*: 'Areas with a comparatively higher degree of naturalness as a result of the lack of or low level of human-induced disturbance or degradation.'

For the three latter criteria, all relative rather than absolute, the intent again is to allow their application at a variety of scales. Whether at the local, regional, basin or global scale, the areas of *relatively* highest productivity and diversity are identified as ecologically significant. Likewise, simply having been spared from past alterations was considered sufficient to have an area considered deserving of enhanced risk aversion in management in the future. The EBSA criteria are applied individually to the information from an area, but the output maps and reports document what areas meet multiple criteria, just one criterion or none. This information is for conservation planning; an area is not necessarily 'more' of an EBSA if it meets several criteria or just one. Conservation priority can be high for an area that meets several criteria, but might also be high for areas meeting a single criterion if harm to such areas was likely to have serious ecological consequences (perhaps a key spawning ground for a forage species).

Initially, a primary motivation for developing and adopting these criteria was to have scientific guidance for creating networks of marine protected areas (MPAs; see http://www.cbd.int/doc/?meeting=PAWG-01). Correspondingly, in addition to the EBSA Criteria in Annex I, the CBD CoP IX/20 decision also has guidance in its annex II on factors to consider in creating such networks. The network guidance begins with inclusion of EBSAs in such networks, but also calls for representativity, connectivity, replication, and adequacy and viability of individual sites in the network. Paragraphs 14 and 18 of the CoP Decision state clearly that EBSAs are 'areas in need of protection'; however MPA networks are only one of several tools that can offer the necessary protection.

The FAO criteria for Vulnerable Marine Ecosystems

The FAO criteria for Vulnerable Marine Ecosystems (VMEs) are an integral part of the International Guidelines for the Management of Deep-Sea Fisheries on the High Seas (DSF Guidelines; FAO, 2009, para. 42). The Guidelines state that vulnerability should be assessed relative to specific threats (para. 15); the intent of the Guidelines is that VMEs are identified based on their ecological features, not the presence of threats. They state clearly that the properties that determine vulnerability to any threat are 'related to the characteristics of the ecosystems themselves, especially biological and structural aspects' (para. 14), and that a 'marine ecosystem should be classified as vulnerable based on the characteristics that it possesses' (preamble to para. 42). Once identified as a VME, the provisions of UNGA resolution 61/105 apply, guided by other provisions of the FAO Guidelines. The specific criteria are not given full definitions in the FAO Guidelines, but associated text clarifies their intended application and the annex to the Guidelines provides examples of potentially vulnerable species groups, communities and habitats, as well as features that potentially support them. Again, the language of the criteria was carefully crafted by both scientific experts and in negotiation by policy-makers.

- *Uniqueness or rarity*: an area or ecosystem that is unique or that contains rare species whose loss could not be compensated for by similar areas or ecosystems. These include: (1) habitats that contain endemic species; (2) habitats of rare, threatened or endangered species that occur only in discrete areas; or (3) nurseries or discrete feeding, breeding or spawning areas. Areas of functional importance to populations including TED species, for which no alternative areas are known to exist, can be captured in this criterion. This gives it application beyond areas with distinct and rare physical attributes.
- *Functional significance of the habitat*: discrete areas or habitats that are necessary for the survival, function, spawning/reproduction or recovery of fish stocks, particular life-history stages (e.g. nursery grounds or rearing areas) or of rare, threatened or endangered marine species. This criterion goes beyond unique and rare habitats, to identify as vulnerable any areas considered necessary for important life history functions. 'Necessary' is an important qualifier here, as merely demonstrating that a species uses an area is not enough to make an area a VME. There has to be some basis to infer impact on the dynamics of the population(s) of interest for an area to qualify as a VME. As with the uniqueness and rarity criterion, again TED species are given explicit mention.
- *Fragility*: an ecosystem that is highly susceptible to degradation by anthropogenic activities. This was considered a straightforward criterion, and readily applied without extra guidance.
- *Life-history traits of component species that make recovery difficult*: ecosystems that are characterized by populations or assemblages of species with one or more of the following characteristics: (1) slow growth rates; (2) late age of maturity; (3) low or unpredictable recruitment; or (4) long-lived. This criterion focuses directly on properties of species and populations, and not of places. To the extent that populations with one or more of these properties are closely associated with specific places, the criterion can be used to identify places that can be considered VMEs. This criterion can also be used to apply directly to populations and communities that are mobile over a large area, however. The entire area in which they might occur does not have to be considered a VME, but they themselves could be a VME when and where they are encountered.
- *Structural complexity*: an ecosystem that is characterized by complex physical structures created by significant concentrations of biotic and abiotic features. In these ecosystems, ecological processes are usually highly dependent on these structured systems. Further, such ecosystems often have high diversity which is dependent on the structuring organisms. As with fragility, this criterion was considered straightforward to apply; the associated text is a rationale for its inclusion more than it is guidance on its application.

As with the EBSA criteria, an area may be considered a VME if it meets even one of the criteria. In assessing VMEs most RFMO/As have mainly been considering subsets of the criteria dealing with corals, sponges and, in some cases, seamounts and vents and dealing with other potentially vulnerable species through general fisheries management measures; there are therefore no precedents to determine if areas receive greater conservation priority when they meet multiple criteria, although UNGA Resolution 61/105 (UNGA, 2006) calls for management action when even one of the criteria is met.

Comparison

Although developed largely independently, the two sets of criteria align well. Some overlap directly including: uniqueness and rarity; functional significance for life history activities; and fragility. The CBD linking of vulnerability with fragility reflects the broader scope of the CBD criteria to all human activities that may alter ecosystems and to the pelagic realm. On the other hand the FAO criteria were specifically developed for deep-sea fishing in the high seas, with paragraph 15 of the Guidelines (FAO, 2009) highlighting that 'vulnerability' should always be assessed relative to specific threats (for VMEs the threat is bottom fishing).

Initially there appear to be some marked differences, such as the CBD criterion of importance for TED species. However, the FAO guidance on uniqueness or rarity and functional significance of the habitat explicitly states that use by TED species is an important consideration in application of those criteria. Likewise the FAO criteria do not explicitly mention either high diversity or high productivity, both included as CBD criteria. This was neither an oversight nor an intentional exclusion, however. Rather, it was decided in the negotiations and made explicit in the FAO guidance on application of the criteria that if the functional significance, life history traits and structural complexity criteria were all applied carefully, then any area of high diversity or high productivity would have a high likelihood of being captured.

In practice, there are three main differences between the CBD and FAO criteria that would be likely to have a major impact on outcomes of their respective applications. Only one relates to places potentially captured by the criteria. FAO does not consider naturalness in itself to be sufficient grounds to consider an area a VME, whereas it is sufficient to qualify an area as an EBSA. On the other hand, the use of the FAO criteria can identify populations and communities as VMEs based on their life-history properties without linking them to any particular site. The CBD criteria only allow particular places to be described as EBSAs, and may have little opportunity to call for enhanced protection to mobile populations or communities with life histories not tied to very specific places. The implications of these differences are discussed in the concluding section of this chapter.

The second relates to geographic scope of application. VME criteria were explicitly developed for deep-sea areas in the high seas where fishing gear may come into contact with the seafloor. The EBSA criteria apply to the entire water column, including the surface pelagic and deep pelagic realms, and are independent of whatever threats may be under consideration.

The other major difference is the consequences triggered by application of the criteria. For the FAO-VME criteria, UNGA Resolution 61/105 requires management actions to be taken to protect an area from significant adverse impacts (FAO, 2009, paragraphs 17–20 and 47 of DSF Guidelines) as soon as a competent RFMO/A or body identifies the area as a VME or likely to contain VMEs. For the CBD-EBSA criteria, describing areas which meet one or more criteria is only one step in a long process possibly leading to designation of the area as an EBSA; even then the actions required by policy-makers and managers are not specified.

Pathways to the separate sets of criteria

If the two sets of criteria for special areas in need of enhanced protection are so similar, why were two developed to begin with? The answer to this question is similar to answers to more general questions about why FAO and CBD are separate and independent IGOs (established in 1946 and 1992, respectively), and why there are both Regional Fisheries Management Organizations/Arrangements (RFMOs/As) and Regional Seas Organizations (RSOs) in many marine areas. The history of their independent origins and recent developments in their functions contain many lessons about coherence, compatibility and, in some cases, competition between the governance streams of fisheries management and of conservation of marine biodiversity.

The common starting point

The World Summit on Sustainable Development (WSSD) included an explicit call for greater protection of special places in the sea. This call, in the WSSD Plan of Implementation[1] paragraph 32(a,c,d) in Box 14.1, does not specify exactly what measures should be taken but places marine conservation clearly in a spatial context.

Although enhanced conservation efforts for corals and other special features of the seafloor (Côté and Reynolds, 2007) and spatial measures for managing some impacts of fisheries (Jennings and Cotter, 1999; MacDonald *et al.*, 1996) both pre-date WSSD, the WSSD Plan of Implementation was a clear trigger for more focused actions in both governance streams.

Box 14.1 Extracts from the WSSD Plan of Implementation

32: In accordance with chapter 17 of Agenda 21, promote the conservation and management of the oceans through actions at all levels, giving due regard to the relevant international instruments to:

(a) Maintain the productivity and biodiversity of important and vulnerable marine and coastal areas, including in areas within and beyond national jurisdiction;
(b) Develop and facilitate the use of diverse approaches and tools, including the ecosystem approach, the elimination of destructive fishing practices, the establishment of marine protected areas … including representative networks by 2012 and time/area closures for the protection of nursery grounds and periods, … and the integration of marine and coastal areas management into key sectors;
(c) Develop national, regional and international programmes for halting the loss of marine biodiversity, including in coral reefs and wetlands.

The CBD pathway

CBD traces its efforts at developing criteria for EBSAs to UNGA Oceans and Law of the Sea Resolution 58/240 (UNGA, 2003). Paragraph 54 was intended to validate the commitments from WSSD and prompt States and IGOs to commence action on enhanced conservation and management of special marine areas, reaffirming nearly *verbatim* the language in paragraph 32c of WSSD quoted above. The time-bound deadline for MPA networks of 2012 focused the CBD/biodiversity community efforts specifically on MPA networks, but the language also referred to the need to maintain 'important and vulnerable marine ecosystems' (from para. 32a) similar to the original EBSA concept as significant areas in need of protection. The CBD/biodiversity pathway was considering many of the same types of areas as the FAO/fishery efforts. The difference was an emphasis on a particular class of tools – MPA networks (which had a cross-sectoral relevance) – rather than on any particular sectoral activity that might threaten the areas.

In 2004 the CBD CoP formed an Ad Hoc Open Ended Working Group on Protected Areas to advance the CBD Programme of Work on protected areas. This working group was only partly a response to UNGA 58/240 but its first meeting (in Montecatini, Italy in 2005) highlighted the importance of marine protected areas in its work. Their report (CBD, 2005) notes 'that the establishment of marine protected areas beyond the limits of national jurisdiction could benefit from ecological criteria for the identification of areas for protection' (para. 3 m) and calls for 'the convening of an inter-sessional meeting of experts to consider ecological criteria for the identification of potential areas for protection in marine areas beyond the limits of national jurisdiction' (para. 59).

At the inter-sessional workshop in 2005, hosted by Canada, invited experts covered a range of scientific disciplines with invitations based on consultations with FAO, the International Seabed Authority (ISA) and particularly the CBD. It was at this workshop that 'ecological criteria' was expanded to criteria for 'ecologically or biologically significant areas' to ensure that areas possibly significant to only a single species were still candidates for protection. They produced a set of criteria covering all the features listed in the final criteria in 'The CBD criteria for Ecologically or Biologically Significant Areas', although with explanatory narrative rather than definitions.

The report of the Canadian workshop was presented as an information document (CBD, 2006) to the CBD CoP VIII which directed that follow-up work be undertaken to 'Refine, consolidate and, where necessary, develop further scientific and ecological criteria for the identification of marine areas in need of protection' (Decision VIII/24, para. 44b). This was implemented through an official CBD Expert Workshop, hosted in the Azores by Portugal in 2007. It completed development and definition of the EBSA criteria, and produced the initial guidance on their application.

The CBD CoP IX in 2008 (CBD, 2008b) received the report of the Azores Workshop (CBD, 2007), adopted the criteria for use and took note of the additional annexes on MPA network properties. In its decision IX/20, the CoP invited States and IGOs to share information on the use of the scientific criteria and the scientific guidance on MPA network establishment, and called for another expert workshop to provide scientific and technical guidance on the description of areas beyond national jurisdiction (ABNJ) which meet the scientific criteria. The explicit mention of ABNJ reflected the CBD's role in provision of scientific and technical advice to UN agencies on biodiversity in ABNJ, and was reinforced by explicit reference to processes within the United Nations on conservation and sustainable use of marine biodiversity in ABNJ as well as the work by FAO and other relevant organizations, showing the awareness of CBD that other groups were working on the same issue in different fora.

The requested Expert Workshop was held in Ottawa in 2009. The experts considered experiences with criterion-based special areas from over 15 Parties, the FAO, ISA, Birdlife International and the International Union for Conservation of Nature (IUCN), and developed further guidance on application of the EBSA criteria. Despite many differences in details of the criteria in the various applications, the workshop concluded that 'there are no inherent incompatibilities between the various sets of criteria that have been applied nationally and by various IGOs' (CBD, 2010a). Consequently, it concluded that most of the scientific and technical lessons learned from applying the various sets of criteria could be generalized, and some sets of criteria could act in complementary ways. The final step in implementation was the CoP X/29 Decision, informed by the results of the above Ottawa Workshop. That Decision encouraged Parties and a long list of IGOs to 'cooperate, on a regional or subregional basis, to identify and adopt, according to their competence, appropriate measures for conservation and sustainable use in relation to ecologically or biologically significant areas ... including by establishing representative networks of marine protected areas' (para. 32). This paragraph acknowledged the relevance of EBSAs to establishing representative networks of marine protected areas, but highlighted their broader relevance in conservation planning. To facilitate implementation the Decision further asked the Secretariat to '... organize a series of regional workshops, ... with a primary objective to facilitate the description of ecologically or biologically significant marine areas through application of [the CBD EBSA] criteria *as well as* other relevant compatible and complementary nationally and inter governmentally agreed scientific criteria ...' (para. 36) and 'establish a repository for scientific and technical information and experience'. The Decision explicitly encouraged the workshops to be inclusive of fisheries RFMO/As and other IGOs, and consider their criteria for special areas (if they had them). However, the fate of the products of such expert workshops was unresolved. CoP Decision X/29 stated that areas meeting each individual criterion would be reported first to the CBD Subsidiary Body on Scientific, Technical and Technological Advice (SBSTTA) for consideration. If SBSTTA considered the reports technically sound they would be transmitted to the next CoP for approval and subsequent inclusion in the repository and submission to the United Nations General Assembly (UNGA), particularly its Ad Hoc Open-ended Informal Working Group on Biodiversity Beyond National Jurisdiction (BBNJ) as well as relevant international organizations, Parties and other Governments. Only the UNGA (or a body it specifically delegates to do so) can take decisions on exactly which areas on the high seas are to have the EBSA label, and was in receipt of the first CBD CoP report on areas meeting the EBSA criteria at the end of 2012. Hence, six years after the UNGA Resolution 58/240, the science-based framework for *description* of areas considered in need of enhanced protection due to their ecological properties was operating. It now falls to competent sectoral policy and regulatory authorities to decide if and how to provide the enhanced protection.

The FAO pathway

The FAO work on VMEs and the VME criteria was also triggered by a 2004 UNGA Resolution in response to WSSD, but in this case it was the annual Sustainable Fisheries Resolution 58/14. It very generally asked the UN Secretary-General to work with FAO, States, RFMO/As and other relevant organizations, to review '... current risks to the marine biodiversity of vulnerable marine ecosystems including, but not limited to, seamounts, coral reefs, including cold water reefs and certain other sensitive underwater features, related to fishing activities' and also asked the review to consider what measures were in place to mitigate such risks. The VME language was already in place, tied to a few types of deep-sea ecosystems that were already considered by much of the marine biodiversity community and some States as at risk from deep-sea fishing. Annual UN Secretary-General reports subsequently had sections on these issues. The primary focus for FAO Fisheries at the time however was the broader challenges of testing, disseminating and building implementation capacity for the scientific and technical guidelines on an Ecosystem Approach to Fisheries (FAO, 2003) including spatial closures, and not on VMEs.

The priority for VME identification changed with UNGA Sustainable Fisheries Resolution 61/105 in 2006.[2] A series of paragraphs gave States and RFMO/As responsibility to: 'sustainably manage fish stocks and protect vulnerable marine ecosystems, including seamounts, hydrothermal vents and cold water corals, from destructive fishing practices' (para. 80) and 'identify vulnerable marine ecosystems and determine whether bottom fishing activities would cause significant adverse impacts to such ecosystems and the long-term sustainability of deep sea fish stocks' (para. 83b). Appearing in the annual Sustainable Fisheries Resolution instead of the Ocean and Law of the Sea Resolution, from the outset VMEs focused on one particular threat: the fisheries. The existence of an overarching framework for managing fisheries under all jurisdictions gave the initiative a narrower and effective starting point for implementation. The Resolution included a number of paragraphs on management measures that were to be taken where VMEs were known or likely to occur, and paragraph 83c directed that where vulnerable marine ecosystems were known or likely to occur, 'to close such areas to bottom fishing and ensure that such activities do not proceed unless conservation and management measures have been established to prevent significant adverse impacts on vulnerable marine ecosystems'. Subsequent paragraphs went further to call on States not to authorize deep-sea-fisheries (DSF) where RFMO/As with competence to implement appropriate management measures did not exist, and to ensure their flag vessels ceased fishing if VMEs were encountered in areas where they had not previously been identified. These measures were also time-bound, for implementation by 31 December 2008.

To assist States and RFMO/As to address these issues, FAO initiated a process to build the science and management framework for sustainable DSF and protecting VMEs on the high seas through the development of the FAO DSF Guidelines.

A series of FAO Expert Consultations on aspects of DSF began and several Consultations were called after Resolution 61/105 was adopted, including one in Bangkok in 2006 (FAO, 2006), that addressed the science basis for VME identification and protection from harm. That report did not list specific candidate criteria for VME identification, but highlighted four categories of impacts that were relevant to the conservation and management of deep-sea fisheries: impacts on target species; bycatch species; impacts on habitats such as cold-water corals and seamounts; and broader food-web/trophodynamic impacts on deep-sea ecosystems. Further consultations included those on: Vulnerable Ecosystems and Destructive Fishing in Deep-sea Fisheries (Rome, 26–29 June 2007); Data and Knowledge in Deep-sea Fisheries in the High Seas (Rome, 5–7 November 2007); and deep-sea fisheries in the high seas: a trawl industry perspective on the International Guidelines for the Management of Deep-sea Fisheries in the High Seas (Cape Town, 26–28 May 2008).

Building on those Expert Consultation reports, FAO hosted a Technical Consultation in 2008, where the DSF Guidelines (FAO, 2009) were negotiated. The DSF Guidelines reaffirmed that vulnerability was to be assessed relative to specific threats, and listed the criteria for VME identification. The Guidelines also lay out the responsibilities of States and RFMO/As for activities to identify VMEs and management measures to protect them where they are known or likely to occur. The Technical Consultation also adopted an annex to the Guidelines listing examples of VMEs that gave prominence to various types of corals, sponges and special seabed features such as hydrothermal vents and seeps, and generic features such as seamounts, canyons and continental slopes. This annex was to have significant impact on the measures taken by RFMO/As and States to identify and protect VMEs.

Given the deadline in Resolution 61/105, all RFMO/As and many States with DSFs immediately commenced efforts to implement the above UNGA resolutions and the DSF guidelines to identify and address the protection of VMEs. The venues for those efforts varied among RFMO/As, but in all cases were science bodies with existing advisory relationships to the RFMO/A Councils. The new requirements led to the forming of new working groups to analyse data relating to the protection of VMEs and the recruitment of new expertise particularly in the area of benthic ecology. In North Atlantic Fisheries Organization (NAFO), North East Atlantic Fisheries Commission (NEAFC) and the South East Atlantic Fisheries Organization (SEAFO), for example, some areas were closed to protect areas assessed as containing VMEs based on the criteria provided in the DSF guidelines.

In 2010, FAO organized an Expert Workshop hosted by the Republic of Korea to analyse the challenges and propose ways forward for the implementation of the DSF Guidelines (FAO, 2011). A discussion paper reviewing the experience of RFMO/As with identifying and protecting VMEs (Rice, 2011) noted that the following points.

- RFMOs/As had all taken some actions to assess the presence of VMEs and protect them from serious adverse impacts.
- Almost all efforts have focused on identifying areas with 'significant concentrations' of corals and sponges and,

in a few cases, seamounts. Very little work has been done with other VME criteria.

- Progress was impeded by the absence of clear standards for how to interpret 'significant concentrations' and by incomplete information on distribution, abundance and species composition of corals and sponges specifically, but more generally of all ecosystem features that may meet the VME criteria.
- In all cases the RFMOs/As have been able to assemble enough information to make at least partial progress, but numerous gaps remain.
- Management measures used to protect VMEs have been almost exclusively closures of areas of DSF.
- There are numerous opportunities for activities to increase progress. Most of these involve collaborative efforts among RFMOs/As, usually with FAO facilitating the collaborations. Development of a global database on known VMEs and sponsoring Expert Meetings for provision of best-practice guidance would be particularly valuable roles for FAO.

Overall the workshop produced a number of important messages relevant to VME identification, including that all RFMO/As (and many States) were taking actions consistent with the requirements of Resolution 61/105. These efforts were however focused on a small subset of the types of VMEs consistent with the full set of criteria; there was mixed experience in the application of the VME criteria to species other than coral and sponges. With regard to mitigation tools, the workshop concluded that closures were the tools most commonly applied for the VMEs that had been identified, although it was noted that there are also many general fisheries management measures in place that would address impacts on some vulnerable species (e.g. international plans of actions on sharks, bycatch restrictions, etc). The need for broadened scientific inputs as well as new tools to address the VME issues was recognized and the specific challenges and needs for support at different levels with regards to VMEs in areas without competent RFMO/As were noted. Priority actions identified included, among others: (1) additional guidance on the interpretation and use of the VME criteria; (2) the need for development of quantitative criteria for clear identification of VMEs; (3) compile and evaluate effectiveness of current management measures for other potentially vulnerable species (e.g. sharks, certain finfish) currently being used by RFMO/As and States; (4) development and training in use of VME species identification guides as well as the need to collect and collate data to support the identification of VMEs and assessments (including by industry); and (5) advice on modelling for VME identification.

Recent status

Developments in the description of areas meeting EBSA criteria

As noted earlier in this Chapter, the CBD CoP Decision X/29 (CBD, 2010b) requested the Executive Secretary to work with Parties and other Governments as well as competent organizations and regional initiatives such as FAO, regional seas conventions and action plans and, where appropriate, RFMO/As to organize, including the setting of terms of reference, a series of regional workshops. The primary objective of the workshops would be to facilitate the description of marine areas meeting the EBSA criteria or other relevant compatible and complementary nationally and intergovernmentally agreed scientific criteria (para. 32) (CBD, 2010a, b). In the same decision (para. 41), the CBD Executive Secretary was asked to collate and make available the scientific and technical data and information and results considered by the workshops prior to the 16th meeting of the CBD SBSTTA.

CoP 10 also requested the Executive Secretary, in collaboration with Parties and other Governments, FAO, UN-DOALOS, UNESCO–IOC – in particular the Ocean Biogeographic Information System (OBIS), the World Conservation Monitoring Centre of the United Nations Environment Programme (UNEP-WCMC) and the Global Ocean Biodiversity Initiative (GOBI) – to establish a repository for scientific and technical information and experience related to the application of the EBSA criteria and other relevant nationally and intergovernmentally agreed scientific criteria, and to develop an information-sharing mechanism be developed with similar initiatives such as the work of FAO on VMEs (para. 39, decision X/29).

Finally, paragraph 42 of decision X/29 requested SBSTTA to prepare reports based on scientific and technical evaluation of information from the workshops, setting out details of areas that meet the EBSA criteria for consideration and endorsement in a transparent manner by the Conference of the Parties to the Convention, with a view to including the endorsed reports in the repository referred to above and to submit the reports to the UNGA (and particularly its Ad Hoc Open-ended Informal Working Group on Biodiversity Beyond National Jurisdiction) as well as relevant international organizations, Parties and other Governments.

Table 14.1 Summary of information on the outcomes of CBD regional workshops held to apply the EBSA criteria in 2001 and 2012.

Title	Information considered and EBSAs described	Venue and reference
The Western South Pacific Regional Workshop to Facilitate the Description of Marine EBSAs	26 EBSAs described based on time series and/or maps of 16 biological datasets and 19 physical datasets.	Nadi, Fiji, 22–25 November 2011; appendix of annex V to UNEP/CBD/SBSTTA/16/INF/6*
The Wider Caribbean and Western Mid-Atlantic Regional Workshop to Facilitate the Description of EBSAs	22 EBSAs described based on 16 biological datasets, 21 physical datasets and scientific information submitted by participants, using the template of the prototype EBSA repository.	Recife, Brazil, 28 February to 2 March 2011; appendix of annex IV to the workshop report UNEP/CBD/SBSTTA/16/INF/7*
The Joint OSPAR/NEAFC/CBD Scientific Workshop on the Identification of EBSAs in the North-East Atlantic	10 EBSAs described based on *pro forma* documents prepared by participants before the workshop, including diverse biotic and abiotic data types in each case.	Hyères, France, 8–9 September 2011; report was withdrawn at CBD SBSTTA 16, pending further review and possible revision.
The UNEP/Mediterranean Action Plan Synthesis Report on the work carried out regarding EBSAs identification in the Mediterranean	10 areas that meet the EBSA criteria based on diverse types of available information, primarily provided by Parties to the Barcelona Convention	17th Ordinary Meeting (Paris, 8–10 February 2012), the Contracting Parties to the Barcelona Convention for the Protection of the Marine Environment and the Coastal Region of the Mediterranean and its Protocols; UNEP/CBD/SBSTTA/16/INF/8*

*See meeting documents, available at http://www.cbd.int/doc/?meeting=SBSTTA-16 (accessed February 2014).

Pursuant to the above requests, a series of regional workshops were convened by the CBD Executive Secretary; results are summarized briefly in Table 14.1. All of these workshops succeeded in describing areas which met the EBSA criteria based on available information, but also concluded that the information was incomplete and further workshops should be held as new information is collected and more archival information can be located and made available for analysis and review. The OSPAR/NEAFC workshop produced a list of lessons learned from their efforts at a cooperative initiative between an RFMO and a Regional Seas Organization, as follows.

- Determination of large areas meeting the EBSAs criteria that favours biogeographic assessments and brings together different scientific assessments provides a logical broader 'envelope' for future nested conservation and management measures.
- Further discussion is needed as to the most appropriate way to describe areas against the EBSA criteria: the use of straight lines versus bathymetric contours or on the basis of natural features versus polygons to be identified/monitored.
- A number of outstanding issues and questions need further consideration, including: the connectivity of populations; what suffices as a reproducing population; the relationship between natural change in communities as against changes resulting from human impacts; linkages of EBSA features within and outside EEZs and between benthic and pelagic and shelf-slope systems; generalizing from scales of marine sampling to scales of policy; and fixed versus mobile EBSA boundaries.
- The results of the workshop may have implications for the direction of future deep-sea research.

This progress was made possible as a result of the revised RFMO mandate, the political will generated within OSPAR towards biodiversity and ecosystems targets and the inclusion of experts nominated by both Contracting Parties and observers from international organizations.

Collaboration to establish EBSA repository and information-sharing mechanism

The CBD Secretariat developed a prototype EBSA repository (http://ebsa.cbd.int/) and an information-sharing mechanism, in collaboration with the partners specified earlier in this section (from para. 39 of Decision X/29). The prototype EBSA repository was developed as a web-based input tool and database to assist countries and competent organizations to submit information and experiences relevant to the description of areas meeting EBSAs criteria, allowing users to share data, information, tools and lessons learned. It has full quality assurance and provides for a place for adding, storing and modifying data and information, with varying degrees of access, visibility and editing Including: (1) data/information private to author; (2) information submitted to the website; (3) information submitted to a regional process; (4) EBSA reports produced by regional workshops; (5) EBSA reports recommended by SBSTTA; and (6) EBSA reports endorsed by the Conference of the Parties.

The repository (accessible at http://ebsa.cbd.int/) can hold many different types of information but there are two main types: (1) reports on descriptions of areas meeting EBSAs criteria; and (2) documents and data supporting description of areas meeting EBSAs criteria. Further developments are underway or planned, including formalization of collaboration with relevant organizations such as FAO (e.g. VMEs database) in terms of sharing data/information, investment of necessary financial resources and strengthening of technical capability and human resources at the Secretariat.

Developments in relation to VMEs

Since the 2010 Busan workshop discussed in 'The FAO pathway' (FAO, 2011), all RFMO/As have continued to make progress on identification and protection of VMEs within their regulatory areas and many of the organizations have proceeded to refine their management measures on reducing significant adverse impacts from bottom-contact fishing gear. A review and workshop organized through UNDOALOS in 2011 also summarizes information on progress made to that date.[3]

Commonly, RFMO/As and States have tended to use their existing scientific advisory bodies to conduct assessments of areas against the VME criteria, relying on associated government scientists. Progress in the identification and management of impacts on VMEs still varies among RFMO/As, and many of the RFMO/As have indicated that they require additional scientific and technical capacities to advance further with data collection and analysis of information on VME distribution and also on the mapping of the fisheries necessary to assess risk.

A full tabulation of VMEs has not been assembled for this paper. VME designation is an ongoing process for many RFMOs, with no periodic common global reporting benchmark such as the CBD CoPs for EBSA description. In the future, the global VME database discussed in the following section will facilitate sharing synthesized relevant information. Readers are therefore referred to the respective RFMOs web pages for further information for the different ocean areas.

In some regions extensive work has been carried out on the identification and management of impacts on VMEs. For example, there has been extensive work on defining risk areas, fine-scale rectangles and registered VMEs in the area covered by the Convention on the Conservation of Antarctic Marine Living Resources (CCAMLR). In the North Atlantic, recent management measures by NEAFC include the extension and continued closure of previously closed areas and the addition of new VME areas closed to bottom fishing. With OSPAR, NEAFC also organized a joint meeting on the application of the EBSA and VME criteria. Within NAFO there has been extensive recent work on mapping of benthic fauna, particularly coral and sponge habitats, around the Grand Bank and Flemish Cap (http://nafo.int/data/frames/data.html). The ICES/NAFO Joint Working Group on Deep-water Ecology (WGDEC) met in 2012 to analyse new information related to VMEs to advise on boundaries, look at the possibility of incorporating a broader set of VMEs and to provide a scientific basis for threshold levels and move-on rules as well as other fisheries measures and tools. ICES has also embarked on the development of an integrated database. The lack of commonly agreed identified VMEs or VME-specific measures in some areas not covered by RFMO/As remains an issue, although in most areas interim arrangements have been discussed and agreed with partners (e.g. in the South Pacific through the South Pacific Regional Fisheries Management Organization or SPRFMO, which came into force in August 2012). The Southern Indian Fisheries Agreement (SIOFA) interim measures were developed early on, but there has not been interim work since the development of the agreement (SIOFA entered into force in June 2012).

FAO has developed a programme on deep-sea high-seas fisheries in support of the implementation of the DSF Guidelines to provide support and capacity development to member states, work welcomed in UNGA Resolution 64/72. This programme encompasses a suite of projects and activities encompassing the development of a VME database, regional workshops on VME-related issues and a global meeting on the VME criteria. As part of the development of the 'Global sustainable fisheries management and biodiversity conservation in the Areas Beyond National Jurisdiction (ABNJ)' programme supported by GEF, FAO is leading the development of a major initiative on 'Sustainable fisheries management and biodiversity conservation of deep-sea living marine resources and ecosystems in the Areas Beyond National Jurisdiction (ABNJ)' with partners from UNEP, the deep-sea RFMO/As, the fishing industry, IUCN and selected regional seas programmes. This initiative has one component that specifically addresses both VME and EBSA issues. Working with CBD on this initiative, there are favourable conditions for effective collaboration in the future. In addition, CBD and FAO held back-to-back regional workshops for application of the EBSA and VME criteria in the western Indian Ocean in 2012, and further meetings are planned for 2013. The two Secretariats are currently investigating ways to create synergies by sharing scientific information as well as expertise.

The VME database

The first VME database meeting was organized with partners in December 2011 to discuss needs, requirements and expectations for such a global VME database, including considerations of possible synergies with other initiatives such as the CBD EBSA repository (FAO, 2013).

The global VME database will contain information related to VMEs, related processes and measures and will serve to facilitate lesson learning and awareness raising for various user groups. The database will be searchable on a number of criteria, and it is expected that the output will include a map interface and a text interface containing information on the biological and physical properties of the VME, historical fishing in the area, management measures, details of the management body, and links to relevant sections of the various management and scientific reports of the RFMO/As. The database is being developed in close collaboration with the RFMO/As and the fishing industry. An initial prototype will be available this year (2014).

Conclusion

Two governance processes dealing with the high seas independently ended up with two sets of very similar criteria to address closely related issues: (1) identify areas of general biological and ecological significance in need of enhanced protection; and (2) identify areas vulnerable to fisheries impact that warrant protection from damage by fisheries. The needs of the two processes were not identical; however, the CBD-led biodiversity process was intended to focus solely on ecological properties of all marine areas from seabed to surface, with threats and management responses outside its mandate. In contrast, the FAO-led fisheries process was intended to deal solely with the deep sea in ABNJ and the threat of damage from bottom fishing, with highly prescriptive management responses required when areas were identified as VMEs. The similarities in their pathways and basic criteria as well as differences in processes and outcomes provide a number of important lessons.

The conservation of marine biodiversity and the fisheries management governance streams basically consider the same factors, but they give different weights to some considerations. The conservation of marine biodiversity stream gives great weight to TED species and their habitats and included biodiversity and productivity as separate criteria, even though they often co-vary (Stachowicz et al., 2007), to avoid missing any candidate areas in the uncommon situation where one of the factors may be high and the other low or average. The fisheries perspective was much more concerned with avoiding any criteria that might be considered to be partially redundant and hence might give conflicting directions. TED species, productivity, and diversity were therefore all explicitly discussed during development of the VME criteria. The discussions concluded that TED species are just special cases of the criteria 'on functional significance of the habitat' and 'uniqueness and rarity', and productivity and diversity are both subsumed under the criteria on 'structural complexity' and 'functional significance of the habitat', and thus there would be no gaps in coverage of areas that were ecologically special and potentially threatened by bottom fishing.

Although the CBD Ottawa workshop in 2009 concluded that there could be very few differences in areas that were found to meet a variety of criteria, there have not yet been careful tests of what types of cases may qualify under one set of criteria and not the other. Such experiments may not add much value or insight to the processes used by either type of institution however, since its goal would be to seek minor differences to amplify rather than seek major similarities on which to build greater coherence. The more constructive developments would be to facilitate more complete use of both sets of criteria, and seek ways for efficient and coherent management responses to the outcomes of applying either set of criteria. It is here where the larger differences lie at present.

As noted in the FAO Busan Expert Workshop, most RFMO/As have concentrated their VME efforts on addressing corals, sponge reefs, seamounts and hydrothermal vents ('The FAO pathway'). Functional significance of particular areas and life histories of species are areas that have not yet played important roles in VME identification, but in many RFMOs are considered in general management measures. The regional workshops applying the EBSA criteria have also given priority to the coral–sponge–vent areas, but all the criteria tend to be considered in the workshop products. Another issue related to the application of the VME criteria has been the difficulty which many RFMO/As have had with interpreting the poorly bounded but implicitly absolute standard of *significant concentration* (how dense and large does a concentration have to be to be considered 'significant'?), whereas the workshops applying the EBSA criteria have reported much less problems with the of explicitly comparative standard of *relatively high* (where being higher than neighbouring areas is all that is necessary to be 'relatively high').

There are both advantages and disadvantages to defining criteria on absolute terms versus relative terms. Absolute terms allow all of a large area that consistently shows an important feature to qualify as an EBSA or VME. They do not require that some portion of a large area always be labelled as an EBSA or VME and the rest as not an EBSA or VME, just because the feature is not *relatively* as strong. However, they require some independent benchmark for properties such as 'functional significance', 'complexity' or 'diversity' to qualify as adequate to be an EBSA or VME, and such benchmarks can be strongly contested among different perspectives. Relative criteria reverse the strengths and weaknesses and, even if they do not require absolute benchmarks, they still require some external framework to allow consistency in deciding what portion of a continuum is *relatively* important enough to be called an EBSA or VME. Clarity on these operational issues of criterion application would benefit agencies applying either set of criteria, and probably contribute more to coherence of outcomes.

The most important difference between the VME and EBSA criteria is not what properties they select or how they are applied, but what actions they trigger. The differences between UNGA Resolution 61/105 (which specifically requires States and RFMO/As to place additional constraints on fishing operations in VMEs or else not authorize such fishing) and UNGA Resolution 58/240 (which simply calls for Parties to provide greater protection) clearly influenced the steps leading to the current practices and the practices themselves. The CBD workshop processes appear to view description of EBSAs as the building of relevant strategic information on biodiversity and the starting point for dialogue with States on what measures they might decide to implement, whereas the RFMO/A actions to identify VMEs quasi-automatically attaches a formal label to places where they are committed to take management actions. It is not surprising that a much smaller proportion of areas have been identified by most RFMO/As as VMEs than have been proposed by workshops as being appropriate for EBSAs.

Once more, it is hard to judge either approach as necessarily either more risk prone or risk averse. In the case of VME identification, the required management actions are explicit but most RFMO/As are far from completing the mapping of their regulatory areas using the full set of criteria. In the case of EBSA descriptions, workshops are applying all of the EBSA criteria and large portions of many areas are included as meeting at least one criterion. However, the ensuing steps in officially designating an area as an EBSA and deciding what to do with it are not yet specified, particularly in areas beyond national jurisdiction. In addition, whatever jurisdiction is accountable for taking actions, the nature of the necessary actions is still vague. Notwithstanding the early expectations in some advocacy groups and States that designation as an EBSA inevitably leads to designation as a MPA, Parties have been careful in Decisions and Resolutions to stress that MPAs are only one of many possible tools for providing the necessary risk aversion in management. Whereas the VME approach may be capturing fewer square kilometres of seabed, it is ensuring them protection (usually through closures); the EBSA approach however tends to capture more areas but leaves their eventual protection to be negotiated.

In fact, the CBD SBSTTA 16 was not prepared to accept all the candidate EBSAs proposed by the joint OSPAR/NEAFC workshop, and the report was withdrawn until the larger ones could be reviewed further on a finer scale. Regardless, if these concerns could be resolved, having both streams start with common maps (of areas which need more risk aversion in management for whatever ecological reasons) as well as with agreement on the features to be protected from harm and the degree of threat of various activities (including but not limited to bottom fishing) may increase the likelihood of compatibility between the institutional responses for conservation and sustainable use. Bringing the other IGOs with management authority on the high seas into the process, such as the IMO and ISA, should address other threats and increase the likelihood of genuinely compatible outcomes of all of these processes – compatible management and conservation measures as well as maps.

The approaches have not been in place long enough and do not feed into similar action-processes. It is therefore still impossible to evaluate which will be more effective in ensuring that sustainable uses can be allowed but serious harm to significant seabed ecosystem features will be prevented. Although collaborative efforts exist, there is little reason to expect coherence in final outcomes if the fisheries management stream adheres exclusively to the VME approach and the conservation of biodiversity stream adheres exclusively to the EBSA approach, as currently practised. Opportunities for greater coherence in institutional responses as well as information on experiences shared could be a way forward. A variety of mechanisms for institutional linkages are available for this (see Chapter 11), which provide useful steps to greater coherence in management for conservation and sustainable use.

Notes

1. http://www.un.org/esa/sustdev/documents/WSSD_POI_PD/English/WSSD_PlanImpl.pdf
2. http://www.un.org/en/ga/search/view_doc.asp?symbol=A/RES/61/105
3. http://www.un.org/Depts/los/reference_files/workshop_fisheries_2011.pdf

References

CBD (2005) Ad Hoc Open-Ended Working Group on Protected Areas. Meeting Documents. Available at http://www.cbd.int/doc/?meeting=PAWG-01 (accessed February 2014).

CBD (2006) Report of the Scientific Experts' Workshop on Criteria for Identifying Ecologically or Biologically Significant Areas Beyond National Jurisdiction, 6–8 December 2005. Available at https://www.cbd.int/doc/meetings/cop/cop-08/information/cop-08-inf-39-en.pdf (accessed February 2014).

CBD (2007) Report of the Expert Workshop on Ecological Criteria and Biogeographic Classification Systems for Marine Areas in need of Marine Protection. Available at http://www.cbd.int/doc/meetings/mar/ewsebm-01/official/ewsebm-01-02-en.pdf (accessed February 2014).

CBD (2008a) Decision Adopted by the Conference of the Parties to the Convention on Biological Diversity at its Ninth Meeting. CBD CoP Decision IX/20. Available at http://www.cbd.int/doc/decisions/cop-09/cop-09-dec-20-en.pdf (accessed February 2014).

CBD (2008b) COP 8 Decision VIII/24. Protected Areas. Available at http://www.cbd.int/decision/cop/?id=11038 (accessed February 2014).

CBD (2010a) Report of the Expert Workshop on Scientific and Technical Guidance on the use of Biogeographic Classification Systems and Identification of Marine Areas Beyond National Jurisdiction in Need of Protection. Available at https://www.cbd.int/doc/meetings/sbstta/sbstta-14/information/sbstta-14-inf-04-en.pdf (accessed February 2014).

CBD (2010b) Decision Adopted by the Conference of the Parties to the Convention on Biological Diversity at its Ninth Meeting. CBD CoOP Decision IX/20. Available at https://www.cbd.int/decision/cop/default.shtml?id=12295 (accessed March 2014).

Côté, I.M. and Reynolds, J.D. (eds) (2006) *Coral Reef Conservation*. Cambridge University Press, Cambridge.

FAO (2003) The ecosystem approach to fisheries. FAO Technical Guidelines for Responsible Fisheries No. 4, Suppl. 2. Rome, FAO.

FAO (2006) Report and documentation of the *Expert Consultation on Deep-Sea Fisheries in The High Seas*, Bangkok, Thailand, 21–23 November 2006. FAO Fisheries Technical Report 838. FAO, Rome.

FAO (2009) *International Guidelines for the Management of Deep-Sea Fisheries in the High Seas*. FAO, Rome.

FAO (2011) Report of the *FAO Workshop on the Implementation of the International Guidelines for the Management of Deep-sea Fisheries in the High Seas: Challenges and Ways Forward*, Busan, Republic of Korea, 10–12 May 2010. FAO Fisheries and Aquaculture Report No. 948. FAO, Rome.

FAO (2013) Workshop for the Development of a Global Database for Vulnerable Marine Ecosystems (VMEs). FAO Fisheries and Aquaculture Report. No. 1018. Rome. 41 pp.

Jennings, S. and Cotter, A.J.R. (1999) Fishing effects in northeast Atlantic shelf seas: patterns in fishing effort, diversity and community structure. 1. Introduction. *Fisheries Research* **40**, 103–106.

MacDonald, D.S., Little, M., Eno, N.C. and Hiscock, K. (1996) Disturbance of benthic species by fishing activities: a sensitivity index. *Aquatic Conservation: Marine and Freshwater Ecosystems* **6**, 257–268.

Rice, J. (2011) Review of progress of implementation of the FAO Deep-sea guidelines for the management of deep-sea fisheries in the high seas: Experience of

RFMO/As with identifying and protecting VMEs. In: FAO 2011. *Report of the FAO Workshop on the Implementation of the International Guidelines for the Management of Deep-sea Fisheries in the High Seas: Challenges and Ways Forward*, Busan, Republic of Korea, 10–12 May 2010. FAO Fisheries and Aquaculture Report No. 948. FAO, Rome, pp. 44–61.

Stachowicz, J.J., Bruno, J.F. and Duffy, J.E. (2007) Understanding the effects of marine biodiversity on communities and ecosystems. *Annual Review of Ecology Evolution and Systematics* **38**, 739–766.

UNGA (2003) Resolution adopted by the General Assembly on 23 December 2003. Available at http://www.un.org/ga/search/view_doc.asp?symbol=A/RES/58/240&Lang=E (accessed February 2014).

UNGA (2006) 61/105. Sustainable fisheries: Agreement for the Implementation of the Provisions of the United Nations Convention on the Law of the Sea of 10 December 1982 relating to the Conservation and Management of Straddling Fish Stocks and Highly Migratory Fish Stocks, and related instruments. Available at http://www.un.org/en/ga/search/view_doc.asp?symbol=A/RES/61/105 (accessed March 2014).

Part IV
Regional governance

Chapter 15

Regional governance for fisheries and biodiversity

R. Warner[1], K. Gjerde[2] and D. Freestone[3]

[1]Australian National Centre for Ocean Resources and Security, University of Wollongong, Australia
[2]Global Marine and Polar Programme and World Commission on Protected Areas, IUCN, Gland, Switzerland
[3]The George Washington University Law School, Washington, DC, USA

Abstract: Strong and coherent regional governance is critical for tackling the increasing number and variety of threats to fisheries and biodiversity within and beyond national jurisdiction including overfishing, destructive fishing practices, marine pollution and climate change impacts. This chapter examines the existing legal and institutional framework for fisheries and biodiversity governance at the regional level including key regional organizations such as regional fisheries management organizations, regional seas organizations and non-binding regional initiatives. As well as highlighting the fragmentary nature of regional oceans governance, this analysis discloses the wide variety of approaches to and differing rates of progress in harmonizing fisheries and biodiversity conservation objectives across major oceanic regions.

Keywords: regional; fisheries; biodiversity; conservation; governance; harmonization; transboundary; threats

Introduction

The regional level of governance is critical to the effective implementation of international legal obligations and conservation and management measures for both fisheries and marine biodiversity: fish stocks and marine ecosystems straddle national boundaries and areas within and beyond national jurisdiction. Strong and coherent regional governance is a vital component in combating the rising catalogue of transboundary threats to fisheries and biodiversity including overfishing, destructive fishing practices, pollution and climate change impacts.

This chapter first describes the legal and institutional framework for fisheries and biodiversity governance at the regional level, focusing on some key features of regional fisheries management organizations (RFMOs) and regional seas arrangements (RSAs), and looking also at other more informal arrangements such as the Large Marine Ecosystem (LME) projects. The analysis reveals the fragmentary nature of the regional oceans governance network for fisheries and biodiversity both in terms of its geographic scope and its functional responsibilities. Examples from different regions underscore the wide diversity and varying rates of progress in aligning and harmonizing fisheries and biodiversity conservation objectives. We then examine the efforts of RFMOs and RSAs to collaborate between themselves and with other global and extra-regional organizations and analyse some of the catalysts for such collaboration. We discuss selected initiatives towards comprehensive strategies for alignment and harmonization of fisheries and biodiversity conservation and management. Finally, gaps in regional oceans governance

Governance of Marine Fisheries and Biodiversity Conservation: Interaction and Coevolution, First Edition.
Edited by Serge M. Garcia, Jake Rice and Anthony Charles.

for fisheries and biodiversity conservation and potential pathways to improved collaboration are identified.

Legal and institutional framework for regional fisheries governance

There are 20 existing and prospective RFMOs with mandates to establish fisheries conservation and management measures. There are still significant gaps, for example in the coverage of non-tuna fisheries on highly migratory species in the Arctic and in the Atlantic south of the NEAFC/NAFO areas of responsibility. Until the end of 2009, there were no general fisheries commissions in the Pacific at all to manage non-highly migratory species. The treaty establishing the South Pacific Regional Fisheries Management Organization (SPRFMO) was concluded in November 2009 and entered into force in 2012. Negotiations are still ongoing for a North Pacific RFMO. In the Indian Ocean, the Regional Commission for Fisheries (RECOFI) covers the Gulf area and the Southern Indian Ocean Fisheries Agreement (SIOFA), concluded in July 2006, entered into force in June 2012 (Freestone, 2008).

Fisheries governance arrangements exhibit considerable diversity and varying rates of progress in their approaches to incorporating environmental protection principles and biodiversity conservation objectives into their management regimes. Recent reviews of RFMO practice at the global level reveal several factors that have limited the effectiveness of RFMOs in implementing fisheries conservation and management measures in an ecologically sustainable manner (HSTF, 2006) including the following.

- *Absence of environmental protection principles in the RFMO conventions.* The absence of modern environmental protection principles or guidelines such as the precautionary approach and ecosystem-based management in some RFMO conventions concluded prior to the UN Fish Stocks Agreement means that unless all RFMO members agree, they are not obliged to consider the consequences of uncertainty when adopting conservation and management measures.
- *Ineffective decision-making frameworks.* It is the established practice of RFMOs to take decisions on their conservation and management measures by consensus, even when their instruments may not require it, and to allow for individual objections to conservation and management measures agreed by the majority of member States (McDorman, 2005). This allows objecting RFMO members to take advantage of uncertainties in scientific advice and can lead to a dilution of conservation and management measures even where the precautionary approach and ecosystem-based management requirements exist. Many of the RFMOs that were established prior to the conclusion of the UN Fish Stocks Agreement allow for States to opt out or object to implementing conservation and management measures that have been agreed within the RFMO.
- *Lack of a formal global coordination mechanism.* There is no overarching global coordination mechanism to oversee the conservation and management activities of RFMOs in marine areas beyond national jurisdiction and monitor their performance against best-practice standards and ensure cross-sectoral exchange of information. This makes it difficult to address global problems such as the conservation of highly migratory marine species or illegal, unregulated and unreported (IUU) fishing as fishing vessels may move between regions concentrating their fishing effort in areas where conservation and management measures are lax or non-existent. At the regional level there has been very little consultation and collaboration between RFMOs. The first meeting between the tuna RFMOs, referred to as the 'Kobe Process' after the first location of the first meeting in Japan, occurred in 2007.
- *Participation levels.* In many regions developing States lack the resources and capacity to participate fully in RFMOs and implement their obligations effectively.
- *Failure to deal effectively with non-Parties.* Few RFMOs include all the participants in a regional fishery among their members. An RFMO may have agreed on environmentally sound conservation and management measures for fisheries in high-seas areas but only those States which have agreed to be bound by its agreement are obliged to apply its measures. The failure to deal effectively with non-Parties or 'free riders' undermines the incentives for fishing vessels of RFMO members to adopt restrictive conservation and management measures (McDorman, 2005).

It is beyond the scope of this chapter to review all RFMOs to determine in detail the extent to which environmental protection principles and biodiversity conservation objectives are reflected specifically in their agreements and practice. What is clear however is the importance of an explicit mandate for addressing both issues in the overarching agreement. This is revealed through the performance of two RFMOs highlighted here that are also directly charged with precaution and conservation of associated species and ecosystems: the Commission for the Conservation of Antarctic Marine Living Resources (CCAMLR) and the Western Central Pacific Fisheries Commission (WCPFC).

CCAMLR was established under the 1980 Convention on the Conservation of Antarctic Marine Living Resources as an integral element of the Antarctic Treaty system. It has a specific mandate to conserve and manage all marine living resources (except whales and seals) in the area south of 60 degrees latitude south and in the area between 60 degrees latitude south and the Antarctic Convergence. CCAMLR's conservation and management responsibilities extend beyond fish species to molluscs, crustaceans and birds found south of the Antarctic Convergence. The Convention explicitly adopts a precautionary and ecosystem-based approach to marine living resource management which recognizes the complex interconnections between all parts of the Antarctic ecosystem (Miller et al., 2004). Article II(3) of the Convention sets out the various elements of CCAMLR's conservation and management approach which allows for rational use of marine living resources in accordance with strict conservation principles. The three key conservation principles which apply to harvesting of marine living resources and associated activities are as follows.

(1) Prevention of decrease in the size of any harvested population to levels below those which ensure its stable recruitment. For this purpose its size should not be allowed to fall below a level close to that which ensures the greatest net annual increment.
(2) Maintenance of the ecological relationships between harvested, dependent and related populations of Antarctic marine living resources and the restoration of depleted populations to the levels defined in sub-paragraph (1) above.
(3) Prevention of changes or minimization of the risk of changes in the marine ecosystem which are not potentially reversible over two or three decades, taking into account the state of available knowledge of the direct and indirect impact of harvesting, the effect of the introduction of alien species, the effects of associated activities on the marine ecosystem and the effects of environmental changes, with the aim of making possible the sustained conservation of Antarctic marine living resources.

Since its inception in 1982, CCAMLR has adopted a variety of innovative measures to implement its ecosystem-based approach to conservation. These include banning destructive fisheries practices such as bottom trawling for particular fish species in the CCAMLR area, mandating measures to reduce incidental seabird mortality caused by baited hooks in long-line fishing, monitoring the effects of fishing on non-target species by collection of data on CCAMLR member state fishing vessels and prohibiting fishing for certain species by CCAMLR member State fishing vessels where the risk to bycatch species is thought to be too great. In 2011, they also adopted a conservation measure for rolling out a comprehensive system of marine protected areas.

A potential weakness in the implementation of CCAMLR conservation measures is the requirement for consensus in decisions on matters of substance such as conservation measures, and the availability of the objection procedure for members to object later to measures for which they may have voted. Despite consensus requirements and contentious meetings, CCAMLR has had some success in implementing the most advanced interpretation of an ecosystem-based approach to marine living resource management in its Convention area (Kaye, 2000). However, the effectiveness of CCAMLR's management was almost undone by its inability to regulate the activities of fishing vessels of non-member States (Rayfuse, 2004). CCAMLR adopted a standard suite of fisheries management measures until the mid-1990s, relying on: flag State implementation of conservation and management measures supplemented by fisheries data reporting; at-sea and in-port inspections by member States of fishing vessels and their catch; and tracking the movement of member States' fishing vessels through vessel monitoring systems and notification of vessel movements. These measures proved to be insufficient to stem increases in IUU fishing in the CCAMLR Convention Area, particularly for Patagonian toothfish, from the mid-1990s.

To address the problem of non-members and IUU fishing, CCAMLR had to resort to trade-related sanctions on a global basis. In 2000, CCAMLR introduced a Catch Documentation Scheme (CDS) which prohibited entry into world markets of Patagonian toothfish without verified catch documents. The scheme has attracted the participation of non-member States and applies to toothfish fishing by member States' vessels and non-member States' vessels. In a relatively short period, the CDS has extended its coverage to more than 90% of the world's toothfish trade and reduced the profitability of this type of IUU fishing. It requires flag State authorization for toothfish fishing both within and outside the CCAMLR Convention area. The scheme has had the side-benefit of refining global estimates of toothfish catch to enable better management. The importance of a strong mandate for biodiversity conservation and precaution is also underscored in the track record of the Convention on the Conservation and Management of Highly Migratory Fish Stocks in the Western and Central Pacific Ocean (WCPFC Convention). It was the first comprehensive conservation and management regime for highly migratory fish stocks to reflect the provisions of the 1995 UN Fish Stocks Agreement. The WCPFC entered into force on 19 June 2004.

The objective of the Convention is to ensure, through effective management, the long-term conservation and sustainable use of highly migratory fish stocks in the Western and Central Pacific Ocean in accordance with the United Nations Convention on the Law of the Sea (UNCLOS) and the UN Fish Stocks Agreement. The area of competence of the WCPFC includes a large area of high seas lying outside and between the 200 nautical mile exclusive economic zones of its Parties and its regulatory competence extends to all fish stocks of the species listed in Annex I of UNCLOS. The region covered by the WCPFC Convention is estimated to have 60% of the world's tuna stocks (Aqorau, 2001). The WCPFC is empowered to adopt principles and measures for conservation and management of the highly migratory fish stocks in its area of competence which reflect the key environmental protection principles in the UN Fish Stocks Agreement. These include measures based on the best scientific evidence available to ensure sustainability and promote optimum utilization. The Commission must apply the precautionary principle in accordance with the methodology set out in annex II of the UN Fish Stocks Agreement, determine the impact of fishing activities on non-target and associated or dependent species and their environment and adopt plans, where necessary, to ensure the conservation of species and protect habitats of special concern. The conservation measures to be taken by the Commission also include those which protect biodiversity in the marine environment and those which assess the impact of fishing activities on other species belonging to the same ecosystem.

Since its inception, the WCPFC has introduced over 30 conservation and management measures ranging from those addressing the level of fishing effort for highly migratory species such as big eye, albacore and yellow fin tuna, sharks, marlin and sword fish and the impacts of fishing for highly migratory species on seabirds, turtles and cetaceans to the prohibition of drift net fishing and reciprocal boarding and inspection procedures. Through an innovative measure to surmount the weaknesses of consensus-driven decision-making, when consensus fails decisions on matters of substance are to be taken by a three-quarters majority of the members present. Decisions become binding on parties 60 days after their notification but members voting against the decision or absent may, within 30 days of their adoption, seek a review of the decision. The review process is another innovative element (Aqorau, 2001) as it can overcome the 'opt out' clauses common in many RFMO conventions and which have the potential to sabotage their conservation and management efforts. The extensive list of conservation and management measures of the WCPFC designed to mitigate the impact of fishing for highly migratory species on associated and dependent species in the Convention Area have the *indicia* of a precautionary approach and ecosystem-based considerations being applied in the decision-making process within the WCPFC.

Although somewhat slow and uneven, there has been some progress in implementing these measures for promoting the conservation of vulnerable tuna stocks and managing fishing capacity to mitigate its impacts on associated and dependent ecosystems (Parris, 2010). To address the problem of non-members and IUU fishing, under Article 24 of the WCPFC Convention flag States must ensure that their fishing vessels do not engage in unauthorized fishing for highly migratory fish stocks beyond national jurisdiction and that as flag States they are able to effectively exercise their responsibilities for fishing vessels operating under their flag. Flag States must maintain a record of fishing vessels authorized to fish beyond national jurisdiction and require such flag vessels to use real-time satellite position-fixing transmitters so that they can participate in the vessel monitoring system established by the Commission. The Commission has established measures for Contracting Parties to board and inspect each other's fishing vessels on the high seas in accordance with the UN Fish Stocks Agreement (UNFSA) model.

In contrast, the four other tuna RFMOs are still operating under pre-UNFSA agreements. Attempts to update their charters are not progressing rapidly, although such changes have been called for by performance reviews undertaken by the majority of these bodies as well as during the Kobe Process. For example, the 2009 performance review of the Indian Ocean Tuna Commission Agreement recommended that the IOTC Agreement be either amended or replaced by a new instrument: 'the absence of concepts such as the precautionary approach and an ecosystem based approach to fisheries management are considered to be major weaknesses' (Edeson, 2009). Similarly, in the review of the Commission on the Conservation of Southern Bluefin Tuna (where the global population of bluefin tuna is ranked on the IUCN Red List of threatened species as 'critically endangered', estimated to be 5% of the population size before fishing began), the absence of agreed management principles (e.g. the precautionary and ecosystems approaches, efficient use, best scientific information, maintaining biodiversity and minimizing effects on the marine environment) guiding management decisions was noted. Similarly, an independent review of the Commission for the Conservation of Atlantic Tunas in 2008 observed that its management of bluefin tuna is 'widely regarded as an international disgrace' (Hurry et al., 2008).

Legal and institutional frameworks for regional marine environmental governance

Since the early 1970s, a diverse array of binding and non-binding regional arrangements has been negotiated around the globe to engage States in the collaborative protection of their offshore marine environments. Many of the binding regional seas arrangements were initiated through the United Nations Environment Programme (UNEP) Regional Seas Programme while others are the result of independent agreements between regional partners (Vallega, 2002). They now cover 18 maritime regions which differ markedly in their character and extent. The UNEP regional seas arrangements, together with the non-UNEP regional marine environmental protection arrangements, involve 149 States, approximately 95.5% of the world's States. The areas of responsibility of many of these arrangements are currently limited to waters within national jurisdiction and very few of them make provision for consensual environmental protection measures in high-seas enclaves and high-seas areas adjacent to waters within national jurisdiction. The geographic scope of these arrangements has been determined by political opportunity rather than any systematic scheme to encompass all the oceanic regions of the world (Boyle, 2000; Sand, 1999). No legally binding conventions have yet been developed for the regional arrangements in the East Asian Seas, South Asian Seas, Northwest Pacific, Northeast Pacific or for the Arctic. Moreover, these conventions are primarily groupings of coastal states, and their jurisdiction is generally restricted to their coastal zones or out to 200 nautical miles. The exceptions are: the OSPAR Convention area, which has high-seas areas within its remit; the Mediterranean, where most coastal states have for various reasons not yet claimed EEZs; the South Pacific, which includes within its mandate the 'doughnut' holes between the EEZs of its members; and the Antarctic Treaty System, consisting of both the Antarctic Treaty and its Protocol on Environmental Protection as well as the CCAMLR Convention.

The spread of regional arrangements for marine environmental protection has paralleled the negotiation and entry into force of UNCLOS and has both reflected and advanced the development of modern environmental protection principles (Treves, 2003). The early focus of most regional arrangements such as the OSPAR Convention and the Barcelona Convention in the Mediterranean was the control of marine pollution, but many have since adopted a more integrated approach to the protection of the marine environment including conservation of its biodiversity and the development of systems of marine protected areas.

The broadening of their scope in relation to approaches to conservation and targets for conservation intervention has enabled many regional arrangements to assimilate new developments in international environmental law and policy through mechanisms such as protocols and non-binding documents such as programmes for action and strategic plans. The majority of regional agreements are based on framework conventions which depend on implementation by State Parties in waters within national jurisdiction. These conventions have been supplemented by Protocols, ministerial level agreements and strategy documents which regulate different sources of marine pollution, provide for the protection of threatened and endangered species and the establishment of marine protected areas to preserve rare or fragile ecosystems. In most regions these binding legal instruments and soft-law accords are accompanied by planning documents which define regional priorities for marine environmental protection.

Key factors that have limited the effectiveness of RSAs in implementing fisheries conservation and management measures in an ecologically sustainable manner include the following:

- the limiting of their areas of responsibility to waters under national jurisdiction;
- the lack of reference to sustainable development and use of marine biodiversity in their mandates; and
- the absence of specific collaboration provisions or arrangements and mechanisms between RSAs and RFMOs.

Catalysts for collaboration on fisheries and biodiversity conservation objectives

Although much of the implementation of fisheries conservation and management and biodiversity conservation measures in RFMOs and RSAs has occurred independently, there are some examples where better alignment or even harmonization of fisheries and biodiversity conservation objectives has been triggered by a variety of catalysts. These include: (1) recognition by global and regional organizations of the need to address transboundary threats to marine living resources and marine biodiversity cooperatively; (2) resolutions and decisions of global bodies on conservation of biodiversity in marine areas beyond national jurisdiction; and (3) the incorporation of modern conservation principles, reflecting those in the UNFSA, into the mandates of RFMOs. This section examines some of those examples and the catalysts which have led to varied degrees of alignment and harmonization between fisheries and biodiversity conservation objectives in different regional settings.

Regional responses to transboundary threats

The need to address the transboundary nature of threats facing marine living resources and biodiversity in particular regions both within and beyond national jurisdiction has been a central catalyst for alignment or harmonization of fisheries and biodiversity conservation objectives and collaboration between regional governance organizations and alliances. Among the non-UNEP RSAs, the Northeast Atlantic RSA has arguably demonstrated the capacity to work with one of the relevant RFMOs to harmonize fisheries and biodiversity conservation objectives. The Convention for the Protection of the Marine Environment of the Northeast Atlantic (OSPAR Convention) seeks to regulate a wide spectrum of human activities which might have an adverse effect on the protection and preservation of ecosystems and the biodiversity of the Northeast Atlantic, while recognizing the competence of other organizations for regulating fishing, shipping and seabed mining. Unlike many other RSAs, there is an explicit provision in Annex V, Article 4(1) of the Convention requiring consultation with the responsible fisheries management organization where fisheries issues arise. The OSPAR Convention applies in a maritime area of the Northeast Atlantic defined in Article 1(a) of the Convention to include all the maritime zones within the jurisdiction of the Contracting Parties and maritime areas beyond national jurisdiction. The Convention has 16 Contracting Parties including the European Community. Most of its Parties are developed State members of the European Union and its decisions are heavily influenced by European Union legislation and policy directives on the environment (Lagoni, 2003).

The preamble of the OSPAR Convention reflects the common objective of the Contracting Parties to provide coordinated protection and sustainable management for the whole maritime environment of the Northeast Atlantic so that the marine ecosystems will continue to sustain the legitimate uses of the sea and meet the needs of present and future generations. To achieve this objective the Contracting Parties must, under the two limbs of Article 2(1)(a) of the Convention, take all possible steps to prevent and eliminate pollution from the maritime area and to protect the maritime area against the adverse effects of human activities in order to safeguard human health and conserve marine ecosystems. In implementing programmes and measures under the Convention, the Contracting Parties must apply the precautionary principle, 'the polluter pays' principle and ensure the application of best-available techniques and best environmental practice. As a demonstration of its evolutionary character, in 1998 OSPAR adopted Annex V on the Protection and Conservation of the Ecosystem and Biological Diversity of the Maritime Area, extending the cooperation of the parties to cover 'all human activities that might adversely affect the marine environment of the North East Atlantic' other than fishing and shipping. In 2003 it adopted its Biological Diversity and Ecosystems Strategy, setting forth its commitment to designate a representative network of MPAs by 2010.

OSPAR's cooperation with the principal RFMO responsible for managing fisheries in the region, NEAFC, was enabled in many ways by its parallel evolution, as described in the next section.

The boundaries of the other regional seas program in the Atlantic Ocean do not extend to areas beyond national jurisdiction; there has therefore been to date little interaction with the RFMOs. The CBD-driven process to describe Ecologically or Biologically Significant Areas (EBSAs) both within and beyond national jurisdiction is likely to create a new need, however. The limitations of the current geographical ambits of both the regional environmental and fisheries treaties is highlighted by the initiative, led by the Government of Bermuda, to introduce conservation and management measures for the Sargasso Sea. The Sargasso Sea, named for the accumulations of holopelagic algae contained within the North Atlantic Subtropical Gyre, is a 2 million square nautical mile ecosystem that is primarily high seas. The OSPAR Secretariat and the Sargasso Sea Alliance have established informal research and information exchange systems and have concluded a Collaboration Arrangement. The Alliance is seeking to use existing sectoral organizations such as ICCAT, IMO and ISA to put protection measures in place and to convene an inter-governmental meeting to establish a collaborative but non-legally binding protection regime for the Sargasso Sea (Freestone and Killerlain Morrison, 2012).

The Global Environment Facility (GEF) has funded a number of fisheries and biodiversity conservation projects conducted on the scale of large marine ecosystem (LME). LMEs are defined as 'regions of ocean space encompassing coastal areas from river basins and estuaries out to the seaward boundary of continental shelves and the seaward boundary of coastal current systems' (Alexander, 1993; Sherman, 1994). Sixty-four LMEs around the globe, including many of the world's enclosed and semi-enclosed seas, have been defined as areas of ocean space in which ecosystem-based management can be applied to maintain and restore ecosystem function while allowing sustainable use (Fanning et al., 2007). The LME approach has formed the basis for an array of GEF-funded projects to promote integrated marine ecosystem governance of LMEs (Duda and Sherman, 2002; Sherman, 2006).

One such GEF-initiated LME project, the Agulhas and Somali Current Large Marine Ecosystem (ASCLME), combines the Agulhas Current Large Marine Ecosystem (ACLME) which stretches from the northern end of the Mozambique Channel to Cape Agulhas and the Somali Current Large Marine Ecosystem (SCLME) which extends from the Comoros Islands and the northern tip of Madagascar to the Horn of Africa. The project covers the countries of Comoros, Kenya, Madagascar, Mauritius, Mozambique, Seychelles, Somalia, South Africa and Tanzania. The objectives of the ASCLME include:

- gathering new and important information about ocean currents and how they interact with and influence climate, biodiversity and the economies of the western Indian Ocean region;
- documenting the environmental threats that are faced by the countries of the region in a Transboundary Diagnostic Analysis (TDA);
- developing a Strategic Action Plan (SAP) which sets out a strategy for the countries to collectively deal with transboundary threats; and
- strengthening scientific and management expertise, with a view to introducing an ecosystem-based approach to managing the living marine resources of the western Indian Ocean region.

One of the key achievements of the ASCMLE project has been to initiate the establishment of a regional coordinating mechanism, the Western Indian Ocean Sustainable Ecosystem Alliance (WIOSEA), to implement the SAP for the region using the existing mandated regional bodies which already have responsibility for the management of marine living resources (e.g. the Nairobi Convention and the South West Indian Ocean Fisheries Commission). In September 2011 the ASCLME signed an Aide Memoire for Cooperation and Collaboration with the South West Indian Ocean Fisheries Project (SWIOFP), a GEF-supported project implemented by the World Bank which is also focused on the western Indian Ocean region and covering the same countries as the ASCMLE.

Through their Aide Memoire the ASCLME and SWIOF projects have agreed to share data and information and undertake joint research cruises which enhance understanding of the region's ecosystems and help to set a baseline from which long-term ecosystem monitoring programs can be derived. They will also cooperate in a policy and governance assessment designed to produce agreed recommendations for implementing an ecosystem-based approach to management of marine and coastal resources in the region.

A similar example of a non-binding regional initiative which forms the basis for biodiversity conservation objectives is the Coral Triangle Initiative (CTI) involving countries in Southeast Asia and the Pacific. The Coral Triangle is a region located along the equator at the confluence of the Western Pacific and Indian Oceans which covers all or part of the exclusive economic zones (EEZs) of six countries (Indonesia, Malaysia, the Philippines, Papua New Guinea, the Solomon Islands and Timor Leste). It is regarded by scientists as one of the richest repositories of marine biodiversity on earth, containing 76% of all known coral species, 37% of coral reef fish, 33% of the world's coral reefs and the most prolific location for mangrove forests and spawning and juvenile growth areas for the world's largest tuna fishery. Threats to the Coral Triangle region include overfishing, destructive fisheries practices, land-based sources of marine pollution and the ravages of climate change. The

CTI was proposed by Indonesia in 2007 as a multilateral partnership to protect the region's coastal and marine resources. The member States (Indonesia, the Philippines, Malaysia, Timor Leste, Papua New Guinea and the Solomon Islands) have committed to five overall goals over ten years:

- the designation of priority seascapes;
- implementing an ecosystem-based approach to managing fisheries and other marine resources;
- the establishment of marine protected areas;
- developing strategies to adapt to climate change; and
- the protection of threatened species.

The member States have also committed to guiding principles including the recognition of the transboundary nature of important marine resources and the need to align their activities with existing international law instruments such as UNCLOS, the Convention on Biological Diversity (CBD), regional fisheries management agreements and the UN Framework Convention on Climate Change (UNFCCC). The first five years of this initiative have revealed the considerable challenges involved in managing the heterogeneous, multiscale and interlinked nature of large-scale marine systems, and prompted analysts to recommend dynamic approaches to governance which link diverse institutional arrangements at multiple scales and involve regular adjustments to governance arrangements (Fidelman, 2012).

Resolutions and decisions of global bodies

Another driver for regional harmonization and integration has been resolutions and decisions of global bodies such as the United Nations General Assembly (UNGA), the Convention on Biological Diversity (CBD) and the Food and Agricultural Organization (FAO). On the basis of growing concern about the adverse impacts of high-seas bottom fishing on vulnerable marine ecosystems, in its 2006 Resolution 61/105 the UNGA called upon States

> '...to take action immediately, individually and through regional fisheries organizations and arrangements, and consistent with the precautionary approach and ecosystem approaches, to sustainably manage fish stocks and protect vulnerable marine ecosystems (VMEs), including seamounts, hydrothermal vents and cold water corals, from destructive fishing practices, recognizing the immense importance and value of deep sea ecosystems and the biodiversity they contain (Fidelman, 2012)'.

Under the resolution, Member States and RFMOs were required to manage fisheries to prevent significant adverse impacts to VMEs (cf. Chapter 14). To assist States in implementing the resolution, the FAO produced a set of International Guidelines for the Management of Deep Sea Fisheries in the High Seas (The Deep Sea Fisheries Guidelines). The principal objective of the Guidelines is to ensure the long-term conservation and sustainable use of deep-sea resources and to prevent significant adverse impacts to VMEs from activities related to such exploitation. This is achieved through a requirement to conduct assessments to establish if deep-sea fishing activities are likely to cause significant adverse impacts in a given area, and to adopt management measures to prevent such impacts such as fisheries closures in areas around VMEs. As a last resort, encounter protocols require fishing vessels to move a minimum distance from a location where species indicating the presence of a VME are captured by their gear.

Although there are still significant problems with implementing the Deep Sea Fisheries Guidelines in a sufficiently precautionary manner, the fact that progress in implementation of the UNGA Resolutions and the Guidelines has been reviewed twice by the UNGA may be a significant factor in prompting many RFMOs to close areas to fishing based on the presence or likely presence of VMEs (Auster et al., 2011; Gianni et al., 2011). The CBD has also been active in aligning and harmonizing fisheries and biodiversity conservation objectives at the regional level through the provision of expert advice on describing marine Areas of Ecological or Biological Significance (EBSAs) and in addressing biodiversity concerns in sustainable fisheries. In 2008, the Ninth Meeting of the Conference of Parties (CoP9) of the CBD adopted the following scientific criteria for identifying 'ecologically or biological significant areas in need of protection in open ocean waters and deep sea habitats':

- uniqueness/rarity;
- special importance for life history stages of species;
- importance for threatened, endangered or declining species and/or habitats;
- vulnerability, fragility, sensitivity or slow recovery;
- biological productivity;
- biological diversity; and
- naturalness.

This decision also provided scientific guidance for selecting areas to establish a representative network of marine protected areas including in open-ocean waters and deep-sea habitats. The 10th CBD CoP in 2010 agreed on a process of regional workshops for the description of EBSAs. The workshop outcomes are meant to inform relevant regional and global organizations. The work is premised on recognition that: the application of the EBSA criteria is a scientific and technical exercise; areas found to meet the criteria may require enhanced conservation and management measures; this can be achieved through a variety of means including marine protected areas and impact assessments; and the identification of EBSAs and the selection of conservation and management measures is a matter for States and competent intergovernmental organizations, in accordance with international law, including UNCLOS.

Regional workshops on describing EBSAs have been organized covering the Northeast Atlantic, the Western South Pacific, the Wider Caribbean and Western Mid-Atlantic, the Western Indian Ocean and the Eastern Tropical and Temperate Pacific. In addition, areas meeting EBSA compatible criteria have been described in the Mediterranean. At the CBD CoP 11 in Hyderabad in October 2012, after review and endorsement by CBD SBSTTA it was agreed that the areas described as EBSAs by these workshops and processes should be sent to the UN and relevant international organizations. In this manner the information on EBSAs can provide valuable information to RFMOs and their scientific commissions in the setting of priorities for conservation and management measures for biodiversity conservation purposes.

Another contribution by UNEP and the CBD has been their efforts to convene experts in biodiversity and fisheries to discuss topics of common concern. In September 2009 in Rome, FAO, UNEP and CBD convened an expert meeting on impacts of destructive fishing practices, unsustainable fishing and IUU fishing on marine biodiversity and habitats with the collaboration of the IUCN Fisheries Expert Group. In December 2011 the CBD Secretariat convened a joint expert meeting (in Bergen) with the same partners, to review the extent to which biodiversity concerns are addressed in existing assessments of fisheries stocks and to propose options to address biodiversity concerns. This meeting also brought together representatives of RFMOs and other relevant organizations, processes and scientific groups.

The meeting report noted (CBD, 2012) that there is much attention paid to the major biodiversity considerations in the RFMO conventions and overarching high-level policies, but implementation is often highly variable and inadequate. In some cases, efforts were constrained by: (1) the primacy to single-species Maximum Sustainable Yield (MSY) goals; and (2) limits on knowledge of ecosystems and the fisheries and on capacities for assessments and management. The expert meeting recommended a number of options for enhancing the treatment of biodiversity considerations in fisheries policy, assessment and management including:

- making CBD expertise on biodiversity assessment and selection of practical indicators for status and trends in biodiversity available to fisheries assessment and management agencies;
- contributing CBD expertise to modelling work and other studies of longer-term ecosystem-scale changes due to climate change, aggregate impacts of multiple uses and other scenarios and making the results of such studies available to sectoral regulatory agencies, including fisheries agencies, when they are dealing with factors such as the robustness of harvesting strategies;
- CBD initiating or supporting pilot projects for fisheries and biodiversity conservation communities to work together and illustrate the concrete benefits of collaboration; and
- CBD being a major contributor to capacity-building initiatives for fisheries management agencies dealing with biodiversity considerations.

Updating mandates to include modern conservation principles and the need for collaboration between RFMOs and RSAs

A further catalyst for alignment and harmonization of fisheries and biodiversity conservation objectives is the inclusion of modern conservation principles such as the precautionary and ecosystem-based approaches in the mandate of RFMOs established before the UNFSA was adopted.

The North East Atlantic Fisheries Commission (NEAFC) was originally formed to recommend measures to maintain the rational exploitation of fish stocks in the Atlantic and Arctic Oceans. It has regulatory competence over three large maritime areas beyond national jurisdiction in the Northeast Atlantic Ocean and may recommend conservation and management measures for all fisheries resources within its Convention Area with the exception of sea mammals and sedentary species and tuna or tuna-like species. In 2006 the NEAFC Convention was amended to include a principal objective of ensuring the long-term conservation and optimum utilization of the fishery resources in the Convention Area, providing sustainable, environmental and social benefits (cf. Chapter 16). Under

Article 4 of the amended Convention, NEAFC must conserve marine biological diversity and control the impact of fisheries on other species and marine ecosystems consistent with this objective. The impact of the change in mandate has been demonstrated through agreed conservation and management measures for fisheries with up to 50% of the fishable area closed to fisheries and strict regulations of bottom fishery in the rest of the area.

NEAFC's incorporation of biodiversity considerations into its fisheries conservation and management measures has also been facilitated by its close working relationship with OSPAR. OSPAR and NEAFC signed a memorandum of understanding in 2008 and both organizations use ICES as their scientific advisory body. ICES has recommended that a coordinated approach be taken between the two organizations in the protection of vulnerable marine ecosystems and there has been considerable overlap between areas proposed for protection by OSPAR and those considered for closure to bottom fishing by NEAFC.

Pathways to improved collaboration on regional fisheries and biodiversity conservation governance

Recognizing that there is considerable diversity among regions in progress towards alignment and harmonization of fisheries and biodiversity conservation goals, this section explores potential pathways to improve regional governance for fisheries and biodiversity conservation. The complex challenges involved in implementing an ecosystem-based approach to fisheries management and the variable capacities of particular regions to meet that objective demands that a broad suite of options be examined, including: (1) expanding the inter-regional capacities of fisheries and biodiversity governance bodies to align their objectives; and (2) strengthening their collaboration with relevant global, intra-regional and national level organizations. Different models for cooperation and collaboration on fisheries and biodiversity objectives are described in the following sections.

Intra-regional options

Better inter-regional alignment of fisheries and biodiversity conservation objectives could be achieved by:

- recognition and incorporation of modern conservation principles such as the precautionary and ecosystem-based approaches and agreed global, regional and national biodiversity conservation targets into the mandates and policies of RFMOs, RSAs and LME projects;
- exchange of information on the status of the marine environment, its resources and biodiversity between governance bodies, e.g. regarding biodiversity data and EBSAs on which RFMOs could act;
- consultation and cooperation in implementing conservation and management measures as exemplified by OSPAR and NEAFC as well as ASCLME and SWIOFP;
- instituting collaborative processes such as strategic environmental assessment and marine spatial planning across regions that involve RFMOs and RSAs as well as other sectoral players in the marine environment;
- developing a common scientific baseline for fisheries and biodiversity conservation measures and recourse to a common scientific advisory body such as ICES in a particular region;
- making a commitment to cooperate on specific pilot projects between RFMOs and RSAs to build commitment for resolving broader challenges facing both fisheries and biodiversity conservation agencies, cementing and fostering working relationships between both sectors;
- undertaking direct collaboration between fisheries and biodiversity conservation bodies at the regional level, promoting openness to receiving input from NGOs and IGOs; and
- strengthening collaboration at the national level amongst the responsible ministries for fisheries and biodiversity conservation, local communities and industry players (Gjerde et al., 2010).

In addition, new protocols or agreements to pursue processes such as environmental impact assessment (EIA), strategic environmental assessment (SEA) and marine spatial planning (MSP) at a regional scale could provide a more legally-binding science-based framework for cross-sectoral cooperation and management in fisheries and biodiversity conservation objectives. Eventually States may wish to consider moving towards a cross-sectoral coordinating mechanism or organization with representatives from the adjacent coastal states as well as relevant sectoral organizations, both global and regional. Such an organization could reflect more comprehensive objectives and may be able to implement a genuine marine spatial planning process for ABNJ which could align and harmonize fisheries and biodiversity conservation objectives in cooperation with comparable national initiatives.

The exchange of scientific information on fisheries and biodiversity between regional governance bodies in adjacent or proximate regions can provide a starting point for enhanced cooperation and collaboration in achieving alignment and harmonization of fisheries and biodiversity conservation goals. Adjacent or proximate regions can also collaborate in developing complementary fisheries and biodiversity conservation objectives and action plans, particularly where fish stocks and vulnerable marine ecosystems straddle both regions. At a more advanced stage, adjacent regions could also collaborate in developing strategic action plans to conserve and manage fisheries and biodiversity in the transboundary context and in proximate areas beyond national jurisdiction.

There is an increasing willingness to cooperate across sectors through memoranda of understanding and other agreements, as evidenced in the North East Atlantic and ASCMLE case studies considered above. This has led to joint work programs, joint meetings and participation at each other's meetings and scientific committees. Underpinning such agreements must be an understanding of the benefits such as certainty, prevention of duplication and increase in efficiency in achieving agreed goals and targets, the process for cooperation, and the human and financial resources committed.

Global connections

Developing stronger links between fisheries and biodiversity conservation bodies at the regional level and relevant global organizations (cf. Chapters 10 and 11) has clear and ongoing benefits. The best-practice standards developed by global organizations such as the FAO and CBD have already contributed to the alignment and harmonization of fisheries and biodiversity conservation objectives at the regional level and have the potential to contribute further. The CBD Guidelines on EBSAs and EIAs could further assist in establishing a common approach to identification of areas and management of risks to marine biodiversity while respecting the varying competences of the regional and sectoral bodies (cf. Chapter 14). At the same time, the wealth of new scientific and technical data and information from RFMO efforts to identify VMEs, from the ISA scientific and planning initiatives as well as from the Census of Marine Life and other scientific initiatives, can inform the conduct of EIAs and SEAs and help to identify EBSAs at the regional level. The development of a CBD-based global inventory of EBSAs in marine areas beyond national jurisdiction is providing a process for stimulating and coordinating cross-sectoral cooperation at the regional level and a globally accessible list of recognized EBSAs for conservation and management purposes.

At the regional level, RFMOs and States could be requested to cooperate in the protection of EBSAs vulnerable to deep-sea or pelagic fishing impacts in the high seas, particularly pursuant to the provisions of the UNFSA for highly migratory and straddling fish stocks. This could be achieved through the adoption of binding management measures including spatial or temporal closures (year round, dynamic or seasonal), effort or gear restrictions, catch or bycatch quotas to reduce impacts on EBSAs or VMEs or other areas and the requirement of full reporting of catches validated by some level of independent observer coverage. The FAO has also published many guidelines and international plans of action for the conservation and management of species such as sharks, sea turtles and sea birds, the prevention of illegal, unregulated and unreported fisheries and the management of bycatch and reduction of discards. These guidelines could also include measures to reduce the catch of vulnerable species in areas where they are most at risk, using the EBSA criteria for endangered and threatened species or special importance for the life history of species.

The potential importance of global level stimulus through UNGA resolutions for example or a possible new multilateral agreement under UNCLOS should also be highlighted. Many States are already calling for a new implementing agreement to UNCLOS to address the conservation and sustainable use of marine biodiversity beyond national jurisdiction. A new global agreement could enhance efforts towards regional cooperation based on common goals and objectives. It could also seek to fill the gaps where no RSAs exist, stimulate capacity building and technology transfer and enhance scientific cooperation and coordination (Gjerde and Rulska-Domino, 2012).

In the short term, the Kobe Process and the five tuna RFMOs could give higher priority to issues related to movements of highly migratory species and their implications for management (oceanic MPAs, interactions between fisheries) and the impact of fish aggregating devices (FADs) on oceanic ecosystems. The GEF-funded ABNJ Program may also be important, with the programme goal of promoting efficient and sustainable management of fisheries resources and biodiversity conservation in ABNJ, in accordance with the global targets agreed in international forums. The programme, led by the FAO in close collaboration with partners, comprises four projects: (1) sustainable management of tuna fisheries and biodiversity conservation in the ABNJ; (2) sustainable fisheries management and biodiversity conservation of deep-sea living resources and ecosystems in the ABNJ; (3) oceans partnership fund; and (4) strengthening global capacity to effectively manage ABNJ. The GEF total of $43,547,119 in indicative funding is expected to catalyse $222,741,000 in co-financing (Turner, 2012). If spent wisely, fairly and transparently, this project could go a long way towards stimulating collaboration for sustainable fisheries management and biodiversity conservation in ABNJ.

Conclusions

Alignment of fisheries and biodiversity conservation objectives varies considerably across regions with relatively high levels of integration occurring in some regions, such as the Antarctic and the Northeast Atlantic, and embryonic levels of coherence in the Western Indian Ocean and South East Asia. There are many reasons for this diversity including: the lack of primacy given to modern conservation principles in some RFMO agreements and policies; the lack of technical expertise and resources to implement ecosystem-based conservation and management measures for fisheries and biodiversity at the regional level; the absence of cross-institutional connections and collaboration between fisheries and biodiversity conservation organizations; and slow uptake by regional bodies of global best-practice standards for fisheries and biodiversity conservation. The absence of a global level mandate for cooperation and regular mechanisms for performance review are also important.

Some of the key ingredients for progressing enhanced alignment of fisheries and biodiversity conservation objectives at the regional level could include:

(1) a focus on creating partnerships or platforms for cross-sectoral discussions and collaboration between fisheries and biodiversity conservation organizations within and across regions;
(2) the incorporation of modern conservation principles into RFMO Conventions and a commitment to implement them in fisheries conservation and management;
(3) the incorporation of sustainable-use principles in RSA Conventions and a commitment to address them in marine biodiversity conservation and management;
(4) the incorporation of global best-practice standards for fisheries and biodiversity conservation management in the policies and practices of RFMOs and RSAs and a commitment to implement them;
(5) the development of a common science advisory foundation for fisheries and environmental bodies to use as a starting point for their policies and management measures;
(6) a commitment to share data across fisheries and biodiversity conservation sectors (e.g. biodiversity data can feed into the RFMO consideration of conservation and management measures and, if EBSAs are identified, RFMOs can take that information into account in adopting particular measures);
(7) a clear understanding from all sides of the benefits of cooperation, the processes that will be used and the resources that each side is committing;
(8) the eventual development of a coordinating mechanism and strategic action plan for fisheries and biodiversity conservation in particular regions; and
(9) a global level agreement for conservation and sustainable use of marine biodiversity in ABNJ which enhances mandates for regional cooperation, accompanied by a global mechanism for oversight, coordination and capacity development.

The model for enhanced alignment and harmonization of fisheries and biodiversity conservation will be different for each region but is likely to contain most of the elements listed above in a form tailored to the political and governance characteristics of the particular region. Multiple benefits will flow from the convergence of fisheries and biodiversity conservation objectives at the regional level. Some of these benefits have been highlighted by the collaboration between RFMOs and RSAs such as NEAFC and OSPAR discussed above and by the CBD in their new 10-year Strategic Plan for Biodiversity 2011–2020 and include, in Aichi Target 6:

> '...the sustainable management and harvest of all fish and invertebrate stocks and aquatic plants applying ecosystem based approaches so that overfishing is avoided, recovery plans and measures are in place for all depleted species, fisheries have no significant adverse impacts on threatened species and vulnerable ecosystems and the impacts of fisheries on stocks, species and ecosystems are within safe ecological limits.'

The achievement of the CBD's target will, to a large extent, be dependent upon continuing efforts to align and harmonize fisheries and biodiversity conservation objectives at the regional level.

References

Alexander, L.M. (1993) Large marine ecosystems: a new focus for marine resources management. *Marine Policy* **17**, 186–198.

Aqorau, T. (2001) Tuna fisheries management in the Western and Central Pacific Ocean: a critical analysis of the Convention for the Conservation and management of Highly Migratory Fish Stocks in the Western and Central Pacific Ocean and its implications for the Pacific Island States. *International Journal of Marine and Coastal Law* **16**(3), 379–431.

Auster, P.J., Gjerde, K., Heupel, E. et al. (2011) Definition and detection of vulnerable marine ecosystems on the high seas: problems with the 'move-on' rule. *ICES Journal of Marine Science* **68**(2), 254–264.

Boyle, A. (2000) Globalism and regionalism. In: *Protecting the Polar Marine Environment* (ed. Davor Vidas). Cambridge University Press, Cambridge.

CBD (2012) Subsidiary Body on Scientific Technical and Technological Advice (SBSTA). Report of joint expert meeting on addressing biodiversity concerns in sustainable fisheries, UNEP/CBD/SBSTTA/16/INF/13, 5 March 2012. Available at http://www.cbd.int/doc/meetings/mar/ebsa-briefing/other/ebsa-briefing-oth-01-en.pdf (accessed March 2014).

Duda, A. and Sherman, K. (2002) A new imperative for improving management of large marine ecosystems. *Ocean and Coastal Management* **45**, 797–833.

Edeson, W.R. (2009) Overview of institutional arrangements for fisheries and marine biodiversity in the Indian Ocean. In: *Fisheries Exploitation in the Indian Ocean: Threats and Opportunities* (eds D. Rumley, S. Chaturvedi and V. Sakhuja). Institute of Southeast Asian Studies, Singapore.

Fanning, L., Mahon, R., McConney, et al. (2007) A large marine ecosystem governance framework. *Marine Policy* **31**(4), 434–443.

Fidelman, P., Evans, L., Fabinyi, M. et al. (2012) Governing large scale marine commons: Contextual challenges in the Coral Triangle. *Marine Policy* **36**(1), 42–53.

Freestone, D. (2008) Fisheries Commissions and Organizations. In *The Max Planck Encyclopedia of Public International Law* (ed. R. Wolfrum). Oxford University Press, Oxford. Online edition.

Freestone, D. and Killerlain Morrison, K. (2012) The Sargasso Sea Alliance. *International Journal of Marine and Coastal Law* **27**(3), 647–655.

Gianni, M., Currie, D.E.J., Fuller, S. et al. (2011) Unfinished business: a review of the implementation of the provisions of UNGA resolutions 61/105 and 64/72 related to the management of bottom fisheries in areas beyond national jurisdiction. Deep Sea Conservation Coalition, September. Available at http://www.savethehighseas.org/publicdocs/DSCC_review11.pdf (accessed March 2014).

Gjerde K.M. and Rulska-Domino, A. (2012) Marine Protected Areas Beyond National Jurisdiction: Some practical perspectives for moving ahead. *International Journal of Marine and Coastal Law* **27**(2), 351–273.

Gjerde, K.M., Ardron, J., Gotheil, S. et al. (2010) Modalities for advancing cross-sectoral cooperation in managing marine areas beyond national jurisdiction. Report for UNEP's 12th Global Meeting of the Regional Seas Conventions and Action Plans.

High Seas Task Force (2006) Closing the Net: Stopping Illegal Fishing on the High Seas. Final report of the Ministerially-led Task Force on IUU Fishing on the High Seas. Governments of Australia, Canada, Chile, Namibia, New Zealand and the United Kingdom, WWF, IUCN and the Earth Institute at Columbia University. Available at http://www.oecd.org/sd-roundtable/papersandpublications/39375276.pdf (accessed March 2014).

Hurry, G.D., Hayashi, M. and Maguire, J.J. (2008) Report of the Independent Review, PLE-106/2008, International Convention on the Conservation of Atlantic Tunas, p. 2.

Kaye, S.B. (2000) *International Fisheries Management*. Kluwer Law International, The Hague.

Lagoni, R. (2003) Regional protection of the marine environment of the Northeast Atlantic under the OSPAR Convention of 1992. In *The Stockholm Declaration and the Law of the Marine Environment* (eds M.H. Nordquist, J.N. More and S. Mahmoud). Kluwer Law International, The Hague.

McDorman, T. (2005) Implementing existing tools: turning words into action. Decision-making processes of Regional Fisheries Management Organizations. *International Journal of Marine and Coastal Law* **20**(3–4), 423–457.

Miller, D.G., Sabourenkov, E.N. and Ramm, D.C. (2004) Managing Antarctic marine living resources: the CCAMLR approach. *International Journal of Marine and Coastal Law* **19**(3), 317–363.

Parris, H. (2010) Is the Western and Central Pacific Fisheries Commission meeting its conservation and management objectives? *Ocean and Coastal Management* **53**(1), 10–26.

Rayfuse, R. (2004) *Non Flag State Enforcement in High Seas Fisheries*. Martinus Nijhoff Publishers, Leiden.

Sand, P. (1999) The rise of regional agreements for marine environment protection. In: *Transnational Environmental Law: Lessons in Global Change* (ed. P.H. Sand). Kluwer Law International, The Hague.

Sherman, K. (1994) Sustainability, biomass yields and health of coastal ecosystems: an ecological perspective. *Marine Ecology Progress Series* **112**, 277–301.

Sherman, K. (2006) The Large Marine Ecosystem Network Approach to WSSD Targets. *Ocean and Coastal Management* **49**(9–10), 640–648.

Treves, T. (2003) Regional approaches to the protection of the marine environment. In: *The Stockholm Declaration and the Law of the Marine Environment* (eds M.H. Nordquist, J.N.

Moore and S. Mahmoudi), pp. 137–138. Kluwer Law International, The Hague.

Turner, J. (2012) The GEF/FAO ABNJ Program on Global Sustainable Fisheries Management and Biodiversity Conservation in Areas Beyond National Jurisdiction (ABNJ), Global Ocean Forum Newsletter. Special Issue on Marine Areas Beyond National Jurisdiction (ABNJ), May 4. Available at http://www.globaloceans.org/sites/udel.edu.globaloceans/files/GEF-FAO-ABNJProgram-FAO.pdf (accessed February 2014).

Vallega, A. (2002) The regional seas in the 21st century: an overview. *Ocean and Coastal Management* **45**(11), 925–934.

Chapter 16

Regional governance: The case of NEAFC and OSPAR

K. Hoydal[1,2], D. Johnson[3] and A.H. Hoel[4]

[1] Director of *Skrivarastova – Fish and Film*, Consultants Ltd, Faroe Islands
[2] IUCN Commission on Ecosystem Management (IUCN/CEM/FEG), Gland, Switzerland
[3] Seascape Consultants Ltd and Emeritus Professor of Coastal Management, Southampton Solent University, UK
[4] Institute of Marine Research, Bergen, Norway

Abstract: This chapter describes how the two North East Atlantic Regional Conventions, NEAFC and OSPAR, operate in practice and how there has been a development towards a focus on biodiversity conservation linked to rational and sustainable practices in the various sectors making use of this maritime area (i.e. fisheries), prevention of pollution and controlling adverse impacts of human activities. We identify a number of acknowledged differences, highlight positive signs of regional cooperation and suggest how this might provide a global model for marine governance and management.

Keywords: regional; governance; maritime area; NEAFC; OSPAR; cooperation; biodiversity; fisheries; pollution

Introduction

The North East Atlantic Fisheries Commission (NEAFC) and the Convention for the Protection of the Marine Environment of the North East Atlantic (OSPAR Commission or OSPAR) are very different multilateral environmental agreements (MEAs). NEAFC was established as a sectoral resource management Convention, OSPAR as a Convention to prevent pollution and control the adverse impacts of human activities including, where appropriate, restoration of ecosystems. Traditionally NEAFC has been concerned with regulating the exploitation of fish stocks and OSPAR with regulating the sources, discharges and concentrations of pollutants. In this sense, both organizations initially took a rather narrow, specific and focused remit.

The situation in the Northeast Atlantic is probably unique. The OSPAR and NEAFC Conventions cover the same geographical area (see Figure 16.1) and have the almost same Contracting Parties;[1] furthermore, both Commissions receive scientific advice from the International Council of the Exploration of the Sea (ICES). The countries responsible for the governance of the productive seas of the Northeast Atlantic, which supports a long list of industries and human activities, have through Regional Fisheries Management Organizations (RFMOs)[2] and a Regional Seas Conventions (RSC) agreed governance systems for both the utilization of marine resources and the protection of biodiversity.

Relevant international processes with respect to regional fisheries management and conservation

Rochette and Chabason (2011) explain that all regional MEAs have had to respond to the ongoing development of global Conventions and tackle universal threats such as

Governance of Marine Fisheries and Biodiversity Conservation: Interaction and Coevolution, First Edition.
Edited by Serge M. Garcia, Jake Rice and Anthony Charles.

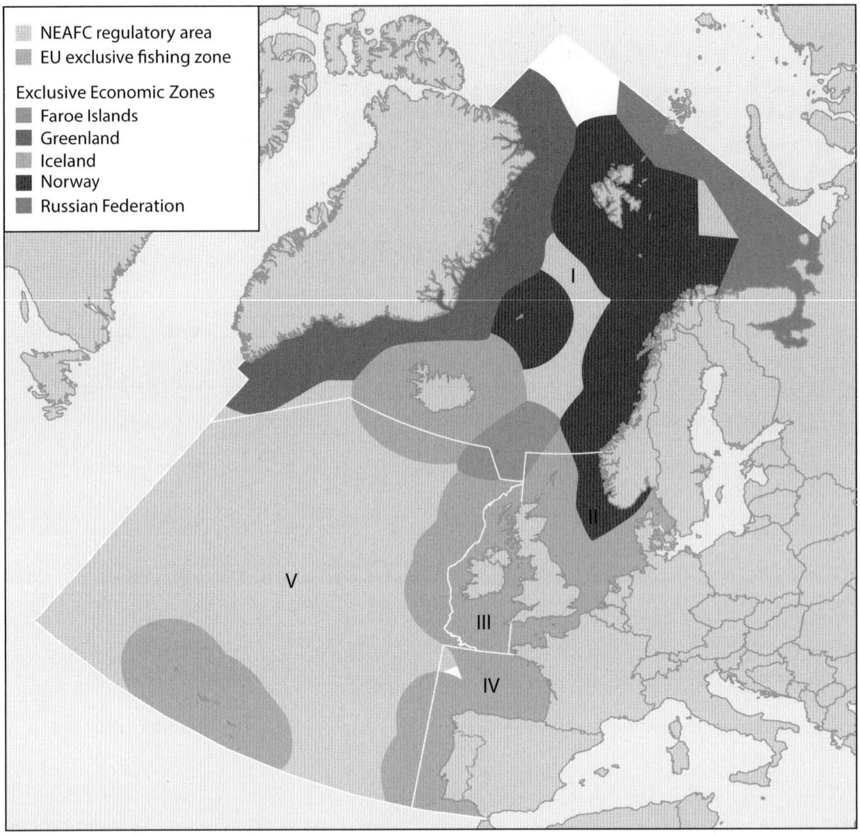

Figure 16.1 Map of NEAFC and OSPAR Convention Areas. The NEAFC and OSPAR Convention Areas are identical. The NEAFC area consists of the Regulatory Area (yellow) and the areas under the management of the EU, the Faroe Islands, Greenland, Iceland, Norway (coastal zone, Jan Mayen and the Svalbard Protection zone) and the Russian Federation. The OSPAR area is divided into Regions I–V (white line boundaries) for the purposes of monitoring and assessment.

climate change. These more recent developments, particularly those stressing the need for biodiversity conservation, have provided NEAFC and OSPAR with a common and unifying concern.

There are a number of international processes at the global level which are important to the structure and function of regional MEAs, particularly RFMO/As and RSCs. Some of these are of a constitutive nature, following up on global legally binding agreements. Others are more process-oriented and soft-law in nature. These processes are described in Chapter 7 together with the development of international law and instruments relevant to the management of living resources and protection of biodiversity since the 1982 UN Law of the Sea Convention (LOSC).

At the global level the Convention on Biological Diversity (CBD) provides a policy framework on the conservation and sustainable use of marine biodiversity, including addressing biodiversity concerns in sustainable fisheries, which is relevant to the work of both FAO and the United Nations Environment Programme (UNEP). This is mirrored regionally and in the Northeast Atlantic, and has provided a further opportunity for synergy.

NEAFC

The UN Fish Stocks Agreement (UNFSA) and its follow-up in the UNFSA Review Conference, as well as the FAO Code of Conduct and Action Plans and the work of the Committee on Fisheries (COFI), including the Regional Fisheries Body Secretariats Network (RSN) meetings, have been of specific relevance to NEAFC. NEAFC has been an active player in the COFI processes and has reported regularly to FAO on progress with respect to implementing the Code of Conduct. Moreover, the processes in the UN General Assembly (UNGA) have played a major role in discussions in NEAFC, and NEAFC has regularly contributed to the annual reports of the UN Secretary-General on Oceans and Sustainable fisheries to the UN General Assembly.

OSPAR

As a Regional Seas Convention, the most relevant international global UN Agency for OSPAR is the United Nations Environment Programme (UNEP). Inspired by the 1972 United Nations Conference on the Human Environment, UNEP launched the Regional Seas Programme in 1974. The aims of this marine and coastal initiative are to address accelerating degradation of the world's oceans and coastal areas by engaging neighbouring countries in comprehensive and specific actions. The network of Regional Seas Programmes comprises 13 organizations established under UNEP auspices, together with 5 partner programmes for the Antarctic, Arctic, Baltic Sea, Caspian Sea and Northeast Atlantic. In most regions an Action Plan (or its equivalent) of jointly coordinated activities is underpinned bya legal framework in the form of a Regional Convention. UNEP considers the strengths of this initiative to include geographic focus, customization to fit threats and priorities of particular regions, cooperation between States, knowledge sharing and flexibility. Annual global meetings of Secretariats work towards delivering a Global Strategy first agreed in 2003. Collectively, the Regional Seas Conventions and Action Plans also contribute to the UN Secretary General's report and the Secretariats meet annually to agree strategic directions and share good practice.

Strengthening the OSPAR convention (annex V)

The origin

A regional-level management framework for protecting the Northeast Atlantic from pollution was formalized by the 1972 Oslo Convention on the Prevention of Marine Pollution by Dumping from Ships and Aircraft and on issues related to pollution from land-based sources with the signing of the 1974 Paris Convention on the Prevention of Marine Pollution from Land-Based Sources.

In 1989/1990 governments actively discussed updating and integrating these original Conventions. Opinion on how to proceed was split; some Parties wished to integrate but to keep the issues covered much the same, while others saw an opportunity to create a more comprehensive marine environmental protection Convention. The compromise was to integrate them, thus streamlining administrative procedures, with the provision to add additional legally binding annexes to the 'new' Convention as required. The opportunity was therefore taken to introduce the possibility of extending the Convention to cover threats to the marine environment from human activities other than pollution.

As a result, the Paris Convention was updated and unified with the London Convention in 1992, forming the OSPAR Convention to be implemented by the OSPAR Commission. Article 1.a of the Convention states that:

> 'The Contracting Parties shall, in accordance with the provisions of the Convention, take all possible steps to prevent and eliminate pollution and shall take the necessary measures to protect the maritime area against the adverse effects of human activities so as to safeguard human health and to conserve marine ecosystems and, when practicable, restore marine areas which have been adversely affected.'

In doing so there is an obligation to apply the precautionary principle, the polluter pays principle and best-available technology/best environmental practice.

Expansion of the mandate to cover biodiversity

In 1992 dumping of radioactive substances at sea was still a major concern. In 1994 this was dealt with when the London Convention agreed to put off any such dumping for 25 years. Globally the focus of international policy-makers had moved to protecting biodiversity and the 1992 World Summit on Sustainable Development and the creation of the Convention of Biological Diversity (CBD) prompted further efforts on biodiversity within OSPAR. A consideration at that time was the opinion that the European Commission Habitats Directive would not apply beyond 12 nautical miles. Scope for OSPAR to protect biodiversity within the maritime area between 12 and 200 NM (nautical miles) therefore started the process, eventually leading to a new annex to the Convention. Some of the OSPAR Contracting Parties were sceptical about this new development. The United Kingdom insisted on keeping fisheries separate and this was also acceptable to Norway.

At that stage the Convention on Biological Diversity was not in force and CITES was only concerned with trade in endangered species. OSPAR's sister Convention in the Baltic, the Helsinki Convention (HELCOM), and non-governmental organizations (NGOs) were supportive.

The possibility to extend the OSPAR Convention was duly exercised in 1998, when the Contracting Parties adopted a new annex V to the Convention (on the Biological Diversity and Ecosystems of the Maritime Area). This authorized the OSPAR Commission to adopt programmes and measures to protect the marine environment relating to all human activities that may adversely affect the ecosystems and their biodiversity, except programmes and measures relating to fisheries management and with a preference for action at the global level on maritime transport.

Annex V came into force on 30 August 2000 for the first seven Contracting Parties (as stipulated in the OSPAR Rules of Procedure). The annex was subsequently ratified by all 16 OSPAR Contracting Parties.[3] Implementing annex V has been guided by the 2003 Biodiversity and Ecosystems Strategy subsequently superseded by the 2010 North-East Atlantic Environment Strategy (OSPAR, 2010b). This Strategy commits OSPAR to halt and prevent by 2020 further loss of biodiversity in the maritime area, to protect and conserve ecosystems and to restore, where practicable, marine areas which have been adversely affected. Two main technical Committees strive to deliver this objective. The Biodiversity Committee seeks to improve the status of threatened and/or declining species and habitats, further the OSPAR network of marine protected areas and monitor and assess ecosystem functioning. The Environmental Impact of Human Activities Committee aims to recognize, reduce and mitigate the adverse effects of human activities and pressures on the marine environment, individually or cumulatively. Particular emphasis is given to marine litter, the introduction of energy including underwater noise and non-indigenous species. At a strategic level the OSPAR Coordination Group seeks to ensure that when environmental impacts affect biodiversity an integrated ecosystem approach is followed.

Strengths of the OSPAR Convention in relation to common issues can be summarized as follows:

- the sea does not respect national boundaries (OSPAR enables States to agree action and cooperate in setting objectives);
- a long-term holistic approach is needed (OSPAR has developed a comprehensive set of strategies to 2020);
- good science, careful monitoring and accurate assessment must underpin policy (OSPAR has developed a Joint Assessment and Monitoring Programme or JAMP);
- an ecosystem approach is needed to ensure holistic solutions (OSPAR Ecological Quality Objectives support this); and
- measures need to be implemented and the implementation needs to be checked (every OSPAR measure has its implementation reporting and assessment procedures).

Progress against the Strategy is evaluated in the Quality Status Report 2010 (OSPAR, 2010a), a major decadal exercise summarizing underpinning contributory assessments and periodic evaluations. OSPAR has been influential in the global debate on protection of marine biodiversity and seeks to draw the attention of other competent authorities to the need to adopt measures when these are not within the OSPAR sphere of competence. Nevertheless, OSPAR has concluded that pressures on biodiversity, including removal of species, loss of and damage to habitats, the introduction of non-indigenous species, obstacles to species migration and poor water quality, continue to be a concern; these can act cumulatively or be exacerbated by climate change. Furthermore, a particular problem frustrating the conservation of deep-sea species and habitats is that marine biodiversity in these waters is still poorly known and understood.

Updating the NEAFC convention

The origin

NEAFC goes back a long time. Negotiations started before the Second World War at the Overfishing Conference in London during 23–28 November 1936. The Convention on Future Multilateral Cooperation in the North-East Atlantic Fisheries, which entered into force on 17 March 1982, replaced the original North-East Atlantic Fisheries Convention of 1959 which in turn had replaced the 1946 Convention for the Regulation of Meshes and Fishing Nets and the Size Limits of Fish. The North-East Atlantic Fisheries Commission (NEAFC) was set up under the framework of the Convention. The text of the 1980 Convention reflects discussions and negotiations leading up to the signing of UNCLOS in 1982.

There are at present five Contracting Parties: the European Union, Denmark (in respect of the Faroe Islands and Greenland), Iceland, Norway and the Russian Federation. NEAFC was something of a dormant Convention before 1995, as it did not adopt any management measures of any significance during its first 13 years.

Strengthening NEAFC mandate and objectives

The signing of the United Nations Fish Stocks Agreement in 1995 sparked life into the organization. From 1995 onwards all major fisheries in the NEAFC Regulatory Area were brought under regulation and NEAFC has systematically adapted the requirements in international law and instruments, that is, science-based management, effective surveillance and inspection, transparency, fast-track dispute settlement and extensive area closures and bottom fishing regulations.

The Contracting Parties of NEAFC updated the 1980 Convention in 2004 and dispute settlement procedures in 2006, bringing the Convention into line with developments in international law since the 1982 Convention was negotiated.

The preamble of the 'new' Convention lists the international instruments of importance (the 1982 LOSC, the 1995 UNFSA and the 1995 CCRF) and stresses that NEAFC – in addition to managing fisheries – has an important role in conservation by integrating environmental concerns into its management.

The new Article 2 of the 2006 Convention amendment states that 'the objective of this Convention is to ensure the long-term conservation and optimum utilization of the fishery resources in the Convention Area, providing sustainable economic, environmental and social benefits'.

Its new Article 4 states that the Commission shall perform its functions in order to fulfil the objective set out in Article 2. When making recommendations in accordance with Article 5 or 6 of the Convention, the Commission shall in particular:

(1) 'Ensure that such recommendations are based on the best scientific evidence available;
(2) Apply the precautionary approach;
(3) Take due account of the impact of fisheries on other species and marine ecosystems, and in doing so adopt, where necessary, conservation and management measures that address the need to minimize harmful impacts on living marine resources and marine ecosystems; and
(4) Take due account of the need to conserve marine biological diversity.

In addition, the Commission shall provide a forum for consultation and exchange of information on the state of the fishery resources in the Convention Area and on the management policies, including examination of the overall effects of such policies on the fishery resources and, as appropriate, other living marine resources and marine ecosystems.'

The 2006 NEAFC Independent Performance Review (NEAFC, 2006a) concluded that the Convention amendments ensured that NEAFC's framework was consistent with modern developments in international instruments and commitments. In the last decade, the NEAFC Secretariat has regularly reported to the Annual Meetings of NEAFC with respect to international developments in fisheries management and ocean governance to assist the continuous review of a wider perspective of ocean governance, especially the ecosystem approach.

Measures implemented individually by NEAFC and OSPAR

NEAFC measures

The previous sections explained the development in the mandates of both NEAFC and OSPAR to consider a holistic ecosystem approach and to conserve biodiversity. This development has been reflected in a number of measures then taken separately by each organization. NEAFC has been fighting IUU fishing with blacklisting since 2005, and had a mature Port State Control system in place in the Northeast Atlantic as early as in 2007.

The pelagic fisheries for redfish, herring, blue whiting and mackerel in the NEAFC Regulatory Area use fishing gear that does not have bottom contact during normal operation. The fishing gear used in deep-sea fisheries (bottom trawl, longline and gillnet) do have bottom contact, as has bottom trawling for Rockall haddock. The pelagic fisheries have been considered clean single-species fisheries where bycatches of non-targeted species are not an issue. This has changed recently and the three main pelagic species – Norwegian Spring Spawning (Atlanto-Scandian) herring, blue whiting and mackerel – are seen by some scientists as a 'pelagic complex' and are in certain areas and seasons caught together. In deep-sea fisheries, bycatches of species not-targeted or of no commercial interest are known to occur.

In 2006 NEAFC prohibited fisheries with gillnets, entangling nets and trammel nets in depths below 200 m, and introduced measures to remove and dispose of unmarked or illegal fixed gear and retrieve lost gear to minimize ghost fishing.

In July 2008 NEAFC adopted additional comprehensive measures on bottom fishing activities in the NEAFC Regulatory Area. These were supplemented at its 2008 Annual Meeting with specific operational procedures. There are now clear procedures and rules in place on how to identify existing bottom fishing areas, exploratory fishing in new bottom fishing areas, how to assess bottom fishing activities, how to deal with encounters with vulnerable marine ecosystems and data collection protocols for observers onboard fishing vessels to increase knowledge of vulnerable habitats in deep waters. These measures are closely related to the implementation of the FAO International Guidelines for the Management of Deep-Sea Fisheries in the High Seas (FAO, 2009).

The need to conserve vulnerable deep-sea habitats and species has been high on the NEAFC agenda in recent years. During this period NEAFC has requested scientific information and advice on vulnerable deep-sea habitats from ICES. This has not been available in a number of instances, pending more research. Consequently, NEAFC has moved forwards in a precautionary and adaptive manner.

The amendments to the NEAFC Convention and the London Declaration in 2006 (NEAFC, 2006b) created the formal basis for taking action; however, NEAFC had already closed a region in the Rockall Area to protect juvenile fish by 2002 and had adopted an interim ban on bottom fishing in a large area on the Reykjanes Ridge (the northern part of the Mid-Atlantic Ridge) and four seamounts adjacent to the Ridge by 2004.

In 2007 NEAFC closed five areas in the Rockall–Hatton Bank area to bottom fishing to protect deep-water corals. These closures were reviewed in 2009 and the measures were extended until 2015. An important element is the scientific evaluation of proposals by NEAFC's provider of scientific advice, the International Council for the Exploration of the Sea (ICES). In 2006 NEAFC decided to set up procedures for reporting scientific information on deep-sea fisheries to ICES, and has a standing request to ICES for more detailed advice on vulnerable habitats and deep-sea fisheries as more information becomes available.

In April 2009, NEAFC decided to close five areas to bottom fisheries on the Mid-Atlantic Ridge in the high seas in the Northeast Atlantic in order to protect Vulnerable Marine Ecosystems (VMEs) from significant adverse impacts. A proposal tabled in June 2008 formed the basis of discussions in NEAFC. NEAFC received advice on this proposal from ICES, which assessed the proposal against the standards and criteria for identifying VMEs, developed by the UN Food and Agriculture Organization (FAO) in response to the United Nations General Assembly resolution 61/105 to support the sustainable management of deep-sea fisheries and responsible fisheries in the marine ecosystem. The NEAFC Secretariat has participated actively in the meetings in FAO on International Guidelines for the Management of Deep-sea Fisheries in the High Seas. Pursuant to the competence of NEAFC, this implies that fishing activities by vessels flying the flags of NEAFC Contracting Parties (or Cooperating non-Contracting Parties) with fishing gear which is likely to come in contact with the seafloor during the normal course of fishing operations are prohibited within these areas.

These new area closures include all five areas which were closed on a temporary basis in 2004. The combined size of the closed areas, covering a large portion of the Mid-Atlantic Ridge, is estimated to be 333,000 km^2 (i.e. larger than the United Kingdom and the Republic of Ireland combined). This measure shall be in force until 31 December 2015. The measure shall be reviewed by the Commission with the intention of extending the period that the recommendation is in force, unless the conclusion of the review is that the continued application of the measure or parts of the measure is not required. If scientific research demonstrates in the future that there are sub-areas where no vulnerable marine ecosystems are found within the areas referred to in this measure, the measure shall be amended in order to exclude those sub-areas from the fishing prohibition.

Most of the remaining part of the NEAFC Regulatory Area is at present subject to the provisions for bottom fishing adopted in 2008. For details, see the map provided in Figure 16.2. NEAFC has analysed the extent of the protection of VMEs in its Regulatory Area. The rough estimates which were arrived at are provided in Table 16.1.

Monitoring and control provisions for fisheries in the NEAFC Convention Area were adopted in 1999 in the NEAFC Scheme of Control and Enforcement and the Non-Contracting Party Scheme. These two schemes have since been integrated. The fully automated vessel monitoring system (VMS) database to support surveillance and inspection is the major element in monitoring and controlling Contracting Party vessels fishing in the Regulatory Area. In addition, IUU fishing by Non-Contracting Party vessels was addressed with A- and B -listings and relevant port control measures since 2005 and Port Control of all vessels in 2007.

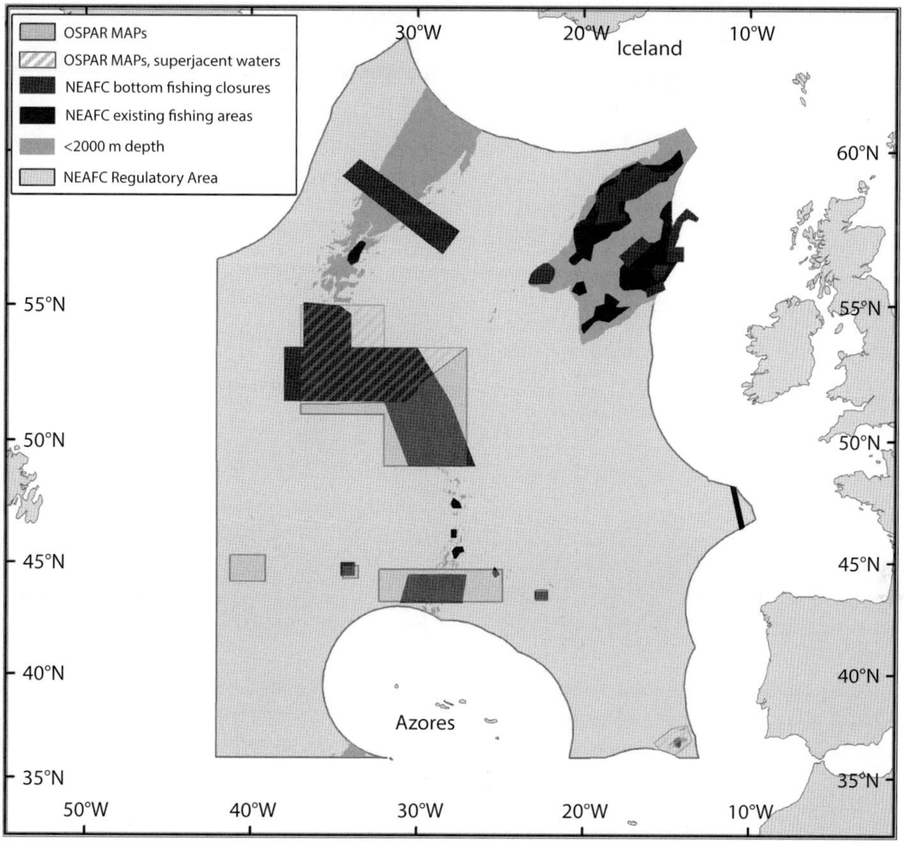

Figure 16.2 Map showing the big area beyond national jurisdiction in the mid-Atlantic, covering the Mid-Atlantic Ridge (MAR) from Iceland to the Azores. It shows areas of less than 2000 m depth, OSPAR MPAs and NEAFC closures and the NEAFC existing fishing areas (areas where Contracting Parties have established fishing activities according to certain criteria in earlier years). The recent OSPAR Charlie-Gibbs North water column closure is not shown. (Source: Adapted from Bergstad and Høines 2011.)

Table 16.1 Fishing and protected areas in the NEAFC regulatory area.

Size of NEAFC Regulatory Area		**Existing fishing areas**		**New fishing areas**		**Closures**	
Areas	**km²**	**km²**	**%**	**km²**	**%**	**km²**	**%**
South of Iceland	4,900,000	42,500	0.9	4,502,200	91.9	355,300	7.3
Norwegian Sea	326,000	0	0	326,000	0	0	0
Barents Sea	71,000	71,000	100	0	0	0	0
Arctic Ocean	275,000	0	0	275,000	0	0	0
Fishable areas and closures							
Areas	**km²**	**%**					
	655,000	100.0					
	355,300	54.0					

It should be noted that all measures adopted by NEAFC are legally binding either: (1) immediately in some Contracting Parties, when entering into force; or (2) after inclusion in domestic law in some Contracting Parties.[4]

An overview of NEAFC measures with detailed descriptions of fisheries management, area management, control and enforcement and stock status can be found in the NEAFC 1998–2007 Fisheries Status Report (NEAFC, 2009).

OSPAR actions and measures

The Rules of Procedure of OSPAR allow for legally binding Decisions, voluntary Recommendations and Other Agreements (such as Guidance or Risk Assessment Frameworks). Decisions and Recommendations can be passed by a three-quarters majority but, being a consensus-driven organization, unanimity is preferred.

Working to protect biodiversity started with establishing an OSPAR List of Threatened and/or Declining Species and Habitats, agreed in 2003 and extended in 2008. The List is based on agreed criteria for decline (expressed in terms of population, distribution and condition of species and distribution, extent and condition of habitats) and threat (expressed in terms of there being a direct or indirect link to human activity). Fishing is seen as the key pressure affecting the species listed and responsible for habitat loss and habitat damage, in many cases through bottom trawling. Collection of background information and mapping of species and habitats has provided OSPAR with the rationale to take action. In 2010 the first formal measures were taken to protect a set of long-lived slow-growing fish species impacted by fishing and deep-sea habitats. Recognizing the competence of fisheries management authorities to deal with questions on the management of fisheries affecting these species and habitats, the measures agreed have been designed to support the management of fisheries. Species concerned are the Angel Shark (*Squatina squatina*), Common skate species complex, Basking Shark (*Cetorhinus maximus*), White Skate (*Rostroraja alba*) and Orange Roughy (*Hosplostethus atlanticus*). OSPAR Recommendations include raising awareness, taking account of these species in environmental impact assessments, protection of breeding sites and action to reduce bycatch. *Lophelia pertusa* reefs, coral gardens and deep-sea sponge aggregations are the habitats selected for protection using a similar precautionary approach.

From 2000 OSPAR has also been working to establish a network of marine protected areas (MPAs). At the 2003 Ministerial Meeting in Bremen, OSPAR Ministers made a commitment (OSPAR Recommendation 2003/3) to create an ecologically coherent network of well-managed MPAs by 2010. Good progress has been made towards establishing this network in coastal seas. From 2005 work progressed to establish MPAs in areas beyond national jurisdiction (ABNJ). In 2010 six areas of outstanding significance were established as the world's first network of High Seas MPAs (see map in Figure 16.2; O'Leary et al., 2012); a seventh MPA, Charlie-Gibbs North (pertaining to the water column only), was added in 2012.

Periodically OSPAR evaluates the quality status of the Northeast Atlantic. OSPAR looks at long-term trends for various pollutant streams (e.g. eutrophication, hazardous substances, radioactive substances) against agreed baselines. According to the OSPAR QSR 2010 (OSPAR, 2010a) there are positive trends (i.e. management policies have resulted in reductions in concentrations and the presence of certain pollutants), although there are areas and sectors where problems persist with human activity impacting biodiversity negatively.

The OSPAR Quality Status Report 2010 (OSPAR 2010a) drew on an assessment of the environmental impact of fishing and a similar ICES assessment. The conclusion was that fishing pressure continues to have a considerable impact on marine ecosystems and many problems remain, despite efforts to improve management. OSPAR recognizes success in reducing IUU fishing, not least as a result of NEAFC initiatives, and OSPAR is committed to supporting improvements in fisheries management.

Comparing the NEAFC areas closed to fisheries and OSPAR MPAs

The map in Figure 16.2 shows clearly that the NEAFC and OSPAR closures, although based on the same ICES advice, do not cover identical areas. No bottom fishing takes place at depths greater than 2000 m, so NEAFC has concentrated on the areas shallower than 2000 m. OSPAR regulates other activities than fisheries, so may need to cover larger areas beyond the 2000 m contour.

Cultural differences between management approaches of the two organizations

In addition to the mandate and actions taken by these two regional organizations to protect marine biodiversity, the nature of the business of the two organizations has implications for how they deliver their responsibilities.

Overall remit

NEAFC is responsible for fisheries management in the NEAFC Regulatory Area. It also has a strong commitment to protection of biodiversity, especially the impact of fisheries on the environment and vulnerable marine areas (VMEs).

OSPAR's remit is much wider and more diverse in that it must deal with a range of different pollutants and a broad range of human activities and uses of the sea. Furthermore, the OSPAR maritime area covers both shallow and deep waters. In addition to addressing the impact of fisheries on biodiversity and ecosystems, OSPAR also deals with eutrophication, hazardous substances, radioactive substances, impacts of the offshore oil and gas industry and a plethora of human activities such as dredging, military activities including dumping of ordnance munitions, artificial reefs, sand and gravel extraction and wind farms.

The role and responsibility of the European Union

The Northeast Atlantic is in many ways a European ocean, albeit that much of the area administered by both NEAFC and OSPAR is in areas beyond national jurisdiction (ABNJ) and thus outside the purview of the EU. The European Commission (EC) is therefore a key player within both organizations, but plays a distinctly different role in each.

In the case of NEAFC, the EC assumes a position for coordination, proposes measures (e.g. many 2011 proposals) and has a right to vote (one of five Contracting Parties; all measures require two-thirds majority), while the EC is one of 16 Contracting Parties within OSPAR. European Member States, 12 of the other 15 Contracting Parties, operate in their own right while respecting relevant European legislation. The EC can exercise the right to determine a coordinated position but this is rarely applied (usually only during a very controversial or crisis situation). The usual pattern in scientific and technical meetings is that EC participants regularly inform meetings and clarify the EU position on specific topics. Over the past 10 years the EC has however passed legislation on a number of subjects pioneered by OSPAR (e.g. control of hazardous substances), causing OSPAR to refocus its efforts and priorities.

The role of observers

Both NEAFC and OSPAR allow Observer organizations to participate in their activities. In both cases Observer organizations, which can be other States, intergovernmental organizations (IGOs) and non-governmental organizations (NGOs), must submit formal applications in line with criteria established in agreed Rules of Procedure. Once admitted, meeting protocol demands that at formal meetings there is a strict 'pecking order' of who may make an intervention, the order being Contracting Parties followed by other States, IGO Observers and then NGO Observers.

Within NEAFC however, Observers observe in the formal sense and only at Commission Meetings. They are all allowed to make interventions at annual meetings (at the discretion of the President). They are not allowed to observe at subsidiary body meetings. By contrast, within the OSPAR community some 50 organizations are recognized as Observers and these organizations participate more openly. Observers are entitled to participate in any expert group. They have a track record of involving stakeholders and promoting issues of concern. Industry groups, including the European Crop Protection Association, European Oilfield Speciality Chemicals Association, International Association of Oil and Gas Producers and World Nuclear Association, contribute expert knowledge and research outputs. With the vested interest of their stakeholders recognized as a central tenet of their mandate, part of their role is also to provide checks and balances, contributing to sustainable development by advising whether proposed regulation is both achievable in practice and equitable. Green NGOs, such as Birdlife International, Friends of the Earth, Greenpeace International, KIMO International, Oceana, Robin des Bois, Seas at Risk and World Wide Fund for Nature, some of whom often work collectively, use the OSPAR platform to highlight concerns and to persuade Contracting Parties to act where they consider this to be appropriate. They also provide an important link to civil society. Active involvement in marine spatial planning issues by several of these NGOs is an important recent consideration (Calado et al., 2012). In this way OSPAR practice is seen to be more open to participation by Observers. Any prospective measure proposed by Observers must be supported by at least one Contracting Party, but Observers are encouraged to draw attention to relevant issues and to submit documents to meetings. OSPAR Observers have initiated work on several key measures and made an invaluable contribution towards achieving the aim of the Convention.

Technical work and ICES advice

Both regional organizations rely on sound science to drive their policy making. However, the approaches are different. NEAFC sends specific requests for advice to ICES every year and refers almost all scientific questions to ICES. OSPAR agrees a Joint Assessment and Monitoring Programme (JAMP), which coordinates thematic initiatives. The JAMP generates the evidence base for periodic (latterly decadal) quality status reports. Each OSPAR subsidiary Committee generates an annual work programme, containing a range of products. Some products reflect the need to evaluate monitoring data, which is collected annually and then compared to previous years, establishing trends against a baseline and allowing for assessments of 'distance to target'. Other products tackle emerging issues (e.g. underwater noise) or react to global concerns (e.g. ocean acidification).

Within OSPAR this work is undertaken on a lead country basis. The lead Contracting Party, supported by contributions from other delegations, is responsible for coordinating work by qualified contributors and delivering the outcome(s). For several data streams (e.g. annual discharge data), the Secretariat collates information for scrutiny by an expert assessment panel and subsequent submission the Committee. Some products require expert intercessional correspondence groups and/or workshops. Committees review the products and recommend them to the annual Commission meeting for endorsement and/or publication. This system manages to leverage input from Contracting Parties on the basis of a modest administrative Secretariat budget. It builds in quality assurance and distributes the work fairly.

ICES advice is requested on the basis of an agreed annual work programme. ICES expertise is used to address information gaps and for a number of cross-cutting topics (e.g. ICES peer-reviewed the QSR; OSPAR 2010a). OSPAR referral to ICES is therefore the exception rather than the rule, unlike NEAFC.

Performance reviews

It is standard practice for many organizations in the private, public and voluntary sectors to undergo some form of periodic performance review. The process is a means of assessing how well any organization is working against its stated aims but it is also a way of enhancing external credibility and transparency.

The Twenty-Sixth Session of the FAO Committee on Fisheries Rome in 2005 called for RFMOs to undergo performance reviews. NEAFC was the first RFMO to undergo such a review in 2006. The outcome was discussed in relevant subsidiary bodies and acted upon. The performance review was a helpful process guiding NEAFC forward and confirmed that NEAFC's overall approach was in accordance with relevant international law and instruments. At the 2011 Annual Meeting of NEAFC, a second performance review was discussed and planning has started.

In 2006 UNEP reviewed the Regional Seas Programmes and Action Plans in general, but to date OSPAR has not undergone any formal performance review. An internal strategic reappraisal of the work of the Commission, resulting in the North-East Atlantic Environment Strategy (OSPAR, 2010b), was conducted during the period 2008–2010.

Secretariat role

For regional multilateral environmental agreements to be successfully implemented, they need an efficient Secretariat whose role is to support Contracting Parties. NEAFC and OSPAR both employ an Executive Secretary who is ultimately responsible for making sure this support function is delivered efficiently and effectively. The relative size of the two Secretariats differs however, OSPAR having three times as many personnel, as does the way they do business.

The operation of NEAFC is not driven by the Secretariat, who is responsible for facilitating the operational implementation of the decisions made by the President and the Commission. Although assuming a similar responsibility the OSPAR Secretariat style is more proactive, producing discussion documents and adopting the standard UN protocol in relation to Contracting Parties (i.e. advice concerning procedure and points of order, but with opportunities for the Executive and his/her Deputy Secretaries to contribute to discussion and, where appropriate, to stimulate consideration of particular topics). For example, the OSPAR Secretariat was instrumental in ensuring that climate change has featured strongly on the Commission's agenda since 2007.

Memorandum of understanding between NEAFC and the OSPAR commissions (2008)

Heads of Delegations of NEAFC and OSPAR met on several occasions at NEAFC headquarters from November 2005 to September 2008 to scope areas of common interest (e.g. best scientific evidence, precautionary approach and conservation of marine biological diversity), recognizing respective complementary competences for multilateral cooperation on fisheries management and environmental protection. The Commissions then formalized arrangements, setting out how these interests and obligations would be jointly recognized and ways in which actions might be taken forward together, by officially signing a Memorandum of Understanding (MoU) (OSPAR Agreement 2008-04; NEAFC, 2008). The MoU promotes mutual cooperation towards the conservation and sustainable use of marine biodiversity through mechanisms such as the free flow of information exchange (including data) and cooperation on marine spatial planning and area management.

NEAFC has also signed terms of Agreement with the International Maritime Organization (IMO) and is considering a MoU with the International Seabed Authority (ISA). In addition to the MoU agreed with NEAFC in 2008, OSPAR has a longstanding Agreement of Cooperation with the IMO (OSPAR Agreement 1999-15) and in 2011 concluded a MoU with the International Seabed Authority (OSPAR Agreement 2010-09). A similar arrangement with the North Atlantic Salmon Conservation Organization is under development and a Collaborative arrangement between the Secretariats of the OSPAR Commission and the Sargasso Sea Alliance was concluded in 2012 (OSPAR Agreement 2012-10). All these bilateral arrangements seek to foster better communication and understanding between the different organizations, recognizing that actions by one affect others.

Practical cooperation between the two organizations

In 2005, 2006, 2007 and 2008 Heads of Delegations of NEAFC and OSPAR met in November at NEAFC Headquarters and discussed issues of common interest. This laid the foundation for the MoU of Understanding between the two organizations signed in 2008. Since the signing of the MoU in 2008, the NEAFC and OSPAR Secretariats have met and reported on the work of both organizations in various fields. For example, in 2011 the following issues were dealt with.

- Setting up the first Regional CBD Workshop to indentify EBSAs in the Northeast Atlantic: it was agreed that, in view of those areas identified by NEAFC as VMEs and by OSPAR as MPAs, the scientific underpinning information was largely in place to inform a Regional Workshop. However, it was also agreed that additional complementary scientific and technical information should be identified and the first possible date for a suitable Workshop, in cooperation with the CBD Secretariat, would be in September 2011. By working together the two organizations were therefore able to organize a Workshop less than a year after the CBD CoP10 in Nagoya, Japan.
- OSPAR was briefed on the UN Regular Process: a regular process for global reporting and assessment of the state of the marine environment, including socio-economic aspects.[5]
- Cooperation with ICES was discussed, as well as the potential for asking for joint ICES advice. Information from bilateral meetings with ICES, where either OSPAR or NEAFC had participated, was presented.
- Area management was discussed, and some parties in OSPAR felt it may be premature to open up discussions on aligning the border of the Mid-Atlantic Ridge closures. There was still the uncertainty stemming from submissions to the Commission on the Limits of the Continental Shelf, pursuant to article 76, paragraph 8, of the United Nations Convention on the Law of the Sea of 10 December 1982 and the timeline for dealing with these submissions.
- OSPAR asked if there were any new developments in NEAFC with respect to area management and drew attention to work in progress with spatial planning.
- OSPAR asked about NEAFC reactions to the 2010 Quality Status Report, in particular the sections dealing with: (1) fisheries (OSPAR, 2010a, chapter 8); (2) trial method for integration (OSPAR, 2010a, chapter 11); and (3) opportunities to communicate results.
- NEAFC agreed to draw the attention of the Commission to the two chapters and the opportunities to communicate common results on these matters.
- The Draft Agreement of Competent Authorities on the management of selected areas in ABNJ within the OSPAR Maritime Area, the Madeira collaborative arrangement, was discussed and it was felt that the collaborative arrangement had been 'parked' for some time. It was perhaps now time to move forward. The important part of the arrangement was the agreement; the principles should not cause any problems for the organizations involved.

Wider sectoral cooperation in the Northeast Atlantic

A first informal meeting between Northeast Atlantic authorities with sectoral competence for aspects of marine environmental protection (i.e. OSPAR, NEAFC, IMO, ISBA, ICCAT, NASCO, IWC) was held in Madeira in March 2010. Participants considered the role and remit of individual competent authorities and discussed a matrix of possible management options for selected areas in the high seas. Technologies available for surveillance and enforcement were discussed and details of the regime put in place by NEAFC to manage fisheries were explained. Drawing on this information, participants proposed a draft *Collective Arrangement*, acknowledging that Competent Authorities should:

(1) inform each other of any updated scientific information and environmental assessment and monitoring data;
(2) notify each other of existing or proposed new human uses;
(3) cooperate on EIAs, SEAs or equivalent instruments;
(4) meet annually to review their respective objectives for the selected areas/status/appropriateness of management measures and any proposals for improvement; and
(5) cooperate to obtain a better knowledge of the areas concerned through, where appropriate, developing exchange of data, sharing of databases and collecting data in standardized formats.

In January 2012 a second of these informal meetings was convened. Participants exchanged information on developments to protect selected areas in the high seas and worked towards generic guidance for developing management plans. There seems to be a consensus that a network of agreements of cooperation and MoUs between existing organizations, with a mandate to regulate human activities in the ocean, can in the short and medium term offer the best opportunity to achieve the correct balance between optimum utilization and conservation in the Northeast Atlantic (Johnson, 2013). This is particularly true considering the absence of an international implementation instrument of the LOSC to set aside specific areas of the high seas in favour of biodiversity conservation.

Examples of converging approaches

Over the past five years NEAFC and OSPAR Secretariats have taken the opportunity to join forces and give a united message. This recognizes the overarching ecosystem approach, the need for robust fisheries management and the need for dialogue between resource managers and conservationists. Presentations have been made to ministers (Seas the Future Conference, held in the Faroes in 2011), stakeholder groups (Long-distance RAC, held in Tallinn in 2010) and the ocean community (Global Forum, held in Parisin 2008).

Concerted efforts on deep-sea species and habitats will also make a difference. For example, the vulnerability of cold-water corals which are easily impacted by fishing gear was highlighted by OSPAR Ministers in 2003. OSPAR raised particular concerns with NEAFC over the protection of corals on the western slopes of the Rockall Bank. Efforts to secure protection include cold-water coral areas within NEAFC Bottom Fisheries closures, within OSPAR MPAs and, following mapping exercises, as specific habitats on the OSPAR List of Threatened and/or Declining Species and Habitats subject to Recommendations. Similarly, both NEAFC and OSPAR have interests in vulnerable species. For example, the Basking Shark (the world's second-largest fish) is vulnerable to accidental bycatch. Both NEAFC and OSPAR have now taken measures to protect this species.

The most notable collaboration between NEAFC and OSPAR to date has been the first joint scientific Regional Workshop, in collaboration with the Secretariat of the Convention on Biological Diversity, to describe areas meeting scientific criteria for Ecologically or Biologically Significant Marine Areas (EBSAs). In its decision on marine and coastal biodiversity (CBD decision X/29), the 10th meeting of the Conference of the Parties to the Convention on Biological Diversity (CBD CoP 10, held in Nagoya, Japan on 18–29 October 2010) sets out the process on the description of areas meeting scientific criteria for EBSAs (annex I to Decision IX/20) through the organization of a series of Regional Workshops. The Northeast Atlantic meeting of scientific and technical experts, held in September 2011, described ten areas considered to meet criteria for EBSAs in the Northeast Atlantic. Work to refine the areas described in 2011 based on additional scientific input and review is underway.

The Workshop concluded that this was a coordinated regional contribution to a global commitment under the framework of CBD made possible as a result of the revised RFMO mandate, the political will generated within OSPAR towards biodiversity and ecosystems targets and the inclusion of experts nominated by both Contracting Parties and observer non-governmental organizations (NGOs). The prototype data visualization tool developed by GRID-Arendal for the workshop was an asset; data layers provided for the tool were instrumental in being able to visualize proposals while they were being discussed, and looking at them in the context of ecological data and other proposed areas for EBSA description. It enabled the sharing of information and map creation, allowing the workshop participants to upload and combine datasets, and reassured participants that access to large datasets can be achieved and managed.

The EBSA process provides a useful scientific foundation towards marine spatial planning for high-seas areas (Ardron et al., 2008). Determination of large areas for meeting criteria for EBSAs, which can facilitate biogeographic assessments and bring together different scientific values, provides a logical broader 'envelope' within which measures might be considered to protect smaller structures and/or determine representative areas in the future. Extensive areas meeting criteria for EBSAs can include areas closed to fishing, MPAs and areas subject to both current and potential future resource extraction (Weaver and Johnson, 2012).

Medium- and long-term expectation

The various global processes relating to fisheries have become increasingly dominated by two sets of issues: the north–south divide in international affairs and a perceived fisheries-versus-environment debate. In recent years, the processes leading to the adoption of the fisheries and oceans resolutions in the UNGA have become increasingly political and an important arena for exerting influence over the future development of fisheries. The meetings relating to the various UN processes mentioned here have, to a growing extent, become arenas for resolving issues that sometimes do not relate to the practical needs for developing fisheries management. In this respect FAO has seen its role as the global norm-setter for fisheries management decline, while the role of UNGA has increased as norm developer of central environmental relevance and has become more political, more cross-sectoral and less technical. Whether or not this trend will continue remains to be seen.

Since their inception the two regional organizations have provided legal frameworks, undertaken scientific assessments and taken measures raised to meet the goals of their respective Conventions. Latterly however, the urgent need to protect marine biodiversity has united NEAFC and OSPAR in a common purpose. Both organizations have awareness and promoted regional cooperation despite political and cultural differences.

This united approach comes at a time when the global context is changing rapidly. Biodiversity conservation is one of the four inter-related crises (the others being economic, climate and food security). Governance of the High Seas is currently the subject of a FAO-GEF project within which the Northeast Atlantic region is perceived as a leading example of bringing relevant stakeholders together. Nevertheless, in other parts of the world regional sea conventions do not extend as significantly into ABNJ, and their overlap with RFMOs is rather limited. For these areas, scenarios favouring an implementing agreement to protect high-seas biodiversity are perhaps more practical (IDDRI, 2011). Finally, it may not be too far fetched to envisage a future renegotiation of an integrated Northeast Atlantic Convention combining not only NEAFC and OSPAR but also the Bonn Agreement – responsible for collective oil spill prevention, preparedness and response in the North Sea and its wider approaches – as a single entity that would potentially be more inclusive, cost-efficient and better able to adapt and react to global change.

Notes

1. The Russian Federation is not a Contracting Party to OSPAR and Switzerland is not a Contracting Party to NEAFC.
2. In addition to NEAFC, the RFMOs NASCO (salmon), ICCAT (tuna) and NAMMCO (marine mammals) also cover the Northeast Atlantic.
3. Contracting Parties to the OSPAR Convention are Belgium, Denmark, European Union, Finland, France, Germany, Iceland, Ireland, Luxembourg, Netherlands, Norway, Portugal, Spain, Sweden, Switzerland and the UK.
4. For example, in the Faroe Islands NEAFC measures become immediately binding while in Norway they have to be included in domestic law.
5. At the World Summit on Sustainable Development held in Johannesburg, South Africa (26 August to 4 September 2002) States agreed, in paragraph 36(b) of the Johannesburg Plan of Implementation, to 'establish by 2004 a regular process under the United Nations for global reporting and assessment of the state of the marine environment, including socio-economic aspects, both current and foreseeable, building on existing regional assessments' (the 'Regular Process').

References

Ardron, J., Gjerde, K., Pullen, S. and Tilot, V. (2008) Marine Spatial Planning in the high seas. *Marine Policy* **32**(5), 832–839.

Calado, H., Bentz, J., Ng, K., Zivian, A., Schaefer, N., Pringle, C., Johnson, D. and Phillips, M. (2012) NGO involvement in marine spatial planning: A way forward? *Marine Policy* **36**, 382–388.

FAO (2009) International Guidelines for the Management of Deep-sea Fisheries in the High Seas. Directives internationales sur la gestion de la pêche profonde en haute mer. Directrices Internacionales para la Ordenación de las Pesquerías de Aguas Profundas en Alta Mar. FAO, Rome.

IDDRI (2011) A legal scenario analysis for marine areas in areas beyond national jurisdiction (eds E. Druel, R. Bille and S. Treyer). Report from the Boulogne-sur-Mer Seminar, 19–21 September 2011. Biodiversity studies 6/11.

Johnson, D.E. (2013) Can competent authorities cooperate for the common good: Towards a collective arrangement in the North-East Atlantic, Chapter 29. In: *Environmental Security in the Arctic Ocean* (eds P. Berkman and A. Vylegzhanin), pp. 333–343. Springer, Dordrecht.

NEAFC (2006a) Performance Review Panel Report of the North East Atlantic Fisheries Commission, NEAFC. Available at http://www.neafc.org/system/files/performance-review-final-edited.pdf (accessed February 2014).

NEAFC (2006b) London Declaration and "New" Convention. Available at http://www.neafc.org/system/files/london-declarlation_and_new_convention.pdf (accessed February 2014).

NEAFC (2008) Memorandum of understanding between NEAFC and OSPAR. Available at http://www.neafc.org/system/files/opsar_mou.pdf (accessed February 2014).

NEAFC (2009) NEAFC Fisheries Status Report 1998–2007. Edited by Kjartan Hoydal, NEAFC Secretariat. Available at http://www.neafc.org/international/fisheries-status-report (accessed March 2014).

O'Leary, B.C., Brown, R.L., Johnson, D.E., von Nordheim, H., Ardron, J. and Packeiser, T. (2012) The first network of marine protected areas (MPAs) in the high seas: The process, the challenges and where next. *Marine Policy* **36**, 598–605.

OSPAR (2010a) *Quality Status Report 2010*. OSPAR Commission, London.

OSPAR (2010b) *North-East Atlantic Environment Strategy*. OSPAR Agreement 2010-03.

Rochette, J. and Chabason, L. (2011) A regional approach to marine environment protection: the 'regional seas experience'. In: *Oceans: The New Frontier* (eds P. Jacquet, R.K. Pachauri and L. Tubiana), pp. 231–241. TERI Press, New Delhi.

Weaver, P. and Johnson, D.E. (2012) Think big for marine conservation. *Nature* **483**, 399.

Chapter 17

Regional governance: The Mediterranean cradle

F. Simard[1,2], M. Camilleri[3] and L. Sbai[4]

[1]IUCN Global Marine and Polar Programme, Gland, Switzerland
[2]IUCN Commission on Ecosystem Management (IUCN/CEM/FEG), Gland, Switzerland
[3]FAO Fisheries and Aquaculture Policy and Economics Division, Rome, Italy
[4]Department of Marine Fisheries, Rabat, Morocco

Abstract: This chapter describes the international governance structure for the Mediterranean Sea, where regional fisheries management organizations (General Fisheries Commission for the Mediterranean or GFCM and International Commission for the Conservation of Atlantic Tunas or ICCAT) and environmental organizations (Barcelona Convention/Mediterranean Action Plan or MAP and the Agreement on the Conservation of Cetaceans in the Black Sea Mediterranean Sea and Contiguous Atlantic Area or ACCOBAMS) have both been established. Despite a long parallel existence, these organizations have set up cooperation arrangements only recently and they are still not yet robust. This paper highlights the cooperation potential among these organizations and focuses on, as an example, the identification and declaration of marine protected areas.

Keywords: Mediterranean; governance; fisheries; biodiversity; cooperation; regional sea

Introduction

The implementation of sound regional governance is particularly important in the Mediterranean region where few States have claimed jurisdiction over areas beyond their territorial waters due to geophysical constraints and politically sensitive issues. Fisheries management organizations and environmental organizations have long been established in the Mediterranean but limited coordination exists between these organizations, resulting in the sectoral management approach over the years even though their Contracting Parties are common. While recognizing that each has its own competencies and mandate in accordance with international maritime law and relevant international fisheries and environmental instruments, in recent years these organizations have promoted the development of cooperation and coordination arrangements within the framework of the ecosystem approach. Furthermore, acknowledging the increasing need for an integrated approach in marine spatial planning which focuses on the management of human activities taking place within an ecosystem, progress has also recently been made in the region on the designation and management of marine protected areas with specific agreements being formulated and cooperative work being undertaken.

Status of marine biodiversity

The Mediterranean Basin is one of the world's richest places in terms of animal and plant diversity. Although the Mediterranean Sea makes up less than 1% of the global ocean surface, up to 18% of the world's macroscopic marine species are found there of which 25–30% are

Governance of Marine Fisheries and Biodiversity Conservation: Interaction and Coevolution, First Edition.
Edited by Serge M. Garcia, Jake Rice and Anthony Charles.

Box 17.1 The Mediterranean Sea

The name Mediterranean is derived from the Latin word *mediterraneus*, meaning 'inland' or 'in the middle of the land'. The Mediterranean Sea is an almost completely enclosed sea covering approximately 2.5 million km^2 and is connected to the Atlantic Ocean through the Strait of Gibraltar (14 km wide). It is surrounded by several Mediterranean coastal States with Europe and Anatolia in the north, North Africa in the south and the Levant in the east. It has an average depth of 1500 m and the deepest recorded point of 5267 m is found in the Calypso Deep in the Ionian Sea.

The Mediterranean Sea was an important route for merchants and travellers of ancient times that allowed for trade and cultural exchange between emergent peoples of the region (the Mesopotamian, Persian, Egyptian, Phoenician, Carthaginian, Iberian, Greek and Byzantine, Roman and Italian, Illyrian and Albanian, Thracian, Levantine, Gallic, Arabic, Berber, Jewish, Aragonese, Nubian, Slavic and Turkish cultures). The history of the Mediterranean region is crucial to understanding the origins and development of many modern societies. Today, the sea suffers from intense pollution and marine debris and is host to some of the world's busiest shipping routes with approximately 220,000 merchant vessels of more than 100 tonnes each crossing the Mediterranean Sea every year.

Mediterranean countries received 275 million international tourists in 2007 and this number is expected to more than double by 2025, half of which visiting the coastal regions alone.

(Source: http://en.wikipedia.org/wiki/Mediterranean_Sea; http://www.planbleu.org/themes/tourismeUk.html)

endemic; this is an incredibly rich biodiversity for such a small area (Cuttelod et al., 2008). Over 20% of the known species are under threat however, and this percentage will likely increase given that currently undescribed species will be added in the future and a large proportion of species are either not assessed or assessed as Data Deficient. This includes emblematic species of conservation concern such as the world's most endangered pinniped, the critically endangered Mediterranean monk seal. Sea turtles, several whales, dolphins, sharks, skates and rays are at risk of extinction or threatened, as well as the overexploited bluefin tuna (Cuttelod et al., 2008; UNEP/MAP-Plan Bleu, 2009).

There are several unique habitats at various levels of risk, including the seagrass meadows of the endemic *Posidonia oceanica* (an important indicator of human impacts and a provider of crucial ecosystem services), vermetid reefs, coralligenic concretions, maerl beds, seamounts and deep-sea coral reefs. Eight hundred marine and coastal protected areas have been established in the Mediterranean so far to protect its marine environment and biota. The current network is however not representative and, excluding the Pelagos Sanctuary (87,000 km^2, the only high-sea MPA), coastal MPAs cover only 0.4% of the Mediterranean Sea (Abdulla et al., 2008).

Status of marine fisheries

Capture fisheries production statistics show that catches in the Mediterranean and Black Sea steadily increased between 1950 and the late 1980s to almost 2 million tonnes, but has decreased since then to tonnages that are, on average, about 23% less than that peak level (Figure 17.1).

The Food and Agriculture Organization of the United Nations (FAO) reported (FAO, 2012) that 33% of assessed stocks in the Mediterranean and Black Sea are fully exploited, 50% overexploited and the remaining 17% non-fully exploited in 2009. All hake (*Merluccius merluccius*) and red mullet (*Mullus barbatus*) stocks in the Mediterranean are considered overexploited, as are probably also the principal stocks of sole and most seabreams. The main stocks of small pelagic fish (sardine and anchovy) are reported to be either fully exploited or overexploited. FAO also informed on a newly identified threat related to the increasing penetration of exotic Red Sea species which in some cases seem to be replacing native species, especially in the Eastern Mediterranean.

With regards to the Black Sea, FAO reported that the situation of small pelagic fish (mainly sprat and anchovy) has recovered somewhat from the drastic decline suffered

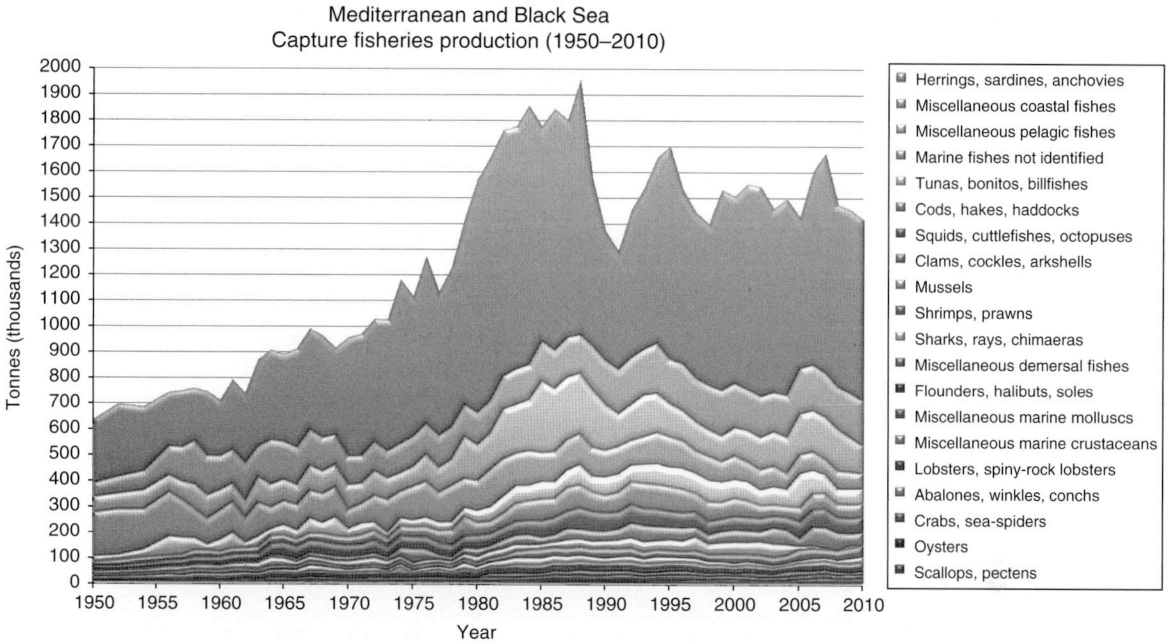

Figure 17.1 Capture fisheries production by groups of species for the period 1950–2010. Species groups are listed in descending order of average production over the entire period. Source of data: FAO Global Statistical Collections (http://www.fao.org/fishery/statistics/en).

in the 1990s, but they are still considered fully exploited to overexploited. This assessment is shared with turbot, while most other stocks are fully exploited to overexploited.

On the bluefin tuna fishery, Bjørndal (2009) (see also FAO, 2010) reported that the number of active states involved in the fishery has declined substantially over the years because the resource has been severely depleted. The spawning stock biomass (SSB) was estimated to be of the order 100,000 tonnes which is the lowest SSB for the resource in recorded history. In 2010, the proposal led by Monaco to ban the international trade in Atlantic bluefin tuna under CITES was ultimately rejected when many parties opposed the listing, stating that the International Commission for the Conservation of Atlantic Tuna (ICCAT) was the appropriate body for management of this species.

Aquaculture

The growth of aquaculture production in the Mediterranean has been uneven among countries and has fluctuated considerably in the region over the last 15 years, albeit with an overall upwards trend (Figure 17.2). Some countries have enjoyed impressive growth rates while others have stagnated or decreased. In particular, Greece has experienced the fastest growth rate overall but has also gone through the classic boom and bust episodes during the 30 years of the industry's existence. Turkey's aquaculture growth halted in recent years, mostly due to the implementation of a spatial planning restructuring programme and as a consequence of the current international economic crisis. The rapid growth of the industry in Spain has been maintained over recent years, while in other European countries such as Italy, France, Cyprus and Malta production has stagnated, stabilized or decreased mainly due to spatial planning issues. Smaller producing countries such as Cyprus, Malta, Croatia and Montenegro have small domestic markets, and strive to be competitive in larger European markets. For southern Mediterranean States the situation is compounded by technical difficulties such as the availability of juveniles, feed and trained technical personnel. Egypt is the exception, with an impressive production growth rate over the past decade and a large and growing domestic market (Barazi-Yeroulanos, 2010).

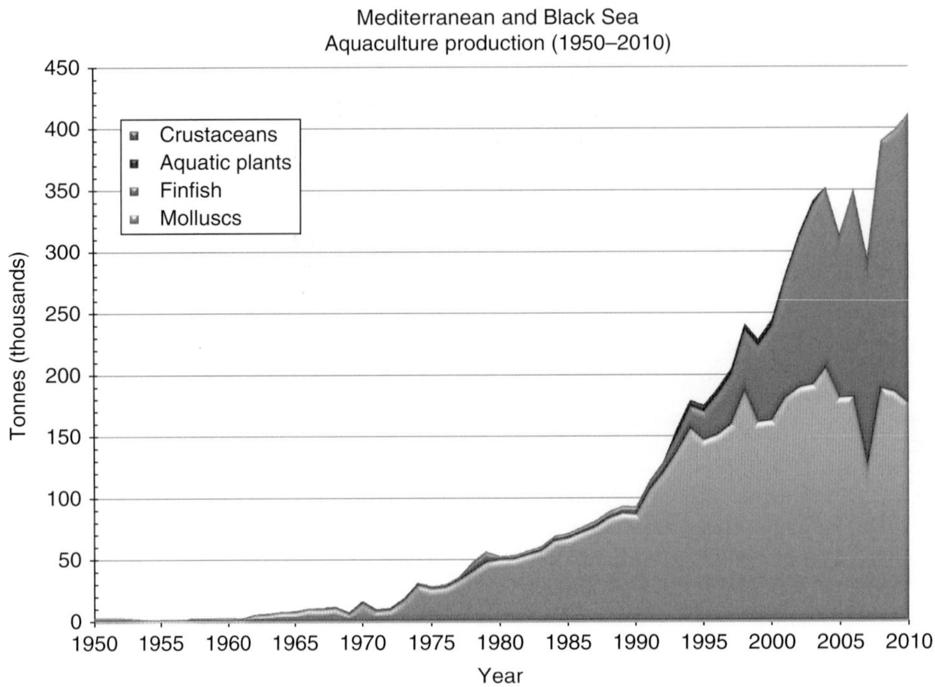

Figure 17.2 Aquaculture production by groups of species for the period 1950–2010. Source of data: FAO Global Statistical Collections (http://www.fao.org/fishery/statistics/en).

Governance framework

Global governance frameworks fully apply in the Mediterranean. The regional governance bodies, which are described in the following sections, refer to global instruments and organizations such as UNCLOS, CBD, CMS, UNFSA and UNEP, FAO and IMO, according to their objectives. Regional governance mobilizes a complex set of institutions and organizations described in the following.

Environmental governance

Barcelona Convention

Following the creation of the Regional Seas Programme in 1974 by UNEP and the adoption of the Mediterranean Action Plan (MAP) in 1975, the Convention for the Protection of the Mediterranean Sea against Pollution (Barcelona Convention) was signed on 16 February 1976 and came into force on 12 February 1978. This Convention was revised in Barcelona, Spain on 10 June 1995 as the Convention for the Protection of the Marine Environment and the Coastal Region of the Mediterranean. The Convention has the following 22 contracting parties: Albania, Algeria, Bosnia and Herzegovina, Croatia, Cyprus, Egypt, European Union, France, Greece, Israel, Italy, Lebanon, Libya, Malta, Monaco, Montenegro, Morocco, Slovenia, Spain, Syria, Tunisia and Turkey. Non-coastal States cannot be party to the Barcelona Convention. However, Article 20 specifies that any State which is not party to the Convention can be granted Observer status, along with any international governmental organization and non-governmental organization which has an interest in the Convention.

The Barcelona Convention does not, unlike the Agreement on the Conservation of Cetaceans in the Black Sea Mediterranean Sea and Contiguous Atlantic Area (ACCOBAMS) and the General Fisheries Commission for the Mediterranean (GFCM), cover the Black Sea which has its own convention (the Convention for the protection of the Black Sea Against Pollution). The Convention's main objectives are to:

- assess and control marine pollution;
- ensure sustainable management of natural marine and coastal resources;

- integrate the environment in social and economic development;
- protect the marine environment and coastal zones through prevention and reduction of pollution and, as far as possible, elimination of pollution whether land or sea-based;
- protect the natural and cultural heritage;
- strengthen solidarity among Mediterranean coastal States; and
- contribute to improvement of the quality of life.

In 2008 in Almeria, Spain the Contracting Parties unanimously decided to create a Compliance Committee to act as an independent body (UNEP(DEPI)/MED IG.17/10 Annex V; Decision IG 17/2: Procedures and mechanisms on compliance under the Barcelona Convention and its Protocols). It comprises seven regular members and seven alternate members, broadly representative of the geographic spread of the countries in the region, all of which are elected by the Contracting Parties. Although the decisions of the Conventions are binding and are to be implemented under legal provisions at national level, in reality no action has yet been taken by the Compliance Committee against a party for non-compliance in the case of non-implementation of decisions.

The following seven Protocols to the Barcelona Convention address specific aspects of Mediterranean environmental conservation.

(1) *Dumping Protocol*: Protocol for the Prevention of Pollution in the Mediterranean Sea by Dumping from Ships and Aircraft. Adopted on 16 February 1976 (Barcelona, Spain). Entered into force on 12 February 1978. The Dumping Protocol was amended and recorded as Protocol for the Prevention and Elimination of Pollution in the Mediterranean Sea by Dumping from Ships and Aircraft or Incineration at Sea. The amendment was adopted on 10 June 1995 (Barcelona, Spain), but has not yet entered into force.

(2) *Prevention and Emergency Protocol*: Protocol concerning the Prevention of Pollution from Ships and, in Cases of Emergency, Combating Pollution of the Mediterranean Sea. Adopted on 25 January 2002 (Malta). Entered into force on 17 March 2004. This Protocol replaced the Protocol Concerning Cooperation in Combating Pollution of the Mediterranean Sea by Oil and other Harmful Substances in Cases of Emergency (adopted on 16 February 1976 in Barcelona, Spain), which had been in force since 12 February 1978.

(3) *Land-Based Sources (LBS) Protocol*: Protocol for the Protection of the Mediterranean Sea against Pollution from Land-based Sources and Activities. Adopted on 7 March 1996 (Syracuse, Italy). Entered into force on 11 May 2008. This Protocol replaced the Protocol for the Protection of the Mediterranean Sea against Pollution from Land-Based Sources (adopted on 17 May 1980 in Athens, Greece), which had been in force since 17 June 1983.

(4) *Specially Protected Areas (SPA) and Biodiversity Protocol*: Protocol concerning Specially Protected Areas and Biological Diversity in the Mediterranean. Adopted on 10 June 1995 (Barcelona, Spain). Entered into force on 12 December 1999. The Annexes to the SPA and Biodiversity Protocol were adopted on 24 November 1996 (Monaco). This Protocol replaced that concerning Mediterranean Specially Protected Areas (adopted on 3 April 1982 in Geneva, Switzerland), which had been in force since 23 March 1986.

(5) *Offshore Protocol*: Protocol for the Protection of the Mediterranean Sea against Pollution Resulting from Exploration and Exploitation of the Continental Shelf and the Seabed and its Subsoil. Adopted on 14 October 1994 (Madrid, Spain). Entered into force on 24 March 2011.

(6) *Hazardous Wastes Protocol*: Protocol on the Prevention of Pollution of the Mediterranean Sea by Transboundary Movements of Hazardous Wastes and their Disposal. Adopted on 1 October 1996 (Izmir, Turkey). Entered into force on 19 January 2008.

(7) *Integrated Coastal Zone Management (ICZM) Protocol*: Protocol on Integrated Coastal Zone Management in the Mediterranean. Adopted on 21 January 2008 (Madrid, Spain). Entered into force on 24 March 2011.

Consistent with these protocols, MAP (http://www.unepmap.org/) priorities for the coming decade are to:

- bring about a massive reduction in pollution from land-based sources;
- protect marine and coastal habitats and threatened species;
- make maritime activities safer and more conscious of the Mediterranean marine environment;
- intensify integrated planning of coastal areas;
- monitor the spreading of invasive species;
- limit and intervene promptly on oil pollution; and
- further promote sustainable development in the Mediterranean region.

MAP's regional influence is very much a product of its close interaction with other United Nations (UN) agencies and other intergovernmental organizations that share MAP's commitment to the welfare of the Mediterranean region and its peoples. It recognizes the expertise and awareness-raising capabilities of NGOs, which are growing in number, supports their initiatives and encourages their participation in MAP activities, especially in relation to their focus on the integration of environment and development.

As the Action Plan for the Barcelona Convention, MAP is implemented on behalf of the Parties to the Convention by UNEP through the Coordinating Unit (MedU), located in Athens. It is responsible for following up and implementing MAP's mandate and activities, and has a diplomatic, political and consultative role. In addition, it supervises MAP subsidiary institutions including the Mediterranean Commission on Sustainable Development (MCSD), the Programme for the Assessment and Control of Marine Pollution in the Mediterranean Region (MedPol) and Regional Activity Centres (RACs).

MCSD is an advisory body to the Contracting Parties. It has a single representative body consisting of the 22 Contracting Parties along with 15 rotating representatives from local authorities, the business community and NGOs. This set-up allows for the development of policies by a multi-stakeholder group related to the promotion of sustainable development in the Mediterranean Basin. The MCSD coordinated the preparation of the Mediterranean Strategy on Sustainable Development (MSSD), which was adopted by the Contracting Parties in 2005.

MedPol is the scientific and technical arm of MAP. It is responsible for the implementation of the LBS, dumping and hazardous wastes protocols. MedPol assists Mediterranean countries in the formulation and implementation of pollution monitoring programmes, including pollution control measures and the drafting of action plans aiming to eliminate pollution from land-based sources.

There are six RACs based in Mediterranean countries, each offering its own environmental and developmental expertise for the benefit of the Mediterranean community in the implementation of MAP activities. One of them is focusing on biodiversity conservation, the RAC on Specially Protected Areas (RAC/SPA). This Centre, based in Tunis, is responsible for the classification of the Specially Protected Areas of Mediterranean Importance (SPAMI).

Decisions are taken by the Conference of the Parties (CoP) which takes place every two years. The CoP adopts the work plan for the next biennium. Decisions are implemented by the MAP components as well as by each Member State. The complexity of the MAP structure and the disparity of the capacities of Mediterranean countries do not facilitate the effective implementation of the processes concerning MAP's priorities. The recently created Compliance Committee has an important role to play in the strengthening of MAP and supporting the implementation of the decisions.

ACCOBAMS

The conservation of cetaceans in the Mediterranean is covered by specific regional agreement under the auspices of the Convention on Migratory Species (CMS), the Agreement on the Conservation of Cetaceans in the Black Sea Mediterranean Sea and Contiguous Atlantic Area (ACCOBAMS), which was signed in 1996.

ACCOBAMS currently has 23 members: Albania, Algeria, Bulgaria, Croatia, Cyprus, Egypt, France, Georgia, Greece, Italy, Lebanon, Libya, Malta, Monaco, Montenegro, Morocco, Portugal, Romania, Slovenia, Spain, Syria, Tunisia and Ukraine. Its scope is to reduce threats to cetaceans in Mediterranean and Black Sea waters as well as to improve the knowledge on these animals. The Meeting of the Parties is the decision-making body of the Agreement and the ACCOBAMS Permanent Secretariat, hosted by the Principality of Monaco, performs the functions listed in Article IV of the Agreement. This includes the organization of the sessions of the Meeting of the Parties, facilitation of co-operation between Parties and non-Party Range States, assistance to the Parties in the implementation of the Agreement and the administration of the budget. ACCOBAMS also has a Scientific Committee which executes its roles and tasks as defined in Article VII of the Agreement, particularly as an advisory body to the Meeting of the Parties. Furthermore, the ACCOBAMS Bureau and two Coordination Units, namely RAC/SPA and the Black Sea Commission, facilitate the implementation of the measures prescribed in the Conservation Plan of the Agreement.

ACCOBAMS also looks at broader conservation issues within the framework of the ecosystem approach and addresses the impact of fisheries on cetaceans. The use of marine protected areas for cetacean conservation is also being explored by ACCOBAMS, following the initiative by France, Monaco and Italy in setting up the Pelagos Marine Mammals Sanctuary in the Ligurian Sea. The Ministers of Italy, France and Monaco met in Rome to sign the final agreement that created the Ligurian Sea Cetacean Sanctuary in 1999. The Sanctuary area of approximately 90,000 km^2 comprises the waters between Toulon (French

Riviera), Capo Falcone (western Sardinia), Capo Ferro (eastern Sardinia) and Fosso Chiarone (Tuscany). In 2001 the Sanctuary became a SPAMI and in 2002 the Sanctuary Agreement entered into force, following ratification by Monaco (2000), France (2001) and Italy (2002).

Fisheries governance

General Fisheries Commission for the Mediterranean

The General Fisheries Commission for the Mediterranean (GFCM) plays a central role in the promotion and implementation of international fisheries instruments such as the 1993 FAO Compliance Agreement,[1] the 1995 United Nations Fish Stocks Agreement[2] and the 1995 FAO Code of Conduct for Responsible Fisheries and related instruments. Its establishment dates back to 1949 when the Conference of the Food and Agriculture Organization (FAO) of the United Nations approved the creation of the General Fisheries Council for the Mediterranean (GFCM) under the provisions of Article XIV of the FAO constitution. The Agreement entered into force in 1952 and was followed by amendments in 1963, 1976 and 1997. The most radical change came about in 1997 when the GFCM's role, as a Regional Fisheries Management Organization (RFMO) for the Mediterranean, Black Sea and connecting waters, was strengthened, bringing about a change in its name to a Commission and the establishment of new obligations for the Contracting Parties. This included their contributions to an autonomous budget, which would alleviate dependence on the support of FAO regular programme resources. These new obligations came into force on 29 April 2004 and today the GFCM enjoys the membership of 23 countries along with the European Union, together with the support of a fully-fledged secretariat which is based at its headquarters in Rome. The current Members are: Albania, Algeria, Bulgaria, Croatia, Cyprus, Egypt, European Union, France, Greece, Israel, Italy, Japan, Lebanon, Libya, Malta, Monaco, Montenegro, Morocco, Romania, Slovenia, Spain, Syria, Tunisia and Turkey. It is worth noting the membership of Japan, being a non-coastal State but with fisheries interests in the region.

Furthermore, it is important to highlight the fact that only half of the six Black Sea coastal states are members of GFCM; however, recent developments indicate that there is a possibility that the others may take up membership in the future. At the 36th session of the GFCM, the Russian Federation and Ukraine participated as observers and a Memorandum of Understanding was signed between the GFCM and the Black Sea Commission.

With a view to effectively promoting the development, conservation, rational management and best utilization of living marine resources, the structure and functioning of the GFCM was revamped immediately following the amendments to the Agreement in 1997. The Commission established the Scientific Advisory Committee (SAC) and four subcommittees (on Stock Assessment; Statistics and Information; Economics and Social Sciences; and Marine Environment and Ecosystems) to undertake the scientific advisory processes and to support it in the formulation of fisheries management measures which consider the four dimensions of sustainability – ecological, economic, social and institutional – identified by FAO within the framework of the 1995 Code of Conduct for Responsible Fisheries. Scientists and experts from Members' national institutions, NGOs and IGOs participate extensively in the activities of the SAC sub-committees and working groups, and contribute to the drawing up of scientific, technical and management advice.

In order to guarantee the sustainability of fisheries in its Convention Area the newly structured Commission decided that the implementation of a fishing effort control regime (Caddy, 1998), as opposed to one focusing on catch limitation, needed to be established. Furthermore, recognizing the dominant multi-species and multi-gear nature of fisheries in the Mediterranean and Black Sea and considering the differences in biological, ecological, physical and socio-economic characteristics in various sub-regions, the logical approach to manage fisheries by sub-area and sub-groups of fleets was studied and deliberated at length. Progress in this regard has been achieved over the years and the management approach has developed into one whereby fishing effort is managed by categories of vessels or Operational Units[3] (Accadia and Franquesa, 2006; Camilleri et al., 2000; GFCM, 2001). Each Operational Unit is associated with certain fishing effort parameters which are regulated accordingly. In this context, the GFCM strives to estimate the relative impact of Operational Units on various resources in order to draw up appropriate effort control measures for each Unit, while considering the ecological, socio-economic and governance dimensions, each having equal prominence, as much as possible in line with the Ecosystem Approach to Fisheries (EAF; Bianchi, 2008). Drawing on the advice from SAC and proposals by Members (individually or jointly), fisheries management, conservation and control measures are adopted by the Commission during its annual Session.

The GFCM also oversees and promotes the sustainable development of aquaculture in the Region. On the basis of the advice from its Committee on Aquaculture (CAQ) and its subsidiary bodies, the Commission also adopts recommendations to support such development and to promote best practices and standards in the growing aquaculture sector. The Compliance Committee (CoC) and the Committee on Administration and Finance (CAF) mainly review the compliance and implementation by Members of GFCM measures (conservation, management, MCS, data submission and enforcement) and administrative, procedural and budgetary affairs, respectively.

During the last decade, the GFCM adopted several management measures related to fleet capacity limits, closed seasons and areas, fishing gear restrictions, minimum landing sizes and effort control. In addition, monitoring, control and surveillance (MCS) requirements, including those related to port State measures and satellite Vessel Monitoring Systems, have been established along with multidisciplinary data submission schemes. The protection of vulnerable marine ecosystems and species of conservation concern has also featured frequently on the GFCM agenda. Binding regulations have been set up in this regard, including the prohibition of bottom trawling in waters deeper than 1000 m and other ecologically sensitive areas (Fisheries Restriction Areas) as well as the protection of marine mammals such as the monk seal.

There is no doubt that the GFCM has evolved as an RFMO and represents the main[4] regional governing body through which international deliberations and negotiations on fisheries are held. However, it still faces considerable challenges in effectively implementing the decisions taken and measures adopted throughout the region, as well as in monitoring the compliance and performance of its Members. Throughout history, the region has been home to many different cultures and political regimes which often makes the implementation processes more complex. While it is customary for the European Union (EU) to mirror GFCM management measures and other decisions in European legislation[5] and to monitor its member States' (currently, nine GFCM Members are EU member States) compliance accordingly, it is largely unclear how and to what extent other GFCM Members are fulfilling their obligations and following GFCM policy and measures. The EU group is likely to increase within the next few years and some of the candidate countries are already receiving considerable assistance from the EU in enhancing their capacity to better monitor, develop and manage their fisheries. However, there is a distinct possibility that the playing field will remain non-level with some of the other countries in the region either not having either the political will or else resources and capacity to fulfil their obligations with respect to GFCM regional policy and measures. Although the latter countries have benefited from FAO sub-regional projects[6] over the years, a more focused and long-term assistance programme would be required to bring them up to scratch with other countries in the region in the development and management of responsible fisheries. Nevertheless, it is important to highlight the fact that over the last 15 years these projects have been instrumental in promoting cooperation among GFCM Members, conducting capacity development programmes and organizing collaborative research activities. This support has generally been considered to be indispensable for several of the GFCM member countries and has contributed significantly to their full integration into the rapidly evolving GFCM mechanisms and governance framework.

While it is the responsibility of GFCM Members to seek to set up mechanisms and develop their capacity to comply with regionally agreed measures, international fisheries instruments and international maritime law, it is equally important for the GFCM to create an operational monitoring and inspection scheme in support of its Compliance Committee, which would enable it to better assess its Members' performance, evaluate the effectiveness of its management measures and be in a better position to oversee fisheries governance in the region. In addition, the GFCM could also continue to rely on the healthy collaboration it has developed with other intergovernmental organizations (IGOs) and non-governmental organizations (NGOs) which, in many cases, could also be instrumental in assisting the GFCM to achieve its goals. As indicated earlier, the GFCM not only manages the exploitation of living marine resources, but is also concerned with the conservation of ecosystems and the control of the impact of fisheries on them. On this issue, it is clear that the GFCM has to collaborate more closely with IGOs dealing specifically with the conservation of the marine environment in order to work towards a common governance framework as the GFCM maintains full competence over governance of fisheries in the region which considers both human and ecosystem wellbeing and equity, in compliance with the EAF.

International Commission for the Conservation of Atlantic Tuna

The International Commission for the Conservation of Atlantic Tuna (ICCAT) is an inter-governmental fishery organization responsible for the conservation of tuna and tuna-like species in the Atlantic Ocean and its adjacent seas, including the Mediterranean Sea (http://www.iccat.es/en/). The Commission was established at a Conference of Plenipotentiaries, which prepared and adopted the International Convention for the Conservation of Atlantic Tuna signed in Rio de Janeiro, Brazil in 1966. After a ratification process, the Convention entered formally into force in 1969. Currently, there are 48 contracting parties, including the European Union Mediterranean States, most of the other Mediterranean States and Japan.

ICCAT compiles fishery statistics from its members and from all entities fishing for tuna and tuna-like species in the Atlantic Ocean, coordinates research, including stock assessment, on behalf of its members, develops scientific-based management advice, provides a mechanism for Contracting Parties to agree on management measures and produces relevant publications. About 30 species (not all of which are present in the Mediterranean) are of direct concern to ICCAT. The Commission also compiles data for other fish species such as sharks that are caught as bycatch during tuna fishing operations in the Convention area.

Other organizations

The Mediterranean Science Commission

The other Mediterranean IGO is the Mediterranean Science Commission (CIESM), more commonly known as the *Commission Intergouvernementale pour l'Exploration Scientifique de la Méditerranée*. It was created in 1919 in Madrid by King Alfonso XIII of Spain and Prince Albert I of Monaco, to promote international research in the Mediterranean Sea and the Black Sea.

The Commission, with headquarters in Monaco, has grown from the 8 founding countries to 22 Member States today. These support a network of several thousand marine researchers, applying the latest scientific tools to better understand, monitor and protect a fast-changing and highly impacted Mediterranean Sea. Structured in six committees and various taskforces, CIESM is governed by a Board and an Advisory Committee and is led by a Director General. It runs expert workshops, collaborative programs and regular congresses, delivering authoritative, independent advice to national and international agencies. CIESM involves researchers from all shores of the Basin in its activities and acts as a focal point for the exchange of ideas, the communication of scientific information and the development of scientific standards across the Basin.

CIESM has established robust networks of scientists in various important fields, but does not play any governance role in the Mediterranean. Furthermore, CIESM does not act as an advisory body to RFMO/As or other Mediterranean international bodies (as the International Council for the Exploration of the Sea or ICES does in the North Atlantic).

The Mediterranean Memorandum of Understanding on Port State Control

Within the international effort to increase the maritime safety and the prevention of pollution in 1995, an agreement for the southern and eastern Mediterranean countries for a Port State Control System was proposed. A Memoranda of Understanding (MoU) was signed in Malta in 1997 under the umbrella of the IMO and ILO to confirm the responsibility of maritime authorities to implement the relevant instruments agreed upon on ships calling at the ports of the MoU to enforce the elimination of sub-standard practices for a safer shipping and cleaner oceans. The MoU includes 10 parties and its secretariat is located in Alexandria, Egypt. The Mediterranean MoU is cooperating with the Paris MoU, Parties to which include European countries and Canada.

The International Union for Conservation of Nature

The IUCN, with a membership including most Mediterranean States as well as most international and national NGOs working in the region, is also highly active in the Mediterranean, particularly through the Centre for Mediterranean Cooperation (Malaga, Spain) which was created in 2001 after a Recommendation adopted at the IUCN Congress in Amman in 2000. Although IUCN does not adopt binding decisions, recommendations and the conclusions emanating from its activities often have great influence on the decisions taken by the States.

The NGOs

In the Mediterranean region, there are a number of Non-Governmental Organizations with marine interest such as the World Wide Fund for Nature (WWF) which has an international office dedicated to Mediterranean affairs, the Mediterranean Programme Office (MedPO). WWF MedPO is especially active in the field of fisheries conservation (office in Barcelona) and for Marine Protected Areas (office in Rome). WWF has also been involved in the origin of the

Mediterranean Protected Areas Network (MedPAN) which links the Mediterranean MPA managers. Other NGOs active in the Mediterranean include Greenpeace and Oceana. The latter has recently been active in marine exploration and the identification of potential marine protected areas.

Cooperation

The relationships among the organizations mentioned in the above sections were weak until the beginning of the 21st century. Progress has been made on this front in recent years however, with several memoranda of understanding having been drawn up in order to undertake common or coordinated activities. It is relevant to mention that the differences in operation of these organizations, including the frequency of convening sessions of the Parties, pose additional constraints in coordinating efforts for the conservation and management of living marine resources and the environment.

GFCM and ICCAT

Being RFMOs, both GFCM and ICCAT cooperate on all issues relevant to tuna and tuna-like species management in the Mediterranean, and enjoy observer status at each other's organization. These RFMOs undertake joint activities as required, such as the convening of meetings of the Ad-Hoc Working Group on Large Pelagic Species in the Mediterranean and the Ad-Hoc Working Group on Sustainable Bluefin Tuna Farming/Fattening Practices in the Mediterranean.

It is customary for ICCAT decisions of relevance to the GFCM convention area to be considered for adoption at every GFCM session; it is interesting to note, however, that this practice is not based on any formal arrangement.

MAP and ACCOBAMS

Similarly, MAP and ACCOBAMS conduct several activities jointly. In order to facilitate the implementation of the measures prescribed in the ACCOBAMS Conservation Plan of the Agreement, the First Meeting of the Parties designated Coordination Units in each sub-region within an existent institution. For this purpose, the UNEP MAP Regional Activity Centre for Specially Protected Areas (RAC/SPA) was set up for the Mediterranean Sea. In 2005, a collaboration framework between ACCOBAMS and the RAC/SPA was agreed upon through which the Mediterranean Regional Coordination Unit (Med/RCU) functions were defined as follows:

- facilitating the implementation in the sub-region of the Conservation Plan of ACCOBAMS;
- collaborate for collecting, assessing and broadcasting the information which facilitate the implementation of the ACCOBAMS;
- preparing a report for the meeting of the Contracting Parties of the ACCOBAMS through the Secretariat of the Agreement on the implementation of the activities carried out by the Med/RCU;
- attending relevant meetings under the frame of ACCOBAMS;
- facilitating the link with the others components of the Mediterranean Action Plan with the view of ensuring synergies between the activities carried out within the ACCOBAMS framework and those carried out in the context of the Barcelona Convention (with special attention to the function devoted by the Agreement to the Barcelona Convention for the establishment of Specially Protected Areas particularly relevant for cetacean conservation);
- providing information about cetacean-related activities developed under the auspices of the Barcelona Convention by non-ACCOBAMS Parties; and
- in consultation with the Scientific Committee and the Permanent Secretariat of the Agreement, preparing a series of international reviews or publications including reports on the status and trends of populations (to be updated regularly) as well as gaps in scientific knowledge; a sub-regional directory of important areas for cetaceans; and a sub-regional directory of national authorities, research and rescue centres, stranding networks, scientists and non-governmental organizations concerned with cetaceans.

Example of cooperation for the conservation of deep sea

There are a few examples of broader cooperation involving both environmental and fisheries organizations. One of these is the cooperation developed between the IUCN Centre for Mediterranean Cooperation and the WWF Mediterranean Programme Office, working with the GFCM. The IUCN and WWF offices have worked together on the conservation of deep sea since 2003 and this venture led to the release of a first document in 2005 entitled 'The Mediterranean deep-sea: highly valuable ecosystems in need of protection'. This document served as the basis for the adoption of a number of measures by the GFCM, namely:

- the prohibition of the use of towed dredges and trawl nets fisheries at depths beyond 1000 m;
- the prohibition of fishing with towed dredges and bottom trawl nets in: (1) deep-sea fisheries restricted area 'Lophelia reef off Capo Santa Maria di Leuca'; (2) deep-sea fisheries restricted area 'the Nile delta area cold hydrocarbon seeps'; and (c) deep-sea fisheries restricted area the 'Eratosthemes Seamount' (South of Cyprus); and
- the obligation of Members to request the appropriate authorities to protect these areas from the impact of any other activity, jeopardizing the conservation of the features that characterize these particular habitats.

In 2009, the area enclosing the submarine canyons of the Gulf of Lions south of Marseille (France) was added to this list of fisheries restricted areas (FRA). These decisions represent an important milestone in the field of marine conservation and fisheries management and their effectiveness would be reinforced if also taken up formally by MAP. In this respect, IUCN and its partners are working closely with GFCM and MAP to strengthen these conservation measures.

In parallel, RAC/SPA is conducting a large-scale project (in collaboration with the ACCOBAMS, GFCM, IUCN and several others partners) on the identification of important areas in the high seas or in deep areas which is expected to lead to the designation of Specially Protected Areas of Mediterranean Importance (SPAMIs) across the Region, including the GFCM FRAs. This project aims to apply the criteria listed in the SPA/BD Protocol to Mediterranean areas beyond national jurisdiction in need of protection, while considering the significant variation between biogeographic regions in terms of their biological and physical aspects as well as their governance.

The use of other criteria already applied by other organizations such as the CBD Ecologically and Biologically Significant Areas (EBSAs) criteria, the FAO Vulnerable Marine Ecosystems (VMEs) criteria, the IMO Particularly Sensitive Sea Areas (PSSAs) criteria and the criteria of the Convention for the Protection of the Marine Environment of the North East Atlantic (OSPAR) has also been explored.

GFCM and MAP

Both GFCM and MAP have drawn up and implemented Memoranda of Understanding (MoUs) with various organizations. At its 36th session recently, with a view to improve cooperation with relevant international organizations (as suggested by the GFCM Performance Review), GFCM signed new MoUs with UNEP-MAP, ICES, ACCOBAMS, Black Sea Commission, MedPAN, RAC-MED and Eurofish; other already existing MoUs include those with IUCN and WWF.

The 17th Meeting of the Contracting Parties of the Barcelona Convention emphasized the need to further reinforce cooperation between all actors in the marine and coastal environment of the Mediterranean and welcomed the ongoing efforts to enhance cooperation with the Secretariats of the Union for the Mediterranean, the General Fisheries Commission for the Mediterranean, the Convention on Biological Diversity, the International Maritime Organization, IUCN and the other regional seas conventions and programmes, and future cooperation with all other relevant organizations. At the first meeting of the Ecosystem Approach (EcAp) Coordination Group held in Athens, Greece on 29–30 May 2012, it was agreed to request the Secretariat to promote the participation of UNEP/MAP partners, relevant projects and the scientific community at all relevant levels.

Issue of national coordination

Both environmental and fisheries bodies consider national coordination and cooperation mechanisms within and among Mediterranean countries. There are however implementation difficulties arising from the fact that different ministries represent the same States in these organizations. The ministries of environment represent the Member State in fora of the Barcelona Convention and ACCOBAMS while it is the ministry responsible for fisheries which represents the member States in the meetings of the RFMOs GFCM and ICCAT. It is widely recognized that both inter-ministerial and intra-ministerial coordination are frequently lacking, resulting in the absence of an agreed national policy and common stance on fisheries and environmental issues during international negotiation fora. This an issue for the Barcelona Convention in particular because, being an environmental convention, national coordination with all technical departments is required for the implementation of decisions taken under the Convention. This problem has led to extreme cases whereby ministries have taken up completely opposite positions publicly on issues of prime importance related to the conservation of the environment and fisheries management. This situation jeopardises the effective implementation of regional governance; such implementation will not be possible without coordination and cohesion within governments first.

Role of the European Union

Although only 7 of the 22 Mediterranean countries belong to the European Union, the EU clearly plays an important role in the improvement of the Mediterranean governance through its internal organizational mechanisms, coordination among its member States and cooperation programmes with non-EU States. At both GFCM and ICCAT meetings the EU, represented by the DG Mare of the European Commission and the EU Presidency, plays a key role in putting forward a common EU position on management decisions, reflecting any prior negotiations undertaken among its member States and in line with the general framework of the European Common Maritime Policy.

At the Barcelona Convention and at the meetings of the Mediterranean Action Plan, both the EU (represented by the DG Environment of the Commission) and the EU member States can take the floor, with internal discussions and coordination actions being organized by the EC as necessary.

Due to the fact that the European Union represents several developed countries as a member of these regional organizations, it is highly influential in the development of regional policies and management decisions. However, it is important to underline the general tendency for the EU to extend EU directives and regulations to a region-wide level which could provoke negative reactions from non-European Mediterranean member States which may not be in a position to implement them (for example the integrated maritime policy). On the other hand, the EU should be duly recognized for the extensive support it provides to non-European Union developing States for the implementation of decisions by regional organizations.

The EU aid process to the non-European Mediterranean countries, the so-called Barcelona Process (changed to the Union for the Mediterranean under the impetus of the French President Sarkozy), is an important tool for cooperation. It does not however play much of a role in terms of marine management and is still far from meeting the expectations raised by political declarations. The European aid to the non-European Mediterranean countries is also implemented in the framework of the European Neighbourhood Policy.

Role of the civil society

All international organizations in the Mediterranean welcome observers on different bases and according to their statute and policy. For example, MAP considers IUCN as an Intergovernmental Organization while the GFCM considers IUCN as an NGO.

IUCN, WWF, Greenpeace, Oceana and other organizations participate regularly in technical meetings, RFMO annual meetings and the conferences of the parties of environmental organizations. Some of them also occupy official roles within these international organizations.

At its 36th Session, GFCM signed several MoU with various organizations including the Eurofish, the Regional Advisory Council for the Mediterranean and the Mediterranean Network of Marine Protected Areas managers (MedPAN).

At its First Meeting of the UNEP/MAP Ecosystem Approach (EcAp) Coordination Group held in Athens, Greece on 29–30 May 2012, it was agreed that the Secretariat should 'promote the participation of UNEP/MAP partners, relevant projects (PEGASO, PERSEUS and others) and the scientific community, at all relevant levels, such as Correspondence Groups and clusters'. This shows a clear will from the Secretariat as well as from the Parties to work together with the diverse constituents of civil society.

Looking beyond the mere will of organizations to work together and the difficulties arising from the internal structures of each, there is a limitation in the capacity of these organizations to follow the diverse processes and activities both in terms of human and financial resources. Sometimes this constraint is alleviated by the existence of funding mechanisms of international organizations to support the participation of the partner organizations in their activities; however, such mechanisms are absent from some of the environmental and fisheries organizations.

Conclusion and discussion

The ideal framework for integrated maritime governance in the Mediterranean is still far from being established; however, good progress is being recorded through the application by many parties of the ecosystem approach principles. This progress could be further enriched through enhanced regional cooperation among regional and international organizations concerned with the conservation and sustainable use of living marine resources, together with better coordination at national level among the governing authorities to formulate and implement integrated policies which take into consideration the requirements of all stakeholders of the marine ecosystem.

The collaboration among regional organizations dealing with the conservation of biodiversity and marine resources is still in its infancy and needs to be developed further. Mechanisms are in place, but the large numbers of parties within each organization as well as the political difficulties in the region do not favour rapid adoption, implementation and enforcement of conservation measures. Furthermore, regional governance issues and differences between relevant organizations on their respective competencies need to be resolved in order to guarantee a sound integrated approach to the sustainable use and management of living marine resources and the conservation of biodiversity and marine ecosystems. As a major example, it is a possibility for the mandate of the Barcelona Convention to be extended to also encompass the management of marine resources through the 1995 SPA/BD Protocol; however, it is also important to recognize that any fisheries conservation and management measures that may be drawn up in protected areas, established through the framework of the SPA/BD Protocol, must be consistent with the fisheries conservation and management objectives and measures of RFMOs. In this regard, any management scheme within such areas should be developed through a cooperation framework involving all concerned entities.

The only road towards achieving sound regional governance in the Mediterranean region, the cradle of civilization, is through perfectly synchronized objectives, goals and efforts by all relevant national and international bodies. Considering the Mediterranean's richness and diversity in politics, cultures, natural resources and ecosystems, it is also plausible, but perhaps utopian, to consider the founding of a single governing body with inputs from RFMOs, IGOs and NGOs, which would integrate the management of all human activity and the conservation of the marine environment through a spatial and ecosystem-based management approach.

Notes

1. FAO Agreement to Promote Compliance with International Conservation and Management Measures by Fishing Vessels on the High Seas, 1993.
2. UN Agreement for the Implementation of the Provisions of UNCLOS relating to the Conservation and management of Straddling Fish Stocks and Highly Migratory Fish Stocks, 1995.
3. Definition of Operational Units: 'Within the context of managing fishing effort by Geographical Sub-Area(s), an Operational Unit is a group of fishing vessels which are engaged in the same type of fishing operation within the same GSA, targeting the same species or group of species and belonging to the same economic segment. Fishing vessels may belong to more than one Operational Unit and the composition of Operational Units is subject to change over time' (GFCM, 2012).
4. The Convention Area of the ICCAT also extends into the Mediterranean and Black Sea. GFCM and ICCAT collaborate closely on the management of tuna and other large pelagic species, including sharks. It is also customary for GFCM to review and adopt ICCAT recommendations of relevance to the Mediterranean and Black Sea region.
5. Regulation (EU) No. 1343/2011 of the European Parliament and of the council of 13 December 2011 on certain provisions for fishing in the GFCM Agreement area and amending Council Regulation (EC) No. 1967/2006 concerning management measures for the sustainable exploitation of fishery resources in the Mediterranean Sea.
6. FAO trust fund projects include: Adriamed (www.faoadriamed.org), ArtfiMed (www.faoartfimed.org), COPEMED (www.faocopemed.org), EastMed (www.faoeastmed.org), MedFisis (www.faomedfisis.org) and MedSudMed (www.faomedsudmed.org).

References

Abdulla, A., Gomei, M., Maison, E. and Piante, C. (2008) *Status of Marine Protected Areas in the Mediterranean Sea*. IUCN, Malaga and WWF, France.

Accadia, P. and Franquesa, R. (2006) The Operational Units approach for fisheries management in the Mediterranean Sea. General Fisheries Commission for the Mediterranean. Studies and Reviews No. 80. FAO, Rome.

Barazi-Yeroulanos, L. (2010) Synthesis of Mediterranean marine finfish aquaculture - a marketing and promotion strategy. General Fisheries Commission for the Mediterranean. Studies and Reviews No. 88. FAO, Rome.

Bianchi, G. (2008) The concept of the Ecosystem Approach to Fisheries in FAO. In: *The Ecosystem Approach to Fisheries* (eds G. Bianchi and H.R. Skjoldal). FAO, Rome, 363 pp.

Bjørndal, T. (2009) Rent in the Northeast Atlantic and Mediterranean bluefin tuna fishery. Final draft. Prepared for the World Bank PROFISH Program, Washington, DC.

Caddy, J.F. (1998) Issues in Mediterranean fisheries management: geographical units and effort control. General Fisheries Council for the Mediterranean. Studies and Reviews No. 70. FAO, Rome.

Camilleri, M., Coppola, S.R. and de Leiva Moreno, J.I. (2000) Operational Units: a preliminary study. General Fisheries Commission for the Mediterranean, Scientific Advisory Committee, Sub-Committee on Statistics and Information. FOA, Rome.

Cuttelod, A., García, N., Abdul Malak, D., Temple, H. and Katariya, V. (2008) The Mediterranean: a biodiversity hotspot under threat. In: *The 2008 Review of The IUCN Red List of Threatened Species* (eds J.C. Vié, C. Hilton-Taylor and S.N. Stuart). IUCN Gland, Switzerland.

European Union (2011) Regulation (EU) No. 1343/2011 of the European Parliament and of the Council of 13 December 2011 on certain provisions for fishing in the GFCM (General Fisheries Commission for the Mediterranean) Agreement area and amending Council Regulation (EC) No. 1967/2006 concerning management measures for the sustainable exploitation of fishery resources in the Mediterranean Sea. Available at http://eur-lex.europa.eu/LexUriServ/LexUriServ.do?uri=OJ:L:2011:347:0044:0061:EN:PDF (accessed March 2014).

FAO (2010) The State of World Fisheries and Aquaculture 2010. Food and Agriculture Organization of the United Nations, Rome.

FAO (2012) The State of World Fisheries and Aquaculture 2012. Food and Agriculture Organization of the United Nations, Rome.

GFCM (2001) Report of the twenty-sixth session of the General Fisheries Commission for the Mediterranean. Lacco Ameno, Ischia, Italy, 10–13th September 2001. GFCM Report No. 26. FAO, Rome.

GFCM (2012) GFCM glossary. Available at http://www.gfcm.org/gfcm/topic/17107/en (accessed February 2014).

UNEP/MAP-Plan Bleu (2009) State of the Environment and Development in the Mediterranean, UNEP/MAP Plan Bleu. UNEP/MAP-Plan Bleu, Athens.

Chapter 18

CCAMLR and Antarctic conservation: The leader to follow?

D. Miller[1] and N.M. Slicer[2]

[1]Antarctic Tasmanian and Science Research Development, Hobart, Tasmania, Australia
[2]Independent consultant, Sandy Bay, Australia

Abstract: The 1980 Convention on the Conservation of Antarctic Marine Living Resources (CAMLR Convention) was the first international agreement to elaborate a definitional distinction between managing marine life exploitation directly and accounting for specific "principles of conservation" expressly. These principles explicitly account for harvested stock sustainability as well as ecological connectivity of "dependent and related species" in an eco-environmental context. This chapter reappraises the "international best practice standing" for CCAMLR, using specific developments to highlight the organization's advances, challenges and shortcomings. Not only is CCAMLR perceived as a leader in international best practice, it is seen as delivering ecosystem and precautionary approaches essential for strong fisheries and ecosystems outcomes. It also provides a model for other RFMOs with respect to the structures, processes and innovative thinking required to operationalize ecosystem-directed management approaches and concepts.

Keywords: CCAMLR performance; ecosystem precautionary management

Introduction

The 1980 CAMLR Convention ('the Convention')[1] was the first international instrument to provide a distinction between managing marine living resources exploitation directly and accounting for specific 'principles of conservation' generally. In these terms, Article II of the Convention equates 'conservation' with 'rational use'[2] within the context of conservation principles for Antarctic marine living resources[3] that comprise:

- preventing harvested population size decreases to below stable recruitment;
- maintaining ecological relationships between harvested, dependent and related Antarctic marine living resources; and
- preventing ecosystem changes[4] or minimizing the risk of such changes which are not potentially reversible over two or three decades.

Together, these principles make 'possible the sustained conservation of Antarctic marine living resources' subject to the Convention's key objective(s). Under Article II, these principles are characterized as:

- precautionary: CCAMLR collects what data it can, then weighs up the extent and effect(s) of data uncertainties and deficiencies before taking a management decision; and
- based on an ecosystem approach: ideally this accounts for the delicate and complex relationships between organisms and physical processes (such as currents, sea

Governance of Marine Fisheries and Biodiversity Conservation: Interaction and Coevolution, First Edition.
Edited by Serge M. Garcia, Jake Rice and Anthony Charles.

temperature etc.) forming part of, or impacting, the Antarctic marine ecosystem.

Simply put, CCAMLR focuses on regulating fishing of harvest target species while aiming to ensure that harvesting activity does not compromise other species or harm the environment. CCAMLR's meeting of Convention objectives was expressly recognized by Willock and Lack (2006). Compared to 15 other RFMOs reviewed, they saw CCAMLR practices as:

- demonstrating 'that application of the ecosystem approach and the precautionary approach can deliver strong fisheries and ecosystem outcomes'; and
- providing 'clear guidance for other RFMOs on the structures, processes and innovative thinking required to operationalize these concepts'.

This chapter appraises CCAMLR's international best-practice standing using specific examples to highlight advances, as well as contemporary challenges and shortcomings. CCAMLR's objectives and practices over the past 30 years are compared to Willock and Lack's (2006) recommended best practices.

The CAMLR Convention

General provisions

The Convention's definition of 'Antarctic marine living resources' south of the APF (Figure 18.1) serves to identify the Antarctic marine ecosystem as 'the complex of relationships of Antarctic marine living resources with each other and with their physical environment'. Despite implicit recognition of seals and whales as important Antarctic marine ecosystem components, these fall outside the Convention's remit.[5] Furthermore, the Convention's area of application in Article I recognizes the biogeographical integrity of marine living resources south of the APF. Linkages with the Antarctic Treaty mean that the Convention acknowledges 'the importance of safeguarding the environment and protecting the integrity of the ecosystem of the seas surrounding Antarctica'.

Key features of Convention Article II's ecosystem-based and precautionary management approach include: (1) setting target levels for sustainable fishing; (2) accounting for ecologically related and harvested species; (3) restoring depleted populations; (4) minimizing risks of irreversible change(s); and (5) accounting for various natural, environmental or human-induced effects other than fishing. The approach had no precedents, and is a unique example of sustainable and precautionary resource management. It also distinguishes the Convention from more traditional fisheries management practices based on MSY principles and single stock management.

Finally, the Convention acknowledges UNCLOS Articles 116 to 119 rights and obligations, in that the right to fish the high seas is moderated by the obligation to address CCAMLR conservation needs.

Contracting parties and Commission membership

The Convention legally binds CPs to decisions and measures agreed by CCAMLR. All substantive Commission decisions are taken by consensus (Article XII). 'Consensus' is not specifically defined, but may be considered the absence of substantive objection to an agreement or decision (Turner et al., 2006).

All Convention signatories are CPs (Table 18.1), but only Commission Members participate in decision-making (Article XII) and contribute to the budget.[6] Acceding States interested in research or harvesting of CCAMLR marine living resources are eligible to become Commission Members (Articles VII and XXIX), but only if actively engaged in research or harvesting of the marine living resources to which the Convention applies. Only Commission Members are currently granted fishing opportunities.

Implementation

The area south of 60°S (including all ice shelves) falls under the Antarctica Treaty, where Article IV effectively 'freezes' all claims to territorial sovereignty in the Treaty Area. This implies that there are no coastal Antarctic claimant States to exercise national sovereignty in waters adjacent to their territories south of 60°S. Despite this situation, Treaty Article VI clearly indicates that nothing in the Treaty shall 'affect the rights, or the exercise of the rights, of any State under international law with regards to the high seas within the area'. These provisions mean that CCAMLR measures apply throughout the Treaty Area, on the high seas and elsewhere, particularly in that Treaty Article IX(1)(f) implies that all Convention CPs are bound to further the Treaty's objectives regarding 'preservation and conservation of living resources' in the Treaty Area.

Several islands in the CCAMLR Area north of 60°S are subject to undisputed territorial sovereignty. They are therefore essentially 'coastal States' with UNCLOS rights and obligations attached to their 'territorial seas' (UNCLOS Article 2 and 3) and adjacent EEZs under UNCLOS Part V

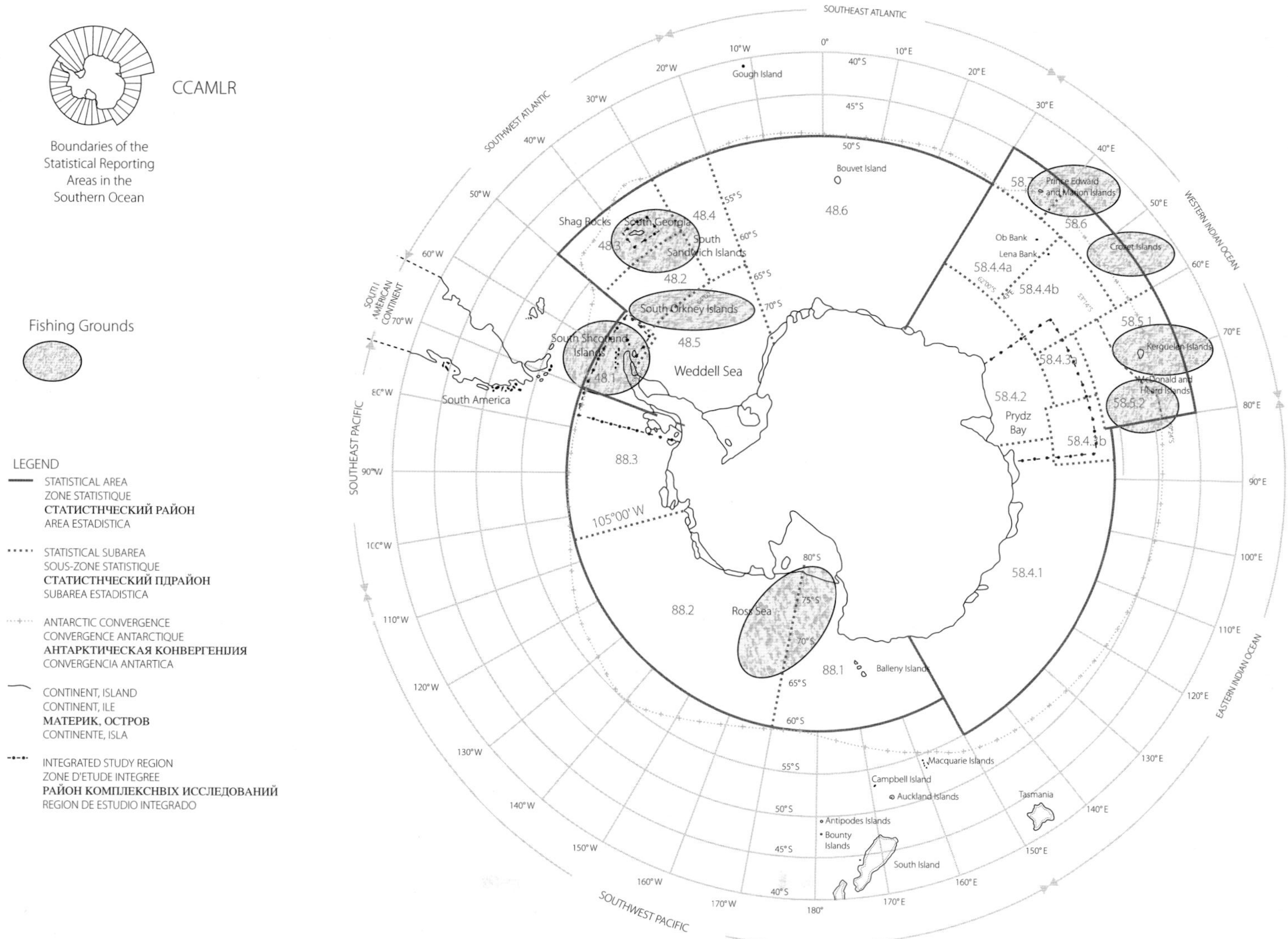

Figure 18.1 The CAMLR Convention Area indicating the location of various fishing grounds. CCAMLR Statistical Subareas and Divisions used for reporting fisheries data. The CAMLR Convention boundary, the position of the Antarctic Convergence (Antarctic Polar Front) and location of various fishing grounds are also shown. (Source: After Miller 2001.)

Table 18.1 CAMLR Convention Contracting Parties

Commission Members		
Argentina	India	Russian Federation
Australia	Italy	South Africa
Belgium	Japan	Spain
Brazil	Republic of Korea	Sweden
People's Republic of China	Namibia	Ukraine
European Community	New Zealand	United Kingdom
France	Norway	USA
Germany	Poland	Uruguay
Non-Commission Members ('Acceding States')		
Bulgaria	Greece	Peru
Canada	Mauritius	Vanuatu
Cook Islands	Netherlands	
Finland	Pakistan	

(Articles 56, 58, 61 and 62 in particular). Consequently, CCAMLR coastal States have the right to determine allowable catches (Article 61), promote optimum utilization of living resources and regulate access to living resources (Article 62) in their EEZs. The CCAMLR Chairman's Statement (http://www.ccamlr.org/en/publications/basic-documents) provides the legal framework for this arrangement.

On the Convention's entry into force, the ecosystem and precautionary approaches outlined in Article II directed CCAMLR's conservation efforts (Constable et al., 2000; Miller, 2002). Ecological uncertainties associated with effective implementation of the Convention's provisions have therefore necessitated innovative thinking to develop a holistic, scientific and ecologically based approach to regulate fishing on target resources, and to minimize indirect harvesting effects on the Antarctic marine ecosystem.

International cooperation

General

Apart from Articles outlining the Convention's relationship to the Antarctic Treaty (Articles III to V), Article XXIII addresses CCAMLR's relations with other international organizations including ATCPs. Both CCAMLR and SC-CAMLR (established under Convention Article XIV, functions outlined in Article XV) are mandated to cooperate with the FAO and other specialized agencies as appropriate. They are also urged to develop cooperative working relationships with relevant IGOs and NGOs contributing to the Commission's work. While only the Scientific Committee on Antarctic Research (SCAR), Scientific Committee on Oceanic Research (SCOR) and the IWC are named in Article XXIII, CCAMLR has established strong working relationships with the environmental NGO ASOC. It also has ties with industry NGOs (the Coalition of Legal Toothfish Operators or COLTO and the Association of Responsible Krill Operators or ARK). These organizations attend Commission and SC-CAMLR meetings as observers in accordance with relevant Rules of Procedure.[7]

Various international ATS and other agreements[8,9] (e.g. SCAR) applicable in the Southern Ocean are potentially important to CCAMLR. These address environmental protection, environmental management, mutual security, scientific research, scientific information exchange and regional governance issues (FAO Compliance Agreement, UNFSA, UNGA Resolution 61/105 and 2009 FAO Port State Measures Agreement). All such instruments strive to promote 'responsible' and 'sustainable' exploitation of marine living resources in general. Preserving the marine environment and protecting biodiversity are specifically mandated by the Madrid Protocol and the CBD.

As a consensus-based organization, CCAMLR has evolved a delicate and inclusive decision-making process (Turner et al., 2008). This binds all CCAMLR Contracting Parties (Article XXI) to the Convention's provisions, in accordance with Convention Article IX. The consequent applicability of adopted CMs is reinforced by various measures[10] where associated cooperative requirements are identified or implied. In particular, some measures require CPs to cooperate with each other and with NCPs to augment the Convention's regulatory provisions and enhance CM compliance. These co-operation polices (see section 5.3.2 of the Convention) encourage countries to work with CCAMLR and include: (1) exchanging information with CCAMLR on IUU fishing; (2) implementing relevant CMs, particularly the toothfish CDS; and (3) acceding to the Convention or joining the Commission.

Best practice assessment

Willock and Lack (2006) identify six RFMO international cooperation best practices. Table 18.2 provides examples of CCAMLR's performance against these practices. Additional examples of other CCAMLR specific practices are included. The results indicate that CCAMLR falls within 'equivalent to international best practice' on average (3.2; see Table 18.2), with three attached considerations.

Table 18.2 CCAMLR international cooperation practices compared with Willock and Lack (2006). Additional CCAMLR practices are included and practices are rated from 1 ('well below international best practice') through 3 ('Equivalent to international best practice') to 5 ('well above international best practice').

Best practice (Willock and Lack, 2006)	CCAMLR practices	Rating
Flexible membership reflecting range of states influencing measures implementation (especially port and market states)	Contracting party conditions flexible for potential accession (e.g. Vanuatu and Cook islands) States (Article XXIX). Not so for membership (Article VII)	2
Targeted capacity building for prospective developing state members	'Policy to encourage NCP cooperation' (see text) with Namibia (Member) and Mauritius (Contracting Party) joining	4
Product trade tracking schemes aimed at influencing CMs effectiveness	Toothfish CDS (CM 10-05): non-discriminatory measure open to Contracting and non-Contracting Parties alike	4
Co-operating state non-member status only as interim step towards full membership	Advanced by China and Namibia becoming Commission Members a short time period after acceding to convention. Other outreach activities	3
Creative solutions to address issues associated with Chinese Taipei through suitable mechanisms	No engagement other than requests for toothfish import data	1
Effect compliance and enforcement measures to counter IUU fishing to preserve member benefits and willingness to comply	Suite of CMs: marking of fishing vessels and gear (CM 10-01), fisheries licensing (CM 10-02), VMS deployment (CM 10-04), port inspection (CM 10-03), CDS (CM 10-05), CCAMLR inspection scheme	5
Other practices		
Sourcing expert and contemporary scientific advice through cooperative mechanisms	Associations with BIOMASS, IWC, SCAR, SCOR, FAO, SCAR Bird Biology Sub-Group, SCAR Group of Specialists on Seals	4
Cooperation with other RFMOs and parallel interest organizations	Member FAO RFSN. MOUs with ACAP and WCPFC	4

The first recognizes Chinese Taipei's minimal involvement in the Antarctic generally and in the CCAMLR Area in particular. Consequently, eliminating Willock and Lack's best practice category for Chinese Taipei raises the CCAMLR grading to 3.6, slightly above prevailing best RFMO practice globally.

Second, CCAMLR and its Members actively reach out to NCPs allegedly fishing in the CCAMLR Area. Diplomatic 'demarches' seek to engage, inform and gather information from these countries on broad application of, and compliance with, CCAMLR CMs. The practice has encouraged cooperation with the Commission and provides information on CCAMLR's objectives.

Third, a number of specific CCAMLR attributes and developments underscore the importance of international cooperation to the Commission's work. Notably, provisions attached to various CMs (Table 18.2) mandate cooperative relations. Furthermore, CCAMLR has established good working relationships on matters of common interest with international bodies such as various ATS arrangements, the BIOMASS Program,[11] SCAR, SCAR Bird Biology Sub-Group, SCAR Group of Specialists on Seals and the CEP.[12] These associations ascribe to the 'freedom of scientific investigation' imperatives detailed in Treaty Article II. Furthermore, CCAMLR has established scientific affiliations with FAO, SCOR[13] and IWC.[14]

Recent international scientific cooperation has been fostered with CCAMLR's establishment of processes to promote scientific burden sharing and capacity building in interested States. This culminated with the 2009 'General Science Capacity Special Fund'[15] which aims to secure wider participation in SC-CAMLR's work, particularly by young scientists from developing countries. With the fund promoting burden sharing and capacity building within SC-CAMLR, it facilitates the collection, study and exchange of relevant information. Other CCAMLR cooperative arrangements involve the RFSN and MOUs with ACAP and WCPFC on matters of mutual interest.

CCAMLR's relationships with certain invited observers[16] have been strained from time to time. Prior to 2004 for example, NGO observers were often excluded from SCIC meetings as well as consultations on developing CMs. These practices have since been discontinued. Nonetheless, meetings of some SC-CAMLR subsidiary bodies (i.e. WGs) fall back on the option outlined in procedural Rule 19(e), where NGO participation is limited by not inviting NGOs to WG meetings. Alternatively, invitations have only been provided in the absence of any Member objection. Nevertheless, NGOs with specific expertise have attended WG meetings on Member delegations, or as 'invited experts' under SC-CAMLR Rule 2. Other recent developments, such as broadening document archives on the CCAMLR website for Members, are aimed at improving transparency and accessibility.

Finally, arrangements such as the scientific observation scheme require bilateral cooperation between Members, while others promote cooperation between Contracting Parties (Conservation Measures agreed by CCAMLR 10-06, or CM 10-06).

Conservation and management

General background

Convention Article IX details functions giving effect to Article II objectives and principles. Broadly, these entail facilitating relevant research, compiling essential data, ensuring collection of relevant fisheries statistics, exchanging essential information, identifying conservation needs and formulating appropriate CMs. Once in place, CMs are legally binding for Convention CPs. CCAMLR also adopts non-legally binding Resolutions that are salutatory in nature and serve as codes of conduct.

CMs include catch limits, effort controls, closed areas and seasons, compliance and trade-related processes and environmental protection provisions. The Schedule of CMs in Force is reviewed annually and takes account of relevant decisions by other Treaty instruments (e.g. CCAS) or fisheries arrangements. CM formulation includes a key requirement that measures are formulated, adopted and revised on the basis of the 'best scientific evidence' available (Article IX(1)(f)) in the form of SC-CAMLR advice, research results and data analyses.

SC-CAMLR undertakes regular (usually annual) analyses of both historical and contemporary fishery data. Standardized, analytical assessment techniques[17] have been developed and data collections methods improved.[18] Additional research aims to improve understanding of Antarctic ecosystem dynamics, biological productivity and functioning. In practical terms, CCAMLR pursues four key actions to address Article II conservation principles: (1) determining the management status for relevant (target and non-target) species and/or ecosystem qualities; (2) assessing ecosystem status in terms of perceived 'health';[19] (3) implementing harvest controls to address differences between the assessed status of exploited stocks and agreed conservation objectives; and (4) striving to reach scientific consensus on key CM formulation advice offered to the Commission.

The management paradigms associated with these actions are not trivial; CCAMLR therefore does not rely on a single management approach alone. Rather, its decisions are scientifically driven, iterative, responsive and ongoing. This has promoted the recognized innovation and flexibility for which CCAMLR is renown in addressing its conservation challenges. For such challenges, Willock and Lack's (2006) three areas of best conservation and management practice (Table 18.3) comprise use of the precautionary and ecosystem approaches to fisheries management, as well as provision of contemporary scientific advice.

CCAMLR case studies

Various case studies illustrate the extent to which CCAMLR's management regime has evolved. To address its conservation principles, CCAMLR has 'operationalized' four essential management requirements: (1) minimizing the risk(s) of irreversible ecosystem change(s); (2) monitoring harvest controls to ensure sustainable exploitation; (3) minimizing potential harvest effects (direct or indirect) on dependent and related species; and (4) refining assessments to account for uncertainties in available information and/or concerning stock status. The latter includes potential responses of non-harvested species[20] and ecosystem function(s). This management approach has therefore come to integrate scientific antecedents with CCAMLR management decisions (Figure 18.2). The underlying drivers strive to: (1) apply correct/timely decisions consistent with Article II conservation principles; (2) undertake monitoring of sufficient power to ensure that dependent predators are not affected by harvest controls; (3) allow sufficient time to detect/rectify harvest-induced ecosystem changes within two to three decades; and (4) refine precautionary assessment

Table 18.3 CCAMLR conservation and management practices compared with Willock and Lack (2006). Ratings as per Table 18.2.

Best practice (Willock and Lack, 2006)	CCAMLR practices	Rating
Precautionary Approach (PA)		
PA application integral to RFMO Convention	Article II Provisions	5
Institutionalize PA use	SC-CAMLR advice and subsidiary working group approaches	5
Management strategies with precautionary reference points for sustainability of all stocks. Explicit rebuilding stocks for overfished stocks and measures to monitor progress	SC-CAMLR PA-based decision rules/ assessments for krill, absence of specific rebuilding strategies except for areal/ species closures to fishing	3
Absence of management strategies; precautionary management measures based on best scientific advice available	Krill fishery, new and exploratory fisheries	4
Ensure resources in new and exploratory high-seas fisheries are subject to precautionary management	Relevant CMs and other approaches	4
Ecosystem Approach to Fisheries Management (EAFM)		
EAFM integral part of convention requiring fisheries management conducted in context of ecosystem rather than target stocks	Article II Provisions	5
Institutionalization of EAFM use by explicit recognition	Operational goals, krill and other EAF-directed management	5
EAFM incorporation into advisory structures supported by adequate data collection, research and analyses	SC-CAMLR, CEMP, WG-EMM	5
Acknowledgement PA integral to EAFM with target stock precautionary management as cost-effective step towards ecosystem impact mitigation	SC-CAMLR 'work in progress'	5
Impacts assessed and mitigation measures applied (including monitoring of measures' effects) for new fisheries/fishing method proposals	CMs on new and exploratory fisheries as wells attached SC-CAMLR assessments	5
Collaboration between RFMOs with responsibilities for ecosystems and/or species to maximize research returns and monitoring fishing ecosystem impacts	MOUs with various relevant institutions, CCAMLR science and SC-CAMLR member's scientific standing	4
Scientific Advice		
Provision independent scientific advice or regular peer review of advice to commission	2008 performance review and peer-reviewed scientific journal	4.5
Subsidiary bodies established to develop options for cms based on scientific advice	SC-CAMLR working groups and SCIC subsidiary groups. Also SC-CAMLR-SCIC joint assessment group (JAG). Other consultations.	4
Application of data collection, verification and exchange standards consistent with UNFSA Annex 1	All data formally verified and exchanged commensurate with UNFSA standards	4
Broadened scope for data collection and scientific advice, including mandatory accounting for non-target and associated dependent species, and broader ecosystem impacts	CEMP, bycatch reporting, VME reporting, fisheries notifications, new and exploratory fishery conditions	5
Appropriate bodies to ensure advice on fishing ecological impacts	SC-CAMLR and WG-EMM	5
Develop management strategies aligned with PA to provide context for considering scientific advice and decision making	SC-CAMLR sanctioned work in progress	4

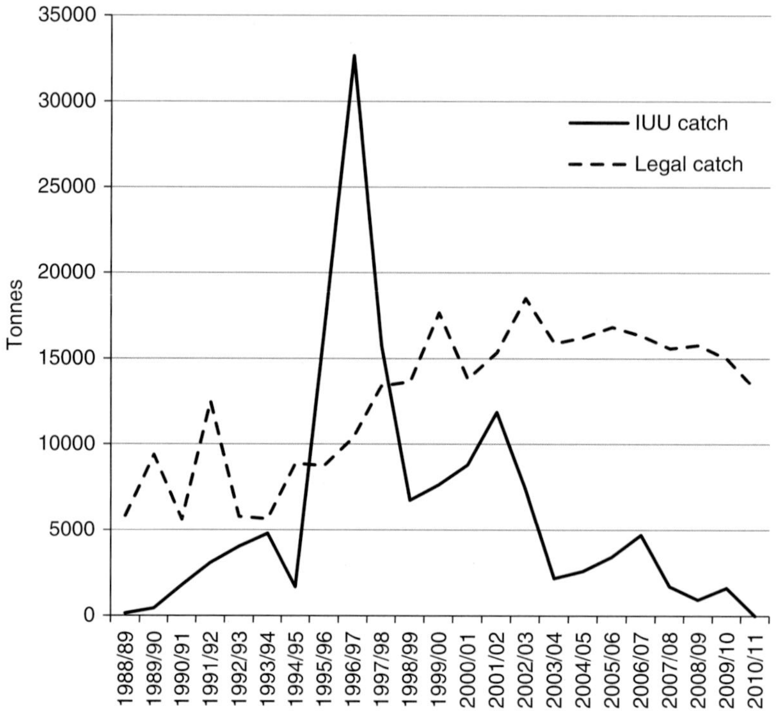

Figure 18.2 CCAMLR's management approach to address *Convention* Article II objectives. Acronyms: CEMP, CCAMLR Ecosystem Monitoring Programme; KYM, krill yield model; GYM, general yield model; VMEs, vulnerable marine ecosystems; MPAs, marine protected areas.

of harvested stock yield to revise key demographic parameter estimates.

Precautionary approach (PA)

Convention Article II outlines the PA for which CCAMLR is recognized (Willock and Lack, 2006). It is unique to the extent that it codifies many principles, later used more widely (Constable et al., 2000) in instruments such as UNFSA and the FAO Technical Guidelines for Responsible Fisheries. Furthermore, CCAMLR has agreed that 'management should adhere to a precautionary approach, according to which decisions taken should have a low risk of long-term adverse effects'. This has important implications for dealing with informational uncertainty where the actual size of exploited stocks is not known precisely, and especially when new stocks are targeted.[21] CCAMLR formally incorporates the PA into assessments of harvested stocks, as well as into decision-making associated with their management (Constable, 2011). Two examples illustrate this practice.

Precautionary catch limits and decisions rules

Precautionary catch limits have been developed for various harvested stocks consistent with SC-CCAMLR advice. In 1991, the CCAMLR-regulated krill fishery became the first fishery globally to be ascribed a 'precautionary status', with catch limits (e.g. CM 51-01) based on direct acoustic surveys of stock biomass. Initially, four management principles were applied to account for sustained consistency of catch levels with time, as well as for uncertainties associated with the place of krill in the ecosystem (i.e. as a predator food source) and its demographic (e.g. recruitment rates) variability (Butterworth et al., 1994). The stock yield is calculated using a stochastic stimulation, where survey-derived biomass ('unexploited biomass') is combined with a proportionality coefficient (λ) to represent estimates of krill recruitment variability, growth and natural mortality (Miller and Agnew, 2000). A three-part decision rule is applied where λ is a 10% probability of the krill spawning stock biomass falling below 20% of the median unexploited biomass level over a 20-year period or the median krill escapement (for predator consumption) over a 20-year period is 75%, or the lesser of these two values. This basic KYM evolved into the more generally structured GYM (Constable and de la Mare, 1996) to provide flexibility for assessing krill growth patterns, natural mortality, spawning and fishing. The decision rules were recast as general principles so that krill spawning stock escapement is sufficient to avoid any likelihood of declining recruitment, and reserves of exploited harvest stock abundance are sufficient to fulfil krill-dependent predator needs. Stock trajectories can then be calculated from estimated levels of absolute recruitment (*R*) in relation to fishing mortality (*F*). The GYM and modified decision rules have since been applied as the IAM to estimate Antarctic toothfish, Patagonian toothfish and Mackerel icefish stock yields and catch limits (Miller, 2011).

New and exploratory fisheries

In the early 1990s, CCAMLR recognized that developing fisheries may become fully operational without essential data for their management being available. To ensure that sufficient 'precaution' is applied to such fisheries consistent with the Convention's objectives, CCAMLR addressed developing fisheries management in 1989. The process recognizes that fishery development is 'directly linked with elaborating timely scientific advice and management measures', with the collection of essential data for management being mandated.

CCAMLR adopted CM 21-01 on 'new fisheries' in 1991.[22] The measure lays out conditions for a new fishery's impending implementation, recognizing that such fisheries must be managed from commencement. A key condition is that pre-notification is essential, particularly when targeting species and/or ground(s) not previously fished. Similar considerations apply if a novel fishing technique is to be used. CM 21-01 aims to collect essential information on target, as well as dependent and related, species to facilitate development of new fishery regulatory controls. It limits catch or effort while allowing fishing to occur,[23] thereby maximizing accrual of essential data for managing fishery sustainability from the onset of harvesting. Along with regulatory provisions, CCAMLR outlined conditions attached to a 'new fishery' becoming an 'exploratory fishery'[24] in 1993.

Both 'new' and 'exploratory' fisheries require pre-notification and review before full-scale fishing is authorized. CCAMLR fisheries are no longer classed as exploratory once the necessary management and catch/effort information is available for their full assessment.[25] The most notable exploratory fisheries to date include toothfish fisheries in parts of the Indian Ocean and in the Ross Sea. All other fisheries are subject to full assessments (i.e. 'developed'), or have yet to be opened as new fisheries. Some fisheries have also been specifically prohibited. Although the krill fishery is considered a developed fishery, it is subject to participatory notification requirements (CM 21-03) given the species' ecological importance and the less-than-full knowledge of its ecosystem interactions.

Ecosystem Approach to Fisheries Management (EAFM)

Krill fishery

The PA to set krill catch limits (section 4.2.2.1 of the Convention) allows for fisheries expansion while minimizing potential catch impacts on the ecosystem in general and on krill predators in particular. As a management tool, krill precautionary catch limits aim to ensure stock sustainability without compromising the species' vital ecosystem role. To augment this strategy, CCAMLR agreed that krill precautionary catch limits should be spread between geographically defined SSMUs in 2002 (Miller, 2011). This would reduce the risks of fishing becoming concentrated in small areas with attached localized damage to key ecosystem components such as land-based krill predators. However, CCAMLR has not yet agreed on how krill precautionary catch limits may be apportioned within current regionally defined fishing grounds, either in relation to SSMUs or any other criteria. This could be perceived as a failure to fully implement a krill fishery PA, although current catches are well below the prevailing precautionary limits and are not yet highly concentrated in localized areas.

Bycatch mitigation

Bycatch manifests direct interaction(s) between fishery and non-target species. CCAMLR has implemented three CMs targeted at avoiding, directly limiting and monitoring fisheries bycatch. For example, CM 33-03 links bycatch measures for selected species with 'move-on' rules to displace fishing when such species are encountered, or when trigger catch levels are exceeded. Bycatch reporting provisions also apply.

Other measures (CMs 32-02 to 32-08) limit directed fishing on certain species, including common bycatch species. CM 33-01 closes areas to fishing when stipulated bycatch limits are exceeded, and CM 32-04 closes the South Shetlands region to directed fishing for Antarctic cod to allow stock rebuilding and to avoid bycatch of other finfish species.

CCAMLR's campaign 'Fish the Sea Not the Sky' (http://www.ccamlr.org/en/document/publications/fish-sea-not-sky) is a notable symbol of success for countering incidental seabird bycatch. Following introduction of mitigation measures in the early 1990s, CMs 25-02 and 25-03 resulted in dramatic seabird bycatch decreases during longline fishing. These measures provided input into the 1999 IPOA-Seabirds (http://www.fao.org/fishery/ipoa-seabirds/legal-text/en) and set a global benchmark for developing successful and cost-effective measures for mitigating seabird bycatch.

Dependent and related species

The primary objective of the 1985 CCAMLR Ecosystem Monitoring Group (CEMP) is to 'distinguish between changes due to harvesting of commercial species and changes due to environmental variability, both physical and biological' (Constable et al., 2000). CEMP monitors key life-history parameters of selected dependent species ('indicator species') likely to respond to changes in harvested species availability. Despite significant and historically relevant time-series CEMP data, it has not been possible to elaborate functional relationships between krill depletion or disruption of aggregations and localized predators (Mangel and Switzer, 1998). Therefore, identification of specific conservation objectives for selected krill predators has not been forthcoming at either a local or metapopulation level (Constable, 2011). As Constable (2005) suggests, such objectives may be unattainable 'not least because of the dynamic nature of ecosystems and the unlikely event that [conservation] "targets" can be met for all populations simultaneously'. Although this has serious implications for EAF management strategies as a whole, it should not impede implementation of EAFM principles (Constable, 2011). The Working Group for Ecosystem Monitoring and Measurement (WG-EMM) continues to explore how ecosystem information (i.e. CEMP data) can be explicitly subsumed into management advice.[26] It is the latter for which CCAMLR is recognized (Willock and Lack, 2006), and it is here that CCAMLR 'work in progress' to formally utilize CEMP data in management decisions retains a high level of anticipated benefit. The recent endorsement of a CEMP Special Fund[27] will augment such work.

Environmental protection

Apart from SSMUs and SSRUs, CCAMLR is initiating environmentally focused measures to address precautionary and ecosystem-directed elements of Article II. For example, CM 26-01 aims to minimize risks associated with fishing-vessel-based alien species contamination and pollution. Other measures protect CEMP monitoring sites (e.g. CM 91-01), or address general safety-at-sea issues (resolutions 20/XXII, 23/XXIII, 28/XXVII and 29/XXVII). The latter aims to ensure that fishing vessels operate in an 'environmentally friendly manner' within the Convention Area.

Apart from interim prohibition of deep-sea gillnetting (CM 22-04), CCAMLR implements spatial management measures to facilitate biodiversity conservation directly. These are consistent with WSSD and UNGA Resolution 61/105 targets over the past decade. For the former, benthic and pelagic bioregionalizations form a basis for a representative MPA network (Miller, 2011). Although no agreement has yet been reached on the specifics for an MPA network within the CCAMLR Area, a general MPA CM was finalized in 2012 (CM 91-04). Anticipating such a network, CCAMLR formulated CMs 22-06 and 22-07 in 2007 before the relevant UNGA deadline. CM 22-06 'freezes' the CCAMLR bottom fishing footprint to areas approved in the 2006/07 fishing season, while CM 22-07 outlines a format for identifying VMEs 'encountered' during fishing operations and the resulting action to be taken by affected vessels (Reid, 2011). The approach remains under consideration with more work scheduled in 2013, including a review of CM 22-07. Notably, CCAMLR was among the first RFMOs to declare (CM 91-03) areal protection through a fishing prohibition around the southern Orkney islands in 2009.

Scientific advice

Reid (2007) emphasized the inherent difficulties of defining operationally based Antarctic conservation objectives, particularly for human impacts in what are largely open-access fisheries. The attached complexities compound incrementally when ecosystem 'health' issues are added, and where potential analogues for target setting are insufficiently precautionary or uncertain. The challenge for CCAMLR scientists and policy-makers alike has therefore been to develop conservation strategies with scientifically defensible management triggers to counter potentially negative human impacts.

This nexus of science and 'policy' is an evolutionary process and one with which CCAMLR has continuously grappled. As early as 1987, SC-CAMLR provided advice reflecting the entire Committee's views. It also outlined the predicted consequences of particular management options when scientific consensus could not be reached. In turn, the Commission endorsed the provenance of SC-CAMLR in 1999 as the sole scientific advice provider. While the Committee's high-quality scientific advice is often provided from the scientific contributions of a proportionately small number of SC-CAMLR Members,[28] the citation record (scientific journal rating SJR = 0.12)[29] of CCAMLR's peer-reviewed scientific journal, *CCAMLR Science*, indicates that SC-CAMLR sponsored science is internationally competitive. Nonetheless, the Commission has occasionally ignored SC-CAMLR advice, most notably in discounting SC-CAMLR-recommended Ross Sea Antarctic Toothfish precautionary catch limit by 50% in 1996 and 1997.[30]

Assessment

For the categories addressed in this section, CCAMLR posts a close-to-perfect mean score (5.0) (Table 18.3) for PA (4.2), EAF (4.8) and scientific advice (4.5). An obvious black mark is attached to the complexities of explicitly linking ecosystem status with fisheries management. Despite more than ten years of effort, model development of functional relationships remains relatively rudimentary. In the latter case, Constable (2011) states 'CCAMLR is demonstrating progress in satisfying the conservation and management requirements identified by Lodge et al. (2007) and Pitcher et al. (2009)'. These requirements aim to maintain universal and orderly fisheries development, so that exploitation does not outpace CCAMLR's ability to achieve the primary objective of Article II of conserving harvested stock sustainability and Southern Ocean biodiversity, subject to the 'best scientific evidence available'.

Compliance and enforcement

Table 18.4 highlights Willock and Lack's (2006) recommended RFMO best compliance and enforcement practices. The attached perspective underwrites activities essential for ensuring that Convention objectives are adequately addressed through CCAMLR fulfilling its governance obligations.

Cooperative mechanisms to detect non-compliance

Like many RFMOs, CCAMLR has no collective enforcement capability *per se*. Rather it relies on flag State control and enforcement by its Contracting Parties to ensure CM compliance. Member States are therefore obligated to enforce CCAMLR compliance-related measures. To optimize enforcement, CMs include provisions for sharing MCS and, in some cases, enforcement among CPs.[31]

System of inspection

The CCAMLR System of Inspection was established under Article XXIV in 1988. It provides for designated CCAMLR Inspectors to undertake at-sea boarding as well as in-port inspection of Member and Contracting Party vessels. Under the System, several CCAMLR Members conduct regular surveillance in the CCAMLR Area. These include Australia and France in the Indian Ocean Sector, the United Kingdom in

Table 18.4 CCAMLR compliance and enforcement practices compared with Willock and Lack (2006). Ratings as per Table 18.2.

Best practice (Willock and Lack, 2006)	CCAMLR practices	Rating
Drawing on UNFSA and IPOA-IUU measures; compliance measures applied and enforced throughout organization as a whole rather than rely on flag state enforcement	Invented 'IUU' as term. Virtually all IUU measures in advance of international practices and basis for IPOA-IUU	4
Urgent need for information exchange within and between RFMOs	IUU Vessel Lists in public domain. exchanges of information with CCSBT, WCPFC, SEAFO, SPRFMO, NAFO and ACAP	3
Standardized RFMO compliance and enforcement measures, including common VMS standards, reporting formats, black and white listing inclusion criteria and trade documentation schemes	All compliance and enforcement measures in keeping with, or in advance of, other RFMOs	4
Strengthening institutional capacity to analyse compliance and enforcement information, including technical subsidiary body establishment and resourcing	SCIC work and developing compliance evaluation procedures. Institutional and member sanctioned training and outreach capacity building	4

the South Atlantic Sector and New Zealand in the Ross Sea. Australia, France and South Africa have established bilateral arrangements for joint patrols in Indian Ocean high-seas areas and in their respective EEZs.

Centralized Vessel Monitoring System (VMS)

VMS has been a requirement since 1988 for all finfish (predominantly toothfish) harvesting vessels active inside the Convention Area. Responding to concerns about the integrity of some flag-State-reported VMS data in 1999, CCAMLR adopted centralized VMS (c-VMS) reporting to the CCAMLR Secretariat. The reporting procedures (CM 10-04) relay VMS data to the CCAMLR Secretariat via the flag State. In the interests of expediency however, CCAMLR allows flag State vessels to report directly to the Secretariat (a practice which most vessels now employ). The USA has adopted a unilateral requirement for centralized port-to-port VMS reporting as a pre-condition for toothfish import approval.[32]

Under the CCAMLR VMS, flag States may also voluntarily report c-VMS data to the Secretariat for vessels fishing outside the Convention Area. As toothfish imports into the USA are linked with port-to-port c-VMS reporting, there is a strong incentive for vessels fishing outside the Convention Area to voluntarily report VMS data. VMS reporting became a requirement for all krill vessels in 2010.[33]

Catch Documentation Scheme (CDS)

Established in 2000, the CCAMLR CDS requires a high level of communication and cooperation between parties. Most toothfish traded under the CDS involves flag States, port States and usually more than one market State. Consequently, CDS operations are subject to multi-party review, with individual catches usually being cross-checked and authorized by three or more parties. The CDS applies to all CPs as well as NCPs that voluntarily assume its application (Sabourenkov and Miller, 2003). CDS-accredited toothfish often fetch higher prices than unaccredited fish (Miller et al., 2010).

Initially, the CDS was hampered by catch misreporting. This was rectified by: (1) ensuring that only Port State officials authorize landings; (2) a unilateral trade ban by the USA on all high-seas catches outside the CCAMLR Area; and (3) linking the CDS with the IUU Vessel Lists and c-VMS reporting. For the latter, port and import officials are able to request c-VMS data for individual vessels prior to authorizing a toothfish catch landing or import.

In 2004, CCAMLR adopted an electronic web-based e-CDS format to provide real-time information exchange. The e-CDS has led to an informal global network of compliance-related contacts, demonstrated by COLTO's unprecedented involvement in the Commission's work. CDS-garnered information has also played a prominent role in various high–profile, multi-jurisdictional and trans-national legal actions.[34]

Port inspection

Over the years, CCAMLR has developed CMs (CM 10-03 and 10-05) to enhance the effectiveness of port inspections in promoting compliance, particularly within the toothfish fishery and CDS. These measures pre-empted the PSMA[35] (2009 FAO Port State Measures Agreement) and are generally more stringent.

Scheme of international scientific observation

The CCAMLR Scheme of International Scientific Observation[36] requires all finfish vessels to carry an international observer. CCAMLR observers have no enforcement powers, with their role being to 'observe and report on the operation of fishing activities in the Convention Area'. The presence of observers plays an important role in reporting IUU activity, documenting incidental mortality of non-target species and detecting non-compliance by licensed vessels. Many vessels also carry national observers who may be vested with enforcement powers. CCAMLR adopted a 50% observer coverage requirement for krill vessels in 2010.[37]

Cooperative measures to deter non-compliance

IUU fishing[38] remains CCAMLR's most significant compliance challenge. IUU fishing was first observed around South Georgia in the mid-1990s. Enforcement action by the United Kingdom pushed IUU fishers east into the Indian Ocean sector, particularly into the South African, French and Australian EEZs in the CCAMLR Area. Strong enforcement by Australia and France eventually displaced IUU fishers onto the high-seas areas further south (Miller, 2007). To date, the Ross Sea remains largely unaffected given the extreme ice conditions for much of the year and the presence of licensed vessels on the relatively restricted fishing grounds. To date, toothfish is the only species targeted by IUU fishing in the CCAMLR Area and the attached trends are highlighted in Figure 18.3. These intimate a recent decline in IUU activities and catches compared to legitimate fishing. In the late 1990s, many IUU vessels were

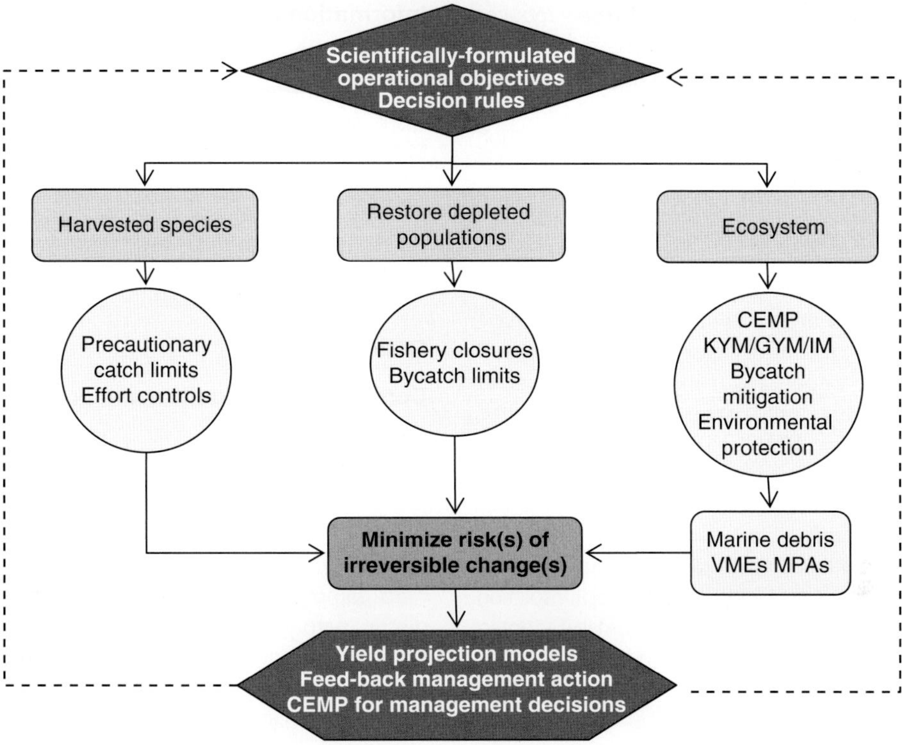

Figure 18.3 Annual estimates of IUU toothfish catches in the CCAMLR Area compared to legitimately reported catches. Data are in tonnes and have been derived from annual CCAMLR reports or estimates. Annual point estimates for IUU catches were discontinued in 2011.

reportedly flagged to CCAMLR Members (Miller et al., 2010). Sanctions for CP-flagged vessels engaged in IUU fishing now include confiscation of vessels and catches, imposed fines, withdrawing licenses and preventing access to future CCAMLR fisheries. Between 2008 and 2011, a number of vessels were denied participation in the forthcoming season's toothfish exploratory fisheries due to repeated incidents of non-compliance.[39] Generally however, the overall compliance record of CP-flagged vessels is high and no cases of IUU fishing by CP-flagged vessels have been reported since the early 2000s. This is evidenced by the CCAMLR IUU Vessel List established under CMs 10-06 and 10-07. From 2011, the List has exclusively comprised NCP-flagged vessels.

All IUU activity in the CCAMLR Area now occurs on the high seas and is confined to NCP-flagged vessels, making it unregulated rather than illegal.[40] CCAMLR's dominant compliance challenge therefore remains the prevention of NCP IUU fishing. While no at-sea enforcement action can be taken against such vessels without flag State consent, all currently active IUU vessels are included on the CCAMLR IUU Vessel List. By default, these are denied port and market access by CCAMLR Members. Several attempts to access Member ports and markets have resulted in catch seizures and the prosecution of individuals.

As CCAMLR has no arrangements for granting fishing access to NCPs, it remains to be seen how this affected, or will affect, IUU fishing trends in the past and in the future. Nonetheless, and despite the successes outlined above, concern still remains concerning emergence of a future IUU krill fishery (Miller, 2011).

IUU vessel lists

In 2001 and 2002 respectively, CMs 10-06 and 10-07 created IUU vessel lists for both CPs and NCPs. Inclusion on the IUU List is itself a sanction as it prevents access to the CDS, as well as to ports and valuable markets.

Furthermore, inclusion on the List considerably reduces a vessel's resale value.

However, an IUU vessel listing does not directly prevent continued engagement in IUU activities. In such cases, the vessels' flag State will receive demarches from the Commission and individual Members, urging it to exert control over vessel activities in accordance with UNCLOS Articles 118 and 119. In some cases this has been successful, with several NCP flag States allowing vessels to be boarded and inspected and/or applying sanctions and de-flagging vessels. In other cases (Equatorial Guinea, North Korea and Togo), flag States have failed to respond. However, decreased profitability for IUU fishers attributable to port and market controls, together with low stock levels in some high-seas areas, has resulted in many vessels ceasing CCAMLR IUU activities.

Since the IUU Vessel Lists have been in place, IUU fleets have demonstrated noticeable mobility with vessels rapidly changing names and flags to avoid detection/listing. This highlights the importance of up-to-date information and information exchanges to avoid name, flag or jurisdiction 'hopping'. Although IUU Vessel Lists are formally adopted annually, CCAMLR makes provisions for updating the Lists inter-sessionally to reflect changes in vessel details. The Lists are publicly available on the CCAMLR website and are circulated to all CCAMLR CPs, RFMOs and the FAO.

Control of nationals

In 2006, CCAMLR become one of the few RFMOs to adopt a formal measure (CM 10-08) controlling Member nationals engaged in IUU activities. The relatively recent CM 10-08 has been successfully applied by Spain to address its nationals' associations with NCP-flagged vessels.[41] However, the use of offshore management companies by IUU operators often means that complex investigations are necessary to acquire sufficient evidence for successful prosecution. Several potential cases have therefore been unsuccessful due to insufficient evidence.

Other measures

Since 2007, the EU has been proposing a CCAMLR market-related measure aimed at imposing trade sanctions on flag States considered to have repeatedly failed to exercise control over the IUU activities of their flag vessels. The proposal is largely compatible with EU IUU Regulation 1005/2008[42] objectives. A lengthy CCAMLR debate has focused on whether the proposal would achieve results above those of the CDS, and hence substantially enhance the effectiveness of the Scheme. The issue remains unresolved.

Information exchange

In 1987, CCAMLR established SCIC[43] as a forum for reviewing CM compliance. A key component of SCIC's work shares compliance-related information among Members. CCAMLR has also established other *ad hoc* information-sharing arrangements with co-operating NCPs (e.g. Indonesia and Mozambique), denying port access to IUU toothfish vessels and providing ancillary information to CCAMLR.

Other issues

Compliance, evaluation and promotion

CCAMLR is considering a formal compliance evaluation process.[44] It is one of the few RFMOs to tackle this difficult problem and the proposed process includes potential ecosystem impact(s).[45] Meanwhile, SCIC continues to review compliance-related information.

To comply with WTO requirements, the CDS is non-discriminatory and any NCP is eligible to participate. However, NCP involvement has often been more problematic than successful. In some cases, CDS-facilitated laundering of IUU toothfish 'legitimized' landings in 'ports of convenience', so boosting illicit trade.

CCAMLR has adopted a formal policy to enhance NCP[46] cooperation as well as CDS implementation (CM 10-05, Annex 10-05/C). The former sets out the requirements for NCPs to achieve and maintain what is effectively a cooperating NCP status. Some NCPs now implement the CDS as port and market States, while there has been progress in broadening port controls by non-cooperating NCPs. Furthermore, CDS training and capacity-building programs are expected to continue into the future and boost implementation of the Scheme.[47]

Assessment

At the height of IUU fishing in the mid-1990s to early 2000s, 55 longline vessels were estimated to be active in the CCAMLR Area with the 1998 estimated IUU toothfish catch being about 30,000 tonnes.[48] An estimated annual incidental mortality of around 7000–20,500 seabirds was attributed to IUU fishing between 1993 and 2001.[49]

IUU fishing in the CCAMLR Area has significantly dropped over the past decade, particularly from 2004 onwards, to around 1000 tonnes in 2010 (Figure 18.3). A corresponding reduction in estimated incidental mortality of seabirds has also occurred.[50] In 2011, only four NCP-flagged IUU fishing vessels were reported to be active and these were included in the CCAMLR IUU Vessel List.

While the need to address the small IUU fleet active in the Convention Area remains, CCAMLR's experiences indicate some success in combating IUU fishing via international cooperation. Nonetheless, counteracting IUU fishing was probably at the detriment to other CCAMLR conservation efforts (Miller, 2011).

CCAMLR's mean compliance grading (4.2) reinforces the notion that the Commission is within the upper bounds of international best practice. It also suggests that the necessary governance structures are in place to ensure conservation measure compliance. Considering that some CMs are directed at maintaining Antarctic marine ecosystem health, then CCAMLR may be considered to be performing well in addressing marine biodiversity and ecosystem conservation requirement.

Performance review and evaluation

The 2008 CCAMLR Performance Review[51] examined the organization's standing as well as its overall performance, achievements and shortcomings in meeting Convention obligations. In terms of Willock and Lack's (2006, chapter IV) best practices, the Review was transparent with external participants being involved. It provided an analysis against international obligations as well as Convention objectives, using pre-set criteria to identify consistent evaluations and benchmarks for improvement, as well as for future review. CCAMLR responded to the Review by maintaining standing items on the Commission, SC-CAMLR and SCIC agendas. It also provides an updated web-based synopsis of the Review's recommendations to track their progress.[52]

In conclusion, it is probably true to say that the CCAMLR Performance Review has largely met all of Willock and Lack's (2006) best-practice requirements. It also highlights the profound effect that IUU fishing has had on the Commission's work, and strongly emphasizes the urgent need to bring unregulated fishing under control or risk significantly undermining CCAMLR's work.

Conclusions

Clarke and Harris (2003) identify unsustainable, unregulated fishing as a key threat to Antarctica's future preservation and conservation. The anticipated large-scale expansion of the krill fishery (Nicol et al., 2011) is obviously critical here, given the species' ecosystem standing. While MSC Certification of the Norwegian CCAMLR krill fishery (Anonymous, 2010) may suggest that it is not all 'doom and gloom' for the fishery's future management, the attached uncertainties remain equal to those when the Convention was first negotiated (Edwards and Heap, 1981). Such uncertainties would be compounded by development of an IUU krill fishery.

More positively, Constable (2011) indicates that 'CCAMLR is undoubtedly able to absorb modern norms for conservation and sustainability because of its regional, broader-than-fisheries remit, combined with a rich pedigree of environmental management and cooperation within the Antarctic Treaty System'. As already indicated, and despite many inherent complexities, CCAMLR continues to demonstrate progress in addressing its conservation and management objectives.

However, the above is not a panacea for CCAMLR's success. Only with continued Member commitment and the involvement of a dedicated and highly professional cadre of scientists[53] will CCAMLR sustain its place at the cutting-edge of best international RFMO practice. In this regard, the fragility of CCAMLR's consensus-based decision-making is both a strength and weakness. It demonstrates remarkable flexibility in dealing with complex challenges posed by marine resource governance, as well as marine biodiversity and ecosystem conservation. However, it runs the risk of falling victim to the 'power of one' dissenting party that consensus decisions imply; a real and present danger if international cooperation becomes undermined by national interests. The conservation outcomes of CCAMLR's next 30 years are therefore eagerly anticipated.

Notes

1. The CAMLR Convention's acronym differs from the Commission's (CCAMLR) established under Convention Article VII. The Commission is the Organization's executive arm serving the collective interests of all CPs.
2. CCAMLR agrees that 'rational use' implies: '(a) harvesting is on a sustainable basis, (b) harvesting on a sustainable basis means that harvesting is conducted to ensure the highest possible long-term yield for the resource subject to the general conservation principles outlined in paragraph 3 of Article II'.
3. Convention Article I(2) defines Antarctic marine living resources as 'populations of fin fish, molluscs, crustaceans and all other species of living organisms, including birds, found south of the Antarctic Convergence'. The Convergence (APF) is the Southern Ocean boundary (±45°S) where cold Antarctic waters meet warmer waters to the north.

4. Potential changes include direct and indirect harvesting impacts and the effects of associated activities, alien (i.e. introduced) species and environmental change(s).
5. Whales and seals are specifically addressed by the 1946 International Convention for the Regulation of Whaling (ICRW) and the 1972 Convention for the Conservation of Antarctic Seals (CCAS), respectively.
6. The budget supports CCAMLRS's day-to-day activities as well as a full-time Secretariat (Convention Articles XIII and XVIII) in Hobart, Australia.
7. See http://www.ccamlr.org/en/document/publications/rules-procedure-commission and http://www.ccamlr.org/en/document/publications/rules-procedure-scientific-committee
8. The most relevant agreements include the ICRW, MARPOL (Annex 5), UNCLOS, CBD, FAO Compliance Agreement, UNFSA and 2007 UNGA Resolution 61/105. Relevant 'soft' (i.e. non-legally binding) instruments include the Code of Conduct and IPOAs for Sharks, Seabirds and IUU.
9. The ATS comprises the international arrangements underpinning relations among States in the Antarctic. It includes the Antarctic Treaty, CCAS, CAMLR Convention and Madrid Protocol. It also includes ATCM Decisions and Recommendations, SCAR and COMNAP.
10. See http://www.ccamlr.org/en/document/publications/schedule-conservation-measures-force-2012/13
11. CCAMLR's initial precautionary catch limits for krill were based on acoustic assessments undertaken during the First International Biomass Expedition (FIBEX).
12. The CEP provides advice and formulates recommendations relevant to the Protocol's implementation. Its specific activities include state-of-environment reporting as well as environmental impact management in the Antarctic Treaty Area. The 1991 ATCM Protocol Final Act (paragraph 7) notes that 'nothing in the Protocol shall derogate from the right and obligations of a Party under the Convention on the Conservation of Antarctic Marine Living Resources, the Convention for the Conservation of Antarctic Seals and the Convention for the Regulation of Whaling'.
13. CCAMLR associations with SCOR include contributions to the planning of and participation by CCAMLR Contracting Parties in the JGOFS and SO-GLOBEC.
14. In particular, a 2008 joint SC-CAMLR and IWC Scientific Committee workshop broadened consideration of Antarctic marine ecosystem model data. Annex 12, Part II at: http://www.ccamlr.org/en/SC-CAMLR-XXVII
15. Paragraph 16.8 at http://www.ccamlr.org/en/CCAMLR-XXVIII
16. Non-Commission Member Contracting Parties are automatically invited to attend Commission and SC-CAMLR meetings as Observers under the respective Rules of Procedure (Rule 30 and Rule 19). Other categories of Observers (NGO and/or IGO) require invitations to meetings, while organizations named in Convention Article XXIII(2) and (3) are eligible to be invited to Commission (Rule 30(d)) and SC-CAMLR meetings (Rule 19(d)).
17. These include the krill yield, general yield and integrated assessment models (KYM, GYM and IAM) where relevant fishery data are combined with computer simulations to account for harvested stock dynamics, model parameter uncertainties and ecosystem considerations. See archive.ccamlr.org/pu/E/e_pubs/am/am-all.pdf
18. Most notably through the 1992 CCAMLR Scheme of International Scientific Observation. Part 10 of CCAMLR's Basic Documents at http://www.ccamlr.org/en/document/publications/basic-documents
19. CCAMLR considers 'ecosystem health' as a safeguard for harvested species, where harvesting does not prejudice the long-term future of dependent species. An 'ecosystem assessment' ensures that the Convention's management requirements are operationally met.
20. Subject to Article II, such species are 'dependent' (e.g. as predators) on harvested species or are ecologically connected to them in some way (e.g. occupy the same geographical space at particular times).
21. Pages 9 and 28 at archive.ccamlr.org/pu/E/e_pubs/am/am-all.pdf
22. CM 21–01 defines a 'new' fishery as a fishery on a species using a particular fishing method in a CCAMLR Statistical Subarea for which: (1) information on distribution, abundance, demography, potential yield and stock identity from comprehensive research/surveys or exploratory fishing have not been submitted to CCAMLR; (2) catch and effort data have never been submitted to CCAMLR; or (3) catch and effort data from the two most recent seasons in which fishing occurred have not been submitted to CCAMLR.
23. To facilitate information gathering and avoid potential overconcentration of exploratory fishing catch or effort, such fisheries are sanctioned in SSRUs or geographically spread areas and are subject to certain conditions including catch limits (see table 1 and figure 1 in CM 41-01).
24. An 'exploratory' fishery is previously defined as 'new' for one year under CM 21-01.
25. This information includes: (1) an evaluation of the distribution, abundance and demography of the target species, culminating in an estimate of the fishery's potential yield; (2) a review the fishery's potential impacts on dependent and related species; and (3) formulation and provision of advice to the Commission by SC-CAMLR to allow appropriate harvest catch levels, fishery effort levels and fishing gear. The recent development of OMPs (Brandão and Butterworth, 2009) for setting allowable toothfish catches at the Prince Edward Islands takes account of appreciable uncertainties about stock status.

SC-CAMLR has encouraged such developments to improve application of precaution and address assessment uncertainty more broadly.

26. WG-EMM assesses Antarctic marine ecosystem 'health' and encourage ways to improve Antarctic ecosystem models in evaluating krill management procedures (Constable 2005). Development of CEMP indices for use in krill fisheries management (de la Mare and Constable, 2000) has also been encouraged.
27. Paragraph 5.72 at: http://www.ccamlr.org/en/ccamlr-xxxi
28. CCAMLR Performance Review Report at http://www.ccamlr.org/en/publications/ccamlr-performance-review
29. The average SJR value (O.12) places *CCAMLR Science* in the 25 top aquatic science journals globally between 1996 and 2010 (http://www.scimagojr.com/journalrank.php)
30. Paragraphs 9.40 to 9.42 at http://www.ccamlr.org/en/SC-CAMLR-XVII
31. Part 9 at http://www.ccamlr.org/en/document/publications/basic-documents
32. See http://www.nmfs.noaa.gov/sfa/domes_fish/FRNotices/CCAMLR_FR_4-9-10.pdf
33. Paragraph 12.9 at http://www.ccamlr.org/en/CCAMLR-XXIX
34. These include the 2004 prosecutions of Hout Bay Fishing and A. Bengis under the USA Lacey Act and Antonio Vidal in 2006; see http://www.dawsons.co.za/july-2013-united-states-vs-bengis-damages-for-illegal-fishing/ and http://www.colto.org/news-archive/
35. At http://www.fao.org/Legal/treaties/037s-e.htm
36. At http://www.ccamlr.org/en/document/publications/scientific-observers-manual-%E2%80%93-2011
37. Paragraphs 4.19 to 4.22 at http://www.ccamlr.org/en/CCAMLR-XXIX and relevant CMs at http://www.ccamlr.org/en/document/publications/schedule-conservation-measures-force-2012/13
38. CCAMLR is contributed with coining the term IUU fishing in 1996, the definition of which followed in the IPOA-IUU. For a full account of IUU fishing in the CCAMLR Area, see Baird (2006).
39. For example, in the case of Korean vessels from the Insung Corporation; Paragraphs 9.17 to 9.18 at http://www.ccamlr.org/en/CCAMLR-XXX
40. This would be illegal fishing if a CP-flagged vessel was involved, although the consent of the flag State would still be required in order to apply enforcement or sanction
41. Notably Spanish domestic legislation on the control of nationals was put in place prior to the adoption of CM 10-08
42. See http://eurlex.europa.eu/LexUriServ/LexUriServ.do?uri=CELEX:32008R1005:EN:NOT
43. Initially named the Standing Committee for Observation and Inspection (SCOI)
44. Paragraph 8.7 at http://www.ccamlr.org/en/CCAMLR-XXX and CM 10-10 at http://www.ccamlr.org/en/document/publications/schedule-conservation-measures-force-2012/13
45. Paragraph 2.42 of Annex 6 at http://www.ccamlr.org/en/CCAMLR-XXIX
46. Page 211 http://www.ccamlr.org/en/document/publications/schedule-conservation-measures-force-2012/13
47. Paragraphs 8.18 to 8.21 at http://www.ccamlr.org/en/CCAMLR-XXIX
48. Table 5 in annex 5 at http://www.ccamlr.org/en/SC-CAMLR-XVII
49. Table 6 at http://www.marinemegafauna.org/wp-content/uploads/2013/02/IUU-Fishing.pdf
50. Paragraph 7.34 in annex 5 at http://www.ccamlr.org/en/SC-CAMLR-XXV
51. At http://www.ccamlr.org/en/publications/ccamlr-performance-review
52. See http://www.ccamlr.org/en/publications/combined-commission-scientific-committee-scic-and-scaf-responses-performance-review
53. As noted, CCAMLR has recognized the need for wider scientific burden-sharing by setting up the General Science Capacity Special Fund.

References

Anonymous (2010) MSC Certification for Antarctic krill fishery. *Fishing News International* August, 10 pp.

Baird, R.J. (2006) *Aspects of Illegal, Unreported and Unregulated Fishing in the Southern Ocean*. Dordrecht, Springer.

Brandão, A. and Butterworth, D.S. (2009) A proposed management procedure for the toothfish (*Dissostichus eleginoides*) resource in the Prince Edward Islands vicinity. *CCAMLR Science* **16**, 33–69.

Butterworth, D.S., Gluckman, G.R., Thomson, R.B., Challis, S., Hiramatsu, K. and Agnew, D.J. (1994) Further computations of the consequences of setting the annual krill catch limit to a fixed fraction of the estimate of krill biomass from a survey. *CCAMLR Science* **1**, 81–106.

Clarke, A.C. and Harris, C.M. (2003) Polar marine ecosystems: Major threats and future change. *Environmental Conservation* **30**(1), 1–25.

Constable, A.J. (2005) A possible framework in which to consider plausible models of the Antarctic marine ecosystem for evaluating krill management procedures. *CCAMLR Science* **12**, 99–115.

Constable, A.J. (2011) Lessons learned from CCAMLR on the implementation of the ecosystem approach to managing fisheries. *Fish and Fisheries* **12**, 138–151.

Constable, A.C. and de la Mare, W.K. (1996) A generalised yield model for evaluating yield and the long-term status of fish stocks under conditions of uncertainty. *CCAMLR Science* **3**, 31–54.

Constable, A.J., de la Mare, W.K., Agnew, D.J., Everson, I. and Miller, D.G.M. (2000) Managing fisheries to conserve the Antarctic marine ecosystem: Practical implementation of the Convention on the Conservation of Antarctic Marine Living Resources. *ICES Journal of Marine Science* **57**, 778–791.

de la Mare, W.K. and Constable, A. (2000) Utilising data from ecosystem monitoring for managing fisheries: Development of statistical summaries of indices arising from the CCAMLR ecosystem monitoring programme. *CCAMLR Science* **7**, 101–117.

Edwards, D.M. and Heap, J.A. (1981) Convention on the Conservation of Antarctic Marine Living Resources. *Polar Record* **20**, 353–362.

Lodge, M., Anderson, D., Løbach, T., Munro, G., Sainsbury, K. and Willock, A. (2007) Recommended best practices for regional fisheries management organisations: Report of an independent panel to develop a model for improved governance by regional fisheries management organisations. Chatham House, Royal Institute of International Affairs, London.

Mangel, M. and Switzer, P.V. (1998) A model at the level of the foraging trip for the indirect effects of krill (Euphausia superba) fisheries on krill predators. *Ecological Modelling* **105**, 235–256.

Miller, D.G.M. (2002) Antarctic krill and ecosystem management: From Seattle to Sienna. *CCAMLR Science* **9**, 175–212.

Miller, D.G.M. (2007) Managing fishing in the Sub-Antarctic. *Papers and Proceedings of the Royal Society of Tasmania* **141**, 121–140.

Miller, D.G.M. (2011) Sustainable management in the Southern Ocean: CCAMLR science. In: *Science Diplomacy: Antarctica, Science, and the Governance of International Spaces* (eds P.A. Berkman, M.A. Lang, D.W.H. Walton and O.R. Young), pp. 103–121. Smithsonian Institution, Washington.

Miller, D.G.M. and Agnew, D.J. (2000) Krill fisheries in the Southern Ocean. In: *Krill: Biology, Ecology and Fisheries* (ed. L. Everson), pp. 300–337. Blackwell Science, Oxford, Fisheries and Aquatic Resources Series 6.

Miller, D.G.M., Slicer, N.M. and Sabourenkov, E.N. (2010) IUU fishing in Antarctic Waters: CCAMLR actions and regulations. In: *Law, Technology and Science for Oceans in Globalisation* (ed. D. Vidas), pp. 175–196. Martinus Nijhoff, Leiden and Boston.

Nicol, S., Foster, J. and Kawaguchi, S. (2011) The fishery for Antarctic krill – recent developments. *Fish and Fisheries* **13**, 30–40.

Pitcher, T.J., Kalikoski, D., Short, K., Varkey, D. and Pramod, G. (2009) An evaluation of progress in implementing ecosystem-based management of fisheries in 33 countries. *Marine Policy* **33**, 223–232.

Reid, K. (2007) Monitoring and management in the Antarctic – making the link between science and policy. *Antarctic Science* **19**, 267–270.

Reid, K. (2011) Conserving Antarctica from the bottom up: Implementing UN General Assembly Resolution 61/105 in the Commission for the Conservation of Antarctic Marine Living Resources. *Ocean Yearbook* **25**, 131–139.

Sabourenkov, E.N. and Miller, D.G.M. (2004) The management of transboundary stocks of Toothfish, *Dissostichus* spp., under the Convention on the Conservation of Antarctic Marine Living Resources. In: *Management of Shared Fish Stocks* (eds A.I.L. Payne, C.M. O'Brien and S.I. Rogers), pp. 68–94. Blackwell Science, Oxford.

Turner, J., Jabour, J. and Miller, D. (2008) Consensus or not Consensus: That is the CCAMLR question. *Ocean Yearbook* **22**, 117–158.

Willock, A. and Lack, M. (2006) Follow the Leader: Learning from Experience and Best Practice in Regional Fisheries Management Organizations. Available at http://wwf.panda.org/about_our_earth/blue_planet/publications/?69480/Follow-the-leader-Learning-from-experience-and-best-practice-in-regional-fisheries-management-organizations (accessed February 2014).

Chapter 19

Implementation of the Ecosystem Approach to Fisheries in the Benguela Current LME area

J. Augustyn[1,2], S. Petersen[3], L. Shannon[4] and H. Hamukuaya[5]

[1]Department of Agriculture, Forestry and Fisheries, Cape Town, South Africa
[2]IUCN Commission on Ecosystem Management (IUCN/CEM/FEG), Gland, Switzerland
[3]WWF Marine Programme, WWF-South Africa
[4]Marine Research Institute and Department of Biological Sciences, University of Cape Town, South Africa
[5]Benguela Current Commission (BCC), Swakopmund, Namibia

Abstract: We explore the successes and challenges of implementing EAF and governance issues at the regional and national levels. Authorities of the Benguela Current region of South Africa, Namibia and Angola are exploring the implementation of an Ecosystem Approach to Fisheries (EAF). Regionally EAF is viewed as a means to integrate fisheries and conservation management. A feasibility study under the Benguela Current Large Marine Ecosystem (LME) programme allowed development of systems and structures to translate the best-available science into management advice. The fisheries institutions in the three countries have begun to implement EAF with varying degrees of success and political buy-in. New initiatives at the country and regional level are allowing EAF implementation and tracking, integrating the underlying science with international initiatives. Under the Benguela Current Commission, a portfolio of EAF projects investigates and informs a more holistic and integrated approach to regional fisheries management and conservation. Institutional developments, mostly in South Africa, are also allowing a more comprehensive and integrated approach.

Keywords: Benguela Current; ecosystem approach to fisheries; tracking tool for EAF; integrated approach to regional management and conservation

Introduction

The Benguela Current Large Marine Ecosystem (BCLME) is a physically and ecologically complex eastern boundary upwelling system. The three developing countries of South Africa, Namibia and Angola exploit the fisheries and conduct marine mining and oil and gas extraction on the coast and offshore. The ecosystem impacts resulting from these activities are considerable and have been addressed in a regional context since the mid-1990s through various regional initiatives.

This chapter focuses on how the region and the individual countries are dealing with fisheries governance and conservation interactions in the context of an Ecosystem Approach to Fisheries (EAF), defined according to FAO (2003): 'The ecosystem approach to fisheries strives to balance diverse societal objectives, by taking into account the knowledge and uncertainties about biotic, abiotic and human components of ecosystems and their interactions and applying an integrated approach to fisheries within ecologically meaningful boundaries'. The implementation of an EAF has, in a sense, been considered by managers in the region as a holistic means of integrating many marine environmental and fisheries governance issues, sometimes in the absence (or as a result of weak implementation) of institutional governance structures. Much of what is described in this chapter with regard to governance issues has not yet

Governance of Marine Fisheries and Biodiversity Conservation: Interaction and Coevolution, First Edition.
Edited by Serge M. Garcia, Jake Rice and Anthony Charles.

been published in formal literature, but reflects the collective experiences of the authors.

In the region, an EAF has been considered to be implemented when the following objectives have been met (Paterson and Peterson, 2010; and see 'Institutional structures, science and management…'):

(1) the managing authority has a good understanding of the ecosystem impacts of fisheries including target, non-target and general ecosystem impacts;
(2) ecosystem impacts of fisheries are included into management;
(3) the social wellbeing of dependent fishing communities is accounted for in management;
(4) the economic wellbeing of the fishing industry is maintained;
(5) the managing authority has transparent and participatory management structures that ensure good communication and information sharing locally and regionally;
(6) management plans incorporate EAF considerations;
(7) compliance to regulations reduces ecosystem impacts of fisheries;
(8) sufficient capacity, skills, equipment and funding exist to support the implementation of an EAF;
(9) good data and procedures exist to support EAF implementation; and
(10) external impacts of fisheries are addressed (e.g. the effect of other sectors, other industries, climate change, etc.).

The above have been used as criteria with which to measure the success of the implementation of EAF in the region.

This chapter reviews the historical development of EAF in the region, from early national efforts through the development of ecosystem-based regional programmes (Benguela Environment Fisheries and Training or BENEFIT and BCLME) to the establishment and implementation of the Benguela Current Commission, a regional advisory body. It reviews the current implementation of EAF (citing examples of progress in science and management) and associated institutional structures and considers the successes, challenges and opportunities of implementing EAF to address integration of fisheries and marine conservation, and relevant governance issues at the regional and national levels. Obstacles and steps that need to be taken to improve governance are identified. EAF has, to a large extent, been used as an integrative process for conservation and fisheries management in the region and particularly in South Africa, largely because of the history of its development and the absence of other effective fora or institutions that are directly addressing it.

A pivotal challenge for the successful implementation of an EAF is that of translating the best-available science into management action as well as the associated systems and structures to facilitate this transition. This has been a focal area of work within the Benguela Current region. Management frameworks are reviewed, including the development of an EAF tracking tool to identify, prioritize and provide a road map for EAF implementation with stakeholder buy-in and support. On the ground, personnel including fishery managers, compliance officers, fisheries observers and fishers have been equipped with the skills and understanding of an EAF through the development of the WWF Responsible Fisheries Training (Okes et al., 2012) which facilitates the sharing of information to build a common understanding of an EAF.

Historical perspective

Although regional issues related to fisheries and ecosystem governance are rooted in international initiatives and agreements culminating in the WSSD JPOI (World Summit for Sustainable Development, Johannesburg Plan of Implementation) of 2002, the region (particularly South Africa and Namibia) was an early leader in managing fisheries based on improved understanding of a complex ecosystem structure and interactions.

The Benguela Ecology Programme (BEP, 1981–2005) was considered a highly successful and international leading programme of its kind (Moloney et al., 2004). Multidisciplinary and multi-institutional, it integrated marine biology, ecology, fisheries management and oceanography in advancing knowledge about the functioning of the Benguela ecosystem. Four major volumes (Payne et al., 1987, 1992; Pillar et al., 1998; Shannon et al., 2004a) summarize this work.

After independence in 1990, Namibia established its own fisheries and environmental Ministries and research institution, the National Marine and Information Research Centre (NATMIRC). Namibia's former Minister of Fisheries and Marine Resources, the late Abram Iyambo, played a role in mainstreaming EAF in important international instruments such as the Reykjavik Declaration on Responsible Fisheries in Marine Ecosystems.

In Angola, the civil war was winding down by the mid-1990s and the country was establishing new regional partnerships to develop its fisheries science capacity. Norway and Germany were already supporting science development in the region. With their support, the Benguela Environment Fisheries and Training (BENEFIT) programme made it possible to take the ecosystem-based approach to a regional level. In 2007 the establishment of the Global Environmental Facility funded the Benguela Current Large Marine Ecosystem (BCLME) programme, laying the basis for EAF. The BCLME's three focus areas were the ecosystem impacts of human activities on fisheries, environmental variability impacting on the ecosystem (the Benguela is one of the most variable marine ecosystems in the world; Shannon et al., 2006b) and ecosystem health (mining and pollution effects on biodiversity).

A key BCLME project investigated the feasibility of EAF management in the region, examining the existing issues, problems and needs related to EAF and developing different management options to achieve sustainable management of resources at an ecosystem level (Cochrane et al., 2009). It laid the formal foundation for EAF in the region under the successor of the BCLME, the Benguela Current Commission (BCC), and allowed the three countries to further their scientific and management support bases for EAF and begin its implementation.

Implementation of EAF at the regional level

Institutional structures, science and management, capacity-building, compliance and incentives

A report commissioned by the BCC (Staples, 2010) compared current institutional arrangements in all three countries with best practice, concluding that institutional arrangements are faring well apart from two major aspects: (1) institutional structure and (2) the legislation, policies and planning category. With the exception of South Africa, there are no dedicated fishery management 'units' or staff. There are no fishery-specific EAF Management Plans that set out operational objectives for each fishery, the strategies of how the objectives will be met or the measures needed to evaluate their performance. A third gap (not mentioned by the report) is the absence of an overarching supporting structure that would facilitate the merging of scientific information and the balancing of management objectives of fisheries (and other resource users) and conservation.

According to the report, these gaps need to be filled at the national level and broader integrated policies developed such as an 'Oceans Policy' (which has now been addressed by South Africa; South African Government Gazette, 2012a). It also recommended the establishment of a structure that facilitates EAF implementation through joint private–government structural arrangements and that the BCC should function as a Regional Fisheries Management Organization, rather than being just an advisory body.

Bringing the fishing industry on board is an important aspect of rolling out an EAF. In South Africa, a Responsible Fisheries Alliance (RFA) between the Worldwide Fund for Nature (WWF) and four major fishing companies in collaboration with other NGOs and government has been a successful initiative. Key results from a recent review of the RFA include six EAF-related research projects completed, over 1000 fishers trained in the Responsible Fisheries Training course and one regulation influenced (WWF, 2012).

With regard to capacity building for EAF, fishers, fisheries observers and compliance officers must have the adequate and appropriate skills to implement changes such as those for the management of bycatch, maintenance of food webs and protection of vulnerable marine habitats. This is critical to translating policy and science into effective action.

Voluntary compliance and incentives have also been important. We believe that the limited success of previous compliance initiatives was the result of an over-reliance on the regulatory system. Given that an EAF is likely to increase the complexity of regulations and management requirements through the additional consideration of broader ecosystem, social and economic issues, it is important to ensure the regulatory framework is adequately supported by incentives and voluntary compliance mechanisms. Locally, three initiatives have incentivized implementation, namely the Marine Stewardship Council (MSC) certification of the South African hake trawl fishery, WWF's Southern African Sustainable Seafood Initiative (WWF-SASSI) and the development of the Responsible Fisheries Alliance (RFA). These initiatives have shifted the burden of responsibility for the management of responsible fisheries from government only to one of responsibility shared by industry and government. The certification of the South African hake trawl fishery has increased ecosystem research and bycatch mitigation while SASSI, through local seafood supply chains, has facilitated market-driven incentives to increase compliance with regulations (e.g. implementation of tori lines to decrease the bycatch of endangered seabirds in the hake trawl fishery). The RFA has provided a dynamic mechanism to undertake key EAF-related research and lobbied government and the broader fishing industry for improved EAF management while voluntarily implementing agreed-upon recommendations to improve the long-term sustainability of the fishery. Thus, engaging rational NGOs has proven useful in facilitating the implementation of an EAF via their unique position at the interface between government, civil society and business.

Capacity building is important. If fishers, fisheries observers or compliance officers do not understand how and why regulations or management measures are in place, there will be poor implementation. The WWF Responsible Fisheries Training (Okes et al., 2012) was developed to empower fishers to understand the environment within which they work, concepts that underpin EAF and processes behind management decisions. It equips fisheries observers with appropriate skills to collect verifiable data on which management decisions are based and inspire them to raise awareness and understanding among fishers. Compliance officers are equipped with an understanding of fisheries regulations and their implementation. It has been highly successful in equipping fishers with the skills and expertise required to implement EAF measures at sea (for example the implementation of bycatch mitigation).

How science feeds into management in the BCC region

The BCC was established in 2007 as an outcome of the BCLME Programme. It is currently an advisory body and does not enforce regulation at the regional level.

Figure 19.1 depicts an organogram of the BCC, showing how scientific advice feeds into management at the regional level. The Ecosystem Advisory Committee (EAC) does not at this stage recommend TACs. Its main role has been to generate and approve scientific projects relevant to management, including for EAF.

International collaboration has enhanced the BCC scientific programme. The BCLME SAP-Imp (Strategic Action Plan Implementation) project continues to support it, as does Norwegian and Icelandic funding (about US$ 8 million) to support the science and capacity building aspects, respectively. The FAO Nansen EAF programme has funded projects on institutional arrangements supporting EAF, integrating the human dimension of EAF into fisheries management; incorporation of stakeholders' knowledge in data collection; and analysis, auditing and tracking of EAF in management. An EU-funded partnership supports projects on the development of ecologically sustainable fisheries practices in the BCLME, promoting EAF through adaptation of state-of-art assessment methods, Marine Protected Areas (MPAs) planning tools and validation or modification of current assessment practices based on spatially explicit analyses.

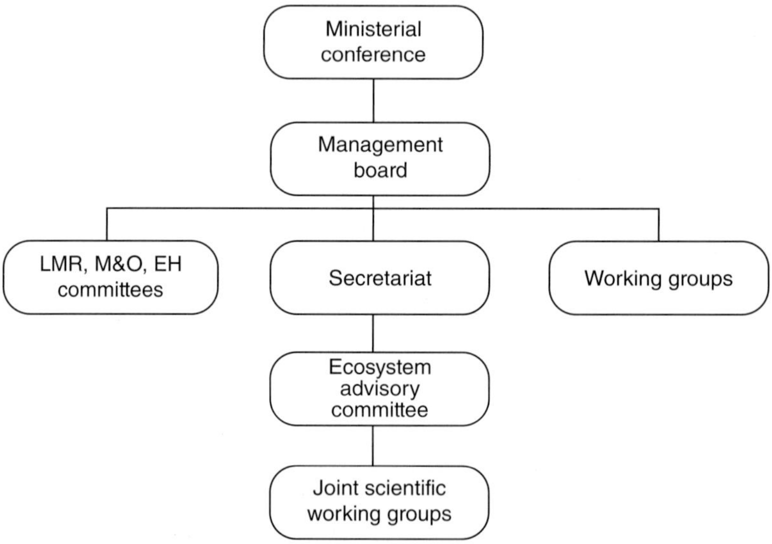

Figure 19.1 Organogram of the Benguela Current Commission, showing how scientific advice feeds into management at the regional level.

The science base for EAF in South Africa

Implementation of EAF requires a solid scientific basis from which management measures can be considered and implemented (Shannon et al., 2010a). Progress has been made in addressing several key issues (Shannon et al., 2006a, 2010a) but the challenge remains to diffuse scientific information through fisheries management frameworks. The EAF research in the Benguela region described below focuses on work undertaken subsequent to earlier overviews (Shannon et al. 2004b, 2006a, 2010a).

Trophic models of the Benguela have been used to assess changes in ecosystem structure and function under fishing and environmental drivers (Shannon et al., 2003, 2008; Watermeyer et al., 2008a,b) and are being considered in support of fisheries management (Shannon et al., 2010a; Travers et al., 2010). Osman (2010) constructed an updated trophic model of the Southern Benguela for the period 2004–2008 and compared a set of model-derived indicators across several periods.

Trophic models have been consulted for management of low trophic level (LTL) fisheries across the world (Smith et al., 2011). This global study found that fishing LTL species at maximum sustainable yield levels may have significant impacts on other components of the ecosystem, particularly when the LTL species constituted a large proportion of overall fish biomass and/or when the species was well connected in the food web. Model results suggested that adopting exploitation rates of half of MSY levels led to significantly lower impacts on other ecosystem components while still maintaining catches at 80% of MSY. This may be considered for adoption in the BCLME.

The needs of top predators are being considered. Using a global database including African Penguin and Cape Gannet (preying on anchovy and sardine), Cury et al. (2011) showed that seabird breeding success declines dramatically and becomes significantly more variable when prey stocks drop below one-third of their maximum observed abundances. This global pattern has been proposed as an indicator of prey needed to sustain productivity of seabirds on long timescales and is consistent with Smith et al.'s (2011) proposed LTL reference level.

A system model has been developed to assess the relative importance of pressures imposed on different breeding colonies of the endangered African penguin (*Spheniscus demersus*), aiming to provide a tool for developing the most strategic approach to curb the severe decline in African penguins (Cecchini, 2012; Ludynia et al., in review). The approach is unique in inclusion of expert knowledge and guidance from the outset, and throughout the modelling process.

The Ma-Re BASICS project (Marine Research in the Benguela and Agulhas Systems for supporting Interdisciplinary Climate-Change Science) provides a research framework for marine data collection and integration into ocean models, projecting physical and biological conditions on seasonal scales under a changing climate. Indicator frameworks are being developed for ocean forecasting and fisheries management.

Indicators of changes in the physical environment of the Southern Benguela shelf and inshore regions were matched to biological indicators of the west coast rock lobster and bank cormorants feeding on lobsters as an early warning system for long-term ecosystem-scale change (Blamey et al., 2012). Temporal shifts have been identified in demersal fish populations off western South Africa in the early–mid-1990s and possibly mid-2000s (Atkinson, 2010; Atkinson et al., 2012). Atkinson et al. (2011) sampled benthic fauna along the South African and Namibian coasts, concluding that trawling intensity at least partly accounts for the noticeable differences in benthic epifauna in lightly versus more heavily trawled areas.

In the global context, the Southern Benguela case study is important for examining the usefulness of simple, easily estimated and understood indicators of fishing impacts (Shannon et al., 2010b). The IndiSeas project (www.IndiSeas.org) focuses on assessing fishing impacts in marine ecosystems through comparative analyses of ecosystem indicators in support of fisheries management decisions (Shin and Shannon, 2010; Shin et al., 2010, 2012). In support of the European Commission Marine Strategy Framework Directive (MSFD), the Convention on Biological Diversity (CBD) and Aichi Targets (calling for effective ecosystem-based fisheries management by 2020), IndiSeas has been expanded to explore biodiversity, environmental and human dimension indicators for EAF (Shin et al., 2012).

Key immediate research challenges

Despite progress with respect to a variety of indicators for EAF, including environmental knowledge, actual management processes require further effort. Bridging the gap between classical single-species approaches and broader ecosystem approaches necessitates respectful collaboration across and within multiple disciplines (Shannon et al., 2010a). Stakeholder relationships and buy-in has been strengthened and collaboration now spans multiple disciplines. However, increased focus on and strengthening of the human dimension and transdisciplinary approaches is also required. The frame-based modelling approach has been proposed to investigate how socio-ecological systems change state over time (Starfield and Jarre, 2011).

To address several ecosystem-level issues (e.g. Shannon et al., 2006a), spatial aspects require further research effort (e.g. in the South African sardine fishery; Van der Lingen, 2011).

Tracking the implementation of an ecosystem approach to fisheries management in the region

Ecological Risk Assessments: A mechanism to review EAF implementation

Insufficient practical guidance currently exists for fisheries managers to implement an EAF. Based on the FAO code of conduct for responsible fisheries (FAO, 1995) and the Australian ecological sustainable framework (Fletcher et al., 2002), the BCLME ran a series of locally adapted Ecological Risk Assessments (ERA; Petersen et al., 2010) workshops to test the feasibility of implementing the approach as well as to investigate the use of ERAs as a tool to facilitate implementation. ERAs provide a structured and transparent process which helps to develop a shared vision among stakeholders and clarifies what is required to implement an EAF. An ERA identifies issues and risks facing a fishery, guided by a holistic framework addressing ecological, social, economic and governance issues. Secondly, it prioritizes these issues using the product of consequence (of the issue for the ecosystem or the community) and likelihood (of a risk occurring). Thirdly, it seeks to gain consensus on actions required to address priority issues. Each ERA provides a snapshot of the current state of a fishery relative to overarching ecosystem objectives. To address the need to track and stimulate EAF implementation a tracking tool was developed (Paterson and Petersen, 2010), reviewing progress against the ten objectives listed in the introduction.

The tool divides general objectives into more specific operational objectives. For each operational objective, a seven-step implementation process has been developed to pinpoint the status of a fishery in EAF terms. Each step reflects the degree of progress towards achieving the objective, allowing for consistency and transparency in decision-making. Identification of the current step allows the status quo for each sector to be determined relatively simply and facilitates discussion and identification of concrete actions needed to move ahead. The EAF tracking tool is now widely used for all major fishing sectors in the region (Paterson and Petersen, 2010; Paterson et al. 2010; Figure 19.2).

Conclusions and lessons learnt through the tracking tool approach

Stakeholder participation is critical to the successful implementation of an EAF. Complexity creates confusion, frustration and reduces the chances of success. A structured approach provides a platform for views to be aired, broadens perspectives and improves understanding of the issues. It helps to simplify the complexities inherent to EAF and supports the building of a common understanding, implementation requirements and implications and benefits for stakeholders. All views must be represented and no group or individual is allowed to dominate.

The tracking tool applies a generic set of objectives across fisheries. A disadvantage of this approach is the potential loss of fishery-specific detail, highlighting the importance of conducting a fishery-specific ERA as a first step, allowing for identification of fishery-specific issues and ensuring that they are addressed in the generic review process. However, the tool is simply a means to structure and facilitate discussion. Its true value emanates from the detailed discussion around each fishery. The advantage of a generic approach is that it allows for comparison, interrogation and reporting at any level. For instance, operational managers can track progress of management actions in a participatory and transparent manner to develop a work plan to address issues. A middle manager can use the tool to compare progress at a sector or fishery level or compare progress between fisheries. Senior managers can use it to track EAF implementation between fisheries, compare implementation of various objectives or investigate progress in overarching issues that could not be tackled by a single sector. Wise use of a limited resource is thus enhanced through improved information, and gaps in progress identified or rewarded. At a policy level, including reporting on WSSD commitments, EAF implementation can be tracked and

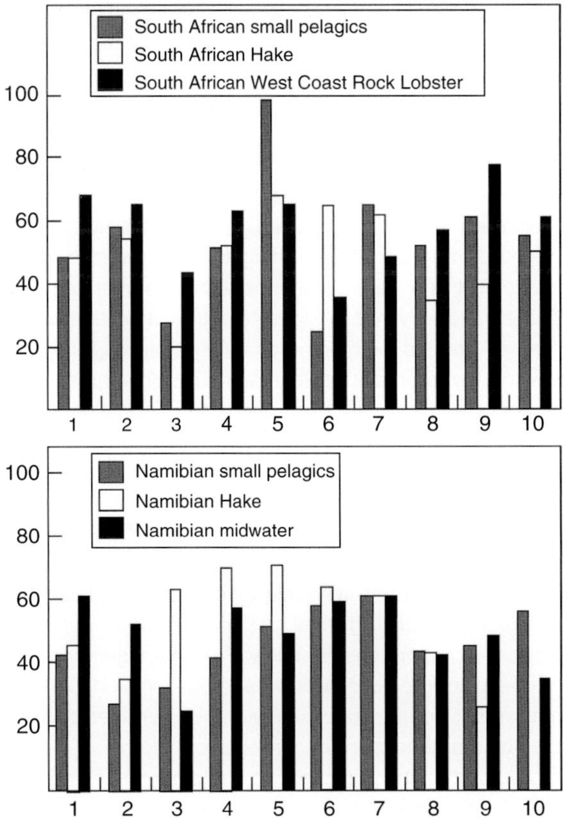

Figure 19.2 EAF implementation shown as percentage and summarized by 10 objectives of an EAF for three South African and Namibian fisheries (note: objective 10 was not discussed at the Namibian Hake ERA review workshop due to time constraints). (Source: Adapted from Paterson and Petersen 2010.)

reported using simple effective diagrams without placing additional burden on managers.

Institutional structures and interactions between biodiversity conservation and fisheries management in the marine environment

At the national level

At the national level, the institutional arrangements and levels of interaction between environmental and fisheries management agencies vary significantly.

South Africa

Until 2010, the Department of Environmental Affairs and Tourism (DEAT) was responsible for both marine fisheries and conservation governance, and issues of integration and/or conflict could be relatively easily dealt with. A further step forward was the establishment of a Chief Directorate for Marine Resource Management in 2007 with the responsibility for EAF included in its mandate. Integration has become more challenging since 2010, since the environmental management functions, including the management of MPAs, top predators and non-consumptive marine activities such as boat-based whale-watching, shark cage diving and SCUBA diving (now in Environmental Affairs, DEA) were split from the fisheries management functions in Agriculture, Forestry and Fisheries (DAFF). Stakeholder interactions are now more challenging, since stakeholders find the difference in emphasis by the two departments to be confusing. It has taken three years to reach agreement at the political level on what the role of each department should be with respect to the management and respective roles in MPAs. In the meantime, it has resulted in some conflicting management decisions, especially with regard to fisheries activities in MPAs.

On the positive side, DAFF has enhanced environmental/fisheries governance through implementation of Ecosystem Risk Assessments (ERAs) and regular reviews involving the key stakeholders. As a result, many regulations and permit conditions have been applied in several fisheries, notably in the demersal and pelagic sectors, and these are revised annually to take account of any new developments. DEA has recently recognized that there was a need for it to clearly define its role with regard to Ocean Governance and has published a Green Paper for public comment. NGOs such as WWF have played an important role in assisting the implementation of EAF and environmental initiatives.

In terms of ICZM, South Africa has promulgated the Coastal Management Act and has ratified both the Nairobi and Abidjan Conventions, regional framework agreements for African east and west coast marine environmental management. South Africa has an active EAF Scientific Working Group (EAF-SWG) which communicates with sectoral SWGs, and annual TAC/TAE recommendations include advice on ecosystem issues. There is also interaction between fisheries scientists and environmental scientists on most SWGs. This is a model that could be considered in regional bodies like the BCC. Consultation with industry scientists and managers also happens formally at the SWG and MWG (Management Working Group) level.

Policy development has partly been addressed in South Africa through a Small-Scale Fisheries Policy (South African Government Gazette, 2012b), and general and sectoral policies that emphasize sustainable solutions to transformation and restructuring the fishing industry are being revised. South Africa has begun developing management plans, including EAF issues in the various sectors.

The development of a new small-scale fisheries policy offers opportunities to address some important socio-economic and governance issues if wisely implemented and properly supported, but also presents dangers if it leads to expansion of fishing effort through political pressure on already stressed coastal resources. An idealistic vision of engendering community stewardship is a major feature of the new policy, but this is threatened in some areas by out-of-control poaching driven by crime syndicates. The re-allocation of fishing rights between 2013 and 2020 could also allow for a more equitable solution in terms of access to fisheries which could reduce political pressure. In MPA management, a shift to sustainable multiple use rather than exclusive conservation may be the only way to maintain their function in future. The two Departments involved (DAFF and DEA) need to formalize integration of decision-making to a greater extent than is presently the case. There is good use of market mechanisms to support EAF implementation: South Africa has a MSC-certified hake fishery and the SASSI programme described above.

Lack of compliance with fisheries regulations is the biggest challenge; organized crime and a culture of impunity are threatening to destroy some inshore fisheries, especially the abalone and rock lobster sectors.

Namibia

In Namibia, the Fisheries and Environmental Ministries have been separated since independence in 1990. The Ministry of Fisheries and Marine Resources has in principle been a strong proponent of ecosystem-based management. However, ecosystem concerns have taken a back seat to exploration and mining development in Namibia. This has been illustrated by insufficient consultation of either Namibian fisheries stakeholders or the BCC when mining exploration applications were recently approved by the Namibian Ministry of Mines and Energy. Socio-economic considerations sometimes override scientific advice on TACs, since final decisions on TACs are taken in consultation with the Fisheries Advisory Council (composed of government managers, industry representative, Labour Unions and NGOs).

A better balance is needed between sustainable use in fisheries and the impact of marine mining activities; more effective, expanded protected areas could contribute more to long-term sustainability. Marine diamond mining began there decades ago and there has been a relatively effective co-management with the fishing industry, particularly the lobster industry. Recently a huge deposit of marine phosphate was discovered in central Namibia with experimental licenses granted before the completion and approval of environmental impact assessment. The government has set up an inter-ministerial committee to evaluate and make recommendations on the issues pertaining to cross-sectoral environmental management.

In Namibia, the EAF WG is largely ineffective and needs to be fully incorporated into the structure of the Ministry to be effective. Namibia conducts periodic reviews of its fisheries with wide stakeholder participation; industry are consulted and play a formal role in the annual TAC setting process of major fisheries, but so far a management plan has been adopted only for hake and one initiated for horse mackerel. Scientists often include EAF considerations in their recommendations of conservation and management measures, but there is a lack of implementation because there is no policy backing.

Namibia has declared one of the largest coastal MPAs in sub-Sahara Africa that include all its islands, but MPAs generally need to be expanded to be in line with the WSSD targets and be supported with a monitoring and evaluation programme.

Angola

Angola has a national EAF committee that consists of small-scale fishers, industrial fisheries and the oil industry, coordinated by the national fisheries agency INIP. Some ecosystem advice is occasionally given (e.g. on turtle conservation) based on results from oil-industry-financed projects and fisheries data are collected from the small-scale and commercial sectors, but the effectiveness of this committee is still to be realized.

The oil industry continues to be the backbone of the national economy and intense oil exploration and production now impact almost the entire coast, at a time when the abundance of commercially important pelagic species like horse mackerel are at historically low levels. A large refinery is being built in Lobito. In some areas like Cabinda, there is conflict between the oil industry and local fishers because of limitations and exclusions placed on fishing areas, and because of pollution incidents that occur from time to time. A proportion of oil revenue is reportedly set aside for development and alternate livelihoods for coastal communities who have for centuries been overwhelmingly dependent on fisheries. A small-scale aquaculture project for artisanal fishers in Cabinda is being funded by the Chevron Texaco Oil Company (K. Kilongo, National Fisheries Agency, personal communication, 2012).

Management plans for EAF implementation exist for the demersal, pelagics and shrimps fisheries, although these have not been fully implemented. Further, fisheries management measures with EAF relevance and the main tools for ecosystem-based management include a closed season for the protection of recruitment (demersal, pelagics, shrimps), closed areas and a minimum distance for industrialized fishing off the coast. There are also gear regulations for each sector (e.g. midwater trawls are banned for horse mackerel) and ban of beach seines for small-scale fisheries. An effort has been made to reduce the number of vessels targeting pelagic fisheries. Although exclusion devices in the shrimp fisheries and bycatch regulations are in place, they are not yet included in permit conditions. So far there are no declared MPAs or regulations broadly relating to marine biodiversity (N. Luyeye, National Fisheries Agency, personal communication, 2012), although some endangered species, mainly cetacean species and turtles, are protected and large-scale fishing is prohibited in bays. TACs were recently introduced to limit shark catches.

In terms of governance structures for integration of conservation and fisheries, there exists a National Department of Fisheries and Ecosystem Protection (DNPPE) and there is awareness of EAF as a result of the regional initiatives, but it is not clear how these issues are integrated with environmental management and implementation is weak (K. Kilongo, personal communication, 2012).

All countries

Compliance with fisheries regulations have been a general problem. The introduction of vessel monitoring systems (VMS) in Namibia in 2004 and more recently in Angola has been negatively affected by ongoing technical problems and have not fulfilled their intended purpose. Using patrol vessels and aircraft in Namibia and South Africa has been successful in combating IUU, but is a hugely expensive undertaking. National institutions are poorly capacitated: the high rate of staff turnover undermines effective capacity of the research institutions in all three countries.

Staff retention is often a challenge when the demand for university graduates in natural sciences is at a peak, due in part to the ever-growing mining and oil and gas industry in the region, especially in Namibia and Angola. Due to competing national priorities, research and training budgets are often under pressure. There are shortages of skills in critical areas, especially in stock assessment, ecosystem modelling, fisheries economics and socio-economics and legislative development and biodiversity planning at the national level.

Institutionally the fisheries departments are largely structured on an outdated model that does not consider EAF. As a result, there are no dedicated EAF managers and EAF is still driven by scientists rather than managers. Full implementation of national-level multi-stakeholder EAF Steering Committees are required to set EAF priorities, advised by EAF SWG. In order to maintain quality and improve the objectivity of scientific inputs into management, scientific institutions related to fisheries and conservation may need to be more independent and better integrated with country science systems. At the same time scientists need to make greater efforts to communicate ecosystem issues more effectively to politicians. In all three countries, involvement of a wider range of stakeholders is needed with respect to the management of fisheries.

Political aspects affecting interactions between conservation and fisheries management at the national level

Although in principle support for the BCC and for the implementation of ecosystem considerations in management has been expressed at the political level in the Benguela region, political will to implement EAF measures at national level is not always strong. For example, in South Africa the split of fisheries and environmental departments have been detrimental to the full implementation of an EAF, and conflicting objectives (particularly with regard to MPAs) have sometimes not been resolved at a high level. Populist politics and bad strategic management sometimes leads to the undermining of good governance. Political leaders are poorly informed and do not adequately engage on ecosystem and sustainability issues with technical managers, and vice versa. A lack of engagement with regional fisheries bodies such as the BCC may lead to transboundary conflicts if not addressed.

In Namibia, an inter-ministerial forum among ministries responsible for ocean governance is often not working effectively. This has been particularly evident in the granting of marine experimental mining licenses, as mentioned above.

In Angola, the government supports small-scale fisheries through investment in infrastructure, gear, vessels, extensive services and value chain marketing but this does not match the scale of resources available, or take into account their biological functioning. Currently, some fisheries (including the most important local fisheries for sardinella and horse mackerel) have been at dangerously low levels as a result of overfishing. It is necessary to limit effort in most fisheries by establishing a centralized register for all fishing vessels and for new effort in fisheries to be managed.

At the regional level

In Southern Africa, there are two major initiatives contributing to improved integration of environmental and fisheries management. We have described the evolution of the three-country BCC on the West Coast. There is also a nine-country Agulhas Somali Current Large Marine Ecosystem (ASCLME) programme, a South West Indian Ocean Fisheries Programme (SWIOFP) and a Project on Land-based Activities in the Western Indian Ocean (WIO-LaB) on the east coast. These Global Environmental Facility (GEF) -backed LME programmes have used an approach of carrying out Transboundary Diagnostic Analyses (TDAs) to determine the causes of ecosystem degradation, followed by the development of a Strategic Action Programme (SAP) to address these causes. All have identified EAF as a key strategy to address environmental governance issues related to fisheries, hence the emphasis on EAF at the regional level. The recently signed BCC Convention strongly emphasizes it. These regional EAF initiatives were also envisaged to be implemented at country level, since most day-to-day governance takes place nationally and is obviously not always related to transboundary issues. In this paper, the focus has been on the BCC since it is further advanced and also pioneered a feasibility study on the implementation of EAF.

As discussed in 'Historical perspective', there has been strong international support for the development of science and BCC is benefiting from the improved capacity to carry out biological and environmental surveys and assessments. Activities will need to be expanded to include RFMO-like processes to implement binding regional decisions such as setting transboundary TACs informed by regional assessments and adopting relevant conservation measures. The foundation for such regional assessments has been laid by annual transboundary surveys which have been conducted for the last decade. These surveys have included intercalibration between research vessels in order to merge longer-term datasets useful for stock assessment from South Africa and Namibia. The work includes resource surveys on the abundance and distribution of commercially important shared fish stocks, notably hake and horse mackerel and for stock assessment. Under both BENEFIT and BCLME, projects focused especially on the transboundary resources, ecosystem variability and developing tools useful for management in the region. A full list of projects and outcomes is available on the BCC website (http://www.benguelacc.org).

Some management measures to address important ecosystem issues have been applied both regionally and nationally, for example management of species caught as bycatch in fisheries (mainly in South Africa and Namibia) and seabird bycatch mitigation in longline and trawl fisheries (all three countries). South Africa has developed a National Plan of Action (NPOA) for Seabirds which includes implementation of bycatch regulations in demersal fisheries, sharks and, to a limited extent, turtles, management of beach-seining for small-scale fisheries and mandatory gear exclusion devices in the shrimp fisheries to reduce bycatch and discards (Angola and SA). A pioneering measure related to food-web considerations is implementation of closed areas for seabird protection. South Africa is exploring, by means of a feasibility study, closed areas around some penguin breeding islands to enhance food availability and is attempting to include top predators in Operational Management Procedures in the small pelagics fishery (Robinson et al., 2010). Implications for future trends in penguin numbers at island colonies of reducing future TAC for sardine and anchovy were considered.

Elsewhere in Africa UNEP Regional Seas Organizations play an important role in conservation governance, for example the Nairobi Convention on the East Coast of Africa. The Convention for Co-operation in the Protection and Development of the Marine and Coastal Environment of the West and Central African Region (Abidjan Convention) is the appropriate UNEP body for the Benguela region and South Africa is a member (Angola and Namibia are not), but it has not been very active in the region despite having been in effect since 1981. Some protocols that have regional application potential have been developed, such as an oil spill contingency plan. Environmental governance issues which could potentially be beneficial are beginning to be considered, but there have not been any real benefits for the region to date. A Memorandum of Cooperation between the BCC and the Abidjan Convention has been developed and is currently being reviewed by the two organizations. The purpose and scope of this Memorandum is to promote the cooperation and complementarity of operations of the Contracting Parties of the Convention and the BCC. The Conference agreed to initiate the development of a MPA Protocol for the effective implementation of these articles in line with the objectives of the Convention on Biological Diversity. While these processes have yet to bear fruit, there appears to be political support for more action.

There is good regional collaboration across fisheries institutions, such as DAFF in South Africa, NATMIRC in Namibia and INIP in Angola, and this extends to BCC and its high-seas counterpart SEAFO (South East Atlantic Fisheries Organization). BCC and SEAFO accommodate each other during key annual meetings. SEAFO itself has played an important role in the high-seas area and made ecosystem-based recommendations on the banning of deep-water shark catches and gillnets. It has also adopted conservation measures on sharks, establishing lists of IUU vessels, limiting trans-shipments at sea, reducing incidental catches of seabirds and turtles, management of deep-water habitats and ecosystems, retrieval of lost gear, setting TACs for deep-sea species in the SEAFO Convention Area and introducing demersal trawling regulations and Port State Measures.

There have also been challenges at the regional level and in all three countries: the ecosystem impacts of marine mining and oil and gas exploration have been of great concern, but it has been a weakness of BCC that there has not yet been a full buy-in and transparency by the departments responsible in each country for management of these activities at the regional level.

While science, observer programmes and MCS are supported by international partners, the BCC is still underfunded. Several environmental projects that could have provided important information on the functioning of the ecosystem and sub-ecosystems have had to be shelved. These include modelling work on the predictability of environmental events such as the Benguela Niños and the intrusion of low dissolved oxygenated water masses into BCLME from the tropics, which have negative impacts on the living marine resources of the region. Long-term sustainability of funding of the BCC Science Plan is not ensured, posing a challenge to maintaining regional research.

On the data and information front, there are other issues related to interactions between environmental and fisheries governance: for example, data collected by the observers at sea mostly exclude discards, an important component of EAF. The challenge is to expand observer training programmes to include identification of all species caught in trawls. There is poor understanding and appreciation of underlying social and economic issues in fisheries and lack of integration into management. Indigenous ecological knowledge is not currently taken into account at the regional or national levels. Only South Africa has made good progress in biodiversity spatial mapping, but BCC has currently launched a region-wide project.

At a regional level, BCC structures are not yet fully elaborated to allow a fuller range of ecosystem issues to be addressed at the management level. If joint obligatory TACs and country species allocations are to be determined for the rational management of transboundary fish stocks, scientific and management committees to deliberate and advise on sustainable levels of fisheries and consider specific ecosystem issues will need to be set up and meet annually.

Nationally and regionally, involvement of a wider range of stakeholders is needed with respect to the management of fisheries. There tends to be a lack of effective reconciliation of objectives with respect to conservation and fisheries.

Despite some of these shortcomings, there are opportunities that have arisen from regional developments. The general approach applied to develop an EAF and some of the successes achieved at regional level can be an example to other developing countries, especially those who are involved in LME-type regional programmes. National institutions have benefited significantly from lessons gained at the regional level; these need to be consolidated. In particular, with an expanded mandate to cover areas such as socio-economics, law and governance, BCC can act as an excellent platform for entrenching EAF in the region. If the basis for EAF can be effectively implemented, as envisaged at regional and national levels, there is reason for optimism. Addressing some of the institutional arrangement issues could lead to great improvement in fisheries management and better reconciliation with conservation objectives. However, the negative influence of populist politics and occasionally corruption can undo years of progress. Managers need to work harder to communicate sustainable management solutions and their economic benefits to political principals.

Other components of ecosystem-based management are under development. For example, functioning MPA networks are in place in two countries and Namibia and SA each have several seabird sanctuaries, including a major transboundary sanctuary at the Gariep River estuary.

Regionally, meeting the objectives of the BCLME SAP and the BCC Convention which leads to binding agreements on the setting of TACs and other ecosystem-based approaches is critical. Full implementation of BCC structures is required in order to allow a wider range of ecosystem issues to be addressed.

Finally, the opportunity is ripe for development and exploration of further frameworks and platforms for communicating EAF science and its implications into enhanced integration of fisheries and conservation management action, with the BCC playing a key role to influence its members and increase the chances of success.

Summary and conclusions

The Benguela Current region has been a laboratory for experimentation with the implementation of EAF, allowing the development of systems and structures required to translate the best available science into management and facilitating the implementation of an EAF in a context-specific manner. The implementation of EAF is to an extent considered as a means of integrating marine environmental and fisheries governance issues, sometimes in the absence (or as a result of weak implementation) of institutional governance structures. The fisheries institutions in the countries have begun to implement EAF management with varying degrees of success. A number of new initiatives are allowing EAF implementation and tracking with some success and the underlying science is well integrated with international initiatives. Under the BCC a portfolio of EAF projects is underway to investigate and advise on key issues that need to be understood in order to adopt a more holistic and integrated approach to fisheries management and conservation in the region. There are also several examples of institutional developments, mostly in South Africa, that are leading the way to allow a more comprehensive and integrated approach. There have been some notable successes, especially with respect to scientific collaboration, management and capacity building; there have also been some serious shortcomings in terms of funding, institutional capabilities and structure, political will and ability to regulate fisheries. There are however opportunities to make further progress through greater political support, policy development and institutional changes as well as through the constructive role of the BCC.

References

Atkinson, L.J. (2010) Effects of demersal trawling on marine infaunal, epifaunal and fish assemblages: studies in the southern Benguela and Oslofjord. PhD thesis, University of Cape Town.

Atkinson, L.J., Leslie, R.W., Field, J.G. and Jarre, A. (2011) Changes in demersal fish assemblages on the west coast of South Africa, 1986–2009. *African Journal of Marine Science* **33**, 157–170.

Atkinson, L.J., Jarre, A., Shannon, L.J. and Field, J.G. (2012) Evidence for shifts in demersal fish populations on the west coast of South Africa: 1986 to 2009. In: *Global Progress in Ecosystem-Based Fisheries Management* (eds G.H. Kruse, H.I. Browman, K.L. Cochrane, D. Evans, G.S. Jamieson, P.A. Livingston, D. Woodby and C.I. Zhang), pp. 45–64. Alaska Sea Grant, University of Alaska Fairbanks.

Blamey, L.K., Howard, J.A., Agenbag, J. and Jarre, A. (2012) Regime-shifts in the southern Benguela shelf and inshore region. *Progress in Oceanography* **106**, 80–95.

Cecchini, L.A. (2012) Robben Island Penguin Pressure Model: A Decision Support Tool for an Ecosystems Approach to Fisheries Management. Master of Science thesis, Department of Statistical Sciences, University of Cape Town.

Cochrane, K.L., Augustyn, C.J., Fairweather, T., Japp, D., Kilongo, K., Iitembu, J., Moroff, N., Roux, J.P., Shannon, L., Van Zyl, B. and Vaz Velho, F. (2009) Benguela current large marine ecosystem—governance and management for an ecosystem approach to fisheries in the region. *Coastal Management* **37**, 235–254.

Cury, P.M., Boyd, I.L., Bonhommeau, S., Anker-Nilssen, T., Crawford, R.J.M., Furness, R.W., Mills, J.A., Murphy, E.J., Österblom, H., Paleczny, M., Piatt, J.F., Roux, J.P., Shannon, L.J. and Sydeman, W.J. (2011) Global seabird response to forage fish depletion – one-third for the birds. *Science* **334**, 1703–1706.

FAO (1995) Code of Conduct for Responsible Fisheries. Rome, FAO.

FAO (2003) Fisheries Management 2. The Ecosystem Approach to Fisheries. FAO Technical Guidelines for Responsible Fisheries, 4 (Suppl. 2), Rome, FAO.

Fletcher, W.J., Chesson, J., Fisher, M., Sainsbury, K.J., Hundloe, T., Smith, A.D.M. and Whitworth, B. (2002) National ESD Reporting Framework for Australian Fisheries: the 'How To' Guide for Wild Capture Fisheries. FRDC Project 2000/ 145, Canberra, Australia.

Ludynia, K., Waller, L.J., Shirley, R.B., Abadi, F., Galada, Y., Geldenhuys, D., Crawford, R.J.M., Shannon, L.J., and Jarre, A. African Penguin population dynamics on Dyer Island, South Africa, and identification of processes relevant for conservation and fisheries management. Paper in press. *African Journal of Marine Science*.

Moloney, C.L., van der Lingen, C.D., Hutchings, L. and Field, J.G. (2004) Contributions of the Benguela ecology programme to pelagic fisheries management in South Africa. *African Journal of Marine Science* **26**, 37–51.

Okes, N.C., Petersen, S.L., McDaid, L.A. and Basson, J. (2012) Enabling people to create change: Capacity building for Ecosystem Approach to Fisheries (EAF) implementation in Southern Africa. *Marine Policy* **36**(1), 286–296.

Osman, W. (2010) Trophic model-generated indicators of the southern Benguela ecostystem for communicating with fisheries managers. Masters thesis. Unversity of Cape Town, South Africa.

Paterson, B. and Petersen, S.L. (2010) EAF implementation in Southern Africa: lessons learnt. *Marine Policy* **34**, 276–292.

Paterson, B., Isaacs, M., Hara, M., Jarre, A. and Moloney, C.L. (2010) Achieving transdisciplinary co-operation for EAF: a South African case study. Proceedings of the GLOBEC/FAO/EUR-OCEANS Symposium on 'Coping with Change in Marine Social–Ecological Systems'. *Marine Policy* **34**(4), 782–794.

Payne, A.I.L., Gulland, J.A. and Brink, K.H. (eds) (1987) The Benguela and Comparable Ecosystems. *South African Journal of Marine Science* **5**, 957 pp.

Payne, A.I.L., Brink, K.H., Mann, K.H., Hilborn, R. (eds) (1992) Benguela Trophic Functioning. *South African Journal of Marine Science* **12**, 1108.

Petersen, S.L., Paterson, B., Basson, J., Moroff, N., Roux, J.P., Augustyn, J. and D'Almeida, G. (eds) (2010) Tracking the Implementation of an Ecosystem Approach to Fisheries in Southern Africa. WWF South Africa Report Series, Cape Town.

Pillar, S.C., Moloney, C.L., Payne, A.I.L. and Shillington, F.A. (1998) Benguela dynamics: impacts of variability on shelf-sea environments and their living resources. *South African Journal of Marine Science* **19**, 512.

Robinson, W., Plagányi, E, Butterworth, D. and de Moor, C. (2010) Summary of penguin–pelagic fish interaction modelling during 2008. Pelagic Scientific Working Group, Marine and Coastal Management, Cape Town. MCM/2010/SWG_PEL/35. 45 pp.

Shannon, L.J., Moloney, C.L., Jarre, A. and Field, J.G. (2003) Trophic flows in the southern Benguela during the 1980s and 1990s. *Journal of Marine Systems* **39**, 83–116.

Shannon, L.J., Cochrane, K.L. and Pillar, S.C. (eds) (2004a) Ecosystem approaches to fisheries in the southern Benguela. *African Journal of Marine Science* **26**, 328 pp.

Shannon, L.J., Cochrane, K.L., Moloney, C.L. and Fréon, P. (2004b) Ecosystem approach to fisheries management in the southern Benguela: a workshop overview. *African Journal of Marine Science* **26**, 1–8.

Shannon, L.J., Cury, P.M., Nel, D., van der Lingen, C.D., Leslie, R.W., Brouwer, S.L., Cockcroft, A.C. and Hutchings, L. (2006a) How can science contribute to an ecosystem approach to pelagic, demersal and rock lobster fisheries in South Africa? *African Journal of Marine Science* **28**, 115–157.

Shannon, L.V., Hempel, G., Moloney, C., Woods, J.D. and Malanotte, P. (2006b) *Benguela: Predicting a Large Marine Ecosystem*. Elsevier, Amsterdam.

Shannon, L.J., Neira, S. and Taylor, M. (2008) Comparing internal and external drivers in the southern Benguela and the southern and northern Humboldt upwelling ecosystems. *African Journal of Marine Science* **30**, 63–84.

Shannon, L.J., Jarre, A.C. and Petersen, S.L. (2010a) Developing a science base for implementation of the ecosystem approach to fisheries in South Africa. *GLOBEC Special Issue, Progress in Oceanography* **87**, 289–303.

Shannon, L.J., Coll, M., Yemane, D., Jouffre, D., Neira, S., Bertrand, A., Diaz, E. and Shin, Y.J. (2010b) Comparing data-based indicators across upwelling and comparable systems for communicating ecosystem states and trends. *ICES Journal of Marine Science* **67**, 807–832.

Shin, Y.J. and Shannon, L.J. (2010) Using indicators for evaluating, comparing and communicating the ecological status of exploited marine ecosystems. 1. The IndiSeas Project. *ICES Journal of Marine Science* **67**, 686–691.

Shin, Y.J., Shannon, L.J., Bundy, A., Coll, M., Aydin, K., Bez, N., Blanchard, J.L. et al. (2010) Using indicators for evaluating, comparing and communicating the ecological status of exploited marine ecosystems. 2. Setting the scene. *ICES Journal of Marine Science* **67**, 692–716.

Shin, Y.J., Bundy, A., Shannon, L.J., Blanchard, J.L., Chuenpagdee, R., Coll, M., Knight, B., Lynam, C., Piet, G. and Richardson, A. J. (2012) Global in scope and regionally rich: an IndiSeas workshop helps shape the future of marine ecosystem indicators. *Reviews in Fish Biology and Fisheries* **22**(3), 835–845.

Smith, A.D.M., Brown, C.J., Bulman, C.M., Fulton, E.A., Johnson, P., Kaplan, I.C., Lozano-Montes, H. et al. (2011) Impacts of fishing low-trophic level species on marine ecosystems. *Science* **333**(6046), 1147–1150.

South African Government Gazette (2012) The South African Policy on Ocean Environmental Management. South African Government Gazette, Volume 568(35783), 88 pp.

South African Government Gazette (2012) Policy for the Small Scale Fisheries Sector in South Africa. South African Government Gazette, Volume 564(35455), 64 pp.

Staples, D. (2010) Final Report on institutional arrangements for implementing an ecosystem approach to fisheries (EAF) management in Benguela Current Commission countries. Benguela Current Commission, Namibia, Doc/25/MB/Meeting/3.

Starfield, A.M. and Jarre, A. (2011) Interdisciplinary modelling for an ecosystem approach to management in marine socio-ecological systems. Chapter 6. In: *World Fisheries: A Social-Ecological Analysis* (eds R.E. Ommer, R.I. Perry, K. Cochrane and P. Cur), pp 105–119. Wiley-Blackwell, Oxford, Fish and Aquatic Resources Book Series 14.

Travers, M., Watermeyer, K., Shannon, L.J. and Shin, Y.J. (2010) Changes in food web structure under scenarios of overfishing in the southern Benguela: Comparison of the Ecosim and OSMOSE modelling approaches. *Journal of Marine Systems* **79**, 101–111.

Van der Lingen, C.D. (2011) The biological basis for hypothesizing multiple stocks in South African sardine *Sardinops sagax*. Marine Resource Assessment and Management Group, University of Cape Town, Rondebosch, South Africa, IWS/DEC11/P/OMP/P7: 1-10 pp.

Watermeyer, K., Shannon, L.J. and Griffiths, C.L. (2008a) Changes in the trophic structure of the southern Benguela before and after the onset of industrial fishing. *African Journal of Marine Science* **30**, 351–382.

Watermeyer, K., Shannon, L.J., Roux, J.P. and Griffiths, C.L. (2008b) Changes in the trophic structure of the nouthern Benguela before and after the onset of industrial fishing. *African Journal of Marine Science* **30**, 383–403.

WWF (2012) Responsible Fisheries alliance: 1st Phase Review. WWF South Africa Report.

Chapter 20

Governance of marine fisheries and conservation in the context of the European Union

S. Beslier[1] and B. Drobenko[2]

[1]IUCN Commission on Ecosystem Management (IUCN/CEM/FEG), Gland, Switzerland
[2]French Université du Littoral de la Côte d'Opale, Lille, France

Abstract: The governance of the European Union (EU) rests on a context-dependent system of delegation of sovereignty by its members and on an original decision-making process. The Common Fishery Policy (CFP) is one of the most integrated policies of the EU and its environmental protection policy is one of its most important. The interactions between those two policies are evident. The resolution of eventual contradictions is more difficult. This chapter demonstrates how the EU institutional and decisional frameworks may lead to segmentation of the policies and indicates the elements of solution needed to achieve the necessary coherence.

Keywords: sovereignty; institutions; framework; Treaty, competence; common fisheries policy; relative stability principle; environmental policy

The institutional framework and decision making of the European Union

The powers

The European Union (EU) is, in a traditional sense, neither a state nor an international organization. Its competences rest in delegation of sovereignty in areas that its Member States (MSs) have agreed to jointly manage when they felt it was in their interest.

The Treaty on European Union (TEU, 2008) and the Treaty on the Functioning of the European Union (TFEU, 2008), which constitute the Lisbon Treaty and entered into force on 1 December 2009, have at the same time codified the earlier treaties and attempted to clarify the competences conferred on the EU and the governance modes thereof.

- 'Competences not conferred upon the Union in the Treaties remain with the Members States' (Article 4 TEU).
- 'The limits of Union competences are governed by the principal of conferral. The use of Union competences is governed by the principles of subsidiarity and proportionality' (Article 5 TEU).

For fishing, TFEU Article 3 indicates that the Union has exclusive competence[1] in the area of conservation of marine biological resources under the Common Fisheries Policy (CFP). Article 4 stipulates that the Union shall share competence with MSs in the area of fisheries, excluding the conservation of marine biological resources. The objectives of the Common Fisheries Policy (CFP) are defined in TFEU Title III (Articles 38 to 44), and are common to fishing and agriculture.

For the environment, TFEU Article 4 indicates that this is a shared competence[2] between the EU and Member States

Governance of Marine Fisheries and Biodiversity Conservation: Interaction and Coevolution, First Edition.
Edited by Serge M. Garcia, Jake Rice and Anthony Charles.

(MSs). Being a transversal policy, Article 11 of the TFEU states that: 'environmental protection requirements must be integrated into the definition and implementation of the Union's policies and activities, in particular with a view to promoting sustainable development'. The objectives of the Union policy on the environment are set out in TFEU Part XX (Articles 191 to 193). Note that, notwithstanding the lack of mention of environmental law in the 1957 Treaty founding the European Economic Community, the Commission initiated a policy in this area, based on a succession of European programmes for the environment prior to its formal recognition by The Hague Treaty (Single European Act in 1986) in 1986.

Member States retain a monopoly on the implementation and execution of measures adopted at EU level, whether under the exclusive competence or as part of a shared competence.

The European institutions and decision making

Primarily, there are three European institutions involved in EU decision-making: the European Parliament (EP), the Council and the European Commission. Other institutions are not involved (or only involved to a small degree) in this process. The European Central Bank is marginally involved in the decision-making process as it may give opinions to the Council on Commission proposals in the financial field. The other institutions, the Court of Justice of the European Union (ECJ) and the European Court of Auditors (ECA), are not involved in that process. Rather, they exercise a control function on the legislator and/or the institutions that implement legislation (see Chapter 28).

The European Parliament (EP)

In the field of fisheries policy, the EP has been exercising legislative functions (co-decide) jointly with the Council since the entry into force of the Lisbon Treaty in 2009. Before that, it only had an advisory role. The environmental policy has however been co-decided by the two institutions since the Amsterdam Treaty in 1997.

Decisions are prepared by the EP Committees. There is a Committee on Fisheries and one on the Environment. The EP Committee, within which the report (on either environment or fisheries) is prepared for the vote by the EP in plenary, must consult with other committees concerned with or affected by the proposal.

The EP is different from national parliaments in terms of its greater openness to the influence of lobbyists. Non-governmental organizations (NGOs) are involved in organizing numerous events and debates in its premises. This is an advantage in terms of enrichment of the debates upstream of decisions, but it also brings a risk associated with the greater or lesser capacity of these groups to influence decisions in favour of partisan interests.

The Council

The Council of Ministers of the EU is unique but, for reasons related to the organization of its work, it meets in different formations corresponding to different sectoral policies. There are currently 9 of these formations, including one on fishing (grouped with agriculture) and one on the environment. Council decisions are prepared by working groups and committees (about 250 of such committees and working groups are active) consisting of delegates from MSs. The Council decision-making process is relatively fragmented. At its bottom, the preparation and coordination of decisions at the technical level by Council groups of experts depends largely on the quality of the coordination achieved at MS level. At intermediate level, the Committee of Permanent Representatives (COREPER) plays a very important role in coordinating different policies. It is the only body where proposals in all different fields are discussed in the same committee, while the Councils only see the proposals which concern their particular field. Arbitrations may be made by the 'General Affairs' Council, which is composed of foreign affairs ministers of MSs or exceptionally by the European Council of Heads of State and Government itself.

From the standpoint of consistency, the EU system is affected by contradictory forces. On the one hand, Ministers serving in a specialized instance of the Council (whether on fishing or the environment) sometimes tend to favour their own sectoral options when negotiating a compromise. On the other hand, the confrontation at EU level of views which are sometimes contradictory because they reflect different MSs interests may lead to more balanced cross-sectoral outcomes than those elaborated at the level of a single State.

The Commission

The European Commission is the executive body of the European Union responsible for proposing legislation, implementing decisions, upholding the Union's treaties and day-to-day running of the EU. The Commission services are organized using the same principles as those governing ministerial departments within a State. The role of Commissioner at EU level is analogous to that of Minister at national level.

The Commission is organized in Directorates-General (DGs). The DG Fisheries or DGXIV, created in 1976, was the 14th in the order of creation of branches of the Commission. Previously fishery products were managed by DG Agriculture. This branch has seen its range of competences expanded in 2005 with the launch of the integrated maritime policy (IMP) and was renamed as Directorate General for Maritime Affairs and Fisheries (DG MARE) in 2007.

The Directorate-General for Environment (DG ENV) was established in 1973 (as DG XI). Most ministries of environment in the MSs (e.g. in France in 1971) were created around the same period.

The Commission has three core functions:

- it has the power of initiative in all legislative proposals;
- it contributes to legislation implementation through the powers delegated to it under the Treaty; and
- it is the guardian of the Treaties and secondary legislation.

The decision-making process within the Commission Services is based on the same basic principles as those encountered in a State. There is a lead-Service which has the obligation to consult with other Commission Services on the proposals to be submitted to the legislative power. The inter-services consultation procedure (ISC) is closely controlled and codified by the General Secretariat of the Commission. Compared to a State system, the main difference lies (in theory) in the arbitration procedure used in case of discrepancy between two irreducible DGs. While national systems are generally pyramidal in nature and the arbitrations are made by the head of state or government, the Commission process is, in principle, more 'collegial' and arbitrations are made by vote of the College of Commissioners. However, a vote in the College is very exceptional. In practice, it is very often the President who makes the arbitrations.

Before tabling a proposal the Commission consults stakeholders, either at the European level (e.g. through the Advisory Committee for Fisheries and Aquaculture, ACFA) or at a sea basin level (Regional Advisory Councils, RACs). These bodies are composed of industry representatives and NGOs. It consults scientific experts through the Scientific, Technical and Economic Committee for Fisheries (STECF). The Commission can also launch broader and open consultations through Green Papers or Blue Papers to which anybody (e.g. ordinary citizens, governments from EU MSs or from non-members of the EU and private companies) can contribute.

Policy development

The Common Fisheries Policy

Fishing, conventionally considered as a sub-sector of agriculture, was included among the areas of Community competence since the Treaty of Rome (Article 38). The first two Community regulations concerning it were adopted in 1970 by the Council, not long before the beginning of the enlargement negotiations with the United Kingdom, Ireland, Denmark and Norway and a few months before the launch of the third UN Conference on the Law of the Sea (UNCLOS). This was not just a coincidence

(1) The first regulation (Council Regulation No. 2141 of 20 October 1970) set out the principle of freedom and equal access to Community waters under a more general principle of the Treaty: that of equal treatment. It aimed to anticipate: (1) the evolution of the law of the sea that eventually led to the appropriation of more than 90% of fish stocks by coastal States due to the creation of Exclusive Economic Zones (EEZ); and (2) the UK negotiating position in the UNCLOS process which challenged the possibility of extending the jurisdiction of coastal States offshore beyond 12 nautical miles.

(2) The second regulation (Council Regulation No. 2142 of 20 October 1970) intended to settle a common organization of markets to anticipate the very liberal concepts supported by the British for the organization of competition on fish market and for trade matters.

From the beginning of the CFP, the link was established between fishing and its environment. Article 102 of the Accession Treaty of the United Kingdom indicates that 'from the sixth year after accession at the latest, the Council, acting on a proposal from the Commission, shall determine conditions for fishing with a view to ensuring protection of the fishing grounds and conservation of the biological resources of the sea'.

The first elements of the EU fishery resources conservation policy were adopted by the Council in 1983 (Regulation 170/83). This policy was based on a definition of the rate of fishing mortality or total allowable catch (TAC) for each stock for which conservation was regulated. This was not original to the extent that it was already the basis of many fisheries management systems, particularly in the North Sea, seeking to ensure a sustainable use of stocks while protecting their environment.

The Relative Stability Principle (RSP)

The main originality of this policy, compared to a national policy of conservation of fishery resources, is based on the introduction of the Relative Stability Principle (RSP) that meets the specific system of community governance. It is often presented as the cornerstone of the CFP. It allows the TAC to be distributed between MSs in a stable manner over the years, on the basis of allocation keys established for the first time in 1983 according to prior fishing patterns (using 1976 as a reference period, the year of creation of Exclusive Economic Zones EEZs by the MSs).

The principle determines the limits within which each MS exercises its responsibility for the implementation and monitoring of catch limits decided by the Council. It also stabilizes the fishing activities in areas dependent on fisheries and thus contributes to steady coastal areas management policies.

It results in allocating fishing opportunities by nationality and is therefore an exception to the principle of free movement of goods. On the other hand, the CFP does not prevent the transfer of capital. A national of one MS may acquire a fishing vessel in another MS, but she must keep its flag of origin if wanting to exploit the fishing opportunities attached to that flag. This practice is generally referred to as 'quota hopping'.

The Green Paper on the reform of the CFP, adopted by the Commission in April 2009, opened the debate on the possibility of replacing the RSP with a system of allocation of individual transferable quotas (ITQs). This debate dwindled rapidly because of the opposition of the majority of MSs to the demise of the principle.

The question of improving or adding flexibility to the RSP has not been raised so far. However, this debate seems all the more necessary as the distribution criteria used to give effect to the principle rest on a snapshot of economic activities of 30 years ago.

The lack of flexibility of the system induces adverse effects when setting TACs for certain stocks. It tends to accentuate the tendency of MSs, when determining the amount of TAC, to select the higher end of the range of the scientific recommendations. As a consequence, when their present needs are higher than their historical rights, some ministers are tempted to push the TAC beyond the upper limit recommended by scientific advice in order to secure enough fishing opportunities for their nationals, increasing risks for the resources.

The control of fishing activities

The control system of fishing activities is the second main original aspect of the Community system of fisheries governance. It illustrates both the particularity of the construction of the European Community and the difficulties resulting from a complex system of shared responsibility.

Control of fishing activities is the weak point of many fisheries management systems. For a long time, the fishing industry has been seen by most national politicians as a socially important, difficult and turbulent system, deserving some particular indulgence. The NGOs have only recently begun to attack this capital of sympathy, in the public opinion, and to denounce a risk-prone behaviour. The phenomena of globalization and development of flags of convenience and illegal fishing, as well as a growing awareness of issues related to destructive fishing practices and the protection of marine biodiversity, have helped focus public attention on these issues and persuaded governments to respond. This realization has led to the adoption of a number of international agreements in an attempt to correct the situation, revealing the main weaknesses of fisheries management systems. These instruments (described in more detail in Chapter 7) include the following agreements

- *The 1993 Compliance agreement*: The FAO Agreement to promote compliance with international conservation and management measures by fishing vessels on the high seas.
- *The 1995 UNFSA agreement*: The UN Agreement for the implementation of the UN Convention on the Law of the Sea of 10 December 1982, relating to the conservation and management of straddling fish stocks and highly migratory fish stocks.
- *The 1995 CCRF*: The FAO Code of Conduct for Responsible Fisheries developed a 'code of ethics' for fisheries, embedding elements of environmental policy emerging from UNCED in UNCLOS-related provisions (e.g. regarding the precautionary approach and references to protection and rebuilding of living aquatic resources and biodiversity; environmentally safe fishing gear; and protection of critical habitats), beyond the conventional reference made in the LOSC to dependent and associated species.
- *The FAO International Plans of Action (IPOAs)* such as the 1999 IPOA for reducing incidental catch of seabirds on longlines; the 1999 IPOA for the conservation and management of sharks; the 2001 IPOA to prevent, deter and eliminate illegal, unreported and unregulated (IUU) fishing.
- *The 2009 FAO port States measures agreement* to prevent, deter and eliminate illegal, unreported and unregulated (IUU) fishing.

This situation and the reactions it induces are also encountered at the European level.

In the EU system, responsibility for implementation, control and enforcement of the decisions relating to the conservation of fishery resources adopted by the European Parliament and the Council lies first with the MSs. The efficiency of the implementation depends therefore on the zeal deployed by MSs in undertaking these tasks. There has also been a temptation for some MSs interacting with their professionals to blame 'Brussels' for the measures and the related hardship, forgetting that they participated in the decision they complained about. European authorities themselves are however rather influential in the system of community control.

As guardian of the Treaty and secondary legislation, the Commission must remind MSs of their obligations and, after giving notice by issuing reasoned opinions, refer as appropriate to the ECJ. This procedure is cumbersome and politically difficult to implement, but it can produce effective results.

For example, France was condemned for its failures in fisheries control on two occasions by the ECJ (see Chapter 28). It was first condemned in 1991 for failing from 1984 to 1987 to comply with all its obligations of supervision to ensure compliance with the fisheries conservation measures of the Community. In 2005, France was sentenced to pay a lump sum of 20 million Euros and a penalty of 57.8 million Euros for each period of six months from delivery of the judgment, until the full implementation of the first decision of 1991. This financial penalty was a first in the history of the Court. It has attracted the attention of other departments that support fisheries and led the French government to deeply reform its organization of fisheries control.

In a special report on the effective implementation of the CFP, adopted in 2007,[3] the European Court of Auditors was highly critical of how it was controlled by MSs. It commented that the data collection mechanisms, which are the responsibility of MSs, showed sufficient weaknesses to compromise a fisheries management policy based on catch limitations.

The Heads of State and Government had already recognized this weakness and had invited the Commission, at the European Council of December 2003, to make a proposal to set up a European Fisheries Control Agency. This agency was created in 2005 and has been operational since 2007. It aims to establish a level playing field and develop a culture of compliance by assisting the MSs and coordinating their monitoring of fishing activities. Although it intervenes within an area which is under the jurisdiction of MSs, it has resulted in an enhanced role for the Commission in the field of control because of the financial supervision and administration that the Commission exercises on the Agency.

In this context, on a proposal from the Commission the EU has recently strengthened its legal arsenal by adopting two regulations: (1) Regulation 'IUU' in 2008;[4] and (2) Regulation 'control' in 2009.[5]

The exercise of the exclusive competence

While the TFEU indicates that the 'conservation of marine biological resources' falls under the exclusive competence of the EU, the basic regulation of the CFP states that:

> 'The Common Fisheries Policy shall cover conservation, management and exploitation of living aquatic resources, aquaculture and the processing and marketing of fishery products and aquaculture, where such activities are practiced on the territory of Member States or in Community waters or by Community fishing vessels or without prejudice to the primary responsibility of the flag State, by nationals of Member States.'[6]

The definition of the scope of the CFP in the basic regulation (2002) is different from that of the Treaty. The Lisbon Treaty (2007) builds upon the formulation of the accession Treaty of the UK (1972), while the basic regulation takes (with one exception[7]) the same formulation as the two previous regulations (1983 and 1992). This formulation was used again, in the same terms, in the proposed reform of the CFP by the Commission in 2011.[8] In this formulation, the use of the term 'living aquatic resources' instead of 'marine biological resources' allows an extension of the Community competence to aquaculture practised in the marine waters as well as the terrestrial territory of the MSs. The addition of the management and exploitation considerably extends the scope of Community action in the areas falling within the shared competence. The extension of the EU legislation scope to 'nationals of the MSs',[9] introduced during the 2002 reform, allowed such legislation to cover previously scarcely 'communitized' fields such as penal prosecution as well as a field that is not within the judicial traditions of most MSs, the extraterritorial application of their jurisdiction.

This acceptance of a broad application of the scope of Community action reflects the characteristics of fishing activity, which takes place within an environment with little tolerance for physical boundaries, and makes the CFP one of the most politically integrated EU policies. However, this tendency to broadly interpret the scope of Community competence has two notable exceptions – marine mammals and genetic resources – as described in the following sections.

Marine mammals

This exception is the result of a deliberate policy and concern. Marine mammals are living resources of the sea, but they are not covered by the CFP. European interventions are affecting them in the context of EU environmental policy, for which EU does not have exclusive competence.

In the legal framework for the CFP, the Commission proposed in 1992 a negotiating mandate for a Community accession to the International Whaling Commission (IWC).[10] The proposal was original enough in that it proposed a membership of the EU and EU MSs in the IWC under a shared competence. A membership based on the exclusive competence of EU on whales would have resulted in a necessary withdrawal of the EU MSs, reducing the number of votes available to the Community (12 at the time and now 28) to only one, risking a shift in the balance between 'conservationists' and 'managers' in the IWC. In order to avoid putting the Community in a delicate political situation at an international level, the College of Commissioners decided to waive the principle of exclusive competence; the Council never endorsed this decision, however.

The Habitats' Directive[11] invited the MSs to take the necessary steps to establish a system of strict protection of marine mammals in EU waters and to prohibit related trade in their territory. The international trade in these species is also prohibited by the regulation[12] implementing the recommendations of CITES[13] in EU law.

It was not until 2007 that the Commission, at the initiative of DG Environment, began to participate actively in the work of the IWC as an observer. It had however relied on the exclusive competence[14] of the CFP in order to overcome the reluctance of MSs. It threatened the MSs with going to the ECJ to force them to accept its presence at the IWC and EU coordination. This remedy would have undoubtedly lead to a result contrary to that sought as the Court would certainly have considered law and, ignoring the political context, the exclusive competence which involves a single EU representation in international organizations such as Regional Fisheries Management Organizations.

The Commission proposed a new mandate for EU accession to the IWC in 2011.[15] This proposal is based on TFEU Article 191 defining the objectives of EU environmental policy and rules out any reference to texts on the CFP (Common Fisheries Policy), stating that this is not a precedent for future negotiations on the conservations and management living aquatic resources under this regulation.

Genetic resources

This other exception is the result of the evolution of scientific knowledge concerning the marine environment and the legal definition of biological resources of the sea.

The CFP applies from its origin of biological resources of the sea and only reflects the state of international law of the sea as it existed at the time of its development.

Concerning biological resources, the 1982 UN Convention on the Law of the Sea (UNCLOS) consisted only of fishery resources and hence related to populations and species. The importance of genetic resources (e.g. of abyssal marine life resulting from processes other than photosynthesis) emerged as the Convention was being drafted and could not be taken into account.

The international debate on the protection and the exploitation of genetic resources of the oceans (it sustainability and equitable sharing of benefit) in the high seas began at the beginning of this century from the problem of trawling species in the deep seabed. This debate is far from closed, either at global or at EU level. The EU has however played a major role in the development of this issue at the global level in fishery institutions, FAO and RFMOs, and in the United Nations where it was the first to defend the idea of an implementing agreement to UNCLOS Part XII for protection of marine biodiversity.

The ecosystem approach

The systematic consideration of the impact of fishing activities on the environment is a relatively recent phenomenon. Protection of the seabed has long been regarded more as a means of preserving stocks themselves rather than part of protecting the marine environment and its biodiversity. The word 'ecosystem' was formally introduced in the vocabulary used for the management of fisheries by the Code of Conduct for Responsible Fisheries (CCRF) adopted by FAO in 1995 (section 7.2.2.d).

The concept of Ecosystem Approach to Fisheries (EAF) was developed by FAO (2002, 2003; see also Garcia et al., 2003) as a result of the Reykjavik Conference on Responsible Fisheries in the Marine Ecosystem organized by Iceland and FAO in 2001. The development of tools allow its implementation is ongoing (e.g. http://www.fao.org/fishery/eaf-net/en).

The Commission launched the debate on 'the role of the CFP in implementing an ecosystem approach to marine management'[16] in 2008. The CFP reform, proposed by the Commission in 2011, reaffirms the need to include European fisheries in a sustainable development and therefore environmentally friendly perspective, particularly with regard to bycatch and protection of vulnerable habitats.

EAF allows for better integration of fishing activities in their environment. It remains a sectoral approach to Ecosystem-Based Management (EBM) and does not act on parameters other than those relating to the management of fish stocks and fisheries environmental impacts.

Although fishing is a maritime industry that has a significant and long-term impact on the marine environment, it is not possible to reduce the issue of management and protection of marine ecosystems to the relation between fishing and the environment.

As recognized in the 2008 Communication (referred to above), the overall strategic objective of an ecosystem approach to marine management cannot be achieved only through a sectoral policy such as the CFP. Achieving this objective requires the establishment of a more comprehensive approach to ocean governance that involves, upstream of decision making, the organization of scientific research that falls within the jurisdiction of MSs, conditioning to a significant extent the effectiveness of the EU policy in that regard.

Environmental policy of the European Union

The EU environmental policy has several objectives (TFEU Article 191), including 'preserving, protecting and improving the quality of the environment, protecting human health and the prudent and rational utilization of natural resources'. It seeks a high level of protection, being based on the 'precautionary principle and on the principle that preventive action should be taken, that environmental damage should as a priority be rectified at source and that the polluter should pay'. The EU's action on the international scene is guided by the promotion of measures, which contribute to 'preserve and improve the quality of the environment and the sustainable management of global resources, in order to ensure sustainable development' (TUE Article 21 f).

Environmental policy is therefore a major parameter of fisheries governance. Fishing can be regarded as disrupting the marine ecosystem; conversely, it is negatively and significantly affected by environmental changes caused by other human activities on land or at sea.

Under its competences the European Union has also ratified a set of international conventions which commit the EU countries, although individual MSs are also independently Contracting Parties to these Conventions (e.g. the Barcelona Convention and its Protocols, the OSPAR Convention and its Annexes).

The environmental policy impacts upon other EU policies, as is the case for the CFP. This European policy has been driven, from a political point of view, by programmes that were subsequently formalized from a legal point of view by secondary legislation (e.g. regulations, directives or recommendations) in order to achieve the objectives. The environmental law has also led the EU to internationally coordinate its policies, contributing to the evolution and strengthening of international conventions on environment (e.g. on climate change, waste and international shipping).

The European programme

Inspired by the Stockholm Conference in 1972, the European Community has subsequently elaborated a policy in the environmental field. Notwithstanding its shortcomings on the subject, the 1957 Treaty lays the foundations of this policy with the adoption of the first European programme for the environment in 1973.[17] The first programme, essentially based on qualitative perspectives, was followed in 1977, 1982 and 1987 by three others pursuing a similar goal of environmental quality.[18]

The Rio conference in 1992 then stimulated a renewed political process with the perspective of sustainable development and the implementation of an integrated approach that affects all the various public policies. The sustainable development concept and the integrated approach characterized the fifth and sixth programmes, until today.[19]

Indeed, from the 5th programme onwards, the European Union has been concerned (in chapter 1 of this programme) about land-based pollution in the light of international commitments, including the Barcelona and OSPAR conventions. The maintenance of biological diversity determines the consideration of impacts of human activities including fishing. The objective of water quality also affects marine waters.

The 6th programme[20] relates directly to fishing and the marine environment andpursues the goal of:

> 'Promoting sustainable use of seas and conservation of marine ecosystems, including the seabed, estuaries and coastal areas, with particular attention to areas of high value in terms of biodiversity, by:
>
> - Promoting greater integration of environmental considerations into the Common Fisheries Policy, on the occasion of its review in 2002;
> - Defining a thematic strategy for the protection and conservation of the marine environment taking into account, inter alia, the conditions and implementing obligations stipulated in conventions related to the sea, and the need to reduce emissions and the impacts of marine transportation and other activities located at sea or ashore; and
> - Promoting integrated management of coastal areas'.

These programmes created the foundation for the adoption of major texts in the field of environment, which also affect marine waters. While the EU adopted some regulations (such as REACH[21]) or recommendations (e.g. on Integrated Coastal Zone Management),[22] the bulk of environmental law falls within the scope of the Directives (see next section).

The Directives

In the environmental field, the EU legislates by means of Directives. These do not apply directly to the Member States' territory, contrary to the regulations used in the case of CFP, but are meant for MSs that are bound by the result to be achieved but retain the choice of form and methods. The Directives must be transposed into national law to be applied, which generates many disputes between the Commission and the MSs when they do not meet agreed deadlines for implementation or fail to reach the objectives.

The Birds Directive

The Birds Directive (Council Directive 79/409/EEC of 2 April 1979 on the conservation of wild birds) aims to protect birds in the wild in the European territory. Although it does not aim specifically at fishing activities[23] and is best known for its impact on hunting, it illustrates how relationships develop between the two policies in the system of EU governance.

The scope of this Directive, limited to the territory of the MSs, is smaller than that of the CFP. DG MARE launched an open consultation in 2010 on ways to reduce the incidental catch of seabirds during fishing operations conducted by the fleet of the EU in and outside the waters of the European Union. The results of this study were published in January 2011[24] and should lead to a European Action Plan in accordance with the International Plan of Action for Reducing Incidental Catch of Seabirds in Longline Fisheries, adopted by the FAO in 1999.

This shows that even though the policy adjustments take time, the system of governance of the EU ensures the coherence of its policies when there is political will to achieve this goal.

The Habitats Directive

The Habitats Directive (Council Directive 92/43/EEC of 21 May 1992 on the conservation of natural habitats and of wild fauna and flora) aims to help to preserve biodiversity through the conservation of natural habitats and of wild fauna and flora in the territory of the MS. For this purpose, Special Areas of Conservation (SACs) shall be designated which, together with the Special Protection Areas (SPAs) established pursuant to the Birds Directive constitutes the European ecological network known as 'Natura 2000'.

This Directive applies to the 'European territory of the MSs where the Treaty applies' (Article 2). It therefore covers the Exclusive Economic Zone (EEZ) of MSs where most of the fishing activities of the EU fleets are performed.

The Water Framework Directive

The Water Framework Directive (Directive 2000/60/EC of the EP and the Council establishing a framework for the Community action in the field of water policy) on water has been developed to protect waters against pollution resulting from discharges, emissions and losses of hazardous substances. It seeks the protection of territorial and maritime waters without indication of legal or physical boundaries. However, it is designed more for the protection of inland and coastal waters than offshore waters.

Water quality, particularly in coastal waters where the spawning grounds and nurseries are mostly located, is an important element for the preservation of biological resources. It is therefore an element of environmental policy particularly crucial for the future of fishing, often little known to fisheries professionals.

The Marine Strategy Framework Directive (MSFD)

The Marine Strategy Framework Directive (Directive 2008/56/EC of the EP and the Council of 17 June 2008 establishing a framework for community action in the field of marine environmental policy) sets up a common framework and objectives for the protection and conservation of the marine environment by MSs. It concerns the management of human activities using an ecosystem-based approach. It is a tool for integrated management of maritime areas under the jurisdiction of MSs and aims to ensure the integration of environmental concerns, under the Treaty, in every policy area. In terms of governance, it requires MSs to designate an implementing authority for each marine region or sub-region concerned.

Among other tools intended to achieve 'Good Environmental Status' the MSFD aims to establish a coherent and representative network of marine protected areas (MPAs) responding satisfactorily to the existing diversity of special habitat types, especially those recognized or identified under EU legislation (Habitats and Birds Directives).

Fishing activities are directly affected by the creation of MPAs as such areas require supervision and restrictions on the activities exerted there. Fishermen are therefore frequently prejudiced against MPAs and may confront radical NGOs and positions about nature conservation. The role of public institutions is to arbitrate to ensure sustainable development of fishing in a protected environment. The position taken by the French government in their memorandum on CFP reform in 2011 illustrates what can be done without becoming antagonistic: 'Marine protected areas are indeed not intended as sacrosanct areas where no fishing is possible, but as management of ocean space of a set of activities. This management fits with the overall goal of ecosystem management of fishing areas with the view to foster the maintenance of activities that are sustainable in their three components: social, economic and environmental.'

International agreements

The complexity of the EU governance system is reflected in its external action, in its participation in international fora as well as in the management of its competences, shared or exclusive, and in the management of international agreements.

The logic of shared sovereignty between the MSs and the EU, which is the foundation of European integration, is at the basis of EU action in the definition and management of its environmental policies and fisheries that are, by their nature, largely internationalized.

The resulting complexity leads to high expenditures of energy in MSs and the Commission for the definition of coherent negotiating positions and may nonetheless lead to inconsistencies when the internal coordination mechanisms do not correctly fulfil their role. This was the case with regard to the management of Bluefin tuna in the International Commission for the Conservation of Atlantic Tuna (ICCAT) and CITES in 2010 (Beslier, 2010). It also causes a lack of clarity or divergences in the positions of the Commission and its MSs that certain third countries may take opportunity of to advance their own interests.

Without going into details of all agreements in which the EU has been involved, it may be noted that many international agreements concern both the fisheries and the environment. This involvement is not systematic however, and this increases to a certain extent the complexity of the EU Governance system. Cases of non-participation of the EU in international environmental agreements have been due to either internal conflicts between the Commission and MSs,[25] the absence in some treaties of provisions for the accession of regional economic integration organizations or else non-ratification of these provisions by the other states.[26]

UNCLOS is the basis of international law for fishing activities as well as for the marine environment. The EU is part of it under exclusive competence with regard to aspects relating to conservation and management of fishery resources, and those concerning the provisions of Parts X (right of transit for landlocked states) and Part XI (operating of seabed mineral resources) within the trade policy. It is also part of it under shared competence of Part XII (Protection and preservation of the marine environment).[27] It participates in the work of the Conference of Parties that is limited to procedural issues mainly due to lack of membership of the United States of America (USA) to the Convention. As a consequence, the substantial discussion is taking place in another forum set up by UNGA, the Informal Consultative Process (ICP) on the Law of the Sea (United Nations Informal Consultative Process on the Law of the Sea or UNICPOLOS). This process is essential for the preparation of resolutions of the United Nations General Assembly (UNGA) regarding the future of fishing and its relationship to the environment, such as the issue of deep-sea fishing and its impact on fragile habitats, which is at the origin of the debate on the protection of high-seas biodiversity and sustainable use of genetic resources. However, the EU is not a UN member. Therefore, while the

United States are fully engaged in the ICP, the EU has experienced serious procedural difficulties to express itself in this process until 2011 when the UNGA[28] granted the EU a privileged observer status to participate in its work and its working groups.

The EU is a member of the Convention on Biological Diversity 1992 (CBD), which is an important tool for biodiversity conservation and particularly for Marine Genetic Resources (MGRs). This convention considers 'protected areas' as a valuable tool for the protection of biodiversity (Article 8) and, at the meeting of the Parties in Nagoya in 2010, adopted the establishment target of an MPA network covering 10% of the oceans by 2020. Although the CBD refers to rights and obligations under the Law of the Sea (Article 22) for the marine environment,[29] its work can be transposed to the high seas, more under good governance than strictly legal terms.

The EU is a member of regional sea conventions such as the Barcelona Convention in 1976 and its protocols, or the OSPAR Convention in 1992 and its annexes. It is, at the same time, a member of the RFMOs covering the same geographical areas: the General Fisheries Commission for the Mediterranean (GFCM) and the North East Atlantic Fisheries Commission (NEAFC). The parallel decisions adopted by NEAFC from 2004 to 2008[30] and OSPAR in 2010 for the creation of MPAs to protect the biodiversity of some seamounts in the Northeast Atlantic owes much to the intervention of the EU (and its MSs) which, because of its location, play a major role in both organizations and is endeavouring to implement a coherent policy for oceans management.

Conclusion

The system of EU governance rests on a context-dependent system of delegations of sovereignty, which would increase the difficulty in efficiently managing the marine environment and its resources if it was only operated by law or regulation.

Experience has shown that coordination mechanisms are far from working with equal efficiency, particularly when it comes to ensure consistency of the CFP with the environmental objectives of the EU and its international commitments. The search for balance between the economic, social and environmental criteria for sustainable development of fishing activities is sometimes hindered by the differing approaches of MSs, not all of whom have the same interests in this area.

The problem of integrated oceans management is not, however, a specifically European problem. The United Nations Convention on the Law of the Sea (1982) has stated that 'the problems of ocean space are closely interrelated and need to be considered as a whole'. Its implementation has however remained largely segmented and different human activities, governed by its various parties, are managed by different international organizations.

Since the Rio conference in 1992 and the World Summit on Sustainable Development held in Johannesburg in 2002, many efforts have been undertaken to develop the tools needed to manage oceans globally. Many States have engaged in efforts to establish integrated maritime policies.

In this context, the Commission decided on 2 March 2005 to launch a broad consultation process on the possibility of an EU maritime policy that transcends questions of jurisdiction and policies segmentation. This consultation, involving both the States, business communities and NGOs, led to the adoption of a Green Paper by the Commission in 2006[31]. This document helped to frame the debate on the issue. It highlights Europe's maritime identity and leadership, which is worth preserving at a time when environmental pressures are threatening the future of maritime activities. Consequently, the maritime policy must aim to promote a maritime industry that is innovative, competitive and environmentally friendly. In addition to maritime activities, the Green Paper proposes that the approach includes the issue of quality of life in coastal regions. With this aim in mind, the Green Paper considers what new tools and modes of maritime governance should be developed.

After a further and very open consultation exercise in 2007, it resulted in proposals for concrete actions summarized in the Blue Book,[32] accompanied by a detailed action plan, some elements of which aim to improve the relationship between fishing and environmental protection. The process focuses more on the preparatory work and the organization of debates upstream of the decision than on the decision itself, which remains subject to the formal rules of procedure of the Treaty. This is particularly important for the implementation of the ecosystem approach which, because of its complexity, calls on non-conventional modes of governance.

Three elements of this Action Plan contribute to achieving this objective and are substantially overlapping: (1) coastal area management; (2) planning of maritime space; and (3) implementation of the ecosystem approach in the fishery sector.

The European Union also urges this process to ensure the place of the European Union in global competition. With the Lisbon Strategy, the EU favours the economic approach as part of a global competition in which they participate. The recent result of Rio+20 confirms that the balance between the three elements of ecological, economical and social goals is of growing importance.

This objective is repeatedly reiterated: 'the Green Paper is grounded on the existing Community policies and initiatives and is in line with the Lisbon Strategy'. It intends to ensure 'the preservation of Europe's leadership in sustainable development of maritime activities'. Among the areas most characterized, the Green Paper refers to the maritime industry, transport, port activities, the growing appeal of coastal areas, coastal tourism and spatial planning for the expanding maritime economy. As stressed by one author, the strategy barely disguises priorities (Billet 2006).

This orientation is characterized by a Commission's communication[33] which aims to: (1) improve knowledge of the seas and oceans, as a cross-cutting tool of the EU's integrated maritime policy; and (2) contribute to a smart and sustainable growth of maritime industries in a better respect of marine environment.

The fisheries policy is one component of this set, the stakes of which lie at the crossroads of fundamental questions: biodiversity, food and economy. In a sustainable perspective it is their balance that will be critical, which implies that the European Commission, Parliament and the Council will have to strengthen their coordination procedures.

Although these elements are largely under the competence of MSs, particularly as regards land use, the development of a common approach and better coordination between MSs enables development of common principles, strengthening of transboundary cooperation and creation of synergies between the policies, whether European or national.

The CFP has been reformed every ten years after its adoption in 1983. The 2002 reform introduced commitment to ensure the integration of environmental concerns into fisheries management. In spite of that, European fish stocks continue to be overfished (approximately 75% of them). Fragile fish stocks lead to suboptimal catches under the Maximum Sustainable Yield (MSY). The last proposal of the Commission for a new reform of the CFP proposes to introduce a deadline to achieve MSY for all European stocks by 2015. The EU subscribed to this objective almost 30 years ago in UNCLOS (Articles 61.3 and 119.1a) and reiterated it in the 1995 UN Fish Stock Agreement (UNFSA Article 5b), in 2002 in the Johannesburg Declaration[34] and in 2010 in Nagoya.[35] The best indicator with which to judge the success of this policy would be the realization that all European stocks are exploited at the level of MSY. The next challenge for the CFP would be to develop tools for the implementation of the ecosystem approach to fisheries. All these reforms will however remain just words if the control system of EU fisheries does not function correctly.

Notes

1. 'When the Treaties confer on the Union exclusive competence in a specific area, only the Union may legislate and adopt legally binding acts, the Member States being able to do so themselves only if so empowered by the Union or for the implementation of Union acts' (Article 2.1 TFEU)
2. 'When the Treaties confer on the Union a competence shared with the Member States in a specific area, the Union and the Member States may legislate and adopt legally binding acts in that area. The Member States shall exercise their competence to the extent that the Union has not exercised its competence. The Member States shall again exercise their competence to the extent that the Union has decided to cease exercising its competence.' (Article 2.2 TFUE)
3. The European Court of Auditors' Special Report No. 7/2007 of 4 December 2007, on the control, inspection and sanction system relating to the rules on conservation of Community resources.
4. Council Regulation (EC) No. 1005/2008 of 29 September 2008 establishing a Community system to prevent, deter and eliminate illegal, unreported and unregulated fishing (IUU).
5. Council Regulation (EC) No. 1224/2009 of 20 November 2009 establishing a Community system for ensuring compliance with the rules of the common fisheries policy.
6. Council Regulation (EC) No. 2371/2002 of 20 December 2002 on the conservation of fisheries resources under the Common Fisheries Policy.
7. The reference to the 'nationals' to strengthen the fight against illegal fishing was introduced in the 2002 reform of the CFP; it does not correspond to most of the member States' traditions of not applying extraterritoriality in their jurisdictional system.
8. Proposal for a Regulation of the EP and of the Council on the Common Fisheries Policy. COM (2011) 425 of 13 July 2011.
9. In line with the FAO IPOA/IUU quoted supra § 2.1.2.
10. Draft Council Decision authorizing the Commission to negotiate on behalf of the Community, a Protocol amending the international convention regulating the Whaling signed at Washington on 2 December 1946 [COM (92) 316].

11. Council Directive 92/43/EEC of 21 May 1992 on the conservation of natural habitats and of wild fauna and flora.
12. Council Regulation (EC) No. 338/97 of 9 December 1996 on the protection of species of wild fauna and flora by regulating trade therein.
13. Convention on International Trade in Endangered Species of Wild Fauna and Flora (in Washington in 1973).
14. Proposal for a Council decision establishing the position to be adopted on behalf of the European Community with regard to proposals for amendment to the Schedule of the International Convention on the Regulation of Whaling. COM (2007) 821 of 19 December 2007.
15. Proposal for a Council decision establishing the position to be adopted on behalf of the EU for amendments to the IWC. COM (2011) 0495 of 25 August 2011.
16. Communication from the Commission to the Council and the EP. COM (2008) 187 of 11 April 2008.
17. The first European programme for environment was adopted on the 22nd November 1973: 'Programme of action of the European Community on the environment, 1973–1976', OJCE No. X112 of 20 December 1973.
18. The 2nd programme (1977–1981) was published on the 17th May 1977 (OJEC No. C139 of 13 June 1977), in which water appears as a priority with air and noise. The 3rd programme (1982–1986) was published on the 7th February 1683 (OJCE No. C46 of 17 February 1983) and the 4th programme (1987–1992) on the 19th of October 1987 (OJCE No. C328 of 17 December 1987).
19. The 5th programme, adopted on the 15th December 1992, is entitled 'Towards sustainability for the period 1992–2000' and was published on the 1st February 21993 (OJEC No. C138 of 17 May 1993). The 6th programme results from Decision 1600/2002 adopted on 22nd July 2002, entitled 'Environment 2010: our future, our choice' (OJCE L 242 of 10 September 2002).
20. Article 6 of the Decision 1600/2002 entitled 'Objectives and priority areas for action on nature and biodiversity'.
21. Regulation (EC) No 1907/2006 of the EP and of the Council of 18 December 2006 concerning the Registration, Evaluation, Authorization and Restriction of Chemicals (REACH), establishing a European Chemicals Agency, amending Directive 1999/45/EC and repealing Council Regulation (EEC) No. 793/93 and Commission Regulation (EC) No. 1488/94 as well as Council Directive 76/769/EEC and Commission Directives 91/155/EEC, 93/67/EEC, 93/105/EC and 2000/21/EC.
22. Recommendation 2002/413/EC of the EP and of the Council of 30 May 2002 concerning the implementation of Integrated Coastal Management in Europe (OJEC L 148 of 06/06/2002).
23. Fishing activities can impact bird populations either through bycatch (longlines, driftnets) or through competition for feeding resources (in the case of fish-meal fisheries).
24. MRAG study available on the DG MARE's website.
25. Case of the IWC, pending since 1992 (cf. supra).
26. Case of the CITES, of which the EU is not a member even if its area of activity falls under a community competence, the trade policy. The Gaboronne protocol which would allow its membership was adopted in 1983 and received only 40 ratifications on the 60 needed in 2011.
27. Declaration of competence provided by the EU at the time of its approval of UNCLOS. Council Decision of 23 May 1998 (OJEC No. 179 of 23 June 1998).
28. Resolution A/65276 of 3 May 2011 on the participation of the EU in the work of the UN.
29. Even if not specified, it must be understood as the high sea, EEZ and territorial waters are under the jurisdiction of the coastal States where the CBD is applicable.
30. For more details see the 'Response of the North East Atlantic Fisheries Commission, NEAFC, to the Secretary-General of the UN on actions taken pursuant to paragraphs 83-84 of resolution 61/105', available on the NEAFC website.
31. Communication from the Commission to the Council, the EP, the European Economic and Social committee and the Committee of the regions: 'Towards a future Maritime Policy for the Union: A European vision for the oceans and seas' (COM/2006/0275 of 7 June 2006).
32. Communication from the Commission to the Council, the EP, the European Economic and Social committee and the Committee of the regions – An integrated maritime policy for the EU. COM (2007) 575 of 10 October 2007.
33. Communication from the Commission to the EP and the Council: 'Marine knowledge 2020: marine data and observation for smart and sustainable growth' (COM 2010 461 of 8 September 2010).
34. Johannesburg Declaration on Sustainable Development (4 September 2002). Annex: Plan of Implementation of the World Summit on Sustainable Development IV.31.a.
35. CBD 10th Conference of Parties: Strategic Plan for Biodiversity 2011–2020. Target 6.

References

Beslier, S. (2010) Pêche et environnement : complémentarité ou concurrence au sein de l'ordre juridique international? Le cas du thon rouge. *Annuaire du Droit de la Mer* **XV**, 165.

Billet, P. (2006) Towards a maritime policy of the European Union. Talking of that the Commission's green paper on oceans and seas. *LexisNexis Jurisclasseur Environnement* **Août–Septembre**, 2.

FAO (2002) Report of the Reykjavik Conference on Responsible Fisheries in the Marine Ecosystem, Reykjavik,

Iceland, 1–4 October 2001. FAO, Rome, FAO Fisheries Report No. 658.

FAO (2003) Fisheries management: The ecosystem approach to fisheries. FAO, Rome, FAO Technical Guidelines for Responsible Fisheries No. 4 (suppl. 2), 112.

Garcia, S.M., Zerbi, A., Alliaume, C., DoChi, T. and Lasserre, G. (2003) The ecosystem approach to fisheries. Issues, terminology, principles, institutional foundations, implementation and outlook. FAO, Rome, FAO Fisheries Technical Paper No. 443.

TEU (2008) Treaty on European Union (consolidated version). *Official Journal of the European Union* C115/13.

TFEU (2008) Treaty on the Functioning of the European Union (consolidated version). *Official Journal of the European Union* C115/47.

Part V
National governance

Chapter 21

The use of national frameworks for sustainable development of marine fisheries and conservation, ecosystem-based management and integrated ocean management

K. Sainsbury[1], P. Gullestad[2] and J. Rice[3,4]

[1] Institute of Marine and Antarctic Studies, University of Tasmania, Australia
[2] The Norwegian Directorate of Fisheries, Bergen, Norway
[3] Fisheries and Oceans Canada, Ottawa, Canada
[4] IUCN Commission on Ecosystem Management (IUCN/CEM/FEG), Gland, Switzerland

Abstract: International agreements and guidelines provide the overall goals of sustainable development and healthy ecosystems, but it is at the national level that these must be implemented while addressing the interests of the nation and its citizens with decisions that affect people as well as the environment. The national frameworks are 'where the rubber hits the road', that is, where the practical instruments and methods to address these challenges must be developed and implemented. Here we summarize the approaches taken by Australia, Norway and Canada during the past 1–2 decades to meet the challenges of ecosystem-based management of fisheries and biodiversity, looking at: (1) developments in legislation and policy, and convergence between fisheries and biodiversity management; (2) methods to address and prioritize issues for ecosystem-based fisheries management; and (3) integrated cross-sectoral management. There are notable similarities in the approaches of the three countries: change of objectives in legislation and policy; greater clarity and agreement about targets, limits and acceptable/unacceptable impacts; methods to comprehensively identify then prioritize management issues; use of risk-based assessment and management methods that can operate at different information levels; significant restructuring and reduction in fleet capacity; and increasing use of fisher rights. An encouraging conclusion is that there are clear examples of approaches, methods and frameworks for sustainable development of fisheries and for using ecosystem-based fishery management that can be applied, in the first instance, without significant new knowledge. There has also been significant effort to address integrated cross-fishery and cross-sectoral management even though all countries are still struggling somewhat with this issue.

Keywords: ecosystem-based fishery management (EBFM); integrated oceans management; marine legislation; marine policy

Introduction

International agreements and guidelines help frame the overall intentions and goals of Sustainable Development, but it is at the national level that these must be implemented and made practical while addressing the interests of the nation and its citizens. It is through the national legal, policy and management frameworks that the high-level intentions and goals of sustainable fisheries in a healthy ecosystem are turned into decisions that affect people as well as the environment.

Governance of Marine Fisheries and Biodiversity Conservation: Interaction and Coevolution, First Edition.
Edited by Serge M. Garcia, Jake Rice and Anthony Charles.

Recognizing this, the key international documents on sustainable fishery development such as the Code of Conduct for Responsible Fisheries and associated guidelines (FAO, 1995, 1996, 2003; UNCED, 1992; UNCLOS, 1982; UNFSA, 1995) and UN General Assembly Resolutions on sustainable fisheries (UNGA, 2013) all presuppose a strong and effective national management framework, and provide guidance on what it should achieve. Where national frameworks are not strong and effective much of the international funding for capacity building to enhance sustainable fisheries (e.g. GEF, 2013) is targeted at improving the legal, policy, management and scientific capabilities of those frameworks. The goal of the capacity building is to enhance the effectiveness of the national frameworks to deliver balanced and sustainable development as envisaged in the international documents. Where co-management arrangements occur, some aspects of national management responsibility are delegated or shared with local organizations (usually a fishing or community organization). Co-management can be an effective component of national frameworks but the nature, role and outcomes of co-management must be consistent with the goals in these international documents (Evans et al., 2011; Gutiérrez et al., 2011).

The national frameworks are 'where the rubber hits the road', that is, where the practical instruments and methods to address these challenges must be developed and implemented. This invariably involves choices and tradeoffs among interest groups, competing users and short- versus long-term outcomes. Further, most jurisdictions have different legislation and institutions for managing the different industry sectors that affect the marine environment and for policies and measures aimed specifically at conservation of the marine environment and its biodiversity. These are referred to through this book as the fisheries (or other sectoral) governance stream and the biodiversity conservation stream, respectively. Even with the best of intentions, sectorally and nationally, this arrangement could give inconsistent policy and management, unintended and/or unconsidered consequences and conflict.

In the last 10–20 years many countries have explored ways to better deliver sustainable fisheries and to integrate management of marine biodiversity across the various industry sectors that all operate in the same ecosystem. These efforts have variously been labelled integrated ocean management and ecosystem-based management. Notwithstanding substantial overlap in the concepts and practical actions involved with definitions and intentions, approaches differ in detail among all jurisdictions. Consequently, experiences have been different in different countries and there is much to be learned from comparison. Here we summarize the approaches taken in three countries: Australia, Norway and Canada. These are all developed countries with strong central governments and significant capacity in marine science, policy and management. Not all elements of the approaches used in these three countries are expected to readily transfer to other contexts, and other chapters specifically address the developing world (cf. Chapters 22–25). These three countries have put significant effort into addressing sustainable fishery development and integrated oceans management however, with policies explicitly endorsing ecosystem-based approaches. They provide good examples of the approaches and outcomes in situations where most of the capacities and frameworks envisaged by the international agreements for sustainable fishery development are in place. Comparison of their experiences may highlight the common elements used to address an ecosystem-based approach to fishery management (which includes impacts of fisheries on biodiversity), provide a suite of potentially different ways to address the elements and identify some remaining challenges. Furthermore, while the technical and management capability in these countries is high, in each country there is still incomplete understanding of the marine ecosystems and human impacts on them and many (mostly small) fisheries that are poorly monitored and assessed. Consequently, even these comparatively rich countries must address questions of how to achieve sustainable development and a healthy ecosystem with incomplete information, and the approaches taken to this are likely to be useful more generally.

National framework for fishery and conservation management in Australia

Australia is a federation with state governments and a central Commonwealth government. Broadly the states are responsible for management of land, water and coasts to three nautical miles while the Commonwealth government is responsible for matters relating to international commitments, from 3 nautical miles out to the boundary of the Exclusive Economic Zone, and for several classes of activity that cross jurisdictions. McPhee (2008) provides an excellent overview of fisheries management in Australia, while Gullett et al. (2011) address some specific aspects including coastal zone management and integrated oceans management.

Developments in legislation and policy, and convergence between fisheries and biodiversity management

An early and important step in the development of a coherent approach to sustainable fishery and conservation management was adoption of the principles of Ecologically Sustainable Development (ESD; see Anonymous, 2012), initially through the 1992 Intergovernmental Agreement on the Environment and later into relevant government Acts and regulation. These principles derived from and resemble the internationally recognized principles of Sustainable Development (i.e. UNCED, 1992). They include:

- integration of economic, social and environmental considerations;
- the precautionary principle and intergenerational equity; and
- use of valuation, pricing and incentive mechanisms.

A National Strategy for ESD was also developed in 1992 that included tailored strategies for each of the main management sectors (e.g. agriculture, fisheries, tourism) and inter-sectoral management (e.g. biological diversity, water and pricing and taxation). The strategies varied greatly in their detail, and the inter-sectoral strategies were generally brief.

ESD as applied to fisheries management is very similar to Ecosystem-Based Fishery Management (EBFM; Pikitch et al., 2004; WWF, 2002) and the Ecosystem Approach to Fishery Management (EAFM; FAO, 2003), particularly with regard to objectives and principles.

The principles of ESD were successively introduced into the main fisheries management and environmental management legislation of all Australian governments during the 1990s. By the early 2000s the same definitions of ESD, and related elements such as precaution, were in the Fishery Management Act and the Environment Protection and Biodiversity Conservation (EPBC) Act. The EPBC provides integrated environmental legislation for, among other things, the promotion of ESD. Two important mechanisms for integration between environmental and fishery management are as follows.

- The Minister for environment must approve all Commonwealth-managed fisheries, and any State-managed fishery that exports its products, as being ecologically sustainable. The assessment is conducted by the environment department (Anonymous, 2007). The assessment criteria are mainly high level and operational interpretation is on a case-by-case basis. These assessments must be repeated at least every 5 years. Approval can be given with conditions, and progress against these is periodically assessed.
- The criteria for EPBC recognition of commercially fished species that are threatened are linked to the limit reference point specified in the Harvest Strategy Policy. Species that are above the biomass limit reference point in the HSP will not be listed under any category of threat. Species that are below the limit reference point could be considered for listing. If a species is below the limit reference point and subject to a rebuilding plan of the Australian Fishery Management Authority (AFMA), the species could be listed as Conservation Dependent and the plan must be jointly agreed and monitored by AFMA and the Minister for environment. Higher-threat categories (vulnerable, endangered, etc.) can apply to species that are more substantially depleted; for these species the Minister for environment is then solely responsible for the recovery plan.

These mechanisms have helped to keep the principles of ESD high on the agenda of fishery management agencies. However, a recent review of the EPBC (Anonymous, 2011a) recognized that current processes are administratively inefficient (e.g. duplication between fishery agencies and the environment department) and have under-specified standards (e.g. no explicit agreement or necessary consistency between the 'acceptable impact' or precautionary benchmarks used by the fishery and environment agencies). An option being explored is for the environment department to audit the performance of fishery agencies rather than provide separate assessments, which will require greater clarity in the performance standard for both fishery management and its audit.

Methods to address and prioritize issues for ecosystem-based fisheries management

By the early 2000s ESD provided Australia with a good policy and legal basis for EBFM and for coherence between fishery management and conservation of biodiversity. However the ESD principles are 'high level' and the challenge since then has primarily been about how to operationalize them in fishery management. This operalization was delivered through three sets of developments that were broadly sequential and that continue to evolve: (1) the Fishery ESD reporting framework; (2) rules for setting exploitation levels for target species and methods for ecological risk assessment/risk management for non-target species; and (3) methods for integrated assessment and prioritized management of all the fisheries operating in the same ecosystem.

The Fishery ESD reporting framework

This provides a method to systematically identify the ESD-related issues for a fishery and risk-based methods to prioritize them (Fletcher et al. 2002, 2005; Fletcher, 2005). It helps to ensure that all relevant issues are taken into account and that consideration is not overwhelmed by the number of potentially relevant issues. It also provides a hierarchical link between high-level objectives and the tools of operational fishery management: indicators, performance benchmarks (e.g. target and limit reference points) and triggered management responses. The methodology is very well suited to workshop applications, and this is mainly how the framework has been used.

The framework begins with the three high-level components of Fisheries ESD: 'ecological wellbeing' (e.g. target species, bycatch and the broader ecosystem); 'human wellbeing' (e.g. economic and social considerations); and the 'ability to achieve' the other two components (e.g. governance and management arrangements). These components are then hierarchically elaborated through layers of subcomponents that are increasingly specific and operational. The terminal subcomponents are each expressed in the form of an operational objective that is practically assessable. These operational objectives are then prioritized by a qualitative likelihood-consequence form of risk assessment.

This methodology was applied to many Australian fisheries during the early 2000s, and continues to be used explicitly in Western Australia. Related approaches are used in other Australian jurisdictions (e.g. Hobday et al., 2011; Smith et al., 2007). It has also been applied in some non-Australian fisheries (Fletcher, 2008) and it influenced the FAO guidelines on the Ecosystem Approach to Fishery Management (FAO, 2003).

Rules for setting exploitation levels and methods for ecological risk assessment and management

The fisheries managed by the Commonwealth government of Australia illustrate several features that have supported practical Fisheries ESD and EBM. The governance model has been important in this, as have some specific approaches to support decision making.

Domestic fisheries in the Commonwealth jurisdiction are managed by AFMA which was established in 1991 by legislation to provide operational fishery management at arm's length from the Minister and other government processes. AFMA fisheries have been limited entry since its inception (limited entry was introduced into most Australian fisheries in the 1970s) and are mostly controlled through individually owned and tradable statutory fishing rights on either catch quota (i.e. individual transferable quota or ITQ, the preferred option) or unitized fishing effort units. AFMA's consultation processes have extensive participation from fishing, scientific and environmental interests. The Commonwealth department of fisheries is responsible for auditing the performance of AFMA and, since 1992, has produced an increasingly comprehensive annual report on the status of AFMA managed fisheries.

This stakeholder engagement and external assessment has helped AFMA make many of the management changes necessary to address Fisheries ESD. Three important developments are described in the following.

(1) Adoption of a specific HSP that sets targets and limits for exploitation of target species and provides a default 'decision rule' or 'control rule' relating harvest levels to current stock status (DAFF, 2007). The target is maximum economic yield. The decision rule linearly reduces fishing mortality as the current biomass decreases below the biomass estimated to produce maximum sustainable yield (MSY) and requires zero-targeted exploitation if current biomass is below half the MSY level. Alternative decision rules and fishery controls can be used if the alternatives can be shown to meet the performance requirements of the HSP. This has led to a wide range of equivalent harvest strategies being developed for 'data-poor' fisheries (Dowling et al., 2008; Prince et al., 2011).
(2) AFMA uses formal Ecological Risk Assessment (ERA) and Ecological Risk Management processes to assess and manage fishery impacts on the ecosystem including bycatch and habitats (AFMA, 2010). Risk levels are used to prioritize investment and allocation of management effort. The risk assessments use a range of methods (Hobday et al., 2011), from qualitative through semi-quantitative to quantitative (e.g. Zhou and Griffith, 2007). This allows assessments with different levels of background knowledge. There is more precaution built into the risk assessment methodology

at low information levels. Risk management can be particularly effective if it uses measures that simultaneously address the risk to many ecosystem properties (e.g. spatial management, fishing gear including exclusion devices).

(3) A major structural adjustment in 2005 removed about one-third of fishing capacity and increased profitability for the remaining participants (Vieira et al., 2010). This was an intervention intended to broadly align fishing capacity with fishery productivity, with autonomous adjustment mechanisms (e.g. individual transferable catch entitlements) expected to maintain that situation in the longer term.

Integrated assessment and prioritized management of all the fisheries in the same ecosystem

The state government of Western Australia has further developed the ESD reporting framework to address all fisheries (recreational and commercial) in the same ecosystem and to include non-fishery activities (Fletcher et al., 2010, 2012). This is a step towards taking the more ecosystem-centric view of EBFM (Pikitch et al., 2004) and towards integrated ecosystem-based management.

This uses the same hierarchical approach as the Fisheries ESD reporting framework. The scope is wider however and so the number and range of potentially important issues is correspondingly greater. The risk assessment and prioritization methods were enhanced to address this increased complexity by (1) consolidating the individual risks into regional-level risks and (2) using multi-criteria analysis to integrate the ecological, social and economic risks into fishery management priorities. The multi-criteria analysis includes a weight for the extent of control possible by the fishery agency. This approach has successfully produced a manageable number of prioritized issues that are then used to allocate agency investment, all on the basis of existing information. It has also supported broader policy efforts to integrate commercial and recreational fishery management.

Integrated cross-sectoral management

The Commonwealth government's Oceans Policy (Anonymous, 1998) was a significant attempt at applying the principles of ESD for integrated management at the scale of regional ecosystems. Oceans Policy envisaged 'whole of government' support for implementation, including enhanced use of existing mechanisms for cross-sectoral and cross-jurisdictional coordination. It was motivated by the need for more integrated planning and management of the various human activities that use or impact the same marine ecosystem (i.e. commercial fishing, recreational fishing, aquaculture, oil and gas extraction, conservation, coastal zone infrastructure, land-based pollution, shipping, irrigation and water diversion, etc.), for better dispute resolution mechanisms when management of these various human activities is conflicting and to establish a National Representative System of Marine Protected Areas (NRSMPA).

Implementation of the Oceans Policy did not occur as initially envisaged, however. It was not possible to achieve the necessary cross-sectoral and cross-jurisdictional coordination by policy alone. Even across sectors managed by the Commonwealth there was very limited success; there was even less success in integrating across state and Commonwealth management. So in 2004 Marine Regional Planning under the Oceans Policy was replaced by Bioregional Planning under a section of the EPBC Act. This allowed regional planning to continue but it changed the nature of the process and the objectives. The decisions are ultimately made by one sector in one jurisdiction under its relevant legislation (the Commonwealth jurisdiction and the environmental management sector), rather than as shared decisions across sectors and jurisdictions. The focus of planning changed from multiple use management to be primarily about identifying and protecting environmental values. The main planning outcome became introduction of the NRSMPA. The Marine Protected Area network is an important tool in biodiversity management, but it is only one aspect of multiple use management to achieve regional ESD outcomes. It is likely that enabling legislation will be necessary to achieve the originally intended integration.

National framework for fishery and conservation management in Norway

Norway is a constitutional monarchy where the State is responsible for the management of both environment and resources, including fish, in waters inside 200 nautical miles. The State is responsible for matters relating to international commitments. Norway's marine fisheries, generally considered to be well managed (e.g. Pitcher et al., 2009), produce about 2.5 million tonnes per year while salmon-based aquaculture production exceeds one million tonnes. Gullestad et al. (2014) describes how the Norwegian fisheries management framework has evolved over the last 40 years.

Developments in legislation and policy, and convergence between fisheries and biodiversity management

Conservation and sustainable use are integrated principles in both the Marine Resources Act (MRA, 2009) and the Nature Diversity Act (NDA, 2009). The previous fisheries Act focused mainly on the exploitation of commercial stocks whereas the new Act applies to all living marine resources. The previous nature management Act focused solely on conservation and protection, whereas the NDA states that its purpose is to conserve biodiversity through conservation and sustainable use. The MRA states that its purpose is to ensure sustainable and economically profitable management of the resources, and several provisions describe conservation of biodiversity as an integral part of sustainable management. Both acts reflect recent developments in international law with regard to biodiversity conservation and fisheries. By integrating conservation and sustainable use as basic principles, the two new laws represent a regime shift in the Norwegian regulatory system.

All human activity has an impact on nature and biodiversity (a footprint). The MRA addresses those caused by the fisheries sector. If all or some of the impacts of human activities stem from sectors other than fishing, and the effects threaten biodiversity, then measures in the NDA may be applied, limited however to the area inside 12 nautical miles from the coastal baseline. It is stated in the act that measures taken in such cases shall be weighed against other important public considerations. Social and economic sustainability are among the relevant public considerations.

The legal basis for good governance of marine resources in an ecosystem context is therefore in place. This does not preclude disagreements between biodiversity conservation and the fisheries sector on what acceptable footprints are. If not resolved at a lower level, such disagreements are to be addressed in consultations between the Ministry of Fisheries and Coastal Affairs and the Ministry of the Environment, and final decisions may eventually have to be taken by the government.

The previous (traditional) fisheries legislation did not prevent substantial progress and improvements in fisheries management. Over the last 20–30 years there has been a dramatic change in the way the economically most important Norwegian marine resources are managed. The major driver for this change was a growing political recognition in the late 1980s that long-term social sustainability could not be achieved at the expense of ecological and economic sustainability. At that time important stocks were depleted and fisheries were heavily subsidized. Later international commitments also gradually became an important driver for change.

Methods to address and prioritize issues for ecosystem-based fisheries management

Management of economically important marine resources

The management of the 'most important' resources, accounting for approximately 85% of total Norwegian first landed value, is supported by international scientific advice from ICES (the International Council for the Exploration of the Sea), based on extensive effort in population monitoring and analytical stock assessments. This advice is used by Norway to guide comprehensive fishery management and control decisions.

Through the closure of the commons by successive introduction of limited entry schemes, the termination of subsidies and the introduction of pervasive fleet structural adjustment measures Norway succeeded in reducing the fishing fleet and halting the growth in fishing capacity (Directorate of Fisheries or DoF, 2010). The structural adjustment measures were initially vessel scrapping schemes but have gradually been replaced by license aggregation schemes. The offshore fisheries were limited access before 1985, and the major coastal fisheries have gradually been made limited access between 1990 and 2004. Remaining open access opportunities in certain coastal fisheries are small scale, and are thought to cause only a minor impact on stocks and on investment behaviour.

Access rights were allocated to those participating in the fishery at the time when open access was closed. An access right is simultaneously connected to both the fisher (company) and the vessel, and gives the holder the right to fish a certain share of the annual national quota of one or more stocks with a particular vessel. The individual quotas are not tradeable but within fleet groups, and subject to regional restrictions, access rights may be aggregated on the condition that the submitting vessel, or a similar vessel, is removed from the fishing vessel register and scrapped. The receiving vessel will hold the extra quotas for a period of 20 years before that access right is returned to the State. It is a politically expressed condition by Parliament that such rights at that time will be shared among all remaining vessels in that particular fleet group.

Management strategies and Harvest Control Rules (HCR) for quota setting based on the Precautionary Approach, along with improved technical measures and an extensive and strict enforcement regime, have contributed to the rebuilding of depleted stocks (DoF, 2010) and laid the foundation for improved profitability in fisheries. So far the management strategies are based to a large extent on single-species analyses. A management strategy begins with the overall management objectives for the stock, including maximization of long-term sustainable yield and minimization of annual variation in quotas. Common elements of HCRs are a target fishing mortality or *F*-value, mechanisms to reduce *F* when the spawning stock falls below a predetermined precautionary limit and, in many cases, mechanisms to curb annual variation in Total Allowable Catch (TAC). All (exept one) of the economically important stocks are shared stocks where the HCRs are negotiated internationally, bilaterally or multilaterally, and ICES is asked to undertake simulation tests to examine whether negotiated outcomes are in accordance with the precautionary approach.

The reduction in number of fishermen and vessels has facilitated increased productivity and profitability for those remaining (DoF, 2010). The industry's economic sustainability is thus considerably strengthened; fewer vessels and fishermen have reduced the industry's role in the maintenance of regional communities and employment. Departure from fishing as a livelihood has however occurred during a period of generally low unemployment, and so good alternative employment opportunities were available.

Ecological sustainability has been radically improved. The aggregate spawning stock of the ten most economically important stocks for Norwegian fisheries has more than tripled since the late 1980s (ICES, 2012). As most of these stocks are transboundary, Norwegian efforts have in particular focused on reaching agreement on sustainable long-term management plans and efficient HCRs for the most important resources, improved technical regulations (including a ban on discards which presently encompasses 55 managed as well as bycatch species and real-time closures for the protection of juveniles) and enforcement and surveillance issues (e.g. illegal, unreported or unregulated or IUU fishing, port State control, vessel monitoring systems, electronic recording and reporting systems, etc.). Further development to optimize the management of the economically important stocks in an ecosystem-based context will proceed along three parallel tracks:

- increase the economic output through improvements in exploitation patterns and reduction in all forms of incidental and unwanted mortality on target species from fishing;
- further optimize the long-term economic yield through possible revisions of management strategies and harvest control rules; and
- as new scientific knowledge becomes available, additional ecosystem considerations is gradually incorporated into management measures including: multispecies interactions; effects of fishing on benthic habitats; and the effects of bycatch of fish, seabirds and marine mammals.

These three bullet points represent the practical approach to EBFM for the economically most important resources for the Norwegian fishing industry. Consistent with the third bullet point, the new Marine Resource Act enables fisheries management to particularly consider possible impact from fisheries on non-commercial species and bottom habitats when regulating fisheries.

What about the economically less-important resources?

In the last decades the Norwegian goal and focus has been on the rebuilding of the economically most important fish resources. Resources of minor economic significance have not been subject to the same research and management efforts. Some of these resources are in a depleted state and are included on the official Norwegian Red List of species at risk (Anonymous, 2010). The Red List, issued for the first time in 2006, is managed by the Norwegian Biodiversity Information Centre and listing is based on the IUCN criteria. The assessment of species is conducted by groups of experts recruited from relevant scientific institutions. As part of the development of ecosystem-based fisheries management, more attention is in recent years being paid to resources of less economic significance. This is in particular the case for Norwegian Red-Listed species for which fisheries management now has a legal obligation to assess appropriate management measures.

However, for at least two reasons, management of these species differs from that of the resources of greater national economic importance. First, applying the same level of research, monitoring, management and control costs would not be cost effective. Second, these are in many cases coastal resources with a large and unknown number of recreational fishers contributing significantly to the exploitation. For these fisheries, restrictions related to gear, fishing period or fishing area are in most cases the preferred management tools.

Different management objectives for large resources, small resources and non-commercial (bycatch) species

Stocks and species accounting for close to 90% of total Norwegian first-landed value may be managed with the objective to optimize long-term economic yield. Stocks with limited information, but still of some economic importance, account for about another 5% of the total first-landed value. These will be managed with the objective to secure a high and, if possible, stable long-term yield. Catches may however, from time to time, be higher or lower than what with more knowledge would have been regarded as optimal.

For the many species that constitute the last 5% of landed value such ambitious objectives are not set. Similarly, such objectives do not apply to non-commercial species, including incidental bycatch of seabirds and marine mammals for which there is no intended catch. Rather, a general and absolute minimum objective for all species, stated both in the Marine Resources Act and the Nature Diversity Act, is the protection of biological diversity. More specifically the objective is to ensure that fishing does not threaten either species or the functioning of ecosystems. The term 'threaten' does not in this context have a strict legal interpretation. Management measures for these species are applied on a case-by-case basis when considered necessary based on the Norwegian Red List, ICES advice and input from the Institute of Marine Research.

Decisions on management objectives for the various species and stocks have turned out to be an important and integral part of the development of EBFM. When this process started in Norway in 2009, it revealed unclear management objectives for several species and stocks. After consultation with stakeholders in 2009 and 2010 this is now rectified (DoF, 2012) although further revision of objectives in some cases can be anticipated.

The need to obtain an overview and to prioritize: The Stock Table and the Fisheries Table

To provide a relatively simple but systematic and updated overview of what might be the relevant issues, the Directorate of Fisheries has developed two checklists – the Stock Table and the Fisheries Table (DoF, 2012) – that provide an overview of issues related to all stocks and fisheries relevant to Norwegian management. So far, 80 species/stocks and 57 fisheries have been included, with extension possible for new stocks, fisheries or emerging issues.

The Stock Table, introduced in 2009, includes information on the status of the stock, exploitation level, management objectives and priorities for action. The Fisheries Table, introduced in 2011, includes information for each fishery such as size selectivity of the gear, discard issues, incidental mortality, habitat impacts and priorities for action.

The elements of the two tables are qualitatively graded or prioritized according to impact or importance, usually based on expert judgment and harmonized within each individual column for consistency and objectivity between stocks or fisheries. Input from science and stakeholders have contributed to this process and explanation/justification for the grading is recorded (DoF, 2012). Because the grading is explicitly expressed as just providing an overview of a complex reality, with no direct regulatory consequences, the process with science and stakeholders has been constructive and non-conflicting.

Both tables are now updated each spring and priorities are discussed with stakeholders. The outcome of these discussions feeds into preparation of the following year's budget proposal by the Ministry of Fisheries and Coastal Affairs to the Parliament, and influences the Ministry's Letter of Expectations each December to the Directorate of Fisheries (DoF) and the Institute of Marine Research.

A major objective of the two tables is to achieve an integration of previous 'target-species focused' approaches to fisheries issues with the emerging and broader conservation issues, into a single prioritization process. The two integrated tables are intended to avoid two different and separate management streams with different sets of players and agendas, resulting in conflicting objectives and measures.

The dialogue and planning associated with the two tables does not resolve the social and political questions about what are acceptable footprints of each fishery. They do clarify the issues and concerns however, and give stakeholders and government an annual opportunity to prioritize where solutions are most needed. In this way they help the process of understanding and achieving the practical actions and balances needed for ecosystem-based fisheries management, and may contribute to reducing the traditional distrust and mutual lack of understanding between conservation and fisheries participants.The tables are not designed to cover cross-sectoral issues related to competing use or impact on fishing from other industries such as oil, shipping, offshore wind energy and aquaculture.

Integrated cross-sectoral management

During the 1990s, the conflict of interest between an expanding offshore oil and gas industry and the traditional fisheries sector regarding use of specific areas at sea became pronounced. At the same time, environmental concerns were raised about effects of both pollution and traditional fisheries on vulnerable marine biota, including overfished fish stocks.

The concept of Integrated Ocean Management Plans for the Norwegian waters was therefore launched in 2002 and the Lofoten–Barents Sea chosen as the first plan area. The Ocean Management Plans were expected to have a cross-sector holistic approach. Representatives of all Ministries and their underlying directorates, as well as institutions that had responsibilities within the ocean areas, participated in the development of the plans. Given its overall responsibility for the environment in areas under Norwegian domain, the Ministry of Environment led the planning process (MoE, 2006, 2009). Reports on the marine environment and on living marine resources, on fisheries, shipping, offshore oil and gas activity and on vulnerable and particularly valuable areas were prepared. Based on these reports, the government presented the first Lofoten–Barents Sea ocean management plan as a white paper to Parliament in March 2006 (Anonymous, 2006; Olsen et al., 2007; Winsnes and Skjoldal, 2008). A revised Lofoten–Barents Sea plan was presented to Parliament in March 2011 (Anonymous, 2011b). A similar plan for the Norwegian Sea was presented to Parliament in 2009 (Anonymous, 2009; Ottersen et al., 2011) and one for the Norwegian sector of the North Sea is expected in 2013. In general these plans do not raise major or particular biodiversity concerns with regard to fishing, although impacts of trawling are discussed as an issue in relation to vulnerable habitats.

So far these plans have had an important impact on decisions with regard to allowing for oil and gas extraction in vulnerable areas.Their direct impact on fisheries management has been limited, but they have contributed to a general political appreciation of marine management that has been helpful in the evolution of an ecosystem-based fisheries management. At the same time, to be successful these cross-sectoral plans wil be completely dependent on comprehensive and effective ecosystem-based management of all relevant sectors, including fisheries.

National framework for fishery and conservation management in Canada

Canada is a constitutional monarchy with the federal government having responsibility for all living aquatic resources and their habitats. Management of most freshwater fisheries and varying components of freshwater habitat protection are formally delegated to the provincial or territorial governments, with case-by-case differences. In addition Aboriginal Bands have many special rights of accesses to resources under a complex set of agreements and judicial rulings. Several aspects of habitat protection, particularly with regard to pollution and contaminants, rest with Environment Canada. Provinces and Territories also have jurisdiction over aquaculture within 3 miles of the high tide mark, and a small number of inter-tidal and near-shore fisheries.

Developments in legislation and policy, and convergence between fisheries and biodiversity management

The primary legislation in Canada for the management of fisheries is the Fisheries Act. First adopted in 1858 following Confederation, it has undergone a number of revisions with the most recent major overhaul in 1985 and revisions to the habitat provisions in 2012. It provides a comprehensive foundation for the management of commercial fisheries and for the conservation and sustainable use of fish stocks (defined to include invertebrates and marine mammals) and their habitats. However, few provisions in the Fisheries Act explicitly address the responsibilities of fisheries management relative to biodiversity more generally, beyond the provisions to protect productivity of aquatic habitats directly supporting fisheries. Canada's Oceans Act (1997) and the Species at Risk Act or SARA (2002) provide the legal basis for broader responsibility for biodiversity. The Oceans Act includes guidance on

Integrated Management, the Ecosystem Approach and the Precautionary Approach in decision-making, as well as provisions for the establishment of networks of Marine Protected Areas (MPAs). SARA specifies processes to identify species at risk of extinction or extirpation and for listing them for protection and recovery. It provides mandatory prohibitions on harming or killing individuals of listed species or damaging their critical habitats, unless the activities causing the harm are explicitly authorized by the Minister of the responsible jurisdiction.

Under the Fisheries Act authority for most decision-making is vested in the Minster of Fisheries and Oceans. The Minister must consult widely prior to decision-making, including with other levels of government, other federal departments when relevant and the public. A number of court decisions have clarified what constitutes 'consultation', particularly with regard to Aboriginal Bands. The Minister is routinely provided with scientific advice to inform the decisions, generally through formal structured science advisory processes. However, 'Ministerial Discretion' is not required to follow the input from the consultations and scientific advice as long as the Minister's decisions are consistent with the Act and relevant judicial interpretations.

Fisheries management and strategic policy are provided by the Sustainable Fisheries Framework (DFO, 2009; SFF, 2009). The SFF is 'comprised of policies and tools designed to help ensure that Canada's fisheries are environmentally sustainable, while supporting economic prosperity'. The SFF establishes the policy basis for 'implementing an ecosystem approach to fisheries management' and moving from the historical single-species approach. Although developed independently from the Integrated Management approaches under the Oceans Act, the SSF complements those approaches.

The central pillar of the SFF is a Fishery Decision-Making Framework Incorporating the Precautionary Approach, intended to guide the management of all targeted stocks in commercial fisheries, and provides the conceptual basis for managing bycatches and fisheries effects generally. It has the classic structure of the 'three-zone' FAO Precautionary Approach framework (FAO, 1996), but it also meets principles for the application of precaution to science-based decision-making in the federal government that were provided by the Canadian Privy Council Office (Government of Canada, 2003). Implementation is aided by Harvest Control Rules although these have only been developed for a minority of managed stocks.

The Policy on New Fisheries on Forage species was originally developed in 1998, but was revised before adoption in 2007 as an early part of the SSF. The Forage Fisheries policy includes objectives for status of target, bycatch and ecologically dependent species and for ecological relationships, making these biodiversity considerations part of the SSF. It also includes guidance on risk management and conditions necessary for sustainable fisheries. The policy has been widely supported regionally, due in part to broad consultation before adoption and the clarity of the prerequisites specified before opening new fisheries on forage species.

The SSF also includes a Policy for Managing the Impacts of Fishing on Sensitive Benthic Areas (PMIFSBA, 2010). It comprises a list of familiar principles including application of the Ecosystem Approach and Precautionary Approach, Integrated Management, risk assessment, shared stewardship and graded protection of areas. From these principles it identifies a series of processes to be applied in guiding fisheries management. They cover information collection on benthic habitats and fishing patterns, risk analysis by regional fisheries managers, selection and implementation of management measures and monitoring and periodic evaluation of experience with the Policy (DFO, 2010). Implementation has been largely independent of the Ecologically and Biologically Significant Area approaches of the Oceans Act, keeping those two streams of policy and management separate.

A Policy on Managing Bycatch and Discards, triggered by the FAO Voluntary Guidelines for Bycatch Management and Discard Reduction (FAO, 2010) is under development within the SFF. Key objectives of the draft Policy include commitments to sustainable harvesting of all aquatic species and minimizaton of waste and risk of harm to resources or ecosystems. The policy addresses data collection and monitoring systems, evaluation of the impact of fishing on bycatch species, minimization of the capture of species and specimens that are not going to be used, maximization of the potential for live release and post-release survival and implementation and regular evaluation of appropriate management measures. The science basis for its implementation is partially in place, with advice provided on potential bycatch in various types of fishing gear and on appropriate methods for setting reference points for bycatch levels in specific fisheries (DFO, 2012a).

The Minister of Fisheries and Oceans is also responsible for implementing the provisions of the Oceans Act. This includes working closely with Ministers of other federal departments, Provinces and Territories to establish networks of MPAs and pursue integrated management. Administration of SARA is complex. The assessment process is under the jurisdiction of the Minster of Environment, the consultation process following a recommendation for listing of any aquatic species is conducted by the Department of Fisheries and Oceans (DFO), the final listing decision is made by the federal Cabinet and, for species that are listed, development and implementation of recovery plans for marine species (except seabirds) is the responsibility of DFO. Because of the mandatory prohibitions on harm or death to species listed under SARA, fisheries management has become increasingly active in recovery planning for listed aquatic species. In practice implementing SARA has proved problematic, leading DFO to develop plans for protection and rebuilding of many depleted species without formally listing them under SARA.

Methods to address and prioritize issues for ecosystem-based fisheries management

The SFF will be implemented by augmenting the existing Integrated Fisheries Management Planning Process. This long-standing process provides a stable forum for government–industry consultations on management options for specific fisheries, taking into account science advice on the status of the stocks, economic analyses of the fisheries and markets and relevant domestic and international policies and agreements.

Through this process Integrated Fisheries Management Plans (IFMPs) are developed regionally according to national templates, with the template including greater consideration of ecosystem issues and more frequent multi-year application over the past decade. The IFMPs specify short- and longer-term objectives for stock and ecosystem conservation, socio-economic outcomes, shared stewardship and management arrangements, licensing practices and harvest allocations among gear sectors or areas. They also specify the management measures to be used such as quotas, spatial and temporal closures and monitoring and surveillance standards.

The IFMP consultation and planning process is the major vehicle for consolidating views on research and management priorities. Because the IFMPs are organized around specified fisheries the discussions and recommendations are often multispecies, but rarely holistic of all ecosystem considerations. The growing number of ecosystem-oriented policies under the SSF has led to increased focus on ecosystem issues in the IFMPs and their consultation processes. This has increased the demands of ENGOs to be full participants in the IFMP planning processes, a demand treated in a variety of ways across fisheries and regions.

In addition to the IFMPs, three major new tools for implementation of the SFF have been introduced: the 'fisheries checklist'; increased allocation of user rights; and an ecological risk assessment framework. The fisheries checklist is a self-diagnostic tool completed annually by fisheries managers, usually in consultation with assessment scientists and industry participants, to qualitatively evaluate the performance of each fishery relative to its objectives and provisions of the corresponding IFMP. The checklist includes sets of questions on the target species in the fishery, its ecosystem effects, state of knowledge for management and the governance, policies and management of the fishery. The checklist is expected to produce a standard by which improvement of a fishery can be monitored over time, and is a core input to the IFMP planning process.

The ecological risk assessment framework (DFO, 2012b) is similar to the ERAF described for Australia (see 'Developments in legislation and policy…' in the second section). Originally developed for implementation of the Sensitive Benthic Habitats Policy, its use is spreading to other planning aspects of fisheries management, particularly with regard to the revised mandate to protect habitats of fish that are part of fisheries.

The other major recent developments in Canadian commercial fisheries are increased numbers of fisheries seeking ecological certification under MSC and other frameworks (cf. Chapter 5) and increased utilization of catch share systems in fisheries (not always in the form of individual transferable quotas). These initiatives usually come from industry, with the Department typically playing a supporting role by planning its own economic strategies.

Under the Oceans Act MPA network, planning is done by a formal working group led by DFO but with representatives of all Provincial and Territorial governments, Environment Canada and Parks Canada. Protection of biological diversity and representativeness of all major marine bioregions are core objectives of Canada's MPA network planning process. Other Oceans Act initiatives for integrated management and health of the oceans are also planned and delivered by federal–provincial–territorial and interdepartmental working groups. The approaches of the different groups vary in detail, but all operate by consensus and include extensive consultion with environmental and industry groups and Aboriginal Bands, as relevant to the issue. All produce priority action lists, including specific requests for scientific advice, which become research priorities for the Department.

For SARA, the Committee on the Status of Endangered Wildlife in Canada (COSEWIC) includes members from all Provinces and Territories, the federal Departments of Fisheries and Oceans, Environment and Parks, First Nations members, and a number of academic and other experts (that is always greater than the number of government members). It sets its own priorities for species to assess, drawing input from a number of supporting specialist groups. Once their priority list for future assessments is released, accountable jurisdictions, including DFO for all aquatic species, makes support for the COSEWIC assessment a science priority. For all species recommended by COSEWIC for a threatened status, SARA requires the jurisdiction to undertake extensive and time-bound consultations regarding both potential impacts of listing and measures needed for recovery. Although there are no structrured processes to coordinate prioritization across demands for implementation of Fisheries Act, Oceans Act and SARA, the mandatory prohibitions and timelines specified by SARA have resulted in SARA-related activities typically receiving special priority.

Integrated cross-sectoral management

Through the early 2000s the DFO invested heavily in building the science foundations and governance processes for integrated planning and decision-making in policy and management. Although momentum for integrated management has waned, there is a significant science legacy of these efforts, including:

- integrated Ecosystem Overview and Assessment Reports for each of the five initial Large Ocean Management Areas (DFO, 2005);
- objective criteria for identifying Ecologically and Biologically Significant Areas (EBSAs, cf. Chapter 14) and Ecologically Significant Species and Community Properties (ESSCPs; DFO, 2004, 2006); and
- guidelines for setting and prioritizing Conservation Objectives from EBSAs and ESSCPs (DFO, 2007).

A recent review of experience with the framework concluded that the EBSA and EBSSCP criteria had performed well and, with minor modifications to the guidance, recommended their continued use in marine areas and expansion of their use to near-coastal areas, estuaries and freshwater systems (DFO, 2011). The review reported uptake of the approach in MPA planning and habitat management, but noted that fisheries management had rarely been engaged in their identification and (with a few exceptions) had shown little uptake of any of the oceans approaches. Rather, fisheries management and strategic policy had embarked on the SFF initiative independently of the Integrated Management processes.

Overall, the last decade has been a time of significant change in the Canadian management of fisheries and in approaches to the conservation of biodiversity. Fisheries management has embraced many aspects of the ecosystem approach and broadened its consultation base, but not to an extent that satisfies many biodiversity conservation interests, while moving at a pace that is considered threatening by some historical fisheries interests. Initial plans for action under the Oceans Act were very ambitious, and did not gain traction wjth many other jurisdictions that would have been affected. More recently focus has narrowed to the MPA network planning and health of the oceans reporting, with full integrated planning and management a distant vision. Actions under SARA have arguably caused the greatest changes because of the more prescriptive nature of the Act, and also triggered some of the strongest opposition. As the strategies and tactics for its implementation become more established the governance of fisheries and oceans management is expected to continue to converge, but not always smoothly.

Discussion

There are several strong and notable similarities in the methods and processes that Australia, Norway and Canada have used in the past 10–20 years to pursue sustainable development and management of marine fisheries and conservation.

There has been a formal change of objectives in legislation and policy. This has progressed from separate specification of fishery objectives (that were typically focused on target species and maximizing yields) and environmental objectives (that were typically focused on conservation and protection) to include wider environmental objectives in fishery management and to recognize sustainable development in environmental management. In different ways this has been a key determining feature of change in all three countries.

Related to this, there has been greater clarity and agreement about targets, limits and acceptable/unacceptable impacts both within fisheries management and between fishery management and management for conservation of biodiversity. This clarity and consistency is not complete in any of the three countries, especially in relation to acceptable/unacceptable impacts on bycatch, habitats and threatened/protected species. But progress has been helped by making the objectives with respect to sustainable development more consistent in the relevant legislation and policy.

The overwhelmingly large number of potential issues that could be considered and the limitations (or cost) of the information needed to understand and manage all of these issues is a recognized challenge. This is being addressed through: (1) methods and processes to comprehensively identify and then prioritize the issues; and (2) use of risk-based assessment and management that can operate at different information levels and that automatically incorporate more precaution when information is limited. All three countries provide different ways forward with the information that is at hand, whether a little or a lot, rather than having to wait for new information.

Two of the countries (Australia and Norway) underwent processes that resulted in significant reduction in fleet capacity, in both cases primarily to achieve economic sustainability. Canada has also undertaken fleet reduction, but as part of major efforts to restructure particular fisheries such as those following the collapses of major Atlantic groundfish stocks in the 1990s. As more fisheries choose to adopt rights-based management approaches many choose to adjust fleet size, but these are generally industry-driven initiatives with government oversight to ensure equitable treatment of participants.

In all three countries there is increasing use of fisher rights, such as individual and transferrable entitlements or catch shares, in addition to more general mechanisms to limited access to fisheries.

All three counties are exploring ways to integrate management of all fisheries in the same ecosystem. This includes the integration of recreational and commercial fishery management where these are exploiting the same stock, although all are struggling somewhat with effective management mechanisms for this.

All three countries have attempted to address integrated cross-sectoral management of multiple human activities occurring in regional ecosystems. In all three cases there has been significant effort put into this, and they have all encountered challenges in gaining cooperation of all necessary regulatory departments and levels of government. Progress has been limited and significant new thinking is needed. The spread of Marine Protected Areas and networks of MPAs has required fisheries management in all three countries to adapt to sharing space in new ways, but MPA planners have also adapted to accommodate more uses, including commercial and recreational fisheries in MPA network design. Finally, the growing number of marine species listed under national protected species legislation is also causing fisheries to alter practices to accommodate biodiversity conservation, with particularly large impacts when legislation regarding protected species is stringent.

Comparison across these three countries is encouraging. There are clear examples of approaches, methods and frameworks for sustainable development of fisheries and for using ecosystem-based management. These can be applied in the first instance without significant new knowledge. There are several outstanding issues still to be fully resolved, for example continuing to build greater clarity and agreement about targets, limits and acceptable impacts across fishery management and environmental management. Fully integrated management of all human uses of the same marine ecosystems remains a recognized but currently distant goal.

References

AFMA (2010) Ecological risk management. Available at http://www.afma.gov.au/managing-our-fisheries/environment-and-sustainability/Ecological-Risk-Management/ (accessed February 2014).

Anonymous (1998) Australia's Oceans Policy. Available at http://www.environment.gov.au/archive/coasts/oceans-policy/publications/policy-v1.html (accessed March 2014).

Anonymous (2006) White paper to the Norwegian Parliament no. 8 (2005–2006) Management of the marine environment of the Barents Sea and the areas off Lofoten. Ministry of Environment, Oslo (in Norwegian).

Anonymous (2007) Guidelines for the ecologically sustainable management of fisheries, 2nd edition. Available at http://www.environment.gov.au/resource/guidelines-ecologically-sustainable-management-fisheries (accessed March 2014).

Anonymous (2009) White paper to the Norwegian Parliament no. 37 (2008–2009) Management of the marine environment of the Norwegian Sea. Ministry of Environment, Oslo (in Norwegian).

Anonymous (2010) The 2010 Norwegian Red List for Species. Available at http://www.biodiversity.no/ThemeArticle.aspx?m=156 (accessed March 2013).

Anonymous (2011a) Independent review of the Environment Protection and Biodiversity Conservation Act 1999. Available at http://www.environment.gov.au/epbc/review/index.html (accessed February 2014).

Anonymous (2011b) White paper to the Norwegian Parliament no. 10. (2010–2011). Revision of the Management Plan for the marine environment of the Barents Sea and the areas off Lofoten. Ministry of Environment, Oslo (in Norwegian).

Anonymous (2012) Ecologically Sustainable Development. Available at http://www.environment.gov.au/about/esd/index.html (accessed February 2014).

DAFF (2007) Commonwealth Fisheries Harvest Policy. Department of Agriculture, Fisheries and Forestry, Canberra, Australia. Available at http://www.daff.gov.au/fisheries/domestic/harvest_strategy_policy (accessed March 2014).

DFO (2004) Identification of Ecologically and Biologically Significant Areas. DFO CSAS. Ecosystem Status Rep. 2004/006. Available at http://www.dfo-mpo.gc.ca/csas/Csas/status/2004/ESR2004_006_e.pdf (accessed February 2014).

DFO (2005) Guidelines on Evaluating Ecosystem Overviews and Assessments: Necessary Documentation. DFO CSAS. Science Advisory Report 2005/026. Available at http://www.dfo-mpo.gc.ca/csas/Csas/status/2005/SAR-AS2005_026_E.pdf (accessed February 2014).

DFO (2006) Identification of Ecologically Significant Species and Community Properties. DFO CSAS. Science Advisosry Report 2006/041. Available at http://www.dfo-mpo.gc.ca/csas/Csas/status/2006/SAR-AS2006_041_E.pdf (accessed February 2014).

DFO (2007) Guidance Document on Identifying Conservation Priorities and Phrasing Conservation Objectives for Large Ocean Management Areas. DFO CSAS. Science Advisory Report 2007/010. Available at http://www.dfo-mpo.gc.ca/csas/Csas/status/2007/SAR-AS2007_010_E.pdf (accessed February 2014).

DFO (2009) The sustainable fisheries framework for Canada. Available at http://www.dfo-mpo.gc.ca/fm-gp/peches-fisheries/fish-ren-peche/sff-cpd/ifmp-pgip-back-fiche-eng.htm (accessed February 2014).

DFO (2010) Occurrence, susceptibility to fishing, and ecological function of corals, sponges, and hydrothermal vents in Canadian waters. DFO CSAS. Science Advisory Report 2010/041. Available at http://www.dfo-mpo.gc.ca/CSAS/Csas/publications/sar-as/2010/2010_041_e.pdf (accessed February 2014).

DFO (2011) Ecologically and Biologically Significant Areas – Lessons Learned. DFO CSAS. Science Advisory Report 2011/049. Available at http://www.dfo-mpo.gc.ca/csas-sccs/Publications/SAR-AS/2011/2011_049-eng.pdf (accessed February 2014).

DFO (2012a) Guidance related to bycatch and discards in Canadian commercial fisheries. DFO CSAS. Science Advisory Report 2012/022. Available at http://www.dfo-mpo.gc.ca/csas-sccs/Publications/SAR-AS/2012/2012_022-eng.html (accessed March 2014).

DFO (2012b) Ecological Risk Analysis Framework (ERAF) for Coldwater corals and Sponge Dominated Communities. Available at http://www.dfo-mpo.gc.ca/fm-gp/peches-fisheries/fish-ren-peche/sff-cpd/risk-ecolo-risque-eng.htm (accessed March 2014).

DoF (2010) Economic and biological key figures. In: *The Directorate of Fisheries*. Available at http://www.fiskeridir.no/english/statistics/booklets/fishery-booklets (accessed February 2014).

DoF (2012) For key economic and biological indicators and an excerpt in English of the two tables included, see 'National framework for fisheries and conservation management in Norway'. Available at http://www.fiskeridir.no/fiske-og-fangst/forvaltningsprinsippet (accessed February 2014) (in Norwegian).

Dowling, N.A., Smith, D.C., Knuckey, I., Smith, A.D.M., Domaschenz, P., Patterson, H.M. and Whitelaw, W. (2008) Developing harvest strategies for low-value and data-poor fisheries: Case studies from three Australian fisheries. *Fisheries Research* **94**, 380–390.

Evans, L., Cherrett, N. and Pemsl, D. (2011) Assessing the impact of fisheries co-management interventions in developing countries: A meta-analysis. *Journal of Environmental Management* **92**(8), 1938–1949.

FAO (1995) Code of Conduct for responsible Fisheries. FAO, Rome. Available at http://www.fao.org/DOCREP/005/v9878e/v9878e00.htm (accessed February 2014).

FAO (1996) Technical Guidelines for Responsible Fisheries. No 2. Precautionary approach to capture fisheries and species introductions. Available at ftp://ftp.fao.org/docrep/fao/003/W3592e/W3592e00.pdf (accessed February 2014).

FAO (2003) The ecosystem approach to fisheries. FAO Technical Guidelines for Responsible Fisheries No. 4 (Suppl. 2). Rome, FAO. Available at ftp://ftp.fao.org/docrep/fao/005/y4470e/y4470e00.pdf (accessed February 2014).

FAO (2010) Report of the Technical Consultation to Develop International Guidelines on Bycatch Management and Reduction of Discards. Rome, 6–10 December 2010. Rome, FAO Fisheries and Aquaculture Report No. 957.

Fletcher, W.J. (2005) The application of qualitative risk assessment methodology to prioritise issues for fisheries management. *ICES Journal of Marine Science* **62**, 1576–1587.

Fletcher, W.J. (2008) Implementing an ecosystem approach to fisheries management: lessons learned from applying a practical EAFM framework in Australia and the Pacific. Chapter 8 Proceedings of CIEAF Conference, November 2006, Bergen, Norway.

Fletcher, W.J., Chesson, J., Fisher, M., Sainsbury, K.J., Hundloe, T., Smith, A.D.M. and Whitworth, B. (2002) National ESD Reporting Framework for Australian Fisheries: The 'How To' Guide for Wild Capture Fisheries. FRDC Project 2000/145, Canberra, Australia. Available at http://www.fisheries-esd.com/c/pubs/index.cfm (accessed February 2014).

Fletcher W.J., Chesson J., Sainsbury K.J., Hundloe T.J. and Fisher, M. (2005) A flexible and practical framework for reporting on ecologically sustainable development for wild capture fisheries. *Fisheries Research* **71**(2), 175–183.

Fletcher, W.J., Shaw, J., Metcalf S.J. and Gaughan, D.J. (2010) An Ecosystem Based Fisheries Management framework: the efficient, regional-level planning tool for management agencies. *Marine Policy* **34**, 1226–1238.

Fletcher, W.J., Gaughan, D.J., Metcalfe, S.J. and Shaw, J. (2012) Using a regional level, risk based framework to cost effectively implement Ecosystem Based Fisheries Management (EBFM). In: *Global Progress on Ecosystem-Based Fisheries Management* (eds G.H. Kruse, H.I. Browman, K.L. Cochrane, D. Evans, G.S. Jamieson, P.A. Livingston, D. Woodby and C.I. Zhang). Alaska Sea Grant College Program.

GEF (2013) Global Environment Facility. Available at http://www.thegef.org/gef/ (accessed February 2014).

Government of Canada (2003) A Framework for the Application of Precaution in Science-based Decision Making about Risk. Available at http://www.pco-bcp.gc.ca/docs/information/publications/precaution/Precaution-eng.pdf (accessed February 2014).

Gullestad, P., Aglen, A., Bjordal, Å., Blom, G., Johansen, S., Krog, J., Misund, O. and Røttingen, I. (2014) Changing attitudes 1970–2012. Evolution of the Norwegian management framework to prevent overfishing and to secure long-term sustainability. *ICES Journal of Marine Science* **71**(2), 173–182.

Gullett, W., Schofield, C. and Vince, J. (eds) (2011) *Marine Resources Management*. LexisNexis Butterworth, Australia.

Gutiérrez, N.L., Hilborn, R. and Defeo, O. (2011) Leadership, social capital and incentives promote successful fisheries. *Nature* **470** (7334), 386–389.

Hobday, A.J., Smith, A.D.M., Stobutzki, I. et al. (2011) Ecological risk assessment for the effects of fishing. *Fisheries Research* **108**, 372–384.

ICES (2012) ICES Advisory Process. Available at http://www.ices.dk/community/Advisory-process/Pages/default.aspx (accessed March 2014).

McPhee, D. (2008) *Fisheries Management in Australia*. Federation Press, Sydney, Australia.

MoE (2006) Ministry of the Environment Report No. 8 (2005–2006) to the Storting. Available at http://www.regjeringen.no/en/dep/md/documents-and-publications/government-propositions-and-reports-/Reports-to-the-Storting-white-papers-2.html?id=701 (accessed February 2014).

MoE (2009) Ministry of the Environment Report No. 37 (2008–2009) to the Storting. Available at http://www.regjeringen.no/en/dep/md/documents-and-publications/government-propositions-and-reports-/Reports-to-the-Storting-white-papers-2.html?id=701 (accessed February 2014).

MRA (2009) Act relating to the management of wild living marine resources. Available at http://www.regjeringen.no/upload/FKD/Vedlegg/Diverse/2010/MarineResourcesAct.pdf (accessed February 2014).

NDA (2009) Act of 19 June 2009 No. 100 relating to management of biological, geological and landscape diversity. Available at http://www.regjeringen.no/en/doc/laws/Acts/nature-diversity-act.html?id=570549 (accessed February 2014).

Olsen, E., Gjøsæter, H., Dommasnes, A., Fossum, P. and Sandberg, P. (2007) The Norwegian ecosystem-based management plan for the Barents Sea. *ICES Journal of Marine Science* **64**, 599–602.

Ottersen, G., Olsen, E., Meeren, G.I., Dommasnes, A. and Loeng, H.O. (2011) The Norwegian plan for integrated ecosystem-based management of the marine environment in the Norwegian Sea. *Marine Policy* **35**(3), 389–398.

Pikitch, E.K., Santora, C., Babcock, E.A. et al. (2004) Ecosystem-based fishery management. *Science* **305**, 346–347.

Pitcher, T., Kalikoski, D., Pramod, G. and Short, K. (2009) Not honouring the code. *Nature* **457**, 658–659.

PMIFSBA (2010) Policy for Managing the Impacts of Fishing on Sensitive Benthic Areas. http://www.dfo-mpo.gc.ca/fm-gp/peches-fisheries/fish-ren-peche/sff-cpd/benthi-back-fiche-eng.htm (accessed February 2014).

Prince, J.D., Dowling, N.A., Davies, C.R., Campbell, R.A. and Kolody, D.S. (2011) A simple cost-effective and scale-less empirical approach to harvest strategies. *ICES Journal of Marine Science* **68**, 947–960.

SFF (2009) Sustainable Fisheries Framework. Available at http://www.dfo-mpo.gc.ca/fm-gp/peches-fisheries/fish-ren-peche/sff-cpd/overview-cadre-eng.htm (accessed February 2014).

Smith, A.D.M., Fulton, E.J., Hobday, A.J., Smith, D.C. and Shoulder, P. (2007) Scientific tools to support practical implementation of ecosystem based fisheries management. *ICES Journal of Marine Science* **64**, 633–639.

UNCED (1992) United Nations Conference on Environment and Development. Available at http://www.un.org/documents/ga/conf151/aconf15126-1annex1.htm (accessed February 2014).

UNCLOS (1982) United Nations Convention on the Law of the Sea. Available at http://www.un.org/Depts/los/index.htm (accessed February 2014).

UNFSA (1995) UN Fish Stocks Agreement (Agreement for the Implementation of the Provisions of the United Nations Law of the Sea). UN Document A/Conf./164/37. Available at http://www.un.org/Depts/los/convention_agreements/texts/fish_stocks_agreement/CONF164_37.htm (accessed February 2014).

UNGA (2013) UN General Assembly Resolutions. Available at http://www.un.org/depts/los/general_assembly/general_assembly_resolutions.htm (accessed February 2014).

Vieira, S., Perks, C., Mazur, K., Curtotti, R. and Li, M. (2010) Impact of the structural adjustment package on the profitability of Commonwealth fisheries. ABARE research report 10.01, Canberra. Available at http://adl.brs.gov.au/data/warehouse/

pe_abarebrs99014396/fisheries_adjustment.pdf (accessed February 2014).

Winsnes, I. and Skjoldal, H.R. (2008) Management plans for the Norwegian part of the Barents Sea ecosystem. In: *The Ecosystem Approach to Fisheries* (eds G. Bianchi and H.R. Skjoldal), pp. 228–246. CAB International and FAO, Rome.

WWF (2002) Policy proposals and operational guidance for ecosystem-based management of marine capture fisheries. World Wide Fund for Nature. Available at http://wwf.panda.org/about_our_earth/blue_planet/publications/?3718/Ecosystem-Based-Management-of-Marine-Capture-Fisheries-Full-version (accessed February 2014).

Zhou, S. and Griffith, S.P. (2007) Sustainability assessment for fishing effects (SAFE): a new quantitative ecological risk assessment method and its application to elasmobranch by-catch in an Australian trawl fishery. *Fisheries Research* **91**, 56–68.

Chapter 22

Small-scale fisheries: Importance, vulnerability and deficient knowledge

J. Kolding[1,2], C. Béné[3] and M. Bavinck[4]

[1] Department of Biology, University of Bergen, Norway
[2] IUCN Commission on Ecosystem Management (IUCN/CEM/FEG), Gland, Switzerland
[3] Institute of Development Studies (IDS), University of Sussex, Brighton, UK
[4] Department of Human Geography, Planning and International Development Studies, University of Amsterdam, Netherlands

Abstract: This chapter explores the relationship between small-scale fisheries and resource conservation from a governance perspective. We outline first the significance of small-scale fisheries, particularly in developing countries. We highlight the serious lack of data, resulting in narratives inherited from industrial fisheries often with very poor relevance for small-scale fisheries. Current governance structures are then described. Many small-scale fisheries are characterized by self-governing mechanisms that have significant conservation value. The sector is vulnerable however, and under increasing pressure. The push for governance reforms, entailing a two-pronged approach of area closures and effort control, usually results in further marginalization and increased conflicts. The reluctance to conform to Western-inspired management models makes small-scale fishers subject to a negative image, particularly for using indiscriminate fishing methods. Recent findings however challenge the mainstream fisheries narratives and suggest that small-scale fishers' exploitation pattern might often be closer to an ecosystem approach than conventional selective fisheries.

Keywords: environment; poverty; welfare; rent; risks; reform; co-management; misperceptions; MPAs; data

Introduction

This chapter and Chapters 23–25 highlight the global sector known as 'small-scale fisheries' (SSFs; see also Chapter 6). Small-scale or artisanal fisheries date back to the dawn of human history, and still constitute a widespread activity and occupation. On the other hand, large-scale or industrial fisheries have evolved during the past century and have now come to dominate the realms of policy and science. Due to their increased marginalization or evolution into recreational activities in many parts of the industrial world, SSFs are now more and more associated with developing countries (Misund et al., 2002), where the vast majority are located.

In this chapter we will primarily focus on linking issues of resource conservation in SSFs with governance,[1] recognizing the difference between traditional knowledge and scientific research but also identifying serious gaps. Conservation is only one of the objectives that fisheries actors try to ensure. Consequently, governance of SSFs is normally shaped by considerations of many kinds such as food security, employment, social justice, economic viability and welfare, in addition to ecosystem health (Béné et al., 2007; Kooiman et al., 2005; Chapter 24). The choices made with regard to these considerations are reflected in contrasting governance approaches, with radically different consequences for SSFs.

Governance of Marine Fisheries and Biodiversity Conservation: Interaction and Coevolution, First Edition.
Edited by Serge M. Garcia, Jake Rice and Anthony Charles.

We will emphasize the societal contributions of SSFs and their relative neglect in the hands of science and policymakers. We also highlight the qualities of SSFs institutions and their relevance for governance. In addition to considering internal aspects of SSFs governance, we also touch on the way in which external actors, concepts and dynamics affect the sector. After all, with perhaps the notable exception of small islands states (cf. Chapter 25), SSFs are usually a tiny player in much larger societies which include many powerful – if not overpowering – interests. Modern industrial fisheries have captured the limelight and shaped most fisheries paradigms, relegating the traditional, 'primitive' and 'inefficient' forms of fishing activities that habitually characterize SSFs to the margins (Bavinck, 2011; Platteau, 1989). Recently however, SSFs have gained renewed recognition for their role in securing livelihoods for hundreds of millions of people (Béné et al., 2007; Heck et al., 2007; Mills et al., 2011).

The chapter is organized as follows. The following section discusses the nature and the condition of SSFs in the world; the third section ('Governance and conservation in SSFs') presents an overview of current governance efforts in connection with conservation and calls for global reform; and the fourth section ('Misperceptions in science-based governance') highlights the consequences of the poor quality of scientific data and limited research for SSFs. The final section summarizes and wraps up the arguments.

Small-scale fisheries: An overview

John Cordell (1989) captured the nature of SSFs in the title of his book *A Sea of Small Boats*. This heading highlights two of the key attributes of SSFs: the small size of the units operating in the sector and also their sheer number and variety. These aspects colour the views on what constitutes the sector (Chuenpagdee et al., 2005; Mills et al., 2011; Misund et al., 2002). We will use the FAO (2012a) definition of small-scale and artisanal fishing: 'fishing households [as opposed to commercial companies], using relatively small amount of capital and energy, relatively small fishing vessels [if any], making short fishing trips, close to shore'.

The most common attributes are therefore households with limited range of operation,[2] dependency on local resources and low capital investment. In view of their numbers and distribution throughout the world, SSFs are by no means 'small' however.[3] Nearly 90% of the world's 120 million full-time or part-time fishers are estimated to derive their livelihood from the small-scale sector. They are estimated to contribute 70% of the total world catch (inland fisheries included), which is used primarily for domestic human consumption (BNP, 2008; Mills et al., 2011). An additional 200–300 million people, many of whom are women, are projected to be employed in the value chain, mostly through informal arrangements. These figures are most likely underestimates however, as they derive from official statistics where SSFs are rarely well accounted for (Mills et al., 2011). SSFs are therefore an important, but underrated, source of employment, food security and income (Béné et al., 2010; Jentoft and Eide, 2011), particularly in the developing world and in rural areas (Béné, 2006; Béné et al., 2009a).

Following demographic growth, coastal migration and forced relocation (e.g. consequences of climatic events and wars), small-scale fishing effort in many developing regions is steadily growing with consistently more people joining the sector and fishing technologies improving (Bavinck, 2014; Jul-Larsen et al., 2003; Salayo et al., 2008). While the world's population increased by 75% between 1970 and 2005, the global number of fishers increased by 178% (Eide et al., 2011). These figures suggest that SSF make important contributions to employment and food security (also see 'Governance and conservation in SSFs'). On the other hand, these developments have also given rise to concern about destructive side effects (Mathew, 2003) as well as governance failures (Symes, 1996). To some however, and perhaps a large extent, these governance failures can be explained as the result of other developments in the coastal zone. In many countries exploitation of inshore waters is amplified by the incursions of industrial trawler fleets (Bavinck, 2005, 2011; Platteau, 1989) and habitat destruction and pollution may pose an even bigger threat than fishing. Governmental fisheries policy (subsidies and technology schemes), constitutional provisions of open access for all citizens and broader economic developments have all contributed to the increased environmental problems. As such, small-scale fishers and their fishing grounds are also among the casualties of development and degradation, rather than the perpetrators. Chuenpagdee et al., (2005, p. 27) note with regard to the governance failures of SSF: 'new driving forces have developed, surpassing the capacity of the old management systems and putting new pressures on the natural and social systems'.

Environmental impact

A comparative examination contrasting small-scale with large-scale fishing (Figure 22.1) provides ammunition for those advocating different views.

	Large-scale company-owned	Small-scale artisanal
Number of fishermen employed	Around 2 million	Over 30 million
Marine fish caught for human consumption	Around 40 million tonnes annually	Around 40 million tonnes annually
Capital cost of each job on fishing vessel	$$$$$$$$$ $$$$$$$$ $$$$$$$$ $ 30 000 to $ 300 000	$ $ 100 to $ 5000
Bycatch discarded at sea	Around 8 million tonnes annually	Around 2 million tonnes annually
Marine fish caught for industrial reduction to meal and oil, etc.	Around 15 million tonnes annually	Almost none
Fuel oil consumption	30 to 40 million tonnes annually	3 to 15 million tonnes annually
Fish landed per tonne of fuel consumed	= 1 to 2 tonnes	= 3 to 15 tonnes
Fishermen employed for each $1 million invested in fishing vessels	3 to 30	200 to 10 000

Figure 22.1 Comparison between large-scale and small-scale fishing (updated from BNP, 2008; FAO, 2012b; Pauly, 2006; Thomson, 1980). (Source: Misund et al. 2002. Reproduced with permission of John Wiley & Sons.)

According to some observers, SSFs are less efficient economically than larger-scale fishing fleets. They are associated with 'Malthusian overfishing' (Pauly, 1990), poverty traps, indiscriminate and unselective fishing methods, illegal and destructive gears (e.g. use of poison and dynamite), overcapacity and resource depletion (Burke et al., 2011; Jentoft and Eide, 2011; Misund et al., 2002; Pomeroy, 2011; Worms and Branch, 2013). In a recent global meta-analysis (Gutiérrez et al., 2011) artisanal multispecies fisheries are consistently considered the least successfully managed on both ecological and socio-economic scores.

Other observers highlight the relative benefits of SSF from both an ecological and social perspective (Béné et al., 2010; Kolding and van Zwieten, 2011). SSFs yield more jobs, generally at the lower end of the labour market where they are needed most. The fact that small-scale fishers generally make use of passive rather than active fishing gears is also considered energy efficient and reduces damage to the marine ecosystem. In terms of fuel consumption, discards and landings per tonne of fuel, SSFs are considerably more efficient than large-scale fisheries (Figure 22.1).

An important characteristic of SSFs is their extraordinary diversity and adaptability to different resources, environments and seasons, which in combination with the almost total absence of discards and notion of bycatch often make them more attuned to the character of the local ecosystem. Some authors (Garcia et al., 2012; Kolding and van Zwieten, 2011) even argue that the overall unselective fishing pattern of SSFs, in some ecosystems, is close to an ecological optimal exploitation and may be considered an operational example of the 'ecosystem approach to fisheries' (EAF) in that it effectively maintains species and size structures. Finally, the constrained mobility of small-scale fishers and ensuing dependency on adjacent fishing grounds is viewed as an incentive for proper management (McGoodwin, 1990). Nevertheless, recognizing that large numbers of small impacts can also generate a large and negative environmental effect, there is a general support for effort control in SSFs (e.g. Berkes et al., 2001; Pomeroy, 2011), even if this is still rarely implemented by governments. The technical modernizations that have and are taking place in SSFs are also a source of concern (Johnson, 2006; Mathew, 2003).

Vulnerability and risk perception

Social scientists have recently drawn attention to another aspect of SSFs that has a bearing on conservation, namely that many small-scale fishers lead vulnerable lives and are extremely susceptible to misfortune (Allison et al., 2011; Béné and Friend, 2011; Jentoft and Eide, 2011). The origins of these vulnerabilities are often found outside the fishery itself however, and are related to basic human needs such as access to drinking water, health facilities or schools for their children, or simply need for political recognition. For these people, poor or degrading environmental conditions are not the primary *cause* of 'poverty'. Confronted with the perils of disease, accidents, death, theft or loss of fishing gear, unavailability of alternative productive assets (such as land) or basic human rights, local communities often do not give priority to the risk of resource degradation. The overall outcome of this situation is that some small-scale fishing people currently lack both the incentive and capacity to engage in collective action or institutional frameworks that aim to conserve stocks. The lesson is that incentives for these fishing communities to invest in resource sustainability will increase only after vulnerabilities that they rank higher have been addressed, and the perceived risk of fishery decline comes to the fore. It follows that the most productive interventions to promote sustainable resource use and conservation in many communities may lie outside the 'natural resource management system' in domains such as education and health services, or institutions and governance to address the lack of secure access to alternative productive assets (including land) and to basic human needs (Allison et al., 2011; Béné and Friend, 2011; Mills et al., 2011).

Governance and conservation in SSFs

The current governance of SSFs

In the developing world where dependence on SSFs is currently high the historical, and in many cases still prevalent, form of governance is embedded in the fishing communities themselves. Fisheries scholars have frequently documented the various forms of self-governance in terms of sea tenure, Territorial Use Rights in Fisheries (TURF; Christy, 1982), or community management. We prefer a perspective of legal pluralism, which suggests that the management efforts of small-scale fishers constitute informal 'legal' frameworks that are often at variance with modern fisheries law (Bavinck, 2005). Starting our discussion at this level, we subsequently consider the ways in which self-governance has meshed with other governance efforts.

The FAO definition quoted above refers to SSFs as activities of 'households'. This view combines well with the sustainable livelihood approach to fisheries, which considers fishing in the context of households' overall income strategies (Allison and Ellis, 2001). We will take a step further by emphasizing the cultural features and social bonds that characterize most small-scale fishing communities (McGoodwin, 2001). Everyone who has ever lived in a village knows what it means to be part of a community: it provides a sense of security, belonging and identity, but also a context of strong social control. In communities that depend on natural resources these normative elements usually extend to the regulation of resource usage. With increased commercialization and urbanization however, these control mechanisms easily break down as is observed in Lake Victoria, for example (Medard, 2012).

As pointed out by many (e.g. Andrew et al., 2007; Béné et al., 2003; Kurien, 2007; Neiland et al., 2005), common property regimes, often in the form of TURFS, are dominant in the developing world. Communities can limit access to fishing grounds by defining who can fish and where. Social boundaries play important roles in allowing certain people into the occupation, and closing the ground for others. The caste system in India or ethnic or tribal organization structures in the rest of the world often limit access to the occupation to certain groups (Bavinck, 2014; Kremer, 1994; Thomas, 1996). Even open-access fisheries such as African lakes have local access-limiting mechanisms that are strongly related to residential rights (Béné et al., 2003; Jul-Larsen et al., 2003).

So, to what extent is the self-governance of fishing communities directed towards conservation? Many observers have noted the prevalence of social rather than biological objectives in community governance, frequently related to social equity, redistribution and/or reciprocity (e.g. Bavinck and Karunaharan, 2006). Others have noted the incidence of religious elements that have little connection to modern conservation objectives. A significant body of authors, starting with Johannes (1978, 1981), have however pointed out the ecological sensibility of the 'words of the lagoon' and their associated rule systems (Berkes et al., 2001; Caddy and Cochrane, 2001; Flood, 1994). The documented evidence of the mechanisms of 'customary management' has therefore increased since the 1980s, particularly with regard to SSFs in small island states (Novaczek et al., 2001; Ruddle, 1988). Although less wide ranging, some work has also been done on self-governance among 'larger' SSFs (Basurto et al., 2012; Bavinck, 2001; Begossi, 2006).

We argue that self-governance is potentially of great value for future conservation for several reasons. The first is that self-governance activities generally enjoy greater legitimacy than dictated policies, particularly in weak political systems, and therefore stand a better chance of success. A second motive is that, although current self-governance efforts do not always pay palpable attention to the needs of conservation, there is no reason to expect that they are inherently incapable of doing so. After all, people are known to adapt institutions in response to new challenges (Jentoft, 2004) and some studies demonstrate that self-governance institutions also change over time (Bavinck, 2001; McEvoy, 1986; Van Ginkel, 2009). Perhap the relatively low attention to conservation in current self-governance is therefore mostly a question of perceived urgency of the challenge at hand.

From the perspective of central government agencies, SSFs have always proved particularly difficult to manage. This results partly from the large number of people involved and their spatial distribution over large and often isolated areas, in combination with governments' limited personnel and financial constraints (Mahon, 1997; Misund et al., 2002). Government agencies also have difficulty in choosing the right management instruments. Output controls (e.g. TACs) are difficult to implement due to lack of regular assessments, the many species (in tropical areas) and the many dispersed landing sites. Input controls (effort) are in principle easier but have barely been introduced in developing countries, perhaps for political reasons. One exception is the nearly universal choice of gear or mesh-size regulations which are largely met with non-compliance (see 'Misperceptions in science-based governance below').

The recent co-management movement in fisheries and in conservation (cf. Chapter 2) has endeavoured to harness and enhance the vitality and the conservation-ethos of traditional small-scale fishers and their self-governance systems, and to combine them with governmental regulation. Such attempts have been based on the assumption that governments and communities are truly interested in joint management as, after all, 'it takes two to tango' (Pomeroy and Berkes, 1997). Nevertheless, the establishment of effective co-management systems has proven to be a more arduous process than anticipated, with substantial risks of breakdown and leakage. Co-management in SSFs has a mixed record in many respects (e.g. Béné et al., 2009b), and in particular with regard to conservation (Berkes, 2006; Evans et al., 2011; Pomeroy et al., 2010). In some places, the lack of success in creating well-functioning co-management institutions has resulted in renewed call for top-down government enforcement, such as for example in Lake Victoria (Kolding et al., 2014). In many cases co-management arrangements are top-down, created by governments with non-negotiable objectives (Nielsen et al., 2004).

In the following sections we shall investigate the possible mismatch in the perceptions about threats and solutions which appear to be entrenched in the governance of SSFs.

The call for governance reform and its implications

The widely heralded global fisheries crisis (see 'Misperceptions in science-based governance') and evidence of a continuing decline of marine ecosystem health (MEA, 2005; UNEP, 2012) have triggered recurrent calls for a global fishery reform. These calls are based on contrasting environmental discourses, i.e. perceptions of problems and their respective solutions (Dryzek, 2005). In the following sections, we discuss only the most important approaches to fisheries reform: the so-called wealth-based approach and its counter-narrative, the welfare-based approach. We also touch briefly on marine protected areas.

Wealth-based approach

In the dominating paradigm of liberal economy, 'privatization' of the resources is often presented as the most direct way to achieve economic efficiency and presumably resource conservation.[4] This school of thought subscribes to the 'tragedy of the commons' narrative, as enshrined by Garrett Hardin (Kolding and van Zwieten, 2011). In industrial fisheries 'property' rights are usually introduced through individually transferable quotas (ITQs), even though they are not *primarily* designed to address specific sustainability issues (Bromley, 2009) other than ending 'derby fishing' (Hilborn et al., 2005) and facilitating market-based capacity optimization. However, as with any other output control, ITQs are particularly difficult to implement in SSFs for reasons explained above (e.g. poor statistics and control over landings). At the helm of the wealth-based approach stand a large number of economists. The problem, these experts explain, comes from the failure to recognize how much wealth (understand 'rent') could be generated by the resource instead of the ubiquitous overcapacity, rent dissipation and subsequent economic and biological over-exploitation. They argue that the only solution is to instigate a drastic fisheries governance reform, the main objective of which is to establish the 'correct' institutions that should capture the huge wealth that the resource can generate (e.g. GPO, 2012; Leal, 2010). For this purpose access to the resource should be given to a limited number of operators through an efficient system of fishing rights. This reform is expected to maximize the sector's economic rent, increase the profits for those who remain in the fishery and, theoretically, lead to the redistribution of benefits to the rest of the society (see also Chapter 6). Many agencies (including FAO, the World Bank and some major NGOs) now see this model as the best way to generate economic growth, reduce poverty and stop overfishing.

How this reform process will deal with the people who become excluded from the fisheries is not clear however, and the subsequent costs to society are rarely considered or estimated. It is also not clear for instance what would be the reduction in labour if we were to apply the 'wealth-based' governance reform. However, experience from elsewhere can provide some rough indications. Béné et al., (2010) showed that the restructuring of the Norwegian fisheries (that supported more than a hundred thousand small-scale operators until the Second World War) into the current wealth-generating, highly capitalized and efficient fishery has required a 90% reduction in the number of fishers. The overall annual catches have however remained fairly constant at around 3 million tons; capacity has simply been substituted by efficiency. Extrapolating to the global scale, this wealth-based reform would mean that more than 95 million people that currently directly depend on SSFs would have to leave the sector to let the remaining 15 million create wealth and accumulate rent.[5] Based on these figures, the value of the poverty prevention function carried out by SSFs (that is maintaining the current people involved above the US$ 1.25 poverty line) is estimated to be around US$ 61 billion per year (Béné et al., 2010). In comparison, the rent dissipated every year in the world's fisheries is equivalent to US$ 50 billion (World Bank, 2008). Considering the approximations involved, these results suggest that even if all the wealth generated by the global reform were to be channelled back to those potentially excluded (an unlikely scenario), it would barely compensate for the poverty prevention function that SSF currently fulfils across the world.

Welfare-based approach

Béné et al., (2010) therefore argue for a more progressive and sequential reform in the case of SSFs, particularly by recognizing and preserving the small-scale fisheries' social and pro-poor (welfare) functions. In addition to being the main source of animal protein and an important source of vital micronutrients for millions of poor across the world (Kawarazuka and Béné, 2010) fisheries, and in particular SSFs, play a remarkable role in poverty alleviation through their capacities to absorb surplus labour (Béné et al., 2007, 2010). As illustrated in Figure 22.1 the SSF sector, often denigrated for its backwardness and inability to generate wealth, is remarkably efficient (in the economic sense) in absorbing the excess of unskilled labour in the developing world.

An additional argument is that of social justice. An efficiency-based approach to fisheries will concentrate returns into the hands of a few whereas a welfare-based approach, which does not exclude conservation measures, will distribute them over a larger number of needful households with few other alternatives (Eide et al., 2011). The difference therefore is not necessarily in the choice of tools (although clearly some specific tools such as ITQs would not be conceivable under a welfare approach), but rather in the general essence/objective of the management: one is geared towards maximization of the rent for the benefit of a few highly capitalistic entrepreneurs, the other is aimed at ensuring that the largest number of poor households can still live from their activities without jeopardizing the ecological sustainability of the resource-base.

Marine protected areas

Marine protected areas (MPAs) represent a form of spatial management that is increasingly implemented in tropical and subtropical SSF regions, particularly for the purpose of conservation and rebuilding of depleted stocks. As Chapter 24 of this volume is devoted specifically to the relationship between spatial management and SSFs, our discussion here is brief.

Many MPAs are frequently established in densely populated and low-income areas, where the impacts of human activities on natural ecosystems are considered among the most severe (Worms and Branch, 2012). This, combined with the strong inverse correlation between biodiversity and latitude and the near-shore location of high biodiversity hotspots of high aesthetic or conservation values (e.g. coral reefs or mangroves), has resulted in most of the world's MPAs being located in developing tropical countries in close proximity to, or right in the middle of, SSFs territories. Consequently, while the overall world target of protected areas by 2010 was set at 10% of the global marine area (CBD, 2004), present coverage has reached 2.3% (Spalding et al., 2013); 25% of the world's coral reefs (Burke et al., 2011) and 27% of mangrove areas (Spalding et al., 2010) have already been designated as MPAs, affecting a large population of SSFs actors.

Whether MPAs are effective in relieving fishing pressure, overall or in SSFs, is still an open question (Thorpe et al., 2011). Nonetheless, their rapid development in the coastal area has an obvious impact upon SSFs, frequently diminishing their opportunities for livelihood and income (Kolding, 2014) and creating significant opportunity costs for populations with extremely low economic resilience. We will briefly return to the interaction between MPAs and SSFs from a biological perspective below.

Misperceptions in science-based governance

Fishery science has emerged over the past century as a necessary response to, and as a result of, increased industrialization and fishing power (see also Chapter 9). Indeed, only the large-scale fisheries were, and still are, valuable enough to afford dedicated research and comprehensive monitoring and enforcement schemes. Consequently fisheries research, as an input to modern governance, is mainly rooted in the notions and paradigms developed in large-scale fisheries from the Northern hemisphere; these are subsequently applied to SSFs in so-called data-poor situations, without considerations of their fundamental differences (Kolding and van Zwieten, 2011). This has important implications for our understanding of the situation and for policy recommendations both in terms of fisheries optimization and resource conservation.

The lack of representative data

When appropriate actions are to be taken for governing fisheries and conservation, there must be a common perception of the problem among the main stakeholders and an agreed approach for solutions. This is where reliable data, analytical capacity and relevant advice are needed in order to diagnose situations and suggest mitigating answers. Good governance requires informed understanding of biological and socio-economic subsystems and how they react and respond to different actions. Unfortunately, one of the most conspicuous features of SSFs is the general lack of quantitative long-term datasets (Mills et al., 2011; Misund et al., 2002). This represents not only a serious constraint for making meaningful suggestions for management, but may also be one of the reasons for the many contradicting opinions that exist about the sector.

When little knowledge exists and decisions need to be made, assumptions and conjectures are often extrapolated from other information sources. The ubiquitous reference to the state of the world fisheries in the global literature and policy documents is the FAO report on the state of fisheries and aquaculture (SOFIA), which reports that about 25% of the stocks are underexploited, 50% fully exploited and 25% are overexploited, depleted or recovering (SOFIA, 2012). These statistics can be interpreted in two ways. On the one hand, they show that 75% of the world stocks are exploited according to international law (i.e. at or below maximum sustainable yield). On the other hand, they also show that 75% of the world stocks are fished at their maximum production level or beyond, reflecting intense fishing pressure. The second interpretation prevails in global communication and conservation advocacy as justification for concern and calls for increased management action, thus contributing to the negative global perception many people (governments, civil society and academics) have of the world's environmental trends in general and fisheries trends in particular.

There are, however, three major problems in using the FAO landing statistics as representative of the global picture and SSFs in particular, as follows.

(1) *Quality of the data*: the landings formally reported by states to FAO are affected by measurement errors, underestimates of SSFs landings and ignorance of unreported fishing catches as well as of the amounts discarded.
(2) *Appraisal methodology*: only a minor proportion of the world's marine resources are formally assessed (16% of the harvested fish taxa according to Ricard et al., 2011). Most of the stocks, particularly from tropical and subtropical fisheries, are assessed only on the basis of catch trends (Costello et al., 2012; FAO, 2011). A large debate is currently taking place about the reliability of that approach in the absence of additional contextual data (Branch et al., 2011; Daan et al., 2011; Froese et al., 2012, 2013). There are therefore serious doubts over how depleted non-assessed stocks actually are.
(3) *Relevance of the FAO global dataset*: of the 441 regional resources for which FAO give assessments (FAO, 2011, tables D1–D19), only about 5% (22 stocks) can be considered representative[6] of SSFs in tropical or subtropical countries (FAO statistical areas No. 31, 34, 51, 57, 71 and 77). Of these 22 stocks, only 4 are classified as overexploited; all comprise sessile highly commercialized export species such as abalone and rock lobsters.

In short, the FAO statistics on which most of the assessments of fisheries resources at the global level are based are virtually unrepresentative for the vast majority of SSFs, of which we often know practically nothing with respect to their long-term exploitation status or their share of the sector impact on a resource. The only 'assessment' metric regularly reported in the SSF literature is the common declaration of 'declining catches' or 'decrease in mean sizes'. Unfortunately, such indicators do not necessarily point towards overfishing. In many instances they may only show the natural consequence of fishing, as is elaborated upon in the following section.

The decline in the fisheries

A decrease in catches, whether observed or inferred from interviews, together with the observation of unrestricted entry (open access) are among the primary justifications for governance interventions. In SSFs nearly every document or investigation focusing on the need for management or conservation action uses the omnipresent 'decrease in stocks' or 'decline in the fisheries' as the first, and often only, argument. The question however is whether this indicator refers to the catch of individual fishers (i.e. catch/fisher/day) or to that of the fishery as a whole. A bit of theory follows.

As a fishery develops and effort expands on a given population, the biomass of fish will necessarily decline as removals increase. Fish density will decrease proportionally and individual fishers will correspondingly experience a reduction in their catch rates, unless they fish harder and adapt their fishing strategies. Usually, fishers can only observe the change in their own daily catch (or that of their peers), but will rarely be able to perceive – or appreciate – that of the accumulated catch of the whole fishery. In that sense a fisher, in an expanding fishery, generally lives in a world of ever-declining personal catch rates unless he innovates constantly (shifting places, gears and targets) to try and maintain its take-home lot. When asked, he will often state 'there is less fish than before'.

However, the individual catch decline (CPUE or density) for each resource component is not only expected as removals increase, it is also an initial prerequisite for increasing stock productivity (net added annual biomass) and thereby generating a sustainable yield. Indeed, according to general production theory, it is appropriate to bring the stock sizes down to around half their virgin level in order to maximize the sustainable yield (MSY, Figure 22.2).

The decrease in fish density (or corresponding CPUE), often erroneously interpreted as an indicator of 'overfishing', is also the general justification for the creation of MPAs in fisheries (Kolding, 2014; Russ et al., 2004; Thorpe et al., 2011) in order to rebuild stocks. In this context it is interesting to notice that the rapid increase (doubling) of fish densities observed inside tropical reserves (Halpern, 2003; Lester et al., 2009), broadly matches the decrease (halving) in fish density levels that would be necessary to produced MSY (Figure 22.2). If the objectives of these MPAs were to rebuild stocks in order to produce a sustainable livelihood for the fishers, these areas were therefore, on average, in no need of protection as they were already producing close to MSY.

Only a decrease in total catches without a decrease in effort is a possible indicator of overfishing. In SSFs however, with their multitude of landing sites and informal character, good long-term series of total catch statistics are invariably missing and literature consistently emphasizes their systematic under-evaluation and underestimation (Mills et al., 2011). Using a 'decline in catches' as an indicator of the status of the stocks may therefore be misleading or even incorrect without specifying whether this concerns the total catch or each individual's catches. Unfortunately this distinction is very seldom made in the SSFs or MPA literature.

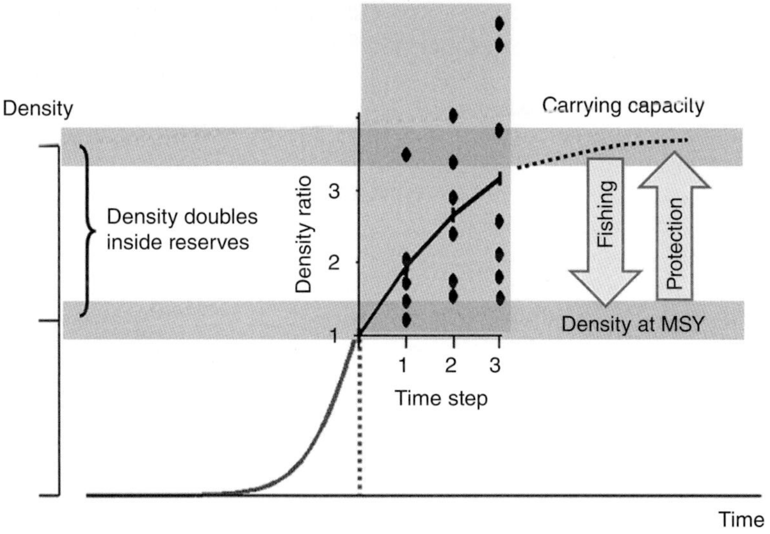

Figure 22.2 Density-dependent sigmoid growth trajectory of a population or community over time. Standard logistic theory predicts the highest growth rate at a density around half the asymptotic carrying capacity, equivalent to the density that returns the highest sustainable yield per time unit (MSY). Meta-analyses (Halpern, 2003; Lester et al., 2009) show that the average density doubles inside tropical reserves after fishing ceases. The grey area is redrawn from Halpern and Warner (2002). It shows density increases in 7 marine reserves for which successive temporal data were collected. Only the first step is significantly different from the previous. These results indicate that rebuilding to carrying capacity is very fast.

The decrease of size

The second most common indicator of unsustainable fishing is the observation of a decrease of large-sized fish or a decrease in the mean size of the fish in the catch (e.g. Worm et al., 2009). Such changes should however be used with great caution as a resource health indicator, as they are highly dependent on climatic conditions, fishing strategies, markets and prevailing management paradigm. For many SSFs, gear and mesh regulations are often the only regulations in place and unselective methods or indiscriminate fishing are by default considered synonymous with destructive or unsustainable fishing (Kolding and van Zwieten, 2011). Non-selective fishing methods are therefore usually banned, and minimum landing sizes or minimum mesh sizes are promoted to protect small sizes. However, when fishing removes part of the population, reducing lifespan, it should not be a surprise that a decrease in mean size is observed, particularly if fishing focuses on the larger individuals. It is a great paradox in fisheries governance that the expected result of fishing within legal requirements is used as a diagnostic of unsustainability and depletion. A decrease in sizes, in the same way as a decrease in catch rates, is not a sign of overfishing; it may only be a sign of fishing activity taking place.

It is also informative to observe that, in many SSFs, fishers do not comply with gear and size regulations because they do not receive their alleged benefits. Fishers know by experience that the highest catches by volume are obtained by targeting juvenile fish or small species located in the lower part of the trophic food chain. This contradiction between theory and practice explains why co-management initiatives, which are often implemented only as a way to better enforce conventional regulations and in particular mesh regulation as in many parts of Africa, fail (Jul-Larsen et al., 2003).

A greater paradox however is that those small-scale fishers that either defy the selectivity regulations, or are left alone without outside interventions, often develop a fishing pattern which, by continuous trial and error, appears finely adapted to match the productivity of individual stocks. In African lakes such a fishing pattern resulted in a fish community structure that did not deviate significantly from the species and size-structure of an unfished situation, while returning significantly higher sustainable yields than in areas managed in a more conventional way (Jul-Larsen et al., 2003; Kolding and van Zwieten, 2011; Law et al., 2012). Unfortunately, the standard response from managers or agencies to this sort of opportunistic adaptive fishing is renewed calls for compliance and enforcement of the conventional paradigm. The preferred fishing practices versus legal frameworks may easily become not only an irresolvable tug-of-war, as seen often when fishers resist the implementation of gear regulations, but also a futile debate as seen from the perspective of ecosystem conservation (Kolding and van Zwieten, 2011; Misund et al., 2002). Sadly, these ongoing conflicts have contributed to the general public image of these fishers to be unruly members of society on the road to self-destruction, and are perceived as just another example of the general governance failure which permeates the view of global fisheries.

Precaution

The above suggests that the data and indicators used for assessing SSFs, as well as justifying conservation measures such as MPAs or indeed conventional fishery management measures for these fisheries, are too often based on inappropriate metrics and/or imperfect understanding of ordinary and expected consequences of sustainable fishing. There seems to be a fundamental problem in many policy and advocacy documents in distinguishing between signs of fishing and signs of overfishing. The question is who should benefit from the doubt? When knowledge is low, then risks are generally considered high. This is a typical situation for applying the so-called precautionary approach (PA). The PA as applied in environmental risk assessment is designed and meant as an insurance policy to give the benefit of the doubt to the environment, thereby trumping human activities. In the present context of environmental awareness and media storming, few governance bodies are willing to take uncertain chances. When such natural reluctance is combined with the unavailability of clear stock status indicators, it becomes particularly tempting and convenient to call on the PA. Like any other model for regulating or conserving fisheries, the PA was developed in the industrial world and promoted by powerful international organizations. The question is whether the same principles can be uncritically applied to SSFs in developing countries where vulnerable livelihoods are at stake. The FAO Guidelines for the Ecosystem Approach to Fisheries (FAO, 2003, para. 147) stipulate that the precautionary principle should apply not only to environmental degradation but also 'to any undesirable outcome (ecological, social, or economic)'. This provision might therefore also be used to protect vulnerable small-scale fishers as well as ecosystems.

Conclusions

In this chapter we have sketched the characteristics of small-scale fisheries and investigated their relationship with conservation and governance. We pointed out the substantial importance of the sector and its continued numerical growth. We also highlighted the lack of data on SSFs, and the fact that in those circumstances many of the recommendations made for SSFs are blindly inherited from the large-scale sector.

In discussing contemporary governance arrangements, we emphasized that many SSFs possess their own self-governance structures that show significant conservation value (cf. Chapter 23). We argued that these institutions should not only be taken at their face value – they can also be seen as 'learning organizations' that adapt to new circumstances. As most SSFs typically depend on their immediate environments and the resources they contain, the incentive to 'learn' is great. It is for this reason that participants in the 2008 FAO conference on SSFs concluded 'small-scale fisheries have yet to fully realize their potential to significantly contribute to sustainable development and the attaining of the United Nations Millennium Development Goals (MDGs)' (FAO, 2009, p. iv).

Significant attention was then paid to the partialities created by the realm of fisheries science. The gloomy assessments of large-scale industrial fisheries were argued to disproportionally dominate the world view, agenda, research, financial resources, knowledge and perception of the fisheries as a whole. There are fundamental differences between the two sectors in terms of impact, resilience and socio-economic role, however.

Due to deficient knowledge, SSFs are surrounded by myths and misconceptions. Small-scale fishers suffer from a negative public image, in particular for a supposed lack of governance and for using indiscriminate and unselective methods, even if there is a dawning recognition that multispecies multi-gear traditional fishing might in fact be closer to an ecosystem approach to fisheries than highly selective gears (Garcia et al., 2012; Kolding and van Zwieten, 2011).

The biggest constraint however remains the general lack of data, insight and knowledge on which to build good and fair governance. Policy-makers have discriminated against SSFs as impediments to modernization and industrialization. Fisheries managers have neglected and underestimated the SSF sector because of its perceived complexity and resistance to top-down technocratic management and partly due to its low economic value. Fishery scientists have ignored SSFs because of the difficulty in assessing their coastal resources with modern fishery science methods and the reluctance to deal with the multidisciplinary approaches required, with strong economic and social components. Instead of remedying the current knowledge shortage by reallocating part of the financial and research resources in this direction, they perpetuate the narrative of resource stocks depleted from open-access situations and therefore in need of stronger regimes of fishing rights. Overall, the governance and management of SSFs have been conceived and implemented mainly as addendums to, or replicas of, the principles developed for industrial fisheries, often with poor results. Continuous biomass decline and overfishing are serious problems in some parts of the world, but very little robust evidence exist on which to claim the same for SSFs (e.g. Worms and Branch, 2012) although we recognize that SSFs (like any fishery) can be overfished if pushed too far. In this context, however, the sweeping extrapolations derived from large-scale fisheries are often detrimental for SSFs.

The lack of data also applies to the human dimension. The variety of SSFs and their societal contributions has only scarcely been documented, as have their own systems of law and governance. All these deficits call for major long-term research efforts, such as recently initiated in the 'Too Big to Ignore' project (http://toobigtoignore.net/).

Acknowledgements

We are grateful to Paul van Zwieten and the editors of this volume for their critical review and constructive comments.

Notes

1. We consider governing as a process rather than a state-of-being, and a condition involving numerous actors and constituencies at multiple nodes and levels (cf. Bavinck et al., 2013; Torfing et al., 2012). In many coastal regions of the world governance proves to be embedded in situations of legal pluralism, in which parties refer to differing legal realms and cultures.
2. There are notable exceptions such as the Senegalese canoe fleet, seasonally active between Mauritania and Sierra Leone and staying at sea for many consecutive days.
3. A recent Canadian-funded project on small-scale fisheries therefore bears the provocative title: *Too Big to Ignore* (www.toobigtoignore.net).
4. Although Clark (1973) has actually demonstrated in theory that it may be the opposite. See also Bromley (2009).
5. This figure does not account for the additional (probably 500+ million) fisher-farmers, the resource-poor who engage in fishing as part of a diversified livelihood strategy.
6. Splitting LSF and SSF landings in FAO statistics is virtually impossible. Countries do not provide separate data and the FAO assessments are given by resources (each containing potentially many stocks in possibly different states). Most of the coastal resources in tropical countries (except some strictly inshore and estuarine species) are jointly exploited by both SSFs and LSFs during their life cycle, with different impacts. With such data, it is difficult to allocate responsibility for overfishing, when it exists, to either of the two subsectors. The stocks we consider 'representative' for SSFs in this section are therefore those inshore demersal species that are predominantly targeted by SSFs only.

References

Allison, E.H. and Ellis, F. (2001) The livelihoods approach and management of small-scale fisheries. *Marine Policy* **25**, 377–388.

Allison E.H., Béné C., and Andrew N.L. 2011. Poverty reduction as a means to enhance resilience in small-scale fisheries. In Pomeroy R., and Andrew, N.H. (eds.) Small-scale fisheries management: frameworks and approaches for the developing world, CABI International, Cambridge, pp.216–237

Andrew, N.A., Béné, C., Hall, S.J., Allison, E.H., Heck, S. and Ratner, B.D. (2007) Diagnosis and management of small-scale fisheries in developing countries. *Fish and Fisheries* **8**, 227–240.

Basurto, X., Cinti, A., Bourillón, L., Rojo, M., Torre, J. and Hudson Weaver, A. (2012) The Emergence of access controls in small-scale fishing commons: A comparative analysis of individual licenses and common property-rights in two Mexican communities. *Human Ecology* **40**, 597–609.

Bavinck, M. (2001) *Marine Resource Management: Conflict and Regulation in the Fisheries of the Coromandel Coast.* Sage, New Delhi.

Bavinck, M. (2005) Understanding fisheries conflicts in the South – a legal pluralist perspective. *Society and Natural Resources* **18**, 805–820.

Bavinck, M. (2011) The megaengineering of ocean fisheries: a century of expansion and rapidly closing frontiers. In: *Engineering Earth: The Impacts Of Megaengineering Projects* (ed. S.D. Brunn), pp. 257–273. Kluwer, Dordrecht.

Bavinck, M. (2014) Investigating poverty through the lens of riches – immigration and segregation in Indian capture fisheries. *Development Policy Review* **32**(1), 33–52.

Bavinck, M. and Karunaharan, K. (2006) A history of nets and bans: restrictions on technical innovation along the Coromandel Coast of India. *Maritime Studies–MAST* **5**, 45–59.

Bavinck, M., Chuenpagdee, R., Jentoft, S. and Kooiman, J. (2013) *Governability of Fisheries: Theories and Applications*. Springer, Dordrecht.

Begossi, A. (2006) Temporal stability in fishing spots: conservation and co-management in Brazilian artisanal coastal fisheries. *Ecology and Society* **11**(5). Available at http://www.ecologyandsociety.org/vol11/iss1/art5/ (accessed February 2014).

Béné, C. (2006) Small-scale fisheries: assessing their contribution to rural livelihoods in developing countries. FAO Fisheries Circular No. 1008. Food and Agriculture Organization, Rome.

Béné, C. and Friend, R. (2011) Poverty in small-scale inland fisheries: old issues, new analysis. *Progress in Development Studies* **11**, 119–144.

Béné, C. Neiland, A., Jolley, T., Ladu, B., Ovie, S., Sule, O., Baba, O., Belal E., Mindjimba, K., Tiotsop, F., Dara, L., Zakara, A. and Quensiere, J. (2003) Natural-resource institutions and property rights in Inland African fisheries, the case of the Lake Chad Basin region. *International Journal of Social Economics* **30**, 275–301.

Béné, C., Macfadyen, G. and Allison, E.H. (2007) Increasing the contribution of small-scale fisheries to poverty alleviation and food security. FAO Fisheries Technical Paper No. 481. FAO, Rome.

Béné, C., Steel, E., Kambala Luadia, B. and Gordon, A. (2009a) Fish as the "bank in the water": Evidence from chronic-poor communities in Congo. *Food Policy* **34**, 104–118.

Béné, C., Belal, E., Baba, M. O., Ovie, S., Raji, A., Malasha, I., Njaya, F., Na Andi, M., Russell, A. and Neiland, A. (2009b) Power struggle, dispute and alliance over local resources: analyzing 'democratic' decentralization of natural resource through the lenses of Africa inland fisheries. *World Development* **37**, 1935–1950.

Béné, C., Hersoug, B. and Allison, E.H. (2010) "Not by rent alone": analysing the pro-poor functions of small-scale fisheries in developing countries. *Development Policy Review* **28**, 325–358.

Berkes, F. (2006) From community-based resource management to complex systems. *Ecology and Society* **11**(45). Available at http://www.ecologyandsociety.org/vol11/iss1/art45/ (accessed February 2014).

Berkes, F., Mahon, R., McConney, P., Pollnac, R. and Pomeroy, R. (eds) (2001) *Managing Small-Scale Fisheries: Alternative Directions and Methods*. International Development Research Centre, Ottawa.

BNP (2008) Small-scale capture fisheries: A global overview with emphasis on developing countries. Big Numbers Project. FAO, World Fish Center and PROFISH – World Bank. Available at http://www.worldfishcenter.org/resource_centre/Big_Numbers_Project_Preliminary_Report.pdf (accessed February 2014).

Branch, T.A., Jensen, O.P., Ricard, D., Ye, Y. and Hilborn, R. (2011) Contrasting global trends in marine fishery status obtained from catches and from stock assessments. *Conservation Biology* **25**, 777–786.

Bromley, D.W. (2009) Abdicating responsibility: the deceits of fisheries policy. *Fisheries* **34**, 280–290.

Burke, L., Reytar, K., Spalding, M. and Perry, A. (2011) *Reefs at Risk Revisited*. World Resources Institute, Washington, DC.

Caddy, J.F. and Cochrane, K.L. (2001) A review of fisheries management past and present and some future perspectives for the third millennium. *Ocean and Coastal Management* **44**, 653–682.

CBD (2004) Draft global outcome-oriented targets for the programme of work on marine and coastal biological diversity. Convention on Biological Diversity. UNEP/CBD/SBSTTA/10/8/Add.1, 5 November 2004.

Christy, F.T. (1982) Territorial use rights in marine fisheries: definitions and conditions. FAO Fisheries Technical Paper No. 227. FAO, Rome.

Chuenpagdee, R., Degnbol, P., Bavinck, M., Jentoft, S., Johnson, D., Pullin, R. and Williams, S. (2005) Challenges and concerns in capture fisheries and aquaculture. In: *Fish for Life: Interactive Governance for Fisheries* (eds J. Kooiman, M. Bavinck, S. Jentoft and R. Pullin), pp. 25–37. MARE Publication Series, Amsterdam University Press, Amsterdam.

Clark, C.W. (1973) The economics of overexploitation. *Science* **181**, 630–634.

Cordell, J. (ed.) (1989) *A Sea of Small Boats*. Serial: Cultural Survival Report No. 26. Cultural Survival Inc., Cambridge, MA.

Costello, C., Ovando, D., Hilborn, R., Gaines, S.D., Deschenes, O. and Lester, S.E. (2012) Status and solutions for the world's unassessed fisheries. *Science* **338**(6106), 517–520.

Daan, N., Gislason, H., Pope, J.G. and Rice, J.C. (2011) Apocalypse in world fisheries? The reports of their death are greatly exaggerated. *ICES Journal of Marine Science* **68**, 1375–1378.

Dryzek, J.S. (2005) *The Politics of the Earth: Environmental Discourses*. Oxford University Press, Oxford.

Eide, A., Bavinck, M. and Raakjær, J. (2011) Avoiding poverty: distributing wealth in fisheries. In: *Poverty Mosaics: Realities and Prospects in Small-Scale Fisheries* (eds S. Jentoft and A. Eide), pp. 13–26. Springer, Dordrecht.

Evans, L., Cherrett, N. and Pemsl, D. (2011) Assessing the impact of fisheries co-management interventions in developing countries: A meta-analysis. *Journal of Environmental Management* **92**, 1938–1949.

FAO (2003) The ecosystem approach to fisheries. FAO Technical Guidelines for Responsible Fisheries No. 4, Suppl. 2. FAO, Rome.

FAO (2009) Report of the global conference on small-scale fisheries. Securing sustainable small-scale fisheries:

Bringing together responsible fisheries and social development. FAO Fisheries and Aquaculture Report No. 911. FAO, Rome.

FAO (2011) Review of the state of world marine fishery resources. FAO Fisheries and Aquaculture Technical Paper No. 569. FAO, Rome.

FAO (2012a) Glossary. Available at http://www.fao.org/fi/glossary/ (accessed February 2014).

FAO (2012b) The State of World Fisheries and Aquaculture. Fisheries and Aquaculture Department, Food and Agriculture Organization of the United Nations, Rome.

Flood, R.C. (1994) An annotated bibliography of community-based and traditional coastal fishery management practices. FAO Fisheries Circular No. 876. FAO, Rome.

Froese, R., Zeller, D., Kleisner, K. and Pauly, D. (2012) What catch data can tell us about the status of global fisheries. *Marine Biology* **159**, 1283–1292.

Froese, R., Zeller, D., Kleisner, K. and Pauly, D. (2013) Worrisome trends in global stock status continue unabated: a response to a comment by R.M. Cook on 'What catch data can tell us about the status of global fisheries'. *Marine Biology*. Available at http://www.seaaroundus.org/researcher/dpauly/PDF/2013/Journal%20Articles/WorrisomeTrendsinGlobalStockStatus.pdf (accessed February 2014).

Garcia, S.E., Kolding, J., Rice, J., Rochet, M.-J., Zhou, S., Arimoto, T., Beyer, J.E., Borges, L., Bundy, A., Dunn, D., Fulton, E.A., Hall, M., Heino, M., Law, R., Makino, M., Rijnsdorp, A., Simard, F. and Smith, A.T. (2012) Reconsidering the consequences of selective fisheries. *Science* **335**, 1045–1047.

GPO (2012) Draft framework document for a Global Partnership for Oceans (GPO). Available at http://www.globalpartnershipforoceans.org/ (accessed February 2014).

Gutiérrez, N.L., Hilborn, R. and Defeo, O. (2011) Leadership, social capital and incentives promote successful fisheries. *Nature* **470**, 386–389.

Halpern, B.S. (2003) The impact of marine reserves: do reserves work and does reserve size matter? *Ecological Applications* **13**(Supplement), 117–137.

Halpern, B.S. and Warner, R. (2002) Marine reserves have rapid and lasting effects. *Ecology Letters* **5**, 361–366.

Heck S., Béné, C. and Reyes-Gaskin, R. (2007) Investing in African fisheries: building links to the millennium development goals. *Fish and Fisheries* **8**, 211–226.

Hilborn, R., Parrish, J.K. and Litle, K. (2005) Fishing rights or fishing wrongs? *Reviews in Fish Biology and Fisheries* **15**, 191–199.

Jentoft, S. (2004) Institutions in fisheries: what they are, what they do, and how they change. *Marine Policy* **28**, 137–149.

Jentoft, S. and Eide, A. (eds) (2011) *Poverty Mosaics: Realities and Prospects in Small-Scale Fisheries*. Springer, Dordrecht.

Johannes, R.E. (1978) Traditional marine conservation methods in Oceania and their demise. *Annual Review of Ecology and Systematics* **9**, 349–364.

Johannes, R.E. (1981) *Words of the Lagoon: Fishing and Marine Lore in the Palau District of Micronesia*. University of California Press, Berkeley, Los Angeles and London.

Johnson, D.S. (2006) Category, narrative, and value in the governance of small-scale fisheries. *Marine Policy* **30**, 747–756.

Jul-Larsen, E., Kolding, J., Nielsen, J.R., Overa, R. and van Zwieten, P.A.M. (2003) Management, co-management or no management? Major dilemmas in southern African freshwater fisheries. Part 1: Synthesis Report. FAO Fisheries Technical Paper 426/1. FAO, Rome.

Kawarazuka, N. and Béné, C. (2010) Linking small-scale fisheries and aquaculture to household nutritional security: an overview. *Food Security* **2**, 343–357.

Kolding, J. (2014) MPAs in relation to fisheries – what are the biological and fish stock implications? In: *Marine Protected Areas (MPA): A Useful Tool in Fisheries Management? Challenges and Experiences from Developing Countries* (eds I. Bryceson and J. Francis). Zanzibar, Tanzania, WIOMSA Book Series No. 9.

Kolding, J. and van Zwieten, P.A.M. (2011) The tragedy of our legacy: how do global management discourses affect small-scale fisheries in the South? *Forum for Development Studies* **38**, 267–297.

Kolding, J., Medard, M., Mkumbo, O. and van Zwieten, P.A.M. (2014) Status, trends and management of the Lake Victoria Fisheries. In: *Inland Fisheries Evolution and Management: Case Studies from Four Continents* (eds R.L. Welcomme, J. Valbo-Jorgensen and A.S. Halls). FAO Fisheries and Aquaculture Technical Paper No. 579. FAO, Rome.

Kooiman, J., Bavinck, M., Jentoft, S. and Pullin, R. (eds) (2005) *Fish for Life: Interactive Governance for Fisheries*. MARE Publication Series, Amsterdam University Press, Amsterdam.

Kremer, A. (1994) Equity in the fishery: a floodplain in N.E. Bangladesh. Research paper R94E. Centre for Development Studies, University of Bath, Bath, UK.

Kurien, J. (2007) The blessing of the commons: small-scale fisheries, community property rights and coastal natural assets. In: *Reclaiming Nature: Environmental Justice And Ecological Restoration* (eds J.K. Boyce, S. Narain and E.A. Stanton), pp. 23–52. Anthem Press, London.

Law, R., Plank, M.J. and Kolding, J. (2012) On balanced exploitation of marine ecosystems: results from dynamic size spectra. *ICES Journal Marine Sciences* **69**, 602–614.

Leal, D. (2010) Political economy of natural resource use: lessons for fisheries reform. Prepared for the Global Program on Fisheries (PROFISH). Agriculture and Rural Development Department World Bank, Washington, DC.

Lester, S.E., Halpern, B.S. Grorud-Colvert, K., Lubchenco, J., Ruttenberg, B.I., Gaines, S.D., Airame, S. and Warner, R.R.

(2009) Biological effects within no-take marine reserves: a global synthesis. *Marine Ecology Progress Series* **384**, 33–46.

Mahon, R. (1997) Does fisheries science serve the needs of managers of small stocks in developing countries? *Canadian Journal of Fisheries and Aquatic Sciences* **54**, 2207–2213.

Mathew, S. (2003) Small-scale fisheries perspectives on an ecosystem-based approach to fisheries management. In: *Responsible Fisheries in the Marine Ecosystem* (eds M. Sinclair and G. Valdimarsson), pp. 47–63. CaBI Publishing, Wallingford, Cambridge, MA.

McEvoy, A.F. (1986) *The Fisherman's Problem: Ecology and Law in the California Fisheries 1850–1980*. Cambridge University Press, Cambridge.

McGoodwin, J.R. (1990) *Crisis in the World's Fisheries: People, Problems, and Policies*. Stanford University Press, Stanford.

McGoodwin, J.R. (2001) Understanding the cultures of fishing communities: a key to fisheries management and food security. FAO Fisheries Technical Paper No. 401. FAO, Rome.

MEA (2005) *Millennium Ecosystem Assessment. Ecosystems and Human Well-Being: Synthesis*. Island Press, Washington, DC.

Medard, M. (2012) Relations between people, relations about things: gendered investment and the case of the Lake Victoria Fishery, Tanzania. *Signs* **37**, 555–566.

Mills, D.J., Westlund, L., de Graaf, G., Willmann, R., Kura, Y. and Kelleher, K. (2011) Underreported and undervalued: Small-scale fisheries in the developing world. In: *Small-scale Fisheries Management: Frameworks and Approaches for the Developing World* (eds N.L. Andrew and R. Pomeroy), pp. 1–15. CABI, Wallingford.

Misund, O.A., Kolding, J. and Fréon, P. (2002) Fish capture devices in industrial and artisanal fisheries and their influence on management. In: *Handbook of Fish Biology and Fisheries*, vol. II (eds P.J.B. Hart and J.D. Reynolds), pp. 13–36. Blackwell Science, London.

Neiland, A.E., Madakan, S.P. and Béné, C. (2005) Traditional Management Systems, Poverty and change in the arid zone fisheries of Northern Nigeria. *Journal of Agrarian Change* **5**, 117–148.

Nielsen, J.R., Degnbol, P., Viswanathan, K.K., Ahmed, M., Hara, M. and Abdullah, N.M.R. (2004) Fisheries co-management-an institutional innovation? Lessons from South East Asia and Southern Africa. *Marine Policy* **28**, 151–160.

Novaczek, I., Harkes, H.T., Sopacua, J. and Tatuhey, M.D.D. (2001) An institutional analysis of sasi laut in Maluku, Indonesia, The World Fish Center. Available at http://EconPapers.repec.org/RePEc:wfi:wfbook:14142 (accessed February 2014).

Pauly, D. (1990) On Malthusian overfishing. *Naga, the ICLARM Quarterly* **13**, 3–4.

Pauly, D. (2006) Major trends in small-scale marine fisheries, with emphasis on developing countries, and some implications for the social sciences. *Maritime Studies (MAST)* **4**, 7–22.

Platteau, J.P. (1989) The dynamics of fisheries development in developing countries: A general overview. *Development Change* **20**, 565–597.

Pomeroy, R. (2011) Managing overcapacity in small-scale fisheries. In: *Small-scale Fisheries Management: Frameworks and Approaches for the Developing World* (eds N.L. Andrew and R. Pomeroy), pp. 75–92. CABI, Wallingford.

Pomeroy, R. and Berkes, F. (1997) Two to tango: the role of government in fisheries co-management. *Marine Policy* **21**, 465–480.

Pomeroy, R., Garces, L., Pido, M. and Silvestre, G. (2010) Ecosystem-based fisheries management in small-scale tropical marine fisheries: emerging models of governance arrangements in the Philippines. *Marine Policy* **34**, 298–308.

Ricard, D., Minto, C., Jensen, O.P. and Baum, J.K. (2011) Examining the knowledge base and status of commercially exploited marine species with the RAM Legacy Stock Assessment Database. *Fish and Fisheries* **13**, 380–398.

Ruddle, K. (1988) Social principles underlying traditional inshore fishery management systems in the Pacific Basin. *Marine Resource Economics* **5**, 351–363.

Russ, G.R., Alcala, A.C., Maypa, A.P., Calumpong, H.P. and White, A.T. (2004) Marine reserve benefits local fisheries. *Ecological Applications* **14**, 597–606.

Salayo, N., Garces, L., Pido, M., Viswanathan, K., Pomeroy, R., Ahmed, M., Siason, I., Seng, K. and Masae, A. (2008) Managing excess capacity in small-scale fisheries: Perspectives from stakeholders in three Southeast Asian countries. *Marine Policy* **32**, 692–700.

Spalding, M.D., Kainuma, M. and Collins, L. (2010) *World Atlas of Mangroves*. Earthscan, London.

Spalding, M.D., Meliane, I., Milam, A., Fitzgerald, C. and Hale, L. Z. (in press) Protecting Marine Spaces: global targets and changing approaches. *Ocean Yearbook* **27**.

SOFIA (2012) The State of the World Fisheries and Aquaculture. FAO, Rome. Available at http://www.fao.org/docrep/016/i2727e/i2727e00.htm (accessed February 2014).

Symes, D. (1996) Fishing in troubled waters. In *Fisheries Management in Crisis* (eds K. Crean and D. Symes), pp. 3-16. Blackwell, London.

Thomas, D. (1996) Fisheries tenure in an African floodplain village and the implications for management. *Human Ecology* **24**, 287–313.

Thomson, D. (1980) Conflict within the fishing industry. *ICLARM Newsletter* **3**, 3–4.

Thorpe, A., Bavinck, M. and Coulthard, S. (2011) Tracking the debate around Marine Pro-tected Areas: key issues and the BEG framework. *Environmental Management* **47**, 546–563.

Torfing, J., Peters, B.G., Pierre, J. and Sørensen, E. (2012) *Interactive Governance: Advancing the Paradigm*. Oxford University Press, Oxford/New York.

UNEP (2012) GEO 5. Global Environment Outlook. Environment for the future we want. UNEP, Nairobi.

Van Ginkel, R. (2009) *Braving Troubled Waters: Sea Change in a Dutch Fishing Community*. Amsterdam University Press, Amsterdam.

World Bank (2008) *The Sunken Billions: The Economic Justification for Fisheries Reform*. Published with FAO, Agriculture and Rural Development Department, Washington, DC.

Worm, B. and Branch, T.A. (2012) The future of fish. *Trends in Ecology and Evolution* **27**, 594–599.

Worm, B., Hilborn, R., Baum, J.K., Branch, T.A., Collie, J.S., Costello, C., Fogarty, M.J., Fulton, E.A., Hutchings, J.A., Jennings, S., Jensen, O.P., Lotze, H.K., Mace, P.M., McClanahan, T.R., Minto, C., Palumbi, S.R., Parma, A.M., Ricard, D., Rosenberg, A.A., Watson, R. and Zeller, D. (2009) Rebuilding global fisheries. *Science* **325**, 578–585.

Chapter 23

Stewardship in tropical small-scale fisheries: Community and national perspectives

P. Christie[1], L.M. Campbell[2] and N. Armada[3]

[1]School of Marine and Environmental Affairs and Jackson School of International Studies, University of Washington, USA
[2]Marine Affairs and Policy, Nicholas School of Environment and Duke University Marine Laboratory, Beaufort, NC, USA
[3]Ecosystems Improved for Sustainable Fisheries (ECOFISH) Project of USAID, Cebu, Philippines

Abstract: Collaboration between policy makers and resource users in the coastal and marine realm are evolving rapidly. This analysis of two cases of collaborative stewardship in the Turks and Caicos Islands and the Philippines, using process and outcomes metrics for ecosystem-based management, demonstrate that sound policy which balances social and environmental goals is attainable. Analysing local or community-level processes within a wider ecosystem and governance context is essential to attain a clear understanding of drivers affecting fisheries and policy options.

Keywords: collaboration; participation; stewardship; fisheries management; evaluation metrics

Introduction

This analysis starts from the premise that well-managed small-scale tropical fisheries are a high priority for both fisheries and conservation agendas. They are challenging to manage well, but lessons are emerging from around the tropical world. The implementation of holistic policies at multiple governance levels is essential to success. Governance can be defined as 'formal and informal arrangements, institutions and mores which determine how resources or an environment are utilized; how problems and opportunities are evaluated and analyzed, what behavior is deemed acceptable or forbidden, and what rules and sanctions are applied to affect the pattern of resource and environmental use' (Juda, 1999, p. 91). Small-scale fisheries vary by context, but share certain attributes: they use small boats with little mechanization; fishing trips are short and to nearby locations; and markets are mainly local or national (Berkes et al., 2001; FAO and WFC, 2008; Pomeroy and Rivera-Guieb, 2005).

Tropical small-scale fisheries rarely constitute a large portion of national economic output, but they provide essential social and economic services. More than half of the fish capture in developing countries is by small-scale fisheries and they directly employ 25–27 million people (FAO and WFC, 2008). The role of small-scale fishing communities in the planning and implementation of sustainable fisheries management and marine conservation initiatives is central to success. Comparative and detailed site-level research demonstrates that without fishing community support, marine fisheries and conservation measures are more likely to fail (Cinner et al., 2012; Gutierrez et al., 2011; Pollnac et al., 2010).

Governance of Marine Fisheries and Biodiversity Conservation: Interaction and Coevolution, First Edition.
Edited by Serge M. Garcia, Jake Rice and Anthony Charles.

The challenges faced by tropical small-scale fishing communities are tremendous (Allison et al., 2012; Andrew et al., 2007; Béné et al., 2010). Their high reliance on marine resources varies seasonally and is frequently in a state of decline over large areas, placing them in a vulnerable state. Additional stresses such as climate change and pandemics such as HIV AIDS disproportionately affect poor and vulnerable coastal fishing communities.

The role of tropical fishing communities, while central to successful fisheries and environmental management, is not assured in many planning processes. How fisheries problems are framed, and therefore responded to, underlies current policy shortcomings. The primacy of fishing rights and economic efficiency in fisheries management policy making does not adequately respond to the wide range of challenges faced by small-scale fishers, including education, health and labour rights (Allison et al., 2012). Linking to broader frameworks such as the human rights framework allows for more holistic strategies in line with the interests of small-scale fishing communities.

A central challenge is the establishment of robust and meaningful decision-making processes that allow for context-relevant policies. Many tropical fisheries are managed in a top-down autocratic manner, with little concern by national policy makers for the impacts on local communities (Christie et al., 2000; Pomeroy and Rivera-Guieb, 2005). In other cases, as in the case described in this chapter of the Philippines, formal laws mandate participatory planning processes. These formal governance processes are complemented by decades of experience with participatory planning, a tool that has been used with mixed success in political and environmental social movements. In other cases, such as the case also described in this chapter of Turks and Caicos Islands (TCI), governments are turning to participatory planning voluntarily in recognition of the limitations of top-down governance. The case of the TCI sea turtle fishery described here similarly demonstrates how international researcher activity and national interests may work towards sustained sea turtle fisheries, which are often under tremendous international pressure to close.

General agreement has emerged based on comparative research and practical experience that participatory, information-rich, sustained planning processes can support the development of sustainable small-scale fisheries with considerable benefits to reliant communities. Considering these emergent lessons and a review of various guidelines for ecosystem-based fisheries management and marine ecosystem-based management, Christie et al. (2007) developed a series of process and output criteria for assessment of fisheries management efforts in the Philippines (Table 23.1). These were refined in Christie et al. (2009b) and

Table 23.1 Marine ecosystem-based management evaluative criteria.

Process
• Planning processes and policies consider local context • Transparent and participatory decision-making processes used for program planning and evaluation • Social- and natural-science-generated information and local knowledge influence planning • The Ecosystem-Based Management (EBM) program area and goals defined to consider ecological scale and interactions, while considering governance feasibility (e.g. management areas represent ecological boundaries while considering institutional jurisdictions and capacities) • Planning processes and policies evolve based on monitoring information and experience • Education program developed to raise awareness about ocean conditions, consider tradeoffs and disseminate lessons among practitioners and appropriate constituencies • Sustained commitment to EBM is fostered • Planning processes consider tradeoffs and establish means to equitably distribute costs and benefits while establishing conflict resolution mechanisms
Output
• Reference points for resource extraction (e.g. catch-per-unit-effort or biomass) and environmental integrity (e.g. biodiversity, habitat condition) established at a precautionary level • A suite of management tools including (but not limited to) MPAs, employed to address resource and habitat goals • Multi-sectoral planning and implementation organizations, which are responsive to ecological scales, established and supported • Formal legal and policy frameworks established to foster EBM • Human and institutional capacity increased to respond to demands of EBM

Christie (2011) and applied as a means to assess marine ecosystem-based management efforts around the tropical world. The criteria suggest means by which planning processes can link fishery-dependent communities to planning processes. The criteria provide a consistent and holist, but not exhaustive, suite of empirically grounded criteria to assess management efforts and will be applied to and refined based on two cases presented within this chapter. The analysis highlights that resource management challenges and related policies are best conceptualized within a broader governance system that links communities with governments, non-governmental organizations, markets and other entities. National policies, especially as related to governance decentralization, clearly influence the feasibility and success of community stewardship.

The following Caribbean and Philippine case studies present examples of small-scale fisheries management in distinct contexts. Both cases describe examples in which fisheries management intersect with high-priority conservation goals, namely sea turtle conservation in the Caribbean and coral reef conservation in the Philippines. The analysis provides examples of management within two distinct governance contexts in which the above criteria are relevant. The cases suggest the need for expansion of the above criteria for effective marine ecosystem-based management.

Case study: Towards co-management of sea turtle fisheries in Turks and Caicos Islands

Although sea turtles are listed on the IUCN's Red List and international trade in sea turtles is prohibited by signatories to the Convention on International Trade in Endangered Species (CITES), sea turtle fisheries continue in many places, illegally and legally (Campbell, 2003). The Caribbean is no exception, and legal sea turtles fisheries operate in several Caribbean countries including Turks and Caicos Islands (TCI) (Campbell et al., 2009; Richardson et al., 2009). Many of these fisheries are neither well regulated nor documented and are under pressure to close, either due to commitments made under international agreements such as the Inter-American Convention for the Conservation of Sea Turtles, or to lobbying pressure from environmental groups.

The rational for closure is ecological and economic. Ecologically, sea turtles are both migratory and long-lived, which poses challenges for assessing the impacts of any management regime on populations. In addition, difficulties associated with studying sea turtles across their range and developmental stages results in a limited understanding of sea turtle populations, their size and dynamics. Given such uncertainty, a precautionary approach is therefore often promoted and equated with eliminating direct use (see Campbell, 2002, 2007). This argument is further supported by a political one; the migratory nature of sea turtles makes them a shared resource, requiring regional management (Campbell et al., 2009; Campbell and Godfrey, 2010). The costs of such closures for local communities are often assumed to be minimal: the fisheries are seen as lacking economic importance and alternative uses of turtles as tourist attractions are promoted as having greater economic value. But what is the value of these fisheries? Because they mostly serve local markets and lack commercial value relative to export-oriented fisheries, there are few data available on their characteristics even when they are legal. Who fishes for turtles? Why and when do they catch turtles? What do they do with them? How does turtle fishing relate to other fishing and livelihood activities? Without answers to these types of questions, implementing any kind of management intervention is challenging.

Since 2008, the government of TCI has been working with a group of researchers from the UK and the USA (Marine Turtle Research Group at the University of Exeter, UK, the Marine Conservation Society, UK and Duke University, USA) on the Turks and Caicos Islands Turtle Project. The project aims to better understand the status of sea turtles and related fisheries for green and hawksbill turtles, with an overall goal of improving fisheries management. The project has collected biological and socio-economic data relevant to the fishery and, based on that data, will recommend amendments to current fishery legislation and a new fishery management plan. Because sea turtles are a migratory resource, 'shared' among nations that host them at various life history stages, one objective of the biological monitoring has been to determine where turtles landed in the TCI fisheries originate from. For example, green turtles (*Chelonia mydas*) likely come from the nesting stock at in Costa Rica, Mexico and Florida (Richardson et al., 2009). Hawksbills (*Eretmochelys imbricata*) are harvested from populations nesting in Mexico, US Virgin Islands (USVI), Antigua, Puerto Rico and Cuba (Richardson et al., 2009). Overall, green turtle populations are listed globally as endangered and hawksbills as critically endangered on the IUCN's Red List, but Caribbean green turtles are increasing at many nesting beaches in the region (Seminoff, 2004) and the rookeries producing green turtles harvested in TCI are large and increasing (particularly in Costa Rica; Bjorndal et al., 1999). These larger-scale issues are considered in the management plan. For example, as part of the management plan the project will recommend changes to TCI's Fisheries Protection Ordinance (introduced in 1998) and the draft recommendations propose more conservative regulations for hawksbills than for green turtles.

The project has adopted some innovative participatory methods, designed to engage local fishers throughout various stages of the research. The overall method of 'Community Voice' was developed to test the premise that a well-designed public participation process could better engage local people in resolving conflicts over environment and resource use (www.communityvoicemethod.org). By combining research with compelling presentation (e.g. documentary film) and structured public discussion of results, the method directs people to identify ways in which conflicts can be resolved (Cumming and Norwood, 2012). Use of this approach in TCI reflects both a research philosophy – those who will be affected by management changes should have some say in them – and the realities of the fishery, where fisher buy-in will be critical to compliance with existing and potential management measures (Campbell et al., 2009; Silver and Campbell, 2005). Although many Caribbean nations have banned sea turtle fisheries in recent years, there is little support for such measures in TCI. The hope is that via extensive consultation throughout the project, fishers will feel ownership of the new fishery management regime, including a fishery management plan and amended regulations designed to enhance both biological and socio-economic sustainability.

The project's efforts were supported by the full-time presence of a project officer in TCI's main fishing community of South Caicos for 2 years. The project officer's presence allowed him to collect extensive data on the fishery and, more importantly, establish relationships with fishers and the broader community. The project officer visited the three fishing landing docks in South Caicos daily when fishers returned home in the afternoon to record, measure and sample turtles landed. During data collection, he conducted short dockside interviews with fishers to probe reasons for turtle capture (e.g. targeted vs. opportunistic), what they were fishing for that day (e.g. conch, lobster, finfish or turtle) and the intended use of the turtle (e.g. sale, consumption, gift). Within days of the dockside interview, the project officer followed up with individual fishers to further understand fisher decision-making about turtle capture and to trace the actual fate of landed turtles (whether they were kept, gifted or sold). This allowed him to identify turtle consumers, who were also interviewed about their use of turtle. In this way, detailed 'dock to plate' commodity chains have been constructed for each landed turtle and the wider impact of the fishery in the community, where a few turtles can circulate among many more users with meat either sold or given as gifts (Figure 23.1), has been characterized. The dockside monitoring and related modelling work has produced a more reliable estimate of the size of the fishery than in the past and includes information on the species, number and sizes of turtles landed (Stringell et al., 2013). For example, modelling suggests that 176–324 green and 114–277 hawksbill turtles are harvested annually; in addition, hawksbill capture is often coincident with lobster fishing. These and other data inform recommendations for legislative amendments and provide a baseline for future monitoring of the harvest, protocols for which are in the draft management plan.

The research has provided insight into the nature of the fishery in terms of who participates, the reasons fishers target turtles, how turtle fishing relates to other fishing activities, what fishers do with captured turtles and the role and importance of the turtle fishery in the wider community. Results in full are too extensive for review here, but were critical to informing the recommendations for the management plan and, more specifically, for amending regulations in the Fisheries Protection Ordinance. For example, the ordinance currently protects small and juvenile turtles smaller than 20 inches shell length from harvest but allows capture of anything larger, including adults at sea (see Richardson et al., 2006). This is contrary to what current sea turtle population modelling suggests is appropriate, and changing the size limit was one of the obvious improvements to be made. Research results from dockside monitoring and interviews with fishers and consumers suggested such change would be accepted, even preferred. In a second example, research showed that the largely opportunistic hawksbill harvest was closely associated with the commercial spiny lobster fishery; fishers looking for lobster regularly encounter hawksbills and sometimes capture them, mostly for personal consumption. Dockside monitoring revealed that many landed hawksbills were sub-adult and adult turtles, often taken during peak hawksbill turtle nesting season. Rather than recommend an unpopular, outright ban on hawksbills, the project married the biological and socio-economic data to recommend that hawksbill fishing be allowed only when the lobster fishery is closed, and that the hawksbill fishery be closed for an 8-month season that coincides with the lucrative lobster open season (and, importantly, the hawksbill nesting season).

Working with the Community Voice Method (see Cumming and Norwood, 2012), 33 stakeholders in the sea

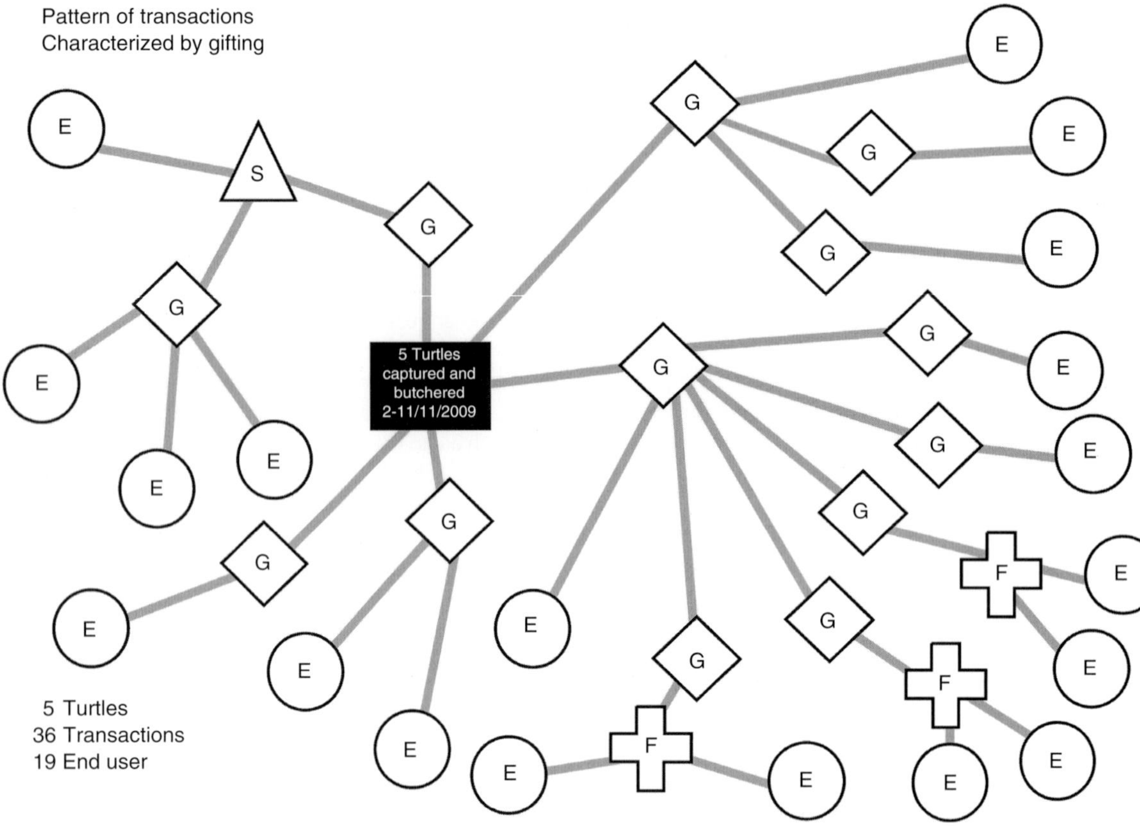

Figure 23.1 Commodity chain associated with five turtles captured over several days in November of 2009. G indicates a gift, F indicates food preparation, S indicates sale of meat to another person, and E indicates eating.

turtle fishery were interviewed resulting in a documentary film constructed around the analysis of the interviews. The documentary is designed to both share research results and spark conversation about what results suggest for future management of the fishery. The film was shown in a variety of venues in TCI and audience feedback was recorded. Twenty-two workshops were also held after some screenings, where the potential turtle fishery management measures discussed by interviewees in the film were debated with the approximately 270 stakeholders who attended. Based on these workshops, amendments to regulations in the Fisheries Protection Ordinance were drafted. A total of 75 turtle fishers were subsequently interviewed individually to record their opinions on these draft recommendations.

Overall, the suggested changes to the Fisheries Protection Ordinance take into account results from the biological and socio-economic research and the feedback received during extensive fisher engagement. Each recommendation is a complex mix of these inputs and, although most measures gained widespread fisher support during interviews, a few were more contentious. For example, the establishment of a hawksbill closed season during the lobster open season was supported by 84% ($n=63$) of interviewees. Similarly, a proposed and novel 24 inches maximum size limited for green and hawksbill turtles received broad support (88%, $n=66$), while the proposed and lowered minimum size limit of 12 inches was rejected by 52% ($n=39$) of the interviewees. These fishers all suggested higher minimum size limits as they felt 12 inches was too low and did not protect enough smaller turtles. Consequently, the final recommendations will include an 18 inches minimum size limit, the mode suggestion from the 39 interviewees. Other supported and thus recommended measures include the ongoing protection of nesting females and nests on the nesting beaches; some gear restrictions; obligatory live landing of turtles for inspection; prohibition of export; and prohibition of take of other turtle species found in the wider Caribbean region. The fishery management plan will seek to establish these amended regulations, promote them among fisher communities and develop a culture of compliance, while developing a regime of fishery enforcement and regular monitoring among Department of Environment and Maritime Affairs (DEMA) officers so that adaptive action can be taken if significant changes to the fishery occurs. The proposed changes to the Fisheries Protection Ordinance were approved by DEMA in October 2012, and the management plan is currently being drafted for approval in 2013 by the appropriate Minister of TCI Government (re-elected in November 2012 after several years of temporary, direct rule by the UK Government).

Members within the TCI national government and its management agencies are aware of the pressure in the region to prohibit turtle fisheries and of the limits on the government's inclination and capacity to follow suit. By instituting new, more biologically robust management measures, the government may be able to better withstand such pressures for complete closures that are socially and institutionally infeasible. The balancing of economic development, cultural tradition, food security and conservation goals has resulted in the development of a management regime that recommends increased regulation of the fishery to meet identified conservation needs, but still allows considerable access to this traditional resource.

Case study: Ubay, Bohol, Philippines

The following describes a fisheries resource management process and outcomes in the Municipality of Ubay, Bohol, Philippines. The initiative was aimed at improving the production potential of a degraded resource base by first ensuring the local capability to enforce already existing fishery laws and regulation, enhancing management activities by cumulatively establishing fisheries control mechanisms and finally enabling management schemes to ensure the effective implementation of activities.

The fishing ground of the Municipality of Ubay is located in Danajon Bank, a double barrier coral reef system in Central Philippines. Ubay has about 60 km of coastline and a body of water of area about 210 km^2 under its jurisdiction. Danajon Bank is known to be one of the most heavily fished and degraded fishing grounds in the Philippines where, for the last four decades, a significant segment of the fishing population was engaged in illegal and destructive fishing activities (Armada et al., 2009). These include dynamite fishing, use of poison in catching coral reef fishes, use of fine meshed nets, coral extraction, cutting of mangroves, destruction of seagrass beds and intrusion of commercial boats into municipal waters.

An environmental summit in 1997 highlighted these illegal activities and a fishing community in Ubay was identified as one of the dynamite and cyanide fishing hotspots in the country. The municipality responded by creating an enforcement team ('bantay dagat' or sea guardian) and establishing a coastal resource management technical working group under the Office of the Mayor. Little materialized out of these initiatives and some members of the enforcement team were embroiled in allegations of incompetence and extortion. Illegal fishing activities continued unabated and the municipality even became the fishing port for illegal commercial fishing boats such as trawls and Danish seines. These fishing gears have significant impact since they are able to catch fishes, including juveniles, in significant quantities depriving small-scale fishers, particularly the hook and line fishers, of their catch. Table 23.2 shows the catch by Danish Seine in relation to other fishing gears in Danajon Bank.

From 2003 to 2010, with support from the US Agency for International Development-funded Fisheries Improved for Sustainable Harvest (FISH) project, the municipality of Ubay started to make serious efforts to overhaul its coastal resource management initiatives. Funding was used partly to hire a multidisciplinary and experienced staff that included fishery scientists, media and communication experts, planners and community organizers. The initiative came at a crucial period. Initial assessment at that time indicated that the standing biomass of Danajon Bank (0.45 tons per km^2) was among the lowest in the country and catch rates were at an all-time low (about 3 kg per fishing operation). The catches were composed mainly of low trophic level and low-value species such as slipmouths (*Leiognathidae*), pufferfishes (*Tetraodontidae*), flat heads (*Paltycephalidae*), goatfishes (*Upeneidae*) and cardinal fishes (*Apogonidae*), mostly considered as trash fish in more productive fishing grounds. These findings prompted stakeholders and resource users to work together to effect change. Initial activities include participatory coastal resource assessment and coastal resource management (CRM) planning. From there, an iterative planning process was used to expand the program, create policies and take action on pressing resource issues

Table 23.2 Catch (kg) of Danish Seine in relation to other fishing gears during a three-month catch monitoring survey in Danajon Bank in 2004.

Species	Danish seine	Bottomset gillnet	Bottomset longline	Drift gillnet	Hook and line (simple)	Roundhaul seine
Nemipterus spp.	1981	16	23	2	61	
Carangidae (jacks)	1857	27	14	173	3	59
Scolopsis spp.	1504	4	2	1	51	1
Leiognathus spp.	1446	24		19	2	242
Sphyraena spp.	1443	59	3	114	9	14
Selaroides leptolepsis	1211	1343		32	9	265
Lolingo spp.	709	4	8	7	3	276
Upeneus sulphureus	649	6		26		
Scomberomorus spp.	540			826	23	
Plectorhinchus spp.	532	2	12		7	
Upeneus spp.	429	55		78	55	13
Megalaspis cordyla	394	30		16		209
Gnathanodon speciosus	342	2	1	3		55
Gerres spp.	265	51	47	14	145	
Leiognathus splendens	234	6		3	1	869
Serranidae (groupers)	193	1	28		2	
Lutjanus chrysotaenia	182	2	50		3	
Dasyatidae	133	2	206		3	
Sardinella spp.	122	23		2850	3	515
Sillago spp.	101	2	4	2	44	
Terapon jarbua	90	3	54	51	18	76
Choerodon spp.	62	2			61	
Alepes sp.	57	9		45		10
Atule mate	55	2		4	1	24
Tastrelliger spp.	51	371		427	3	2

that interested local policy makers and resource users. The CRM plan included the creation of a CRM Office to address illegal fishing and other coastal and marine issues in a more holistic manner as well as the immediate rehabilitation of the enforcement team.

Because 2004 was an election year the initiatives were put on hold for a while. The election resulted in a change of administration and initially nobody could predict how the new mayor would deal with CRM initiatives (most people involved during the previous administration belong to the opposing political party). Surprisingly however, the new mayor retained the entire personnel involved in CRM, pushed for the formal establishment of the CRM Office and for the appropriation of a separate budget for CRM activities. Key to the entire initiative was the overhaul of the enforcement team. An effectiveness assessment indicated that the existing enforcement team was a failure because it lacked the skills to address the complexity of fishery law violation, lacked the legal authority and accountability, and was not properly compensated leading some members to engage in extortion. The overhaul included purging of guilty members, recruitment of new members and subjecting members to a series of trainings to improve skills in prevention, apprehension and prosecution of illegal fishing violations. They were likewise trained for land-based enforcement support activities including mechanisms to deny marketing of illegally captured fish and working with the public to report fishery law violations. The municipal government provided team members the necessary financial compensation, allowances and support during litigations.

As enforcement capability was strengthened, the fisheries management initiatives crafted and implemented by the municipality of Ubay became more and more sophisticated. It started with the establishment of marine protected areas (MPAs) and was followed by establishment of a closed season for rabbit fish (*Siganus canaliculatus*), a ban on catching of undersized (12 cm carapace length and smaller) and berried blue crabs (*Portunus pelagicus*), a ban on the use of beach seine due to its fine-meshed net and limits to the number and deployment of stationary gears. These were supported through enabling management schemes such as zoning of fishery uses, registration of fishers and fishing boats and licensing of fishing gears and operations.

The body of water of Ubay is part of a larger fisheries management system and, together with eight other adjacent municipalities, they form the Danajon Bank marine ecosystem. Following an ecosystem-based management approach, Ubay's coastal and fisheries management initiatives are linked to initiatives of these neighbouring municipalities (McLeod and Leslie, 2009). Their individual MPAs together form a network of MPAs, fisheries control mechanisms were harmonized and uniformly implemented, fisheries zoning schemes were harmonized and made compatible across water boundaries and fishery law enforcement is governed through a coastal law enforcement council (CLEC) supported by the nine municipalities. An inter-municipality agreement was forged by the nine local governments to strengthen the CLEC, as well as enable the council to perform as a policy-making body on coastal and fisheries management.

Various sectors of the local society were engaged and fisheries resource management activities were implemented by a variety of institutions at various governance levels in addition to the local government. MPAs were directly managed by communities or 'barangays' and supported by the municipal government through legislation and training of MPA managers and personnel on monitoring and enforcement. Habitat assessments were conducted in collaboration with the FISH project and non-government organizations (NGOs) operating in the area. Together with the Department of Education, youths were engaged through youth camps, stage plays and mural paintings to increase student interest in coastal and fishery issues. Information, education and communication (IEC) campaigns were conducted in conjunction with medical missions by the Philippine Navy and Health Department, as well as 'ecological evangelization' by selected members of the Catholic clergy.

Monitoring over a seven-year period showed that average catch rates of four fishing gears monitored in fisheries-independent surveys increased by 9.8% (ranging from −75% for bottom-set longline to 109% for trawl), reef fish biomass by about 12% and reef species richness by 30% (FISH, 2010). Catch rates of gears monitored via fisheries-dependent surveys showed varied results. The average catch rates from fisheries-dependent data were observed to have generally increased but are not statistically significant because of high variance and many extreme values. These monitoring results were regularly communicated to the local government, stakeholders and resource users in various fora. Capacity review of municipal government units showed significant improvements in their capacity to conduct CRM planning, law enforcement, IEC and legislative and technical support to MPA establishment. Municipal investment in CRM activities has significantly increased over the years. CRM budget of Ubay increased from practically zero in 2003 to 2 million pesos ($40,000) in 2006 and almost 3 million pesos ($60,000) in 2010. It set the stage as a pilot-scale economic analysis of CRM investment, showing that the LGU's annual expenditure on CRM can be offset by the potential annual income from resource rent if properly implemented (Rosales, 2008). The municipality also became the core of CRM and fisheries management destinations for study tours from all over the country and abroad.

As ongoing activities, the municipality of Ubay is trying to find ways to limit fishing effort to optimize harvest and is also looking into the administrative processes to expedite prosecution of apprehended illegal fishers. The Ubay CRM Office is trying to work with their counterparts in other municipalities of Danajon Bank towards a unified or, at least, harmonized modes of development and implementation of future initiatives.

The Philippine marine governance context is highly decentralized, with municipal governments maintaining jurisdiction out to 15 km offshore (Eisma et al., 2005). Such decentralization can make institutional coordination to scale up to ecosystem levels challenging but, as demonstrated here, not impossible. National laws, namely the Local Government Code of 1991 (Republic Act 7160) and the Philippine Fisheries Code (Republic Act 8550) recognize and encourage the creation of clusters of municipal and community coordination resulting in ecosystem-level management processes (Christie et al., 2009a; Eisma et al., 2009).

Analysis and discussion

Comparative studies (Cinner et al., 2012; Gutierrez et al., 2011; Pollnac et al., 2010) and these cases studies of progress towards sustainable small-scale fisheries demonstrate that success is possible and that elements of success are consistent, but require local adaptation. Attention to local dynamics and engagement of resource managers as co-managers are clearly elements of successful management that balances local needs for employment and food with conservation goals. The definition and implementation of meaningful participation is context specific and framed by legal and cultural specifics. There is no substitute for lasting and painstaking engagement over decades to ensure that management models, scientific and non-scientific knowledge and implementation know-how are incorporated at all relevant governance levels. The influence of decentralized Philippine marine governance system has been well documented (e.g. Christie and White, 2007). TCI legal context provides less support for decentralized governance of marine resources. Nonetheless, this case demonstrates that participatory planning and implementation are still possible and results in impressive outcomes.

The TCI and Philippine cases represent ambitious and interesting examples of simultaneous engagement of coastal and fishing communities in conservation and fisheries management. These efforts, which use a highly participatory process, naturally integrate both conservation and fisheries management goals. Perhaps this is because many tropical coastal communities are 'natural integrators'; they do not commonly separate their activities and concerns by ecosystem boundaries and they do not commonly dichotomize resource management and environmental stewardship. Their livelihoods and activities move between ecosystems. Their livelihoods and cultures are also dependent on healthy ecosystems, which have sometimes evolved over long periods of time (Aswani and Hamilton, 2004). They quickly feel the consequences of unsustainable behaviours and policies, and are motivated to take action based on their interests to maintain livelihoods, resources and ecosystems. For example, the sense of threat from illegal fishing or poaching by commercial vessels has been determined by two distinct studies involving more than 500 informants to support, implement and protect marine protected areas in which they have sacrificed fishing grounds for long-term interests (Christie et al., 2009a; Pollnac et al., 2010). By using participatory methods which foster community fishery and environmental stewardship, the historic dichotomy of fisheries and biodiversity management is more likely resolved.

The process and output criteria for marine EBM help with the systematic analysis of these cases. Table 23.3 provides summary information about how each criterion was met in both contexts. The case studies strongly suggest that successful small-scale fisheries management and marine EBM will require analysis and complementary policy at multiple governance levels, including international, national and local levels. Markets for fisheries products are frequently international. Fishers move between national boundaries. International marine conservation and fisheries management initiatives are increasingly multinational. As such, and in line with Ostrom (1990), the criteria for EBM evaluation should include the following additional element: linkages between local, national and international governance are considered and result in nested governance arrangements.

The decentralized governance context of the Philippines has encouraged the proliferation of community-based and co-management approaches. In that context, scaling up to robust ecosystem-level management will require innovative, multi-institutional approaches so that scarce management resources (e.g. enforcement boats, environmental monitoring equipment) can be shared and regulations harmonized (Christie et al., 2009b). In other tropical (and temperate) contexts, 'scaling down' to local levels is in need of development (Cochrane et al., 2009). Strong institutions and laws do not ensure successful compliance with fisheries laws. Self-monitoring and self-compliance with rules were identified as essential fisheries mechanisms decades ago (McCay and Jentoft, 1996). Experiments to link large-scale management processes such as multi-national large-marine ecosystem (LME) are underway in the Caribbean and have the potential to support local management initiatives as described in this case study (Fanning et al., 2009). Given the life history of marine turtles, linked local and regional management is essential for turtle sustainability.

The collection of social ecological information at relevant scales remains an ongoing challenge in many tropical contexts. The TCI and Philippine case both collected management-relevant fisheries and biological data. The collection of social data was incomplete in the Philippine case, as is common in many such management programs (Christie, 2011). The use of spatially explicit databases to organize management data is an essential tool which can be made particularly relevant at local levels when local knowledge is gathered and incorporated. Increasingly, such data are amassed at higher levels to create compelling representations of ecosystem health at regional and global levels (WRI, 2011). Maintaining such databases remains expensive and the linkages to local management are not always apparent (De Leon and White, 2005). Development of datasets which respond to specific management needs and a process to incorporate such data in policy-making exercises are obvious – but frequently overlooked – elements of planning programs (Christie, 2011).

Table 23.3 Assessment of TCI and Philippine cases using marine ecosystem-based management evaluative criteria.

Process	Turks and Caicos Islands	Ubay, Philippines
Planning processes and policies consider local context	The sea turtle management plan aims to manage the fishery in context, as part of livelihoods. Community Voice approach adopted because of need to engage local stakeholders, and reality of limits on fisheries enforcement. Community buy-in will be essential, and project designed to increase it.	Local history, management capacity, ecology are considered in planning and implementation process. Fisheries management plans tailored to fisheries and governance capacities of local government institutions.
Transparent and participatory decision-making processes used for program planning and evaluation	Participatory techniques used for data collection, design of management plan and ultimate approval. Fisher feedback solicited at all stages. Regular presentation and discussion of research results to community.	Participatory techniques used for resource monitoring, management plan design and implementation. Philippines decentralized governance system requires participatory planning.
Social and natural-science-generated information and local knowledge influence planning	Biological and ecological data on sea turtles and socio-economic and cultural data on the fishery collected. Local fishers participated in biological data collection, based on knowledge of fishing grounds and capture techniques.	Emphasis on natural science and fisheries information, but basic socio-economic information collected. Local knowledge of ecosystem, fisheries practices and management feasibility influenced policy design and implementation.
The EBM program area and goals defined to consider ecological scale and interactions, while considering governance feasibility (e.g. management areas represent ecological boundaries while considering institutional jurisdictions and capacities)	Caribbean wide data on status of sea turtles and the origins of those coming to TCI accounted for in management recommendations (e.g. hawksbill use more limited than green turtle use).	FISH program designed to manage Danajon Bank ecosystem by linking relevant municipal and national institutions with responsibilities over political jurisdictions.
Planning processes and policies evolve based on monitoring information and experience	Baseline and ongoing monitoring information were used to determine minimum and maximum allowable turtle size, closed seasons and social feasibility (e.g. tolerance or lack thereof for certain measures).	Initial management interventions focused on immediate actions and attainable goals, and then expanded to more ambitious goals. Management interventions were reviewed and adapted regularly based on feedback from stakeholders and in some cases from monitoring information.
Education program developed to raise awareness about ocean conditions, consider tradeoffs and disseminate lessons among practitioners and appropriate constituencies	Several satellite tracking projects undertaken with fisher participation, to illustrate migratory routes. Regular updates on location were posted, and one turtle in particularly that travelled several hundred miles before returning to TCI generated considerable local interest (Richardson et al., 2010).	Education campaigns developed for general public using various educational techniques. Capacity development program including planning, implementation and enforcement skills implemented for policy makers.
Sustained commitment to EBM is fostered	Although the project was funded by UK donors, an officer in the Department of Environment and Maritime Affairs was assigned to the project. DEMA staff will be responsible for monitoring and enforcement of the new management plan.	Participatory processes, involvement of local and external partners, hiring of local personnel and local funding by municipalities fostered sustained commitment.

(continued)

Table 23.3 (Cont'd)

Process	Turks and Caicos Islands	Ubay, Philippines
Planning processes consider tradeoffs and establish means to equitably distribute costs and benefits while establishing conflict resolution mechanisms	The draft management plan takes into account results from the biological and socio-economic research and the feedback received during extensive fisher engagement. Each recommendation is a complex mix of these inputs.	Fishery limits implemented for various fisheries resulting in distributed costs and benefits for large- and small-scale fisheries. Improved fisheries management benefits fishery sector.
Output		
Reference points for resource extraction (e.g. catch-per-unit-effort or biomass) and environmental integrity (e.g. biodiversity, habitat condition) established at a precautionary level	Data collected provide reference points for sea turtle harvest regarding landings, seasonality and catch composition. Project did not assess biodiversity or habitat condition.	Fisheries-dependent, fisheries-independent and ecosystem baselines established prior to project implementation and used to establish realistic targets (approved by project sponsor) for fishery recovery. Most fishery recovery goals met.
A suite of management tools including (but not limited to) MPAs, employed to address resource and habitat goals	Recommended changes to Fisheries Protection Ordinance include multiple measures, i.e. 'slot' size limits, hawkbill fishery closed season, gear restrictions. Management plan identifies need to look at other measures (e.g. guidelines for coastal development) to protect habitat.	Various fisheries management techniques used in concert with MPA networks.
Multi-sectoral planning and implementation organizations, which are responsive to ecological scales, established and supported	Multi-sectoral agency exists and new management plan responsive to ecological, biological, economic and social aspects of the sea turtle fishery drafted.	Community, municipal and Danajon-Bank-wide management organizations established for input into planning and implementation, especially for regulation enforcement.
Formal legal and policy frameworks established to foster EBM	Management plan for sea turtles, in agreement with CITES obligations.	Coastal and fisheries management plans developed by each municipality. Plans and regulations rationalized between municipalities. Resources, especially for enforcement, shared between municipalities.
Human and institutional capacity increased to respond to demands of EBM	DEMA partnered with project throughout. Individual offices trained in research and monitoring skills.	Human and implementation capacity for fisheries and coastal management developed. Environmental and fishery management capacity enhanced, although dependent on external skills and resources. Municipal government commitment solidified and ongoing.

Linkages between research and education are fundamental to long-term success. Documentary film making which captures local voices (as in TCI) and educational programs linked to youth and religious movements (Philippines) help to make fisheries management relevant to larger societal concerns. Increasingly, social and learning networks, which complement formal educational processes, are becoming part of the management toolkit.

Conclusions

In summary, successful small-scale fisheries management in challenging, tropical contexts is possible as demonstrated by these and other cases. Successful programs have multiple resource management and environmental benefits. The application of a suite of basic design principles to fisheries and conservation programs can improve success and long-term feasibility of management initiatives. These cases have incorporated most of the EBM design principles and suggested at least one other: the need for integration across levels of governance. They clearly highlight the importance of participatory and decentralized decision making. They also highlight the importance of how tropical small-scale fishery issues and management practices are analysed. Small-scale tropical fisheries are complex systems which function within and are influenced by local, national and international policies and institutions.

Acknowledgements

The socio-economic research and participatory engagement of fishers in the Turks and Caicos turtle project was led by Amdeep Sanghera, assisted by Lisa Campbell, Gabriel Cumming, Quentin Phillips, Sue Ranger and Peter Richardson. Research in the Philippines was supported by the Pew Charitable Trust Pew Fellows program (Christie). The FISH program was supported by USAID.

References

Allison, E.H., Ratner, B.D., Asgard, B.A., Willmann, R., Pomeroy, R.D. and Kurien, J. (2012) Rights-based fisheries governance: from fishing rights to human rights. *Fish and Fisheries* **13**, 14–29.

Andrew, N.L., Béné, C., Hall, S.J., Allison, E.H., Heck, S. and Ratner, B.D. (2007) Diagnosis and management of small-scale fisheries in developing countries. *Fish and Fisheries* **8**, 227– 240.

Armada, N., White, A.T. and Christie, P. (2009) Managing fisheries resources in Danajon Bank, Bohol, Philippines: An ecosystem-based approach. *Coastal Management* **37**, 275–299.

Aswani, S. and Hamilton, R. (2004) Integrating indigenous ecological knowledge and customary sea tenure with marine and social science for conservation of bumphead parrotfish (Bolbometopon muricatum) in the Roviana Lagoon Solomon Islands. *Environmental Conservation* **31**, 1–15.

Béné, C., Hersoug, B. and Allison, E.H. (2010) Not by rent alone: analysing the pro-poor functions of small-scale fisheries in developing countries. *Development Policy Review* **28**, 325–358.

Berkes, F., Mahon, R., McConney, P., Pollnac, R. and Pomeroy, R. (2001) *Managing Small-Scale Fisheries, Alternative Directions and Methods*. International Development Research Centre, Ottawa.

Bjorndal, K.A., Wetherall, J.A., Bolten, A.B. and Mortimer, J.A. (1999) Twenty-six years of green turtle nesting at Tortuguero, Costa Rica: an encouraging trend. *Conservation Biology* **13**, 126–134.

Campbell, L.M. (2002) Science and sustainable use: views of conservation experts. *Ecological Applications* **12**, 1229– 1246.

Campbell, L.M. (2003) Contemporary culture, use, and conservation of sea turtles. In: *The Biology of Sea Turtles*, Volume II (eds P.L. Lutz, J.A. Musick and J. Wyneken), pp. 307–338. CRC Press, Boca Raton.

Campbell, L.M. (2007) Reconciling local conservation practice with global discourse: a political ecology of sea turtle conservation. *Annals of the Association of American Geographers* **97**, 313–334.

Campbell, L.M. and Godfrey, M.H. (2010) Geopolitical genetics: claiming the commons through species mapping. *Geoforum* **41**, 897–907.

Campbell, L.M., Silver, J.J.,Gray, N.J., Ranger, S., Broderick, A., Fisher, T., Godfrey, M.H., Gore, S., Jeffers, J., Martin, C., McGowan, A., Richardson, P., Sasso, C., Slade, L. and Godley, B. (2009) Co-management of sea turtle fisheries: Biogeography versus geopolitics. *Marine Policy* **33**, 137–145.

Christie, P. (2011) Creating space for interdisciplinary marine and coastal research: Five dilemmas and suggested resolutions. *Environmental Conservation* **38**, 172–186.

Christie, P. and White, A.T. (2007) Best practices for improved governance of coral reef marine protected areas. *Coral Reefs* **26**, 1047–1056.

Christie, P., Bradford, D., Garth, R., Gonzalez, B., Hostetler, M., Morales, O., Rigby, R., Simmons, B., Tinkam, E., Vega, G.,

Vernooy, R. and White, N. (2000) *'Taking Care of What We Have': Participatory Natural Resource Management on the Atlantic Coast of Nicaragua*. University of Central America Center for Research and Documentation of the Atlantic Coast and the International Development Research Centre, Managua and Ottawa.

Christie, P., Fluharty, D.L., White, A.T., Eisma-Osorio, R.L. and Jatulan, W. (2007) Assessing the feasibility of ecosystem-based fisheries management in tropical contexts. *Marine Policy* **31**, 239–250.

Christie, P., Pollnac, R.B., Oracion, E.G., Sabonsolin, A., Diaz, R. and Pietri, D. (2009a) Back to basics: An empirical study demonstrating the importance of local-level dynamics for the success of tropical marine ecosystem-based management. *Coastal Management* **37**, 349–373.

Christie, P., Pollnac, R.B., Fluharty, D.L., Hixon, M.A., Lowry, G.K., Mahon, R., Pietri, D., Tissot, B.N., White, A.T., Armada, N. and Eisma-Osorio, R.L. (2009b) Tropical marine EBM feasibility: A synthesis of case studies and comparative analyses. *Coastal Management* **37**, 374–385.

Cinner, J.E., McClanahan, T.R., MacNeil, M.A., Graham, N.A.J., Daw, T.M., Mukminin, A., Feary, D.A., Rabearisoa, A.L., Wamukota, A., Jiddawi, N., Campbell, S.J., Baird, A.H., Januchowski-Hartley, F.A., Hamed, S., Lahari, R., Morove, T. and Kuange, J. (2012) Comanagement of coral reef social-ecological systems. *Proceedings of the National Academy of Sciences of the United States of America (PNAS)* **109**, 5219–5222.

Cochrane, K.L., Augustyn, C.J., Fairweather, T., Japp, D., Kilongo, K., Iitembu, J., Moroff, N., Roux, J.P., Shannon, L., Van Zyl, B. and Vaz Velho, F. (2009) Benguela Current Large Marine Ecosystem: Governance and management for an ecosystem approach to fisheries in the region. *Coastal Management* **37**, 235–254.

Cumming, G. and Norwood, C. (2012) The Community Voice Method: using participatory research and filmmaking to foster dialog about changing landscapes. *Landscape and Urban Planning* **105**, 434–444.

De Leon, R.O. and White, A.T. (2005) Do biophysical studies and coastal databases enhance coastal management sustainability? Several Philippine cases. *Ocean and Coastal Management* **48**, 411–426.

Eisma, R.L., Christie, P. and Hershman, M.J. (2005) Legal issues affecting sustainability of integrated coastal management in the Philippines. *Ocean and Coastal Management* **48**, 336–359.

Eisma, R.L., Amolo, R.C., Maypa, A.P., White, A.T. and Christie, P. (2009) Scaling-up local government initiatives towards ecosystem-based fisheries management in Southeast Cebu Island, Philippines. *Coastal Management* **37**, 291– 307.

Fanning, L., Mahon, R. and McConney, P. (2009) Focusing on living marine resource governance: The Caribbean Large Marine Ecosystem and adjacent areas project. *Coastal Management* **37**, 219–234.

FAO and WFC (2008) Small-scale capture fisheries: A global overview with emphasis on developing countries. A preliminary report of the Big Numbers Project. Available at http://www.worldfishcenter.org/resource_centre/WF_1085.pdf (accessed February 2014).

FISH (Fisheries Improved for Sustainable Harvest) Project (2010) FISH Document No. 53-FISH/(2010) Completion report. Tetra Tech EM, Inc., Pasig City, Philippines. 252p. Available at http://www.oneocean.org (accessed February 2014).

Gutierrez, N.L., Hilborn, R. and Defeo, O. (2011) Leadership, social capital and incentives promote successful fisheries. *Nature* **470**, 386–389.

Juda, L. (1999) Considerations in the development of a functional approach to the governance of large marine ecosystems. *Ocean Development and International Law* **30**, 89–125.

McCay B.J. and Jentoft, S. (1996) From the bottom up: Participatory issues in fisheries management. *Society and Natural Resources* **9**, 237–250.

McLeod, K. and Leslie, H. (2009) *Ecosystem-Based Management for the Oceans*. Island Press, Washington DC.

Ostrom E. (1990) *Governing the Commons the Evolution of Institutions for Collective Actions*. Cambridge University Press, New York.

Pollnac, R., Christie, P., Cinner, J.E., Dalton, T., Daw, T.M., Forrester, G.E., Graham, N.A.J. and McClanahan, T.R. (2010) Marine reserves as linked social–ecological systems. *Proceedings of the National Academy of Sciences of the United States of America (PNAS)* **107**(43), 18262–18265.

Pomeroy, R.S. and Rivera-Guieb, R. (2005) *Fishery Co-Management: A Practical Handbook*. International Development Research Centre, Ottawa.

Richardson, P.B., Broderick, A.C., Campbell, L.M., Godley, B.J. and Rangerab, S. (2006) Marine turtle fisheries in the UK overseas territories of the Caribbean: domestic legislation and the requirements of multilateral agreements. *Journal of International Wildlife Law & Policy* **9**, 223–246.

Richardson, P.B., Broderick, A.C., Bruford, M., Campbell, L.M., Clerveaux, W., Formia, A., Henderson, A., McClellan, K., Newman, S., Pepper, M., Ranger, S., Silver, J.J., Slade, L., and Godley, B.J. (2009) Marine turtles in the Turks and Caicos Islands: remnant rookeries, regionally significant foraging stocks and a major turtle fishery. *Chelonian Conservation and Biology* **8**, 192–207.

Richardson, P.B., Calosso, M.C., Claydon, J., Clerveaux, W., Godley, B.J., Phillips, Q., Ranger, S., Sanghera, A., Stringell, T.B. and Broderick, A.C. (2010) Suzie the Green Turtle: 6,000 kilometres for one clutch of eggs? *Marine Turtle Newsletter* **127**, 26–27.

Rosales, R. (2008) Developing a framework for economic analysis of CRM investments: the case of Ubay, Bohol.

Fisheries improved for Sustainable Harvest (FISH) Project. Tetra Tech EM, Inc., Pasig City, Philippines. 26p. Available at http://www.oneocean.org (accessed February 2014).

Seminoff, J.A. (2004) IUCN Redlist report, *Chelonia mydas*. Available at http://www.iucnredlist.org/details/4615/0 (accessed February 2014).

Silver, J.J. and Campbell, L.M. (2005) Fisher participation in research: Dilemmas with the use of fisher knowledge. *Ocean & Coastal Management* **48**, 721–741.

Stringell, T.B., Calloso, M.C., Claydon, J.A.B., Clerveaux, W., Godley, B.J., Lockhart, K.J., Phillips, Q., Ranger, S., Richardson, P.B., Sanghera, A. and Broderick, A.C. (2013) Marine turtle harvest in a mixed small-scale fishery: evidence for revised fishery management legislation. *Ocean and Coastal Management* **82**, 34–42.

World Resources Institute (2011) *Reefs at Risk Revisited*. Washington, DC.

Chapter 24

Making space for small-scale fishing communities: Use and misuse of spatial management instruments

M.R. Sowman[1], R. Rajagopalan[2], C. Sharma[2,†] and J. Sunde[1]

[1]University of Cape Town, Rondebosch, South Africa
[2]International Collective in Support of Fishworkers (ICSF), Chennai, India

Abstract: This chapter considers the impacts of conservation and conventional fisheries management approaches and practices, especially the implementation of marine protected areas (MPAs) and other spatially based management measures, on the socio-economic and customary rights of coastal fishing communities. Despite a plethora of international instruments recognizing the need to respect the rights of local fisher communities, research suggests that communities continue to be impacted. Case studies in India and South Africa demonstrate how MPAs have resulted in major impacts on fishers' access to food and livelihoods as well as loss of customary rights and cultural practices. The disjuncture between policy rhetoric and implementation practices is evident and, on this basis, adoption and mainstreaming of a human-rights-based approach from policy level through to management action is proposed. Governance mechanisms include recognition of small-scale fishing communities' human rights, integration of social and cultural dimensions in planning and management and involvement of fishers in decision-making processes.

Keywords: small-scale fisheries; governance; marine protected areas; spatial management instruments; human rights; customary rights; South Africa; India

Introduction

During the past 15–20 years there has been a significant shift in thinking about and approaches to fisheries and conservation management. Calls for the adoption of a more participatory, integrated, systems-orientated and, more recently, a human-rights-based approach to resource management are being articulated in various international instruments as well as in country policies and legislation. The concept of an 'ecosystems approach', first articulated in the United Nations Convention of Biodiversity (UN CBD) in 1992, has permeated all resource management sectors including fisheries and conservation. Methods and tools to give effect to this all-embracing concept, which places people and their needs squarely on the agenda, are being pursued in most countries. The implementation of this approach is proving challenging however, especially in terms of embracing the various sectors in planning and decision-making, enabling fishing communities to participate fully and effectively in the design of appropriate measures and ensuring that the needs of the poor and marginalized are addressed. In the fisheries sector, the focus on industrial-scale operations (both in terms of development and management), as discussed in Chapter 22, and the neglect of traditional and local small-scale resource users remains an issue of concern. Although the notion of human dimensions is now part of the lexicon in the fisheries management and conservation arena, managers and

† Deceased.

Governance of Marine Fisheries and Biodiversity Conservation: Interaction and Coevolution, First Edition.
Edited by Serge M. Garcia, Jake Rice and Anthony Charles.

natural scientists, who have traditionally driven management decisions, are grappling with how to embrace human dimensions and give effect to human rights in practice.

Spatial management tools including Marine Protected Areas (MPAs[1]), zonation, closed areas and 'no-take' reserves are increasingly being used by fisheries and conservation managers to achieve their objectives; in some cases, these tools serve the objectives of both sectors. Although there is still some scepticism regarding the value of protected areas as a fisheries management tool (Hilborn, 2003; Hilborn et al., 2004; Kolding, 2011), there is increasing evidence that MPAs do have a role to play in fisheries management and can yield benefits for both conservation and fisheries (FAO, 2011; Mascia, 2010). However, the costs to small-scale coastal fishing communities associated with these interventions are in some cases very significant, raising questions about the principles and approaches guiding their establishment and ongoing management.

This chapter focuses on the implementation of MPAs and other spatial management measures, and associated impacts on the socio-economic and customary rights of coastal fishing communities. This is examined against the backdrop of an increase in the scope and number of international instruments that call for states to implement measures to give effect to human rights in all planning and decision-making. It explores the underlying reasons for ongoing tensions between fisheries management and conservation initiatives on the one hand, and socio-economic rights of traditional fishing communities on the other, by drawing on two very different case studies (one from India and one from South Africa). We provide a review of historical and current approaches and practices to MPAs and other spatial management measures in both countries, before outlining some of the key impacts and outcomes of these measures on coastal resource users. The final section explores why there continues to be a disjuncture between policy rhetoric and implementation practices and proposes a number of processes and mechanisms to improve practice.

International instruments informing the development of spatial management tools

This section focuses on selected international instruments that highlight a more people-centred and human-rights-based approach to fisheries management and conservation. In addition to protection of the marine environment, the United Nations Convention on the Law of the Sea (UNCLOS, held in 1994) requires states to consider the economic needs of coastal fishing communities as well as the nutritional needs of populations of landlocked and disadvantaged states. The UN Fish Stocks Agreement (UNFSA, held in 1995) similarly recognizes the needs of poorer countries and the need to protect vulnerable groups. The Convention on Biodiversity (launched in 1992), which is primarily concerned with the conservation of biological diversity, contains several articles that require attention to human dimensions.

The UN FAO Code of Conduct (FAO, 1995), especially article 6.18, stresses the need to protect the rights of fishers and fishworkers, particularly those engaged in subsistence, small-scale and artisanal fisheries, to a secure and just livelihood as well as preferential access, where appropriate, to traditional fishing grounds and resources in the waters under their national jurisdiction. Article 10.1.2 stresses the importance of engaging fishing communities in management and decision-making. Article 7.6.6 of the Code of Conduct for Responsible Fisheries (CCRF) asks States to give due recognition to traditional practices, needs and interests of indigenous people and local fishing communities which are highly dependent on fishery resources for their livelihood, when deciding on the use, conservation and management of fisheries resources.

In addition, there are various international human rights instruments that should guide all management actions and decisions that may affect the lives and livelihoods of resource users and communities. These include, among others, the Universal Declaration on Human Rights (UDHR in 1948) and the International Covenant on Economic, Social and Cultural Rights (ESCR Covenant in 1976). Special instruments such as the Convention on the Elimination of all forms of Discrimination against Women (CEDAW in 1979) seek to give special focus to certain rights articulated in the main covenants, in this case the protection of the rights of women.

A key message emanating from the above instruments is that human rights, including social, economic and cultural rights and needs, must be taken into account by all organs of state in decision-making, including decisions relevant to conservation and fisheries management. However, while governments including South Africa and India and key conservation organizations have endorsed these principles and provisions, many struggle to marry conservation and human rights principles in practice. Although the adoption of these provisions into national policies appears to be a relatively simple next step, translation of these principles into management and decision-making, especially at the local level, is proving difficult.

South Africa

Historical perspective on fisheries management and marine conservation

South Africa has a lengthy history of spatially based fisheries and conservation management measures in the coastal zone with the first of such measures dating back as early as 1908. Over the past century, these measures have taken various forms including declaration of marine reserves, protected areas, 'no-take' reserves, national parks, biosphere reserves and species-specific sanctuaries. However, few of these areas were specifically declared for fisheries management purposes; the primary motivation in many cases was conservation, although in some cases local resource conflicts prompted the establishment of protected areas (Sowman et al., 2011; van Sittert, 1992, 2003a). It was only after the 1950s, when several marine and coastal resources were considered to be in a state of crisis, that the state, with support from marine and fisheries scientists, began to look at a range of management tools including MPAs to rebuild dwindling marine resources.

The use of spatially based management measures in South Africa needs to be located within the context of the political economy of access to natural resources in South Africa (Sunde and Isaacs, 2008). As early as during the colonial period, the perceived threat to 'nature' by the actions of indigenous peoples prompted urgent state intervention to protect and scientifically manage the 'threatened species' through the creation of protected areas such as nature reserves (Beinart, 2003; van Sittert, 2003a). During apartheid, the rationale for declaring many of the protected areas was to further support the apartheid governments' segregated conservation and land development policies (Carruthers, 1989; van Sittert, 2003b). The impacts on local communities in terms of loss of access to land, resources and cultural heritage were enormous. Conservation therefore became a highly politicized arena and local resource-dependent communities became hostile towards conservation; in many areas, the state-imposed rules were ignored (Fabricius and de Wet, 2002).

Current policy and institutional framework

Following the democratic elections in 1994, the country embarked on a massive law reform process guided by the Constitution of the Republic of South Africa, Act 108 of 1996, which provided a new human rights framework for all policy and law-making in South Africa. The Constitution sought to redress past injustices and promote substantive equality in all sectors and spheres of society. The environmental right (section 24 of the Constitution) guarantees everyone the right to an environment that is not harmful to their health and wellbeing. It requires government to secure ecologically sustainable development and, at the same time, promote justifiable economic and social development. Balancing these often competing rights and interests is proving one of the greatest challenges in all sectors.

The Marine Living Resources Act 18 of 1998 (MLRA) replaced previous legislation governing marine resources management. In addition to ensuring the long-term sustainable use of marine resources, the Act is concerned with addressing past inequities, promoting equitable access to resources, transforming the fishing industry and securing economic benefits for coastal communities. With regard to the implementation of spatial tools, the MLRA empowers the Minister to declare special management areas for the exclusive use of a designated community or group of fishers as well as the declaration of MPAs (section 43 of the MLRA).

During the past 10–15 years, MPAs have become an important tool for both marine scientists and conservation agencies in their efforts to meet CBD targets. The approach to MPA identification and management has been largely top-down, science-based, resource-orientated and regulatory. This is in contrast to the more people-centred approach being followed in the governance of terrestrial conservation; the latter is very explicit with regard to consideration of equity, participation and benefit sharing as required by the National Environmental Management: Biodiversity Act 10 of 2004 (Sunde and Isaacs, 2008).

Currently, MPAs in South Africa cover 21.5% (785 km) of South Africa's 3000 km coastline (DEA, 2012). Of this, 9.1 per cent (344 km) comprises 'no-take' zones while only 0.4 per cent of the economic exclusive zone (EEZ) is under MPA status (DEA, 2012) (Figure 24.1). The 21 MPAs within the inshore zone consist of a continuum of categories of MPAs ranging from complete 'no-take' MPAs (three nationally) to those with multiple zoning including sanctuary no-take areas, areas reserved for restricted sustainable use by commercial rights holders only and areas reserved for recreational and sustainable use by commercial rights and permit holders. South Africa has developed a National Protected Areas Expansion Strategy (NPAES) (DEAT, 2008) and has set ambitious targets for MPAs. This strategy aims

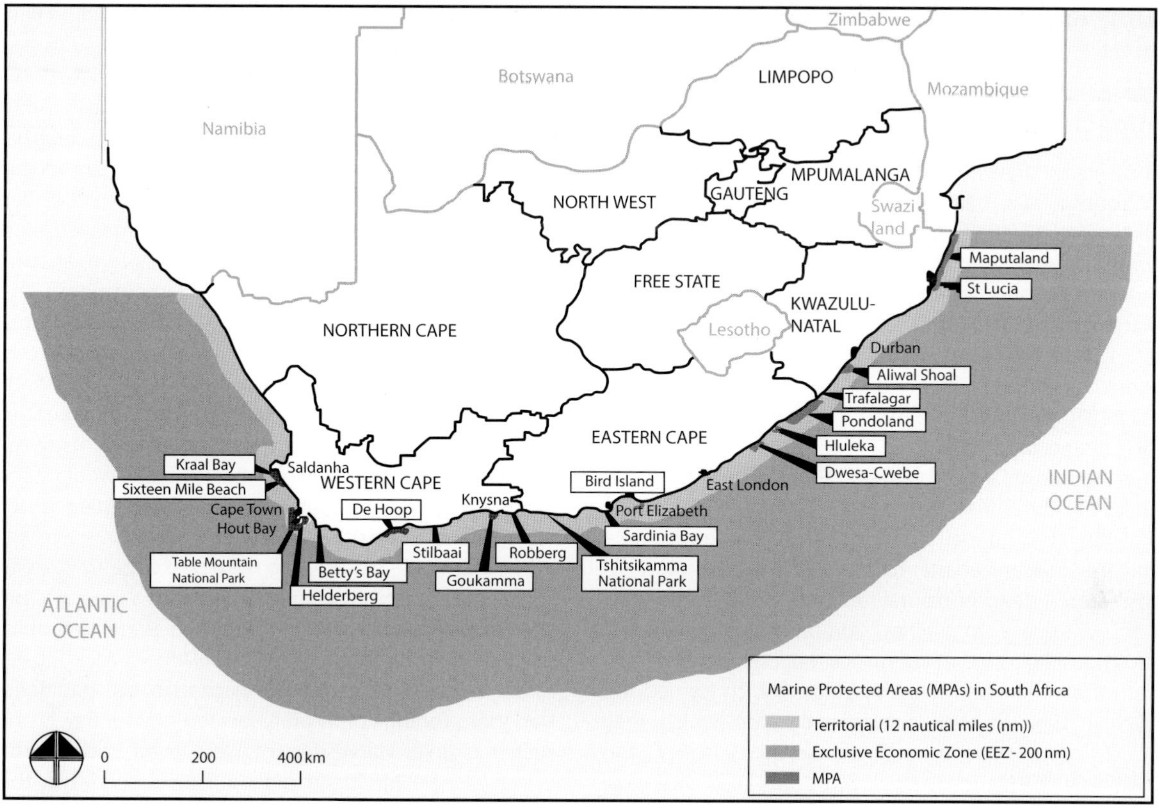

Figure 24.1 Map of MPAs in South Africa.

to have 25% of the coastline declared MPAs by 2028, of which 15% of the inshore coastal area would be declared 'no-take' (DEAT, 2008).

Recently, the Department of Agriculture, Fisheries and Forestry (DAFF) gazetted a new small-scale fisheries policy (DAFF, 2012). This was in response to extensive advocacy actions by small-scale fishers and associated court action,[2] resulting in recognition that the post-1994 fisheries management framework did not cater for the thousands of traditional small-scale fishers operating along the coast (Masifundise, 2010). This policy makes provision for the establishment of small-scale fishing community management areas and for small-scale fishing communities to be given preferential access to resources in these areas. This new policy is ground-breaking in its commitment to recognizing the rights of small-scale fishing communities, including the customary rights of these communities in so far as they are compatible with the Bill of Rights (DAFF, 2012). It is not yet clear how this new small-scale policy will articulate with existing MPAs where such MPAs prohibit access to, and use of, marine resources by local, traditional small-scale fishing communities that live adjacent to these MPAs. Most of the traditional communities living adjacent to the current MPAs have historically fished in these areas and many argue customary rights to access and use of resources in these areas (Sunde and Isaacs, 2008). This issue is currently the subject of both legal and advocacy actions in South Africa (Sunde et al., 2013). While these customary communities acknowledge, in line with international precedent (the Australian Government's Native Title Act, 1993), that their customary rights are subject to regulation, they are arguing that their customary fishing rights are not extinguished by the MLRA. In accordance with the protection afforded customary law in terms of section 211 of the Constitution, these rights must be recognized. While MPAs have enabled South Africa to respond to its international commitments to expand its network of MPAs, the approach to establishing, planning and managing MPAs in South Africa has had significant negative impacts on coastal fishing communities living adjacent to these protected areas.

Impacts and outcomes

This section draws on research conducted in two MPAs in South Africa, namely the Table Mountain National Park (TMNP) MPA and the Dwesa-Cwebe MPA (see Figure 24.1). As these cases illustrate, the impacts and outcomes of both area-based conservation and fisheries management measures on small-scale fishing communities in South Africa are closely tied to the political economy of the country and, in particular, the historical marginalization of these communities under both colonialism and apartheid.

In the case of Dwesa-Cwebe, an MPA on the eastern seaboard of South Africa (Figure 24.1), the local communities that historically depended on the coastline of the MPA and the adjacent coastal forest have a well-established customary system of natural resource use. Over the past century these communities, comprising approximately 2300 households, were systematically deprived of access to both the coastal and forest resources which formed the material basis of their culture and upon which they depended for their livelihoods (Sunde, 2011). In 1991, the Dwesa-Cwebe Marine Reserve was established. Inter-tidal harvesting was prohibited but shore-based angling was permitted under strict regulations. In 1996, after the democratic elections and the introduction of land restitution legislation, the communities' land claim was recognized and negotiations on how they would access the natural resources began. However, despite the fact that negotiations leading up to the signing of the Settlement Agreement confirmed recognition of the communities' right to participate in the co-management of natural resources in the reserve, in 2000 the Dwesa-Cwebe MPA was promulgated as a complete 'no-take' MPA just six months prior to the signing of the land Settlement Agreement. There was no consultation with the affected communities. The community understood that their right to access marine resources would continue to be negotiated, however. To date the fisheries and the conservation authorities have failed to recognize the communities' claim to their customary fishing rights along this coastline or to establish appropriate co-management arrangements.

These local communities, located in the poorest district in the country (Shackleton et al., 2007), have continued to harvest both marine and forest resources, despite running the risk of prosecution (Sunde, 2011). Both fishermen and women inter-tidal harvesters have experienced high fines (up to US$ 200 or 3 months in prison) when they cannot afford to pay their fines. During the past two years, two persons have been killed by conservation rangers while harvesting resources in the reserve. The community has recently submitted a complaint to the South African Human Rights Commission about a number of alleged human rights abuses and have also commenced legal action in support of their customary rights to their traditional fishing grounds, citing both international and regional fisheries, environmental and human rights instruments in their defence (Sunde et al., 2013). In a recent court judgement, the Magistrate confirmed that the community had a customary fishing system, that they had a right to have their culture and customs protected in terms of the Constitution and it was unlikely that the MLRA would stand up to Constitutional scrutiny in this regard (State versus Gongqose and two others, 2012).

The TMNP-MPA adjacent to the city of Cape Town (refer Figure 24.1) comprises a combination of area-, species- and gear-based measures to promote and protect the biodiversity of the region. The Karbonkelberg rock lobster sanctuary within the TMNP, adjacent to the traditional Hangberg fishing community located in Hout Bay, is one such example. Harvesting of west coast rock lobster (*Jasus lalandii*) has been practised for centuries in this area, with strong customary use rights evolving from the 19th century (van Sittert, 1993, 1994). With the growth of the export-led rock lobster industry from the early to mid-1900s, restrictions on access to the lobster resource increased, severely limiting customary fishing practices. Nevertheless traditional fishing continued, often 'illegally', as a means to supply food and basic income (Hauck, 2009; van Sittert, 1994).

Although the TMNP-MPA was only declared in 2004, the Karbonkelberg 'lobster sanctuary', which was declared as a no-take zone in 1934 (van Sittert, 1994), was simply incorporated into the MPA. The declaration of this MPA in 2004 completely ignored the historical rights of the local fishing community. In view of the basic gear used by these fishers, it is difficult for them to access fishing grounds outside the sanctuary. The fishers' sense of injustice is further exacerbated by the fact that industry has an experimental permit to harvest lobster for scientific purposes in the Karbonkelberg Sanctuary during March every year (Hauck, 2009). The consequence has been increased illegal harvesting and fishing in dangerous conditions (Hauck, 2009). The response of the fishery authority has been to increase law enforcement efforts and administer fines and confiscate boats, gear and catches. Recently, after extensive lobbying, the authorities have agreed to withdraw the experimental quota.

The above case studies reveal that despite South Africa's adoption of a host of international instruments relevant to 'people and protected areas', and the promulgation of legislation to give effect to these instruments, declaration and management of MPAs in South Africa are characterized by state-driven, science-based and top-down decision-making processes. Local community needs and interests have been given scant attention. Inability to access food (especially in times of crises), loss of livelihoods, alienation of fishers from their environment, loss of cultural identity, erosion of customary governance systems, increasing conflicts among resource-dependent communities and claims of harassment and human rights abuses are all impacts associated with MPAs. Failure to recognize and take account of these human dimensions have undermined the legitimacy of MPAs and curtailed achievement of management objectives.

India

Historical perspective

India has a long history of 'conservation', characterized by a centralized and commerce-oriented system of natural resource management put in place by its British colonial powers. The colonial state saw in the exploitation of forests the opportunity for wealth; laws were therefore created not to preserve ecological integrity but to protect the state's monopolistic access to forest resources (Kothari and Pathak, 1998). Large areas of community-used lands and forests were appropriated and, in the process, rights of local communities were extinguished. The period after the independence of India in 1947 saw a continuation of such policies. The same mindset characterizes the Wild Life (Protection) Act of 1972 (WLPA), as amended in 2002 and 2006, that provides the legislative framework for the establishment of national parks and sanctuaries.

The trajectory within fisheries is markedly different. The focus after Independence was primarily on development, modernization and expanding production to enhance income and employment opportunities, to meet protein and food security needs and to increase exports, rather than on management and conservation. The need for improving conservation and management of fisheries resources to ensure long-term sustainability is, to some extent, now being recognized.

With a long coastline of 8118 km and an EEZ of 2.02 million km^2, fisheries have traditionally been a very important source of livelihoods in India. Fishing communities living along the coast have long histories of fishing and associated governance and traditional knowledge systems, and fisheries are as much a cultural as a social and economic activity. Communities commonly hold a spatial perception of their 'rights' where fishing by outsiders, or the use of certain gear, is regulated (Koshy, 2007; Vivekanandan, 2009). Customary institutions, although important, have never been officially recognized and/or incorporated into resource management, either by fisheries or conservation agencies.

Current policy and institutional framework

Conservation of coastal and marine biodiversity

As mentioned, spatial management approaches for conservation of marine biodiversity are being implemented through the designation of sanctuaries and national parks in coastal and marine areas under the WLPA. Reports by the Government of India (GoI) to the CBD refer to these designated areas as MPAs although, in an Indian context, marine and coastal protected areas (MCPAs) or MPAs are not a specific legal category. According to the WLPA, national parks are afforded the highest degree of protection and all human activities are banned unless they contribute to conservation. Regulations in the case of sanctuaries are less stringent, and fishing activities may be regulated or prohibited. The 2002 and 2006 amendments to the WLPA created new protected area categories: community reserves, conservation reserves and tiger reserves, all of which have participatory provisions and acknowledge the principle of 'sustainable use'. Although these new categories have been introduced, they have yet to be applied in the marine space.

The WLPA, essentially a terrestrial framework, has some provisions that are specific to the coastal and marine context and have relevance for protecting the interests of local fishing communities. For example, the Act requires that if any part of the territorial waters is to be included within a sanctuary or national park, adequate measures must be taken to protect the occupational interests of local fishers. The right of innocent passage of any vessel or boat travelling through territorial waters should also not be affected by the designation of a sanctuary. The Ministry of Environment and Forests (MoEF) is the main agency at the Central level responsible for the conservation of biodiversity and wildlife and for the implementation of environmental legislation, such as the WLPA. At the State level, the Department of Forests is the main agency under the MoEF, responsible for managing protected areas.

Marine and coastal resource protection measures were first initiated in 1967 in India, pre-dating the existence of any specific legal frameworks for protected areas. According to the GoI's third national report to the SCBD (2006), there are 31 MCPAs designated in the country, of which 18 are in the marine environment (Figure 24.2). The remaining 13 have both terrestrial and marine components (SCBD, 2006).

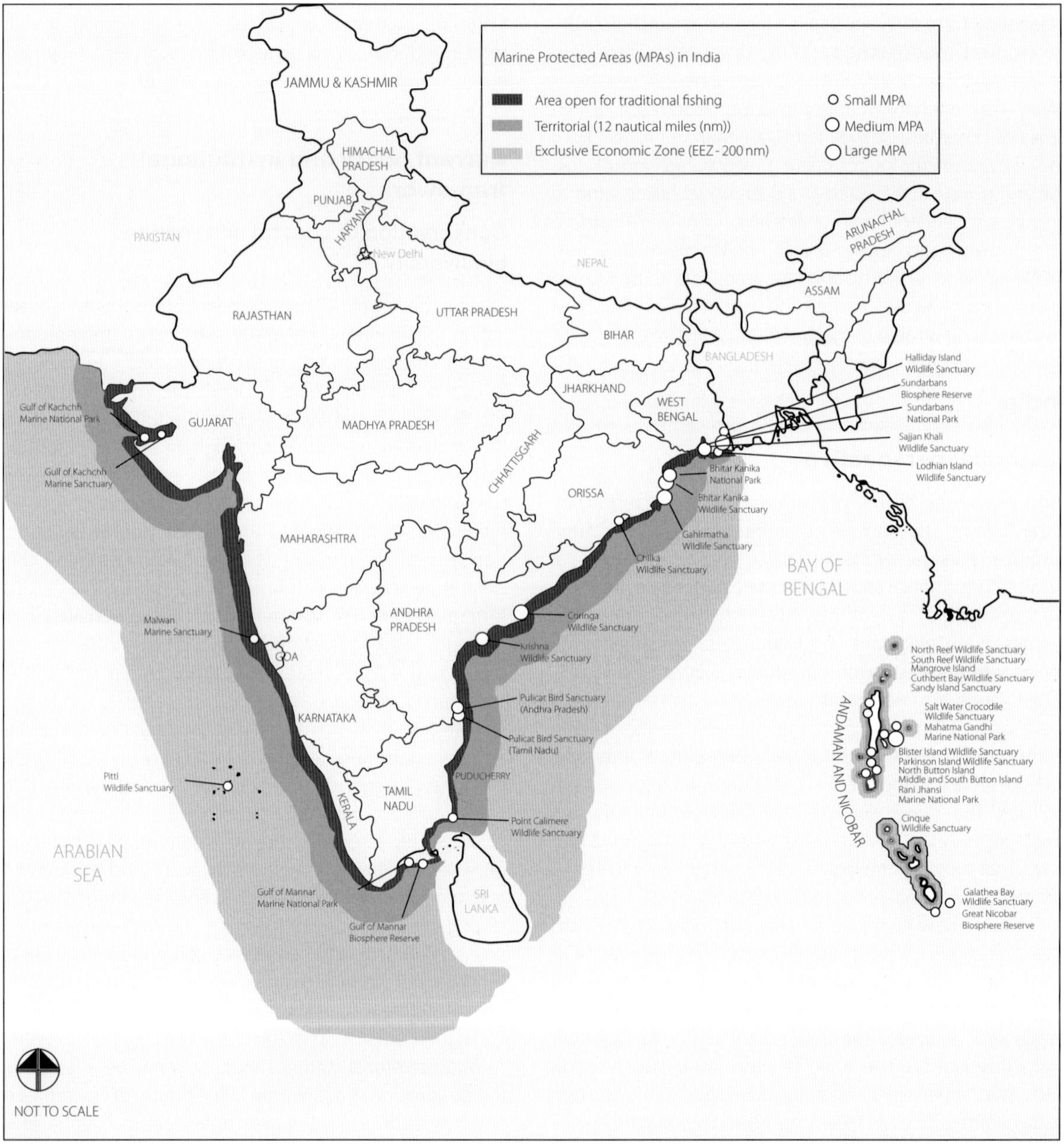

Figure 24.2 Map of MPAs in India.

All 31 MCPAs are located within territorial waters and cover an area of 6271.21 km^2. The lack of a precise definition for MPAs/MCPAs, combined with no specific legal designation, makes it difficult to determine the actual number of designated MCPAs. India plans to increase the area under MCPAs from 18.5% of the inland area to 36.14% and, in the case of coastal biogeographic zones, from 6.16% to 7.12%. The Eleventh Five Year Plan of the GoI (2007–2012) specifically calls for more protected areas for the conservation of coral reefs (GoI, 2008). It is notable that most of the MPAs/MCPAs were established for habitat or species protection, not as a mechanism for fisheries management.

Fisheries management

In India both the Union (federal) government and State (provincial) governments are involved in fisheries management. While the State governments have the power to legislate on fisheries issues in territorial waters, the Union government has the authority to legislate on fisheries issues in the EEZ. Fisheries within the 12 mile territorial limits are managed under the state-level Marine Fishing Regulation Acts (MFRA). The Act is based on a model piece of legislation prepared by the Indian Ministry of Agriculture in 1979 in response to demands from fishers operating unpowered fishing vessels, seeking to protect their fishing space and equipment from bottom trawlers. It was drawn up at a time of tremendous conflict between the two sub-sectors over access to fishing space and resources, sometimes even leading to destruction of life and property (Kurien and Mathew, 1982). Fisheries management is undertaken mainly through licensing, prohibitions on certain fishing gear, regulations on mesh size and the establishment of closed seasons and areas.

Spatial management tools in fisheries include permanent or temporary bans on fishing in certain areas, through measures such as declaring no-fishing areas and demarcating fishing zones for mechanized and non-mechanized vessels. Under the MFRAs, zones are demarcated by each State based on distance from the shoreline (5–10 km) or on depth. These inshore zones, where trawling and other forms of mechanized fishing are not permitted, are perhaps the most important spatial management measures in place. The closed season, or 'monsoon fishing ban', is another important 'temporal-spatial' management measure implemented on both the east and west coasts of India for a period of 47 and 65 days, respectively (coinciding with what is considered to be the spawning and breeding season). Notably, both the above management measures were put in place as a result of the struggles, campaigns and demands of small-scale fishing communities. There are also several state-specific management initiatives that adopt spatial management measures to regulate fishing.

Impacts and outcomes for coastal communities

Impacts and outcomes of spatial management measures used in conservation

Various studies have been undertaken by the International Collective in Support of Fishworkers (ICSF) to understand the socio-economic impacts of protected area management in the five main MCPAs in mainland India. MCPAs, as implemented under the WLPA, have typically been implemented in a top-down fashion with a focus on keeping people out. In most cases, regulations and prohibitions imposed on fishing have affected livelihoods of local communities in a context where coastal fishing villages are generally located in remote areas and characterized by high levels of poverty and poor access to basic services.

In the Gahirmatha (Marine) Wildlife Sanctuary in Orissa, for example, regulations on fishing have not only reduced access to fishing grounds for mechanized and motorized vessels by almost 50%, they have also cut the number of fishing days from 240 to fewer than 100 (Mathew, 2004; Rajagopalan, 2009). The livelihoods of women, who are involved in crab collection and fishing in creeks, have also been affected (ICSF, 2009). About 43% (around 11,000) of the affected fishers are estimated to live below the poverty line.[3] This may represent an underestimate as accurate estimates of the total number of people affected, such as workers on vessels, are lacking. The absence of basic facilities in remote areas hinders the development of alternative livelihood options (Aide et Action, 2008). Further challenges are posed by the proximity of certain landing centres to the boundaries of a sanctuary. Fishers passing through the sanctuary to access fishing grounds are often arrested and fined, and their boats and catch are confiscated. Such harsh restrictions have reduced incomes, leading to livelihood crises and high levels of indebtedness. Fishers are no longer able to repay loans, often issued at high rates of interest. This situation has led to some fishers committing suicide (Rajagopalan, 2009).

In the Gulf of Mannar National Park, restrictions on access have affected the livelihoods of about 5000 women seaweed harvesters. Fishers in this area express concern over arbitrary confiscation of vessels, gear and catch; levying of fines; arrest of fishers and lengthy legal processes with associated financial burdens; as well as violence and harassment (Rajagopalan, 2009). It is often difficult for such fishers to establish their innocence with respect to their actual fishing location, given the lack of boundary lines delimiting the National Park. The situation is similar in other protected areas.

Few systematic initiatives have been implemented by State governments to compensate active fishers and others who depend on the fisheries for loss of livelihoods. In all protected area sites, communities point out that small-scale traditional fishers operating small non-motorized boats are often the worst affected by protected area (no-take) regulations, even though their fishing activities tend to be of low impact. Larger vessels, by contrast, are typically able to shift their operations to more distant waters. Fishing communities tend to perceive MCPAs as tools to 'keep people out'. They are sceptical of conservation measures that ban all types of fishing activities, while not regulating in any way industrial and development activities within or adjacent to MCPAs with large impacts on the ecosystem. A lack of consultation with communities, combined with a lack of recognition of communities as 'rights holders' in decision-making processes, threatens the legitimacy and outcomes of the protected area management. Although recent amendments to the WLPA require community participation in the designation and management of protected areas, and the settling of rights of affected individuals within a two-year period, they are yet to be implemented in any meaningful way. Other provisions under the WLPA that protect the occupational and other interests of fishing communities also remain poorly implemented, and are differently interpreted by implementing officials (ICSF, 2012).

It is evident that the implementation of MCPAs in India has had severe socio-economic implications, fostering a sense of alienation among fishing communities living adjacent to, and customarily fishing in, such areas. The extent to which such protected areas are achieving their environmental objectives, in a context where community support for enforcement of regulations is negligible and where efforts to regulate the impact of industrial and other developments are entirely lacking, also remains a moot issue.

Impacts and outcomes of spatial management measures used in fisheries

The main spatial management tools used in fisheries management are the artisanal fishing zones and the seasonal ban on fishing (monsoon ban). The enforcement of artisanal zones remains weak, however. Small-scale fishers consistently demand better enforcement of these zones and maintain that if prohibitions on trawling and destructive gear in inshore zones were in fact enforced, ecological and social benefits would be significant and there would be little need for other conservation measures. The monsoon fishing ban is perhaps the most effectively implemented spatio-temporal management measure. Such seasonal fishing bans that prohibit fishing by mechanized vessels during specified periods have been strongly supported by small-scale fishing communities. In recent years there have even been calls to extend the period of such fishing bans. Scientific studies also indicate positive impacts on fisheries resources due to this measure on both the east and west coasts (GoI, 2007; Vijayan et al., 2000).

Discussion

Drawing on insights and lessons from South Africa and India, this final section focuses on why there continues to be a disjuncture between policy rhetoric that appears to be converging in terms of conservation, fisheries management and human rights and wellbeing goals on the one hand, and implementation practices on the other. Evidence from both countries highlights how the declaration and management of MPAs (in various forms) has resulted in major impacts on fishers' access to food and livelihoods as well as loss of customary rights and cultural practices. In both countries, failure to ensure a 'bottom-up' approach to the planning of spatially based measures, whether primarily for fisheries management or conservation objectives, has undermined not only the policy objectives of such measures but, most importantly, the human rights and livelihoods of local communities, thereby compromising the sustainability of these interventions in the longer term. In the case of India where fishing communities have motivated for spatial management measures, lack of enforcement undermines the potential benefits of these measures.

Different values and goals

The promotion of a conservation agenda in South Africa and by the Forestry Department in India without giving due attention to human dimensions is underpinned by deep-rooted values and beliefs that conservation of nature, protection of biodiversity, restoration of degraded habitats and rebuilding of dwindling stocks are overriding goals. Long-term resource sustainability goals and consideration of the national and public good override local rights, needs and interests. In India, a production-orientated approach to fisheries management rests side by side with this conservation agenda; economic efficiency goals therefore trump the rights of local communities in this case. In contrast, for many traditional and local resource users, environment, livelihoods and culture are indivisible and damage to the environment or loss of access threatens not only their livelihoods but their cultural identity. These differences in construction of place (Nursey-Bray, 2011) as well as in values and goals, mitigate against formulating a shared vision and goals for the area. Given the historic power of the conservation and marine science community in fisheries management and conservation in South Africa, and the power of the market in India, it is the values of these prevailing interests that have dominated planning and decision-making.

On the other hand, local resource users, development NGOs and social scientists, although concerned about long-term sustainability of resources, have tended to focus on provisions in the international instruments that require protection of human rights in all aspects of planning and decision-making. Creating space for local community-based planning processes, recognizing and respecting local rights and needs and accepting that there are different values and goals across different stakeholders engaged in marine and coastal governance are necessary initial steps to any negotiations on implementation of spatial management measures including the declaration of protected areas. This has been highlighted in several studies and reports (Borrini-Feyerabend, 2008; Charles and Wilson, 2009; Christie and White, 2007; Cinner and Aswani, 2007; FAO, 2011; ICSF, 2009, 2012; Mascia, 2010; Nursey-Bray, 2011; Pomeroy et al., 2007; Rajagopalan, 2009; Sowman et al., 2011) and reinforced in the Global Protected Areas Management Evaluation (PAME) Study (IUCN-TILCEPA, 2010). The process of finding a solution acceptable to all parties is likely to require considerable time and effort and will inevitably involve compromises, tradeoffs and mutual adjustment. However, a decision negotiated in this way is likely to be more respected and sustainable than one imposed by government.

Institutional limitations: MPAs are created in isolation of broader marine governance processes

MPAs are one of several spatial mechanisms employed by government and other coastal stakeholders to manage the myriad of activities taking place in coastal and marine environments. As such, MPAs are embedded in a larger social, cultural, economic, institutional and political context and cannot be planned and implemented in isolation of the broader planning and governance processes relevant to the particular area (Charles, 2011; Charles and Wilson, 2009; Jentoft et al., 2003; Pomeroy et al., 2007).

In both India and South Africa, the research indicates that implementation of spatial management tools (especially no-take MPAs) leads to loss of access to food and livelihoods and undermines customary rights. It is therefore imperative that other government departments at different levels of government be integrally involved in discussions regarding identification and planning of MPAs. In practice however, due to the sectoral nature of government, engaging with this wide array of actors when dealing with a conservation or fisheries matter is generally considered beyond the ambit of their mandate. These various agencies should be part of the process from the outset so that the feasibility of proposals and consideration of alternative or supplemental livelihoods can be ascertained during negotiation, and further investigations to inform ongoing negotiations can be commissioned if required. The involvement of other relevant government departments and agencies in conservation planning and management is important as fishers are often overlooked due to their marginalized status in the community or a perception that they already have a livelihood, even though access to fishing grounds may have been restricted.

Inequitable distribution of costs and benefits

In most cases, it is local resource-dependent communities that experience the greatest hardships from MPAs while benefits accrue to other fishery sectors (recreational and commercial) as well as non-fishery-dependent sectors (tourism, conservation). These benefits may also occur over various geographic scales and different time frames. As Jentoft et al. (2007, p. 617) notes, one cannot expect the poor to 'postpone their needs in anticipation of an MPA augmenting the resource base'. Perceptions among local resource users regarding fairness and justice with respect to the impacts and benefits associated with the declaration of MPAs can also influence willingness to adhere to rules and regulations relevant to the MPA. Frequently, government's response to an escalation of 'illegal' activity is to enhance law enforcement efforts (Hauck, 2011; ICSF, 2009). However, this approach has not proved to be effective or sustainable.

There is currently no comprehensive methodology for calculating compensation for either the impact of evictions or the ongoing loss of food and livelihoods, or impact on culture and spiritual life. There is also a tendency to view the benefits of MPAs in terms of their benefits for the commercial and recreational fisheries sector, with little attention paid to the fact that local small-scale fishing communities seldom benefit from the predicted 'spill-over effects' of MPAs due to limitations on mobility and the existing inequities in the commercial sector.

Weak governance processes and practices

Despite commitments to a host of international instruments calling for inclusive and meaningful participation of civil society, and traditional and local resource users in planning and decision-making, evidence in both South Africa and India suggests that small-scale fishers and coastal communities are seldom adequately involved in such decisions. Local resource users feel aggrieved at the lack of consultation with respect to the establishment or expansion of MPAs. The lack of appropriate and meaningful participation is frequently the root cause of the grievances, misunderstandings and conflicts occurring in coastal contexts where local resource users compete for resources with other fishery and non-fishery sectors. Data from the Global PAME study (IUCN-TILCEPA, 2010) confirms that effective public participation is one of the most significant indicators of successful PAs.

Determining appropriate governance arrangements for a particular local context requires participation of local resource users, relevant government agencies and other relevant stakeholders. However, the process of developing a shared vision, setting objectives, exploring alternative management tools, agreeing on rules and regulations and clarifying roles and responsibilities is time consuming and requires willingness from all parties to learn and adapt. There is therefore a need for careful consideration of what governance type would be most appropriate in a particular context and how the principles of good governance (Borrini-Feyerabend et al., 2011) can be embraced and implemented by the governing institutions.

Processes and mechanisms to address disjuncture between policy rhetoric and local realities

Moving beyond rhetoric and achieving a just balance between human rights and ecological and fisheries management objectives requires the mainstreaming of a human-rights-based approach into policies, legislation, plans, assessments and management activities at all levels. However, mainstreaming will only be achieved when there is awareness and understanding among all stakeholders – traditional and local resources users as well as government functionaries and other stakeholders – of the nature of these rights and the duties that such rights impose on the individual and broader society. It requires that users are fully involved from the outset of the process, that their values and interests are identified and addressed and that their knowledge systems as well as best-available scientific information inform decision-making. Of crucial importance is developing capacity of all stakeholders to participate as equal partners in deliberations regarding appropriate spatial management tools and governance models that are suited to the particular context. In order to move forward however, the starting point has to shift from ecological and fisheries science imperatives dictating the location, size and category of protected areas, to joint assessments which includes the clarification of rights, claims and duties (Greiber et al., 2009) and deliberations on what objectives and outcomes are desired and what spatial or other management tools would be most appropriate in a given context. This requires a significant shift in attitude and recognition that local communities can be powerful allies in marine and fisheries conservation and management.

Conclusions

The findings of this chapter suggest that despite significant differences between the past and present governance frameworks relevant to conservation and fisheries management in South Africa and India, the imposition of spatial management measures to address conservation and/or fisheries management objectives has led to significant negative socio-economic and cultural impacts in both countries. These findings are cause for concern given the commitment by both countries to a plethora of international instruments that (1) increasingly recognize the need to respect the rights of traditional and local small-scale fisher communities and (2) require adoption of good governance principles in fisheries and conservation management. In particular, the current target-driven approach to expand areas under marine protection without active involvement of resource users in decision-making is exacerbating the vulnerability of these resource-dependent communities, eroding socio-cultural systems, alienating them from mainstream conservation efforts and promoting illegal fishing activities.

Clearly, the principles articulated in these global agreements are being interpreted and implemented in different ways by the range of actors involved in these processes and decisions and are having little impact at the local level. This research suggests that different values and goals, institutional blockages, sectoral agendas, the challenge of reconciling global imperatives with local realities, inequitable distribution of costs and benefits, weak governance and the challenge of balancing human rights and environmental principles are some of the key factors undermining alignment between policy and practice in both South Africa and India.

Adoption and mainstreaming of a human-rights-based approach from policy level through to management action is offered as a practical way forward. This requires raising awareness of rights articulated in the various international instruments and building capacity among all stakeholders to articulate their rights. Furthermore, it requires undertaking joint assessments in a manner that integrates local and scientific knowledge and recognizes legitimate local governance systems as the starting point for determining appropriate governance arrangements and management regimes that address the unique constellation of ecological, social and cultural needs of each context.

Acknowledgements

The South African authors would like to extend thanks to the National Research Foundation and WWF SA for their financial support for this research.

Notes

1. Throughout the paper we use the term MPAs in a generic sense to refer to a range of spatial management tools that seek to achieve protection of ecosystems and/or species and impose some level of restriction on human activities in these areas. However, we recognize that different countries use different terms. For example in Brazil, MPAs that permit sustainable use are referred to as Marine Extractive Reserves whereas in Costa Rica they are referred to as Marine Areas for Responsible Fishing.
2. Kenneth George versus the Minister 2007 (EC 1/05).
3. In India, BPL is estimated using 13 scoreable socio-economic parameters: operational holdings of land, housing, clothing, sanitation, ownership of consumer durables, literacy, status of households in labour force, means of livelihood, status of children of age 5–14 years, type of indebtedness and migration.

References

Aide et Action (2008) Seeking substitutes: Study report on "Alternative livelihood vocation for the marine fishermen affected due to turtle conservation measure, in Orissa Coast". Aide et Action.

Australian Government (1993) Native Title Act No. 10 of 1993. Office of Legislative Drafting and Publishing. Attorney-General's Department, Canberra, Australia.

Beinart, W. (2003) *The Rise of Conservation in South Africa: Settlers, Livestock and the Environment*. Oxford University Press, Oxford.

Borrini-Feyerabend, G. (2008) Governance as key for effective and equitable protected area systems. IUCN Commission on Environmental, Economic and Social Policy (CEESP), Briefing Note 8.

Borrini-Feyerabend, G., Chatelain, C. and Hosch, G. (2011) *Sharing Governance! A Practical Guide for Marine Protected Areas in West Africa*. PRCM, IUCN and CEESP, Dakar.

Carruthers, J. (1989) Creating a national park, 1910 to 1926. *Journal of Southern African Studies* **15**, 188-216.

Charles, A. (2011) Human rights and fishery rights in small-scale fisheries management. In: *Small Scale Fisheries Management: Frameworks and Approaches for the Developing World* (eds R. Pomeroy and N.L. Andrew), pp. 59–74. Centre for Agricultural Bioscience International, London.

Charles, A. and Wilson, L. (2009) Human dimensions of Marine Protected Areas. *ICES Journal of Marine Science* **66**, 6–15.

Christie, P. and White, A.T. (2007) Best practices in governance and enforcement of marine protected areas: An overview. In: *FAO Expert Workshop on Marine Protected Areas and Fisheries Management: Review of Issues and Considerations*. FAO, Rome, 12–14 June 2006, pp. 183–220.

Cinner, J. and Aswani, S. (2007) Integrating customary management into marine conservation. *Biological Conservation* **140**, 201–216.

DAFF (2012) *Policy for the Small-scale Fisheries Sector in South Africa*. DAFF, Pretoria.

DEA (2012) Chapter 5: Oceans and coasts. In: *South African Environmental Outlook (SAEO) 2012: A Report on the State of Environment*. Draft 1. DEA, Pretoria.

Department of Environmental Affairs and Tourism (DEAT) (2008) National Protected Area Expansion Strategy for South Africa (NPAES). Priorities for expanding the protected area network for ecological sustainability and climate change adaptation. DEAT, Pretoria, Republic of South Africa.

Fabricius, C. and de Wet, C. (2002) The influence of forced removals and land restitution on conservation in South Africa. In: *Conservation and Mobile Indigenous Peoples. Displacement, Forced Settlement and Sustainable Development* (eds D. Chatty and M. Colchester), pp. 142–157. Berghan Books, New York.

FAO (1995) *Code of Conduct for Responsible Fisheries*. FAO, Rome.

FAO (2011) Fisheries management 4: Marine protected areas and fisheries. FAO Technical Guidelines for Responsible Fisheries No. 4 (Suppl. 4), FAO, Rome.

GoI (2007) Report of the Committee to Study the Impact of the Closed Season on the Fisheries. GoI, New Delhi.

GoI (2008) Eleventh Five Year Plan (2007–2012). GoI Planning Commission. Oxford University Press, New Delhi.

Greiber, T., Janki, M., Orellana, M., Savaresi, A. and Shelton, D. (2009) Conservation with Justice: A Rights-based Approach. IUCN Environmental Law and Policy Paper 71. IUCN, Gland, Switzerland.

Hauck, M.B. (2009) Rethinking small-scale fisheries compliance. *Marine Policy* **32**, 635–642.

Hauck, M.B. (2011) Small-scale fisheries compliance: Integrating social justice, legitimacy and deterrence. In: *Small-Scale Fisheries Management: Frameworks and Approaches for the Developing World* (eds R.S. Pomeroy and N.L. Andrew), pp. 196–215. Cabi, Wallingford.

Hilborn, R. (2003) State of the world's fisheries. *Annual Review of Environment and Resources* **28**, 359–399.

Hilborn, R., Stokes, K., Maguire, J.J., Smith, T., Botsford, L.W., Mangel, M., Orensanz, J., Parma, A., Rice, J., Bell, J., Cochrane, K.L., Garcia, S., Hall, S.J., Kirkwood, J.P., Sainsbury, K., Stefansson, G. and Walters, C. (2004) When can marine reserves improve fisheries management? *Ocean & Coastal Management* **47**, 197–205.

ICSF (2009) Proceedings of the Indian MPA Workshop, Social Dimension of Marine Protected Areas Implementation in India: Do Fishing Communities Benefit? IMAGE Auditorium, Chennai, India, 21–22 January.

ICSF (2012) Workshop on fishery-dependant livelihoods, conservation and sustainable use of biodiversity: the case of marine and coastal protected areas in India. India International Centre, New Delhi India, 1–2 March.

IUCN-TILCEPA (2010) Joint PAEL-TILCEPA Workshop on Protected Areas Management Evaluation and Social Assessment of Protected Areas. IUCN, Gland, Switzerland.

Jentoft, S., Minde, H. and Nilsen, R. (eds) (2003) *Indigenous Peoples: Resource Management and Global Rights*. Eburon, Delft.

Jentoft, S., van Son, T.C. and Bjorkon, M. (2007) Marine protected areas: A governance system analysis. *Human Ecology* **35**, 611–622.

Kolding, J. (2011) MPAs in relation to fisheries – what are the biological and fish stocks implications? Department of Biology, University of Bergen, Bergen, Norway.

Koshy, N. (2007) Orissa Case Study. Presented at ICSF Workshop, Asserting Rights, Defining Responsibilities: Perspectives from Small-scale Fishing Communities on Coastal and Fisheries Management in Asia. Siem Reap, Cambodia, 3–5 May. Available at http://rights.icsf.net/images/stories/Asserting-Rights/case_orissa_summary.doc (accessed March 2014).

Kothari, A. and Pathak, N. (1998) Sharing benefits of wildlife conservation with local communities: Legal implications. *Economic and Political Weekly* **XXXIII**(**40**), 4 July.

Kurien, J. and Mathew, S. (1982) *Technological Change in Fishing: Its Impact on Fishermen*. ICSSR Monograph. Centre for Development Studies, Thiruvananthapuram, India.

Mascia, M.B. (2010) Impacts of marine protected areas on fishing communities. *Conservation Biology* **24**, 1424–1429.

Masifundise (MDT) (2010) *A Handbook Towards Sustainable Small-Scale Fisheries in South Africa*. MDT, Cape Town.

Mathew, S. (2004) Socio-economic aspects of management measures aimed at controlling sea turtle mortality: a case study of Orissa, India. Presented at the Expert Consultation on Interactions between Sea Turtles and Fisheries within an Ecosystem Context. Rome, 9–12 March. In: FAO Fisheries Report 738, FAO, Rome. Available at http://www.fao.org/docrep/007/y5750e/y5750e0d.htm (accessed March 2014).

Nursey-Bray, M. (2011) Social contexts and customary fisheries: marine protected areas and indigenous use, Australia. *Environmental Management* **47**, 671–683.

Pomeroy, R., Mascia, M. and Pollnac, R. (2007) Marine protected areas: the social dimension. In: *FAO Expert Workshop on Marine Protected Areas and Fisheries Management: Review of Issues and Considerations*, FAO, Rome, 12–14 June 2006, pp. 49–275.

Rajagopalan, R. (2009) Social dimensions of sea turtle protection in Orissa: Case study of Gahirmatha (marine) wildlife sanctuary and the nesting beaches of Rushikulya and Debi. International Collective in Support of Fishworkers (ICSF), Chennai, India.

SCBD (2006) India's third national report to the CBD. Available at http://www.cbd.int/doc/world/in/in-nr-03-en.doc (accessed March 2014).

Shackleton, C.M., Timmermans, H.G., Nongwe, N., Hamer, N. and Palmer, R. (2007) Direct-use values of non-timber forest products from two areas on the Transkei Wild Coast. *Agrekon* **46**, 135–156.

Sowman, M., Hauck, M., van Sittert, L. and Sunde, J. (2011) Marine protected area management in South Africa: new policies, old paradigms. *Environmental Management* **47**(4), 573–583.

State vs Malibongwe David Gongqose and two others (2012) Case No: E382/10. Available at http://joasa.org.za/aricles/S v D Gongqose and others.pdf (accessed March 2014).

Sunde, J. (2011) Dwesa-Cwebe Scoping Report. Environmental Evaluation Unit, University of Cape Town, Cape Town.

Sunde, J. and Isaacs, M. (2008) Marine Conservation and Coastal Communities: Who Carries the Costs? A Study of Marine Protected Areas and Their Impact on Traditional Small-Scale Fishing Communities in South Africa. International Collective in Support of Fishworkers (ICSF), Chennai, India.

Sunde, J., Sowman, M., Smith, H. and Wicomb, W. (2013) Emerging proposals for tenure governance in small-scale fisheries in South Africa. *Land Tenure Journal* **1**(13), 117–146.

van Sittert, L. (1992) Labour, capital and the state in the St Helena Bay fisheries c. 1856–c. 1956. PhD thesis, University of Cape Town, Cape Town, South Africa.

van Sittert, L. (1993) 'More in the breach than the observance': crayfish conservation and capitalism. *Environmental History Review* **17**(4), 20–46.

van Sittert, L. (1994) Red gold and black markets: the political economy of the illegal crayfish trade c. 1890–c. 1990. Department of History, University of Cape Town, Cape Town, South Africa.

van Sittert, L. (2003a) The tyranny of the past. Why local histories matter in the South African fisheries. *Oceans & Coastal Management* **46**, 199–219.

van Sittert, L. (2003b) Making the Cape floral kingdom: the discovery and defence of indigenous flora at the Cape c. 1890–1939. *Landscape Research* **28**, 113–129.

Vijayan, V., Edwin, L. and Ravindran, K. (2000) Conservation and management of marine fishery resources of Kerala State, India. *Naga, the ICLARM Quarterly* **23**, 6–9.

Vivekanandan, V. (2009) Fishing community institutions: role in conserving marine living resources. Presented at the ICSF workshop on Social dimensions of Marine Protected Area (MPA) implementation in India: Do fishing communities benefit? 21–22 January, Chennai, India. Available at http://mpa.icsf.net (accessed March 2014).

Chapter 25

ENGOs and SIDS: Environmental interventions in small island developing states

P. McConney[1], R. Pomeroy[2] and Z. Khan[3]

[1]Centre for Resource Management and Environmental Studies (CERMES), University of the West Indies (UWI), Barbados, West Indies
[2]University of Connecticut, Avery Point, Groton, CT, USA
[3]Foundation for the Peoples of the South Pacific International (FSPI), Suva, Fiji

Abstract: Fisheries in small island developing states (SIDS) are governed and operated in ways that may either complement, or compete and conflict with, other coastal and marine uses and users. Environmentally oriented interventions by non-governmental organizations (NGOs) impact upon fisheries harvest, post-harvest and supporting activities from policy to practical levels. Environmental actors, interests and issues may either enable or constrain fisheries in SIDS. Environmental interventions are of increasing importance as ecosystem approaches, climate change, depletion of major fish stocks and sustainable development occupy global centre stage. The complex interactions associated with these interventions are poorly documented and understood, however. This chapter examines the concepts, interactions, outcomes and lessons related to the environmental interventions by NGOs in the fisheries of Caribbean and Pacific SIDS. We highlight the synergies, solutions, challenges and constraints. From a governance perspective we suggest how NGO interventions may enhance the benefits derived mainly from small-scale fisheries in SIDS.

Keywords: environment; fisheries; non-governmental organizations; small island developing states

Introduction

The United Nations (UN) monitors the progress towards sustainable development of 52 nation-states and territories that are categorized as small island developing states (SIDS). Due to their small geographic size, the use of coastal and marine space, resources and facilities in SIDS can be intense. Fisheries-related activities in SIDS are governed, planned, managed and operated in ways that either complement, or compete and conflict with, other coastal and marine uses and users such as those involved in tourism, transportation, biodiversity conservation and aquaculture. Uses and users may interact negatively to contribute to overfishing, pollution and habitat degradation, or positively for food security, alternative or supplementary livelihoods and increased well-being if assisted by ecosystem-based management (EBM) and integrated coastal management (ICM) (Fanning et al., 2011).

Environmental non-governmental organizations (ENGOs) and multi-agenda NGOs of different types and sizes impact upon how fisheries-related activities are accommodated in SIDS. Biophysical, socio-economic and governance issues are the contexts for interactions among fisheries and

Governance of Marine Fisheries and Biodiversity Conservation: Interaction and Coevolution, First Edition.
Edited by Serge M. Garcia, Jake Rice and Anthony Charles.

environmental actors along the fish chain. NGO interventions may especially enable or constrain small-scale fisheries (SSF) since actors in these fisheries may not be well organized, have adequate capacity or be particularly powerful. Where active fisherfolk organizations exist, they can be allies to environmental interests. Fisheries-NGO interactions, although poorly documented and understood, are increasing in importance as ecosystem approaches, climate change, poverty alleviation, depletion of major fish stocks and sustainable development all receive global attention (Fanning et al., 2011).

This chapter is mainly about tropical marine small-scale capture fisheries in SIDS. Freshwater inland fisheries are minor in most SIDS except for the largest islands such as Papua New Guinea and coastal states with extensive aquatic systems such as Guyana. Small-scale fisheries predominate in SIDS, but some (especially in the Pacific) have large-scale fisheries that are either domestic or foreign-operated. Marine protected areas (MPAs) and coastal aquaculture are important in many SIDS but are addressed to a greater extent in other chapters, as are small-scale fisheries more generally.

This chapter examines links between fisheries and environmental conservation in SIDS. It highlights synergies and solutions as well as challenges and constraints in such interventions. We suggest how interactions among fisheries and environmental actors may be enhanced in SIDS to improve the governance of small-scale fisheries and benefit the societies in which they are embedded

What are SIDS?

Of the 52 small island developing states (SIDS) listed by the UN, the few that are not islands are small, coastal developing states such as Belize and Guyana (http://sustainabledevelopment.un.org). SIDS share similar sustainable development challenges and prospects, and are often categorized into three geographic regions: Caribbean; Pacific; and Africa, Indian Ocean, Mediterranean and South China Sea (AIMS) (UN, 2011). In this chapter we focus mainly on Caribbean and Pacific SIDS.

SIDS typically have small populations, limited natural resources, high external transportation costs, expensive public administration, costly infrastructure and high dependence on international trade. They are susceptible to disasters and vulnerable to external shocks. Their small size and internal markets limit economic activities as well as the development of human capacity (UN, 1994). Some SIDS are among the world's least-developed countries with few opportunities to create economies of scale except through collaboration (http://www.sidsnet.org). SIDS work together through the Alliance of Small Island States (AOSIS) to tackle environmental issues and sustainable development (http://aosis.info).

Environment and fisheries

Fisheries–NGO environmental interactions are diverse and complex. Big ENGOs with conservationist stances may promote coastal and marine management measures such as MPAs that displace fishing and threaten livelihoods. However, these ENGOs may also work with fisheries elsewhere to build capacity, empower fisherfolk and develop alternative or supplementary livelihoods. Fisheries and environmental actors are working in partnership from international to local levels in subjects as diverse as poverty, gender, quality assurance, trade, fisheries management and governance. Global policy arenas set the stage for such interactions.

Understanding the global policy underpinnings of the environment in which the SIDS fisheries–NGO interactions take place is important. The Barbados Programme of Action for the Sustainable Development of SIDS (SIDS-POA) was adopted at the United Nations Global Conference on Sustainable Development of Small Island Developing States held in Barbados in 1994. The SIDS-POA (UN, 1994) addresses many coastal and marine issues, including fisheries. Several multinational environmental agreements and instruments recognize the special circumstances of sustainable development in SIDS, and accord them preferential treatment.

Most ENGOs intervene with the aim of benefiting society at national, regional or global levels. Some ENGOs cast the negative impacts of conservation, on fisheries for example, merely as acceptable collateral damage. This is the arena of the thorniest fisheries–ENGO interactions. Thorpe (2005) suggests that only a few SIDS have managed to integrate fisheries into their sustainable development plans. This leaves room for fisheries NGOs, SIDS multi-agenda NGOs and big ENGOs to fill gaps or to agitate for interventions where government action is slow or absent. Where fisheries are integrated into sustainable development or biodiversity conservation, strategies mutually beneficial to fisheries–NGO interactions are more likely to occur.

Patterns of SIDS fisheries–NGO interactions are dynamic, changing with fisheries and environmental agendas. Important trends are embedded within these dynamics such as fisheries environmental issues becoming less the province of fisheries authorities due to the ascendance of large ENGOs wielding significant financial and political power. This is especially the case where there is a lack of political will or capacity in small, weak or failing states to tackle critical environmental issues. Big ENGOs may overwhelm smaller indigenous SIDS NGOs and fisheries NGOs even in collaboration. They may also simply ignore and marginalize them. It is worth considering what these ENGOs and SIDS NGOs are before we examine what they do. Size, scale, levels, power, networks, learning, adaptive capacity, self-organization, vulnerability and resilience are all concerns in these interactions.

NGOs and environmental and fisheries interactions

NGOs are typically self-governing civil society organizations that assign themselves specific roles and goals within defined issue areas (Shariff, 1996). Not all NGOs that have environmental agendas are labelled ENGOs, and not many ENGOs address fisheries issues. SIDS NGOs that have environment or fisheries in their multi-agenda portfolios must be distinguished from the global big ENGOs. Figure 25.1 sets out a typology that locates NGOs along ranges of geographic and environmental scope.

The number of regional NGOs has recently declined in the Caribbean (Potter and Parsram, 2011). These have been displaced from below by local and national NGOs and from above by big international ENGOs. There are few regional SIDS NGOs in the Pacific (Lam, 2006). International ENGOs, such as The Nature Conservancy (TNC), Conservation International (CI) and the World Wildlife Fund (WWF), have their own agendas and significant financing that can sometimes influence entire regions (Potter and Parsram, 2011). Governance and accountability of environmental NGOs is a current topic of considerable importance for sustainable development (Jepson, 2005).

ENGOs became more powerful in world politics after the 1992 United Nations Conference on the Environment and Development (UNCED) (Princen and Finger, 1994). Big ENGOs compete with each other and smaller NGOs for public sector and private foundation funds (Chapin, 2004). Multilateral environmental agreements are proliferating at a faster rate than resources can be allocated for their implementation, and this intensifies competition. Big ENGO capacity, based on project experience, generally makes donor agencies more comfortable with them than with lesser-known indigenous SIDS NGOs. Big ENGOs can grow larger and carve out more territory at the expense of SIDS NGOs unless partnerships for the sustainability of the latter are formed, or the SIDS NGOs become internationally competitive despite their small size (e.g. CANARI in the Caribbean and FSPI in the Pacific). Dynamics of competition and collaboration need to be appreciated in NGO interventions.

Fisheries NGOs (fisherfolk organizations) are shown in a multi-level network typology (Figure 25.2) based on geographic scale and jurisdiction (McConney and Phillips, 2011; Parsram and McConney, 2011). Fisherfolk organizations operate at local, national, regional or global levels, or combinations of these via formal or informal networking. Networking is critical in SIDS NGO transboundary collective action.

Fisheries NGOs focus on sustainable use and livelihoods. These areas intersect with most issues that concern SIDS NGOs and big ENGOs. Interactions among the organizations can be cross-level where, for example, a big ENGO partners with a fisheries NGO to achieve national penetration or conversely a fisheries NGO allies with a big ENGO for funds. In

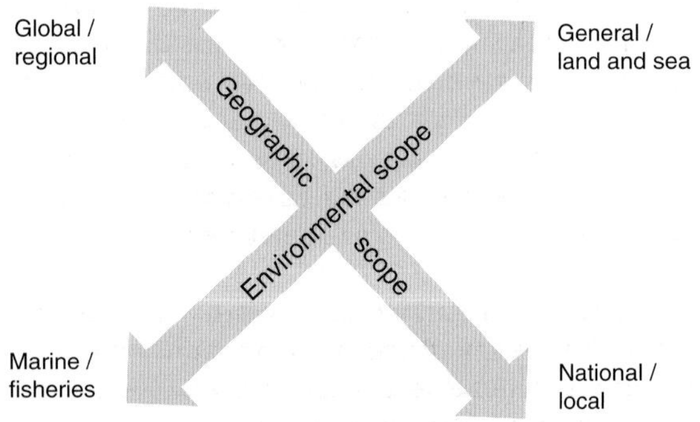

Figure 25.1 Typology of environmental non-governmental organizations.

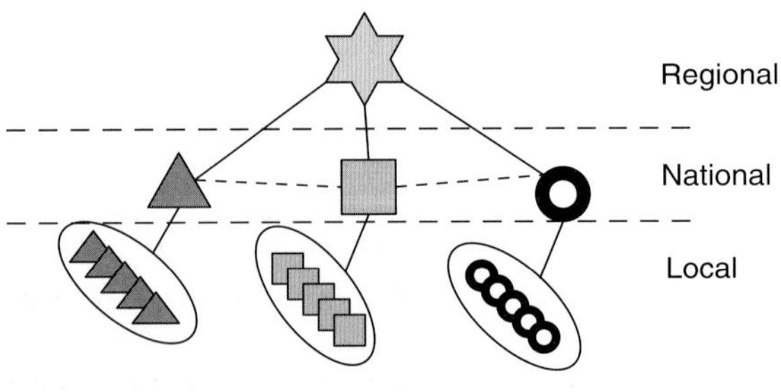

Figure 25.2 Fisherfolk organization multi-level structure.

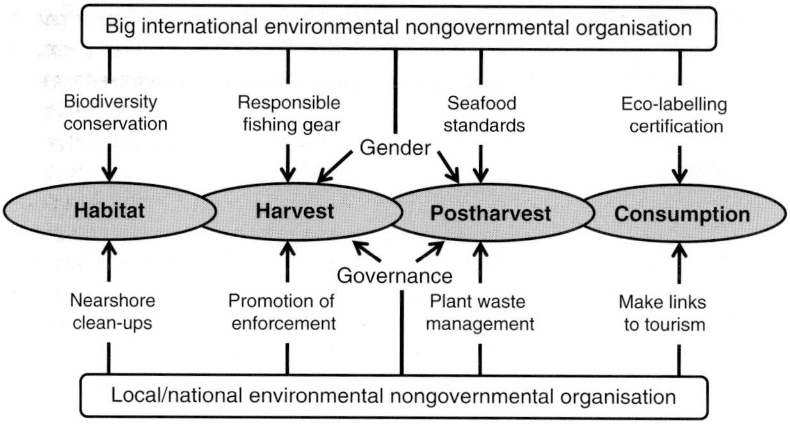

Figure 25.3 Examples of ENGO-fisheries interactions along the fish chain.

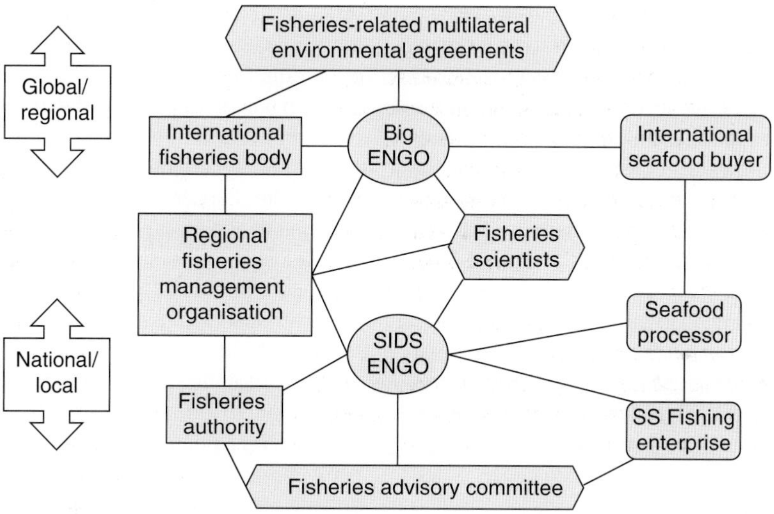

Figure 25.4 ENGOs are often brokers in multi-level networks.

reviewing NGOs roles to support ecosystem-based management (EBM) in Caribbean fisheries, Potter and Parsram (2011) note that they can provide resources and capabilities that range from networking, public awareness, education and project management to scientific research, advocacy and funding. Figure 25.3 illustrates interactions along the fish chain, distinguishing the roles of different NGOs.

Although big ENGOs, SIDS NGOs and fisheries NGOs may make separate interventions along the fish value chain there is a trend towards multi-faceted and collaborative interventions being integrated into both the ecological and social dimensions of SIDS fisheries. Figure 25.4 illustrates the strategic positions that NGOs occupy in SIDS fisheries systems as they interact with many actors on several issues at multiple levels. It is a simplified representation of highly complex and fluid relationships.

Big ENGOs and SIDS NGOs often broker connections to key actors such as government, research, donor and civil society agencies across boundaries and disciplines. Fisheries NGOs may have many ties with the potential for integration. However, big ENGOs are usually able to network with actors and to exercise power, more effectively than fisheries NGOs. Shariff (1996) notes that NGOs of all types are often regarded with suspicion and as potential competitors by governments, but they also generate information valuable to policy advisers and policy makers. NGOs intervene in fisheries and other economic sectors at multiple levels of governance in the ecosystem approach to fisheries (EAF). Chapin (2004) points out that NGOs serve agendas such as conservation more than they solve issues of sustainable livelihoods and community wellbeing. Benefits from NGO interventions into fisheries are highly variable. In the following sections we provide examples of NGO-fisheries interactions in the Caribbean and Pacific. These are followed by a summary of lessons learned from the outcomes of such interactions.

Caribbean

There are over 20 SIDS in the Caribbean ranging from tiny island nations to coastal continental states. They are diverse in geography, resources, population, politics, economies, development, culture and language. Fanning et al. (2011) review the region in the context of marine EBM, noting the importance of fisheries and tourism. Below are examples of fisheries and environmental conservation interactions among big ENGOs (transboundary and linked mainly to a developed country), SIDS NGOs (indigenous regional, national or local level) and fisherfolk NGOs (mainly national and local).

The Caribbean Challenge Initiative was launched in 2008 by The Nature Conservancy (TNC) (www.nature.org/ourinitiatives/regions/caribbean/caribbean-challenge.xml). Under this initiative, eight Caribbean SIDS have committed to protecting 20% of their marine and coastal habitat by 2020 to help meet global biodiversity conservation targets similar to the Coral Triangle Initiative in the Pacific. The participating SIDS receive financial and technical assistance from TNC and other international agencies to establish new MPAs and strengthen existing MPAs. Fisheries conservation interactions occur since many MPAs in the Caribbean are managed either exclusively by government fisheries authorities or partially through co-management agreements. Fisheries authorities balance conservation commitments to the Caribbean Challenge with fisheries management and development. Tradeoffs among coastal uses increasingly relate to ecosystem-based management (EBM), ecosystem approach to fisheries (EAF), integrated coastal management (ICM) and marine spatial planning (MSP) as frameworks which can bring fisheries into the wider policy-making arena with a voice at the table. The role of ENGOs in marine spatial planning is increasing globally (Calado et al., 2012). At times however, fishers are at risk of being excluded or marginalized due to the NGO-driven conservation focus and their lack of capacity to participate (McConney et al., 2007).

Most Caribbean MPAs are multi-zoned rather than entirely no-take. Fishing is allowed in some zones. Securing funding for fisheries-related initiatives is often difficult unless tied to MPAs through NGO-led conservation interventions. MPA funding accommodates sustainable livelihoods interventions to reduce commercial fishing effort. It is unusual for Caribbean fisheries NGOs to oppose MPAs in the Caribbean Challenge countries, but they are concerned about the threat to livelihoods (McConney and Phillips, 2011). Co-management involving NGOs and fisheries stakeholders have helped to allay fears at some MPAs (Pomeroy and Goetze, 2003).

Grant funding may not adequately cover SIDS NGO core costs such as organizational management costs and infrastructure. Collaboration between big ENGOs, SIDS NGOs and fisheries NGOs is most often aimed at enhancing the delivery of outputs through cost and capacity sharing. Unless outputs and outcomes feature in the strategic and management plans of SIDS and fisheries NGOs, they may find themselves chasing project funds to remain viable in the short term while core capacity remains under-developed despite implementation collaboration with big ENGOs. This has been addressed by ensuring that big ENGOs are aware of the SIDS and fisheries NGOs' plans (Pomeroy and Goetze, 2003) and institutionalizing a culture of planning with partner NGOs (Fernandes and TIDE, 2005).

TNC is promoting the monitoring, control and surveillance (MCS) to reduce illegal, unreported and unregulated (IUU) fishing by nationals and non-nationals on the Pedro Bank, Jamaica, along with providing facilities for fishers, enforcement officers and researchers. The ENGO is also assisting fishers to become better organized and institutionalize conservation ethics. It is a multi-faceted approach that addresses sustainable resource use and poverty alleviation (TNC, 2011).

Conservation capacity and awareness are being built by big ENGOs and SIDS NGOs through stakeholder workshops on conservation in an era of climate change. ENGO videos feature fishers and fisheries officers promoting no-take MPAs and responsible fisheries as a result of NGO-organized exchange visits to MPAs where fish are large and plentiful (TNC, 2011). Such capacity-building events and communication reach target groups within and outside fisheries, but often ignore livelihoods and offer the illusion that small-scale fishers and their households have many alternatives to resource use.

Sea turtles are fully protected by law in several Caribbean islands; harvest is allowed in some outside of closed seasons. Where sea turtle conservation impacts livelihoods, the Wider Caribbean Sea Turtle Conservation (WIDECAST) Network promotes alternative or supplemental livelihoods as part of its approach. WIDECAST is an ENGO formed to conserve sea turtles in the Wider Caribbean Region. Its objectives are to facilitate the exchange of information and best practices, implement research and recovery plans, improve knowledge of turtle migrations and nesting and develop a regional database on sea turtles (http://www.widecast.org). SIDS NGOs are also concerned with turtle conservation and livelihoods. Nature Seekers in Trinidad is known for community-based conservation of leatherbacks at Matura Bay in association with the government's Forestry Department. The coastal community is involved in conservation and turtle-watching tours for locals and tourists. Nature Seekers successfully lobbied for sea turtle conservation measures, bringing together actors at several levels of governance (James et al., 2006).

In Belize, some alternative livelihood programmes of local NGOs are aimed at reducing fishing effort on threatened resources. For example, the Toledo Institute for Development and Environment (TIDE) trained over 15 fly fishing tour guides who later earned more from guiding than commercial fishing, but return to commercial fishing during the low season for tourism (Fernandes and TIDE, 2005). However, tourism earnings are re-invested into commercial fishing if no lucrative, less seasonal, alternatives are available. Even with viable alternatives the strong cultural and lifestyle attractions of fishing are difficult to overcome. For example, farming and processing sea moss (algae) was spearheaded in St Lucia by the Caribbean Natural Resources Institute (CANARI), a regional SIDS NGO (HTTP://www.canari.org). This may be successful for household livelihood diversification where women get involved in aquaculture, but seldom do the men stop fishing entirely as a result of alternative livelihood interventions.

Examples of NGO influence on fisheries conservation come from Belize. In December 2010 trawling was banned in the exclusive economic zone (EEZ). The big ENGO Oceana gave tangible support to the proposal to ban shrimp trawling along the Belize Barrier Reef that has World Heritage Site status. It negotiated the buyout of shrimp trawlers in collaboration with the government and Belize Fishermen Cooperative Association (BFCA), a fisheries-NGO. The BFCA is one of the most powerful fisherfolk organizations in the insular Caribbean. It has had a seat on the country's Fisheries Advisory Board, which has advised successive fisheries ministers on policy for over 30 years (McConney et al., 2007).

Conservation of spawning aggregations of groupers and snappers is high priority for fisheries NGOs and fisheries authorities concerned with commercial fisheries as well as for big ENGOs and SIDS NGOs. Belize declared 11 protected spawning aggregation sites and formed a National Spawning Aggregation Working Group that includes NGOs. The goal of the group is 'to manage spawning aggregation sites and gather data that can be used to develop strategies to maintain these as viable sites for the protection, conservation and sustainable use of the fishery' (http://www.tidebelize.org).

Rights-based fisheries management is uncommon in the Caribbean where open access, or technical measures and input controls predominate. The Environmental Defense Fund (EDF) teamed up with the Wildlife Conservation Society (WCS) to introduce 'catch shares' to Belize in specific fisheries in 2008 as a pilot programme (http://www.edf.org/oceans/how-catch-shares-work-promising-solution). There is also collaboration with TNC and a SIDS NGO (TIDE) to do this in a co-managed marine reserve. It is unlikely that this conservation experiment would have taken place without big ENGO intervention.

Exchanging information among the many stakeholders and projects across the Caribbean is a major challenge, but necessary to encourage collaboration, the sharing or scaling up of successful initiatives and the conservation of transboundary marine resources. The Gulf and Caribbean Fisheries Institute (GCFI) has as its goal the exchange of information among diverse fisheries interests (http://www.gcfi.org). GCFI hosts an annual scientific meeting that covers fisheries biology, socio-economics and management plus recreational fisheries, aquaculture, MPAs and special topics. It brings together scientists, students, managers, fishers, big ENGOs, SIDS NGOs, fisheries NGOs and others. There is an annual award for fishers who lead sustainable resource use. Big ENGOs sponsor fishers to attend GCFI meetings, providing a space in which fisheries interests can be articulated and collaboration among stakeholders strengthened. ENGOs are important for communicating marine conservation science to policy advisers and makers in Caribbean SIDS, often in partnership with fisheries NGOs (Deane and McConney, 2011).

The Caribbean Network of Fisherfolk Organizations (CNFO) was created in 2007 out of a Caribbean Regional Fisheries Mechanism (CRFM) project to strengthen and network national fisherfolk bodies (mostly cooperatives and associations) of Caribbean countries (McIntosh et al., 2010).

Although it is an informal network it has influenced regional fisheries policy via participation in the CRFM Fisheries Forum and Ministerial Council meetings in part due to assistance from SIDS NGOs (McConney and Phillips, 2011). Sustainable Grenadines (a SIDS ENGO) has worked with the CNFO and big ENGOs on fisherfolk organizations, MPAs and network development in the Grenadine islands.

Legislative reform is a recent area of intervention for big ENGOs and SIDS NGOs in the Caribbean. The WCS is assisting the government of Belize in a participatory process for updating its legislation to operationalize EAF. Fisheries NGOs are part of the steering group and they, as well as SIDS NGOs, are vocal at community consultations. This NGO-led initiative is significant given the typically slow pace of legal reform and low priority of revisions to fisheries legislation.

Big ENGOs, SIDS NGOs and fisherfolk NGOs have promoted EBM and EAF more than fisheries authorities. TIDE, which co-manages the Port Honduras Marine Reserve (PHMR) in Belize, has integrated watershed management with fisheries management (Fernandes and TIDE, 2005). The PHMR protects lobster and finfish fisheries as well as key habitat for the endangered West Indian Manatee. TIDE's programmes are assisted by several big ENGOs. Sustainable Grenadines, another SIDS NGO, is working in close association with TNC on sustainable inter-sectoral use of coral reef resources on the transboundary Grenada Bank. The Caribbean Network of Fisherfolk Organizations (CNFO) has included EBM/EAF in its vision and mission (McIntosh et al., 2010). It has undertaken initiatives ranging from workshops to posters to get messages across to its members as well as fisheries authorities. All three categories of NGO are catalysts for cutting-edge change in fisheries.

Although big ENGOs and SIDS NGOs focused on conservation may exclude or marginalize fisheries interests, most often these organizations are vehicles for meaningful participation. Finch (2006) discusses how fisheries interests in Belize are integrated into the structure and functions of indigenous ENGOs. They have boards that include tourism, fisheries, environmental conservation interests, local government and nearby communities (Pomeroy and Goetze, 2003). In the PHMR fishers were originally opposed to the ban on using gill nets in the reserve. Now fishers work with TIDE to report and prosecute offenders. Acceptance of the gill net ban only came after TIDE educated fishers and provided financial incentives such as scholarships and employment for those who depended upon gill nets for their livelihoods (Fernandes and TIDE, 2005). The employment of fishers as SIDS NGO staff (e.g. as rangers or wardens) helped improve environmental conservation capacity.

Pacific

The 22 Pacific Island Countries and Territories (PICTs) span much of the tropical and subtropical Pacific Ocean. The area encompassed by the exclusive economic zones of these countries and territories exceeds 27 million km^2. Besides being more than an order of magnitude larger than the Wider Caribbean Region, there are numerous socio-ecological system differences between the Caribbean Sea and the Pacific Ocean. PICTs are more heavily dependent than Caribbean SIDS on marine resources for economic development, government revenue, food security, livelihoods and culture (Petersen, 2002). Pacific SIDS are highly vulnerable to adverse changes in the status of fisheries resources. Much attention is paid to impacts of climate change on fisheries, aquaculture and habitats (Bell et al., 2012).

The Pacific is known for biological diversity and species richness. Its waters hold the world's largest stocks of tuna and related pelagic species. Offshore tuna fisheries contribute to the economies of Pacific SIDS through access agreements, processing and exports while inshore, mainly reef and lagoon, fisheries contribute mainly to food security for local populations. A point of contention, however, is that Pacific SIDS access agreements typically yield in total less than US$ 100 million annually compared to the US$ 2 billion market worth of the tuna caught mainly by distant water fishing nations (DWFN) through agreements and otherwise. Plans to increase benefits from improving access agreements, from expanding PICT domestic fisheries offshore or from both often involve ENGOs (Cartwright, 2007). We examine Pacific SIDS fisheries interactions with conservation NGOs not in direct comparison to the Caribbean, but to further expand and diversify the analysis. Pacific big ENGO interventions are often large scale and programmatic in structure (Lam, 2006). Because of this the examples are set out mainly around the programmes of the NGOs, in contrast to the Caribbean section. Environmental conservation interactions between fisheries and NGOs primarily focus on tuna management at national and regional levels, and inshore fisheries management at community or local level through the establishment of MPAs and locally managed marine areas (LMMAs) (Govan, 2009).

The geographic scale of big ENGO initiatives can be immense. Conservation International (CI) spearheads work on the Pacific Oceanscape, the Phoenix Island

Protected Area in Kiribati and on establishing high seas MPAs for resource management and protection in areas beyond national jurisdiction (http://www.conservation.org/where/asia-pacific/pacific_islands/pages/overview.aspx). These areas range from millions to hundreds of thousands of square kilometres in size and are dedicated to innovative marine spatial planning and resource conservation. In order to conserve vast areas, CI and its partner ENGOs engage high-level policy-makers. Similarly, the Pew Environment Group's Global Ocean Legacy (http://www.pewenvironment.org/campaigns/global-ocean-legacy/id/8589941025) is a large-scale high-level initiative. Such grand schemes to conserve global biodiversity transcend fisheries and demonstrate the power of big ENGOs to a much greater extent than seen in the Caribbean.

The Nature Conservancy (TNC) also has entry points at the highest levels of government to create and support the commitment of countries to conservation. This big ENGO develops conservation tools and capacity, raises environmental awareness and aims to increase the pace of achieving national and regional implementation of global (e.g. WSSD and CBD) conservation targets. TNC works in the Coral Triangle and with the Micronesia Challenge to set aside large underwater tracts or 'seascapes' as formal MPAs, linking them into a region-wide network and equipping local partners to successfully manage them (http://www.nature.org/ourinitiatives/regions/asiaandthepacific/index.htm). TNC supports the development of community-based ecosystem approaches for managing nearshore fisheries, sustaining livelihoods and piloting conservation sustainable finance mechanisms (SPC, 2010).

Several big ENGOs assist Pacific SIDS to enhance sustainable development through improved management of transboundary oceanic fisheries resources, still in the context of marine biodiversity conservation but with multi-level policy engagement. Prominent among these are the International Union for the Conservation of Nature (IUCN) -led Pacific Islands Oceanic Fisheries Management Project and Pacific Roundtable (http://www.iucn.org/about/union/secretariat/offices/oceania/roundtable/). The latter is a network of NGOs, donors and other organizations working in nature conservation in the region. The coordination is important for the success of fisheries initiatives. For example, Pacific bigeye tuna is listed as endangered on the IUCN Red List and other species may follow if the fisheries are not properly managed. IUCN is therefore also interested in building capacity and links between experts to develop policy on access arrangements (Cartwright, 2007).

Fisheries management is a focus of World Wildlife Fund (WWF) programmes. The focus of WWF national marine programmes is the establishment of marine managed areas following a process of stakeholder consultations and scientific assessment of selected areas (Cartwright, 2007). WWF has been working with communities, local councils of chiefs, provincial administrations, national government authorities and NGO partners in the Macuata province of Fiji. This was one of the founding sites for the Fijian Locally Managed Marine Area (LMMA) network, which currently covers hundreds of mainly community protected areas.

The WWF offshore fisheries programme addresses advocacy, public awareness, research and policy in conjunction with partners such as Greenpeace to improve the health and management of tuna fisheries and reduce negative climate change impacts in the Pacific Islands. These transboundary functions are often collaborative. WWF was given the responsibility by the Secretariat of the Pacific Community (SPC) Oceanic Fisheries Management Programme and the Forum Fisheries Agency (FFA) to carry out awareness and advocacy on Western and Central Pacific Fisheries Commission (WCPFC) tuna fisheries issues. WWF is part of the Pacific Islands Roundtable for Nature Conservation Offshore Fisheries Working Group that also includes Greenpeace, IUCN Oceania, SeaWeb, the University of South Pacific and the US Regional Environment Advisor. WWF's role in sustainable fisheries management (for tuna) and species conservation (for turtles and whales) is also growing, particularly at policy level. It plays a facilitative role in ensuring that national and regional capacities are enhanced and it is expected to be a major big ENGO asset in the future (Lam, 2006).

Among the big ENGOs taking a scientific research approach to establish conservation best practices are the Wildlife Conservation Society (WCS) and Birdlife International. The WCS has been piloting the demonstration of EBM and EAF based on global research and practical experience tailored to the context and requirements of implementation in the western Pacific. It works with IUCN to promote high-seas MPAs (Cartwright, 2007). WCS also supports the LMMA initiative by complementing the network's practical approach with scientific research (Govan, 2009). For example, the WCS Fiji programme works with fishing communities and the Fiji government to protect biodiversity and natural resources by supporting the ridge-to-reef protected area networks designed to maintain ecological connectivity and improve resilience to climate change (http://wcsfiji.com/).

Since consumer markets, particularly in Europe and the USA, are starting to restrict trade in species known to come from unsustainable fisheries, Greenpeace is concerned with the market-based aspects of conservation and the need for fair trade in fish (Cartwright, 2007). Markets for bigeye and yellowfin tunas are among those of interest. Greenpeace supports traceability and raising public awareness in DWFNs on fisheries trade and sustainability issues. This big ENGO works on increasing awareness of IUU problems within Pacific SIDS government agencies and on strengthening port state control. When Greenpeace conducted compliance patrols in some PICT EEZs it found irregularities on most vessels boarded, including lack of valid licenses and numerous non-operational VMS (Cartwright, 2007). This information can assist the Pacific SIDS to improve conservation via their foreign fishing regimes.

Reducing and managing the impacts of oceanic tuna longline fishing on seabirds, such as threatened species of albatross taken as bycatch, is the main aim of Birdlife International. Birdlife implements this regionally through RFMOs such as the WCPFC and IATTC (Cartwright, 2007). This big ENGO also promotes use of standardized methods for recording seabird bycatch and is lobbying for increasing the levels of observer coverage.

Big ENGOs may implement multi-country fisheries-related programmes in collaboration with inter-governmental bodies such as the FFA, SPC, WCPFC and South Pacific Regional Environment Program (SPREP). For example, the SPC has collaborated with TNC and FAO to develop community ecosystem-based approach fisheries management guidelines for Pacific Island countries (SPC, 2010). Efforts have been made to strengthen fishing industry associations, including artisanal (community) fishers' associations, through training, advice and support to build and improve their capacity to engage in dialogue effectively and meaningfully with other stakeholders in marine resource use, and especially with governments regarding fisheries resource management and development policies and strategies (Carleton and Vuki, 2011).

A key SIDS NGO is the Foundation for the Peoples of the South Pacific International (FSPI). This network links civil society development NGOs from nine Pacific Island countries (http://www.fspi.org.fj). The FSPI Communities and Coasts Programme works in partnership with local CBOs, national NGOs and government authorities to build capacity for community-based coastal management that includes nearshore fisheries. The FSPI programme expands the LMMA approach to coastal fisheries, which originated in Fiji, to other Pacific island countries, and promotes the use of traditional sanctuaries and LMMAs as the foundation of national MPA systems. In the Solomon Islands the LMMA consortium includes TNC, FSPI, WWF, Solomon Islands Development Trust (SIDT) and WorldFish Centre (Fa'asili, 2006). The LMMA approach recognizes the initiatives of indigenous peoples and local communities to sustainably manage marine resources based on their traditional and local knowledge. The programme engages policy to seek widespread recognition of the benefits of LMMAs. Its research component focuses on assessing resource management effectiveness and supplementary livelihoods in situations of scarce options (Cartwright, 2007). A growing area of cooperation is for NGOs to engage other organizations in interdisciplinary fisheries science (Hastings et al., 2012).

Taking an alternative approach to turtle conservation, the Wan Smolbag Theatre is a SIDS NGO based in Vanuatu that raises environmental awareness in South Pacific countries by using drama to inform and encourage public discussion on environmental conservation, fisheries resource and governance issues (http://www.wansmolbag.org). The Wan Smolbag environmental programme initiated the Turtle Monitors Network (TMN) that comprises village-based turtle monitors throughout Vanuatu. The network adopted the LMMA approach and has led MPA management at sub-national and national levels by replicating interventions and creating case studies for learning.

Fisherfolk NGOs in the Pacific Islands are few in number and low in capacity. Engagement with them has generally been inadequate (Carleton and Vuki, 2011). The major regional fisherfolk body is the Pacific Islands Tuna Industry Association (PITIA) with membership of national industry associations in 14 SIDS. PITIA informs and encourages its members to engage in policy decisions affecting their businesses. It facilitates and encourages sustainable use of tuna and tuna-related resources. It was set up in 2004 to co-ordinate or undertake negotiations with national, regional and international entities having interests in, or impacts on, fishing or associated industries in the island countries of the FFA. National level fisherfolk bodies include the Cook Islands Fishermen Association, Fiji Tuna Boat Owners Association, Samoa Fish Exporters Association, Nauru Fishermen's Association and others. Some of these PICT fisherfolk NGOs are also member s of the Organization for the Promotion of Responsible Tuna Fisheries (OPRT), an industry-based international NGO established in Tokyo for the purpose of promoting the sustainable use of tuna resources (http://oprt.or.jp). PICT organizations also support the sustainable expansion of domestic tuna fisheries.

In a series of consultations (Cartwright, 2007, 2009a, b, 2011) suggestions were made by fisheries stakeholders for improving the roles of all types of NGOs in the management of oceanic species in the Western Central Pacific Ocean. These suggestions included NGOs assisting by:

- combating illegal, unregulated and unreported (IUU) fishing;
- maintaining accessible databases on foreign fishing fleets;
- developing and disseminating non-technical fisheries information;
- acting as independent 'watchdogs' to ensure good governance;
- reminding States of national, regional and international responsibilities;
- facilitating the timeliness and quality of evidence-based decision-making;
- coordinating civil society collective action and voice on issues;
- promoting the ecosystem approach to fisheries and resilience;
- lobbying and campaigning to mobilize resources from donors;
- building (adaptive) capacity among all types of fisheries stakeholders;
- reviewing access agreements to inform NGO understanding of issues;
- briefing SIDS NGOs prior to RFMO meetings and debriefing afterwards;
- developing communication strategies to better reach fisheries audiences; and
- actively networking and helping to manage conflicts among stakeholders.

Outcomes and lessons learned

The examples of fisheries–environmental conservation NGO interactions in the Caribbean and Pacific reflect the changing role of NGOs in fisheries governance. The FAO Fisheries and Aquaculture Department notes that on a global scale the role of all types of NGOs in the governance of fisheries, as policy-influencing and implementing actors, has increased at local, national and international levels (http://www.fao.org/fishery/topic/13310/en). The prominence of NGOs is linked mainly to social justice and sustainable development. NGOs address biophysical, socio-economic and governance issues within the socio-ecological systems of fisheries at multiple levels. FAO perceives involvement of NGOs to be generally constructive (and sometimes indispensable) for achieving effective fisheries management within some governance arrangements such as co-management.

Since UNCED, NGOs have taken stronger interest in fisheries issues. According to FAO, issues of most interest have included: 'excess fleet capacities and harmful subsidies in world fisheries, impacts of fishing agreements and fish trade on the environment, food security and the socio-economic conditions of small-scale and indigenous fishers, and environmental and social effects of certain types of aquaculture (e.g. shrimp and salmon culture)' (http://www.fao.org/fishery/topic/13310/en). Big fisheries NGOs such as the International Collective in Support of Fishworkers (ICSF) are rare. It is mainly the big environmental NGOs, and to a lesser extent the SIDS NGOs, that have become channels for information exchange between the fisheries authorities, fishing industry, other stakeholders and general public. In developed countries the ENGO role has grown as environmental conservation dimensions of fisheries become better understood and accepted by fishery managers and the fishing industry. ENGO roles have changed from 'problem identification' to 'problem-solving'. ENGOs are pursuing deeper involvement in fisheries governance in developed countries (Dunn, 2005) as well as SIDS (Potter and Parsram, 2011).

Table 25.1 summarizes some of the outcomes and lessons learned from the Caribbean and Pacific. The remainder of the section considers, from a governance perspective, how to enhance the benefits to be derived mainly from small-scale fisheries in SIDS. This does not discount the importance of large-scale fisheries but acknowledges that SSF are prevalent in SIDS at national and local levels.

The global trend for NGOs to become more involved in fisheries matters is evident in Caribbean and Pacific SIDS. Their attention focuses more upon environmental conservation than on sustainable small-scale fisheries. The bias varies considerably by region, country, type of NGO and specific issue. Big ENGOs operating at very large geographic scales such as in the Pacific tend to be more conservation oriented; negative impacts on commercial fisheries tend to be viewed as short-term collateral damage in view of longer-term conservation goals. SIDS NGOs and fisheries NGOs attempt to balance conservation with livelihood and development objectives. They serve a different, less global, constituency but cannot avoid operating in the context of global targets and agendas driven largely by the big ENGOs. Large-scale initiatives such as the Caribbean Challenge can significantly change the context of regional and national coastal and marine management. Such initiatives may absorb much of the limited capacity in SIDS for ICM and fisheries

Table 25.1 Major fisheries interactions, outcomes and lessons learned.

Fisheries–ENGO interactions	Outcomes experienced and key lessons learned
Promoting direct biodiversity conservation (e.g. via MPAs)	A key area for ENGOs that can either integrate or exclude fisheries interests depending upon the approach adopted
Strengthening monitoring, control and surveillance	Most common in relation to IUU fishing and often about fishers violating MPA rules. Focus on both enforcement and compliance
Developing sustainable or alternative livelihoods	Difficult in SIDS with strong fishing culture and few occupational alternatives. Often best tackled at household or community level
Empowering fisherfolk NGOs or groups to influence policy	Requires an appreciation of networking. Even groups with low technical capacity can become influential under good leadership
Developing capacity within fisherfolk organizations	Capacity is multidimensional and several facets need to be developed simultaneously to achieve effective self-organization
Promoting or catalysing the implementation of EBM/EAF	Resource user groups often have an implicit understanding of EBM/EAF which does not need to be overly complicated
Implementing fisheries management measures	Many fisheries interests in SIDS support fisheries management that is equitable and compatible with fisheries livelihoods
Reforming or amending fisheries legislation	Legislative changes typically proceed slowly unless there is a crisis or catalyst, but an accelerated pace must be participatory
Introducing rights-based fisheries management	Experience and traditions of rights-based fisheries vary greatly in SIDS. Consider local customary tenure and property regimes
Facilitating information exchange and communication	Small islands are often isolated and much attention must be paid to communication if this disadvantage is to be overcome

management. This may not be detrimental if NGO interventions are well integrated into the sustainable development plans of SIDS along with fisheries. However, Thorpe (2005) found that the integration of fisheries into sustainable development was not common in SIDS. It is critical that this is addressed as a priority and is assisted by ENGO–fisheries interactions becoming more based on people-centred long-term programmes rather than short-term projects. Framework funding for ENGO activities in SIDS can contribute to this adaptive approach by reducing the competition for short-term funding and enhancing the conditions under which durable partnerships can develop between big ENGOs, SIDS NGOs and fisheries NGOs.

Similar to the Caribbean, several big ENGOs and SIDS NGOs with marine interests operate in the Pacific region, but not all of their initiatives demonstrate strong or appropriate links to PICT national policy (Govan, 2009). He notes for example that many NGO activities do not support indigenous approaches such as the development of the LMMA network. LMMAs have been judged the most promising type of marine management regime for western Pacific communities (Govan, 2009). Some big ENGO interventions may have peripheral benefits at best, and may actually undermine benefits from SIDS NGO and governmental initiatives. Powerful actors such as the big ENGOs (despite their good intentions) can often exacerbate local fisheries management problems by driving environmental conservation agendas via their own well-funded but narrow strategies and programmes. Government fisheries authorities often look on helplessly with inadequate funding, unable to influence or direct the host of big ENGO conservation activities in their countries which may nonetheless engage their limited human resources (Fa'asili, 2006).

Key contributors to loss of biodiversity and deterioration of marine environmental resources in most Caribbean and Pacific SIDS include inadequate government and NGO leadership, and consequent low levels of capacity for formal marine resource governance (Carleton and Vuki, 2011; Fanning et al., 2011). Although the Pacific has a long and well-documented history of customary tenure and culturally embedded marine conservation for sustainable use, the decline of local ownership of environmental and fisheries problems and their solutions calls for more voice in formal arrangements, especially for women and disadvantaged groups (Novaczek et al., 2005). Partially linking local action with national policy is a multitude of big ENGOs and smaller SIDS NGOs with marine interests and mandates. NGOs in both the Caribbean and Pacific provide fisheries-related capacity building and technical and policy assistance using donor funds. However, opportunities for coordination of interventions and collaboration are enhanced by the Pacific Islands Roundtable for Nature Conservation whereas institutions in the Caribbean are less well developed. Governance arrangements for coordinating fisheries-related NGO interventions can greatly improve the development of adaptive capacity and resilience in fisheries socio-ecological systems at multiple levels.

Environmental conservationists must build adaptive capacity and resilience in socio-ecological fisheries systems. Current approaches to building capacity tend to foster dependence upon ENGOs, especially those that can access funding not available to fisheries authorities and fisheries NGOs. This is evident in the Caribbean where little is implemented without ENGO intervention and where programmes may fail when ENGO-funded or implemented projects come to an end. Instead, ENGO efforts should be aimed at building adaptive capacity in concert with enabling self-organization (Mahon et al., 2008). In SIDS with limited capacity this may only be achieved through national and transboundary networking. We have shown examples of this from the Grenadines Islands. For greater self-organization however, there must also be an enabling policy environment that relinquishes command and control approaches for flexible co-management arrangements with strong stakeholder ownership (McConney et al., 2007; McConney and Charles, 2010). In the Caribbean, trust funds for sustainable financing may help to address some MPA issues and consequently also benefit coastal fisheries.

NGOs demonstrate the ability to incorporate new and emerging concepts and practical methodologies ranging from EBM/EAF to marine spatial planning (Calado et al., 2012). With NGO assistance, stakeholders may be better able to keep up with trends in fisheries governance. In an increasingly globalized world this is critical since change is occurring rapidly, whether in climate or technology. The new and emerging concepts and practical methodologies that NGOs can bring to fisheries must be feasible. The environmental conservation agendas of big ENGOs may be incompatible with the ecological, cultural, social, economic and political realities of SIDS fisheries. SIDS need to be more proactive in setting their own broad strategic agendas for fisheries through institutions such as AOSIS, RFMOs or their regional bodies for social and economic integration, such as the SPC or CARICOM in our two study areas. If issues such as climate variability and change are truly mainstreamed, then they need not divert funds from fisheries management but will instead enhance initiatives to cope with uncertainty. Introducing or strengthening EAF/EBM is also useful in this regard (Fanning et al., 2011).

Fisheries NGOs need to be strengthened and not overlooked in pursuit of environmental conservation that often drives ENGO fisheries interventions. Networking can assist as described in this chapter, but bringing fisheries NGOs more fully into the policy domain, as seen in the case of the CNFO and which is being pursued with PITIA, is also essential (McIntosh et al., 2010). Big ENGOs and SIDS NGOs can prepare the fisherfolk organizations for this transition into collaborative decision-making (McConney and Phillips, 2011). If environmental conservation NGOs partner with fisheries NGOs at multiple levels, there are likely to be mutual benefits to all stakeholders and ultimately outcomes reflecting improved wellbeing. The role of civil society in fisheries governance is on the ascendant (Mikalsen et al., 2007).

References

Bell, J.D., Johnson, J.E. and Hobday, A.J. (eds) (2011) *Vulnerability of Tropical Pacific Fisheries and Aquaculture to Climate Change*. Secretariat of the Pacific Community, Noumea, New Caledonia.

Calado, H., Bentz, J., Ng, K., Zivian, A., Schaefer, N., Pringle, C., Johnson, D. and Phillips, M. (2012) NGO involvement in marine spatial planning: A way forward? *Marine Policy* **36**, 382–388.

Carleton, C. and Vuki, V. (2011) The GEF/UNDP Oceanic Fisheries Management Project Terminal Evaluation. Report prepared for the United Nations Development Program, Fiji. Available at http://www.ffa.int/gef/files/gef/RSC6_WP%206%20Terminal%20Evaluation_Attachment%20A.pdf (accessed March 2014).

Cartwright, I. (2007) NGO and civil society workshop on oceanic fisheries management in the Western and Central Pacific Fisheries Convention Area. Tanoa Plaza Hotel, Suva, Fiji, 24–25 April 2007. Summary record and outcomes.

Cartwright, I. (2009a) Oceanic Fisheries Management in the Western and Central Pacific Fisheries Convention Area: Roles for ENGOs and CSOs. Quality Inn, Honiara, Solomon Islands, 26–27 January 2009. Summary record and outcomes. Available at http://www.ffa.int/gef/files/gef/SI%20OFM%20Workshop-%20Summary%20Record_FINAL.pdf (accessed March 2014).

Cartwright, I. (2009b) Oceanic Fisheries Management in the Western and Central Pacific Fisheries Convention Area: Roles for ENGOs and CSOs. Ela Beach Hotel, Port Moresby, Papua New Guinea, 29–30 January 2009. Summary record and outcomes. Available at http://www.ffa.int/gef/files/gef/PNG%20OFM%20Workshop-%20Summary%20Record_FINAL.pdf (accessed March 2014).

Cartwright, I. (2011) OFM In-Country Workshop Report, Tuvalu 19–22 November 2010. Report to World Wildlife Fund.

Available at http://www.ffa.int/gef/files/gef/OFM%20Tuvalu%20Report%20Part_1.pdf (accessed March 2014).

Chapin, M. (2004) A challenge to conservationists. *World Watch Magazine* **Nov/Dec**, 17–31.

Deane, L. and McConney, P. (2011) Communication between marine science and policy in the eastern Caribbean. *Proceedings of the Gulf and Caribbean Fisheries Institute* **63**, 406–410.

Dunn, E. (2005) The role of environmental NGOs in fisheries governance. In: *Participation in Fisheries Governance* (ed. T. Gray), pp. 209–218. Springer, Dordrecht.

Fa'asili, U. (2006) Establishment of Solomon Islands National Community-based Fisheries Management Programme. Preliminary study. Available at https://www.spc.int/coastfish/Sections/CFMS/Publications/CountryReports/Solomon_Islands/Prelim_study_Solomon.pdf (accessed March 2013).

Fanning, L., Mahon, R. and McConney, P. (eds) (2011) *Towards Marine Ecosystem-Based Management in the Wider Caribbean*. Amsterdam University Press, Netherlands.

FAO (2005–2012) World inventory of fisheries. The role of NGOs in the governance of fisheries. Issues Fact Sheets. Text by Peter Manning. In: *FAO Fisheries and Aquaculture Department*. Available at http://www.fao.org/fishery/topic/13310/en (accessed March 2014).

Fernandes, D. and TIDE (2005) Lessons from the Equator Initiative: Community-based Management of the Port Honduras Marine Reserve, Belize. Natural Resources Institute, University of Manitoba, Canada.

Finch, J. (2006) Negotiating influence: the experience of three local NGOs in marine reserve co-management in southern Belize. Master of Science Thesis, University of Montana, USA.

Govan, H. (2009) Status and potential of locally-managed marine areas in the South Pacific: meeting nature conservation and sustainable livelihood targets through wide-spread implementation of LMMAs. SPREP/WWF/WorldFish-Reefbase/CRISP. Secretariat of the Pacific Regional Environment Programme, Apia, Samoa.

Hastings, J.G., Gruby, R. and Sievanen, L. (2012) Science-based coastal management in Fiji: Two case studies from the NGO sector. *Marine Policy* **36**, 907–914.

James, C., McConney, P. and Pomeroy, R. (2006) Case study analysis of coastal resources comanagement in the Caribbean. Centre for Resource Management and Environmental Studies, University of the West Indies, Barbados. CERMES Technical Report No. 2.

Jepson, P. (2005) Governance and accountability of environmental NGOs. *Environmental Science and Policy* **8**, 515–524.

Lam, M. (2006) Strategy to promote and strengthen Environmental NGOs stakeholder participation and public awareness of oceanic fisheries management issues. Pacific Islands Oceanic Fisheries Management (OFM) Project Report. Pacific Islands Oceanic Fisheries Management Project, Honiara, Solomon Islands.

Mahon, R., McConney, P. and Roy, R. (2008) Governing fisheries as complex adaptive systems. *Marine Policy* **32**, 104–112.

McConney, P. and Charles, A.T. (2010) Managing small-scale fisheries: moving toward people-centered perspectives. In: *Handbook of Marine Fisheries Conservation and Management* (eds R.Q. Grafton, R. Hilborn, D. Squires, M. Tait and M. Williams), pp. 532–545. Oxford University Press, New York.

McConney, P. and Phillips, T. (2011) Collaborative planning to create a network of fisherfolk organisations in the Caribbean. In: *Collaborative Resilience: Moving Through Crisis to Opportunity* (ed. B. Goldstein), pp. 207–230. MIT Press, Cambridge, Massachusetts.

McConney, P., Mahon, R. and Pomeroy, R. (2007) Challenges facing coastal resource co-management in the Caribbean. In: *Adaptive Co-management: Collaboration, Learning and Multi-Level Governance* (eds D. Armitage, F. Berkes and N. Doubleday), pp. 105–124. UBC Press, Vancouver, Canada.

McIntosh, S., Lay, M., McConney, P. and Phillips, T. (2010) The development of a Caribbean regional network of fisherfolk organisations and its role in influencing fisheries policy. *Proceedings of the Gulf and Caribbean Fisheries Institute* **62**, 298–305.

Mikalsen, K., Hernes, H. and Jentoft, S. (2007) Leaning on user-groups: The role of civil society in fisheries governance. *Marine Policy* **31**, 201–209.

Novaczek, I., Mitchell, J. and Veitayaki, J. (eds) (2005) *Pacific Voices: Equity and Sustainability in Pacific Islands Fisheries*. Institute of Pacific Studies, Suva, Fiji.

Parsram, K. and McConney, P. (2011) A network approach to understanding coastal management and governance of small scale fisheries in the eastern Caribbean. In: *World Fisheries: A Social-Ecological Analysis* (eds R. Ommer, R.I. Perry, P. Cury and K. Cochrane), pp. 334–350. Wiley-Blackwell, Oxford.

Pena, M. and McConney, P. (2011) Institutional arrangements for local management of marine areas in the eastern Caribbean. *Proceedings of the Gulf and Caribbean Fisheries Institute* **63**, 267–273.

Petersen, E.H. (2002) Economic policy, institutions and fisheries development in the Pacific. *Marine Policy* **26**, 315–324.

Pomeroy, R. and Goetze, T. (2003) Belize case study: marine protected areas co-managed by Friends of Nature. Caribbean Conservation Association (CCA), Garrison, Barbados.

Potter, B. and Parsram, K. (2011) Roles for non-governmental organizations in EBM of marine and coastal areas of the wider Caribbean. In: *Towards marine Ecosystem-Based Management in the Wider Caribbean* (eds L. Fanning, R. Mahon and P. McConney), pp. 307–317. Amsterdam University Press, Netherlands.

Princen, T. and Finger, M. (1994) *Environmental NGOs in World Politics*. Routledge, London.

Secretariat of the Pacific Community (2010) A community-based ecosystem approach to fisheries management: guidelines for Pacific Island Countries. Secretariat of the Pacific Community, Noumea, New Caledonia.

Shariff, M. (1996) Role of non-governmental organizations (NGOs) in fish health management in the Asia-Pacific. In: *Health Management in Asian Aquaculture. Proceedings of the Regional Expert Consultation on Aquaculture Health Management in Asia and the Pacific* (eds R.P. Subasinghe, J.R. Arthur and M. Shariff), pp. 51–59. FAO Fisheries Technical Paper No. 360, Rome, FAO.

Thorpe, A. (2005) Integrating fisheries into the national development plans of Small Island Developing States (SIDS): ten years on from Barbados. *Natural Resources Forum* **29**, 51–69.

TNC (2011) Fishy Times Pedro Bank Management Project Newsletter No. 4, January 2011.

United Nations (1994) Programme of Action for Small Island States: Agreed at the Global Conference on Sustainable Development of Small Island Developing States, Barbados. The Nature Conservancy, Kingston, Jamaica.

United Nations (2011) *World Statistics Pocketbook 2010: Small Island Developing States*. United Nations, New York.

Chapter 26

The role of capacity building for improving governance of fisheries and conservation of marine ecosystems

J.C. Seijo[1] and S. Salas[2]

[1] Marist University of Mérida, Yucatan, Mexico
[2] Marine Resources Department, Cinvestav Unidad Merida, Mexico

Abstract: This chapter identifies current training and capacity building to facilitate the process of shifting from fisheries governance of targeted single-species stock assessment and management to an ecosystem approach to fisheries governance considering ecological, economic and social interdependencies. It also identifies and discusses linkages of capacity building for governance and conservation of ecosystems hosting fisheries, in relevant topic areas such as: (1) joint capacity building for understanding and implementing the ecosystem approach to fisheries management; (2) educational efforts to increase awareness of potential impacts of alternative fishing technologies on ecosystem health and for adopting species- and habitat-friendly fishing methods and gears; (3) modelling and analysis of vulnerable sedentary and low-mobility marine species and the ecosystems in which they live; (4) capacity building for using quantitative and qualitative criteria and methods to aid decision-making in an environment of risk and uncertainty; and (5) research questions to identify possible changes in ecosystem performance and species distribution patterns resulting from climate change. The chapter also reports on capacity-building and training programs for understanding and establishing an ecosystem approach to fisheries management and for conserving ecosystems through marine protected areas (MPAs).

Keywords: capacity building; governance coordination; conservation of ecosystems; EAF; training for spatial management; MPA; risk and uncertainty

Introduction

It has been acknowledged that fisheries can offer multiple economic and social opportunities to address development needs of coastal communities in many countries (Berkes et al. 2002; Béné 2009), but capacity building is needed in order to improve fisheries governance sustained by healthy ecosystems (Cicin-Sain et al., 2000). Training and capacity building for fisheries governance and ecosystem conservation is usually fragmented and uncoordinated. It tends to be undertaken by different branches of governments and private organizations with diverse agendas. This fragmentation may also include conflicting sectoral governance objectives. There is a need to reconcile, through training and capacity building, responsible harvest of marine species and ecosystem biodiversity conservation (Brander, 2010).

Consistent capacity building is needed for fishery managers, conservation officials, fishers, local scientists and other stakeholders concerning principles, concepts, methods of analysis and instruments for marine responsible

Governance of Marine Fisheries and Biodiversity Conservation: Interaction and Coevolution, First Edition.
Edited by Serge Garcia, Jake Rice and Anthony Charles.

fisheries management and for conservation of the associated marine ecosystems. Education to internalize attitudes of cooperation instead of conflict in shared management are also considered to mitigate emerging issues in fisheries governance and conservation of transboundary resources and of shared ecosystems as a result of climate change and anthropogenic activities.

Capacity-building process needs to be focused, iterative and adaptive and needs to be set up within a well-integrated framework and coherent strategies that favour multiplier effects. To achieve these effects, commitment is required from local authorities and users of the marine ecosystem. Commitment can be shown when the benefit of participation can be perceived (Gatewood, 1984; Guttman, 1996). The benefits of capacity building for cooperation can be many, but they have to be spelled out and the scales of impact defined. Those effects could include: (1) increased fisheries monitoring, control, surveillance and enforcement capabilities of those involved; (2) increased opportunities (economic, reduction of social conflicts, conservation) for the actors, arising from sustainably managed fisheries and ecosystems; and (3) improved ability of countries, regions or communities to manage fisheries sustainably, strengthening governance of fisheries and the ecosystems sustaining them to achieve shared objectives.

Capacity building and training for governance

Capacity building should take place on individual fishers, fishing community and institutional levels. On an individual level, capacity building requires the development of conditions that allow individual participants in the fishery to build and enhance existing knowledge and skills for their responsible fishing. It also calls for the establishment of conditions that will allow individuals to engage in the process of learning and adapting to changing ecosystem and market conditions. Fishing community capacity building should support the establishment interactive fisheries management that learns from its actions and from feedback it receives from fisheries stakeholders participating responsibly within the community. On an institutional level, capacity building should involve strengthening pre-existing fisheries sector institutions in coastal States. It should not necessarily involve creating new institutions, rather modernizing existing fisheries sector institutions and supporting them in forming sound policies, organizational structures and effective methods of fisheries management with ecosystem considerations.

Applying the United Nations Development Program approach of capacity building to fisheries governance and ecosystem conservation involves five major considerations (UNDP, 2008):

(1) *Engage fishery stakeholders on capacity development.* An effective capacity-building process for effective fishery governance and ecosystem conservation must encourage participation by all those involved. If stakeholders are involved and share ownership of the fish resource and its ecosystem they will feel more responsible for the outcome and sustainability of ecosystem use and conservation. Engaging ecosystem stakeholders who are directly affected by the fishery management strategy allows for more effective decision-making; it also makes fishery governance and ecosystem conservation more transparent.

(2) *Assess capacity needs and assets of the fishery/ecosystem to be managed and conserved.* Assessing pre-existing fisheries sector capacities through engagement with stakeholders allows capacity builders to see what areas require additional training, what areas should be prioritized and in what ways capacity building can be incorporated into local and institutional development strategies.

(3) *Formulate a capacity-building response.* Once a fisheries governance and ecosystem conservation assessment has been completed, a capacity-building response must be created based on four core issues: institutional arrangements, leadership, knowledge and accountability.

(4) *Implement a capacity-building response.* Implementing a capacity-building program should involve the inclusion of multiple governance systems (national, local and institutional). It should involve continual reassessment and expect change depending on changing natural and anthropogenic factors and situations. It should include evaluative indicators to measure the effectiveness of fisheries governance and ecosystem conservation of initiated programs.

(5) *Evaluate capacity development.* Evaluation of capacity building promotes accountability. Measurements should be based on changes in an institutions performance. Evaluations should be based on changes in performance based around the four main issues: institutional arrangements, leadership, knowledge and accountability.

Concerning the identification of specific needs for capacity building for governance of fisheries and marine ecosystem conservation, a four-tier hierarchy of capacity-building needs could include: (1) structures, systems and roles; (2) staff and facilities; (3) skills; and (4) tools (Potter and Brough, 2004).

The training dimension of capacity building for fisheries governance and marine ecosystem conservation should, as suggested by Bavinck et al. (2005), accomplish the following:

- demonstrate the diversity, complexity and dynamics of both the natural and human systems involved and the associated difficulties for predicting and controlling fisheries management impacts;
- involve in the training process the multiplicity of stakeholders/governors in fisheries governance, as their influence would be felt even if they were excluded;
- foster awareness of challenges faced by fisheries governance that exceed the carrying capacity of the ecosystem;
- raise concerns of what fisheries governors have to take into consideration, such as ecosystem health, livelihood and employment, social justice and food safety and security, and underline the difficult tradeoffs among them;
- propose governance systems that address (1) diversity with inclusiveness; (2) complexity with holistic and integrative approaches; and (3) dynamics through interactive learning.

Fisheries governance and marine ecosystem conservation

Swan (2000) defines governance as:

> 'A continuing process through which governments, institutions and stakeholders of the fishery sector – administrators, politicians, fishers and those in affiliated sectors – elaborate, adopt and implement appropriate policies, plans and management strategies to ensure resources are utilized in a sustainable and responsible manner. It could be at global, regional, sub-regional, national or local levels. In the process, conflicting or diverse interests may be accommodated and cooperative action may be taken.'

On the other hand, marine ecosystem conservation is understood as the study of conserving physical and biological marine resources and ecosystem functions. Marine conservation involves a combination of scientific principles derived from oceanography, marine biology and fisheries science, with human factors such as fishers' behaviour studied by fisheries economics and marine policy and law.

Governance of fisheries and conservation of the marine ecosystems sustaining them could be understood as a process through which governments, institutions and stakeholders of the fishery and the ecosystem in which it operates elaborate, adopt and implement appropriate policies, plans and management strategies to ensure that resources are utilized in a sustainable and responsible manner to conserve the functional integrity of marine ecosystems. The main characteristics of good governance include it being participatory, transparent, equitable, accountable, consensus-oriented and effective. Alternative management approaches have been proposed over time to improve fisheries governance. These approaches could vary from prescribed command-and-control centralized systems to more decentralized, adaptive, participatory systems involving co-management. It is critical to understand the nature and possible performance of alternative fisheries management systems in different fishery contexts. It therefore seems appropriate to undertake training on documented lessons of experiences of alternative fisheries management designs and implementation schemes with their corresponding fishery performance and ecosystem effects. In this capacity-building effort strengths and weaknesses of possible management systems should be discussed properly.

The analytical or technical approaches to deal with marine conservation issues will depend on the scale of the system of interest, but also on the goals expressed by managers and other stakeholders. It is necessary to understand the socio-economic context and the dynamics of institutional arrangements (legislation, policy processes, mechanisms for cooperation, information collection) for managing fisheries systems and conserving the ecosystems sustaining them (Smith, 2002). The identification of these components, context and its interactions can be among the first steps of moving towards adequately building capacities to improve fishery governance with healthy ecosystems (Charles 2001; Seijo et al., 2011).

Capacity building for understanding the ecosystem approach to fisheries

One basic component of capacity building for governance of fisheries and conservation of the marine ecosystems in which they occur is the understanding of the ecosystem approach to fisheries. The need to apply an ecosystem approach to fisheries management is now globally accepted. The ecosystems approach to fisheries 'strives to balance diverse societal objectives, by taking account of the knowledge and uncertainties about biotic, abiotic and human components of ecosystems and their interactions and applying an integrated approach within ecological meaningful boundaries' (FAO, 2003). Also, as pointed out by NOAA (2009), 'an important aspect of an ecosystem approach is balancing the needs of the environment with the economic, social, and recreational needs of the people who use the environment'. It is further acknowledged that this approach involves making decisions using a comprehensive and inclusive framework for all living resources, rather than reacting to the status of a single stock of fish. It represents a move away from fisheries management systems that focus only on the sustainable harvest of target species and towards systems and decision-making processes that balance environmental wellbeing with human and social wellbeing within improved governance frameworks.

It should be mentioned however that in many fishing communities of coastal States this approach is still an abstraction. As pointed out by FAO (2011), the adoption of the ecosystem approach to fisheries management has changed the requirements for technical skills in a number of ways. Firstly it has widened the skills and knowledge required to include ocean and climate sciences, species and habitat ecology and ecosystem modelling. Secondly it has increased the level of information and analysis required through expanded and costly data collection systems (Seijo, 2007).

To mitigate the above-mentioned challenges for understanding and implementing the EAF paradigm, there have been a number training and capacity-building courses and workshops on ecosystem approach to fisheries management. Very recently, Okes et al. (2012) documented success stories in their implementation stages. Table 26.1 provides an illustration of the worldwide training and capacity-building efforts on the ecosystem approach to fisheries.

Environmental education of potential ecosystem impacts of alternative fishing technologies and methods

The functional integrity of marine ecosystems have been jeopardized not only through the use of different kinds of illegal fishing methods including the use of toxic products (cyanide) or dynamite fishing methods, but also from the use of habitat-disrupting fishing technology (e.g. bottom trawling). The impact of such elements on marine ecosystems over time has been highlighted by several authors

Table 26.1 An illustration of some capacity-building and training programs for understanding and establishing an ecosystem approach to fisheries management.

Year	Capacity building and training program	Place	Institutions
2011	Ecosystem approach to fisheries	Rhodes, South Africa	FAO, EAF-NACEN Project, Rhodes University
2011	Ecosystem indicators for ecosystem approach to fisheries	Rome, Italy	FAO
2011	Ecosystem approach to fisheries	Agadir, Morocco	FAO, EAF-NACEN Project, Government of Morocco
2011	Regional exchange and policy workshop on ecosystem approach to fisheries management	Kota Kinibalu Sabah, Malaysia	NOAA, WWF, USAID-Asia
2009	Practical implementation of the ecosystem approach to fisheries and aquaculture	Colombo, Sri Lanka	APFIC, FAO
2009	Adapting to climate change: The ecosystem approach to fisheries and aquaculture in the Near East and North Africa Region.	Abbassa, Egypt	Worldfish Center, FAO
2009	Regional workshop on the application of the ecosystem approach to shrimp and lobster fisheries of the Centroamerican region.	San Salvador, El Salvador	Fisheries and Aquaculture Organization of the Centroamerican Isthmus (OSPESCA)
2007	Sustainable use and management of sea cucumber fisheries with an ecosystem approach.	Puerto Ayora, Galápagos Islands, Ecuador	FAO

(Caddy and Seijo, 2011; Chuenpagdee et al., 2003; Collie et al., 2000; Langton and Auster, 1999; Pauly et al., 2002; Pet-Soede et al., 1999).

In the last decade, efforts have been made to promote reduction of fishing pressure and elimination of non-selective gears that can generate externalities between fisheries and to other users of the marine ecosystem (Pauly et al., 2002; Zhoua et al., 2010). In these efforts, NGOs and academics have been more proactive; enforcement at local level is however limited, especially in nations where poverty limits fishers' engagement in these types of initiatives (Béné, 2009; Olmos et al., 2008; Pet-Soede et al., 1999). People need to internalize, through education and training, the benefits of responsible harvesting practices which can be reflected on healthy ecosystems and, in turn, provide multiple services for direct and indirect users over time. It is necessary to offer opportunities for governments and the communities to seek approaches for sharing responsibilities according to their capacities. These capacities can be very limited in some cases, such as Latin America and the Caribbean region, either in terms of knowledge, technical abilities or financial and personnel support (Salas et al., 2007). Training and education for adopting species and habitat-friendly fishing methods and gears therefore seems critical for a sustained effort towards a more responsible use of marine ecosystems without compromising their functional integrity. A number of fishing gears operate in the benthic ecosystem, targeting a number of high-value species with considerable vulnerability because their sedentary and low-mobility nature facilitates harvesting and increases the propensity for resource exhaustion.

A simple approach for mitigating capacity-building fragmentation

As mentioned in the introduction, training and capacity building for fisheries governance and ecosystem conservation is usually fragmented and uncoordinated. It tends to be undertaken by different branches of governments and private organizations with diverse agendas. This fragmentation may also include conflicting sectoral governance objectives. A simple approach for mitigating this fragmentation and sometimes inconsistent capacity building is to migrate from governance through a single-species fisheries approach to a coordinated governance approach involving fisheries and conservation of ecosystem functions (Figure 26.1).

The former has concentrated its training and capacity building in the assessment and management of: (1) single-species/single-fleet fisheries; (2) regional fisheries of trans-boundary stocks; and (3) highly migratory stocks. The latter is concerned with joint capacity building and training of fishery managers, conservation officials, fishers, local scientists and other stakeholders for understanding and acquiring skills for modelling and analysis of: (1) oceanographic patterns of fishery ecosystems; (2) marine ecosystem dynamics; (3) ecological and technological interdependencies in fisheries; (4) space-time stock and fishers dynamics; (5) potential impacts of fishing technologies and methods on ecosystem health and functions; (6) fisher behaviour and dynamics of coastal communities; and (7) cooperation for governance and conservation of shared stocks and their ecosystems.

One critical aspect that needs to be reinforced through training and capacity building is to facilitate dealing with the growing uncertainties and risks associated with including ecosystems dimensions in fishery data collection and analysis, rather than just the uncertainties inherent in single species targeted by one or more fleets in a given space. This could be done by enhancing the capacity of fisher/ecosystem decision-makers through the use of quantitative and qualitative risk analysis tools and other relevant decision theory tools.

The process of developing viable governance schemes and effective tools for management, to the best of our current scientific knowledge and capabilities, requires identification of gaps in local knowledge, technology and capacities. It is important to evaluate carefully and rigorously those gaps according to context, in order to develop capacity-building strategies at different scales. Training and capacity building in institutions at local, regional and international scales are essential for sound governance of marine fisheries and the ecosystems in which they occur.

The analysis and transferring of information generated by bio-ecologic, socio-economic and governance indicators to management actions within an ecosystem framework is confronted with fishery/ecosystem heterogeneous scenarios involving varying levels of: (1) fishery data availability; (2) existing capacity for indicators estimation and analysis; (3) existing capacity to apply available decision-making tools and methods; (4) existing capacity of decision-makers, and other stakeholders to make systematic decisions in an environment of risk and uncertainty.

Once the specific scenario is characterized in terms of data availability and local analytical capacity, it is important to consider alternative and complementary decision-making tools to use and analyse fisheries and ecosystem

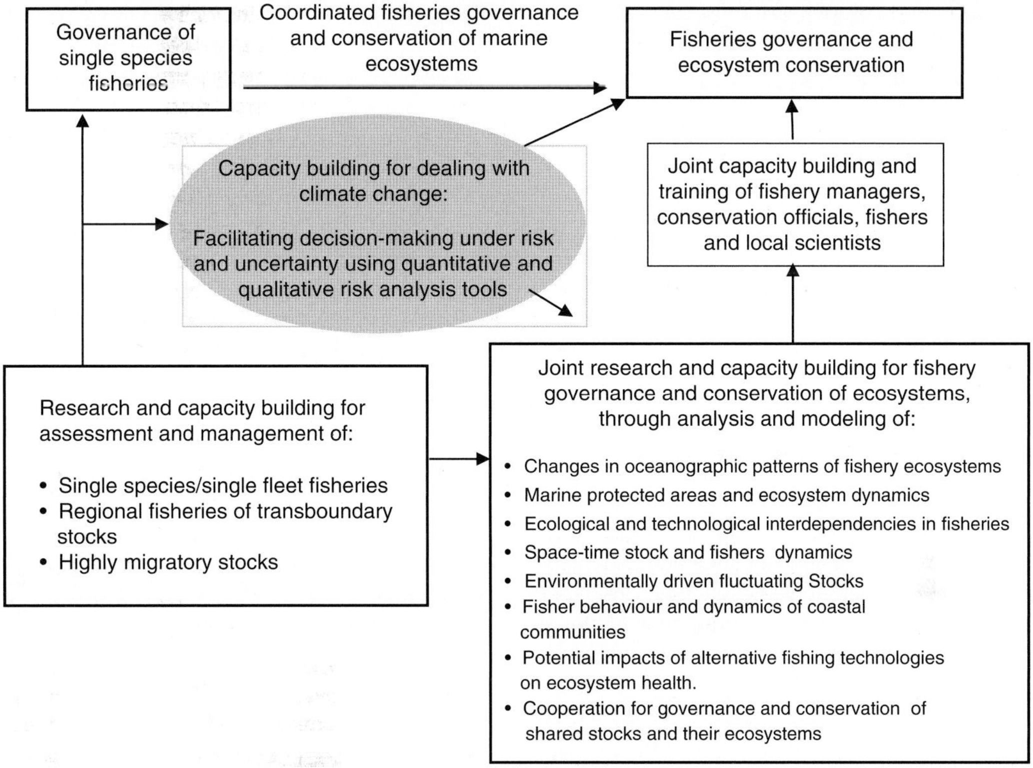

Figure 26.1 Towards capacity building and research for governance and conservation of fisheries and the ecosystems sustaining them.

management options under varying degrees of uncertainty and associated risks.

To build the capacity for effective governance of fisheries within an ecosystem context, it is vital to develop sufficient knowledge of the fishery system to be managed. To foster commitment, the expansion of the knowledge base from single-species fishery assessment to ecosystem assessment where the fishery occurs requires a participatory governance process.

Capacity building in modelling and analysis: Sedentary and low-mobility marine species with MPAs

Increasingly, there is a need for capacity building in terms of fishery analysis and modelling. An example of this arises in meeting the challenge of managing highly vulnerable sedentary and low-mobility resources, with their heterogeneous patchy distributions within fragile ecosystems (e.g. coral reefs). The Schaefer–Gordon and Beverton–Holt models laid the foundations for such analysis. Nevertheless, these models are based on dynamic pool assumptions, which establish that: (1) the resource is homogeneously distributed in space; (2) ages are perfectly mixed; and either (3) fishing effort is applied uniformly over the range of resource distribution; or (4) after fishing effort has been applied, the resource is able to redistribute itself according to (1) and (2).

Not surprisingly, for sedentary and low-mobility resources models based on dynamic pool assumptions are inadequate and can result in serious model error. Local habitats are unequal in quality and holding capacity throughout the stock range. The spatial distribution of these resources is patchy in terms of size, density and age structure. As a result, fishing intensity is also spatially heterogeneous responding to spatial patchiness of resource abundance. The principal consequence of this spatial heterogeneity is that, under dynamic pool assumptions, the productive potential of the stock is overestimated, increasing the risk of over-exploitation and collapse of the fishery with severe effect in the benthic ecosystem (Seijo and Caddy, 2008).

To effectively cope with these complexities, responsible spatial management and conservation of fisheries and marine ecosystems require capacity building for understanding: (1) the heterogeneous spatial distribution of sedentary benthic organisms and the corresponding spatial behaviour of fishing intensity over time; (2) the recognition of the implications of dynamic pool assumptions in overestimating stock abundance of sedentary species and many low-mobility demersal resources; and (3) the ecological interdependencies of benthic communities within an ecosystem framework.

Models and marine protected areas

Training and capacity building for applying and developing adequate spatial models for analysis of alternative management strategies for fisheries targeting sedentary species is another area that should be seriously considered to aid decision-making in preventing exhaustion of highly vulnerable stocks. Marine protected areas have been one approach in attempting to avoid collapse of low-mobility resources and protect the fragile ecosystems in which they live, but before establishing a marine protected area ecosystem managers should be aware of the potential costs and benefits involved. An MPA will generate desired effects only if it is located within the source area of the population to be protected. If located outside the main source area or in the sink area, the performance of the fishery will likely be worse off than without the MPA (Seijo and Caddy, 2008). For the same fishing effort level, fishing intensity (effort per unit area) will increase in the source area as a result of the reduction of the overall fishing area. Good governance and conservation of marine ecosystems require clear understanding of these complexities to avoid compromising the sustainability of sedentary or low-mobility stocks and the ecosystems in which they live.

Management and conservation of metapopulations with source–sink configurations

Metapopulations are composed of linked subpopulations distributed over space with interrelationships. They are fairly common, especially for sedentary or territorial species (Sinclair, 1987). Metapopulation theory has become an important focus for fisheries studies, but until recently few studies examined its practical implications for fisheries management. The effective reproductive component of a population may be much smaller than its total population size, and reproductive age groups and spawning sites may not be equally successful in their reproductive activities throughout the species range (e.g. Smedbol et al., 2003). Source–sink resources follow a specific metapopulation configuration where one or more local groups showing a positive rate of population growth, referred to as sources, also contribute to other local groups (sinks) on sub-optimal habitats which do not contribute significantly to the next generation. That disruption of unrecognized source–sink configurations may lead to stock collapse. This calls for coordinated research and capacity building for fisheries management and conservation of the ecosystem sustaining it.

For metapopulations with a source–sink configuration, in addition to the size of the marine protected area its location is mainly relevant with respect to source areas and the position of fishing ports used by fishermen targeting a benthic resource. Management decisions regarding location, size and linkages between MPAs and other components of ecosystems must be clearly understood by fishers, managers and the scientists providing the best science available for responsible ecosystem and fisheries management. Adopting MPAs as a major tool of management will require a shift in management emphasis from single-species management to spatial management of ecosystems. Oceanographic features, bathymetry, hydrography and the transport of organisms into or out of MPAs can be critical factors in MPA design. Training and capacity building for understanding these complexities when making spatial management decisions seems critical. Recent capacity building and training programs for marine protected areas worldwide are reported in Table 26.2.

Capacity building for marine protected areas

Concerning guidelines for capacity building for marine protected areas, a guide for conducting marine ecological gap assessments was developed by TNC, WWF, USAID and Parks in Peril. The Convention on Biological Diversity published a brochure on criteria for identifying ecologically or biologically significant marine areas in need of protection and the scientific guidance networks of MPAs. IUCN (2004) developed a toolkit for managing marine protected areas in the Western Indian Ocean. The FAO Technical Guidelines on MPAs as a Fisheries Management Tool (the MPA Guidelines) are aimed at providing information and advice on MPAs in the context of fisheries management (FAO, 2011). FAO

Table 26.2 An illustration of some capacity-building and training programs for conserving ecosystems through marine protected areas (MPAs).

Year	Capacity building and training program	Place	Institution
2005	Management capacity training in support of developing a marine protected area (MPA) network for South China Sea Region	Nha Trang, Vietnam	NOAA, IUCN, DANIDA, WWF, CI
2007	Development, implementation and management of marine protected areas.	Honiara, Solomon Islands	Division for Ocean Affairs and the Law of the Sea (DOALOS)
2008	Marine protected areas	Libreville, Gabon Francophone Africa regional workshop	CBD-PoWPA
2008	Marine protected areas	Salinas, Ecuador Latin America regional workshop	CBD-PoWPA
2008	Marine protected areas	Guinée, Bissau General Assembly of the Network of MPAs of West Africa	CBD-PoWPA
2009	Marine expert workshop on scientific guidance on the use of biogeographic classification systems for identification of marine areas in need of protection	Otawa, Canada	Convention on Biological Diversity (CBD)
2009	Marine protected areas	Nadi, Fiji Pacific Island workshop	CBD-PoWPA
2012	Putting into practice the FAO Technical Guidelines on MPAs and Fisheries: MPAs as a potential management tool for sustainable fisheries in South and South East Asia	Bangkok, Thailand	Southeast Asian Fisheries Development Center (SEAFDEC)

indicates that the purpose is not only to address MPAs when used directly for fisheries management, but also to discuss the implementation of MPAs with multiple objectives. The MPA guidelines seek to clarify the potential effects of MPAs on fisheries, the fishery resource and on the ecosystem including biological, physical and socio-economic aspects. More recently a set of case studies for assessing the performance of MPAs considering policy, governance and institutional issues of marine protected areas in Brazil, India, Palau and Senegal has been reported by Sanders et al. (2011). It should be pointed out that training could also be needed for modelling the potential bio-economic effects of location and size of marine protected areas.

Capacity building in relation to MPAs will need to facilitate improvements in addressing possible spatial changes in the location of source areas with climate change and corresponding implication on proper location of MPAs. Some of the questions to be addressed include the following. (1) Are source areas changing over space and time as a result of climate change? (2) What is an adequate location of an MPA for responsible management and conservation of metapopulations? (3) Given a specific location, what is the bio-economic optimum size of the MPA? (4) How are fishers responding to spatial management (source area closures) of metapopulations? (5) What is the rate of compliance with MPA management approaches? (6) Is it economically feasible to have adequate surveillance to avoid or mitigate illegal fishing in established MPAs?

Steps for implementing joint capacity-building and training for spatial management

Some of the steps for establishing joint training and capacity-building programs for MPA management are as follows (NOAA, 2011):

- working with local and regional partners, a need and interest for MPA management capacity building is identified;

- a legal and management framework is in place, basic infrastructure needs are met and the MPA authority is willing to support skill and knowledge-based development for the purpose of moving towards an effectively managed network of MPAs;
- a site-specific management capacity training needs assessment is conducted;
- the structure of the training program and design of curriculum is specifically tailored to regional needs;
- trainers are selected and comprise local, regional and international expertise in MPA management;
- ongoing support is provided until local and regional MPA management capacity is secured;
- milestones and performance indicators are built into the program to ensure MPA sites are moving towards more effective management and conservation goals are met; and
- communication and coordination plans are designed and implemented among the participating MPAs to provide support for scaling up from individual sites to networks of MPAs.

Capacity building to cope with risks and uncertainties (notably climate change)

Training and capacity building are needed to support responsible fisheries management through acknowledging and coping with possible sources of uncertainty, for example in terms of long-term fish stock fluctuations resulting from changes in the carrying capacity of the ecosystems driving the abundance over space and time of target pelagic fish resources. Uncertainties also arise in fishery and ecosystem indicators due to data collection procedures and sampling design, indicator estimation procedures, environmental changes, un-reporting and under-reporting of catches, illegal fishing and other sources of uncertainty such as climate change.

Specifically, coping with the latter impacts of climate change (modifying the size and distribution of marine and freshwater species; Cochrane et al. 2009) calls for training to help those throughout the fisheries. It is important to provide fishery actors with the rapidly evolving knowledge on displacement of species towards the poles and changes in the size and productivity of their habitats. There are also anticipated changes in ecosystem productivity, with decreases at tropical and sub-tropical ocean areas and increases at higher latitudes. Habitat effects of changes in water temperature are also critical, as this can significantly influence fish metabolism, growth rate, productivity, seasonal reproduction and susceptibility to diseases and toxins. Higher temperatures in marine and freshwater ecosystems are likely to affect physiological processes of species, causing positive as well as negative effects in fisheries; this will depend on species sensitivity to changes in temperature and salinity as well as their mobility capacity to migrate to more suitable environments. Changes in species abundance may arise (Cochrane et al. 2009) due to effects on reproduction and recruitment patterns, individual growth and productivity of ecosystems sustaining the fishery, among other factors. There may also be changes in species availability including effects on spatial distribution of species and of fishing intensity, as well as changes in species catch composition over space and time.

Overall, fishery stakeholders need to understand how climate change will increase uncertainties associated with fish harvest and fishery management. Such uncertainties create new challenges to risk assessments which are usually based on knowledge of probabilities of occurrence of past events. Data to determine effects of previous climate changes in the best of cases cover some decades and would not be an appropriate guide for future expectations. To estimate such risks, quantitative methods such as Monte Carlo analysis can be used to calculate crucial probabilities under a range of alternative management strategies in order to mitigate the effects of climate change (Seijo and Caddy, 2000).

Concluding remarks

(1) It seems essential that future governance of fisheries and the ecosystems that sustain them undertake joint training and capacity building of conservation officials, fishery managers, fishers, local scientists and other stakeholders.

(2) The adoption of the ecosystem approach to fisheries management has changed the requirements for technical skills in a number of ways. Firstly it has widened the skills and knowledge required to include ocean and climate sciences, species and habitat ecology and ecosystem modelling. Secondly it has increased the level of information and analysis required through expanded and costly data collection systems. Training and capacity building to account for these requirements of technical skills and adequate ecosystem data collection systems should be a priority.

(3) It seems essential that training for spatial management of fisheries within an ecosystem framework should provide adequate modelling approaches and tools to properly explore the size and location effects of MPAs.
(4) Training and capacity building for conservation of marine ecosystems hosting fisheries targeting sedentary and low-mobility resources should enhance the understanding of modelling and analysis of source–sink dynamics of metapopulations.
(5) Increasing uncertainties involved in the process of shifting from single-species management to an ecosystem approach to fisheries require training in quantitative and qualitative risk analysis to properly apply the precautionary approach to ecosystem management.

References

Bavinck, M., Chuenpagdee, R., Diallo, M., van der Heijden, P., Kooiman, J., Mahon, R. and Williams, S. (2005) *Interactive Fisheries Governance: A Guide to Better Practice*. Eburon Publishers, Delft.

Béné, Ch. (2009) Are fishers poor or vulnerable? Assessing economic vulnerability in small-scale fishing communities. *Journal of Development Studies* **45**(6), 911–933.

Berkes F., Mahon, R., McConney, P., Pollnac, R. and Pomeroy, R. (2002) *Managing Small-scale Fisheries. Alternative Directions and Methods*. International Development Research Centre, Ottawa, ON.

Brander, K. (2010) Reconciling biodiversity conservation and marine capture fisheries. *Current Opinion in Environmental Sustainability* **2**(5–6), 416–421.

Caddy, J.F. and Seijo, J.C. (2011) Destructive fishing practices by bottom gears: A broad review of research and practice. *Ciencia Pesquera* **19**, 5–58.

Charles A. (2001) *Sustainable Fisheries Systems. Fish and Aquatic Resource Series 5*. Blackwell Science, Oxford.

Chuenpagdee, R., Morgan, L.E., Maxwell, S.M., Norse, E.A. and Pauly, D. (2003) Shifting gears: assessing collateral impacts of fishing methods in the US waters. *Frontiers in Ecology and the Environment* **1**(10), 517–524.

Cicin-Sain, B., Knecht, R.W., Vallega, A. and Harakunarak, A. (2000) Education and training in integrated coastal management: lessons from the international arena. *Ocean and Coastal Management* **43**(4–5), 291–330.

Cochrane, K., De Young, C., Soto, D. and Bahri, T. (2009) Climate change implications for fisheries and aquaculture: overview of current scientific knowledge. FAO Fisheries and Aquaculture Technical Paper, No. 530. FAO, Rome.

Collie, J.S., Hall, S.J., Kaiser, M.J. and Poiner, I.R. (2000) A quantitative analysis of fishing impacts on shelf-sea benthos. *Journal of Animal Ecology* **69**(5), 785–798.

FAO (2003) The ecosystems approach to fisheries. FAO Technical Guidelines for Responsible Fisheries, No. 4(2). FAO, Rome.

FAO (2011) Fisheries Management 4: Marine Protected Areas and Fisheries. FAO Technical Guidelines for Responsible Fisheries No. 4(4). FAO, Rome.

Gatewood, J. (1984) Cooperation, competition, and synergy: information-sharing groups among Southeast Alaskan salmon seiners. *American Ethnologist* **11**(2), 350–370.

Guttman J.M. (1996) Rational actors, tit-for-tat, and the evolution of cooperation. *Journal Ecological Behavior Organization* **29**, 27–56.

IUCN (2004) Managing marine protected areas: A Toolkit for the Western Indian Ocean. IUCN Eastern African Regional Programme, Nairobi, Kenya.

Langton, R.W. and Auster, P.J. (1999) Marine fishery and habitat interactions: to what extent are fisheries and habitat interdependent? *Fisheries* **24**(6), 14–21.

NOAA (2009) Fisheries management: building a sustainable future for America's future. National Oceanic and Atmospheric Administration, Washington.

NOAA (2011) MPA Management Capacity Building. National Marine Sanctuary Program. Available at http://www.sanctuaries.noaa.gov/management/pdfs/handout.pdf (accessed March 2014).

Okes, N.C., Petersen, S., McDaid, L. and Basson, J. (2012) Enabling people people to create change: Capacity building for the Ecosystem Approach to Fisheries (EAF) in Southern Africa. *Marine Policy* **36**(1), 286–296.

Olmos, M.E., Salas, S., Beltrán-Morales, L.F., Preceda-Solís, A. and Ortega-Rubio, A. (2008) Riqueza ecológica y pobreza económica en un área natural protegida en Baja California Sur. *Región y Sociedad* **42**, 133–164.

Pauly, D., Christensen, V., Guénette, S., Pitcher, T.J., Sumaila, U.R., Walters, C.J., Watson, R. and Zeller, D. (2002) Towards sustainability in world fisheries. *Nature* **418**, 689–695.

Pet-Soede, C., Cesar, H.S.J. and Pet, J.S. (1999) An economic analysis of blast fishing on Indonesian coral reefs. *Environmental Conservation* **26**, 83–93.

Potter, C. and Brough, R. (2004) Systemic capacity building. *Health Policy and Planning* **19**(5), 336–345.

Salas, S., Chuenpagdee, R., Seijo, J.C. and Charles, A. (2007) Challenges in the assessment and management of small-scale fisheries in Latin America and the Caribbean. *Fisheries Research* **87**, 5–16.

Sanders, J.S., Gréboval, D. and Hjort, A. (compilers) (2011) Marine protected areas: country case studies on policy, governance and institutional issues. FAO Fisheries and Aquaculture Technical Paper No. 556/1. FAO, Rome.

Seijo, J.C. (2007) Considerations for management of metapopulations in small-scale fisheries of the Mesoamerican barrier reef ecosystem. *Fisheries Research* **87**, 86–91.

Seijo, J.C. and Caddy, J.F. (2000) Uncertainty in bio-economic reference points and indicators of marine fisheries. *Marine and Freshwater Research* **51**, 477–483.

Seijo, J.C. and Caddy, J.F. (2008) Port location for inshore fleets affects the sustainability of coastal source-sink resources: implications for spatial management of metapopulations. *Fisheries Research* **91**(2–3), 336–348.

Seijo J.C., Charles, A., Chuenpagdee, R. and Salas, S. (2011) Toward sustainability for coastal fisheries of Latin America and the Caribbean: Effective *Governance and Healthy Ecosystems*. In: *Coastal Fisheries of Latin America and the Caribbean* (eds S. Salas, R. Chuenpagdee, A. Charles and J.C. Seijo), pp. 395–413. FAO Fisheries and Aquaculture Technical Paper No. 544. FAO, Rome.

Sinclair, M. (1987) *Marine Populations*. Washington Sea Grant Program. University of Washington Press, Seattle.

Smedbol, R.S., McPherson, A., Hansen, M.M. and Senchington, E. (2003) Myths and moderation in marine metapopulations. *Fish and Fisheries* **3**, 20–35.

Smith, H.D. (2002) The role of the social sciences in capacity building in ocean and coastal management. *Ocean and Coastal Management* **45**(9–10), 573–582.

Swan, J. (2000) Regional fishery bodies and governance: issues, actions and future directions. FAO Fisheries Circular No. 959. FAO, Rome.

UNDP (2008) Supporting Capacity Development: The UNDP Approach. United Nations Development Programme, New York.

Zhoua, S., Smith, A.D.M., Punt, A.E., Richardson, A.J., Gibbsa, M., Fulton, E.A., Pascoe, S., Bulmanb, C., Baylissa, P. and Sainsbury, K. (2010) Ecosystem-based fisheries management requires a change to the selective fishing philosophy. *PNAS* **107**(21), 9485–9489.

Chapter 27

Fishers' organizations: Their role in decision-making for fisheries and conservation

M. Makino[1,3], A.S. Cabanban[2,3] and S. Jentoft[4]

[1]Fisheries Research Agency, Kanagawa, Japan
[2]PRIMEX, Inc (Philippines) and ANZDEC, Ltd (New Zealand)
[3]IUCN Commission on Ecosystem Management (IUCN/CEM/FEG), Gland, Switzerland
[4]Norwegian College of Fishery Science, Fisheries and Economics, University of Tromsø, Norway

Abstract: Using examples from Norway, Japan and Philippines, this chapter discusses the roles of fishers' organizations (FOs) in relation to the characteristics of the social and ecological systems surrounding their fisheries sectors. Food provision, as one of the ecosystem services from the sea, and job creation are both socially important in all three countries, and FOs are playing a core role in the conservation of these services. FOs can act as representatives of the fisheries sector in fisheries governance and as partners with representatives of other sectors in marine conservation.

Keywords: fishers' organizations; social and ecological factors; sectors; decision-making; marine conservation

Introduction

The principal aims of fishers' organizations (FOs) are to promote fisheries activities and achieve sustainable use of marine living resources, drawing on the provisioning services provided by marine ecosystems; biological diversity conservation or marine ecosystem conservation (referred to simply as marine conservation here) is not the main objective of FOs. Our study aims to demonstrate the roles of FOs in decision-making and describe how FOs have participated in both fisheries governance and marine conservation.

For any given country or region, both fisheries governance and marine conservation should be conducted using measures that match the social, cultural and ecological underpinnings (Jentoft, 2004; Makino, 2011). For this reason, we discuss the social and ecological characteristics which determine the role of FOs in fisheries governance and marine conservation activities. This is accomplished through a comparative approach of three countries from socially and ecologically different areas: Norway, Japan and Philippines.

The following section briefly introduces the profiles of these three countries, including ecological systems, national economy, fisheries sector and food consumption. The roles of FOs in fisheries governance are summarized in the third section, with examples from the three countries. Similarly, 'Roles of FOs in marine conservation' summarizes the roles of FOs in marine conservation activities. Finally, in the last section we discuss the implications of the socio-ecological factors in relation to conservation and the participation and role of FOs in marine conservation.

Governance of Marine Fisheries and Biodiversity Conservation: Interaction and Coevolution, First Edition.
Edited by Serge M. Garcia, Jake Rice and Anthony Charles.

Country profiles of Norway, Japan and Philippines

Ecological systems

Norway is located in the northern part of Europe, surrounded by three Large Marine Ecosystems (LMEs)[1]: Barents Sea, Norwegian Sea and North Sea. Its coast is very large, with 25,148 km of the Norwegian coastline along the mainland and fjords and 58,133 km around its islands (CIA, 2012). The northern part of the coast is within the Arctic Circle and its marine ecosystems are Arctic and sub-Arctic; the biodiversity of the marine area is therefore relatively low (Hillebrand, 2004; Makino and Matsuda, 2011). According to the latest five years of statistics by FAO FISHSTAT PLUS (2006–2010), the top 10 species and 5 families (in terms of volume) constitute 90% of total landings by capture fisheries.

Japan is located in the temperate zone of the north-western Pacific. The Japanese archipelago extends for more than 3000 km with a coastline of 29,751 km along the full range from sub-Arctic to tropical marine ecosystems. It includes five LMEs: East China Sea, Kuroshio Current, Sea of Japan, Oyashio Current and Sea of Okhotsk. Because of the variety in types of ecosystems, its marine biodiversity is high with 31 species and 12 families constituting 90% of the total landings. The land area is mostly rugged and mountainous, and only about 11.6% is arable. All the major cities are in the coastal area (Tokyo, Yokohama, Osaka, Nagoya, etc.); the coasts near the major cities are therefore mostly reclaimed and highly developed.

The Philippines is a tropical and archipelagic country consisting of 7107 islands with a coastline of 36,289 km. It is surrounded by the West Philippines Sea, Sulawesi Sea and the Pacific Ocean. The Sulu Sea and the northern part of (Celebes) Sulawesi Sea are within its territory. The marine biodiversity in these tropical seas is very high; 34 species and 19 families constitute 90% of the total capture landings.[2]

National economy and fisheries sectors

The Norwegian economy is comparatively small (Table 27.1) with the main industries – mining of mineral fuels and oil extraction – constituting 20% of the GDP. The productions of aluminium, silicon, chemical fertilizer, etc. are highly developed industries using the country's rich hydroelectric resources. The total population is also relatively small (about 5 million) and the per capita GDP is one of the highest in the world.

The Japanese economy is the world's third largest in terms of GDP and fourth largest in terms of export value. Although it has a large population (about 130 million, the 10th largest in the world), the per capita GDP is quite high compared to the world average. The main industry is manufacturing, including automobiles, electronics, office machinery, etc.

The Philippines' national economy in terms of GDP is close to the world average, and is growing rapidly with a GDP annual growth rate in 2010 of 7.6%. The total population is large (about 100 million, the world 12th largest) and also growing rapidly, and the per capita GDP is quite low. About 26.5% (2009) of the population is below the poverty line (ADB, 2012). The main industries are agriculture, forestry and fisheries; more than 30% of the nation's total workers are employed in these sectors. The country's economy was dependent on exports in the 1970s but now is no longer heavily reliant.

The capture fisheries landings in Norway, Japan and Philippines are ranked 10th, 4th and 11th in the world, respectively, so all three can be categorized as major fishing countries.

The Norwegian fisheries sector is highly oriented to the international market. It is the second-largest fisheries exporting country in the world, and fisheries products (including aquaculture products) are the second-largest export item for Norway. The total number of capture fishers in Norway has dramatically decreased over the last several decades, and is now the smallest of the three countries. More than half of these are living in Finmark, Troms and Nordland, which are the three most northern counties of Norway (ICFO, 2011). The per-fisher landings and per-vessel landings are very high, and the fisheries sector is highly efficient.

In Japan, more than two hundred thousand fishers are using almost the same number of fishing vessels, largely in small-scale coastal fisheries. There are about 6300 fishing villages along the coastline, half of which are located in remote islands or peninsulas (Fisheries Agency of Japan, 2009). The fisheries sector is highly important as the source of employment in these areas. Japan is one of the world's largest importing countries of fisheries products. The trade deficit of fisheries products was US$ 10.338 million in 2010.

A very large number of fishers are operating in the Philippines, with a large number of vessels, in coastal waters and within the Exclusive Economic Zone. Fishers are found in all regions of the country. The production is 5.3 tons/vessel/year, indicating very small-scale and labour-intensive fisheries. The fisheries production is consumed domestically, with exports of fisheries products by the Philippines below the world average (Table 27.1).

Table 27.1 Major figures relating to the national economy and fisheries sectors in three countries. Landings and values refer to capture fisheries.

	GDP (10^9 US$)[a]	GDP/capita (US$)[a]	Total export value (×10^9 US$)[b]	Fisheries landings (×10^3 tons)[d]	Fisheries values (×10^6 US$)[c]	Export value of fisheries products (×10^6 US$)[b]	N. fishers (× 1000)[c]	N. fishing vessels (× 1000)[c]	Per fisher landing (tons)
Norway	413	84,589	131	2510	1935	8510	13	5	197
Japan	5459	43,141	769	4320	14,080	1285	222	185	19
Philippines	1200	2140	51	2522	3171	338	1614	476	5
World ave.	1075	14,633	75	397	n.a.	383			

[a] United Nations Statistics Division (2012)
[b] United Nations Statistics Division (2011)
[c] International Cooperative Fisheries Organization (2011)
[d] UN FAO FISHSTAT PLUS (2006–2010)

Food consumption

The food intake of Norwegians, Japanese and Filipinos differ considerably. The average person in Norway consumed 3464 kcal of animal protein per day in 2007 (FAO Food Balance Sheet; http://faostat.fao.org/site/368/default.aspx#ancor), about the same as the average in Europe (3406 kcal/day) but higher than the world and Asian averages of 2769 kcal/day and 2668 kcal/day. The Japanese average of 2812 kcal per day is about the same as the world average and slightly higher than the Asian average. The Filipino average food consumption is about 2565 kcal/day, which is slightly lower than the world and Asian averages.

The intake of protein and consumption of fisheries products among the Norwegians, Japanese and Filipinos are also different. The average Norwegian consumes 64.2 grams per day of protein, of which 24.0% is from fisheries products. The protein intake is considerably more than the average European person (57.7 g/day), which is very high compared to the world average (29.8 g/day) and the Asian average (23.4 g/day). The dependence on fisheries products is much higher than the European average (10.9%) and the world average (16.1%). The average Japanese consumes 52.0 g/day of animal proteins, which is more than two times higher than the Asian average but lower than the European average. Japan's reliance on fisheries products is 42.5%, which is very high and about twice the Asian average (22.6%). Filipinos consume 25.3 g/day of animal protein, which is about the Asian average. The main source of protein (41.1%) is from the fisheries products. The self-sufficiency ratios of fisheries consumption in Norway, Japan, Philippines are 154.4%, 50.4% and 98.5%, respectively.[3]

FOs and their roles in fisheries governance

Norway

Norway has traditionally one national-level FO, the Norwegian Fishermen's Association. It represents not only fisher members but also the whole industry in the fisheries governance system, with the central state fisheries ministry as its counterpart. This arrangement dates back to 1926, when the organization was established.[4] This happened largely on the initiative of the government which needed a single corresponding partner to consult in fisheries policy making, an organization which could speak on behalf of the entire industry. For the government this was partly a way of reducing transaction costs; rather than having to relate to many interest groups one by one, it could concentrate on one. For the fishing industry this raised a challenging demand of having to reach agreement and being able to speak with one voice. This created a system of 'centralized consultation' within the Norwegian fisheries sector, which granted the Norwegian Fishermen's Association considerable power both within the industry and in the relationship with the government.

As the government needed this industry representation on its side for the purpose of legitimization and implementation, the organization also obtained considerable leverage in public policy. The government could not easily ignore what the Association demanded when it negotiated on behalf of members and non-members alike, as well as those within the processing and export sectors.

The role of the Norwegian Fishermen's Association should not be underestimated. It has for instance been instrumental in keeping Norway outside the European Union. The fishery issue was always important in determining the outcome of negotiations about Norway's membership. Contrary to their European counterparts, the organization is an observer when the government negotiates quotas with other countries (for example with Russia and the European Union) on shared stocks. The government and the Association may disagree at times, for instance on offshore oil exploration, but over the years they have been able to reach consensus on major fisheries management matters; this has created legitimacy and eased the implementation of fisheries management policies, such as rule enforcement. On some issues, the government has left it to the organization to design the rules (e.g. the so-called 'trawl ladder' which is a key for determining the quota allocation between the offshore trawler and the coastal fleet). The key works so that when the TAC is low, the coastal fleet is granted a higher share than when TAC is high, an arrangement which is meant to secure the livelihood of small-scale fishers in difficult times. When the TAC is high, the catching capacity of small-scale fishers is insufficient to reach the TAC.

It is fair to say that the Norwegian management system is a co-product of this organization and the government. Still, one should not neglect the distributional and other policy conflicts between different gear groups within the industry, which have also affected the organization. In 1987, a considerable section of the small-scale fishing membership broke out of the organization and established 'The Norwegian Coastal Fishermen's Association', and these two organizations differ sharply on various issues such as individual transferable quotas (ITQs) and whether the system should now be extended to included vessels below 11 m in length. The Norwegian Fishermen's Organization still continues to be the most influential interest organization, not least because it also includes the large-scale trawler and purse seine fleets whose economic and political clout in Norwegian fisheries is considerable. The Norwegian Fishermen's Association now seems to have changed its identity and profile from being firmly embedded in civil society with a social responsibility to secure the existence of local communities and a decentralized settlement pattern. In recent years, largely because of the quota rights system and the regional concentration of the fishing industry and down-scaling of the small-scale fishery, the organization has abandoned its traditional role as spokesman for the coast and become an interest organization for quota owners (Mikalsen et al., 2007), thus with a narrower agenda.

In addition to these national-level FOs, fishers formed management committees for the different parcels of the ocean and, largely by themselves, defined the rules for fisheries. One example of such co-management is the area management measures that were established for the Lofoten cod fishery more than a century ago (Jentoft and Kristoffersen, 1989). These area management rules by FOs are primarily established to avoid conflicts between different gear groups however, which has always been an issue in coastal waters where the contenders are many and space is limited. Sometimes these arrangements are also aimed at easing the fishing pressure on spawning grounds or juvenile fish in periods of the year. More than 60 closures, amendments and re-openings are undertaken annually in an effort to protect juvenile fish. Trawling has been banned in areas with coral reefs and more than 150 smaller management areas along the Norwegian coast have been established. An impressive 800,000 km^2 of deep ocean is now closed for regular bottom fishing. As a comparison, the Norwegian mainland covers an area of approximately 324,000 km^2. A recent central government regulation for waters shallower than 1000 m requires every fishing vessel to record and report all contact with vulnerable habitats all the way to shore, and to move two nautical miles away before they can resume their fishing activity.

Japan

The Law of Fisheries Cooperative Associations of 1948 and the Fisheries Law of 1949 govern fisheries management in Japan. Fisheries Organizations are nested systems at various scales, from the local level to the national level (Makino and Matsuda, 2005). The smallest-scale FOs are the local Fisheries Cooperative Associations (FCAs), which are composed of local fishers. They are the core of coastal fisheries governance not only in the planning process but also in rule enforcement, as illustrated in the examples in the following sections.

Sea cucumber

The Kawauchi-machi FCA is located in Aomri Prefecture, the northern part of Japan. Fishing for sea cucumber using a dredge net is the main source of employment for the community in the prefecture and, for FCA members, based on the fishing licenses issued from the prefecture governor. In 2011, the total number of fishers in this FCA was 172. An autonomous panel for sea cucumber fishery governance, established in 1999 in the FCA, is called the 'Council for Promoting Sea Cucumber Resource Utilization'. This is the core FO for sea cucumber fisheries governance and is responsible for implementing a wide range of measures. For example, the Resource Assessment and Harvest Planning Division of the Council conducts resource assessments of sea cucumber populations every year. The results of the assessments are the basis for the setting of the autonomous limit on total catch volume. Official researchers provide support by performing the analytical calculations and formulating advice. As an operational rule, the Division sets the stricter minimum size limits, no-take zones, no-take periods and other measures in addition to the official regulations set by the prefecture government.

The Processing Division of the Council employs local people and operates a drying process at the FCA-direct-sale factory. This division also communicates with the Resource Assessment and Harvest Planning Division regarding fisheries operations in order to optimize the working rate of the processing factory, e.g. the daily catch limit is set to 50–60 kg per vessel per day. The Marketing Division sells the dried cucumber to domestic and Hong Kong markets and also conducts surveys in the targeted market, giving instructions to the drying processes and fishing operations based on the results of the surveys. For example, based on the consumer demand in China, the maximum catch size was set to 300 g per individual sea cucumber. As the result of these well-organized governance measures by the Council, this FCA was awarded the Emperor's Prize in 2004 which has added value to the product in the Hong Kong market. For more details of this fishery, see Makino (2011).

Sand eel

Another example is the sand eel (*Ammodytes personatus*) pelagic trawling in Ise Bay, located in the central part of the Japanese Pacific coast and bordered by Aichi and Mie prefectures. Historically, trawl fishers in Aichi Prefecture mainly target juveniles while the Mie fishers target adult sand eels. About 700 fishing vessels operate through licenses issued from Aichi and Mie prefectures, and belong to 1 of the 12 FCAs established in the surrounding bay. The multiple FCAs are a result of the migratory nature of the sand eel and the fact that a single FCA is not able to effectively conserve the resources.

Due to the introduction of new technologies (e.g. echo-sounders) and the increased demand for sand eel as feed in aquaculture operations, the resource was intensively exploited in the late 1960s and the 1970s and finally collapsed in the early 1980s. After the collapse, the 12 FCAs established the General Assembly of Fisheries Unions for Sand Eel Fishers in Ise Bay. This is a cross-prefecture organization for the governance of the sand eel fishery in the bay, formed of autonomously organized unions of sand eel fishers within each prefecture. Official researchers in the two prefectures scientifically support the Assembly's activities. For example, based on the data gathered by the member fishers, the stock assessment of the sand eel resource is conducted and provided to the Assembly. The representatives of Fisheries Unions then gather together to decide the opening date of the fishing season. In order to protect the spawning stocks, the Assembly also set autonomous no-take zones in summer at the mouth of the bay. The area of the no-take zones is adaptively set according to the scientific results of the annual stock assessment. For more details, see Tomiyama *et al.* (2008).

Philippines

The Philippines is an archipelago with 7000 islands and over 2000 species of marine fishes, with fishing being multi-gear, catching multi-species and landing in numerous sites. As a result, full management of fisheries, especially coastal fisheries, would require great manpower and budgetary allocations; in fact, the annual budget of the Bureau of Fisheries and Aquatic Resources (BFAR), the governmental agency mandated to manage fisheries, is very small. Community-based management in municipal waters and co-management between BFAR and the FOs are management options that circumvent these problems in fisheries (Evans et al., 2011). The Act providing for the Development, Management and Conservation of the Fisheries and Aquatic Resources of 1998 (The Republic Act No. 8550) is the legal basis for FOs to participate in decisions on management regulations (Batongbacal, 2001). Section 2 of the Act states that the '(d) right of the fisherfolk, especially of the local communities with priority to municipal fisherfolk, in the preferential use of the municipal waters, … on the basis of resource and ecological conditions, and shall be consistent with … commitments under international treaties and agreements' and the '(g) grant to the private sector to utilize the fishery resources under the basic concept that (they) shall not only be a privileged beneficiary … but also an active participant and partner of the Government in the sustainable development, management, conservation and protection of the fishery and aquatic resources of the country'. Based on this legal basis, FOs have participated in fisheries governance, as described in the examples below.

Sardines

Production of sardines in Moro Gulf and Visayan Sea has declined considerably in recent years, presumably because of overfishing. In response, the Southern Philippines Fishing Association (SoPHIL), based in Zamboanga City, Mindanao advocated the introduction of a closed season for sardine fisheries to the national government (Bureau of Fisheries and Aquatic Resources or BFAR). The advocacy, with financial allocation for research into the reproductive biology of sardines to determine the timing of the closed-season, took seven years. The Government of the Philippines also funded research to increase the understanding of the biology of sardines and the ecology of Celebes Sea and Sulu Sea, which are the habitats of pelagic sardines. Based on these scientific results and consultations with the fishing sector, the government passed a Joint Administrative Order No. 1 in 2011 establishing a closed season for the conservation of sardines in East Sulu Sea, Basilan Strait and Sibuguey Bay. After the introduction of the closed season, sardine FOs recognized the effectiveness of this measure. The scientific findings also noted that the sardines harvested in southern Philippines, Visayan Sea and Sulu Sea were of the same stock (Wilette and Santos, 2013). The Alliance of the Philippines Fishing Federation, Inc. and the Northern Cebu Commercial Fishing Boat Operators Association therefore supported spatial expansion of the closed season in 2012 to cover the migration of the sardines.

Tuna

Another example of FOs participating in fisheries governance in the Philippines is the large federation of the tuna fishing industry. The Philippines' tuna fishery is the second largest in the world in terms of landing volume. The tuna industry is located around General Santos City, Mindanao where it provides jobs for more than 75,000 people. In 1999, stakeholders of the tuna industry, including tuna FOs, boat owners, landing associations, canners, exporters, etc., established the Federation of Fishing and Allied Industries, Incorporated (SFFAII; http://www.sffaii.org). The Federation sits on advisory councils including the National Tuna Council, National Agricultural and Fishing Commission and National Fishing and Aquatic Resources Management Council. It participates in congresses and has taken part in the passage of relevant fisheries regulations. For example, participation in the National Tuna Congress resulted in resolutions for consideration by BFAR. In addition, it participates in the international governance of the tuna fishery as a member of the Delegation of the Philippines to the West Central Pacific Fisheries Commission.

Roles of FOs in marine conservation

Norway

With the advent of salmon aquaculture in the early 1970s and the enormous growth of this industry in Norway, competition for coastal space has increased. The need for marine spatial management has therefore become urgent, demanding new legislation and active involvement of municipal government and other stakeholders including FOs (Buanes and Jentoft, 2005). The new concept of marine protected areas (MPAs) was introduced in a situation where Norway already has extensive experience of similar management arrangements. In recent years, the government has introduced ambitious plans for the establishment of additional MPAs where the need for special protection of marine bio-diversity has been identified by marine scientists. For instance, as part of the management measures for lobster, four pilot MPAs were established in 2006 prohibiting all gears excepts hook and hand-line (Pettersen et al., 2009).

The definition of an MPA by The International Union for the Conservation of Nature (IUCN) as an area '…which has been reserved by legislation or other effective means, including custom, with the effect that its marine and/or coastal biodiversity enjoys a higher protection than its surroundings' is not seen by government to be particularly novel in the Norwegian context. Nor has it triggered much opposition among Norwegian fishers and their organizations, as they find the concept and practice of MPAs as something they are quite familiar with and can make use of. With salmon aquaculture, fishing tourism and the expansion of offshore oil occurring in traditional fishing areas, they feel the need for protection. They are no longer on the top of the list of stakeholders that municipal governments would care for (Buanes *et al.*, 2005). Rather, municipalities see the new industries as a means to diversify their economic base as the fishing industry is down-scaling and has become a less reliable employer. The potential for further expansion in these new industries is also huge.

With such prospects, fishers have an interest in defending the coastal zone and in controlling their usage. MPAs are therefore seen as a useful tool that may serve them well, but only as long as it does not interfere with their fishing activity too much and they are allowed to be engaged in defining MPA location and rules. They would however be inclined to accept restrictions on bottom trawling which affect local ecosystems, a view largely shared by trawl fishers.

The government recognizes the need to bring fishers on their side. Active cooperation and sharing of responsibility with the FOs is therefore seen as essential for the success of these measures.[5] The Norwegian Fishermen's Association has mixed feelings about local management so far as the municipal governments are setting the stage and other stakeholders become involved. This can for instance be seen in the case of a local initiative by the Tvedestrand municipality, located at the southern end of Norway, about which the Norwegian Fishermen's Association has expressed heavy criticism. Not only does the Association feel excluded but it also sees the initiative, if the government should go along with it, as representing a major policy shift in the Norwegian fisheries management system. The Association insists on the principle that since fisheries resources are a national common property, central government must be at the helm. The organization explains that it is not against marine conservation, but it has problems accepting a decentralized management system (which may in effect end up reducing its hegemony).

How to balance the influence of fishermen in relation to other stakeholder groups is a major governance challenge. If the Tvedestrand model is adopted in other coastal areas, more opposition from fishermen can be expected which will hinder the implementation of MPAs in Norway regardless of what the government may want. Still, both the Norwegian Fishermen's Association and the Norwegian Coastal Fishermen's Association will have problems arguing against MPAs. In fact, given the pressure from the aquaculture industry and the offshore oil industry (which is now eager to drill outside the Lofoten islands, which are Norway's most important spawning grounds of the Arctic cod), their best chance is to make sure they have an influence on MPA organization, location and design. It is therefore essential to have a constructive relationship with government at all levels, including municipalities. The latter are inclined to talk as much to other stakeholders as to fishermen, and central government cannot easily ignore what regional and municipal government want even if the central government is sympathetic to the fishermen's arguments. Regional and local municipalities in the Lofoten area now enthusiastically welcome the oil industry despite vociferous protests from the fishermen's organization, while the central government is waiting to take a stand.

Japan

The Shiretoko peninsula is located in the northern part of Japan. The peninsula and its surrounding sea areas were added to the UNESCO World Natural Heritage List in 2005. Since then, various new measures have been implemented to conserve its exceptional ecosystems. This rich ecosystem has also been a home to humans for thousands of years; as of 2012, a total of 18,474 people in 7768 households are living in the Shiretoko Peninsula, where the major industries are fisheries, agriculture and tourism. The marine governance approach taken here is not to eliminate local fishers from the area, but to place their activities at the core of a conservation scheme (Makino et al., 2009).

There are four main components in this conservation scheme. The Shiretoko World Natural Heritage Site Regional Liaison Committee was established with officers from a wide range of ministries and departments in central and local governments. Fisheries Cooperative Associations (FCAs), the tourism sector, the Scientific Council and Environmental NGOs are members of this liaison committee. The Committee serves as the core arena for policy coordination among administrative bodies. The Shiretoko World Natural Heritage Site Scientific Council provides scientific advice on governance measures and on research and monitoring activities. It is composed of natural scientists and social scientists, and is observed by representatives of ministries and departments in central and local governments, FCAs and Environmental NGOs. The Shiretoko National Park Committee for the Review of Proper Use carries out research and discussions on proper-use rules for tourists. Through these organizations and their interrelationships, stakeholder participation is ensured, information and opinions are shared and consensus between the wide-ranging interests of multiple users of the ecosystem services is built, naturally increasing the legitimacy of the management plans and rules.

Local FCAs have been collecting and compiling catch data for more than 60 years, and these data cover many of the indicator species and other major marine species in the food web. For some species, such as walleye pollock, a great deal of detailed biological information such as size, time and place of catch and maturity has been accumulated on an autonomous basis by FCAs. This information is an important baseline for monitoring changes in the functions and structure of the Shiretoko marine ecosystem. The data provided by FCAs are officially utilized in the governance system. The monitoring activities for non-commercial species, as well as for basic environmental indicators such as weather, water quality, sea ice and plankton, are conducted by governmental agencies. This is therefore a system of role-sharing between government and FOs for sustainable and economically efficient ecosystem monitoring.

Another illustration of how FOs are involved in marine conservation is found in the coastal areas of Yokohama city, located on the western side of Tokyo Bay. Since the 17th century, this area has been famous for its production of high-quality fish for sushi. According to fisheries ground maps in the late 19th century, the majority of the coastal areas around Yokohama were tidal lands and shallow-bottom areas covered by seagrasses. Japan experienced very rapid economic growth after the Second World War and, since the 1970s, seagrass beds (whose existence is crucial for the eggs and juvenile stages of fish and shellfish) have almost entirely disappeared around urban areas such as Yokohama. Today, nearly all of the 140 km of Yokohama coastline has been developed for industrial or residential use; only about 500 m of natural coastline are left.

In 1981, a volunteer group of scuba divers organized a sea bottom clean-up in Yokohama City. This group later became the main organizer of seagrass-replanting activities in the area after a researcher at the prefectural fisheries research station found a local seagrass bed at the southernmost part of the Yokohama coast. The group began to plant the seeds of local seagrass in neighbouring areas. Today, seagrass-recovery activities are enthusiastically conducted by a wide range of participants, including environmental NGOs, local FCAs, local schools, private companies, central/local governments and local research institutions (Makino, 2011). The areas where seagrasses have been re-established are now protected as no-take zones by the local FOs, with part of these activities financially supported by the local and central governments. Their activities have successfully expanded the seagrass-covered areas. As evidence of the success of their activities, spawning by oval squid was observed in 2004 for the first time in 30 years.

Philippines

FOs in the Philippines have been deputized by the municipalities, and enforce some of the marine conservation measures. For example, in Dauin Municipality, Negros Oriental, Negros Island and Central Visayas, local fishers formed the Dauin United Fishermen's Association which protects a small coral reef area (less than 10 ha) as the Dauin Fish Sanctuary, and charges entrance and diving fees. Of the funds collected, 60% are allocated for wages of the fishermen and operational costs, with the remaining 40% constituting a payment for ecosystem services to the Municipality. The latter amount is more than 50% of the total revenue from taxes and service fees. The partnership between the fishermens' association and the municipality, within the context of the Local Government Code, has reduced fishing effort, generated livelihoods and revenue and protected a coral reef area.

Another important institution for marine conservation in the Philippines is the Fisheries and Aquatic Resource Management Councils (FARMC), which is legally based on the Philippines Fisheries Code of 1998. In the Philippines, the management of municipal waters is delegated to the municipal and city governments (Local Government Code of 1991; RA 8550). The municipal governments can pass ordinances on the use of municipal waters provided there is '(c) prior consultation, through public hearing, with the Municipal/City FARMC (M/C-FARMC) among other provisions'. Also, no regulation can be implemented by the national government without the prior consultation of the National FARMC (N-FARMC) and relevant M/C-FARMC. FARMCs are composed of the representatives from municipal (coastal) fishers and commercial (offshore) fishers, youth and women sectors, the accredited non-governmental organizations, the private sectors and the Department of Agriculture. FARMC can therefore be understood as an official body representing a wider range of interests relating to fisheries resource management and marine conservation measures (see also Baticados, 2004).

An example of the FARMC's role in marine conservation can be found in Bolinao, Pangasinan which is known to produce rabbit-fishes (Family Siganidae, living in seagrass meadows), coral reef fishes and cultured milkfish (*Chanos chanos*). The local stakeholders are concerned with overfishing, pollution from mariculture, bleaching (as an impact of climate change), crown-of-thorns starfish explosion and coral diseases. Destruction of seagrass by bottom-trawls and water pollution by increased organic compounds from milkfish farms are relatively new threats (Geronimo et al., 2010). The Global Environment Facility (GEF) provided a grant to demonstrate community-based management of seagrass habitats and resources, building on the achievements in public awareness and in the establishment of protected areas, and monitoring by various groups. A 60 ha seagrass protected area was established in 2007, managed by the Municipality and Bolinao FARMC, formed in addition to the eight MPAs on coral reefs in Bolinao. Monitoring is conducted by the community of the ecology, fisheries and socio-economic benefits from these protected areas, but there also needs to be monitoring conducted by non-government organizations, the local government or scientists to ensure sound and verifiable findings.

Another example of fishers involved in marine conservation arises in southeast Cebu Island in Central Visayas. This area has seven municipalities along its coastline, which in 2007 had a combined population of about 210,000 (of which about 50% lived along the coast). More than 5500 were full-time small-scale fishers, with the coastal capture fisheries contributing 6.5% of the municipal fisheries of the Central Visayas Region in 2003. The current issues and threats faced by the municipalities and fisherfolks are illegal and/or unregulated fishing, intrusion of commercial fishing operations within municipal waters, degradation of mangroves, seagrass, coral reefs, sand extraction, uncontrolled foreshore development and coastal and marine pollution. These threats were addressed by establishing protected areas between 1974 to 2003 covering mangroves, seagrass, macroalgal beds and coral reefs (Diaz et al., 2010), with MPAs ranging in size from 5 to 40 hectares. The results were assessed as ranging from 'good' (MPA is enforced) to 'very good' (MPA is sustained). Thirteen of the marine sanctuaries (covering 139 ha) have core zones that do not allow extraction and any human activity, while eight marine sanctuaries (covering 131 ha) have core zones that do not allow extraction but allow diving for a fee. The seven municipalities, with jurisdictions over these protected areas, formed the Southeast Cebu Coastal Resource Management Council (SCCRMC) to provide protection to a larger area and to regulate transboundary impacts of activities. The SCCRMC works in partnership with FARMCs, FOs, Environmental NGOs, the Provincial Government of Cebu and many other stakeholders to implement coastal resources management.

Discussion

What is conservation and who defines it for whom?

All three countries have long histories of fishing, large fisheries production and people who are considerably dependent on seafood as the source of animal protein. Seafood provision is therefore a marine ecosystem service that is socially important. All three countries have established FOs with a legal basis to conserve and sustain the provision of this ecosystem service, and have also established some types of co-governance regime (community-based or centralized consultation). These reflect various socio-ecological conditions such as long coastlines (Norway), many small islands (Philippines) or large number of small-scale fishers (Japan and Philippines), and can provide a way to reduce transaction costs for fisheries governance. Under these regimes, FOs play important roles in developing operational rules for fisheries management such as coordinated different gear types and spatial uses and gear/season limits, as described in 'FOs and their roles in fisheries governance'.

FOs have advocated and/or introduced various measures to conserve marine living resources and ecosystem structures to which the resources are closely related (seagrasses, water quality, mangroves, coral reefs, etc.) in order to achieve sustainable fisheries. Some of these activities are characterized and motivated by the national economic context in each country. For example, the Philippines' national economy is rapidly growing and coastal areas are rapidly developed and polluted, so FOs represent the interests of coastal marine conservation in order to protect local resources.

The main targets of these FO activities are edible and economically valuable species and, to some extent, the ecosystem that supports the exploited species. In other words, the conservation objective of FOs is limited to the seafood provisioning service rather than other wide-ranging services. Furthermore, marine living resources (and therefore fisheries income based on those resources) often show natural fluctuations and the fishers' intrinsic discount rate for the future tends to be high. This means the timeframe for FO activities can be considerably shorter than that for the ecosystem. These two points reflect the fundamental limitations of FOs in marine conservation.

What, then, is the objective in conserving the ecosystem services of coastal and marine ecosystems for these three countries? In other words, what is a 'good ecosystem' or 'good portfolio of ecosystem services' to be conserved? The objectives of management of land and marine ecosystems and resources are 'a matter of societal choices' (Principle 1 of the CBD Ecosystem Approach). The vision of a 'good ecosystem' in these three countries is the coexistence of a productive ecosystem and the human activities (including fishing and seafood consumption) within it, rather than a wilderness from which people are excluded. The people of these countries also understand that various aspects and objectives of marine conservation, and the relative importance of each ecosystem service, should not and cannot be determined by the fisheries sector alone.

How can FOs participate in and play roles in marine conservation?

FOs can be expected to take on appropriate roles in marine conservation in the three countries, given that they are legally recognized and are considered partners in managing fisheries, as provisioning services of ecosystems. However, the specific roles of FOs in conservation will be determined by the social and ecological factors surrounding the fisheries sector (Table 27.2).

All three cases in this chapter show that, when the fisheries sector has relatively strong power among marine ecosystem

Table 27.2 Implications of social and ecological factors on marine conservation in three countries (Japan, Philippines and Norway).

	Social and ecological factors	Implications on marine conservation
Affinities	Long history of human residence and large fisheries production with high dependence on seafood	Humans are part of the 'good ecosystem' and not to be eliminated from the original 'wilderness'. The seafood provisioning service of ecosystems is highly appreciated by society.
	Legally established FOs	FOs can play the core role in fisheries governance (co-management). FOs can play important roles in the conservation of seafood provisioning services.
	Long coastline with many fishing communities	FOs and local fishing communities can play the principal role in ecosystem monitoring.
Differences	Ecosystems	Area-based management (MPAs) fits well in tropical ecosystems and is therefore is one of the main management tools in municipal waters in Philippines.
		Output control (quota management) fits well to Arctic and sub-Arctic ecosystems, and is therefore one of the main management tools in Norway which dramatically changed the agenda and structure of FOs.
	National economy (development, per capita GDP)	The economy is rapidly growing in the Philippines. Coastal development is increasing and poor enforcement of laws results in marine pollution. Coastal FOs can represent the interests of users of coastal ecosystem services.
		The economy is highly developed in Japan and the coastline around the major cities was reclaimed. FOs actively participate in ecosystem restoration activities.
		Per capita income (GDP) is very high in Norway. Small-scale fishers have been reduced in number in order to increase economic efficiency.
	Relative power of the fisheries sector among marine users	When strong, FOs can play the principal role in the seafood-related part of marine conservation (Japan and Philippines).
		When weak, FOs represent fisheries interests. Conservation measures will be supported by FOs when they protect fisheries interests (Norway).

service users or when government is willing to protect local fisheries, then construction of a new conservation scheme which includes FOs as well as other stakeholders is an option. In this scheme, FOs can play a role in a consultative body or management body (e.g. Bolinao FARMC or SCCRMC in Philippines). Examples also show that FOs can play the principal part in cost-effective ecosystem monitoring, which is an indispensable part of adaptive capacity. In Japan's Shiretoko World Natural Heritage, local FOs have been collecting and compiling catch data of wide-ranging species for many years, and this became the baseline for ecosystem monitoring. Similarly in the Philippines, local FOs and communities are playing the core role in monitoring. From this viewpoint, keeping the scattered settlement of fishing communities along national coastlines is an important option for effective ecosystem monitoring.[6]

When the fisheries sector is not strong compared to other sectors, or when there is no political will to protect local fisheries, then FOs can represent the interest of the fisheries sector in negotiating against or coordinating with other powerful marine sectors. FOs can use conservation measures in order to avoid the degradation of seafood provisioning services by other types of uses (e.g. supporting MPAs to exclude oil drilling and sea farming in Norway). The general rule in such circumstance is that all sectors should achieve accountability about the sustainable use of the ecosystem services.

Challenges for the future

In the mid- to long-term, balancing the interests of the fisheries sector relative to other stakeholders is a big challenge. The principles of the CBD may be referenced in order to provide guidance in coordinating various sectoral interests, and in organizing and managing conservation schemes involving multiple stakeholders. In this process, the desired roles of FOs in marine conservation and the interrelationship with other sectors should be discussed socially and scientifically. One of the most important points in this process is that the discussion should not only be limited to marine ecosystem or biodiversity conservation but should be based on the broader framework of an integrated marine policy, covering all types of marine ecosystem service uses such as food provision, supporting services, regulating services and cultural services.

Another important point in the discussion on the guiding principles is the nature of fisheries rights. Experiences in these three countries show that the role and contribution of FOs to marine conservation are likely to be affected by the nature of rights (individual/private, collective/group/community or state property, etc.) as well as area-based versus catch-based.

In Norway for example, the catch quota system is one of several causes of changes in geographical distribution of fishers and quotas (from north to south) as well as of decreases in the total number of fishers, which subsequently changes the agenda of FOs. In Norway the level of income is very high, the Arctic ecosystem fits well with the quota management scheme (i.e. low biodiversity with large biomass) and there exist alternative sources of local employment (oil development and sea farming). Under these conditions, this policy direction would be the most realistic option to keep the fisheries sector income equal to that of other sectors. On the other hand, Japan is also a high-income country but sea cucumber fishing rights in Mutsu Bay are area-based rights, and the fisheries sector is the major source of employment in this remote community. The feasible approach in this case was to make sea cucumber more profitable by the FOs' processing and marketing activities, thereby also creating local jobs, and generating government support of FO activities via scientific information and social recognition (the Emperor's Prize). Finally, in the Philippines an FO has the legal authority (the right) to collect entrance fees and diving fees from tourists, which helps to reduce fishing pressure while keeping local employment.

As organizations, FOs are likely to be more empowered, internally as well as externally, by collective and area-based rights. These rights give the FOs greater clout against individuals who would prefer to exit, should they feel that the organization does not serve their interest, and against a government whose authority and ability to intervene will then be reduced. Such rights would therefore be expected to lead to a more proactive role in fisheries governance and marine conservation.

Finally, it should be noted that discussions in this chapter are based on the reality of three countries where the fisheries sector is large, the coastline is long and fish is an important component of people's diet. In other countries with a small fisheries sector, little coastline or low fish consumption, the discussion in this chapter cannot be applied. This is because the image of the 'good ecosystem' or 'good portfolio of ecosystem services' is considerably different. Furthermore, there can be great disparities among areas and sectors even within one country, so the scientific clarification of such differences is very important in order to ensure constructive discussion on fisheries governance and marine conservation at both the international and the domestic levels.

Notes

1. LMEs are areas of the ocean characterized by distinct bathymetry, hydrography, productivity and trophic interactions. For more details, see NOAA's LME webpage (http://www.lme.noaa.gov).
2. The number of species is underestimated, due to the different statistical precision of the Philippine's fisheries statistics system compared with the statistics prepared by FAO.
3. In this chapter, the self-sufficiency ratio of fisheries consumption is defined as the domestic fish production volume divided by the domestic supply volume.
4. http://www.regjeringen.no/en/archive/Stoltenbergs-1st-Government/fid/Veiledninger-og-brosjyrer/2000/Facts-about-the-Norwegian-Fisheries-Industry---2000/10.html?id=420994.
5. http://www.fiskeridir.no/fiske-og-fangst/aktuelt/2011/0911/vernar-800-000-kvadratkilometer-havbotn.
6. The local and traditional knowledge of fishers are utilized as the basis to draft the adaptive strategy for climate changes. For example, Makino and Sakurai (2012) in the Shiretoko World Natural Heritage, and Redmap (http://www.redmap.org.au/) in Australia.

References

Asian Development Bank (2012) Philippines Fact sheet. Asian Development Bank, Mandaluyong City.

Baticados, D. (2004) Fishing cooperatives' participation in managing nearshore resources: the case in Capiz, Central Philippines. *Fisheries Research* **67**, 81–91.

Batongbacal, J.L. (2001) The evolution of Philippine Fisheries Legislation. Manila. *Philippine Law Journal* **76**(4), 497–524.

Buanes, A. and Jentoft, S. (2005) Challenges and myths in Norwegian coastal zone management. *Coastal Management* **33**, 153–167.

Buanes, A., Jentoft, S., Karlsen, G. R., Maurstad, A. and Søreng, S. (2005) Stakeholder participation in Norwegian coastal zone management. *Ocean & Coastal Management* **48**, 658–669.

Central Intelligence Agency (2012) *The CIA World Factbook 2011*. Skyhorse Publishing, New York.

Diaz, R.D., Sabonsolin, A.Z. and Apistar, D.R. (2010) Southeast Cebu. In: *Coral Reef Information Network of the Philippines (PhilReefs). State of the Coasts: Promoting the State of the Coast Reporting*, pp. 76–80. Coral Reef Information Network of the Philippines (PhilReefs), MPA Support Network, Marine Environment and Resources Foundation, Inc. and the Marine Science Institute, University of the Philippines, Diliman, Quezon City.

Evans, L., Cherrett, N. and Pemsl, D. (2011) Assessing the impact of fisheries co-management interventions in developing countries: A meta-analysis. *Journal of Environmental Management* **92**(8), 1938–19379.

Fisheries Agency of Japan (2009) White paper on fisheries. Nourin-Toukei-Kyokai, Tokyo (in Japanese).

Geronimo, R.C., Vergara, M.W.B., Alino, P.M. and Echavez, A.S. (2010) Bolinao, Pangsinan. In: *Coral Reef Information Network of the Philippines (PhilReefs). State of the Coasts: Promoting the State of the Coast Reporting*, pp. 21–29. Coral Reef Information Network of the Philippines (PhilReefs), MPA Support Network, Marine Environment and Resources Foundation, Inc. and the Marine Science Institute, University of the Philippines, Diliman, Quezon City.

Hillebrand, H. (2004) Strength, slope and variability of marine latitudinal gradients. *Marine Ecology Progress Series* **237**, 251–267.

International Cooperative Fisheries Organization (2011) *2011 World Fisheries*. International Cooperative Fisheries Organization, Seoul.

Jentoft, S. (2004) Institutions in fisheries: what they are, what they do, and how they change. *Marine Policy* **28**, 137–149.

Jentoft, S. and Kristoffersen, T.I. (1989) Fishermen's co-management: The case of the Lofoten Fishery. *Human Organization* **48**(4), 355–365.

Makino, M. (2011) *Fisheries Management in Japan: Its Institutional Features and Case Studies*. Springer, Dordrecht.

Makino, M. and Matsuda, H. (2005) Co-management in Japanese coastal fishery: its institutional features and transaction cost. *Marine Policy* **29**, 441–450.

Makino, M. and Matsuda, H. (2011) Ecosystem-based management in the Asia-Pacific region. In: *World Fisheries: A Social-Ecological Analysis* (eds R.E. Ommer, R.I. Perry, K. Cochrane and P. Cury), pp. 322–333. Wiley-Blackwell, Oxford.

Makino, M. and Sakurai, Y. (2012) Adaptation to climate change effects on fisheries in the Shiretoko World Natural Heritage area, Japan. *ICES Journal of Marine Science* **69**, 1134–1140.

Makino, M., Matsuda, H. and Sakurai, Y. (2009) Expanding fisheries co-management to ecosystem-based management: a case in the Shiretoko world natural heritage. *Marine Policy* **33**, 207–214.

Mikalsen, K.H., Hernes, H.K. and Jentoft, S. (2007) Leaning on user-groups: The role of civil society in fisheries governance. *Marine Policy* **31**, 210–209.

Pettersen, A.R., Moland, E., Moland Olsen, E. and Knutsen, J.A. (2009) Lobster reserves in coastal Skagerk. An integrated analysis of the implementation process. In: *Integrated Coastal Zone Management* (eds E. Moksness, E. Dahl, and J. Støttrup), pp. 178–188. Wiley-Blackwell, Oxford.

Tomiyama, M., Komatsu, T. and Makino, M. (2008) Sandeel fisheries governance in Ise Bay, Japan. In: *Case Studies in Fisheries Self-governance*, pp. 201–220. FAO Fisheries Technical Paper No. 504. FOA, Rome.

United Nations Statistics Division (2011) International Trade Statistics Yearbook 2010. United Nations Statistics Division, New York.

United Nations Statistics Division (2012) National Accounts Statistics: Main Aggregates and Detailed Tables 2010. United Nations Statistics Division, New York.

Wilette, D.A. and Santos, M.D. (2013) Correcting widespread misidentifications of the highly abundant and commercially important sardine species *Sardinella lemuru* in the Philippines. *Journal of Applied Ichthyology* **29**(4), 881–885.

Chapter 28

The role of courts in fisheries management and marine biodiversity protection: US and EU systems

P. Shelley[1] and T. van Rijn[2]

[1]Conservation Law Foundation, Boston, MA, USA
[2]European Commission, Brussels, Belgium

Abstract: This chapter is a snapshot view of some of the influences that courts in the US and the EU have had in fisheries management with an additional focus, where possible, on the judicial function integrating broader marine resource conservation issues into fisheries management. The impact of the courts is very much determined by the respective legal structures of the US and EU. In the US, where the regulatory environment is more object-oriented and the courts have direct authority over the implementing entities, the courts have played a bigger role integrating marine conservation considerations into fisheries management. In the EU the arbitrage between fisheries management and the conservation of marine resources is made by the legislator and the courts can only influence the achievement of those objectives after rules are adopted.

Keywords: courts; integration; environmental review; regulatory implementation; jurisdiction; competence

Introduction

Courts have played an important role in fisheries management in the United States and the European Union, particularly in their application and interpretation of broader legal and policy objectives to specific real-world fact patterns. In the US in particular, courts have served a critical function by integrating marine biodiversity considerations into the fishery management decision-making process. Nonetheless, the formal literature studying the judicial role remains limited and complicated by the diversity, scope and contextual variation of US and EU court opinions. Comparative analyses of the role courts have played in this field in different governance systems are even more limited since the value of such comparisons are made infinitely more complicated by virtue of the fact that the constitutional governance structure within which a court exists and operates as well as the nature of the substantive provisions in the laws under scrutiny by a court dictate the reach and breadth of any particular court's sphere of influence.

The role of courts in fisheries management and marine biodiversity conservation in the United States

The national policy framework for marine resources conservation and management in the US is characterized by competing national objectives and strategies. There is no

Governance of Marine Fisheries and Biodiversity Conservation: Interaction and Coevolution, First Edition.
Edited by Serge M. Garcia, Jake Rice and Anthony Charles.

overarching ocean policy designed to reconcile or integrate these various and varying policy goals (cf. Chapter 21). To a significant degree, the federal courts have had to play that role.

The US Constitution empowers the Congress to establish national policy through legislating and the executive branch through various departmental agencies under the control of the President with implementing those national policies. Congress often defers to those agencies to amplify, clarify and interpret the broadly stated national policies of Congress. While the executive agencies are less directly accountable to the public than the Congress, they are designed to be politically responsive in their actions through the annual Congressional appropriations process and the quadrennial Presidential elections.

As federal judges enjoy lifetime appointments, they are somewhat immune from electoral politics. The federal judiciary is therefore perceived as being the ultimate backstop in the US constitutional system of checks and balances, charged with adjudicating the law in the context of the resolution of particular cases outside the winds of public opinion. 'It is emphatically the province and duty of the Judicial Department to say what the law is' (Marbury v. Madison, 5 US 137, 177 (1803)).

The court's role is to determine and apply legislative intent. While US courts are not theoretically in the business of writing law or providing advisory opinions, the judicial exercise is hardly mechanical and often produces the functional equivalent of law, particularly in situations where Congress has been silent or vague. This has always been understood: 'Those who apply the rule to particular cases [the judiciary] must, of necessity, expound and interpret that rule.' As a result of its constitutional authority and political insulation, the US federal court system has played a significant role in shaping marine policy by implementing and integrating various fishery management, wildlife management and biodiversity laws that were never integrated by Congress.

US courts have been significant and relevant in two other respects. First, court interpretations of the law are absolutely binding on the executive branch. As agencies do not have the option of ignoring or disregarding a judicial determination, court decisions can produce substantial changes in agency direction and investment of resources. Second, third parties to a particular decision can challenge the agency action in federal court. As the following cases will illustrate, conservation organizations have become adept at using the federal courts to challenge federal fishery management decisions that they believe to be in violation of other more-conservation-oriented laws.

The legislative context of the judicial role

Any examination of the US judiciary's role in marine resources management must naturally begin by examining the relevant statutory frameworks. In the US, there is no overarching marine ecosystem management law or set of laws for the integration of fisheries management and marine conservation. In this respect, US management of the ocean resources is very different from management of national forests, where 'multi-use' concepts and comprehensive and integrated management planning are typically at the core of the process.[1] Complicating the matter further, Congress relies on different agencies[2] that often have significant mission and cultural differences to advance the complex ocean management agenda; these differences often produce both inter-agency tensions as well as competing, if not conflicting, decisions.

The laws regulating living US ocean resources covered by this chapter[3] can be broken down into two broad categories: *substantive laws* and *procedural laws*. Of the three major substantive laws reviewed here, one fundamentally focuses on economic development while the others pertain to marine conservation. Notwithstanding the numerous laws that influence domestic and international ocean policy, most US cases are brought pursuant to one of these three laws.

In regards to fisheries management, US objectives, as codified in the Magnuson-Stevens Fishery Conservation and Management Act (Magnuson-Stevens Act[4]), focus first and foremost on optimizing economic development of the nation's fish and shellfish resources. While the national program for fisheries management seeks 'to prevent overfishing, to rebuild overfished stocks, to insure conservation, to facilitate long-term protection of essential fish habitats, and to realize the full potential of the Nation's fishery resources',[5] the objective is realization of long-term maximum sustainable yields from US fisheries. Through its National Marine Fisheries Service (NMFS), NOAA has primary jurisdiction over the Magnuson–Stevens Act.

With one notable exception – the protection of deep-sea corals[6] – the Magnuson–Stevens Act approaches all marine resource and ecological management issues strictly in terms of optimizing fisheries production. Under Magnuson–Stevens, management actions are primarily determined by regional fishery management councils which comprise politically appointed commercial and recreational fishermen or their representatives as well as the fishery management officials in the states that make up the regional council.[7]

These councils have immense powers and the law intentionally allows them great discretion in shaping regional fisheries policy. For example, the National Marine Fisheries Service (NMFS) has limited authority to overrule or adjust the management actions taken by councils, and any identified deficiencies in a plan have to be resubmitted to the council for action. While NMFS and NOAA are mission-dedicated to the stewardship of living marine resources through science-based conservation and management and the promotion of healthy ecosystems, the fishery management councils they oversee have neither a similar mission nor set of conservation biology expertise.

Regarding broader marine conservation interests, the Endangered Species Act of 1973[8] (ESA) and the Marine Mammals Protection Act of 1972 (MMPA)[9] are the two primary laws that introduce substantive ecosystem and biodiversity elements to marine and fishery resources management. Under the ESA, a marine plant, invertebrate or vertebrate species is considered to be 'Endangered' when there is 'a danger of extinction throughout all or part of its range'[10] and 'Threatened' when it is in danger of becoming extinct.[11] Listing is meant to be exclusively based on scientific considerations. 'Critical habitat' must be designated with endangered species 'to the maximum extent prudent and determinable'[12] and may be designated with threatened species.

The ESA prohibits all persons from taking[13] an endangered or threatened species except under limited circumstances.[14] It also requires that federal agencies 'insure that any action authorized, funded, or carried out by such agency … is not likely to jeopardize the continued existence of any endangered … or threatened species or result in the destruction or adverse modification' of designated critical habitat.[15] All listed endangered species, and some threatened species within the discretion of the agency, are also required to have recovery plans designed to aid in the re-growth of species populations.

The MMPA focuses on the conservation of all marine mammals. Primarily directed at domestic conservation, the MMPA evolved to include an extra-territorial element addressing international marine mammal conservation issues. Under the MMPA, it is illegal to take[16] a marine mammal in US waters without a permit.[17] Despite the stated objective of the MMPA to conserve 'optimum sustainable populations' of all US marine mammals, a 1994 amendment significantly compromised the protective aspects of the MMPA by creating an exception for the taking of marine mammals in connection with commercial fishing. While these changes were not intended to provide a long-term exemption for marine mammal takes by commercial fishing, the reality is that the programs implemented in many fisheries to reduce mammal takes are required 'to take into account the economics of the fishery, [and] the availability of existing technology'[18] and therefore have rarely achieved mortality levels of 'approaching zero' as required.

Interactions and integration of the statutory goals of these substantive laws has been substantially driven by judicial challenges brought under two procedural laws. The National Environmental Policy Act (NEPA)[19] applies to any major federal action affecting the marine environment. NEPA is broadly written to encompass both direct federal actions that affect the environment, as well as indirect actions such as federal permitting of a private activity or the use of federal funding for a non-federal action affecting the marine environment. Any fishery management plan that NMFS approves must be accompanied by a full environmental record review demonstrating that NMFS has considered the management plan's known and potential impacts on protected marine species, important marine habitats, long-term ecological productivity and the short-term and long-term social and economic impacts. While the particular federal action going through the NEPA process might be quite narrow, such as setting harvest levels for a stock of fish, the NEPA analysis is much broader. A proposed federal action may not proceed until the conclusion of the NEPA review.

NEPA is a powerful tool for forcing integration of fisheries management and marine biodiversity conservation because the action agency has to take a hard look at the potential environmental impacts of any action under the full array of federal regulatory and management programs, including programs over which it may not have any direct jurisdiction. With the goal of full disclosure of expected or potential impacts to the public, the agency must review the biological, economic and social impacts of its proposed action. The agency must provide an opportunity and respond to all public comments on the action. Additionally, the action agency must issue a formal record of their decision based on the environmental review, demonstrating that they have taken all practical actions to minimize and mitigate the identified adverse impacts.

NEPA does not compel a particular outcome, just a process, but that process is comprehensive and open to the public and to other federal or state agencies with an interest in the issue. The result of a NEPA review is the production of a publicly available administrative record for any action, and guaranteed agency access for third parties to comment on a proposed action, ultimately influencing its direction. The value of the expert analysis and opinion that is incorporated into a NEPA review and highlights the costs and benefits of a particular proposed action is invaluable to the public, who could only rarely afford to independently commission such a costly analysis. The final agency decision, which has to be based on this NEPA review and supported by the record, can then be challenged in federal court for consistency with the underlying analysis.

The second procedural statute is the Administrative Procedures Act[20] (APA), which is the key to the federal judiciary system for parties seeking to challenge agency actions that are alleged to be in violation of any of the preceding acts. Under the APA, virtually all federal agency decisions are subject to judicial review. The judicial standard of review of the agency action is highly differential in that an agency's action can only be overturned if the decision is determined to be 'arbitrary, capricious, an abuse of discretion, or otherwise not in accordance with law.'[21] The courts are not permitted to substitute their preferences for the agency's preferences as long as the latter is consistent with the law. That deference, however, does not protect against improper or unreasoned decisions and courts routinely rule against agencies under the APA. When courts find that an agency's action has violated the APA, the usual remedy would be to remand the action to the agency for revisions that comply with the court's decision.

Fisheries management and marine conservation in the US courts

The court's role in implementing these various laws will now be highlighted by a few cases that specifically address fisheries and marine resource management issues.

Fisheries management in US courts

There are numerous cases that illustrate the court's role in fisheries management (see the discussion of the role of the courts in US fisheries management in the report by the National Research Council; NRC, 2010). The two cases profiled here were selected because they demonstrate the power of the courts to declare the law on the topic of preventing overfishing, and how those declarations tend to influence later policy.

Conservation Law Foundation (CLF) v. Franklin[22] involved an appeal of an agency decision that resolved litigation brought by a conservation organization challenging the legality of a fishery management plan. The agency had agreed to an interpretation advanced by the conservation group that the Magnuson–Stevens Act required fishery management councils to 'prevent overfishing' and had signed an enforceable judicial decree, called a consent decree, that required NMFS to eliminate the ongoing overfishing. Even though this litigation happened some 15 years after the 'prevent overfishing' requirement became the law,[23] it was the first case to consider what that clause meant in the context of a particular fishery management plan. In the challenge, the fisher organizations were arguing that since the fishery was already subject to overfishing rates, managers could not logically prevent overfishing.

The court upheld the requirement in the consent decree that the agency had to *eliminate* the existing overfishing, even though the controlling statutory term was 'prevent[ing] overfishing'. Focused on legislative intent, the court concluded that the facts established that overfishing was occurring in the managed stocks but that that didn't matter: 'one can't prevent what has already occurred. … Thus, the consent decree establishes that a plan to rebuild will be developed in order to 'eliminate' present overfishing, and 'prevent' future overfishing.'[24] While Congress had been silent on what course of action was required when overfishing was already occurring, the court filled in the gap in order to effect Congressional intent.

As the Magnuson–Stevens Act was under a periodic review by Congress, this judicial decision immediately found its way into the political process. Consistent with the court's view, the Sustainable Fisheries Act of 1996[25] amended the Magnuson–Stevens Act to add mandatory and time-bound rebuilding requirements and procedures for fisheries that were already subject to overfishing or overfished. The court's decision therefore became national law, applicable to all US fisheries. After the Conservation Law Foundation decision, a conservation floor established by the overfished and overfishing limits now backstopped all fishery management planning and economic optimization arguments.

Natural Resources Defense Council v. Daley[26] was a lawsuit brought by a conservation group opposed to an approved fishery management plan. In *NRDC v. Daley*, the conservation group brought claims under the Magnuson–Stevens Act and the APA that the NMFS-approved fishing quota for summer flounder was arbitrary and capricious. The fishery management plan for summer flounder, acknowledged to be one of the most important flounder species in the US, indicated that the species was both overfished and that overfishing continued. The regional management council proposed an annual catch that had a 3% chance of achieving a target fishing mortality rate (target *F*). In its place, NMFS proposed an annual catch that had an 18% probability of achieving the desired target *F*. A lower court upheld the quota and the process leading to it, and NRDC appealed that ruling.

The appeals court reversed the lower court, concluding that 'this case presents a situation in which [NMFS's] quota … so completely diverges from any realistic meaning of the [Magnuson–Stevens Act] that it cannot survive scrutiny….'[27] In phrasing that has permanently entered US fisheries management lore, the court opined that 'only in Superman Comics' Bizarro world, where reality is turned upside down, could [NMFS] conclude that a measure that is almost four times as likely to fail as to succeed offers a "fairly high level of confidence".'[28] The court suggested that the summer flounder quota had to have at least a 50% chance of staying below the target *F*, a probability that had been considered but rejected during the planning process. In response to agency pleas that it defer to NMFS's scientific expertise, the court said simply that courts 'do not hear cases merely to rubber stamp agency actions….'[29] The summer flounder management plan was rejected by the court and the matter was remanded back to the agency to fix. The 50% probability metric was quickly adopted throughout the agency and incorporated into the agency's published fishery management planning guidelines that were applicable to all fishery management councils.

Marine biodiversity protection in the US Courts

The US courts play an equally important role with other marine conservation issues and are often called on to adjudicate fishery management agency actions that raise the ESA and MMPA issues. Judicial review of decisions made pursuant to these statutes is also brought pursuant to the Administrative Procedures Act (APA).

One of the most interesting set of cases showing these statutory interactions involved a decadal series of lawsuits and policy battles between the conservation community and fishing community over the management of walleye pollock in the US North Pacific Ocean.

From a fisheries perspective, the pollock fishery is not only the largest but also among the most significant US fisheries in terms of economic value and employment. It enjoys a reputation as one of the best-managed fisheries in US, in no small part due to the high levels of industry-sponsored research and self-management. From an ecosystem perspective, walleye pollock is a critical forage fish in the North Pacific ecosystem and is particularly important in the diet of the Stellar sea lion, a marine mammal species whose western Pacific sub-population is listed as endangered under the ESA.

In *Greenpeace v. NMFS*,[30] the court was called on to resolve the 'complicated legal dictates' of the Magnuson–Stevens Act, the ESA and NEPA. The fisheries issues in the case were driven by the council process under the Magnuson–Stevens Act while the protected species issues associated with the Stellar sea lion were managed directly by the agency. The agency's conservation perspective on the interactions between the Stellar sea lion and the pollock harvest proposed by the regional council took the form of a formal biological opinion (BiOp), elaborated by the National Marine Fisheries Service, and a NEPA environmental review.

The agency's BiOp concluded that the proposed pollock fishery would likely result in jeopardy to the survival of the Stellar sea lion and adversely affect the animal's critical habitats primarily because of expected local depletions of pollock. As required by the ESA, the agency proposed a number of reasonable and prudent alternatives that the agency felt would resolve any jeopardy issues. The regional fisheries management council then adopted these alternatives as part of its pollock fishery management plan, which NMFS subsequently approved. Conservation groups challenged NMFS' decision as illegal under the ESA and NEPA. The pollock industry also intervened in the case, raising a significant challenge to the underlying science in the BiOp finding of jeopardy.

With respect to the industry's challenge to the ESA, the strength and candour of the BiOp prevailed on all fronts. Pointing to the legal ESA requirement that 'Congress intended that agencies "give the benefit of the doubt to the species,"'[31] the reviewing court concluded that the BiOp reflected the best available science, albeit still 'inconclusive' by the agency's own admission.[32] Unlike the record reviewed by *NRDC v. Daley* court, the Greenpeace court concluded that 'NMFS provided a reasonable interpretation of equivocal evidence, to which this court must defer.'[33]

The conservationists' challenge to the agency's ESA decisions fared much better. The court was clearly influenced by the weakness of the agency record: 'the voluminous record does not contain a single sentence reflecting the *reasons* why NMFS found the final reasonable and prudent alternatives to be adequate.'[34] While the court clearly recognized the complexity of some of the predator/prey and environmental factors at issue in the biological analysis, 'NMFS must comply with the mandates [of the ESA], *including asking the required questions*'.[35] The court concluded by finding the agency's alternatives arbitrary and capricious, not because the particular alternatives the agency looked at were necessarily inadequate or inappropriate but because the agency had completely failed to explain how it reached its decision: 'without some rational explanation, the Court cannot conduct a meaningful review'.[36]

The ESA aspects of the case did not fully resolve the litigation, however. There remained the conservationists' NEPA challenges, a law that the court approvingly referred to as 'our basic national charter for the protection of the environment'.[37] It had been more than twenty years since NMFS had conducted a comprehensive environmental review of the northern Pacific ecosystem and there were significant changes in the region from multiple human and environmental sources. While this baseline review had been updated by the agency in connection with its approval of the pollock management plan, the agency only considered the updated information from the narrow perspective.

Specifically, the agency had limited its review to the actual or probable environmental effects of several different proposed quota levels in the pollock fishery, ignoring a wide range of potential impacts connected to the fishery beyond just the quotas. Issues such as: the ecosystem impacts of fisheries; the abundance of marine mammals, birds, and other marine species in the fishing zones; and a current understanding of the ecosystem in general had been ignored. The court concluded that a 'broad, programmatic [environmental] review was needed in order to fairly evaluate the dramatic and significant changes which have occurred in the … groundfisheries'.[38] The court also condemned the agency's failure to address the critical comments of a separate federal agency, the US Environmental Protection Agency.[39] The public nature of the NEPA debate, coupled with the court oversight, can be seen here as providing a critical opportunity to ensure that conflicting views were not 'swept under the rug'.

The role of courts in fisheries management and marine conservation in the European Union

To understand the role of the European Union courts regarding fisheries management and marine conservation, it is useful to first explain briefly the way marine resources are managed through the common fisheries policy (CFP) in the European Union (Churchill and Owen, 2010). Indeed any role the EU courts have with respect to the preservation of marine biodiversity arises primarily through the court's role in attaining the policy goals of the CFP.

Regulation of fisheries management and conservation of marine resources in the EU

The common fisheries policy (CFP) covers the conservation, management and exploitation of living aquatic resources.[40] Conservation of all biological marine resources is therefore an important objective and functions on the same level as the management and exploitation of such resources. The European Court of Justice (ECJ) has always emphasized this integration of conservation and management of marine resources as part of its judicial task. The focus at this stage of *policy* development at the EU level, however, is on commercially exploited or exploitable fish and shellfish species.

The importance of the conservation of the biological marine resources is recognized in 2009 by Article 3 of the Treaty on the Functioning of the European Union (TFEU), according to which this subject is an *exclusive competence* of the European Union. Accordingly, fisheries conservation and management can only be regulated by the Union institutions. The Member States have in principle no such rule-making power although they enjoy exclusive competence in the critical task of implementing Union rules. This division has consequences for the extent of the competences of the different courts in the EU.

Based on these various competencies, there exist three layers of fisheries conservation and management regulation. The first two layers of regulation are situated at the level of the European Union. One comprises the legislative regulations enacted by the Council (composed of ministers of fisheries of the Member States) and the directly elected European Parliament as co-legislators. Those regulations are directly applicable in the Member States (cf. Chapter 20). The second layer comprises the detailed rules established by the European Commission, an independent institution within the Union, and the regulations of the Council acting alone. The enactment of detailed rules supplementing or implementing the legislative acts has been delegated by the legislator to the Commission. Specific competences, mainly in the area of the fixing and repartition of fishing opportunities, have been attributed to the Council.[41] Apart from its regulating competences, the European Commission has the task to ensure that the Member States respect their obligations imposed by the Union legislation.

To this end, it may launch infringement procedures before the European Court of Justice.[42]

The third layer of regulation occurs at the level of the Member States. It consists of the rules enacted by the Member States relating mainly to their internal control and enforcing competences. While the law-making powers are lodged with the Union institutions, the Member States have the sole authority to apply and enforce the rules established by the Union within their territory. They have to enforce the law on their fishers through their national legal systems, which are often highly variable between the Member States. Such enforcement can be effected through decisions by the Member State administration imposing obligations or penalties on fishers or through the bringing of penal actions by the public prosecutor before the Member States courts.

The rules established by the EU and the Member States institutions determine almost exhaustively the way fisheries are managed in respecting the conservation of marine resources. The courts interpret and apply those rules, but cannot directly set requirements for the exercise of fisheries activities or for the protection of marine biodiversity.

The organization of the courts within the EU

The courts in the European Union are organized at two levels for administering the enacted laws and rules on fisheries conservation and management: the first level comprises the courts of the European Union and a second level comprises the courts of the Member States. The courts of the Member States play the direct role in terms of the integration of fisheries conservation and management in the life of the fishers. Indeed, penal sanctions for infringements of the rules are imposed against fishers exclusively in those courts. Those national courts can also be seized when interested parties believe that the authorities in Member States have not correctly applied the rules set by the European Union, in order to reverse decisions endangering sound fisheries conservation and management or when fishermen contest administrative sanctions and obligations imposed on them.

The courts of the European Union – the General Court and the European Court of Justice (ECJ) – have a different function. They are the courts that determine whether the *acts* of the Union institutions – legislative and other – are *legal*, that is, in conformity with the rules of the TFEU. They are also responsible for determining whether the second-layer regulations adopted by the European Commission and the Council are in conformity with the first-layer legislative acts.

A second and very important function of the ECJ is to ensure *consistency in the interpretation and application of Union law by the courts* of the Member States. It is important for the attainment of the objectives of a common fisheries policy that the rules and regulations are interpreted and applied by the courts of the Member States of the EU in the same manner. To that end, a system of preliminary rulings has been established that allows national courts to seek interpretive guidance from the ECJ on matters within its competence.[43]

Finally, the ECJ plays a role ensuring that the Member States respect Union law when the Member States implement and apply the common fisheries policy. On an application of the European Commission in an infringement procedure, the ECJ may declare that a Member State has failed to respect its obligations under Union law.[44] This is a declaratory judgement, but if the Member State does not comply with the judgement the ECJ may, on an application of the Commission, impose a fine and/or a penalty payment on the Member State in default.[45]

The role of the EU courts in fisheries management and conservation of marine biodiversity

A distinction must be made in the EU between the roles of the courts of the Member States and the roles of the European Union courts. Moreover, as at the level of the EU all marine resource conservation (see Chapter 20) is realized primarily through the operation of the common fisheries policy (CFP), the case law discussed here is limited to situations involving that policy.

Role of the courts of the Member States

The main involvement of the national courts with the rules of the CFP occurs when the courts adjudicate applications by Member State authorities to sanction infringements of those rules or where fishermen contest the legality of administrative decisions imposing penalties or other measures on them because they did not comply with their obligations under the rules. It is difficult to make any generalizations about such decisions because the courts all operate within the different national legal and governance systems. The challenge is further complicated by the variety of different national legal cultures, with differences often existing even within a Member State. Those cultures were – certainly in the last century – not always favourable to environmental issues and therefore not always receptive towards the protection of marine resources.

Since the beginning of this century, the trend has started to change slowly. It must be said however that the courts once were and still are more inclined to see the interests of local fishers as paramount, fishers who are often seen as having to struggle for their livelihood and should therefore not be punished too heavily. Sometimes, this bias prevents achievement of the wider interests of conserving the resources of the sea for future generations embodied in the legislation of the European Union.

In order to try to overcome this problem of widely differing judicial decisions, the Union decided in 1998[46] to identify certain categories of fishing behaviour that were to be considered to constitute serious infringements of the rules of the common fisheries policy. The idea was that pre-identifying such a list would induce the authorities and courts of the Member States to take account of the relative seriousness of those infringements when applying for and imposing sanctions. Indeed, Union law requires such sanctions to be proportionate, dissuasive and effective. If successful, the identification of such serious infringements should bring more unity and uniformity in the application of the sanctions among the Member States.[47]

National administrative courts also have competence in general to adjudicate the legality of decisions by the government administration in applying the rules set by the Union. An interesting case in this regard is a decision by the French Conseil d'Etat of 10 August 2005[48] annulling a ministerial decree of the French government on the use of driftnets. The ministerial decree[49] tried to circumvent an EU prohibition[50] on the use of driftnets, which was designed to protect bluefin tuna, by licensing the use of a similar gear (thonaille). This gear, according to the French government, was different from a driftnet but certainly appeared to have the same negative impacts and effects as a driftnet. The Conseil d'Etat found that the gear was not, in fact, different from a driftnet and that by expressly allowing such gear, the French government had infringed the EU ban on the use of driftnets. The ministerial decree was annulled and French fishers could no longer fish with such gear.

The courts of the European Union

Unlike the national courts, the role of the courts of the European Union with regard to the integration of the principles of marine conservation into fisheries management is much clearer. There are only two judicial bodies: the European Court of Justice and the General Court. Because the common fisheries policy was determined at the level of the EU based on its exclusive competence, it is natural that the courts of the EU play the greater role in the interpretation and development of the rules of that policy. This function can be seen in a few leading cases.

The driftnets cases

The driftnets case from the French Conseil d'Etat referenced above in connection with the role of the courts of the Member States was also subject to parallel proceedings in the European Court of Justice. The European Commission brought infringement proceedings against France reproaching the French government for circumventing the ban on driftnets. After lengthy but ultimately unsuccessful discussions with the French government about the interpretation of the notion of driftnet in the European regulation, the Commission brought the question before the ECJ. In *Commission v. France, C-556/07*,[51] the ECJ supported the Commission's interpretation of the driftnet regulation[52] and declared that France had infringed the driftnet ban by allowing its fishermen to use a type of gear that was similar to a driftnet, and had the same detrimental effect for the selectivity of the catches of bluefin tuna, and by not taking the necessary enforcement measures. The ECJ in substance confirmed the ruling by the Conseil d'Etat.

This case illuminates some of the differences between actions taken in courts of the Member States and the EU courts. The decision of the Conseil d'Etat had the most direct impact on the challenged practice in that the French decree was annulled. The ECJ could not have accomplished this result since the judgement of the ECJ is simply declaratory in nature. Nonetheless, the ECJ's interpretation of the European regulation is more authoritative as it has the final say in this regard.[53]

Italian vessels also used driftnets on a large scale in contravention of the EU regulation. Political pressures from many countries, including the United States, had not resulted in Italy abandoning that practice. The European Commission carried out many inspection missions to Italy that led to discussions that were held with the Italian authorities in order to have them control and enforce the driftnet ban. As this process was without result, the Commission introduced an infringement procedure before the ECJ, leading to a judgment in 2009 declaring that Italy did not respect Union law by not controlling and enforcing properly the prohibition on the use of driftnets.[54] Since then, Italy has taken the necessary measures to comply with Union law.

These driftnet cases demonstrate the importance of the EU infringement procedure whereby the European Commission can sue a Member State for failure to respect Union law for purposes of ensuring the integration of marine conservation and fisheries management in the EU. Often this judicial intervention at the level of the ECJ has been critical in shifting behaviour at a Member State level. However, it must also be noted that the infringement procedure is much slower than a direct action for annulment of a state measure before a Member State's national court.

The case of undersized hake

The importance of the infringement procedure is demonstrated even more clearly by another leading case that also involved a confrontation between the European Commission and France, concerning the minimum size of hake. The EU Regulation on Technical Measures[55] forbids the landing of hake below a certain size. The EU Control Regulation[56] imposes an obligation on the Member States to control the landing operations in order to prevent the landing of undersized fish. The same Control Regulation creates a duty for the Member States to ensure that fishers and traders who land or sell undersized fish are prosecuted and sanctioned in a proportionate, dissuasive and effective manner.

Small hake, however, is very much appreciated by consumers on the French West coast and especially in Brittany (Bretagne). In fact, there was an important market operating in Brittany for undersized hake apparently with the tacit approval of the French authorities which did not control the landings of undersized fish, did not regularly prosecute offenders and, when they did do so, imposed very low sanctions.

Investigations by the inspectors of the European Commission brought out those infringements to the Control Regulation and the Commission brought the case before the ECJ. In *Commission v. France, C-64/88*, the Court, by judgement of 11 June 1991,[57] agreed with the Commission and declared that France had infringed Union law by not controlling the landings of undersized fish and by not pursuing those persons that landed or sold such fish.

Ten years of compliance discussions between the European Commission and the French authorities followed that judgment. Commission inspectors during this period continued to regularly find undersized hake landed in French ports and sold on the French markets. Concluding that the situation did not improve, the European Commission launched a new case with the ECJ in 2002, this time requesting that the ECJ impose a periodic penalty payment on France because of its failure to execute the judgement of the Court.

On 12 July 2005 in *Commission v. France, C-304/02*,[58] the ECJ upheld the Commission's determination that France had not executed the ECJ's 1991 judgement and imposed a fine of €20 million and a penalty payment of €57,761,250 for each future period of six months that France failed to take the necessary corrective actions. The Court asserted that 'it is imperative that the Member States fulfil the obligations incumbent on them under the Community rules in order to ensure the protection of fishing grounds, the conservation of the biological resources of the sea and their exploitation on a sustainable basis in appropriate economic and social conditions'. It was the first time a fine was imposed on a Member State by the ECJ.

After six months of very intense discussions between France and the European Commission, the latter concluded that France had still not taken all the measures needed to ensure there were efficient controls on the landings and sale of undersized hake as well as a system for the proportionate, dissuasive and effective sanctioning of the infringements. The Commission decided in early 2006 to demand the payment of the periodic penalty that was due. This decision was challenged by France before the General Court, but the action was dismissed.[59] In the summer of 2006, France implemented the necessary measures and controls on the undersized hake fishery.

Although this case took 17 years, the catch by French vessels of undersized hake and of undersized fish in general was stopped, the landing controls were improved to ensure effective supervision at the docks and the sanction regime in France was reformed as required by Union law. The case also established a precedent that allowed similar cases in other Member States to be settled with a positive result for fisheries management and conservation in the EU.[60]

The Onkel Sam case

The final case to be considered here is the case of the *Onkel Sam*. This case demonstrates the role of the ECJ in interpreting the extent to which Union law on the conservation of fish species applies to vessels flying the flag of a third country. *Onkel Sam* was a fishing vessel flying the flag of Panama but controlled by Danish interests. It had caught salmon in international waters in the east Atlantic Ocean. On its way to Poland where the fish would be landed, the vessel had to call at a Danish port because of engine problems. In the port, the vessel was inspected by Danish authorities and the salmon were seized, as it was forbidden to retain, transport and store on board salmon caught in the North Atlantic by the EU Regulation on Technical Measures.[61] The master of the vessel (a Danish national) and the owner were prosecuted before a criminal court for infringement of that prohibition. The Danish criminal court wondered whether the EU regulation applied to a vessel flying the flag of a non-EU country or to an EU national on the basis of his status as a crew member of such a vessel. As those questions concerned the interpretation of EU law, the criminal court submitted those questions to the ECJ for a preliminary ruling.

The ECJ[62] started its analysis by placing the relevant provision of Union law within its context, that being the International Convention for the Conservation of Salmon in the North Atlantic[63] which prohibits the fishing for salmon on the high seas. The Court concluded from this context that the provision of Union law has to be interpreted in order to give it the greatest practical effect, within the limits of international law.

With regard to the vessel,[64] the ECJ concluded from relevant provisions of the Law of the Sea that coastal states' legislation cannot be applied to vessels flying the flag of a third country fishing on the high seas, nor to such a vessel when sailing through the economic zone (freedom of navigation) or territorial waters (right of innocent passage) of a Member State. The Court ruled that under the Law of the Sea the situation is however different when such vessel enters the internal waters and ports of the coastal state. Then the law of the coastal state, in this particular situation the relevant provision of Union law, applies with no exceptions for vessels of third countries coming to port in need of repair. Therefore, the ECJ determined that the vessel may be properly held guilty of an infringement with regard to the prohibition against retaining, transporting and storing North Atlantic salmon on board and that the cargo may be confiscated. In this case, the ECJ interpreted the relevant provision of Union law in the widest sense possible within the framework of the rules of international law and thereby tried to give effect to the objective of the rule to conserve an endangered fish species.[65]

Conclusions

The legal regimes and governance structures in the United States and the European Union, and therefore the roles of the respective courts, have some important similarities and some differences. In both the US and EU, judicial review is a fundamental component of interpreting the law. In both systems, the courts not only have the final word on what the law means, but their decisions are often adopted by policy makers to provide for a consistent interpretation and application of the law going forward.

In the US, the ability of third-party conservation organizations to initiate the judicial review process, coupled with implementation of the rigorous analytical requirements of NEPA and the APA, provides a set of judicial tools to those groups to influence the direction of fisheries management and marine conservation. NEPA requirements have also empowered US courts to force an integration of broader ecological and environmental concerns and policy objectives into fisheries management decision making under appropriate circumstances.

The European Union operates under very different jurisdictional principles in terms of the competence of its judicial institutions. While there is not an overarching environmental policy act such as NEPA embodying marine conservation considerations (Chapter 20), conservation of all marine resources is the leading objective of the common fisheries policy. However, unlike in the US, third-party conservation organizations have no extensive legal standing in the EU courts to challenge the implementation of that policy.

Another feature of the EU legal system is that while the common fisheries policy itself is within the exclusive competence of the EU institutions, implementation of the regulations developed pursuant to the CFP occurs at both the EU level and the Member State level and Member States develop their own implementing regulations. The judicial control of those implementing Member State regulations lies with the Member States courts. As the legislation at EU level and Member States level is rather detailed, the role of the EU and Member State courts in many respects is subordinate to that of the institutions that create the regulations. The courts in the US appear to have greater latitude to integrate the national policies laid out generally in different laws while the courts in the EU have to apply and interpret rather precise texts, leaving them less margin of interpretation.

It should therefore be noted that the EU and Member State courts are important in three respects. Firstly, the respect for regulations by the fishing community is secured through a consistent and effective enforcement of the rules. The national courts of the Member States are the guardians of the lawful conduct of the government administration and, through their sanctioning capacities, they can contribute and have contributed to regulatory compliance. In this role, the Member State courts and the US federal courts are functionally similar.

Second, where the provisions of EU regulations are unclear so that their consistent application is endangered, the preliminary reference procedure can give an authoritative and consistent interpretation of the regulation. This is a collaboration procedure between the national courts, which put preliminary questions to the ECJ, and the ECJ, which provides binding interpretations of the EU regulations in order for the national courts to decide the cases before them. In that role, the ECJ gives great weight to marine conservation considerations that may be less compelling at the Member State level. There is not a counterpart to this important judicial efficiency function in the US system.

Lastly, the unique procedure whereby the ECJ can ultimately force Member States to comply with their obligations to correctly apply the tasks imposed on them by Union regulations has proven to be an effective instrument for the EU to integrate considerations of marine conservation in the minds and practices of those who have to apply the rules everywhere within the EU. In comparison with the plenary authority of the US courts to directly impose change on agencies, it is a protracted procedure because of the initially declaratory and only in second instance compelling nature of the Court judgments, but it has been proven to achieve excellent results in the EU context.

Acknowledgements

Peter Shelley gratefully acknowledges the assistance of Ivria Fried, Northeastern Law School, for her significant assistance on the treatment of the role of the United States courts. This chapter expresses only the personal opinions of the authors.

Notes

1. National Forest Management Act of 1976, 16 USC § 1600 et seq.
2. The agency with the most comprehensive ocean mission, including fisheries management, is the National Oceanic and Atmospheric Administration (NOAA). Significantly, NOAA is located within the Department of Commerce.
3. There are, of course, many other US laws that apply to marine conservation and fisheries management both domestically and others that seek to have an international reach. It is not possible to cover them in a condensed manner for this chapter. The author would call particular attention to the Lacey Act, 18 U.S.C. § 42, 16 U.S.C. §§ 3371-3378. The Lacey Act, in theory, is a powerful tool for reducing illegal trade in flora and fauna around the world. It prohibits any import, export, transportation, sale, receipt, acquisition or purchase of wildlife or plants in the US where it is illegal to take, possess, transport or sell such species under foreign law. Stiff criminal penalties and asset forfeitures are available to the government. Recent amendments focus upon controlling the introduction and spread of invasive species. US v. Bengis, 631 F.3d 33, 36 (2d Cir. 2011), is an interesting example of how the US courts have used the Lacey Act to advance domestic and international conservation objectives.
4. Magnuson–Stevens Fishery Conservation and Management Act of 1976, 16 USC §§ 1801 et seq.
5. Magnuson–Stevens Fishery Conservation and Management Act of 1976, 16 USC § 1801(a)(2).
6. Magnuson–Stevens Fishery Conservation and Management Act of 1976, 16 USC § 1884.
7. Representatives of academia and conservation non-governmental organizations are sometimes appointed to the regional councils but this is not common.
8. Endangered Species Act of 1973, 16 USC §§ 1531-44.
9. Marine Mammals Protection Act of 1973, 16 USC §§ 1361 – 1423 h. There are, of course, many other laws that apply to marine conservation such as the National Marine Sanctuaries Act, 16 U.S.C. §§ 1431-34, that are not possible to cover in a condensed manner.
10. Endangered Species Act of 1973, 16 USC § 1532(6).
11. Endangered Species Act of 1973, 16 USC § 1532(20).
12. Endangered Species Act of 1973, 16 USC § 1533(a)(3)(A).
13. The definition of a 'take' is very broad: 'to harass, harm, pursue, hunt, shoot, wound, kill, trap, capture, or collect, or to attempt to engage in any such conduct', Endangered Species Act of 1973, 16 USC § 1532(19).
14. Endangered Species Act of 1973, 16 USC § 1538(a)(1)(B).
15. Endangered Species Act of 1973, 16 USC § 1536(a)(§2).
16. The MMPA definition of a 'take' is similar to that in the ESA: 'to harass, hunt, capture, or kill, or to attempt to harass, hunt, capture, or kill any marine mammal.' Marine Mammals Protection Act of 1973, 16 USC § 1362(13).
17. Marine Mammals Protection Act of 1973, 16 USC § 1372(a)(2)(A).
18. Marine Mammals Protection Act of 1973, 16 USC § 1371(f)(2).
19. National Environmental Policy Act of 1972, 42 USC § 4321 *et seq.*
20. National Forest Management Act of 1976, 16 USC § 701 *et seq.*
21. 5 USC § 706(2)(A).
22. 989 F.2d 54 (1st Cir. 1993).
23. One of the greatest policy drivers in the Magnuson-Stevens Act, National Standard One requires that fishery

management plans 'prevent overfishing while achieving, on a continuing basis, the optimum yield from each fishery for the United States fishing industry.' 18 USC § 1851 (a)(1).
24. *Ibid*, 989 F.2d at 61.
25. Public Law 104–297, 110 Stat. 3559 (1996).
26. *Ibid*, 209 F.3d 747 (D.C. Cir. 2000).
27. *Ibid*, 209 F.3d at 753.
28. *Ibid*, 209 F.3d at 754.
29. *Ibid*, 209 F.3d at 755.
30. 55 F. Spp. 2d 1248 (W.D. Wash. 1999)
31. *Ibid*, 55 F. Supp. at 1262. This language is itself derived from another court decision on the ESA interpreting Congressional intent. See *Conner v. Burford*, 848 F.2d 1441 (9th Cir. 1988)
32. *Ibid*, 55 F. Supp. at 1261-62.
33. *Ibid*, 55 F. Supp. at 1262.
34. *Ibid*, 55 F. Supp. at 1265 (emphasis in original).
35. *Ibid* (emphasis added).
36. *Ibid*, 55 F. Supp. at 1267.
37. *Ibid*, 55 F. Supp. at 1269 (quoting from the NEPA regulations at 40 C.F.R. § 1500.1(a)).
38. *Ibid*, 55 F. Supp. at 1273.
39. *Ibid*, 55 F. Supp. at 1274–75.
40. Article 1 and 2 of Reg. No. 2371/2002, Official Journal of the European Community (OJEC) 2002 No. L 358/59.
41. Treaty of the Functioning of the European Union, available at http://www.eur-lex.europa.eu. Article 43 (3) TFEU.
42. Article 258 TFEU.
43. Article 267 TFEU.
44. Article 258 TFEU.
45. Article 260 TFEU.
46. Regulation (EC) No. 1447/1999, OJEC 1999, No. L 167 p. 5.
47. See now Article 42-47 of Regulation (EC) No. 1005/2008, OJEC 2008, No. L 26 p. 1 and Article 90 of Regulation (EC) No. 1224/2009, OJEC 2009, No. L 343 p. 1.
48. Published on http://www.conseil-etat.fr/fr/base-de-jurisprudence (265034).
49. JORF of 30 August 2003, p. 14829.
50. Regulation (EC) No. 1239/1998, OJEC 1998, No. L 171 p. 1.
51. European Court of Justice, judgement of 5 March 2009 (ECJ 5.3.2009, published on http://eur-lex.europa.eu).
52. The fact that the legislator introduced during the procedure a definition of 'driftnet' along the same line has had no role in the reasoning of the Court. See also ECJ 5.3.2009, *France v. Commission, C-479/07*, published on http://eur-lex.europa.eu.
53. That was the reason for the European Commission to pursue the case although the decree had already been abolished two years before its case was launched.
54. ECJ 29.10.2009, *Commission v. Italy, C-249/08*, published on http://eur-lex.europa.eu.
55. Regulation (EC) No. 850/98, OJEC 1998 No. L 318 p. 1.
56. At the time Regulation (EC) No. 2847/93, OJEC 1993 No. L 261 p. 1; now replaced by Regulation (EC) No. 1224/2009, OJEU 2009, No. L 343 p. 1.
57. European Court Reports (ECR) 1991 p. 1–2727.
58. ECR 2005 p. I-6263.
59. GCEU 19 October 2011, *France v. Commission, T-136/08*, published on http://eur-lex.europa.eu.
60. See e.g. ECJ 22 December 2008, *Spain v. Commission, C-189/07*, published on http://eur-lex.europa.eu.
61. Article 6 (1) of Regulation (EEC) 3094/86, OJEC 1986, No. L 288 p. 1 replaced by Article 26 (1) of Regulation (EC) No. 850/98, OJEC 1998 No. L 318 p. 1.
62. Judgment of 24 November 1992, *Poulsen and Diva Corp., C-286/90*, ECR 1992 p. I-6048.
63. OJEC 1982 L 378 p. 25. Panama is not a Contracting Party
64. Charges against the master were determined to be inappropriate since there were no express provisions in Union law allowing such a charge against the master or other member of the crew on the basis of their nationality. Such changes have now been made so that European nationals are now responsible under the Regulation on illegal fishing anywhere they are fishing and on any vessel. Article 39 of Regulation (EC) No. 1005/2008, OJEC 2008, No. L 286 p. 1.
65. Article 39 of Regulation (EC) No. 1005/2008, OJEC 2008, No. L 286 p. 1.

References

Churchill, R. and Owen, D. (2010) *The EC Common Fisheries Policy*. EC Law Library, Oxford. 574 pp.

NRC (2002) *Science and its Role in the National Marine Fisheries Service*. National Research Council. The National Academies Press, Washington, DC.

Part VI
Conclusion

Chapter 29

A tale of two streams: Synthesizing governance of marine fisheries and biodiversity conservation

A. Charles[1,2], S.M. Garcia[2] and J. Rice[2,3]

[1] Sobey School of Business and Environmental Science Department, Saint Mary's University, Halifax, Canada
[2] IUCN Commission on Ecosystem Management (IUCN/CEM/FEG), Gland, Switzerland
[3] Fisheries and Oceans Canada, Ottawa, Canada

Abstract: Attention to the governance of fisheries and of biodiversity conservation has a long history, arising from common roots. The corresponding governance streams have evolved over the years – in terms of objectives, approaches, nature and intensity of interactions – and their similarities have increased with time through externally driven convergence and interactive coevolution. This chapter examines the manner by which the streams have converged and coevolved, as well as the possible limits to integration, related to risk tolerances and core beliefs in both streams. The chapter also explores how convergence and coevolution appear across different governance levels, scales and issues, how they are affected by the differing but interacting sustainability 'pillars', and how higher-level cross-sectoral integration and internal 'diversity' of views within each stream affect the final outcome. Ultimately, it is concluded that while total overlap of the two governance streams may be unlikely, and even undesirable, there are reasonable avenues leading to further productive integration in the future.

Keywords: governance streams; levels of governance; integrated governance; convergence; coevolution; scales; risk; drivers; sustainable development; win-win-win

Introduction

People have been fishing for millennia, and attempting to manage at least some of those fisheries for centuries. Attention to biodiversity conservation, while perhaps not described as such, has also been present for centuries, albeit more apparent in terrestrial and coastal environments (Chapters 1 and 2). If we think of 'governance' as the process of decision-making, then we can correspondingly consider decision-making over fisheries and over biodiversity as constituting two potential 'streams' of governance. The more differentiated the processes of decision-making about fisheries and about biodiversity, the freer the streams are to follow different pathways to different places. The more interrelated the problems those streams try to address however, the more opportunities for conflicts to arise if the streams are following different pathways to different places.

The streams we have explored in this volume flow through the highest levels of global governance, including United Nations fora of various sorts. They flow through nations of the world, often found within their environmental and fisheries agencies. The streams are also found in thousands of coastal communities, in all parts of the world. These two streams of governance run through time and across scales, within a common institutional landscape but parallel 'watersheds'. They produce legislation, policy and management plans, they affect economies, ecosystems and societies and they generate passions and conflicts. If the streams were actually water flowing downhill, their

Governance of Marine Fisheries and Biodiversity Conservation: Interaction and Coevolution, First Edition.
Edited by Serge M. Garcia, Jake Rice and Anthony Charles.

ultimate confluence would be assured as all rivers flow to the sea. For streams of governance however, that confluence may or may not occur over time. This matter was discussed in detail earlier in the book (Chapter 2) and is explored further in this concluding chapter.

In this book we have explored – for each of the two governance streams – their past evolution, factors of progress, principles, conceptual objectives, policies, processes, instruments and institutions. We have focused on the interface between the two streams, including interactions at global, regional, national and local levels and across science, management and policy. The analysis herein has explored conceptual and operational synergies and conflicts, duplications and gaps, progress and impediments. We have sought to improve understanding of the two governance streams, as well as their commonalities and fundamental differences, seeking to understand how their joint performance in terms of fishery sustainability might be improved.

Throughout, we have looked for the extent and shape of coherence between the streams, their trajectories (and functional distance between them), outcomes (in terms of management performance) and processes. The latter proves to be crucial, since processes are more tractable than their uncertain outcomes and, more fundamentally, the durability of superficially similar outcomes of very distinct processes might depend on whether they result from mere convergence or active coevolution (see 'Evolving approaches').

Roots in conservation

A fundamental premise in modern conservation is the idea of foregoing some degree of immediate gratification in order to save resources, biodiversity and healthy ecosystems for a future time. Both streams agree that this conceptualization of 'conservation' is desirable. The fisheries stream has a long history of and engagement in resource 'conservation' and modern responsible fisheries do not want to damage biodiversity and ecosystems, just as responsible biodiversity conservationists do not want to deprive people of sustainable livelihoods and food from the sea. The differences between the streams, in terms of thinking of 'conservation', lie in exactly what to save for the future and to what extent.

We laid out this core thesis in the opening three chapters of the book. In this chapter, we ask what has been learned through the subsequent chapters about the similarities and differences in how each stream views and acts upon the need for conservation. Such actions are driven by the desire to reach each stream's negotiated objectives, which may include: (1) protecting endangered non-commercial species or ecosystems; (2) protecting ones that are not (yet) endangered; (3) sustaining the production of food and livelihoods; and (4) investing in 'natural capital' and 'living on the interest'.

These objectives reflect a dramatic broadening from the narrow forms of economic valuation of marine resources of the past, focusing exclusively on monetary rewards from extraction and utilization. Historically, this often took place with little thought for the future or with such a high discount rate as to be unsustainable. In recent decades however, modern economic analyses may incorporate sustainability considerations (albeit typically with a limited natural resource focus) and recognition of the importance of broader ecosystem wellbeing has led to increased emphasis on valuation of future ecosystem services from healthy biodiversity. This helps to balance the picture, providing a common currency to contextualize more fully the values of current harvests.

A key message to be explored here however is the need to go beyond biodiversity and economics, producing a comprehensive approach to the valuation of benefits and costs, in order to fully meet the marine governance challenge. Specifically, we argue in this chapter that increasing the coherence of the two governance streams requires a third (social) dimension of valuation, to better assess the role of resource users and their communities in food security and provision of livelihoods in the spirit of sustainable development (see 'The three prongs of sustainable development' below), broadening the idea of 'conservation' to include also the conservation of human communities and social systems (see Chapter 22).

Evolving objectives

Despite linkages through a focus on 'conservation', the two streams have spent much of their history flowing with limited interaction. Indeed, in many chapters we have seen that if we turn the clock back just 20 years there were two solitudes: a fisheries governance world and a biodiversity conservation governance world. Despite some middle-of-the-road voices expressing the need for progress towards meeting the objectives of both streams together, broadly each stream saw the other as a major source of problems. Many in the biodiversity community blamed the fisheries community for both destroying marine ecosystems and disenfranchising anyone except fisheries interests from any meaningful role in decisions about use of the ocean; many in the fisheries community blamed the biodiversity community for turning global public opinion against them by presenting worst-case examples as the norm, and seeking to put fisheries out of business. Although direct confrontations of the governance streams were uncommon, tensions and disagreements were arguably more common than constructive meetings designed to iron out differences.

Over time however, the substance behind the confrontations and the potential impacts became apparent, including a perception of poor performance in both streams, loss of trade opportunities, harm to the image of the various players and issues of possible non-compliance. This forced each stream to broaden its objectives in the direction of the other stream with significant changes from historical 'business as usual', at least at the strategic and policy levels. This trend is clear from the evidence presented in the chapters. Such changes were driven by rising civil society intervention and public awareness, as well as new alternatives in positive interaction and collaboration.

Indeed, it can be argued that each stream now sees a path forward from its current status to a future that at least *partially* reconciles a range of broad objectives which are agreed-upon in very broad terms. The overall aim may be expressed as 'sustainable use' of fishery resources and biodiversity by 'responsible fisheries'. The objectives include: (1) in the ecological dimension, to maintain or rebuild the ecosystem structure and function; (2) in the economic dimension, to have fisheries that remain viable within a long-term framework that internalizes all their costs; and (3) in the social dimension, to maintain and improve the sustainable livelihoods of coastal communities with particular attention to vulnerable groups. Food security, a central concern of fisheries, is being progressively integrated in conservation rhetoric and 'good governance' is recognized by both fields as essential. The general agreement on these high-level objectives is however vulnerable as soon as each stream considers more specific reference values (for targets, acceptable impacts and constraints) and implementation time frames for different resources, and for both natural and human systems.

Efforts to recognize and incorporate a broader range of objectives are seen in both governance streams. In the fisheries stream, perhaps the biggest shift has been in the widespread recognition of the need for, and adoption of, the ecosystem approach to fisheries management. Another important direction has been the incorporation of more social and economic dimensions of fisheries (as described in Chapters 5 and 6) into formal analysis and decision-making. There is also a growing global focus on small-scale fisheries and coastal livelihoods and their contribution to the Millennium Development Goals, particularly in terms of food security and poverty alleviation. In addition, the growing influence of eco-certification brings more attention to broad ecosystem objectives, as well as social and governance aspects of fisheries.

Parallel to this, as it receives increasing attention globally the biodiversity governance stream is also broadening its set of objectives, moving beyond its original preservation orientation and terrestrially based emphasis on protecting areas, habitats and species. As for the fisheries stream, this is seen in a transition over the past decade or so to a fully ecosystem-based focus. There has also been a shift to accommodating sustainable development ideas – including economic and livelihood considerations of biodiversity, linking sustainable use and 'green growth' goals – and ultimately increasing legitimacy in governance of *all* uses of biodiversity.

Evolving approaches

The objectives of the two streams are evolving and broadening while the overlap in the approaches they use is increasing – notably the precautionary, ecosystem and participatory approaches and the integration of economic, social and ecological considerations – which represent crucial avenues toward coherence and coordinated governance.

It is interesting to note that as fisheries interests took on the ecosystem approach, they rapidly recognized the legitimacy of many of the concerns of the biodiversity community (even though they sought to do so on their own terms, through their own institutions and with their own language). In parallel, as biodiversity interests developed their own practical plans in relation to people and industries that use the ocean, they adopted the concept of 'sustainable use' and variants of spatial planning. These included greater diversity of meanings for 'marine protected areas' (MPAs) and implementation of integrated conservation and development projects, also in their own terms, institutions and language.

The broadening of approaches used within the fisheries stream now tends to hold fisheries agencies accountable also for vulnerable marine ecosystems (VMEs) in which they operate, bycatches, impact on habitat, non-commercial endangered species, etc. This implicitly, and increasingly explicitly, reflects biodiversity goals and calls for increased use in fisheries of spatial tools and protection for iconic species – long preferred within the biodiversity stream. Application of precaution and ecosystem approaches makes all this central to core practices of the fisheries stream (Chapter 4). Furthermore, fisheries governance institutions are increasingly granting at least observer status to the biodiversity community at their 'tables'. Biodiversity conservation organizations sometimes have equal status in key bodies, such as the US Fisheries Management Councils and the EU Regional Advisory Committees. A parallel broadening of approaches within the biodiversity stream includes the adoption of ecological valuation and economic analyses of biodiversity[1] as major evolutionary steps to merge economics with a traditional ecological base. This is based on an expectation that an economic analysis would favour better biodiversity outcomes if the true 'value' of biodiversity was factored into the decisions. The stream also considers biodiversity conservation and the 'sustainable development agenda' (*sensu* WWSD and Rio + 20) as a natural marriage (through the concept of 'sustainable use') arguing that, to meet social goals sustainably, healthy ecosystems are fundamental. Altered ecosystems may be seen as acceptable outcomes as long as the alterations are sustainable and readily reversible (and details of reversibility and risk can be explicitly evaluated and agreed upon). The broadened approach of the biodiversity conservation stream has also led to strengthened linkages with industries and the formation within IUCN's Commission on Ecosystem Management of a Fisheries Expert Group. Furthermore, each Marine and Coastal Resolution of CBD since its 9th Conference of Parties (CoP IX) has explicitly directed fisheries-oriented projects, to be developed in collaboration with FAO.

Evolving interactions

As the two streams evolved in terms of objectives and approaches, they could not continue to remain exclusively with their own terms, institutions and language. Economic forces and especially industrialization increased the tensions between the two streams during the 19th and 20th centuries, but also led to some greater coherence of concepts and practices. This has accelerated since UNCED in 1992. The two streams now cannot easily avoid crossing each other constantly, at all levels, because neither group can achieve its own objectives without cooperation from and impact on the other.

Furthermore, societal demands for synergy in governance have pushed the streams towards more intentional planned collaborations, reinforced by hybrid institutions, learning – or at least arguing – that planned interactions with agreed groundrules often produce more constructive dynamics and outcomes than unplanned encounters. These trends have brought greater dialogue while also crystallizing tensions, thus highlighting areas of shared perspectives and of fundamental differences that have been examined in most of the chapters in this volume.

There have also been clear signs of increasing adaptation to accommodate the other stream and some shared changes, such as increased 'good governance' including inclusiveness, stakeholder engagement and accountability. Common concepts have been progressively mainstreamed in each field, and respective concerns are now better understood and taken into account. As an insight from producing this volume we have seen that, strategically, the streams are agreeing on the endpoints they wish to reach (i.e. on outcomes) far faster than they are agreeing on the best tactical pathways to reach those endpoints (i.e. on processes and methods/tools).

Most chapters indicate growing recognition among all parties of the need for more coherence, coordination and harmonization between the two governance streams. Indeed, there may be at times nearly full agreement on the desired endpoints. Progress seems particularly strong when both groups are in terrain relatively new to them, so institutions must be flexible in learning by doing (e.g. dealing with EBSAs and VMEs, or with impact assessments). This is tempered by the reality that each stream has its own long history of how their governance processes do their business. In particular, difficulties arise when either stream believes it can exercise exclusive 'control' over the setting in which the interactions take place (e.g. managing how much a species can decline, managing what areas are closed to most or all uses). Furthermore, when a stream has failed to resolve internal governance issues, these interact dynamically with fisheries–biodiversity tensions, sometimes making it difficult to disentangle them (e.g. when tensions between biodiversity conservation and fishery governance interact with conflicts between large-scale and small-scale fisheries).

The key to coherent governance are common processes in which both streams engage, and agreement on the paths to reach the desired endpoint. We can point to successes in this regard, such as cases of relatively harmonious travel on a common path (as with the coordinated back-to-back meetings to describe EBSAs and identify VMEs; Chapter 14) and the complex but effective methods used by CCAMLR for bringing ecosystem considerations into their fisheries management decisions (Chapter 18). However, there are cases on the other extreme where realities remain largely separate. This includes FAO and CITES still running independent evaluations of risk of extinction of commercially exploited marine species (Chapter 13), and the frequent stresses between expansion of coastal MPA networks and constriction of small-scale fisheries (Chapter 24). The move to common processes is still far from complete but, despite this, there are opportunities arising to form alliances under the right conditions (Chapter 10).

Convergence or coevolution?

The previous three sections provide evidence from this book of growing similarities between the two streams of governance in terms of objectives, approaches and interactions. Understanding the underlying roots of these similarities is important in identifying the potential future dynamics of the two streams. Here we examine this in terms of two concepts from Chapter 2, namely 'convergence' and 'coevolution'. These are defined as describing different mechanisms by which the 'functional distance' between the two streams decreases (and integration increases) with time, as a consequence of either similar responses to external forcing (*convergence*) or interactive and deliberate change to facilitate reciprocal change, reduce tensions, etc. (*coevolution*). This differs from usage in a number of chapters where the term 'convergence' merely indicates that components, processes or trajectories seem to go in the same direction, getting closer and closer with time, without any assumption about the underlying mechanism.

In Chapter 2, it is hypothesized that the two streams of governance are displaying more than just convergence and that there is a degree of coevolution with each stream evolving in ways that are becoming increasingly interdependent. This volume indicates that an important part of the similarities is due to convergence, prompted and facilitated internationally by United Nations cross-sectoral summits and at a national level by integrative frameworks (cf. Chapters 2, 10–12 and 21). However, many chapters also provide reasonable evidence of coevolution across multiple scales, with each stream building new governance features in direct response to developments in the other. Some examples of this dynamic, from nations and regions, are provided here.

At a national level Chapter 19 indicates that for the Benguela Current region of South Africa, Namibia and Angola, 'the institutional arrangements and levels of interaction between environmental and fisheries management agencies vary significantly', but there are positive examples indicating coevolution. The review of experiences in Australia, Norway and Canada (Chapter 21) suggests that in meeting 'the challenges of ecosystem-based management of fisheries and biodiversity' these three industrialized countries demonstrate 'clear examples of approaches, methods and frameworks for sustainable development of fisheries and for using ecosystem-based management'. The comparison of the roles of European and United States court systems (Chapter 28) conclude that in the US 'the courts have played a bigger role integrating marine conservation considerations into fisheries management' while in Europe, 'arbitrage between fisheries management and the conservation of marine resources' has tended to be handled more by legislatures.

At a regional level Chapter 15 concludes that in terms of 'alignment of fisheries and biodiversity conservation objectives' there has been 'relatively high levels of integration occurring in some regions such as the Antarctic and the North East Atlantic'. This is reinforced in Chapter 18 which notes that the Convention on the Conservation of Antarctic Marine Living Resources (CCAMLR) 'is seen as delivering ecosystem and precautionary approaches essential for strong fisheries and ecosystems outcomes'. Similarly, Chapter 16 explores interactions of the North East Atlantic Fisheries Commission, an RFMO, and the OSPAR Commission, a Regional Seas Convention dealing with environmental concerns, noting that as these organizations work to 'meet the goals of their respective Conventions', 'the urgent need to protect marine biodiversity has united NEAFC and OSPAR in a common purpose'. The coevolution may well continue in the future and indeed both Chapter 16 (for the North East Atlantic) and Chapter 17 (for the Mediterranean) speculate on the merging of institutions to integrate across fisheries and biodiversity conservation. The latter refers to 'a single governing body … which would integrate the management of all human activity and the conservation of the marine environment through a spatial and ecosystem-based management approach'. We focus on this matter of future coevolution in the following two sections of this chapter.

Limits to convergence and coevolution

Another hypothesis put forward in Chapters 2 and 3 is that, whether processes underway reflect convergence or coevolution, there are limits to both of these (as a result of rising integration costs) that are inherent in the differing values, goals and risk preferences of the two streams. While this book has documented the increasing interactions and linkages between the two streams, as highlighted in preceding sections of this chapter, it remains an open question as to what extent these are leading to a sense of 'common purpose'. Certainly there remain strong differences between the streams.

In the fisheries governance stream, there is a widespread feeling that (1) it now has the proper tools, and the knowledge to use them, to manage fisheries sustainably in an ecosystem context, so (2) the stream should be left alone to deal with its expanded duties to achieve 'sustainable, responsible management of fisheries'. This, it is felt, should be done in line with agreed international norms but to an extent that is based on the stream's own assessment of 'adequate' implementation in fisheries, without further pressure from the biodiversity stream.

On the other hand, many in the biodiversity stream see too little progress in fisheries (and sectoral management more broadly). Sectoral jurisdictions are seen as protective of their own long-established modes of operation and as exercising delay through 'incrementalism' and lengthy 'studies' at each step. They would argue that despite reasonable concepts for conservation being included in the 1995 Fish Stocks Agreement, the 1995 Code of Conduct and its many guidelines and plans of action and the 2001 Reykjavik Declaration, real change in RFMOs has only come when biodiversity advocates have pressured States into measures with teeth, such as UNGA Resolution 61/105.[2]

With these differences in mind, what is the potential for a full merger of the streams? Two points should be highlighted. First, neither convergence nor coevolution makes a complete merger (or other form of full equivalence) inevitable and little in this book suggests that either stream envisions a future of a full merger in which it has completely lost its identity to some hybrid governance stream. Second, the idea of a full merger of biodiversity governance and fisheries governance, in isolation from other impacting activities such as other economic sectors, would not seem sensible (as suggested in 'Evolving interactions' and discussed in 'A third stream' below).

Underlying these basic points, and the logic that complete fusion of the streams seems unlikely, are fundamental differences between the streams in relation to precaution (Chapter 12). There have been significant debates at all levels of governance over the application of precaution in the form of a 'Precautionary Principle' – considered as more prescriptive and thus preferred by the biodiversity conservation stream – or a 'Precautionary Approach', considered as more cost effective and practical and hence preferred by the fisheries stream. Both streams claim legitimacy by referring to UNCED Principle 15 which indeed calls for precautionary action but also for a precautionary approach in case of uncertainty.

More fundamentally, however, the differences in preferred terminology reflect differences in what kinds and what level of risks the two streams are willing to tolerate. Consider first that in an uncertain world, perfect decisions are impossible and there is always a risk of error. Structured decision-theory highlights two kinds of errors: *misses* (*not* taking actions when, in retrospect, they were needed) and *false alarms* (taking actions when, in retrospect, they were *not* needed) (Chapter 9). There are inescapable tradeoffs in the two types of errors: the lower we want the rate of one type of error, the disproportionately greater we must allow the other error rate to become (Hastie and Dawes, 2001).

To the biodiversity conservation community, the cost of 'misses' in a fisheries–environment decision can be very high: for example, environmental damage that is difficult or impossible to reverse, the cost of which is at least initially wholly borne by the ecosystem. Consequently, their tolerance for misses is much lower than for false alarms, the cost of which is wholly for the industry and which they view as readily reversible and an incentive for industry to improve information. In contrast, the fishery community, bearing the full cost of false alarms, wishes to reduce their occurrences, accepting an increased probability of misses. The difference in the sensitivity to misses and false alarms therefore reflects the fundamental disagreement on how the risk (and the costs of errors) should be distributed between people and nature.

The above discussion supports the hypothesis that there may indeed be limits to reducing the 'functional distance' between the streams. With fundamentally different tolerances for different types of management risks, it seems unlikely to reach total integration and much less a full merger of the streams. The limits are both situation-specific and labile: the acceptable balance of risks that the two streams can negotiate in one place may be unacceptable in other places depending on social or economic needs, available options and the status of marine biodiversity. Moreover, even in the same place, changes in social or economic conditions (possibly completely external to the fisheries) or changes to environmental conditions (either possibly external to any fisheries impacts) may cause either stream to reopen negotiations about the acceptable balance of risks.

Pathways of convergence and coevolution

The previous section demonstrates some of the possible limits to convergence and coevolution, and explains why we should expect enduring differences and tensions between the streams. Given this reality, it is fair to evaluate the hypothesis that reduction of functional distance between the streams (i.e. their integration) will have a net positive impact on both streams, as generally assumed, accompanied by *lower aggregate costs to society* (cf. 'Integration through interaction' in Chapter 3).

As shall be shown, the validity of this hypothesis is not as clear as for the other two discussed above. This is because an outcome that is 'acceptable' from the long-term perspective of both streams must balance the future and immediate human needs of both use and conservation. Many chapters in the book show that the 'right' balance depends on the situation, the trends, the actors, their history, their set of values, their capacity and so on. The best that can be expected are situation-specific negotiations that lead to a suitable level of interaction based on a good understanding of the differing values, goals and risk preferences of the two streams. We examine below some potential pathways of interaction and their respective impacts.

Constructive or defensive responses

In Chapters 2 and 3 we viewed convergence, coevolution and integration as opportunities to make governance more coherent, seeking out the common expected outcomes of both streams and efficiently selecting approaches and tools that work 'best' (as seen by both streams) to obtain them. The MOUs being signed between RFMOs and Regional Seas Organizations (Chapter 16) illustrate the fact that such constructive coevolution is happening even though still limited to cooperation, information sharing, etc. Where the governance streams have long been flowing, as in the Mediterranean (Chapter 17), there seems to be more 'resistance' to reducing functional distance (despite the fact that in the Mediterranean, the environmental governance actively emerged from the fisheries side).

However, differences in preferred evolutionary pathways are also visible. There is evidence that some of the evolution is 'defensive' in nature, i.e. with each stream making moves to protect itself from developments originating in the other stream. If indeed evolution is defensive as often as (or more often than) it is constructive, this challenges the validity of the hypothesis above that coevolution has a net positive impact on both streams. Consider two examples.

First, MPAs had been promoted in the biodiversity stream since the early 1970s (Chapter 1) and were considered for use in fisheries management in the Code of Conduct for Responsible Fisheries (in 1995). It was only after WSSD in 2002, when MPAs gained strong policy momentum in the international biodiversity governance stream (which began to produce its own MPA guidance documents, e.g. in the CBD and IUCN), that the fisheries stream called on FAO to develop guidelines specifically linking MPAs and fisheries management. This enabled fisheries to 'translate' the general guidance specifically for fisheries, thus staying securely within its own stream in terms of the approach taken while using words and concepts of the other stream (cf. Chapter 8).

Second, in response to two decades of stress over criteria for listing fishery species considered at risk of extinction because of international trade, an example of dual 'defensive' coevolution arose as a negotiated compromise. In addition to the existing scientific advisory process used by CITES to assess listing proposals (based exclusively on IUCN's advice), a second scientific process was established under FAO (involving an independent panel of experts established under an MOU between FAO and CITES). The advice is produced for the same institution, and the two streams evaluate the risk of extinction for the same listing proposals with the same public information. Each stream retains some independence however by using its own experts and process and its own risk perceptions (cf. Chapter 13), producing conclusions that may or not coincide.

Institutional coevolution

Consider the current situation of FAO and CBD, each creating and maintaining independent repositories for the areas found to meet VME and EBSA criteria, respectively. Coevolution to date has stopped at sharing of information; common data repositories, shared processes or shared outcomes across the streams have yet to be achieved. There appears to be potential for further coevolution by creating one common site with appropriately labelled contents of all the information on special areas of the high seas (cf. Chapter 14). However, this must deal with the institutional reality that the frameworks under which VMEs and EBSAs are respectively established (RFMO conventions and UN Resolutions) create very different obligations on States with regard to management. This may be seen to justify separate processes, or at least to make more difficult a shift in the institutional arrangements.

A different challenge to institutional coevolution arises when one (or both) of the streams sees too little progress towards needed reforms and resolution of looming crises. This sense of urgency (addressed in 'Limits to convergence and coevolution') can lead to forms of institutional change that challenge the idea of 'coevolution'. It led the biodiversity stream to call for a dedicated Implementing Agreement of UNCLOS, a major step beyond earlier proposals in UNCED and WSSD for more effectively linking biodiversity conservation to broader frameworks of sustainable development. The Implementing Agreement, as proposed in RIO + 20, would have the same legal status as the Fish Stocks Agreement but would place the governance of biodiversity and marine genetic resources at the forefront of spatial and ecosystem-based management. Its advocates envision a model wherein the overall goals and priorities regarding conservation of biodiversity for all sectors would first be set within such a framework based on the concepts of conservation and sustainable use. The separate governance streams of all economic sectors, including fisheries, would thereafter remain responsible for managing their activities on an operational basis, but within the norms set under the Implementing Agreement. In contrast to this outcome, economic sectors including fisheries would probably prefer to operate in a different multi-sectoral spatial planning and management framework, complying with UNCLOS and its Implementation Agreements, but in which the norms, objectives, constraints and strategies and plans could be negotiated.

Moving ahead

Given that no radical take-over of a stream by the other is likely to be viable, practical solutions are needed to accelerate coherence between the two streams and increase the chances that both meet their objectives. As described in Chapter 3, the integration process might be seen as involving several phases: (1) the two streams work alone and are blind to each other; (2) there is a realization, but not an interest, that there is someone 'on the other side'; (3) the two streams realize that they should 'care' about each other but remain firmly on their own side; and (4) a progressively larger portion of actors from each stream 'walk all the way to the other side and see, with their own eyes, what it looks like'. How do we learn together and thereby move to integrating fisheries and biodiversity streams more effectively? What are the affordable mechanisms to bring the streams closer together?

The evidence presented in this volume indicates that integration requires suitable cross-scale processes, together with active consensus building. It is fostered through the use of: (1) common data, tools and processes, including information sharing and joint (integrated) assessments; (2) opportunistic and strategic alliances; and (3) suitable cross-sectoral frameworks (global to national). There are important roles to be played by legal, institutional, procedural and methodological integrative instruments and mechanisms, as discussed in various parts of this volume. These may include common committees and task forces, MOUs and tools such as social and environmental assessment, GIS, modelling and scenario analysis, stakeholder analysis and ecological accounting. Some of these linking mechanisms are indicated in Figure 29.1.

Perhaps a focus on commonalities and common new directions can be useful – including moves to multidisciplinarity, participation and facilitation. For example, as pointed out in Chapter 26, the shared need to function in a world of ecosystem-based management (and the limitation of budgets available) may provide an incentive for those in both streams (from conservation officials to managers, scientists and other stakeholders) to come together for joint training and capacity building. External drivers can also play a role in moving towards integration. Exogenous threats from climate change, food insecurity and globalization may amplify the need for more integration of the two streams, to reduce the risks to both or for emergence of a third path (e.g. *integrative governance*; see below). These pathways would need to give both constituencies legitimacy and credibility, have objectives and priorities that adherents of both streams are willing to accept and

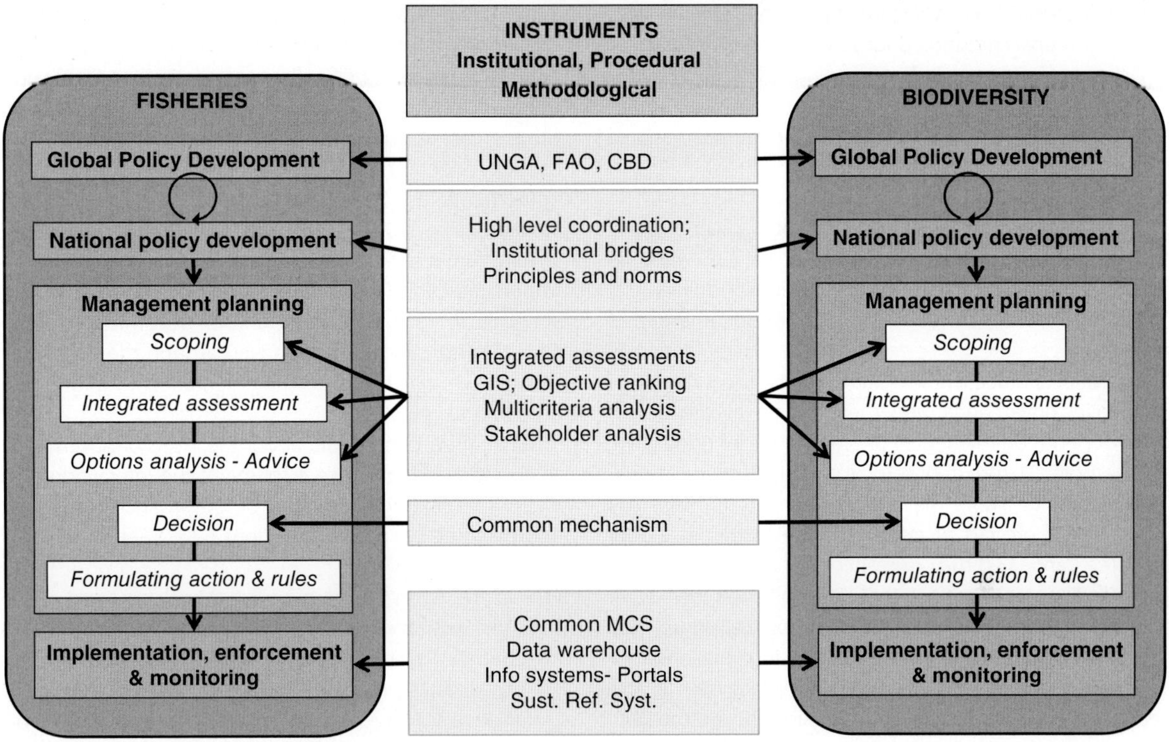

Figure 29.1 Legal, institutional, procedural and methodological linkages between the two governance streams for fisheries and biodiversity conservation (from Chapter 3).

strive to achieve (even though neither considers them perfect) and contain combinations of tools that both streams can accept.

The two streams across scales, levels and issues

This book has covered its core theme (the evolution of the two governance streams) as it arises at many different scales and levels of governance (Cash *et al.*, 2006; Charles, 2012; Gibson et al., 2000). While the book's structure reflects an institutional scale (from global to regional, national, local and community levels of institutions), its content also addresses other interconnected or intersecting scales and levels: spatial (from coasts to open oceans and LMEs to MPAs); jurisdictional (from international, to regional and national); and legal (from overarching frameworks to specific local/sectoral regulations and rules).

Governance faces different issues and is executed differently at these different scales and levels, even though the fundamentals (e.g. good governance principles) may apply throughout. Governing the implementation of a global agreement (at the UN level) requires different actors and processes and has different pertinent time frames than governing a specific fishery or MPA. Issues of coherence and phenomena of convergence and coevolution play out across scales, but progress at one level may not ensure comparable progress at others. It is clear from accumulated evidence that there are crucial differences in what constitutes effective governance at:

(1) *local level*, where all the players can (and often do) know each other and are closely connected by history, culture and common local realities and are directly affected by decisions;
(2) *national level*, too large for 'town hall' approaches but where you have a single government that can elaborate policies, strategies and regulations within and across economic, social and environmental sectors with different degrees of participation of the people concerned; and

(3) *international level*, where multiple governments with different backgrounds, priorities and capacities have to negotiate common policy frameworks and legal instruments and then 'trust' each other to set and implement corresponding policies and regulations at their national and local levels of governance.

Properties of governance such as equity, power balance, transparency and performance are important but emerge differently at different levels or scales. Similar drivers (e.g. of economic, social or societal nature) are documented at all levels.

The effects of *local drivers* are manifest most clearly in local-level governance, and their impact at the global level depends on the degree of 'scaling-up' depending on, for example: (1) the extent of local participation in national policy-making; (2) the perceived usefulness by government officials of reflecting local issues in global debates; and (3) the existence of larger-scale and intermediary organizations that take common local-level effects to higher-level fora. This reality of 'scaling-up' may be seen for example in the increasing presence of small-scale fisheries in debates and processes at the regional and global levels, including Rio + 20 and the FAO. An important role in 'scaling-up' is often played by intermediary organizations and NGOs, taking local innovations and insights to the global arena.

In contrast, the effects of *global drivers* such as demography, climate change and the global economic situation are quite different and tend to be felt directly across all levels, albeit in different ways and with differing power and consequences. For example, at the community level, fishers and fish farmers may feel the impacts of drivers such as climate change and international trade in terms of increasing poverty and diminishing livelihood opportunities, with limited options to address those global drivers. At the global (UN) level, these drivers impact more on process and institutional needs, for example in requiring institutions that are either new (e.g. the International Framework Convention on Climate change, IFCCC) or greatly adapted (World Trade Organization, WTO). This may also have a domino effect in diverting resources from other global concerns in order to address these drivers. There are also challenges in terms of a lack of tools to translate global agreements, if they can be reached, directly into action at the community level.

Overall, it would seem that no level of governance is immune to drivers at different levels, and governance systems need to somehow deal effectively with the challenges the same drivers pose at different levels. Global economic forces directly affect local living conditions and hence local-level governance processes, while local drivers can strongly affect the lower-level buy-in and capacity needed to effectively implement large-scale agreements (e.g. the WSSD commitment to have all fish stocks able to produce MSY by 2015). With respect to the latter point, it is notable that multi-level feedback can be important. High-level mechanisms, such as widespread shifts to more participatory and/or decentralized decision-making, can shift local drivers around the world and this in turn can improve implementation of large-scale agreements.

What, then, can be said about the levels of governance that is relevant to our core theme of the interaction between two differing streams of governance? Here we address three key related questions.

(1) *Do the two streams manifest themselves, and in the same way, at all levels?* There is evidence through the chapters that the two streams are present in governance processes and established institutions at all levels. However, not surprisingly, the issues on which they play out vary considerably. At global and regional levels the streams are interacting on *strategic issues,* principles, global norms and legal and policy frameworks and the outcomes might be binding or voluntary. At the national level, the interaction tends to shift to *tactical issues* related to planning and implementation, and perhaps too rarely on optimal pathways. The more local the level, the more tactical the dynamics. The more consensual, clear and explicit the global outcome (the 'what'), the less stressful its adaptation and implementation (the 'how') at national and local levels. When global outcomes are agreeable but vague platitudes, every actor will take home its own perception of what that means and the lower levels will have to figure out both the 'what' and the 'how', and even 'how far' to go.
(2) *Do the same tensions between the two governance streams arise at all levels?* The chapters show that many tensions exist at all levels, for example those between short-term and long-term thinking, protection and sustainable use, ecological, economic and social outcomes and in competition for political influence and budgets. However, the effects at a local

level are less conceptual and political and more tangible, economic and social. While tensions hinge more on ideologies, concepts and systems of values at global levels, at the local level they hinge on effective or expected impacts on livelihoods, the (re) distribution of costs and benefits (equity), security and the eventual redistribution of decision-making power. The disruptive effects of larger-scale drivers (e.g. conservation strategies, multinational investments and changes in international markets and trade rules) are very clearly felt at the local level, triggering strong tensions. At the global level, we can observe the limitations placed on advancement of progressive policies because of regional differences in how such policies will address their development needs or increase their costs and regulatory burdens, again triggering strong tensions.

(3) *Are some levels of governance doing better than others in resolving tensions?* Tensions may emerge from disagreements regarding the object of a proposed action (e.g. the protection of a species or an area), the approach to that action (participative or not), the instrument considered (top-down MPAs versus community-managed reserves) or the perceived consequences of that action (e.g. on livelihoods or power distribution). The chapters show, as one would expect, that there have been both successes and failures to resolve tension at each level of governance. Notably, failures at the global level often hide local successes, while success at the global level does not automatically imply success at lower levels. This book contains many examples of local successes, such as those in the Philippines (Chapter 27), where fisher organizations are playing a major role in representing the interests of coastal marine conservation in order to protect local resources against rapid coastal development and pollution. There have also been successes at the global level, for example the FAO Code of Conduct and its guidelines and plans of action integrating biodiversity concerns and facilitating that integration in regional and national legal instruments and policies. However, national-level implementation of the Code of Conduct remains rather partial (Pitcher et al., 2008). Similarly, the CBD Marine and Coastal Resolutions *always* direct the CBD Secretariat to work with FAO and RFMOs on all fisheries-related initiatives it undertakes, but this cooperative attitude has been slow to be taken up at national levels (with some exceptions in advanced countries under integrative frameworks; see Chapters 8 and 21).

Just as we can ask how scales and levels of governance affect interactions between the two streams, it is also important to consider how those interactions are affected by the different *issues* arising in fisheries and in biodiversity conservation. Specifically, it was seen earlier in this chapter that the more a given issue touches on vital or fundamental elements of each stream's identity, the more likely adaptation is to be defensive rather than conciliatory. While deliberate coevolution may result in making 'functional distance' very small, resulting in a nearly complete 'meeting of the minds' (e.g. on the need for participatory and adaptive management processes) for some issues, there are others (e.g. determining generally acceptable levels of impact on the ecosystem compatible with acceptable levels of food security) that are so fundamental as to result in a larger distance between the streams. Indeed, given the wide and differing range of forces at play on some issues, the two streams could actually be moving apart rather than together. In such cases, we would argue that these divergent trajectories are unlikely to be sustainable.

Overall, our conclusion from this book is that it is not the thematic issue or approach itself that underlay any of the successes or failures in improving coherence and synergy between the streams. Instead, assessments of risk seem to be a major determinant. Issues with low potential environmental, social or economic costs of poor decisions are probably those where it may be easier to find consensus, but these low-cost issues are also low-value in terms of what is gained by agreeing. In any case, these 'low-hanging fruits' may play an important role in initiating cooperation on more complicated agreements with higher stakes. On the contrary, if an issue is perceived as implying a potentially high cost of management errors to either stream (e.g. risk of extinction vs. risk of unnecessarily losing revenues) and, for that reason, has been contentious for years or even decades, many chapters (and 'Limits to convergence and coevolution') illustrate the fact that there is little flexibility in compromising one's risk tolerances just for the sake of 'sharing' the way forward.

We can therefore conclude that the streams function, and tensions exist, at all scales and levels and across many issues. Governance can however provide more coherent outcomes across streams, although this is more difficult as the decisions posed to the governance process have larger and more immediate real-world consequences (particularly social and economic consequences). There is no cookbook recipe to fix the governance process where it is not working well: the more a decision has real-world and potentially inequitable consequences, the longer the process will have to struggle with it.

The three prongs of sustainable development

The search for and realization of so-called 'win-win' solutions has become ubiquitous in discussions of each of fisheries and biodiversity governance; this is more appealing than the alternative of tradeoffs, 'hard choices' and negotiated settlements, which all imply more arduous processes and less desirable outcomes in which either no-one fully achieves their goals or some 'succeed' while others 'fail'.

With respect to biodiversity conservation and fisheries governance, win-win solutions might look for policy or actions that satisfy both sides in a competition between two groups of actors (e.g. conservationists versus fishers) or two sustainability dimensions (e.g. ecological versus economic performance). For example, the currently popular *blue economy* (a marine version of the *green economy*) intends to simultaneously conserve ocean biodiversity while generating economic wealth. It advocates market-based approaches to conservation, in which markets or market-based mechanisms (e.g. eco-labels) properly value biodiversity and resource sustainability, generating economic and biodiversity benefits simultaneously.

One major problem with such approaches, regardless of the stream, is that they see fisheries only through an economic lens with 'win-win' viewed as achieving fishery-related economic goals (in addition to biodiversity goals), disregarding social and distributional impacts (Chapter 6). However, the global policy agenda, institutions and processes (Chapters 2 and 10–12) by which biodiversity conservation is conditioned places it in the context of sustainable development (SD) as a fundamental overarching objective and calls for balance in its ecological, social and economic dimensions through good governance. Indeed, in Rio + 20 UNCED commitments to develop internationally applicable Sustainable Development Goals (SDGs) have been confirmed, and should provide guidance on more concrete implementation of a balance across the three dimensions.

Taking a SD approach to both streams requires understanding and incorporating the social prong of sustainable development in both. This reality is far from universally recognized, as all our chapters on small-scale fisheries (Chapters 22–25) emphasize in relation to their respective topics. In fact, Chapter 24 suggests that this explains 'why there continues to be a disjuncture between policy rhetoric that appears to be converging in terms of conservation, fisheries management and human rights and wellbeing goals on the one hand, and implementation practices on the other'.

Indeed, the hidden conflict within each stream between those who do or do not acknowledge and incorporate the social dimension is likely as significant and consequential as is the more apparent conflict between streams, which is the focus of this book. An example would certainly arise in the contentious debate over market-focused environmentalism and market-based (individual) fishing rights. A tendency of both streams to ignore or downplay the social prong of sustainable development certainly constitutes a critique of these governance streams. Accordingly, the coverage in this book tries to reflect a reasonable balance of concerns arising as a result of the fishery governance stream (such as the imposition of governmental fishery-access policies that may harm small-scale fishing communities) and as a result of biodiversity conservation policy directions (such as the imposition of MPAs).

Both the SD and the small-scale fisheries perspectives point to the importance of pursuing broader 'win-win-win' outcomes across the three dimensions of sustainable development. This should not be seen as simultaneous achievement of perfect 'maximized' results in each of the ecological, economic and social dimensions, as it is generally impossible to maximize in three dimensions at once. Furthermore, benefits and costs will accrue differentially across the various sectors of society, across fishing communities, across nations and across generations. Instead, we need to define 'win-win-win' outcomes as resulting from negotiated solutions (in most cases) that face the necessity of tradeoffs and hard choices, outcomes and pathways that avoid the least desirable or most unjust results. The goal is for all sides to feel like a winner, or at least feel that the result gave no more to others than to them (in analogy to a Nash Equilibrium) leaving the actors reasonably content that a compromise was reached based on a proper process with a level playing field. The idea of getting the incentives right (Chapter 5) is a major ingredient in achieving this balance. Paradoxically, the resolution of the tensions *within* each stream over the social dimension of sustainability may provide some of the most productive pathways to reducing the functional distance *between* the streams.

The challenge of getting agreement on outcomes and pathways is complex in such a situation, requiring an inclusive approach to sustainable development, accounting for distributional impacts, values and a full-cost accounting of social and opportunity costs. Who gains and who loses are fundamental questions. This need for a three-prong approach and a full-cost analysis applies to each of the two streams individually, as well as to the interaction between them.

In this regard, it is of interest to consider the concepts of 'hard' and 'soft' sustainability. Some would argue that a 'hard' view of ecological sustainability underpins biodiversity conservation, limiting human impacts on the natural world by deeming it unacceptable to irreversibly deplete natural capital while increasing social or economic capital. The contrasting 'soft' view of sustainability often seems to underpin the human side of the fishery system; at several points in the book arguments were made for a 'harder social sustainability' perspective. Within this perspective, it would be considered unacceptable to deplete social capital through measures to increase natural (or financial) capital without compensation or mitigation measures. For example, imposing a marine park around a coastal community would be unacceptable if it jeopardized the viability of that community. This view is reflected in the suggestion (Chapter 22) that the Precautionary Approach could be applied 'to protect vulnerable small-scale fishers as well as ecosystems'. More broadly, this perspective would call for all evaluations of governance measures to take into account the impacts on ecosystems, economic opportunities *and* communities, and the acceptability of those consequences to those experiencing them.

While a three-prong focus on sustainable development can be challenging to implement, there can be 'success stories'[3] at all levels from the local community to international agreements. For example, Chapter 22 shows that 'many SSFs possess their own self-governance structures that show significant conservation value'. Moreover, Chapter 23 documents several 'ambitious and interesting examples of simultaneous engagement of coastal and fishing communities in conservation and fisheries management' that 'naturally integrate both conservation and fisheries management goals'. They suggest that 'this is because many tropical coastal communities are 'natural integrators ... they do not commonly dichotomize resource management and environmental stewardship'. Integrative legal and policy frameworks at global and national levels (cf. Chapters 12 and 21, respectively) facilitate similar outcomes at a larger scale.

The emergent message here is that efforts seeking coherence of the fisheries and biodiversity governance streams must embrace societal choices about wealth distribution, poverty alleviation and food security, ensuring that human activities (including both fishing and biodiversity conservation initiatives) are ecologically, economically and socially sustainable. These are complex questions, and governance challenges in the future are not made any easier by considerations of equity, opportunity and social outcomes, and tactical concerns about compliance not only between the two streams but also across the range of realities 'on the ground'.

A third stream

While this book has been built around a dichotomy between two important streams, it is important to recognize that these do not capture the full range of thought and action on ocean affairs. There are in fact multiple streams of thought, including those oriented to other ocean users such as energy, defence and navigation as well as specific issues such as climate change. Within the range of governance streams, fisheries and biodiversity conservation are but two (albeit major) players.

One major governance stream to be highlighted in this section arises out of the desire, expressed by many individuals and institutions, for integrated multi-sectoral solutions to marine challenges. This multi-sectoral or integrated management stream looks at oceans as a whole and has been at the heart of many institutions and major events at the international level (e.g. UNCLOS, UNEP, DOALOS, UNCHE, UNCED, WSSD and Rio + 20). These broader initiatives have been informed and sometimes led by global institutions such as the UNGA (for its convening power), the Intergovernmental Oceanographic Commission (for its scientific coordination capacity) and UNDP, The World Bank and the Global Environment Facility (for their financial capabilities).

The manner with which both the fisheries and the biodiversity conservation governance streams interact with this third stream is worthy of further analysis. As noted in many chapters, interactions between fisheries and biodiversity conservation have often taken place within the broader context of cross-sectoral integrative discourse, including considerations of integrated coastal and ocean management as well as marine spatial planning, for example in Chapter 8 (on spatially based governance), Chapters 10 and 11 (on global processes) and Chapter 20 (for a discussion of the European experience).

Furthermore, it may well be the case that some concerns fisheries have with the biodiversity conservation stream may be linked just as much with the 'third stream' and, in particular, the move towards multi-sectoral integrated ocean management. In many jurisdictions, fisheries may see this as a threat to their traditional access to aquatic resources, and may be more concerned about the potential effects of the integrated governance stream than of the biodiversity conservation stream. Ultimately, the fisheries sectoral view and concerns may well need to be nested within an emerging 'integrative framework' with its high-level goals, norms and constraints and will need to compete, more than it already does, in a world of integrated governance with other sectors including more powerful players such as the energy sector. While this presents serious challenges, integrated management does provide fisheries with an entry point to potentially meet its long-frustrated hope that land-based pollution and degradation of productive ecosystems could be halted and possibly reversed (FAO, 1996).

For the biodiversity conservation stream, since the 1972 UNCHE there is a natural bond with the integrated governance stream in that both look beyond sectors to a more holistic view of oceans and how they are used by humans. For example, ENGOs may see integrated management as a natural approach to incorporating biodiversity conservation goals in sectoral policies. However, it is not clear whether integrated management and marine spatial planning processes would place equal weight on 'environment' and 'economy' (i.e. putting biodiversity conservation as equal to all economic sectors together) or would consider the environment as merely one voice (or 'stakeholder') among many within a multi-stakeholder process. In any case, key issues such as pollution and climate change require a multi-sectoral approach, which will lead the biodiversity conservation stream to involve itself in the integrative stream regardless of the power dynamics in place.

In the end, interactions of fisheries and biodiversity conservation streams – the focus of this book – will likely take place and be conditioned by the emergence of operationally effective integrative frameworks and the interaction of fisheries with other sectors inside those frameworks. These include national integrative frameworks of ocean and coastal management, marine spatial planning and multi-use MPAs (e.g. Chapters 8 and 21) and the UN framework (e.g. Chapters 10 and 11). The future of the biodiversity conservation stream may therefore be strongly influenced by the power of the integrative stream, while the future of fisheries will be conditioned by interactions not only with the biodiversity conservation stream but also with other, sometimes more economically powerful, sectors competing for political power and space. This further complicates the discussion earlier in this chapter of 'win-win' or 'win-win-win' approaches, as it highlights the issue of who determines the playing field and who decides what is equitable.

Closing thoughts: An empathetic ear

In this book we have sought to present a balanced analysis by a balanced and diverse set of authors, reflecting a comprehensive representation of the forces at play and the behaviours and outcomes observed. While the various chapters in the book deal with a common theme, the book's chapters do not adopt a homogenous perspective. On the contrary, there is a widely varied series of chapters written from many different perspectives.

We must acknowledge that the two-stream approach we have used throughout the book is only a part of a more complex and evolving flow of governance. First, there are other streams that influence thinking, policy and action on oceans – notably the integration stream discussed above. Second, there are other crucial dimensions of conflict, such as between those who incorporate social aspects into decision-making and policy directions and those who do not. Third, the metaphor of 'streams' itself is limited in discussing fisheries and biodiversity conservation, in that the two streams are distinct but neither is homogenous. Each exhibits a complex and evolving distribution of thinking, from radical and often marginal positions (e.g. in favour of outright protection or of total freedom to use) to more moderate and often dominating views that agree on sustainable use (even when disagreeing on what it entails). Indeed, within either stream we can see substantial differences among the set of players. For example, the review in Chapter 25 of the role of environmental NGOs in small island developing states (SIDS) contrasts large ENGOs operating at large geographic scales (which they suggest 'tend to be more conservation oriented') with local SIDS NGOs and fisheries NGOs (which attempt to balance conservation with livelihood and development objectives). The shape of the resulting distribution of thinking for each stream may differ across scales, levels and issues but, in any given situation, we can envision each stream comprising a spectrum of individual perspectives (on values, goals and risk preferences) with some individuals lying close to, and others more distant from, the dominant mode of the other stream.

Considering this, would it be desirable if those in each stream lying closest to the other stream were to hold power and/or dominate thinking within their stream? That scenario may be most likely to accelerate the pace to a harmonious coexistence, but would the resulting 'monoculture of thought' be desirable or is there value in a diversity of views on such core themes as conservation and use? And would such a middle-of-the-road evolution be stable, if those on the 'far end' of the distribution were consistently to see decisions being made that were much further from their preferences than had been the case when decisions were made entirely within their own stream? In any case, even in a democracy recognizing leadership by the majority, there could be value in understanding the full distribution of views within each stream in the hope of facilitating stable coevolution to the extent possible.

The above discussion might be broadened to address a fundamental question arising from our implicit assumption that, starting with common roots, the identified processes of convergence and coevolution have been leading towards the 'right' outcome, albeit slowly and with difficulties. But are convergence and coevolution in fact guiding the two streams in the 'right' direction? Or could coevolution actually produce bad mutations and non-viable adaptations? These questions have not been addressed in detail within this volume, and it would seem that the question may not have an answer since it is clear there is no global consensus on what a 'right' outcome would be. Perhaps a more realistic 'best' outcome would be continual movement of both streams away from the outcomes which most agree are undesirable. This would still mean movement in directions more favoured by some perspectives than others, so progress in the 'right' direction requires a careful balancing of whose benefits are increasing at what rate and whose are decreasing. As with so much concerning fisheries and oceans, adaptive management is needed; since both the end point of the evolutionary process and the process itself are dynamic, there needs to be regular adaptation to changing circumstances as lessons are learned from success and failures.

In closing, despite the nuances underlying the two-stream analysis of this book, it has provided a conceptual framework that has, in our view, proved its value. Within the book it has helped us frame a set of hypotheses, the robustness of which we might not have easily agreed initially. It helped us extract emergent insights from the book's chapters (e.g. on governance scales, the role of risk perceptions and the mechanisms of convergence and coevolution) that were not present or were very elementary previously. In addition, in the real world the two-stream focus allows one to understand better and rationalize retrospectively what has happened at any number of meetings on fisheries and/or ocean issues worldwide, whether at a local community level or at national, regional or global levels. More importantly, the two-stream dynamic conceptualization may help us all prepare better to make institutions, processes and individual meetings achieve more in the future. For example, one sees in the drafting of global agreements or guidelines, on subjects such as marine species at risk, how elusive it can be to find the right interpretational words that reflect real understanding of both streams and the right tone that acknowledges both perspectives as legitimate. Accordingly, we hope that this book has helped not only in providing those holding one perspective with factual knowledge of the other perspective (in terms of its history, events, major documents, etc.) but also in creating a more genuine understanding of the other perspective, and even a certain empathy.

Notes

1. For example, the 'TEEB' initiative on The Economics of Ecosystems and Biodiversity (http://www.teebweb.org/).
2. This resolution, adopted in December 2006, called for a series of specific actions to be taken by States regarding authorization of their vessels to engage in bottom fishing on the high seas.
3. Being cautious with the term, since success today may not continue tomorrow as sustainability calls for dynamic adaptation to changing conditions.

References

Cash, D.W., Adger, W., Berkes, F., Garden, P., Lebel, L., Olsson, P., Pritchard, L. and Young, O. (2006) Scale and cross-scale dynamics: governance and information in a multilevel world. *Ecology and Society* **11**(2), 8.

Charles, A. (2012) People, oceans and scale: governance, livelihoods and climate change adaptation in marine social–ecological systems. *Current Opinion in Environmental Sustainability* **4**, 351–357.

FAO (1996) Integration of fisheries into coastal areas management. FAO Technical Guidelines for Responsible Fisheries, No. 3. FAO, Rome.

Gibson, C., Ostrom, E. and Ahn, T.K. (2000) The concept of scale and the human dimension of global change: a survey. *Ecological Economics* **32**, 217–239.

Hastie, R. and Dawes, R.M. (eds) (2001) *Rational Choice in an Uncertain World: The Psychology of Judgment and Decision Making*. SAGE Publications, Thousand Oaks, California.

Pitcher, T.J., Kalikoski, D., Pramod, G. and Short, K. (2008) Safe conduct? Twelve years fishing under the UN Code. WWF, Gland. 34 pp. Available at http://www.panda.org/safeconduct (accessed March 2014).

Annex 1

History of fisheries and biodiversity conservation: A timeline of key events (1850–2012)

S.M. Garcia

IUCN Commission on Ecosystem Management (IUCN/CEM/FEG), Gland, Switzerland

The origin and historical development of fisheries management and conservation of nature (specifically in the oceans) varies according to regions and countries and with the scale at which these are considered (from local to global). This book focused on the international level and refers to international initiatives when it is felt that they signal the beginning of a significant change at global level. However, national and international initiatives are connected by an action–reaction chain and a number of international events described below have their origin in some national event/initiative and vice versa. Mentioning a subject (e.g. the precautionary approach) for the first time (in 1995 when adopted by FAO) may not always mean that this event is the absolute first in that domain. There were some national predecessors. Similarly, important international initiatives are often preceded by preparatory work (e.g. meetings of like-minded countries, preparatory working groups, early debates at conferences) and followed by review meetings to check on implementation and performance. Unless these have produced something of high relevance for the purpose of the book or serve as examples of implementation, they are not referred to.

The table below refers to events which are believed to have had global repercussions, shaping evolution. It is intended to show key milestones but does not pretend to be exhaustive. In the early years, it follows the developments in the Northern Hemisphere where overfishing, fishery and conservation science, State-led governance and international organizations started shaping governance. While the fisheries part of the table is exclusively focused on marine fisheries, the conservation part includes references to the conservation of nature in general and of terrestrial species and ecosystems, as these were precursors for marine conservation (emerging in the 1960s–1970s), biodiversity conservation (after 1992) and the establishment of the CBD. The cells containing events of a cross-sectoral nature of relevance both for fisheries and biodiversity conservation are shaded for better visualization.

Governance of Marine Fisheries and Biodiversity Conservation: Interaction and Coevolution, First Edition.
Edited by Serge M. Garcia, Jake Rice and Anthony Charles.

Year	Nature/biodiversity conservation	Fisheries management
1850–1900	• **Establishment of first national parks** in the USA (Christophersen, 1997). **Fortress Conservation** (parks without dwellers) was the North American Model. That model was used by colonial powers until the 1970s (Büscher and Diets, 2005). It is still used in many former colonies as the dominant model (Cernea and Schmidt-Soltau, 2003).	
1854		• **Overfishing**: John Cleghorn of Wick introduced the term 'overfishing' in the British Association for the Advancement of Science (BAAS) in relation to herring fisheries (Roswadowski, 2002).
1860–1870	• Acceleration of industrial development. Growing concern for the environment.	• Acceleration of industrial development. Growing concern for resources sustainability.
1860		• Norway establishes a commission to investigate fisheries problems, building the idea that **science** should provide the basis for decisions in fisheries (Roswadowski, 2002).
1868		• The **UK Sea Fisheries Act** deregulates fishing activities repealing more than 50 fisheries-related Acts established over centuries. The action is based on the advice of the 1883–1885 **Royal Commission on Sea fisheries of the United Kingdom** and the influence of T.H. Huxley, member of this Commission.
1870	• **Creation of a seal reserve** and closure of the Pribilof Island fur seal exploitation.	
1871		• Creation of the 1st **US Commission on Fish and Fisheries** because of concern about the state of some fishery resources (Atkinson, 1988).
1872	• Creation of the **Yellowstone National Park** in USA.	• Establishment of the first marine research laboratory in Naples (Stazione Zoologica di Napoli).
1882		• **T.H. Huxley** maintains his position, despite years of criticism, in his opening lecture of the Fisheries Exhibition in London (Roswadowski, 2002, p. 28). Forgetting the qualifiers in his statement, it is often cited as having unleashed **overfishing.**
1883	• **Paris Convention on the protection of seals in the Behring Sea.** One of the first initiatives to protect species in addition to areas.	
1890		• The plaice fishery in the North Sea starts stagnating (Gulland and Carroz, 1968). • Heinke (Germany) works on the increasingly urgent question of **overfishing** of the North Sea (Roswadowski, 2002).

Year	Nature/biodiversity conservation	Fisheries management
1899		• At the Stockhom Conference on fisheries, Delegates affirmed that rational exploitation of the sea should rest as far as possible on **scientific enquiry** (Roswadowski, 2002, p. 34).
1900	• **London Convention on the Protection of Wild Fauna in Africa** (never ratified). It aimed at (1) creating hunting reserves (excluding traditional hunters); and (2) ensuring conservation of wild, useful or inoffensive animal species. It established a list of protected species and species-based hunting regulations.	
1902	• Creation of the **International Council for the Exploration of the Sea** (ICES) following the 1901 **Kristiania Conference** and the **International Fisheries Congresses** of 1896 and 1898. The principle to use **scientific inquiry** as the basis for a rational exploitation of the sea in endorsed and ICES will implement the principle with a mandate covering both environmental and fishery-related science as well as fishery science.	
	• **Paris Convention for the Protection of Birds Useful to Agriculture**. In force but superseded. Essentially limited to the protection of small passerines and nocturnal raptors, it lists 'noxious species', including diurnal birds of prey and fish-eating birds such as herons, pelicans and divers, the destruction of which is encouraged (de Klemm and Shine, 1993).	
1903		• **Establishment of the ICES Overfishing Committee**. Following the Kristiana Conference, ICES establishes two committees dealing with: (1) the annual and decadal fluctuations of stocks; and (2) the effect of trawlers (the Overfishing Committee).
1905		• The haddock fishery in the North Sea starts stagnating (Gulland and Carroz, 1968).
1909	• **The Th. Roosevelt initiative** to convene a 1st International Conservation Conference in The Hague aborted, obstructed by the US Congress. • **North American Conservation Conference** (Washington) recommends nomination of Conservation Commissioners in each country, to promote collaboration (in fugitive resources) (Ciriacy-Wantrup, 1968).	
1910	• **8th International Zoological Congress, Gratz (Switzerland).** The establishment of the *International or* ***World Commission for the Protection of Nature*** is considered for the first time (UNESCO, 1948). Conferences in 1923 and 1923 will follow.	

(continued)

Year	Nature/biodiversity conservation	Fisheries management
1911	• **First national agency for protected areas**: Canada created the world's first park agency, the Dominion Parks (D. Laffoley, personal communication, 2012).	• **Convention for the preservation and protection of fur seal in the North Pacific** (US, Canada, Japan, Russia) (Atkinson, 1988; Gulland and Carroz, 1968). Terminated in 1941. • **The problem of common resources**. Warming publishes (in Danish) the first systematic study of the problem of overuse of common pool fishery resources. His theory (translated in English by Peder Andersen in 1983) was independently rediscovered 43 years later by Scott Gordon in 1954 and 'immortalized' as 'The tragedy of the commons' by Hardin (1968).
1913	• **1st whale sanctuary**. Australia establishes a sanctuary to protect important calving grounds for humpback whales at Norwegian Bay in Western Australia (currently part of the Ningaloo Marine Park). An example of marine protected area focused on the protection of habitats for iconic species, rather than taking an ecosystem-based approach to conservation and management (D. Laffoley, personal communication, 2012).	
	• **World Treaty on Conservation** (Bern). Proposes the establishment of an **International Commission for the protection of Wildlife**. A draft was adopted but the process was aborted by WWI. A new attempt was made in 1923 (UNESCO, 1948).	
1914		• **Natural fluctuations**: Johan Hjort and Einar Lee establish the natural origin of fisheries fluctuations through fluctuation of year-class strength. WWI interrupted progress in that direction (Roswadowski, 2002, p. 85).
1916	• **The US National Park Service** is established (D. Laffoley, personal communication, 2012)	
1919		• Very high post-WWI catch rates demonstrate the reality of the **overfishing** problem in the years before the wars (Gulland and Carroz, 1968). • The **Commission Internationale pour l'Exploration de la Mer Méditerranée (CIESM)** is established (Roswadowski, 2002).
1920s		• Many stocks (Plaice, hake, cod, haddock) are overfished in the North Atlantic (Gulland and Carroz, 1968). • Early 1920s: **ICES struggles** to establish international regulations of the trawling activity, amidst scientific uncertainty, financial difficulties, and industry's opposition (Roswadowski, 2002). • Mid-1920s: **The fishing industry recognizes the role of science** in fisheries regulation. • 1929–1930: general **recognition that stocks and fisheries could be predicted** by monitoring year-class strength (Roswadowski, 2002, p. 85).

Year	Nature/biodiversity conservation	Fisheries management
1923	• **First International Congress on the protection of nature, Paris.** The attempt to establish an international institution failed (UNESCO, 1948). The concept will be revived with the establishment of IUPN in 1948 and discussed at the first UNESCO-IUCN International Technical Conference on Nature Protection at Lake Success (USA) in 1949 (de Klemm and Shine, 1993).	
1925	• **The League of Nations (LON) Expert Committee** identifies two issues related to fisheries worth the LON attention: (1) the territorial waters regime and (2) the exploitation of ocean resources. Two international projects were formulated: (1) an Office for International Waters to deal with coastal States rights, and (2) a central Institute for Ocean Resources that would produce the scientific and economic data to help define the right. The first was too controversial and abandoned in 1930. The second was supposed to define rational exploitation, establish reserves, rotation exploitation schemes, closing seasons, minimum age for capture and efficient methods to control the measures implementation. The project was appreciated by many States and opposed by the big fishing nations (which, together, caught more than 1 million tonnes). A global project was not consensual. The preference of the 'big ones' was a set of regional agreements between interested States and only them. ICES was mentioned. The threatened species were marine mammals. The key word was 'protection' (Savini, 1976).	
1928	• **International Office for the protection of Nature** established in Brussels, for documentation purposes and to facilitate cooperation.	• **ICES is internationally recognized.** The League of Nations recognizes ICES as 'the appropriate expert body from which to seek advice on marine resources such as whales, seals and fisheries' (Roswadowski, 2002, p. 76).
1930	• **The LON Conference** resolution provides that 'protection of the various species of marine fauna should be considered not only in the territorial sea but also out of it'. This question of resources in areas beyond national jurisdiction will keep lawyers, biologists, politicians and lobby groups busy for the following 80 years (at least). The issue will be dropped and taken up again by the newly established United Nations in the 1950s (Savini, 1976) and will be partly resolved by the LOSC in 1982.	
1931		• **Controlling overfishing?** Edward Russel recognized that effort could be adjusted each year if natural stock fluctuations could be accurately predicted.
1932	• **First transboundary national park** between the USA and Canada, and the first bilateral protected area between Poland and Slovakia.	• **RFBs:** establishment of the **International Pacific Halibut Commission** (Atkinson, 1988).
1933	• **London Convention Relative to the Preservation of Fauna and Flora in their Natural State.** This African Convention had similar objectives to the aborted 1900 Convention for Africa (hunting reserves and **protected areas**). Remained in force until 1968 when it was replaced (de Klemm and Shine 1993, p. 141)	• **Optimum yield:** Hjort, Jahn and Ottestad establish that maximum catch would be obtained from a population reduced to half its original size. This was a precursor to the concept of **Maximum Sustainable Yield** (Roswadowski, 2002, p. 91). The concept was further developed by Graham (1935) and Beverton and Holt (1957), and adopted enshrined in the LOSC.

(*continued*)

Year	Nature/biodiversity conservation	Fisheries management
1935	• **The 1st MPA for ecosystem protection** was probably the Fort Jefferson National Monument marine and coastal park in Florida (D. Lafffoley, personal communication, 2012)	• The British Biologist Michael Graham synthesized his **Great Law of Fishing**, predicting that 'all fisheries that are unregulated become unprofitable'. He also articulated the role that **science and scientists** should have in fisheries regulations, 'including through the study of economic and social aspects of the industry' (Roswadowski, 2002, p. 77, 89).
1937		• **London Convention for the Regulation of Meshes for Fishing Nets and the Size Limits for Fish** (Saetersdal, 1992). It is also called the first **London Overfishing Convention** (Roswadowski 2002, p. 77). • **London Convention on Whaling** (Roswadowski, 2002, p. 77) • **International Pacific Salmon Fisheries Commission**
1940	• **Pan-American Convention on Nature Protection and Wildlife Preservation in the Western Hemisphere.** Requires Contracting Parties to adopt appropriate measures to **protect migratory birds** of economic value or aesthetic interest, or to prevent the extinction of a given species which is under threat including in **protected areas** (Ciriacy-Wantrup, 1968). • **President Truman's Declaration** on the Policy of the United States with respect to Natural Resources of the subsoil and the seabed of the continental shelf	
1945	• **H.S. Truman Proclamation**. It set a claim of sovereignty by the USA to the outer continental shelf (OCS) and the resources therein, establishing the right of the US to establish conservations zones 'in areas of the high seas contiguous to the coasts of the United States'. While recognizing some limited sovereignty over an expanded region of the sea, the proclamation was careful to stipulate that the new US policy did not affect 'the right of free and unimpeded navigation' (Truman, 1945).	
		• **Establishment of FAO**. Tensions build up between FAO and ICES. • The **1st FAO Technical Committee on fisheries** identifies all the key issues that will be addressed for the following 6 decades by the FAO programme of work. Bycatch is seen as a 'waste' issue. • **Salmon is overfished** in the USA (Royce, 1988).
1946	• The **Swiss League for the protection of nature** facilitates the adoption of an international resolution stating the desirability of an active **organization for the protection of nature.**	• The very high catch rates obtained when resuming fishing after WWII confirmed the previously overfished status of many important stocks such as plaice, haddock and hake in the North Sea or bream in the Yellow sea (Gulland and Carroz, 1968).

Year	Nature/biodiversity conservation	Fisheries management
1946 (*cont'd*)		• **London Overfishing Conference and Convention** (Gulland and Carroz, 1968; Saetersdal, 1992) calls for regulation of mesh sizes and landing fish sizes and for regulation of fishing fleet capacity. The latter could not be agreed. The Convention became effective in 1954 and was later transformed into NEAFC in 1959 (in force in 1963). • **RFBs-IWC: The International Convention for the Regulation of Whaling** signed in Washington. It adopts MSY as the basis for allocating quotas. • **Overfishing:** ICES express fear that the reconstruction efforts would recreate the pre-war overfishing situation. Proposals by Russel to limit fleet tonnage of countries were turned down. (Roswadowski, 2002, p. 107, 109).
1947	• The **International Conference for the Protection of Nature** organized by the Swiss League set up a **Provisional International Union for the Protection of Nature.** UNESCO is asked to convene a follow-up conference in Paris (UNESCO, 1948).	• Peru and Chile unilaterally extend their jurisdiction to 200 miles.
1948	• Establishment of the **International Union for the Protection of Nature** (IUPN) under the auspices of UNESCO. As reflected by the use of 'protection' in its name, IUPN intended to mobilize international support to **protect** living species and habitats for endangered species from humans (Christophersen, 1997). The name was changed to IUCN in 1956, replacing 'protection' by 'conservation'.	• First issue of the **State of Food and Agriculture (SOFA)** that included small sections on fisheries. In 1997, the State of Fisheries and Aquaculture (SOFIA) took over that function. • **RFBs-APFIC:** Establishment of the Asia-Pacific Fishery Commission.
1949	• **1st UNESCO-IUPN International Technical Conference on Nature Protection** at Lake Success (USA). Discussed again the 1913 concept of a **World Treaty on Conservation**, revived at an IUCN meeting in Fontainebleau in 1948. Disparities between the conditions and legislations of countries made a treaty unworkable. Regional and sectoral treaties were considered more feasible. The momentum for a global treaty was renewed in the early 1980s, culminating in the adoption of the **CBD** in 1992 (de Klemm and Shine, 1993). • **Aldo Leopold (1887–1948)** published *A Sand County Almanac* in USA, developing the idea of a 'land ethic', a responsible relationship between people and the land they inhabit. A landmark in the American conservation movement.	• **RFBs-ICNAF:** Establishment of the International Commission for the Northwest Atlantic Fisheries (ICNAF) (Later NAFO). • **RFBs-GFCM**. Establishment of the General Fisheries Council for the Mediterranean (GFCM). Later renamed 'Commission'. • **High seas fishing regulation** selected at the UN as a topic for codification by the International Law Commission (Gulland and Carroz, 1968). • **Overfishing**: Salmon and halibut (North Pacific) and Plaice (North Atlantic) are considered overfished (Gulland and Carroz, 1968). • **RFBs-IATTC:** Establishment of the Inter-American Tropical Tuna Commission.

(*continued*)

Year	Nature/biodiversity conservation	Fisheries management
Until 1950s		• Fishery regulations were largely designed to promote orderly marketing and orderly fishing, not really for the purpose of conservation in the usual sense that we think of it (Royce, 1988)
1950s		• Emergence of a progressively more relevant **fishery science** for management (Royce, 1988; Roswadowski, 2002). • Expansion of North Atlantic fisheries to the Arctic. Discovery of deep-sea stocks (Sebastes).
1950s–1970s	• **Fortress conservation** (FC) in Africa. The '**fences and fines**' approach, often with exclusion and forced relocation of the poor original communities (Büscher and Dietz, 2005). This approach is deemed to have failed because of a sense of injustice and illegitimacy.	• **1950–1965: Overfishing spreads.** Most high value stocks of the North Atlantic become overfished (Gulland and Carroz, 1968). • **Mid-1950s: Emergence of bio-economic models** to fisheries management with the work of Gordon and Scott.
1951	• **Rachel Carlson (1904–1964)** publishes her famous book *The Sea Around Us*. A fishery biologist by training, she was credited with advancing the environmental movement.	
1952		• **200 miles**: Santiago Declaration on the 200 miles extended jurisdiction (Chile, Peru, Ecuador).
1953	• **200 miles:** A report of the UN proposes: (1) to give to coastal States a management authority on the 200 miles contiguous to its territorial waters in order to establish the prohibitions needed to protect ocean resources from extermination; and (2) to create an International Court of Justice to resolve conflicts. The proposals were opposed as 'favoring too much the coastal nations'.	
1954		• **The 'Fairtry'** industrial factory trawler is launched in the USA (Warner, 1983), initiating the era of long-range factory vessels that will roam the world, discover and deplete resources in just three decades.
1955	• **The UN-FAO Technical Conference on the Conservation of the Living Resources of the Sea** favoured the option of regional agreements (over that of a central institution) based on the geographical distribution of the species concerned. It is interesting to note that a similar process, aiming at the establishment of a global authority for nature conservation (by President Th. Roosevelt in 1909), failed 50 years earlier, ending with the same international preference for regional institutions.	
1956	• **From IUPN to IUCN:** The IUPN Assembly in Edinburgh changed 'International Union for the **Protection** of Nature (IUPN)', to '**International Union for Conservation of Nature and Natural Resources**', reflecting a transition from a so-called 'preservationist' approach towards the utilitarian aspect of 'conservation' as rational long-term resource management, which the 1980 IUCN World Conservation Strategy later codified as '**sustainable development**' (Christophersen, 1997).	

Year	Nature/biodiversity conservation	Fisheries management
1957		• **Beverton and Holt** publish their famous book *On the Dynamics of Exploited Fish Populations*. • The **North Pacific Fur Seal Convention** (replaces the 1911 Agreement).
1958	• **First UN Conference on the Law of the Sea** (UNCLOS I, Geneva).	
		• **Convention on Fishing and Conservation of the Living Resources of the High Seas**. In force in 1966. Never really implemented. • **First 'Cod war'** (1958–1961) between Iceland and Britain. It ended with the recognition of Iceland's 12 miles exclusive zone by Britain.
1959	• **Protected areas.** The United Nations Economic and Social Council (ECOSOC) established a UN List for Nature Parks and Equivalent Reserves and IUCN is asked to prepare and maintain it. • **Antarctic Treaty**, signed in Washington, gave birth to the Convention for the Conservation of Antarctic Seals (in 1972) and the **Convention on the Conservation of Antarctic Marine Living Resources** (CCAMLR in 1980).	• **Development of the RFBs network**: FAO conference authorizes the establishment of regional fishery bodies despite resistance from ICES (Roswadowski, 2002).
1960	• **Second Conference on the Law of the Sea** (UNCLOS II)	
	• **ICNP:** Creation of the IUCN **International Commission on National Parks (CNP)**.	
1960s	• In the 1960s, the **International Biological Programme (IBP**) (of UNESCO) was instrumental in raising the level of awareness about the need for conservation.	• **12 miles fishing zone:** After 1960, several states enacted legislation providing for an exclusive fishing zone beyond 12 miles (Gulland and Carroz, 1968). • **Creeping jurisdiction:** without claiming jurisdiction, some Asian states reserve their rights to regulate fisheries up to 100 miles beyond their territorial sea (Gulland and Carroz, 1968). • During the 1960s, overfishing of whales becomes clear. Pirate whaling develops to evade IWC measures Development of long range fleets to the southern hemisphere. • Strong emphasis on technological development in FAO programme • Development of national statistical systems and fishery research capacity in the developing world. • Large-scale fishery resources discovery supported by UNDP and FAO. Large-scale development of the **research capacity.** • Large-scale development of resources assessment capacity and spreading of the **science-based fishery management paradigm.** • Development of the interest for species interactions (food chain) and **multispecies fisheries** considerations that would develop two decades later (Roswadowski, 2002, p. 247).

(continued)

Year	Nature/biodiversity conservation	Fisheries management
1961	• **WWF:** After years with continuing funding problems in IUCN, members decided to create a parallel but complementary body, the World Wildlife Fund, to focus on fund raising, public relations and large-scale public support. (Christophersen, 1997).	• **RFBs-CARPAS**: Establishment of Regional Fisheries Advisory Commission for the Southwest Atlantic (CARPAS). For local political reasons, it remained practically idle after 1974. It was abolished in 1999.
1962	• **MPAs. The World Parks Congress on National Parks** gives a first international impetus to the MPA movement (D. Laffoley, personal communication, 2012).	
1963	• **Red List.** The establishment of the **IUCN Red List** was a prime example of species-based conservation. CITES was another one key example while RAMSAR was an example of space-based approach to conservation.	• **RFMOs-NEAFC:** Establishment of the Northeast Atlantic Fisheries Commission to replace the 1954 London Convention.
1964	• **Agreed Measures for the Conservation of Antarctic Fauna and Flora** adopted by the Parties to the 1959 Antarctic Treaty, specifically protect a certain number of areas of particular ecological value.	
Mid-1960s	• **Tuna-dolphin issue:** the very high mortality of dolphins in tuna purse seining in the EC Pacific start raising concern. A 'dolphin safe' label is issued under ENGOs pressure.	
		• **By mid-1960s**, management starts focusing on technical aspects of management. Creation of numerous RFBs under FAOs constitution. • **Inadequate management**. ICES scientists warn that mesh-size regulations would soon be inadequate to face the rapid rising of fishing capacity (Roswadowski, 2002, p. 189).
1966		• **1st FAO Committee on Fisheries**: focus on fishery bodies and management while the FAO extra-budgetary Field Programme remains focused on development. • Scott and Christy again address the central issue of **common property** with little or no echo in the decision-making quarters. • **The 1958 Convention on Fishing and Conservation of the Living Resources of the High Seas** (adopted at UNCLOS I) enters into force. • **RFBs-ICCAT:** Establishment of the International Commission for the Conservation of Atlantic Tuna.
1967	• **UNCLOS III preparatory process:** Avid Pardo (Ambassador of Malta) calls on the UNGA to develop 'an effective international regime over the seabed and the ocean floor' that clearly defines national jurisdiction and to apply the concept of **Common Heritage of Mankind** to the resources of the seabed beyond national jurisdiction. One month later, the UNGA resolutions 2467 A (XXIII) and 2750 C (XXV) created the Committee on the Peaceful Uses of the Sea-Bed and the Ocean Floor beyond the Limits of National Jurisdiction and called for the convening of a third Law of the Sea meeting to be held in 1973 (**UNCLOS III**).	

Year	Nature/biodiversity conservation	Fisheries management
1967 (*cont'd*)	• **Wreck of the oil tanker Torrey Canyon.** A catastrophe that triggered an acceleration in the pressure to conserve marine areas.	• **RFBs-CECAF**: Establishment of the Commission for the Eastern Central Atlantic Fisheries (FAO). • **RFBs-IOFC:** Establishment of Indian Ocean Fishery Commission (FAO).
1968	• The so-called **Tragedy of the Commons** (Hardin, 1968) meant to criticize unmanaged use of the decades of misunderstanding about **common property.** The error was acknowledged only in 1998 (Hardin, 1998).	
	• **UNESCO Conference on the scientific basis of the wise use and conservation of biosphere resources**. Stresses environmental degradation and creates the Man and Biosphere Program (MAB) that aims at developing the scientific basis for management of biosphere reserves. • **African Convention on the Conservation of Nature and Natural Resources** (Algiers) signed in the wake of the independence movement (de Klemm and Shine, 1993).	• **Sequential overfishing.** Gulland and Carroz (1968) refer to a 'switch from the preferred [whale] species to the next most preferred species' only after the first has been 'drastically depleted'. • **First ICES Symposium on the food chain** (convener: J. Steele)
Late 1960s	• Late 1960s–early 1970s: **From species to nature conservation**: Existing species-based legislation is deemed to provide insufficient and too narrow protection. Broader and stronger protection is called for through the enactment of specific nature conservation legislation. The legitimacy of this approach is no longer linked to safeguarding certain resources and activities, but derived from the increasingly recognized need to **preserve species and biological diversity for their intrinsic value** and for the use of future generations (de Klemm and Shine, 1993).	• **Catch limitations (TACs) are introduced** in North Atlantic Fisheries on the argument that effort limitations would be impossible to implement (Roswadowski, 2002, p. 190). ICNAF TACs start to be implemented in 1969.
1969	• **GESAMP**. Establishment of the Group of Experts on Scientific Aspects of Marine Pollution (GESAMP) sponsored by IMO, FAO, UNEP, WHO, WMO, IAEA), UNESCO, IOC and the UN.	• **RFBs-ICSEAF:** Establishment of the International Commission for the South East Atlantic Fisheries. ICSEAF closed after the independence of Namibia.
1970–1990	• 1970–1990: Shift to **community-based conservation (CBC)** in Africa but based on western objectives of conservation of aesthetic and intrinsic values of wildlife. The approach met with limited success (Büscher and Diets, 2005).	• Mid to late-1970s**: Unilateral expansion of numerous national jurisdiction** (in fishing zones and then EEZs). Long-range foreign fleets concentrate on Western Sahara, Namibia and the Southwest Atlantic. **Pressure on developing countries** to open their waters builds up, based on the 'surplus' provision of UNCLOS. **Rapid development of national fleets** to 'fill the void' left by excluded foreign fleets.
1970	• **Launch of the Man & Biosphere Programme of UNESCO**	• FAO global **Species Identification Programme** develops in support of research and management.

(*continued*)

<table>
<tr><th>Year</th><th>Nature/biodiversity conservation</th><th>Fisheries management</th></tr>
<tr><td>1971</td><td>• The Ramsar Convention. The Wetlands Convention on Wetlands of International Importance Especially as Waterfowl Habitat (Ramsar, Iran). It represents a space-based approach to conservation and management of wetlands included in the List of Wetlands of International Importance. Parties are required to promote 'wise use' to establishing nature reserves (de Klemm and Shine, 1993). Wise use of wetlands is defined in the Convention as 'the maintenance of their ecological character, achieved through the implementation of ecosystem approaches, within the context of sustainable development'.
• The Man and Biosphere (MAB) programme of Unesco is launched. Biosphere reserves are selected as representing different types of ecosystems. They are new types of protected areas in which human occupation and activities compatible with the purpose of the reserve remain authorized but may be strictly regulated (de Klemm and Shine 1993, p. 153).</td><td>• FAO first estimate of world fisheries potential (100 million tonnes). It warns that many fisheries are not sustainable and recommends redistributing existing effort on underutilized species.</td></tr>
<tr><td>1972</td><td colspan="2">• The United Nations Conference on Human Environment (Stockholm) laid the initial basis for sustainable development. The Stockholm Declaration formed the foundation for a number of international instruments laying down general conservation objectives and sometimes very specific conservation rules. This confirmed the evolution of the principle of national sovereignty over natural resources but also the recognition of the real interest of the international community in the conservation of certain of these resources (de Klemm and Shine, 1993).
• The Marine Mammal Protection Act (MMPA) in USA requires improved gear design, fishing methods and dolphin release methods to reduce dolphin deaths in tuna fisheries (particularly in the EC-Pacific).</td></tr>
<tr><td></td><td>• UNEP is established.
• UNESCO World Heritage Convention for the Protection of the World Cultural and Natural Heritage is adopted in Paris. The Convention considers that there should be: (1) a legally recognized duty on the part of all States to conserve their natural and cultural heritage; and (2) a corresponding duty on the part of the world community to contribute financially to the conservation of that natural heritage in developing countries (de Klemm and Shine, 1993).
• Convention for the Conservation of Antarctic Seals, London (under the Antarctic Treaty).
• Oslo Convention. Convention for the Prevention of Marine Pollution by Dumping from Ships and Aircraft. In force from 1974, it will merge in OSPAR in 1992.</td><td>• Collapse of the Peruvian anchoveta fishery.
• Second Cod war between Iceland and Britain after Iceland declared a 50 miles jurisdiction limit.
• The FAO Advisory Committee on Marine Resources (ACMRR) confirmed the poor state of whales resources.
• A 10-year moratorium on whaling recommended.
• The Assessment of the state of whales (1972–1976) began and confirmed the poor state and the urgency of constraining measures.</td></tr>
</table>

Year	Nature/biodiversity conservation	Fisheries management
1973	• **UNCLOS III starts:** Launching of the Third Conference on the Law of the Sea, in line with UNGA Resolution 3067.	
	• **Endangered Species Convention (CITES)** Convention on International Trade in Endangered Species of Wild Fauna and Flora (Washington) entered into force in 1975. Although CITES does not provide for any possibilities of retaliation, Parties have been asked not to accept imports of specimens from countries not complying with the Convention and these prohibitions have been quite effective (de Klemm and Shine, 1993). • **NGOs:** Technically qualified NGOs first accepted as non-voting observers in the CITES process (Article 11 of CITES convention). • **The Endangered Species Act** in USA provides that the identification of a particular area essential to the survival of an endangered species **automatically sets in motion habitat protection measures**, signalling the transition from species protection to integrated habitat protection *for these species*. This Act also signals a **transition from prohibition measures to positive action** such as the implementation of rebuilding plans.	• **FAO Technical Conference on Fisheries Management and Development (Vancouver)**: recognized resources limitations, overcapitalization, insufficient science, the problem of free access, environmental degradation and the need for precautionary, adaptive, anticipatory management. • FAO experts recommends to **lower fishing objective from F_{MSY} to F0.1 or $F_{2/3MSY}$** • **RFBs-WECAFC:** Establishment of the Western Central Atlantic Fishery Commission (FAO).
1974	• **RSOs-Helsinki Convention on the Protection of the Marine Environment of the Baltic Sea Area**. In force in 1980. Revised in 1992. In force in 2000. • **Paris Convention.** Convention for the Prevention of Marine Pollution from Land-Based Sources was opened for signature in June 1974 and entered into force in 1978. Will merge in OSPAR in 1992.	• First **FAO meeting on climatic oscillations** impacts on resources (Guayaquil) and vulnerable oscillating stocks.
1975	• **The Great Barrier Reef Marine Park** is established in Australia. It is among the largest marine parks and one containing large commercial fisheries. A reference in MPA management. • **CNPPA**. The international Commission on National Parks becomes the i**nternational Commission on National Parks and protected Areas.** • First **International Conference on Marine Parks and Reserves** Tokyo.	• **Third Cod war** between Iceland and Britain after Iceland's unilateral extension of its jurisdiction to 200 miles.

(*continued*)

Year	Nature/biodiversity conservation	Fisheries management
1976	• **RSOs-Convention on the Conservation of Nature in the South Pacific** (in Samoa). Entered into force in 1990. Provides for the protection of species and areas. • **RSOs-Barcelona Convention.** Convention for the Protection of the Marine Environment and the Coastal Region of the Mediterranean. In force in 1978. Revised in 1995.	• GFCM sponsors the establishment of the **Barcelona Convention for the protection of the marine environment.** • The US **Fisheries Conservation and Magnuson Act** represents the first extension of jurisdiction in the developed world with a 200-miles fisheries conservation zone (now the EEZ). It declares the 'prevention of overfishing' as a national goal alongside producing optimal yield out of a fishery.
1977		• **ACFM.** The Advisory Committee on Fisheries Management (ACFM) is established in the ICES framework (Roswadowski, 2002, p. 254) and advises the North Atlantic Fisheries Commission (NEAFC), International Baltic Sea Fishery Commission (IBSFC), NASCO, the European Commission and ICES Member Countries.
1978	• **RSOs-Kuwait Convention. Regional Convention for Co-operation on the Protection of the Marine Environment from Pollution.** In force in 1979.	• **RFBs-NAFO** the North Atlantic Fisheries Organization replaces ICNAF (1949–1978). As ICNAF members are extending their waters to 200 miles, NAFO' s jurisdiction is limited to high seas.
1978–1981		• **Social concerns** are strongly expressed in advance of the catch reductions recommended by ACFM.
1979	• **Bonn Convention on the Conservation of Migratory Species of Wild Animals (CMS, the Bonn Convention)** aims to conserve terrestrial, marine and avian migratory species throughout their range, established under the aegis of UNEP. In force in 1983. • **Berne Convention on the Conservation of European Wildlife and Natural Habitats**, the regional conservation treaty for Europe, signed under the auspices of the Council of Europe. • **The European Community Directive on the Conservation of Wild Birds** (79/409) sets out a list of species in annex I whose **habitats** require special conservation measures.	
1980s	• The **World Conservation Strategy** (IUCN, UNEP, WWF) elaborated in collaboration with FAO and UNESCO coins the term '**Sustainable Development**'. It identified 3 objectives: (1) maintenance of essential ecological processes; (2) preservation of genetic diversity; and (3) **sustainable use** of species and ecosystems. It meets with some serious opposition from inside, particularly from the northern NGOs, and sympathy from southern NGOs. • **RFB-CCAMLR (Convention on the Conservation of Antarctic Marine Living Resources)** is signed in Canberra.	
		• ICES develops a focus on **multispecies modelling, assessment and management.** During the following 2 decades, the models will never be used directly for management (Roswadowski, 2002, p. 263).

Year	Nature/biodiversity conservation	Fisheries management
1980s (*cont'd*)		• FAO pushes a new **focus on multispecies fisheries assessment and management**, bycatch, direct effort control, resources allocation and SSFs. • John Gulland (of FAO) **assesses economic losses** due to overfishing as one billion US$ for the Northeast Atlantic (before the Cod collapse). • **The FAO worldwide resources watch** is strengthened through comprehensive regional reviews. • **The SOFA report** stresses the poor state of many resources and calls for improved management, better control of foreign fleets, adjustment of fishing capacity, determination of EEZ boundaries and management of shared stocks.
1980	• **CCAMLR:** Convention for the Conservation of Antarctic Living Marine Resources is established.	
1981	• **RSOs-Abidjan Convention.** Convention for Co-operation in the Protection and Development of the Marine and Coastal Environment of the West and Central African Region. In force in 1984. • **RSOs-Lima Convention:** Convention for the Protection of the Marine Environment and Coastal Area of the South-East Pacific. In force in 1986.	• **Safe biological limits**. Focusing on environmental concerns, ACFM adopts the concept of 'safe biological limits' leaving to States the decision to select among options designed to keep stocks within these limits. Single conservation recommendations will still be made for endangered stocks (Roswadowski, 2002, p. 255).
1982	• Adoption of the **UN Law of the Sea Convention (LOSC**), after an unprecedented effort of codification of the rules relating to all aspects of the use of the marine environment by man.	
	• **World Charter for Nature** adopted by the UNGA. Based on the work of IUCN. The Charter is a piece of soft law that proclaims principles of conservation 'by which all human conduct affecting nature is to be guided and judged'. It incorporates the Stockholm Principles but also the three objectives of the World Conservation Strategy (de Klemm and Shine, 1993). • **MPAs. The 3rd World Parks Congress on National Parks** calls for the incorporation of marine and coastal (and freshwater) sites into the worldwide network of protected areas (Toropova et al., 2010). • **Jeddah Convention.** Regional Convention for the Conservation of the Red Sea and Gulf of Aden Environment. In force 1985. • **SPA Protocol.** The Protocol Concerning Mediterranean Specially Protected Areas (SPA Protocol). In force in 1986. Revised in 1995 as the Protocol Concerning Specially Protected Areas and Biological Diversity in the Mediterranean (**SPA and Biodiversity Protocol**).	• **UNCLOS III** ends with the adoption of the LOSC in Montego Bay, Jamaica. • **10-year Moratorium on whaling** is adopted by the IWC. Recommended in 1972 in Stockholm, it will really start being implemented in 1988. • **RFBs-NEAFC:** The new NEAFC Convention adopted in 1980 enters into force.

(*continued*)

Year	Nature/biodiversity conservation	Fisheries management
1983	• **CMS:** Convention on the Conservation of Migratory Species of Wild Animals (CMS), 1979, in force 1983. • **RSOs-Cartagena Convention.** Convention for the Protection and Development of the Marine Environment of the Wider Caribbean Region. In force in 1986. • **The World Commission on Sustainable Development** is created by the UN. • **1st Workshop on Biosphere reserves (Minsk USSR)** by UNESCO.	
Mid 1980s	• Introduction of **Integrated Conservation and Development Projects (ICDPs)** by WWF (Hughes and Flintan, 2001). • **The tuna-dolphin issue** develops into a crisis in the Eastern Central Pacific.	
1984		• **The World Commission on Environment and Development** (WCED, Brundtland Commission (1984–87) starts its work as an independent body of the UN. • **FAO Conference on Management and Development of fisheries**: With the LOSC, it signals the opening of the new order of the ocean. Priority is on management. The battle for appropriation of resources turns into a battle for access of the new resource owners to foreign markets. Fish trade takes international political importance. Social and economic issues take an upper end. • **The International Conference of Fishworkers and their Supporters,** organized simultaneously in Rome by ICSF as a form of demonstration, drew attention on the importance and problems of the neglected small-scale fisheries.
1985	• **Agreement on the Conservation of Nature and Natural Resources** (Kuala Lumpur) between the ASEAN nations to achieve global environmental protection, based on the objectives of the World Conservation Strategy. Contains provisions on protected areas (de Klemm and Shine, 1993). • **ROSs-Nairobi Convention.** The Convention for the Protection, Management and Development of the Marine and Coastal Environment of the Eastern African Region. In force in 1996. • **Protected Areas and Wild Fauna and Flora in the Eastern African Region.** The Protocol to the Nairobi Convention is adopted 1985.	• **FAO COFI Sub-Committee on trade** is established.

Year	Nature/biodiversity conservation	Fisheries management
1986	• **RSOs-Noumea Convention.** Convention for the Protection of Natural Resources and Environment of the South Pacific Region. In force 1990. • **The IUCN Commission on National parks and protected areas nominates a Vice-Chair for marine areas** to accelerate the development of MPAs (Toropova et al., 2010).	
1987	• **Brundtland Report**: *Our Common future*. The work of the Commission and the report give content to the concept of '**sustainable development**' leaving it vague enough so that each country could develop its own ways and means to reach this ethical goal. The Conclusions of the Brundtland report were adopted by the UN General Assembly as a framework for future cooperation in the field of environment and development.	
	• The NGO **Conservation International** shifts to **sustainable use.** • **The 4th World Wilderness Congress passes a resolution on MPAs** (Kelleher *et al.*, 1995a–d).	
1988	• IUCN sets up the **World Conservation Monitoring Centre (WCMC),** co-sponsored by UNEP and WWF. • UNEP sets up the ad hoc **Working Group of Experts on Biological Diversity**, starting the process leading to the establishment of the CBD in 1992. • The **Flora and Fauna Guarantee Act** of Victoria (Australia) provides, for the first time, for the listing of potentially threatening processes, specifying further the notion of 'harm' in addition to 'taking'. • **Marine conservation objective:** IUCN primary goal of marine conservation is 'to provide for protection, restoration, wise use, understanding and enjoyment of the marine heritage of the world in perpetuity through the creation of a **global representative network of MPAs** and through the management, in accordance with the principles, of human activities that use or affect the marine environment'. Resolution 17:38, 17th General Assembly, Costa Rica). This puts the management of uses and nature protection as integral part of 'conservation'. • **E.O. Wilson** introduces the term '**biodiversity**' in his book ***Biodiversity***, National Academy Press, Washington DC 1988.	

(*continued*)

Year	Nature/biodiversity conservation	Fisheries management
1989	• **Driftnets: The Wellington Convention for the prohibition of fishing with long driftnets in the South Pacific** requires its Parties to prohibit the use of drift nets longer than 2.5 km in areas under their jurisdiction.	
	• **Protocol to the Noumea convention for the Conservation and Management of Protected Marine and Coastal Areas of the South-East Pacific**. In force 1994. • UNEP establishes the Ad Hoc Working Group of Technical and Legal Experts to prepare an international legal instrument for **the conservation and sustainable use of biological diversity.**	UNGA resolution 44/225 on large-scale pelagic driftnet fishing and its impact on the living marine resources of the world's oceans and seas could be considered a case of radical application of the concept of **precaution**, despite the lack of explicit reference to the Precautionary Principle. It called for a moratorium to be imposed on all driftnet fishing by 30 June 1992.
1990	• **Drift nets:** UN Resolution 44(225) recommends a moratorium on the use of large drift nets in the high seas as from 1 July 1991 unless effective conservation and management measures have been taken. • **Tuna-dolphin**: The Dolphin Protection Consumer Information Act (DPCIA) establishes criteria for labelling canned tuna products as 'dolphin safe' when caught on a trip during which no dolphins were encircled. The crisis is over. • **PICES: Adoption of the Convention for the North Pacific Marine Science Organization.**	
	• **'Return to the barriers'** in Africa. A paradigm develops again arguing that conservation should return to the original values of aesthetic beauty and intrinsic value (Büscher and Diets, 2005). It is said that 'protected areas are the last bastions of biodiversity and wildlife conservation and hence should be protected at all costs and focus on preservation, the strictest form of conservation' (Terborgh and Van Shaik, 2002). • **SPAW. Protocol of the Cartagena Convention concerning Protected Areas and Wildlife.** • **Sustainable use**: The 18th IUCN General Assembly, (Perth, Australia) in Recommendation 18.24 resolved that IUCN should develop guidelines on **sustainable use** of wildlife. It recognized that 'the ethical, wise and sustainable use of some wildlife can provide an alternative or supplementary means of productive land use, and can be consistent with and encourage conservation, where such use is in accordance with appropriate safeguards'. • Establishment of the Specialist Group: Sustainable Use of wild Species (SG-SUWS)	• **Greenpeace International** asks for observer status in ICES (Roswadowski 2002, p. 272). • **ICES Study group on ecosystem effects of fishing** is established. • **Establishment of PICES** the North Pacific Marine Science Organization (the Pacific ICES) by Canada, China, Japan, USSR and USA. Following an early ICES model, PICES was established with no assessment groups and no management-related responsibilities.

<table>
<tr><th>Year</th><th>Nature/biodiversity conservation</th><th>Fisheries management</th></tr>
<tr><td>1991</td><td colspan="2">• ICES Listings: 1st proposal to list Bluefin Tuna in CITES Appendix II
• The UN Moratorium on the use of large-scale driftnets in the high seas is applicable. While not binding, UN resolutions have considerable moral force (de Klemm and Shine, 1993).
• ESD: The Australian Fisheries Management Act introduces Ecologically Sustainable Development (ESD) in fisheries, recognizing their potential impact on the ecosystem (McLoughlin et al., 2008).</td></tr>
<tr><td></td><td>• Caring for the Earth: A Strategy for Sustainable Living developed by IUCN, UNEP and WWF. It lays down basic principles and actions to be taken to adapt the WCS to the user's needs and capabilities.
• GEF: Establishment of the Global Environment Facility partnership.</td><td>• Risk: ACFM reformulates its advice formats to take into account the risk inherent in assessments and explicit the implications for management (Roswadowski, 2002, p. 255–256).
• Responsible fishing. COFI19 invites FAO members to define new concepts to make fisheries more responsible and more sustainable.</td></tr>
<tr><td>1992</td><td colspan="2">• UNCED. UN Conference on Environment and Development (UNCED, Rio Conference). Sets important principles of conservation and development (including Principle 15 of the Rio Declaration on the precautionary approach) and outlines a large diversity of programmes of work (including chapter 17 on oceans and seas and the protection and rational use and development of their living resources). The conference will have a profound influence on policy agendas at all levels.
• The UN Framework Convention on Climate Change (UNFCCC) is adopted at UNCED.
• The Convention on Biological Diversity is opened for signature (in force in 1993). The objectives are conservation, sustainable use and the equitable sharing of benefits arising from the utilization of biodiversity and genetic resources. The Convention reiterates (as in the LOSC) the sovereign rights of coastal States over their biological resources as well as the responsibility for conserving their biological diversity and for using them in a sustainable manner. It calls for integration of conservation and development and recognizes the role of indigenous communities in biodiversity protection. Protected areas are an explicit part of the in situ conservation measures (de Klemm and Shine, 1993).
• CITES Listing: Attempted listing of Tuna species in CITES COP 8. The proposal by Sweden to list Western Atlantic Bluefin Tuna on Appendix I and Eastern Atlantic tropical tunas on Appendix II was turned down (mainly on the arguments that they were managed by ICCAT and were not declared endangered) and withdrawn.
• The Global biodiversity strategy is published by the World Resources Institute (WRI), IUCN and UNEP as a complement of (or as an attempt to condition) the work of the future CBD.</td></tr>
<tr><td></td><td>• CBD: The UNEP Nairobi Conference for the Adoption of the Agreed Text of the Convention on Biological Diversity finalizes the draft text of the Convention that to be opened for signatures at UNCED one month later.
• RSOs-Convention for the Conservation of Biological Diversity and the Protection of Priority Wild Areas in Central America (signed by Costa Rica, El Salvador, Guatemala, Honduras, Nicaragua and Panama) (de Klemm and Shine, 1993).</td><td>• Raising concern: A special fishery section of SOFA on fisheries stresses that UNCLOS objectives were being missed, underlining mismanagement, economic losses, social stress, environmental degradation and the need for integrated management. Bycatch emerges as a threat to conservation in addition to a waste. High-seas management is an emerging issue as fleets expand their operations in that area.
• Cod fishery moratorium: Following the collapse of the Cod stocks of Labrador and Newfoundland, Canada closes the fishery. It is still practically closed today, 30 years later.</td></tr>
</table>

(continued)

Year	Nature/biodiversity conservation	Fisheries management
1992 (*cont'd*)	• **EU Council Directive 92/43** on the **Conservation of Natural Habitats and of Wild Fauna and Flora** to promote the maintenance of biodiversity. The Directive provides for two sets of measures covering the protection of species and species' habitats and the protection of habitat types. It is the only recent international instrument combining comprehensive conservation rules and a financial mechanism to assist in their implementation (de Klemm and Shine, 1993). • **OSPAR:** the Convention for the Protection of the Marine Environment of the North-East Atlantic (OSPAR Convention) is adopted, joining the 1972 Oslo Convention and the 1974 Paris Convention. It will enter into force in 1998 with an annex V that extends its mandate to cover biodiversity concerns.	• **Responsible fishing:** In response to COFI 19, the Cancun International Conference on Responsible Fishing Conference takes place in Mexico.
1993	• **The UN Commission on Sustainable Development** is established at the UN to follow up the implementation of Agenda 21. • **The LME programme** starts, supported by GEF and led by NOAA, to assist developing countries in planning and implementing an ecosystem-based management of coastal resources. Originally focused on general environmental analyses and contamination, the programme will progressively integrate and focus on fisheries problems, particularly after WSSD (Sherman, 2008). • **The implications of UNCED and the Agenda 21** for fisheries, in terms of precaution, environmental impact and management integration are brought to the attention of COFI 20. • **The CBD** comes into force.	
		• **Compliance agreement:** adoption of the FAO Agreement to Promote Compliance with the International Conservation and Management measures by Fishing Vessels in the High Seas. The Agreement applies to all types of fishery resources, not only the straddling and highly migratory as the 1995 UNFSA. • **RFBs-IOTC:** Establishment of the Indian Ocean Tuna Commission.
1994	• **UNCLOS** comes into force. • **The ICES WG on ecosystem effects of fishing** is established. • **ACFM meets with ACME.** The Advisory Committees on Fisheries Management (ACFM) and on the Marine Environment (ACME) have their first joint meeting. • **The IUCN Sustainable use Initiative** starts in IUCN. The concept was Implicit in the 1980 WCS. It was reinforced by the WCED (in 1987), Caring for the Earth (in 1991) and UNCED (in 1992). Reiterated in the 1996 World Conservation Congress (WCC, Montreal, Canada) (Christophersen, 1997). It will be formally adopted by IUCN in 2000 (WCC, Amman, Jordania).	

Year	Nature/biodiversity conservation	Fisheries management
1994 (*cont'd*)	• **ICRI. The International Coral Reef Initiative** is established at CBD CoP 1 by Australia, France, Japan, Jamaica, Philippines, Sweden, USA and UK as well as multilateral development banks and environmental and development al NGOs and the private sector.	• **Social sciences in ICES?** The 1994 ICES Statutory meeting discusses the need to bring in experts in economics, sociology and anthropology. Discussions were pursued for 2 more years. The difficulties of moving to fishery 'systems' analysis and management advice were considered. (Roswadowski, 2002, p. 257).
1995	**The Jakarta Mandate** on marine and coastal biological diversity is adopted by CBD CoP2 and a study is called for on the relationship between the **CBD and UNCLOS** on issues related to conservation and sustainable use of genetic resources of the deep seabed. CoP2 recognizes the use of **Integrated Marine and Coastal Area Management (IMCAM)** as the most suitable framework for addressing human impacts on marine and coastal biodiversity and promoting its conservation and sustainable use. The implementation of its Programme of work calls for the **precautionary** and the **ecosystem approaches**. • **Ecosystem approach** is adopted by CBD COP2 in Jakarta (Decision II/8) (Vierros, 2008).	
	• **Comprehensive inventory of potential Marine Protected Areas** is produced by the Great Barrier Reef Marine Park Authority, the World Bank and IUCN (Kelleher et al., 1995a–d). • **Sustainable use initiative (SUI)**: Launched to improve understanding of factors affecting sustainability. The regionally based Sustainable Use Specialist Group (SUSG) is launched in support of the SUI. The sustainable use initiative was disbanded in 2001 but a Sustainable Use Team was established.	• **FAO Code of Conduct for Responsible Fisheries** (CCRF) is adopted. A cornerstone is the **precautionary approach to fisheries** (PAF). • **UN Fish Stock Agreement (UNFSA),** the Agreement for the Implementation of the Provisions of the UNCLOS of 10 December 1982 relating to the Conservation and Management of Straddling Fish Stocks and Migratory Fish Stocks is adopted. It also reflected the PAF and shifted the status of MSY from a target to a limit for fisheries development and a target for rebuilding depleted stocks. Will come into force in 2001. • The CCRF and the UNFSA integrate the **precautionary approach** in fisheries management. • The FAO Ministerial meeting adopts the **Rome Consensus on World Fisheries** recognizing overfishing, overcapacity and threats to resources, and calling for improved conservation and management. • **Kyoto Declaration and Plan of Action** on the Sustainable Contribution of Fisheries to Food Security (organized by Japan and FAO) called for assessment of stocks and fisheries status on food security, employment, trade and production sustainability. • **Turbot war.** An international dispute between Canada (supported by the United Kingdom and Ireland) and Spain (supported by the European Union) in which Canada captured a Spanish trawler (Estai) and its crew just outside the 200 miles for illegal fishing on Greenland Halibut.

(*continued*)

Year	Nature/biodiversity conservation	Fisheries management
1996	• **WCPA**: The international CNPPA becomes the **World Commission on Protected Areas** (WCPA) at the IUCN Congress in Montreal. The Commission will hold the World Parks Congresses (1962, 1972, 1982). • **Sustainable use:** the IUCN WCC (Montreal) reaffirms the Sustainable Use principles and calls for a Policy Statement on Sustainable Use to be adopted in 2000.	• **The FAO SOFA report** that 60% of world's fish resources are in need of urgent management, 35% are currently overfished and 20 million tons of fish could be added to total world landings through rehabilitation of degraded resources, responsible exploitation of underdeveloped resources and reduction of discards and waste.
1997	• **Creation of the Marine Stewardship council (MSC)** by WWF and Unilever to elaborate an eco-labelling system for fisheries. The MSC will become independent soon after. • **Shrimp-turtle conflict**: India, Malaysia, Pakistan and Thailand brought a joint complaint to WTO against a ban imposed by the US on the importation of certain shrimp and shrimp products on all countries not using Turtle Excluder Devices (TEDs). With decisions made in 1998 and 2001, WTO recognized both the right of any WTO member to apply trade restrictions to protect the environment and endangered species and the need to do so in a non-discriminatory manner. This was an important precedent for the use of trade measures to protect biodiversity from fisheries impacts. (http://www.wto.org/english/tratop_e/envir_e/edis08_e.htm).	
		• **The ICES Multispecies assessment WG is retired** considering the little use made of the assessments and the high cost of the methodology (Roswadowski, 2002, p. 254). • A **Resources Management Committee** is established in ICES to deal with economics, sociology and management advice as well as the effect of natural oscillations and climate change on management decisions (Roswadowski, 2002, p. 257).
1998	• **OSPAR** comes into force. Its **annex V on ecosystems and biodiversity** introduces provisions regarding the 'protection and conservation of the ecosystems and biological diversity of the maritime area'. This will lead to the 2003 Recommendation to establish an OSPAR Network of MPAs and the collaboration with NEAFC in identifying EBSAs. • **Marine genetic resources (MGRs):** the CBD Programme of work on marine and coastal biodiversity (decision IV/5) includes an operational objective related to making available information on marine and coastal **genetic resources,** including **bioprospecting.** • **Malawi Principles**. 12 principles for the **Ecosystem Approach** are elaborated by the Workshop on the Ecosystem Approach held in Lilongwe (Malawi) that will be adopted by the CBD CoPV (Decision V/VI) in 2000.	

Year	Nature/biodiversity conservation	Fisheries management
1999	• **ICP. Establishment of UNICPOLOS** (United Nations Open-ended Informal Consultative Process on Oceans and the Law of the Sea (also called the **Consultative Process**) by the UN General Assembly to facilitate the annual review by the UNGA of developments in ocean affairs and the law of the sea (resolution 54/33). • **Ecosystem impact of fishing**: ICES-SCOR symposium on the ecosystem effects of fisheries in Montpellier.	
	• **IUCN** publishes **Guidelines for Marine Protected Areas.**	• **FAO IPOAs.** FAO adopts International Plans of Action for (1) reducing incidental catch of seabirds in longline fisheries; (2) conservation and management of sharks; and (3) management of fishing capacity. • **Precautionary approach**: The ICES Advisory Committee on Fisheries Management (ACFM) establishes an operational definition for the precautionary approach to fisheries. • **FAO Ministerial meeting on IUU fishing**. Adopted a Declaration expressing concern at growing IUU fishing. The Ministers agreed to develop an international Action Plan within the framework of the Code of Conduct.
2000	• **The UN Millennium Summit** (New York) adopted the eight Millennium Development Goals (MDGs), 18 targets and 48 indicators regarding poverty, hunger, health, education, gender issues, child mortality and environmental sustainability. The MDGs are universally accepted as a framework for measuring progress in development. In 2002, the UN launched the Millennium Project to devise a plan of implementation for enabling developing countries to meet the MDGs by 2015, and to assess progress towards their achievement by 2005. • **Census of Marine Life (CoML) starts.** This 10-year international effort undertook to assess the diversity, distribution and abundance of marine life, at a scale never attempted before. The Census stimulated marine science by tackling these issues globally, and engaging some 2700 scientists from more than 80 nations who participated in 540 expeditions, mobilizing 650 million dollars. It produced more than 2600 scientific publications, identified more than 6000 potential new species and registered more than 30 million biodiversity records in the **Ocean Biogeographic Information System (OBIS).** • **Adoption of the Sustainable Use concept** the IUCN World Conservation Congress (Amman, Resolution 2.29) without any amendment. The links between sustainable use, human livelihoods, poverty, and equity emerges as a common focus across IUCN's programmes and units (Christophersen 1997). • **ICP-1.** Discusses fisheries and the impacts of marine pollution and degradation.	
	• **GMOs. Cartagena Protocol on biosafety** adopted by CBD (entered into force 2003) addresses safe transfer, handling and use of living modified organisms that may have an effect on biodiversity with focus on transboundary movements. • **CBD CoP V endorses the 1998 Malawi Principles** for the **Ecosystem Approach.**	• **RFBs-WCPFC:** The **Convention for the Conservation and Management of Highly Migratory Fish Stocks in the Western and Central Pacific Ocean** is opened for signature. It will enter into force in 2004.

(*continued*)

<table>
<tr><th>Year</th><th>Nature/biodiversity conservation</th><th>Fisheries management</th></tr>
<tr><td>2001</td><td>Sustainable use: Elaboration of the 'White Oak' Principles of sustainable use (IUCN)</td><td>• The 1995 UNFSA comes into force
• EAF: FAO adopts the Ecosystem Approach to Fisheries in Reykjavik.
• FAO IPOA-IUU: to prevent, deter and eliminate Illegal, Unreported and Unregulated Fishing (IUU).
• RFMOs-SEAFO. Establishment of the Southeast Atlantic Fisheries Organization. Replaces ICSEAF.</td></tr>
<tr><td>2002</td><td colspan="2">• WSSD. World Summit on Environment and Development (Johannesburg). It enshrined the Ecosystem Approach, stressed the need to eliminate destructive fishing practices and establish marine protected areas (MPAs) and networks of MPAs by 2012.
• Regular Process: WSSD agreed (para. 36b of the POI) to 'establish by 2004 a regular process under the United Nations for global reporting and assessment of the state of the marine environment, including socio-economic aspects' (Global Marine Assessment, GMA). The UNGA endorsed the proposal (Res. 57/141: para 45).
• CITES listing: Proposal to List Patagonian toothfish in CITES appendices. Proposal by Australia to list the Patagonian toothfish under Appendix II, because many vessels fishing for Patagonian toothfish were non-parties to the CCAMLR, the competent RFMO. The measure was seen as complementing CCAMLR's efforts.
• Tuna-dolphin conflict. US Secretary of Commerce changes the definition of the dolphin-safe label concluding that intentional encirclement of dolphins with purse seine nets to catch tuna was no longer having a 'significant adverse impact' on any depleted dolphin stocks.
• ICP-3. Discussed on the protection and preservation of the marine environment, capacity building, regional cooperation and coordination and integrated oceans management.</td></tr>
<tr><td></td><td>• MGRs: Bonn Guidelines (CBD Decision VI/24): The Bonn guidelines on access to genetic resources and the fair and equitable sharing of the benefits arising from their utilization are adopted. They are a useful first step in the implementation process of the CBD provisions on MGRs.
• Biodiversity targets. CBD CoP 6 members commit to reduce significantly biodiversity loss by 2010.
• NE Pacific Convention. The Convention for Cooperation in the Protection and Sustainable Development of the Marine and Coastal Environment of the Northeast Pacific.
• Establishment of the Pelagos Sanctuary in the Mediterranean (87500 km^2).</td><td></td></tr>
<tr><td>2003</td><td colspan="2">• Sea mounts biodiversity: The UNGA encourages urgent integration and improvement of risk management for marine biodiversity on seamounts (Resolution 57/141).
• Regular Process: The UNGA requests the UNSG (Res. 58/240) to convene a group of experts to define the 'Regular Process'.
• VMEs and EBSAs: the UNGA encourages urgent integration and improvement of risk management in threatened and vulnerable marine ecosystems (VMEs) in ABNJ (Resolution 58/240).
• ICP-4. VMEs. ICP4 adopted recommendations on the protection of vulnerable marine ecosystems.
• IPBES. G8 Summit (Evian) agrees to act to implement MDGs and the Johannesburg Plan of Implementation. Will lead to the 2005 Paris Conference.</td></tr>
</table>

Year	Nature/biodiversity conservation	Fisheries management
2003 (*cont'd*)	• **MGRs and deep seabed.** CBD and UNCLOS secretariats undertake a joint study on conservation and sustainable use of deep seabed **genetic resources** beyond national jurisdiction (ABNJ). • **RSOs-Caspian Convention.** Framework Convention for the Protection of the Marine Environment of the Caspian Sea. • **The 5th World Parks Congress** recognizes the suitability of multiple types of governance for protected areas (State-based, community-based, private, etc.).	• **Deep sea conference**: International Conference on Governance and Management of Deep-Sea Fisheries organized with FAO in Queenstown, New Zealand. • **RFBs-SEAFO**: South East Atlantic Fisheries Organization established. • **High Seas Task Force on IUU Fishing** is established by ministers from Australia, Canada, Chile, Namibia, New Zealand and the UK and international non-governmental organizations (Earth Institute, IUCN, WWF and MSC).
2004	• **Sustainable Use Initiative:** formally endorsed at the World Conservation Congress in Amman (Jordania) (Christophersen, 1997). • **Sustainable use:** CBD COP 7 (Kuala Lumpur) adopts the 14 **Addis Ababa Principles and Guidelines for the Sustainable Use of Biodiversity** (Decision VII/14). • **CBD and the FAO Code of Conduct.** COP7 recognizes and recommends the implementation of the Code and its guidelines as the instruments to implement CBD policies in fisheries (Decision VII/5). Preamble paragraph 7 as well as operational objectives 2.1g and 4.1b. • **MPAs and biodiversity**: CBD COP 7 (Kuala Lumpur) strengthens the work program on MPAs and high seas biodiversity, highlighting the urgent need for better international cooperation in ABNJ to establish more MPAs and **fight against destructive practices.** • **Destructive fishing:** the UNGA calls on States to take interim prohibition measures against destructive fishing (Res 59/25: paras 66, 67). • **VMEs:** the UNGA calls for action to regulate bottom fishing in VMEs and establish RFMOs where there are none (Resolution 59/25: paras 68, 69). • **Regular Process;** The GMA International Workshop was held in conjunction with the 5th ICP meeting. It recommends a Task Force to organize an Assessment of Assessments (**AoA**). • **ICP-5. ABNJ**. Adopted recommendations on new sustainable uses of oceans, including the conservation and management of the **biological diversity of the seabed in areas beyond national jurisdictio**n (ABNJ). • **BBNJ**: Resolution 59/24 (paragraph 73) of the UNGA established the **Ad Hoc Open-ended Informal Working Group** to study issues relating to the conservation and sustainable use of marine biological diversity beyond areas of national jurisdiction (BBNJ) to review the activities of the United Nations and other relevant international organizations with regard to the conservation and sustainable use of marine biological diversity in areas beyond national jurisdiction (ABNJ) and provide guidance.	
	• **MGRs and the deep Sea**. CoP7 in decision VII/5: (1) invited Parties and other States to identify activities and processes under their jurisdiction or control with significant adverse impact on deep seabed ecosystems and species in ABNJ; (2) requested the Secretariat to compile information on the methods for identification, assessment and monitoring of **genetic resources** of the seabed and ocean floor and subsoil thereof, in ABNJ; and (3) compile and synthesize information on MGRs status and trends, identifying threats and technical options for their protection.	• **RFBs-WCPFC The Western Central Pacific Fisheries Commission** enters into force under the Convention for the Conservation and Management of Highly Migratory Fish Stocks in the Western and Central Pacific Ocean.

(*continued*)

Year	Nature/biodiversity conservation	Fisheries management
2004 (*cont'd*)	• **Biodiversity in ABNJ:** the UNGA decides to establish an **Ad Hoc Open-ended Informal Working Group** on issues relating to conservation and sustainable use of marine biological diversity in ABNJ (Res. 59/24). Progress will be followed up in 2006, 2009 and 2011 (61/105, 64/72 and 66/68, respectively).	
2005	• **EBSAs identification criteria are elaborated** at the CBD Scientific Experts Workshop on criteria for identifying ecologically and biologically significant areas beyond national jurisdiction, Ottawa, Canada. • **ICP-6. Marine debris:** adopted recommendations on fisheries and their contribution to sustainable development, and considered the issue of **marine debris**.	
	• **Millennium ecosystem assessment** makes a negative assessment of the state of the global environment and biodiversity. • The **Green Growth** concept is adopted at ESCAP. • **GRAMED**. The Global and Regional Assessments of the Marine Environment Database (GRAMED) starts developing (led by UNEP and IOC/UNESCO) to support the '**Assessment of Assessments**' and provide support to a future Regular Process. • **International Conference on Biodiversity, Science and Governance (Paris**). This independent conference calls for an international multi-stakeholder consultative process towards establishment of an **International Mechanism of Scientific Expertise on Biodiversity (IMoSEB)**. The process (2005–2007) will lead the establishment of IPBES in 2008. • First **International Marine Protected Area Congress** in Australia in 2005. • **IPBES.** The **Paris International Conference** on 'Biodiversity, Science and Governance' organized by France is an important step towards the creation of **IPBES** in 2010.	• **International Conference on the Governance of High Seas Fisheries and the United Nations Fish Agreement – Moving from Words to Action** held in St John's, Newfoundland and Labrador. Launched the discussions on the RFMO reform.
2006	• **FAO-CITES Memorandum of Understanding** is signed. FAO and CITES will review and consult together on the scientific, legal and technical evaluation of commercially exploited aquatic species listed or proposed for listing in the CITES Appendices. Following this agreement, CITES will send to FAO, for examination by the FAO Panel Of Independent experts, the country's proposals for listing fishery species on CITES appendices. • **VMEs:** UNGA elaborates a framework for management of **VMEs** (coining the term) in which States authorizing vessels to trawl on the high seas should: (1) conduct assessments of impacts on VMEs; (2) manage impacting activities including through closed-areas prohibit fishing; (3) establish and implement VMEs 'encounter protocols'; (4) implement the precautionary and ecosystem approaches; and (5) sustainably manage deep-sea fish stocks **Resolution 61/105 of ICP-7.** • Enhanced understanding of **ecosystem-based management (EBM)**, and adopted recommendations on **ecosystem approaches (EA**) and oceans. • **IMCAM.** CBD CoP8 stressed again the importance of Integrated Marine and Coastal Area Management.	

Year	Nature/biodiversity conservation	Fisheries management
2006 (*cont'd*)	• **CBD-UNGA cooperation on MPAs in ABNJ**. CBD COP8 (Curitiba) recognizes the role of the CBD scientific and technical support to the work of the UNGA on MPAs in ABNJ. • **The CBD adopts international targets for MPA coverage** starting both a huge effort to meet the target and some criticism about the real value of such targets for improved conservation, in practice. At least 10% of each of the world marine and coastal ecological regions to be protected by 2012. • **MGRs**. CBD **Decision VIII/21** (1) calls for information on research activities related to deep seabed genetic resources in ABNJ; and (2) indicates options for protection of deep seabed genetic resources in ABNJ; codes of conduct, guidelines and principles; threats reduction and management through permits and EIA, MPAs, prohibition of detrimental and destructive practices in vulnerable areas. • **The 1st UNGA WG on biodiversity in ABNJ** discusses issues behind establishment of MPAs in ABNJ. • The **Northwestern Hawaiian Islands Marine National Monument** is established in USA as a sanctuary and as a World Heritage Site (in 2010).	• **EAF implementation**: The 2006 FAO-Norway Conference on Implementing the Ecosystem Approach to Fisheries is a follow-up to the FAO-Iceland Reykjavik Conference in 2001. • **NEAFC strengthens conservation**: it adopts the London Declaration and amends the 1982 Convention, strengthening commitment to conservation. • **1st Review Conference on the UN Fish Stocks Agreement** to assess the adequacy of the Agreement's provisions and means of implementation. Commitment to: (1) integrate ecosystem considerations in fisheries management; (2) develop a legally binding minimum standards for port State measures; (3) develop a comprehensive global register of fishing vessels; and (4) expand assistance to developing countries. Stresses the urgency of (5) reducing fishing capacity; (6) strengthening RFMO mandates; and (7) reviewing RFMOs performance. • **SIODFA.** The **Southern Indian Ocean Deep-sea Fishers Association** is formed. Registered under the Incorporated Societies Act of the Cook Islands, it promotes responsible management of the deepwater fishery resources of the area for sustained harvests benthic biodiversity conservation. SIODFA voluntarily closed a number of SIO areas to its fleets. • **SIOFA.** The Southern Indian Ocean Fisheries Agreement is signed. It will come into force in 2012.
2007	• **ICP-8. MGRs:** discussed issues particularly related to **marine genetic resources**. Delegates were unable to agree on key language referring to the relevant legal regime for marine genetic resources in areas beyond national jurisdiction and, as a result, no recommendations were adopted. However, a Co-Chairs' summary report was forwarded to the General Assembly for consideration. • **Regular Process. AoA**: 1st Meeting of the Group of Experts for the start-up phase, Assessment of Assessments (AoA) of the Regular process for the global reporting and assessment of the state of the marine environment, including socio-economic aspects.	
		• **RFMOs best practices.** Meeting of the Chatham House Independent Panel to develop a model for improved governance by RFMOs. The Panel recommended a set of Best Practices (Lodge et al., 2007). • **FAO Workshop on Vulnerable Marine Ecosystems and Destructive Fishing in Deep-Sea Fisheries**. Clarifies issues regarding ecosystem vulnerability, destructive fishing and related impacts. These guidelines are important for the implementation of the resolution UNGA 61/105 and to bridge nature conservation and fisheries management in vulnerable marine ecosystems.

(*continued*)

<table>
<tr><th>Year</th><th>Nature/biodiversity conservation</th><th>Fisheries management</th></tr>
<tr><td rowspan="2">2008</td><td colspan="2">• Destructive fishing practices: FAO, UNEP and CBD organize jointly the Expert Meeting on Impacts of Destructive Fishing Practices, Unsustainable Fishing and Illegal, Unreported and Unregulated (IUU) Fishing on Marine Biodiversity and Habitats (Rome, facilitated by the IUCN Fisheries Experts Group).
• ICP-10: discusses the outcomes, achievements and shortcomings of the Consultative Process in its first nine years and forwards its conclusions to the UNGA or consideration.</td></tr>
<tr><td>• IUCN produces a new set of guidelines for protected areas categories, including a new definition replacing the 1994 definition.
• Sustainable use and livelihoods. IUCN 4th WCC establishes the Sustainable Use and Livelihoods Specialist Group (SULI).
• EBSAs: The 3rd UNGA WG on biodiversity in ABNJ calls on States to work towards a common methodology for the identification and selection of marine areas to be protected.
• EBSAs criteria: CBD CoP9 (Bonn) adopts Scientific Criteria for Identifying EBSAs in Need of Protection in Open Ocean Waters and Deep-Sea Habitats, as well as scientific guidelines for selecting areas for representative MPAs networks.
• IPBES. 1st IMoSEB-Millennium Assessment meeting on strengthening the Intergovernmental Science-Policy Interface on Biodiversity and Ecosystem Services, under the auspices of UNEP. A concept note is elaborated about an Intergovernmental Science-Policy platform on Biodiversity and Ecosystem Services (IPBES).
• CBD-SBSTTA 13: looks at options for preventing and mitigating the impact of some activities to selected seabed habitats, and at ecological criteria and biogeographical classification systems of marine areas (and networks of MPAs) in need of protection.
• Green Growth. OECD adopts a Declaration on Green Growth at the Council at Ministerial Level (C/MIN(2009)5/ADD1/FINAL.</td><td>• FAO adopts the International Guidelines for the management of deep-sea fisheries in the high seas after 2 years of preparations. The guidelines clarify the VME concept. 'The most vulnerable marine ecosystems are ones that are both easily disturbed and are very slow to recover, or may never recover. Vulnerable ecosystem features may be physically fragile, but some may be functionally fragile even if physically robust.'
• OSPAR and NEAFC sign a memorandum of understanding in order to facilitate collaboration on high seas MPAs.</td></tr>
</table>

<table>
<tr><th>Year</th><th>Nature/biodiversity conservation</th><th>Fisheries management</th></tr>
<tr><td>2009</td><td colspan="2">• Regular Process: In its Resolution 65/37 A (para 202) the UNGA decide that the Regular Process is accountable to the UNGA and should be an intergovernmental process guided by international law, including the UN LOSC and other applicable international instruments, and take into account relevant UNGA resolutions. The Regular Process would be overseen and guided by an Ad Hoc Working Group of the Whole of the General Assembly composed of Member States (para. 203, 204). The year 2014 was selected as the deadline for the first integrated assessment.
• Biodiversity in ABNJ. GEF initiates a global pilot program focused on the protection of marine biodiversity in Areas Beyond National Jurisdiction (ABNJ).
• CBD Strategic Plan 2011–2020 is adopted by CoP10 (Nagoya), with the objective to protect effectively and equitably, by 2020, at least 10% of coastal and marine areas, noting that the determination of EBSAs and the selection of conservation and management measures is a responsibility of States and competent intergovernmental organizations. Further collaboration with FAO on fishery matters is invited.
• EBSAs and VMEs. SBSTTA deals with EBSAs and VMEs and calls for the establishment of a CBD global inventory of EBSAs, sharing info with the FAO VMEs initiative.</td></tr>
<tr><td></td><td>• IPBES. Following intergovernmental and multi-stakeholders meetings in Malaysia (2008), Kenya (2009) and Korea (Busan, 2010), governments decided that an Intergovernmental Platform on Biodiversity and Ecosystem Services (IPBES) should be established. In 2010, the Busan Outcome was welcomed by CBD CoP10 in Nagoya and considered by UNGA. UNEP was requested to convene a plenary meeting to fully operationalize IPBES.
• CBD Nagoya Protocol on MGRs. (Decision X/1). The Nagoya Protocol on Access to Genetic Resources and the Fair and Equitable Sharing of Benefits Arising from their Utilization is adopted and settles the use of MGRs in EEZs. Article 10 opens the possibility for negotiating a future multilateral benefit-sharing mechanism, possibly under UNCLOS rather than CBD.
• CBD Expert Workshop on Scientific and Technical Guidance on the use of Biogeographic Classification Systems and Identification of Marine Areas beyond national jurisdiction in need of protection (Ottawa, Canada).</td><td>• Port States Agreement: FAO adopts a legally binding Agreement on Port State Measures to Prevent, Deter and Eliminate Illegal, Unreported and Unregulated Fishing (IUU).
• RFBs-SPRFMO. Adoption of the Convention on the Conservation and Management of High Seas Fishery Resources in the South Pacific Ocean. Establishment of the South Pacific RFMO. It will enter into force in 2012.
• Flag State Performance. FAO Expert Consultation on Flag State Performance. Agreed on draft criteria for flag State performance self-assessment and international/multilateral assessment.</td></tr>
</table>

(*continued*)

Year	Nature/biodiversity conservation	Fisheries management
2010	• **Regular Process.** Intense international activity is developed at UN level to try to operationalize the concept. • **Aichi Targets:** In Decision X/2, CBD COP 10 (Nagoya, Aichi Prefecture, Japan) adopted a revised and updated Strategic Plan for Biodiversity, including the **Aichi Biodiversity Targets**, for 2011–2020.	
	• **GEF ABNJ Program**: GEF 41st Council adopts a programme for Global sustainable fisheries management and biodiversity conservation in the areas beyond national jurisdiction (ABNJ) (Program 4580/4581) developed in conjunction with the 5 tuna RFMOs, FAO (implementation leader), UNEP and the World Bank. • **EBSAs identification.** A series of regional meetings are initiated at CBD to identify EBSAs. • **Aichi Targets:** CBD COP 10 adopts the Aichi Targets.	• **2nd Review Conference on the UN Fish Stocks Agreement** to assess the progress made in implementing the 1995 United Nations Convention on the Conservation and Management of Straddling Fish Stocks and Highly Migratory Fish Stocks. Noted progress in relation to RFMOs. Confirmed concern about overfishing. Called for stronger action. • **Deep seas guidelines**. FAO Workshop on the Implementation of the International Guidelines for the Management of Deep-sea Fisheries in the High Seas: Challenges and Ways Forward, Busan, Republic of Korea. • **VMEs:** The UNGA review was preceded by a two-day workshop to discuss implementation of resolutions 61/105 and 64/72, addressing the impacts of bottom fishing on vulnerable marine ecosystems and the long-term sustainability of deep-sea fish stocks (A/66/566). **The next review will take place in 2015**.
2009–2010	• **IPBES establishment**: UNEP Governing Council, based on the 2010 UNGA resolution, decides to establish IPBES as an interface between the scientific community and policy makers that aims to build capacity for and strengthen the use of science in policy making. The process brought together two initiatives: the International Mechanism of Scientific Expertise on Biodiversity (IMoSEB) consultation (which finished in November 2007) and the follow-up processes to the Millennium Ecosystem Assessment (MEA).	
2011		• **RFBs-North Pacific RFMO.** The Convention on the Conservation and Management of High Seas Fisheries Resources in the North Pacific Ocean. Successful negotiations on the draft text. The draft Convention will be opened for signature.
2012	• **Rio + 20.** The new **United Nations Conference on Sustainable Development**, 20 years after UNCED, led to mixed outcomes. It confirmed all previously agreed commitments but did not come up with any new solution to existing problems. The Heads of State simply renewed their political commitment to sustainable development and declared their commitment to the promotion of a sustainable future. The document largely reaffirms previous action plans such as Agenda 21. No decision was made towards establishment of a global environment agency or to improve the high seas biodiversity management framework. The meeting illustrated the tension between the developed and developing worlds on the issue of 'green economy', the new avatar of the tensions between development and conservation. • **Global Partnership on Oceans.** The GPO is a growing alliance of over 100 governments, international organizations, civil society groups, and private sector interests that will mobilize knowledge and financial resources to address threats to ocean health, resilience and productivity. Its three pillars are: (1) protected areas; (2) fishing rights; and (3) wealth accounting.	
		• **RFBs-SIOFA** comes into force.

Acknowledgements

Many thanks to K. Gjerde, K. Hoydal, D. Miller, L. Ridgeway and M. Vierros for their contributions.

References

Andersen, P. (1983) On rent of fishing grounds: a translation of Jens Warming 1911 article with an introduction. ***History of Political Economy* 15**(3), 381–396.

Atkinson, C.E. (1988) Fisheries management: An historical overview. ***Marine Fisheries Review* 50**(4), 111–123.

Beverton, R.J. and Holt, S.J. (1957) On the dynamics of exploited fish populations. ***Fishery Investigations* XIX**(II), 1–533.

Büscher, B. and Dietz, T. (2005) Conjunctions in governance: The state and the conservation-development nexus in Southern Africa. ***Journal of Transdisciplinary Environmental Studies* 4**(2), 1–15.

Cernea, M.M. and Schmidt-Soltau, K. (2003) The end of forcible displacements? Conservation must not impoverish people. ***Policy Matters* 12**, 42–50.

Christophersen, L.E. (1997) IUCN, a bridge builder for Conservation of Nature. ***Green Globe Yearbook* 1997**, 59–69.

Ciriacy-Wantrup, S.V. (1968) *Resource Conservation. Economics and Politics*, 3rd edition. University of California Press.

de Klemm, C. and Shine, C. (1993) Biological diversity conservation and the law. Legal mechanisms for conserving species and ecosystems. Gland (Switzerland). IUCN Environmental Policy and Law Paper, 29.

Graham, G.M. (1935) Modern theory of exploiting a fishery and its application to the North Sea trawling. ***Journal du Conseil/Conseil Permanent International pour l'Exploration de la Mer* 10**(3), 264–274.

Gulland, J.A. and Carroz, J.E. (1968) Management of fishery resources. ***Advances in Marine Biology* 6**, 1–71.

Hardin, G. (1968) The tragedy of the commons. ***Science* 162**, 1243–1248.

Hardin, G. (1998) Extension of 'The tragedy of the commons'. ***Science* 280**(5374), 682–683.

Hughes, R. and Flintan, F. (2001) Integrating conservation and development experience. A review and bibliography of the ICDP literature. IIED, London, Biodiversity and Livelihoods Issues No. 3.

Kelleher, G., Bleakley, C. and Wells, S. (eds) (1995a) *A Global Representative System of Marine Protected Areas. Volume 1: Antarctic, Arctic, Mediterranean, Northwest Atlantic, Northwest Atlantic and Baltic.* Great Barrier Reef Marine Park Authority, the World Bank and the World Conservation Union, Washington, DC.

Kelleher, G., Bleakley, C. and Wells, S. (eds) (1995b) *A Global Representative System of Marine Protected Areas. Volume 2: Wider Caribbean, West Africa and South Atlantic.* Great Barrier Reef Marine Park Authority, the World Bank and the World Conservation Union, Washington, DC.

Kelleher, G., Bleakley, C. and Wells, S. (eds) (1995c) *A Global Representative System of Marine Protected Areas. Volume 3: Central Indian Ocean, Arabian Seas, East Africa and East Asian Seas.* The World Bank, Washington, DC.

Kelleher, G., Bleakley, C. and Wells, S. (eds) (1995d) *A Global Representative System of Marine Protected Areas. Volume 4: South Pacific, Northeast Pacific, Northwest Pacific, Southeast Pacific and Australia/New Zealand.* The World Bank, Washington, DC.

Lodge, M.L., Anderson, D., Lųbach, T., Munro, G., Sainsbury, K. and Willock, A. (2007) Recommended Best Practices for Regional Fisheries Management Organizations. Report of an independent panel to develop a model for improved governance by Regional Fisheries Management Organizations. The Royal Institute of International Affairs, Chatham House, London.

McLoughlin, R., Rayns, N., Smith, A.D.M., Johnson, D., Hobday, A. and Smith, T. (2008) Implementing the ecosystem approach in Australian Commonwealth (federally)-managed fisheries. In *The Ecosystem Approach to Fisheries* (eds G. Bianchi and H.R. Skjoldal), pp. 293–300. CABI, Wallingford and FAO, Rome.

Millennium Ecosystem Assessment (2005) *Ecosystems and Human Well-Being: Synthesis*. Island Press, Washington, DC.

Roswadowski, H.M. (2002) *The Sea Knows No Boundaries: A Century of Marine Science under ICES*. ICES, University of Washington Press, Copenhagen, Seattle, London.

Royce, W.F. (1988) Centennial Lecture IV: The historical development of fishery science and management. ***Marine Fisheries Review* (50th Anniversary Issue: Marine fisheries history) 50**(4), 30–39.

Saetersdal, G. (1992) Fishery resources and their environment, management and development. Paper presented at the *International Conference on Responsible Fishing*, Cancun, Mexico, 6–8 May 1992. FAO, Rome.

Savini, M. (1976) Contribution à l'étude des organisations internationales de pêche et de chasse maritimes. Thèse pour le Doctorat en Droit. Institut du Droit de la Paix et de l'Environnement. Université de Nice, France.

Scott Gordon, H. (1954) The economic theory of a common property resource, the fishery. ***Journal of Political Economy* 62**(2), 124–142.

Sherman, K. (2008) The large marine ecosystem approach to marine resources assessment and management. In: *The Ecosystem Approach to Fisheries* (eds Bianchi, G. and H.R. Skjoldal), pp. 47–75. CABI, Wallingford and FAO, Rome.

Terborgh, J. and Van Shaik, C. (2002) Why the world needs parks. In: *Making Parks Work. Strategies for Preserving Tropical Nature* (eds J. Terborgh, C. Van Shaik, L. Davenport and M. Rao). Island Press, Washington DC.

Toropova, C., Meliane, I., Laffoley, D., Matthews, E. and Spalding, M. (eds) (2010) *Global Ocean Protection. Present Status and Future Possibilities*. Gland (Switzerland). CN/TNC/UNEP/WCMC/UNU-IAS/AAMP.

Truman, H.S. (1945) Proclamations on U.S. Policy Concerning Natural Resources of Sea Bed and Fisheries on High Seas. Available at http://www.ibiblio.org/pha/policy/1945/450928a.html (accessed March 2013).

UNESCO (1948) Conference for the establishment of the International Union for the Protection of Nature. General information. Document NS/UIPN/1 prepared by the UNESCO Secretariat for the French Government.

Vierros, M. (2008) The ecosystem approach of the Convention on Biological Diversity. In: *The Ecosystem Approach to Fisheries* (eds G. Bianchi and H.R. Skjoldal), pp. 39–46. CABI, Wallingford and FAO, Rome.

Warming, J. (1911) Vor grundrenteaf fiskgrunde. ***Nazional Okonmiske Tidsskrift*** **49**, 499–505 (translated by Andersen in 1983).

Warner, W.W. (1983) *Distant Water: The Fate of the North Atlantic Fisherman*. Little, Brown, London.

WCED (1987) *Our Common Future. World Conference on Environment and Development*. Oxford University Press, UK.

Annex 2

Key global institutions, bodies and processes: Roles, participation and main focus[1]

L. Ridgeway
Formerly with Fisheries and Oceans Canada, Ottawa, Canada

This annex describes key global institutions, organizations, bodies and processes relevant to Chapter 11 that oceans governance and, more specifically, fisheries and biodiversity issues. This annex also paints a high-level picture of the nature and diversity of global marine governance issues being addressed. It covers: key UN-based institutions, programs and processes; MEA-established bodies; key non-UN or non-environmental institutions dealing with fisheries or biodiversity; a key global financial mechanism; some coordination mechanisms; and selective NGO organizations that are visible and active in global processes. It also describes typical delegates, how IGOs and NGOs are engaged and primordial areas of focus, that is, prisms through which issues tend to be addressed.

Governance of Marine Fisheries and Biodiversity Conservation: Interaction and Coevolution, First Edition.
Edited by Serge M. Garcia, Jake Rice and Anthony Charles.

		Potential points of functional intersection *(vertical and horizontal)*			Participation		Primordial contextual focus
Organization/ body	**Body/process/ program**	**Fisheries-related topics**	**Marine science**	**Marine biodiversity/ marine environment protection**	**Typical delegates** ***(majority in bold)***	**Role of IGOs/ NGOs/ stakeholders**	**Not necessarily 'mandate', but preoccupation on oceans and fisheries**
INSTITUTIONS OF THE UNITED NATIONS							
United Nations Principal Organ	**UN General Assembly (UNGA)**[2] Annual review of developments relating to ocean affairs and law of the sea Resolutions on oceans and the law of the sea and on sustainable fisheries[3]	Resolutions set goals for Member States on oceans and fisheries and address a wide variety of issues, including oceans governance In fisheries, recent focus on impacts of bottom fisheries on vulnerable marine ecosystems (VMEs)	Resolutions set goals for Member States on a wide variety of marine science issues, including capacity-building and transfer of technology Recent focus on the issue of the implementation of Part XIII of UNCLOS	Resolutions set goals for Member States on a wide variety of marine biodiversity and marine environment issues, including oceans governance Recent focus on oceans governance in ABNJ, legal regime for marine genetic resources, Regular Process, MPA networks and broader capacity issues	UN Permanent Representatives, **Foreign Affairs/ Legal**, Permanent Observers Draft resolutions negotiated in closed informal consultations of Member States	Some IGOs are Permanent Observers and can attend UNGA meetings IGOs and NGOs with consultative status can attend as observers	**Gives political legitimacy and mobilization to diverse actions** Consistency of oceans and fisheries developments with international law Linkages of fisheries and oceans developments with broader UN goals, in particular sustainable development
General Assembly	**Second Committee (Economic and Financial Committee)**[4] Main Committee of the General Assembly dealing with economic questions Draft Resolutions on sustainable development are negotiated for adoption in the General Assembly	Resolutions address oceans and fisheries issues relating to sustainable development (e.g. SIDS, capacity-building and technology transfer, climate change)	Resolutions address marine science issue relating to sustainable development (e.g. capacity-building and technology transfer, cooperation and coordination)	Resolutions address marine biodiversity and marine environment issues relating to sustainable development (e.g. marine ecosystems, corals, marine pollution)	UN Permanent Representatives, **Foreign Affairs/ Legal,** Permanent Observers	Some IGOs are Permanent Observers and can attend UNGA meetings IGOs and NGOs with consultative status can attend as observers	**Gives political legitimacy and mobilization to diverse actions** Increasing role on oceans and fisheries in the context of sustainable development, capacity building, technology transfer

General Assembly	**Open-ended Informal Consultative Process on Oceans and the Law of the Sea (ICP)**[5] Supports annual review by the UNGA of developments in oceans and law of the sea Suggests particular issues to be considered, especially where enhanced coordination and cooperation is needed Experts on panels provide a common basis of understanding for discussions	A wide range of oceans and fisheries issues are discussed, including sustainable fisheries, IUU fishing and oceans and fisheries governance Fisheries issues have been raised in virtually every meeting	A wide range of marine science issues are discussed, including capacity-building and transfer of technology	A wide range of marine biodiversity and marine environment issues are discussed, including marine genetic resources, VMEs, ecosystem approaches, marine biodiversity, integrated assessments and oceans governance	UN Permanent Representatives, **Foreign Affairs/ Legal, IGOs and NGOs**	IGOs and NGOs with consultative status attend meetings and participate fully in panel discussions Science/IGO/NGO experts on panels and host side events Science/IGO/NGO experts on some official delegations, depending on issue, practice and capacity	**Forward-looking: emerging issues** on policy and oceans law needing consideration and clarity **Fostering greater coordination and cooperation on these topics** **Sustainable development,** capacity building, cooperation and coordination
General Assembly	**Ad Hoc Open-ended Informal Working Group to study issues relating to the conservation and sustainable use of marine biological diversity beyond areas of national jurisdiction (BBNJ)**[6] Studies issues and makes recommendations to the General Assembly on the conservation and sustainable use of marine biological diversity in areas beyond national jurisdiction (ABNJ)	Addresses impacts of fisheries in areas beyond fisheries jurisdiction Governance gaps and issues relating to legal framework for marine biodiversity in ABNJ	Need for capacity-building in science and transfer of technology concerning ABNJ Addressing gaps in information and marine science	Governance and legal frameworks and issues relating to marine genetic resources, area-based management tools (including, MPAs) and environmental impact assessments Current focus on legal framework for marine biodiversity in ABNJ (gaps, improved implementation of existing instruments, possible need for new agreement under UNCLOS)	UN Permanent Representatives/ **Foreign Affairs/ Legal, IGOs and NGOs**	IGOs and NGOs with consultative status can attend and participate as observers IGOs and NGOs and policy experts on some official delegations, depending on topic Side events hosted by IGOs and NGOs	**Legal frameworks and governance gaps for marine biodiversity in ABNJ** **Relevant legal regime on marine genetic resources in ABNJ** Some developing States want new legal regime for marine genetic resources in ABNJ, including sharing of benefits Some developed States want new legal regime for biodiversity protection, particularly high seas MPAs

(*Continued*)

		Potential points of functional intersection *(vertical and horizontal)*			Participation		Primordial contextual focus
Organization/ body	**Body/process/ program**	**Fisheries-related topics**	**Marine science**	**Marine biodiversity/ marine environment protection**	**Typical delegates *(majority in bold)***	**Role of IGOs/ NGOs/ stakeholders**	**Not necessarily 'mandate', but preoccupation on oceans and fisheries**
General Assembly	**Regular Process for Global Reporting and Assessment of the State of the Marine Environment, including Socio-economic Aspects**[7] Established to assess the marine environment, building on existing regional assessments, including by improved understanding of marine and coastal ecosystems, increased scientific and technical collaboration and capacity building in marine science, information and management	Fisheries issues considered in context of assessing the state of the marine environment and the impact of anthropogenic activities on the marine environment	Assessment based on marine science research (MSR); techniques for assessments, integrating trends across ecological processes and structures; socio-economic benefits and provision of ecosystem services; capacity-building and technology transfer	Issues relating to biodiversity and protection of the marine environment considered in the context of assessing the state of the marine environment and the impact of anthropogenic activities on the marine environment	UN Permanent Representatives/ **Foreign Affairs/ Legal**, Science experts nominated by States to oversee and carry out assessments	IGOs and NGOs with consultative status can participate as observers Scientific institutions and major groups can participate as observers Science experts on some official delegations; can be nominated by States to carrying out assessments	**Developing institutional arrangements and preparing the First Global Integrated Marine Assessment by 2014** Regional workshops to support the Regular Process and enhance assessment capacity

United Nations Principal Organ	**United Nations Security Council**[8] Primary responsibility under the UN Charter for the maintenance of international peace and security	May consider fisheries issues in the context of its mandate (e.g. illegal fishing and international peace and security, security implications of climate change)	Considers issues relating to science, capacity-building and transfer of technology in the context of its mandate (e.g. security implications of climate change)	Considers issues relating to biodiversity and the protection of the marine environment in the context of its mandate (e.g. impacts of illegal dumping on international peace and security; security implications of climate change)	**UN Permanent Representatives/ Foreign Affairs/ Legal**	Some IGOs are Permanent Observers and can attend meetings IGOs as well as NGOs with consultative status can attend some meetings as observers	**Maintain international peace and security in accordance with the principles and purposes of the United Nations** Identify threats or acts of aggression and recommend action
United Nations Principal Organ	**Economic and Social Council (ECOSOC)**[9] Coordinates economic, social and related work of the UN specialized agencies, functional commissions and five regional commissions Central UN forum for international economic and social issues	Considers governance, food security, including fisheries and sustainable development issues in context of broad development policy and Millennium Development Goals (MDGs)	Considers capacity-building, transfer of technology, science and research issues in the context of broader development policy	Considers governance, biodiversity and sustainable development issues in the context of broader development policy and achieving the MDGs	**UN Permanent Representatives/ Foreign Affairs** Representatives of IGOs, NGOs, the private sector and academia	IGOs and NGOs with consultative states can participate, present written contribution and make statements to the Council and its subsidiary bodies	Focus on **progress in the implementation of the international development goals**, including the MDGs, and coherence and effectiveness of development partners

(Continued)

		Potential points of functional intersection *(vertical and horizontal)*			Participation		Primordial contextual focus
Organization/ body	**Body/process/ program**	**Fisheries-related topics**	**Marine science**	**Marine biodiversity/ marine environment protection**	**Typical delegates** ***(majority in bold)***	**Role of IGOs/ NGOs/ stakeholders**	**Not necessarily 'mandate', but preoccupation on oceans and fisheries**
ECOSOC	**UN Commission on Sustainable Development (UNCSD)**[10] Ensures effective follow-up to 1992 United Nations Conference on Environment and Development (UNCED); reviews progress in implementing Agenda 21 and Rio Declaration on Environment and Development and 2002 Johannesburg Plan of Implementation (JPOI) Two-year cycles on thematic issues. 2014–15 focus is Oceans and Seas, Marine Resources, Small Island Developing States and Disaster Management and Vulnerability To be replaced by an intergovernmental high-level political forum pursuant to the outcome of the 2012 United Nations Conference on Sustainable Development	Various global commitments and targets for coming 5–10 years, launch of new MEAs as appropriate, institutional reform as appropriate	Various global commitments and targets for coming 5–10 years	Various global commitments and targets for coming 5–10 years; launch of new MEAs as appropriate	**UN Permanent Representatives/ Foreign Affairs;** Environment and sectoral specialists depending on State practice and choice	Many preparatory processes (including non-governmental) to set the agenda Broad participation from IGOs and NGOs in UNCSD sessions Supports innovative activities (e.g. Partnerships Fair, Learning Centre and panels, roundtables, side events) High-level dialogue among Ministers, including special session with Major Groups	To monitor developments in sustainable development and global summits; intended originally to focus on implementation but now negotiating forum for multi-stakeholders Broad agenda setting and targets; **subsequent global environment meetings have become more explicitly focused on sustainable development for environmental goals** Past recommendations led to establishment of Open-ended Informal Consultative Process on Oceans and the Law of the Sea (ICP)

United Nations Programme	**UN Development Programme (UNDP)**[11] Coordinates UN development activities at country level; supports national and local capacity development to help achieve the MDGs Leading programming implementer for the Global Environment Facility and other climate-related funds	Issues relating to food security, including fisheries and oceans governance in context of development policy and MDGs (e.g. overfishing, habitat loss and degradation, sustainable fish stocks)	Issues relating to capacity-building and the transfer of technology in the context of broader development policy and MDGs (e.g. integrating science into development plans)	Issues relating to biodiversity, protection of the marine environment and ocean governance in the context of broader development policy and MDGs (e.g. coastal and marine ecosystems, pollution, invasive alien species, marine protected areas)	**UN Permanent Representatives/ Foreign Affairs** participate in meetings of the Executive Board	IGOs as well as NGOs with consultative status may participate in meetings of the Executive Board upon invitation	**Focus on achieving MDGs, strengthening ability of developing States to manage the environment in a sustainable manner and advancing poverty reduction efforts** Provides inter-governmental support to and supervision of UNDP, UNFPA and UNOPS
United Nations Programme	**United Nations Environment Programme (UNEP)**[12] Designated authority in UN on environmental issues; located in developing State (Kenya); intended to act as catalyst and to coordinate on environmental issues in UN; launches MEAS (e.g. CBD) Focus on environmental assessments, instruments/ institutions for environmental management, transfer of	**Marine and Coastal Strategy (2009)** guides marine work **Freshwater and Marine Programme** for coordination with specialized agencies (e.g. FAO). Direct involvement in projects, especially developing countries (see also GEF) Capacity building regarding international fisheries issues	Responsible in UN for environmental assessment and review (e.g. Millennium Ecosystem Assessment, Global Environment Outlook (GEO) and other thematic reports) Jointly sponsors knowledge activities/projects Launched with CBD an **Intergovernmental Science–Policy Platform on Biodiversity and Ecosystem Services (*IPBES*),** addressing both	**Marine and Coastal Strategy (2009)** guides marine work Capacity-building links to Regional Seas Conventions/ Action Plans (see GEF) **Marine and Coastal Strategy – The Other 70%** vision; land-oceans links; ecosystems and human wellbeing; vulnerable people/ places, reconciling use and conservation **Freshwater and Marine Programme**	**Government representatives (Environment)**, IGOs and NGOs and research institutes attend meetings of the GMEF	IGOs and NGOs can participate in meetings of the GMEF as observers Major conduit for GEF funding projects on conservation and sustainable use	**Promoted as lead among UN agencies for integrated planning and assessments** **Focus on environmental governance, management priorities and risks, capacity building for developing States** Marine and coastal program sees coordination as a major function, across global institutions involved in marine issues

(*Continued*)

Organization/ body	Body/process/ program	Potential points of functional intersection *(vertical and horizontal)*			Participation		Primordial contextual focus
		Fisheries-related topics	**Marine science**	**Marine biodiversity/ marine environment protection**	**Typical delegates** ***(majority in bold)***	**Role of IGOs/ NGOs/ stakeholders**	**Not necessarily 'mandate', but preoccupation on oceans and fisheries**
	knowledge/ technology for sustainable development, partnerships/ cooperation with civil society and private sector **UNEP Governing Council/Global Ministerial Environment Forum (GMEF)** is annual environmental ministerial forum for review of policy issues 2012 UN Conference on Sustainable Development invited General Assembly at its 67th session in 2013 to adopt a resolution to strengthen and upgrade UNEP	and negotiations (e.g. WTO subsidies, CITES) Policy briefs (e.g. Fisheries and Aquaculture, 'greening' of Fisheries) 'Emergency Assessment' of emerging issues (non-governmental)	terrestrial and marine biodiversity; issue of relationship to and duplication with UN Regular Process	for strengthened GPA; water quality; Regional Seas Programmes; marine spatial planning; integrated ecosystem assessment; marine habitats and climate change mitigation/ adaptation; ecosystem tools for SIDS; environmental and infrastructure			Provides directly parallel guidance documents on issues where FAO has already issued guidance documents (e.g. ecosystem approach, MPAs and fisheries)

UNEP	**UNEP Regional Seas Programme**[13] Individual Regional Seas Programmes (RSPs) address degradation of global oceans and coastal areas through sustainable management and use Contribute to global goals (e.g. Agenda 21, WSSD Plan of Implementation, MDGs) Function through conventions and action plans; provide comprehensive strategy and framework for protecting the environment/ promoting sustainable development for countries sharing common body of water	RSPs deal with fisheries indirectly by addressing coastal management and development, ecosystems, biodiversity and SIDS Cooperate with FAO and RFMOs to address environmental impacts of fisheries/ ecosystem approach	RSPs focus on marine science and assessments in context of marine and coastal environment (e.g. chemical wastes, oil spill preparedness and response, conservation of marine species and ecosystems) Capacity-building and transfer of marine technology	RSPs directly address protection of marine biodiversity and the marine environment through wide range of monitoring, assessment and management activities/ initiatives (e.g. coastal management and development, marine and coastal protected areas, ecosystems and biodiversity, SIDS) Cooperate with IGOs addressing these issues	**Government representatives (Environment)**, IGOs and NGOs and research institutes Regional Seas Global Meetings attended by IGO Secretariats and NGOs	RSPs develop own rules on participation, but typically allow participation by IGOs and NGOs IGOs and NGOs can attend the Regional Seas Global Meetings	**Sustainable management and use of the marine and coastal environment** Many RSPs have principal **focus on capacity-building**, transfer of marine technology
UNEP	**Global Programme of Action for the Protection of the Marine Environment from Land-based Activities (GPA)**[14] Aims to prevent degradation of marine environment from land-based activities	Addresses fisheries in context of impacts of land-based activities and review of implementation of the GPA	Focus on improving science, assessment of the marine environment and enhancing capacity-building and transfer of technology	Deals with impacts of land-based pollution on marine biodiversity/ environment	**Government representatives (Environment)**, IGOs and NGOs, in particular science and technical experts	IGOs and NGOs can attend meetings as observers	**Coastal-marine interactions and Improving integration of land-based issues** in the work of other organizations Development of national plans of action

(*Continued*)

		Potential points of functional intersection *(vertical and horizontal)*			Participation		Primordial contextual focus
Organization/ body	**Body/process/ program**	**Fisheries-related topics**	**Marine science**	**Marine biodiversity/ marine environment protection**	**Typical delegates** ***(majority in bold)***	**Role of IGOs/ NGOs/ stakeholders**	**Not necessarily 'mandate', but preoccupation on oceans and fisheries**
United Nations Programme	**UN Office on Drugs and Crime (UNODC)**[15] Operates globally through an extensive network of field offices to address illicit drugs and international crime Relies on voluntary contributions, mainly from governments	UN Convention against Transnational Organized Crime (UNTOC), UN Convention against Corruption (UNCAC), addresses international cooperation in criminal matters, including trafficking, smuggling of people/drugs/ weapons on fishing vessels/ IUU links			**Government representatives (Legal)**, IGOs and NGOs	IGOs as well as NGOs with consultative status can participate as observers in meetings of the Commission on Narcotic Drugs (CND) and the Commission on Crime Prevention and Criminal Justice (CCPCJ), as well as meetings of the Crime Congress Individual academic experts can attend	Normative work and field-based technical cooperation **projects to enhance the capacity of Member States to** counteract illicit drugs, crime and terrorism

UNCLOS Treaty Bodies and Institutions	**States Parties to United Nations Convention on the Law of the Sea (SPLOS)**[16] Addresses budgetary and administrative matters relating to UNCLOS Also receives reports from Secretary-General on issues of a general nature, relevant to States Parties, that have arisen with respect to UNCLOS	Fisheries issues raised by some delegations in the context of the reports of the Secretary-General (e.g. IUU fishing, destructive fishing practices, sustainable development)	Marine science issues raised by some delegations in the context of the reports of the Secretary-General (with a current focus on issues related to Commission on the Limits of the Continental Shelf)	Issues of biodiversity and marine protection raised by some delegations in the context of the reports of the Secretary-General (e.g. marine genetic resources, ocean governance, marine pollution, marine biodiversity in areas beyond national jurisdiction)	UN Permanent Representatives/ **Foreign Affairs/ Legal**	Very limited role for IGOs and NGOs on official delegations IGOs as well as NGOs with consultative status can attend as observers	**Budgetary and administrative issues relating to UNCLOS** Significant focus on issues relating to the **workload of the Commission on the Limits of the Continental Shelf**
UNCLOS Treaty Bodies and Institutions	**International Seabed Authority (ISA)**[17] Organizes and controls activities in the seabed and ocean floor/beyond national jurisdiction (the Area), especially administering the resources of the Area Regulates deep seabed mining, ensuring the marine environment is protected from harmful effects of mining activities, including exploration	Issues relating to fisheries only indirectly considered in context of implications of measures protecting marine environment from harmful effects of deep seabed mining	Significant activities regarding marine scientific research, including detailed resource assessments, monitoring the status of scientific knowledge of the deep-sea marine environment and promoting marine scientific research in the Area	Ensuring environmentally sustainable development of seabed mineral resources, including measures to regulate deep seabed mining and ensure the protection of the marine environment from harmful effects Understanding seabed features and developing measures to regulate seabed mining activities and protect the marine environment	**Foreign Affairs/ Legal** Role for scientific and technical advisors on official delegations and in the work of the ISA	Very limited role for IGOs and NGOs on official delegations	**Focused on management of deep seabed mining, spatial planning and ecosystems associated with non-living resources**

(Continued)

		Potential points of functional intersection *(vertical and horizontal)*			Participation		Primordial contextual focus
Organization/ body	**Body/process/ program**	**Fisheries-related topics**	**Marine science**	**Marine biodiversity/ marine environment protection**	**Typical delegates *(majority in bold)***	**Role of IGOs/ NGOs/ stakeholders**	**Not necessarily 'mandate', but preoccupation on oceans and fisheries**
United Nations Fish Stocks Agreement (UNSFA)	**Informal Consultations of States Parties to the United Nations Fish Stocks Agreement (ICSP)**[18] Forum for States Parties to discuss issues relating to implementation of the Agreement, planning of UNFSA Review Conferences	Wide variety of issues relating to oceans and fisheries and the implementation of UNFSA, including impacts on fisheries sustainability	Issues relating to conservation and management of fish stocks and enhancing scientific research and assessments relating to the sustainability of fish stocks	Issues relating to biodiversity, precautionary approach and ecosystem approaches relating to the implementation of UNFSA	UN Permanent Representatives/ **Foreign Affairs/ Legal**/Fisheries policy experts (especially in UNFSA Review Conference)	IGOs as well as NGOs with consultative status can attend as observers	Promoting a wider participation in the Agreement, strengthening implementation of the provisions of the Agreement **Preparations for Review Conference on UNFSA**
UNFSA	**UNFSA Review Conference**[19] Convened to assess the effectiveness of the Agreement by reviewing and assessing the adequacy of its provisions and proposing means of strengthening implementation	Conservation and management of stocks; mechanisms for international cooperation; monitoring, control and surveillance and compliance and developing States and non-parties	Enhancing scientific research and assessments relating to the sustainability of fish stocks	Issues relating to biodiversity, precautionary approach and ecosystem approaches relating to the implementation of UNFSA	UN Permanent Representatives/ **Foreign Affairs/ Legal**	Limited role for IGOs and NGOs on official delegations IGOs as well as NGOs with consultative status can attend as observers	Promoting a wider participation in the Agreement **Strengthening implementation of the provisions of the Agreement**

United Nations Coordinating Mechanism	**UN-Oceans**[20] Created to establish effective, transparent and regular inter-agency coordination on ocean and coastal issues in UN Reviews relevant programmes and activities in UN system, identifies emerging issues, defines joint actions and specific task teams; promotes integrated management; facilitates inputs to Secretary-General's annual report on oceans and the law of the sea Currently under review by the UN Joint Inspection Unit General Assembly to consider reviewing its mandate and enhancing transparency and reporting of its activities to Member States	**Task Force on Global Partnership for Climate, Fisheries and Aquaculture** to strengthen collaboration and coordination regarding oceans and climate change and support for the UNFCCC Strategic approach to health and resilience of global oceans, strengthening the capacity vis-a-vis broader development strategies.	Task Force on Establishing a Regular Process for Global Assessment of the Marine Environment	Task Force on Marine Biodiversity in Areas Beyond National Jurisdiction Task Force on Marine Protected Areas and Other Area-based Tools	Comprising **Secretariats of the UN system** and financial institutions (potentially, UN-DESA, UN-DOALOS, FAO, IOC-UNESCO, UNEP, World Bank (IBRD), IMO, WMO, UNDP, IAEA, CBD, ISA, ILO, UNIDO, WTO, WHO, UNHSP (UN-HABITAT), UNFCCC, Ramsar, UNCTAD, UNU, OECD)	Participation of relevant NGOs and other stakeholders in work of Task Forces encouraged under the responsibility of lead institutions	**Coordination and cooperation of UN activities related to oceans and coastal areas**

(*Continued*)

Organization/ body	Body/process/ program	Potential points of functional intersection *(vertical and horizontal)*			Participation		Primordial contextual focus
		Fisheries-related topics	**Marine science**	**Marine biodiversity/ marine environment protection**	**Typical delegates *(majority in bold)***	**Role of IGOs/ NGOs/ stakeholders**	**Not necessarily 'mandate', but preoccupation on oceans and fisheries**
United Nations Coordinating Mechanism	**Environment Management Group (UNEP-EMG)**[21] Coordination body of specialized agencies, programmes and organs of the UN including the MEAs (e.g. FAO, UNEP, UNESCO-IOC, WMO, CMS, CITES); chaired and supported by UNEP **Issue Management Groups (IMGs)** address specific issues, including background documents/ reports (e.g. sustainability management, green economy and biodiversity)	IMG on Green Economy; report on how the UN system can support, including facilitating common understanding and measures for the transition; addresses fisheries issues in a green economy (e.g. harmful subsidies, full cost pricing, sustainable trade)	Science considerations underlie work of IMGs, in particular preparation of reports	IMG on Biodiversity; report on UN-system wide perspective on the biodiversity challenges and link to human well-being/ socio-economic development goals; biodiversity aspects of IGO activities and potential for collaborative programmes and initiatives UN system-wide approach to environmental and social sustainability and international environmental governance reform	Representatives of specialized agencies, programmes and organs of the UN work through technical meetings, Issue Management Groups and task forces	Representatives of IGOs and NGOs can be invited to contribute.	Identifying environmental issues that warrant cooperation engaging collective capacity in coherent management responses **Providing UN system-wide responses to environmental issues and coordinating inter-agency cooperation**

United Nations Specialized Agency	**Food and Agriculture Organization (FAO)**[22] Forum for 192 member nations to negotiate agreements and debate policy in fisheries and aquaculture in context of food security Four main fisheries bodies: **Committee on Fisheries (COFI)**, **Sub-Committees on Aquaculture and Fish Trade and Coordinating Working Party on Fisheries Statistics** Biennial publication on State of Fisheries and Aquaculture (SOFIA) Jointly runs CODEX with WHO (see below) Many joint activities with other IGOs	Overarching program is **Code of Conduct for Responsible Fisheries and related International Plans of Action (IPOAs)**: sharks, IUU fishing: fishing capacity; seabirds bycatch; and programs supporting conservation and management Normative and field work to improve fisheries sustainability; key databases and programs supporting management; technical assistance (Global program for responsible fisheries)	Secretariat performs specialized independent research/joint activities Knowledge dissemination, global databases, information networks (statistics, fisheries status and trends; aquatic sciences and fisheries abstracts, Fishery Resources Monitoring System (FIRMS); Fisheries Global Information System (FIGIS); Species ID and data) Advice to CITES regarding aquatic species listings	Application of and capacity building in ecosystem approach, including of standards and guidelines for management components (e.g. bycatch, deep sea guidelines/VMEs Marine protected areas as a tool for fisheries management Reduction of environmental impacts of shrimp trawling GEF programme on sustainable fisheries management and biodiversity conservation in ABNJ in partnership with UNEP and the World Bank, among others	**Government representatives (Fisheries**: scientists, managers, policy experts), Foreign Affairs, Legal Industry representatives often on delegations	IGOs including RFMO/As as well as NGOs can participate in FAO meetings as observers Participation in expert processes and collaborative projects among IGOs and Secretariats New frameworks and instruments are developed through expert processes before being considered in State-to-State negotiations and adoption by COFI	**Sustainable management of fisheries; fisheries perspective on emerging tools and frameworks** in support of food security; ground work on normative standards related to fisheries For small-scale fisheries and developing States, focus is development as well as tailored application of norms and standards in small-scale context **Large fieldwork program oriented to capacity building in developing States and States in transition**

(*Continued*)

		Potential points of functional intersection *(vertical and horizontal)*			Participation		Primordial contextual focus
Organization/ body	**Body/process/ program**	**Fisheries-related topics**	**Marine science**	**Marine biodiversity/ marine environment protection**	**Typical delegates *(majority in bold)***	**Role of IGOs/ NGOs/ stakeholders**	**Not necessarily 'mandate', but preoccupation on oceans and fisheries**
FAO/WHO	**Codex Alimentarius Commission and Joint FAO/WHO Food Standards Programme**[23] Develops general standards for food labelling, food hygiene, food additives, pesticide residues; procedures for assessing the safety of foods derived from modern biotechnology; import and export inspection and food certification systems	Standards for fish and fishery products (marine, fresh water and aquaculture) through the **Codex Committee on Fish and Fishery Products** Codex Alimentarius used as basis for dispute resolution in WTO	Standards are based on the best available science and independent risk assessment bodies or ad-hoc consultations organized by FAO and WHO FAO convenes activities for scientific advice for food safety (e.g. biotechnology, nanotechnologies		**Government representatives (Food safety experts, Health), Food safety organizations** All Codex meetings are open to the public to the limits of available space	IGOs and NGOs can become accredited Codex observers to provide expert information, advice and assistance to the Commission	Pre-eminent international food standard setting body working to **protect the health of consumers and ensure fair practices in the food trade**

United Nations Specialized Agency	**United Nations Educational, Scientific and Cultural Organization (UNESCO)**[24] International collaboration through education, science and culture. Pursues its objectives through five major programs: education, natural sciences, social and human sciences, culture, communication and information Pursuant to **World Heritage Convention** 45 world heritage sites are located in marine areas		**Mobilizing science** and policy for sustainable development **World Heritage Marine Program** for analysis of threat and management gaps for marine World Heritage sites, case studies of specific marine sites/ecosystems, cross-cutting research on threats to sensitive/rare ecosystems (e.g. marine invasive species)	**World Heritage Marine Program** for effective conservation of existing and potential marine areas of Outstanding Universal Value Comprehensive planning basis for the identification of new sites Conservation and capacity building for marine spatial planning training and heritage sites; promoting the World Heritage Convention as an instrument for marine conservation	**Government representatives (Environment); Permanent delegates to UNESCO**	IGOs and NGOs with consultative status can attend meetings of the General Conference	**Contribute to the global goals and intercultural dialogue** through education, the sciences, culture, communication and information

(*Continued*)

Organization/ body	Body/process/ program	Potential points of functional intersection *(vertical and horizontal)*			Participation		Primordial contextual focus
		Fisheries-related topics	**Marine science**	**Marine biodiversity/ marine environment protection**	**Typical delegates** ***(majority in bold)***	**Role of IGOs/ NGOs/ stakeholders**	**Not necessarily 'mandate', but preoccupation on oceans and fisheries**
UNESCO	**International Oceanographic Commission (IOC)**[25] United Nations body for ocean science, ocean observatories, ocean data and information exchange and ocean services (e.g. tsunami warning systems) Programs focus on marine environmental protection, ecosystem dynamics, climate change, global observing systems, data and information management, coastal area management and disaster management Active in UN-Oceans and cooperative processes assessing the state of the marine environment, including the Regular Process	Indirectly addresses fishery issues through objective work on oceans observation, climate change, science-related tools and knowledge for decision-making; translation of discovery science into policy-relevant products that can be used for management Focus on establishment of tsunami warning systems, monitoring ocean acidification and climate effects on atmosphere/ ocean	Cooperation and coordination of programmes/data exchange in marine scientific research, ecosystem health, climate change, Global Oceans Observation System (GOOS), hazard mitigation and capacity development Operates the International Oceanographic Data and Information Exchange (IODE) Coordinates/ integrates meteorological observing data with WMO and supports IPCC	Scientific program in aid of useable and integrated oceans knowledge for use in governance management tools and mechanisms, as well as integrated planning (coastal and protection)	**Science experts** from governments and institutions	Many cooperative projects (e.g. through UN-Oceans, Regular Process, WMO on oceans/ atmospheric work, UNFCC on climate monitoring through earth observation and with FAO and CBD on oceans, fisheries and biodiversity issues	Objective science processes, knowledge dissemination and institutional capacity building **Understanding needs for management of resources, oceans and coastal areas and facilitating creation of usable knowledge**

UNESCO/IOC	**Advisory Body of Experts on the Law of the Sea (ABE-LOS)**[26] Experts from 144 countries work to implement Part XIII of UNCLOS and help develop marine scientific research	Sponsors and addresses various aspects of research and management tools relevant to fisheries (e.g. integrated management, MPAs, implementing the ecosystem approach)	Assists States in meeting UNCLOS provisions on MSR; scientific legal and policy aspects of marine bio-prospecting; legislative reviews; scientific research, methods for gathering data, transfer of marine technology	Publications and reports address various aspects of research and management tools (e.g. integrated management, MPAs, implementing the ecosystem approach)	Each IOC Member State can nominate two experts: one with a scientific background and one with legal knowledge	Scientific experts from IGOs and NGOs can be nominated by IOC Member States	**Assisting in implementation of marine scientific research obligations under UNCLOS**
United Nations Specialized Agency	**International Maritime Organization (IMO)**[27] Responsible for maritime safety and preventing pollution from ships (e.g. MARPOL deals with accidental and operational oil pollution, pollution by chemicals, goods in packaged form, sewage, garbage and air pollution; OPRC provides a global framework for cooperation in combating major incidents or threats of marine pollution; HNS Protocol covers marine pollution by hazardous and noxious substances) Also deals with London Convention and the London Protocol on pollution by dumping	Fishing vessels under 24m generally not covered by IMO instruments (i.e. most vessels in developing countries) **Torremolinos International Convention for the Safety of Fishing Vessels** applies to fishing vessels 24 m and over; new Protocol will raise limit to 45 m Addresses fishing vessel safety, port State measures and control, vessel tracking and detection systems, marine pollution from fisheries Joint work with ILO and FAO	Science for assessing impacts of shipping on marine environment (e.g. pollution from vessels, invasive species) Technical assistance; funding provided by UNDP and UNEP and others; projects through the World Maritime University and International Maritime Law Institute Participates in GESAMP; lead agency on ballast water and environmental hazards of harmful substances	**Marine Environment Protection Committee (MEPC)** addresses prevention and control of pollution of the marine environment from ships, adoption of instruments (e.g. pollution from shipping, invasive species, anti-fouling, greenhouse gas emissions, wrecks) Area protection provided by Special Areas and Particularly Sensitive Sea Areas Partnerships with UNDP and GEF to assist developing countries reduce the transfer of harmful aquatic organisms and pathogens in ships' ballast water	IMO Committees, Assembly and Council attended by **government representatives (Transport, Shipping, Environment, Coast guard, Legal)** Formal arrangements for cooperation with more than 40 IGOs More than 60 NGOs have consultative status to participate in the work bodies as observers	IGOs and NGOs with consultative status can participate in meetings and the work of bodies as observers	**Maritime safety and preventing pollution from ships** Increasing coordinating role with other UN agencies (ILO, FAO)

(*Continued*)

Organization/ body	Body/process/ program	Potential points of functional intersection *(vertical and horizontal)*			Participation		Primordial contextual focus
		Fisheries-related topics	Marine science	Marine biodiversity/ marine environment protection	Typical delegates *(majority in bold)*	Role of IGOs/ NGOs/ stakeholders	Not necessarily 'mandate', but preoccupation on oceans and fisheries
United Nations Specialized Agency	**International Labour Organization (ILO)**[28] Promotes rights at work, encourages decent employment opportunities, enhances social protection and strengthens dialogue in handling work-related issues Formulates international labour standards, sets minimum standards of basic labour rights Promotes development of independent employers' and workers' organizations and provides training and advisory services to those organizations	ILO instruments apply to fishing workers (minimum age, medical, competency certificates, accommodation, work hours, training) **Work in Fishing Convention, 2007** deals with work conditions and operation of vessels; Action Plan with IMO/ FAO on implementation Cooperative projects with IMO/ILO/FAO (e.g. codes and guidance on safety of fishermen/ vessels; guidance on training and certification of fishing vessel personnel; joint publications and actions plans)	Marine science dealt with indirectly in the context of job creation and work conditions (e.g. 'green' jobs)	Deals with marine biodiversity indirectly in the context of job creation and work conditions (e.g. 'green' jobs)	**Government representatives (Employment, Labour),** IGOs and NGOs attend meetings of the International Labour Conference Only 'tripartite' UN agency bringing together representatives of governments, employers and workers to jointly shape policies and programmes	IGOs and NGOs with consultative status can participate in meetings and the work of bodies as observers Role for representatives of employers and workers on official delegations	**Enhancing rights of workers and social protections for employees** Promoting social justice and internationally recognized human and labour rights

United Nations Specialized Agency	**World Meteorological Organization (WMO)**[29] Specialized agency for meteorology (weather and climate), operational hydrology and related geophysical sciences Cooperates with other IGOs involved in oceans activities (e.g. IMO, UNESCO/IOC, UNDP), universities, training institutions, schools Joint WMO-IOC Technical Commission for Oceanography and Marine Meteorology (JCOMM) established in 1999	**Marine Meteorology and Oceanography Programme (MMOP**) focus on meteorological forecast and warning services **Marine Broadcast System for the Global Maritime Distress and Safety System (GMDSS)** coordinated with World-Wide Navigational Warning Service operated by the International Hydrographic Organization (IHO)	Ten major scientific and technical programmes on a wide range of meteorological and hydrological services issues JCOMM supports GOOS and the Global Climate Observing System (GCOS) **WMO-IOC-ICSU World Climate Research Programme**-scientific information for predictions of global and regional climatic variations, frequency and severity of extreme events	**Marine Meteorology and Oceanography Programme (MMOP)** focus on quality meteorological forecasts and warning services in support of safety of life and property at sea and the role of the oceans in climate variability and change Understanding, monitoring, mapping and predicting maritime weather and ocean conditions offer the opportunity for adequate planning of coastal zone and marine activities and provide a structure for early detection and warning of marine-related hazards	**Government representatives (Meteorological, Environment, Transport),** IGOs and NGOs with consultative status	Representatives of IGOs can conclude arrangements or agreements providing for representation at the WMO Congress NGOs with consultative status can attend meetings of the WMO Congress	Provision of operational hycrology and related geophysical sciences; **focus on climate change and climate variability**

(*Continued*)

		Potential points of functional intersection *(vertical and horizontal)*			Participation		Primordial contextual focus
Organization/ body	**Body/process/ program**	**Fisheries-related topics**	**Marine science**	**Marine biodiversity/ marine environment protection**	**Typical delegates *(majority in bold)***	**Role of IGOs/ NGOs/ stakeholders**	**Not necessarily 'mandate', but preoccupation on oceans and fisheries**
United Nations Specialized Agency	**World Bank Group**[30] Fights poverty and improves living standards in developing world; loans, policy advice, technical assistance and knowledge sharing to low- and middle-income countries **World Agriculture and Rural Development: Global Program On Fisheries** for policy dialogue and specific investments Action Plan for Agriculture and Rural Development (2009)	**Global Program on Fisheries (PROFISH)** works with client countries to achieve WSSD fisheries goals Strategic vision for fisheries and aquaculture to revitalize fisheries in developing world: governance; poverty alleviation; legal/ regulatory frameworks; management regimes; subsidies reform; capacity building; coastal communities; sustainable aquaculture		**World Agriculture and Rural Development: Global Program On Fisheries** promotes marine reserves and protected areas PROFISH workshops raise the oceans profile, discuss latest developments/ science/role of the World Bank (e.g. Oceans of Carbon: The Role of the World Bank in Oceans and Climate Change)	Program collaborates with **large aid donors, recipients and fisheries experts** in its annual planning meetings Experts participate as relevant in sponsored activities	PROFISH to continue to undertake well-chosen global analyses, in collaboration with agencies such as FAO, IFPRI, OECD, WorldFish Center, national institutions, NGOs and other World Bank units PROFISH is active partner in GEF and new Global Partnership on Oceans (GPO)	**Global effort to ensure fisheries play a stronger role in development and poverty alleviation** Mobilizing knowledge and resources for concrete directions and investments

MULTILATERAL SECRETARIATS AND OTHER BODIES

MEA Secretariats and other bodies	**Convention on Biological Diversity (CBD)**[31] Focus on conservation of biological diversity, the sustainable use of the components of biological diversity, and the fair and equitable sharing of the benefits arising from the utilization of genetic resources Conference of the Parties (COP) reviews progress in implementation of commitments, adopts programmes of work and provides policy guidance Strategic Plan calls for cooperation among all relevant international instruments, MEAs and processes for policy coherence **Jakarta Mandate** focused on integrated marine and coastal area management, the sustainable use of living resources, marine and coastal protected areas, mariculture and alien species	Fisheries issues in the context of addressing conservation and sustainable use of biodiversity components New Aichi targets include sustainable management and harvesting of fish stocks and application of ecosystem-based approaches to fisheries **Global Biodiversity Outlook** on status of biological diversity, includes fisheries Joint activities with CMS on migratory species and their habitat **Jakarta Mandate** for fishing-related targets, protection and recovery of features/habitats; review ecosystem considerations in fisheries assessments and effects of fisheries	**Subsidiary Body on Scientific, Technical, and Technological Advice (SBSTTA)**: recommendations on scientific and technical aspects of implementing the CBD Convention Collaborative science-based tools for ecologically significant area protection, design of management tools and assessments, target setting for biodiversity generally **Jakarta Mandate** for collaborative science-based tools for vulnerable area protection, design of management tools and assessments, target setting for marine biodiversity specifically	Wide range of issues on conservation and sustainable use of marine biodiversity/fair and equitable sharing of marine genetic resources in national jurisdiction **Biodiversity Liaison Group** among six biodiversity conventions for coherence and cooperation in implementation Programme of work on protected areas with UNESCO-WHC, Ramsar and CMS Participates with UNEP in **International Coral Reef Initiative** (ICRI), a partnership among governments, IGOs/NGOs to preserve coral reefs ecosystems **Jakarta Mandate** for assessment and tools related to biodiversity	**Government representatives (Environment)**, IGOs and NGOS, science institutions attend COP **Government representatives (Environment)**, science experts, IGOs, science institutions and NGOs attend SBSTTA	IGOs and NGOs involved in the conservation and sustainable use of biological diversity may participate as observers Secretariat works with diverse partners to increase cooperation and coordination on biodiversity, including Rio Conventions, IGOs, NGOs, indigenous organizations, scientific and technical research and assessment bodies and private sector	Halt the loss of biodiversity, ensure that ecosystems are resilient and provide essential services, contribute to human wellbeing and poverty eradication **Tools and frameworks for biodiversity protection** and sustainable use (terrestrial and marine), **with emphasis on area-based management tools, including MPAs** Biodiversity as a common heritage. **Access and benefit sharing**

(Continued)

		Potential points of functional intersection *(vertical and horizontal)*			Participation		Primordial contextual focus
Organization/ body	Body/process/ program	Fisheries-related topics	Marine science	Marine biodiversity/ marine environment protection	Typical delegates *(majority in bold)*	Role of IGOs/ NGOs/ stakeholders	Not necessarily 'mandate', but preoccupation on oceans and fisheries
		on forage stocks; identification of ecologically and biologically significant areas (EBSAs), regional workshops, ideally with RFMO input re VMEs; promoting environmental impact assessments, including fisheries		protection (e.g. prior environmental assessment, delineation and identification of EBSAs; review of guidelines for considering biodiversity in environmental impact assessments)			
MEA Secretariats and other bodies	**Convention on International Trade in Endangered Species (CITES)**[32] Aims to ensure species are not at risk as a result of international trade Accords varying degrees of protection through controls on trade (e.g. prohibitions, licence requirements) to more than 30,000 species of animals and plants (including	Approximately 100 commercially exploited aquatic species are listed Application of listing criteria to fish species and adequacy of species management CITES/FAO MOU cooperation regarding application of criteria to commercially exploited	Parties must designate one or more scientific authorities to advise them on the effects of trade on the status of the species Scientific issues are dealt with in Standing Committees	Focus on conserving biodiversity and contributing to sustainable use by ensuring against unsustainable exploitation from international trade Addresses MDGs, WSSD biodiversity targets, promotes wider involvement of civil society in the development of conservation policies and	**Government representatives (Environment)** Species experts (e.g. fish) take part in specific Standing Committees on technical issues and join delegations at COP	IGOs may participate in plenary sessions and committees as observers Technically qualified NGOs or others, if approved, can also participate as observers	Species protection where trade is a threat Increasing focus on **promoting trade restrictions to complement other management tools thought to be inadequate,** which can trigger debates over jurisdiction and conflicts with other management regimes

	aquatic) based on the application of certain listing criteria by parties	species, other issues (e.g. 'trade' status of species caught in high seas); CITES considers advice of FAO Expert Advisory Panel		practices, and a coherent and internationally agreed approach based on scientific evidence			
MEA Secretariats and other bodies	**Convention on the Conservation of Migratory Species of Wild Animals (CMS)**[33] Aims to conserve terrestrial, aquatic and avian migratory species throughout their range; migratory species threatened with extinction are listed; protected by coordinated action of Range States, including conserving or restoring habitats, mitigating obstacles to migration Instruments concluded under CMS range from legally binding treaties to less formal instruments such as MOUs	Resolutions and recommendations calling for immediate action to address the problem of bycatch and non- target species Several CMS-related agreements and memoranda of understanding dedicated to species for which bycatch is a major issue	Scientific Council provides scientific advice, identifies research and conservation priorities; recommendations to COP; advice on project funding **Global Register of Migratory Species (GROMS)** has information on species, GIS-maps, population data, bibliography, threat status (2000 IUCN Red List) and other convention protection (e.g. CITES)	Instruments concluded under CMS aim to conserve specific migratory species (e.g. marine turtles, cetaceans) Cooperates with five other MEAs through the **Biodiversity Liaison Group**	**Government representatives (Environment**, Fisheries)	MOUs among Range States for short-term conservation measures for critically endangered migratory species, including with specialized NGOs IGOs and NGOs collaborate in policy, specific projects/ fieldwork IGOs and NGOs can participate in meetings of some agreements and MOUs as observers	**Focus on the conservation of migratory species, their habitats and migration routes through species listing**

(*Continued*)

		Potential points of functional intersection *(vertical and horizontal)*			Participation		Primordial contextual focus
Organization/ body	**Body/process/ program**	**Fisheries-related topics**	**Marine science**	**Marine biodiversity/ marine environment protection**	**Typical delegates** ***(majority in bold)***	**Role of IGOs/ NGOs/ stakeholders**	**Not necessarily 'mandate', but preoccupation on oceans and fisheries**
MEA Secretariats and other bodies	**London Convention and London Protocol**[34] Objective to promote the effective control of all sources of marine pollution and to take practicable steps to prevent pollution from dumping of wastes and other matter Under the Protocol all dumping is prohibited; permits may be issued for certain specified materials, subject to conditions	Specific guidance for assessment of fish waste or organic materials from industrial fish processing; screening focused at potential presence of chemicals used in aquaculture and their residues	Scientific groups provide assessment advice and recommendations (e.g. ocean fertilization activities, sequestration of CO2; development of precautionary protocols on MSR) Framework to assess proposed oceans fertilization activities on case-by-cases basis for conformity as legitimate scientific research	Specific guidelines for assessment of wastes or other matter that may be considered for dumping; guidance on amenities, biological features and uses of the sea affecting location of dump-sites (e.g. areas of special scientific or biological importance sanctuaries, fishing, spawning, nursery and recruitment, migration routes)	**Government representatives (Environment)**; IGOs and NGOs	IGOs and NGOs with consultative status can attend meetings as observers	**Promoting a wider participation in the London Protocol**; prevention of pollution of the sea by dumping of wastes and other matter**; mitigating the impacts of climate change**

MEA Secretariats and other bodies	**United Nations Framework Convention on Climate Change (UNFCCC)**[35] Sets overall framework for intergovernmental efforts on climate change; Kyoto Protocol legally binds developed countries to emission reduction targets during commitment periods	Considers fisheries-related issues in the context of national adaptation plans and programmes; impacts of climates climate change on States and vulnerability, in particular SIDS	**Subsidiary Body for Scientific and Technological Advice (SBSTA)** for work on methodological and scientific matters, including scientific, technical and socio-economic aspects of mitigation; land use, research and observation Research dialogues highlight activities and findings, needs and priorities, including topics related to oceans/ marine environment, in particular ocean acidification	Biodiversity-related issues in context of national adaptation plans/ programmes; impacts of climate change on States, especially SIDS Recommendation of COP to enhance cooperation and action to enable adaptation, reduce vulnerability (including VMEs), build resilience in developing countries IMO provides regular reports on developments on reducing greenhouse gas emissions from ships	**Government representatives (Environment,** Foreign affairs, Meteorological), IGOs and NGOs	Several categories of observer organizations attend sessions of the COP and its subsidiary bodies, including IGOs and NGOs IGO and NGO host range of side events	**Development of new legally binding instruments and binding emission reduction targets** for developed and developing countries; national mitigation and adaptation plans; common but differentiated responsibilities

(*Continued*)

		Potential points of functional intersection *(vertical and horizontal)*			Participation		Primordial contextual focus
Organization/ body	**Body/process/ program**	**Fisheries-related topics**	**Marine science**	**Marine biodiversity/ marine environment protection**	**Typical delegates** ***(majority in bold)***	**Role of IGOs/ NGOs/ stakeholders**	**Not necessarily 'mandate', but preoccupation on oceans and fisheries**
MEA Secretariats and other bodies	**Ramsar Convention**[36] Framework for national action and international cooperation for conservation and use of wetlands and their resources Contributes to achievement of MDGs, 2010 Biodiversity targets, 2012 target for MPAs and responses to climate change	Wetlands include mangroves and coral reefs; specific criteria to identify wetlands based on supporting fish species and contributing to biological diversity, or as important source of food for fishes	**Scientific and Technical Review Panel (STRP)** provides scientific and technical guidance Focused on 'wise use', national policies/ wetland management based on the best available scientific (technical and traditional) knowledge CBD-Ramsar Joint Work Plan	List of Wetlands of International Importance for conservation of global biological diversity Listing criteria based on supporting vulnerable/ threatened/ endangered species, ecological communities; supporting biological diversity of biogeographic regions; supporting species at a critical stage in their life cycles or refuge during adverse conditions	**Government representatives (Environment),** IGOs and NGOs	IGOs and NGOs can participate in COP as observers Certain NGOs have status of 'International Organization Partners' (e.g. IUCN); provide expert advice; full members of STRP Cooperative agreements with MEAs and IGOs; CBD Biodiversity Liaison Group	**Wetlands contribution to biodiversity protection**

FINANCIAL SUPPORT/CAPACITY BUILDING							
Financial support/ capacity building	**Global Environment Facility (GEF)**[37] Unites 182 countries, IGOs, civil society organizations and the private sector on global environmental issues; financial mechanism for the CBD and many MEAs **International Waters** finances activities to collectively manage transboundary water basins and coastal and marine systems, including restoring/ sustaining fish stocks and associated biological diversity, addressing impacts of land-based pollution of coastal waters in Large Marine Ecosystems	**FAO/UNEP/World Bank project** on global sustainable fisheries management and biodiversity in ABNJ; project areas to target measures, including regarding tuna fisheries **Oceans Partnership Fund** with World Bank, UNEP, tuna RFMOS, CBD, UN-DOALOS	Scientific and advisory panel provides technical and scientific advice to GEF partners and projects	**FAO/UNEP/World Bank project** addresses identification of VMEs and EBSAs, development of management strategies; area-based planning tools for deep-seas ecosystems; testing area-based planning methodologies; marine profiles and plans for select seascapes including MPAs; information improvements; cross-sectoral policy coordination for VMEs and EBSAs; new communities of practice for integrated planning	**Aid agencies/ donors**	UNDP, UNEP and World Bank original GEF partners; now a separate institution Partnership includes: UNDP, UNEP, World Bank, FAO, UNIDO, African Development Bank, Asian Development Bank, EBRD, Inter-American Development Bank, International Fund for Agriculture Development	**Only multi-convention financing entity; improves funding stability for environmental governance**

(*Continued*)

Organization/ body	Body/process/ program	Potential points of functional intersection *(vertical and horizontal)*			Participation		Primordial contextual focus
		Fisheries-related topics	**Marine science**	**Marine biodiversity/ marine environment protection**	**Typical delegates** ***(majority in bold)***	**Role of IGOs/ NGOs/ stakeholders**	**Not necessarily 'mandate', but preoccupation on oceans and fisheries**
INTERNATIONAL TRADE							
International trade	**World Trade Organization (WTO)**[38] Forum for negotiating agreements to reduce obstacles to international trade and ensure a level playing field to contribute to economic growth and development; provides a legal and institutional framework for implementation and monitoring of agreements and settling disputes An important source of dispute resolution, especially regarding whether environment or sustainability can be used to restrict trade Important links to CODEX standards and WIPO	Agreements on tariffs, sanitary and phytosanitary (SPS) measures and health protection claims; technical barriers to trade (TBT); subsidies and countervailing measures including reducing fisheries subsidies; anti-dumping; import licensing and quotas; safeguards for domestic industry; intellectual property rights (TRIPs) over plants, animals and technologies Framework for subsidies negotiations based on linking environmental impacts of subsidies to fish stock status and management regimes			**Government representatives (Trade policy, Foreign affairs, Legal)** Most fishery countries are members of WTO and bound by trade agreements, which have a direct impact on fish trade	IGOs with observer status can attend meetings of WTO bodies NGOs can apply to attend Ministerial Conferences; also participate in issue-specific symposia	Consistency of actions with WTO Agreement and free trade, conditions and **justifications for infringement of free trade commitments** Subsidies negotiation is one of the first negotiations of a trade regime for sustainability purposes

MULTI-STAKEHOLDER FORA OPERATING IN GLOBAL PROCESSES

Multi-stakeholder fora operating in global processes	**International Union for Conservation of Nature (IUCN)**[39] Global environmental network with focus on conserving biodiversity through setting global scientific standards, conservation projects on sustainable management of biodiversity and natural resources and participation in international processes **IUCN Global Programme** gives priority to biodiversity, climate change, sustainable energy, human well-being and green economy More than 1000 government and NGO member organizations and 11,000 volunteer scientists in 160 countries	Species-based protections, including **IUCN Red List**; ecosystem management **Global Marine and Polar Program (GMPP)** includes work on fisheries governance; eco-certification; sustainable aquaculture; IUU fishing; corruption in fisheries	Sponsors considerable technical and scientific work in networks and experts groups under marine program Criteria to be adopted in 2012 for assessing and listing ecosystems at risk, to complement species at risk list	**GMPP** focus on ensuring marine and polar ecosystems are maintained and restored in their biodiversity and productivity and any use of resources is sustainable and equitable Projects in ten theme areas: climate change mitigation and adaptation; conserving threatened species; energy and industry; fisheries and aquaculture; invasive species; marine protected areas; oceans governance; securing coastal livelihoods; polar activities; coral reefs Participates in **Global Invasive Species Programme**, to minimizing the spread and impact of invasive species, with CBD mandate	Congress attended by government, public sector, NGOs, business, UN agencies and social organizations Convenes special events with diverse global participation	Unites States and NGOs at IUCN Congress who define global programme of work and elect IUCN Council Member organizations participate in national and regional committees Fosters issues-based networks on a wide range of issues with wide range of experts	**Analysis-based advocacy of networked activities** to help create international framework for marine biodiversity protection **High-level agenda setting, with recent focus on marine biodiversity and fisheries**

(*Continued*)

Organization/ body	Body/process/ program	Potential points of functional intersection *(vertical and horizontal)*			Participation		Primordial contextual focus
		Fisheries-related topics	Marine science	Marine biodiversity/ marine environment protection	Typical delegates *(majority in bold)*	Role of IGOs/ NGOs/ stakeholders	Not necessarily 'mandate', but preoccupation on oceans and fisheries
Multi-stakeholder fora operating in global processes	**Global Forum on Oceans, Coasts and Islands (GOF)**[40] Forum brings together cross-sectoral ocean leaders from over 110 countries to promote implementation of international agreements related to oceans, coasts and SIDS; assess progress, identify obstacles and opportunities for sustainable development; emerging issues (oceans governance in ABNJ); international consensus building on unresolved ocean issues; mobilize knowledge, resources and organizational action Secretariat provided by the **International Coastal and Ocean Organization (ICO)**	Issues-based working groups and items at conferences (e.g. sustainable fisheries, tunas, IUU fishing, links to oceans governance, climate and oceans)	Integrates knowledge into policy-related dialogue; no original research	Expert working groups and multi-stakeholder dialogues on specific implementation gaps (oceans governance, biodiversity, marine genetic resources, climate and oceans) Reports on progress achieved regarding global environmental summits Hosts 'Oceans Day' profile oceans issues at large global meetings (e.g. UNFCCC, COPs, Global Summits)	**Multi stakeholder advocacy;** some governments, academics, lawyers, IGOs; some delegations have fisheries experts	Multi-stakeholder presence at sponsored events including IGO and government representatives, although often few fisheries experts	**Advocacy of specific policy options based on multi-stakeholder dialogue** from special events and at special events regarding implementation of global commitments (e.g. WSSD) **Very focused on sustainable development**, climate change and sharing of benefits across States

Industry organization	**World Oceans Council (WOC)**[41] Cross-sectoral industry leadership alliance on corporate ocean responsibility Collaborates to improve ocean science in support of safe and sustainable operations, educate public and stakeholders on role of responsible companies, engage in ocean policy and planning, and develop science-based solutions to cross-cutting environmental challenges	Working groups convene around platforms to advance WOC vision and mission through priority policy processes, including UNCLOS, CBD, issues, such as invasive species and ocean noise, and regions of concern	Develops science-based solutions to cross-cutting and cross-sectoral environmental challenges, such as invasive species, ocean noise, marine mammal impacts, marine debris and the Arctic	Addresses marine ecosystem footprint of operations, supply chains; business value of collaboration for marine environmental issues; launching collaborative programs on shared priorities **National Business Forum on Marine Spatial Planning (MSP)** cross-sector business involvement in the MSP process	Key companies in shipping, oil and gas, fisheries, aquaculture, tourism, renewable energy, ports, dredging, cables and pipelines, carbon capture and storage, as well as the maritime legal, financial and insurance communities and others	Member companies, associations and organizations can participate in committees and working groups; some participation by ICFA members	Business value of collaboration and branding for oceans sustainability **Ensure business interests find entry into oceans planning processes** Strategic Action Plan to match those in governments and other oceans communities, especially around marine spatial planning
Industry organization	**International Coalition of Fisheries Associations (ICFA)** Coalition of national fish and seafood industry trade associations from the world's major fishing nations Objective is to preserve and maintain the oceans as a major fish source of food	Advocate policies for the long-term sustainable use of living marine resources Facilitates meetings with various UN agencies addressing seafood issues and participates in meetings where fisheries issues are debated			Representatives of national fish and seafood industry trade associations from the world's major fishing nations; representing countries harvest more than 85% of the global fish	Members meet semi-annually to identify trends, develop policies, work planning	To provide decision-makers with a unified voice on global fish and seafood issues **Committed to objective science-based and fully participatory fishery conservation and management processes** Participate in meetings to ensure the voice of the seafood industry is heard

(*Continued*)

Organization/ body	Body/process/ program	Potential points of functional intersection *(vertical and horizontal)*			Participation		Primordial contextual focus
		Fisheries-related topics	Marine science	Marine biodiversity/ marine environment protection	Typical delegates *(majority in bold)*	Role of IGOs/ NGOs/ stakeholders	Not necessarily 'mandate', but preoccupation on oceans and fisheries
UNITED NATIONS SCIENCE, RESEARCH AND EDUCATION FORA							
UN science, research and education	**Joint Group of Experts on the Scientific Aspects of Marine Environmental Protection (GESAMP)**[42] Expert group to advise UN system on scientific aspects of marine environmental protection Targeted scientific studies providing technical advice on scientific aspects of prevention, reduction and control of the degradation of the marine environment Cross-sectoral, interdisciplinary and science-based approach to advise policy development in marine environmental affairs; provides for coordination and cooperation among UN agencies	Programs, working groups and studies on a variety of fisheries-related topics, including fisheries habitat and ecosystems, sustainable coastal aquaculture, assessment and communication of environmental risks in coastal aquaculture, impacts of potentially harmful substances	Integrates and synthesizes regional and thematic assessments for global assessments of marine environment; scientific and technical guidance on design and execution of assessments; advice on specific topics on the marine environment; overviews of marine environment monitoring and assessment in UN agencies; improving coordination/ integration; identify new and emerging issues	Programs, working groups and studies on diverse topics, including impacts of pollution on marine ecosystems, invasive alien species, threats to marine biodiversity Working Group on environmental hazards (EHS) and harmful substances carried by ships **Ballast Water Working Group on Active Substances** reviews proposals for IMO on ballast water management systems **Working Group on Global Trends in Pollution** of coastal ecosystems	Technical Secretariats of the UN agencies	Governments and IGOs can initiate partnerships Individual experts can participate in the Pool of Experts; nominations made by UN organizations, Member States, regional organizations, international scientific bodies and NGOs Specialized experts from sponsoring agencies (IMO, FAO, UNESCO-IOC, WMO, IAEA, UN, UNEP, UNIDO, UNDP)	Technical advice on scientific aspects of prevention, reduction and control of the degradation of the marine environment **Cooperation and coordination of scientific work by UN agencies**

UN science, research and education	**United Nations University (UNU)**[43] Worldwide network of academic institutes and programmes coordinated in Tokyo Contribute through collaborative research and education, dissemination and advisory services, to efforts to resolve global problems of human survival, development and welfare 'Think tank' for the United Nations system and bridge between UN and the international academic and policy communities	Activities are clustered into interlinking thematic areas, including global change and sustainable development Articles and publications on range of fisheries issues, for example persistent organic pollutants in fisheries food chain; no-take fisheries reserves	Thematic areas include science, technology and society Articles and publications on wide range of marine science issues	Articles and publications on biodiversity issues, such as implications of dying coral reefs; true value of ecosystem services; Atoll islands and climate change		Worldwide network of institutes and programmes, organizations and academic institutions on UN topics of interest Members of IGOs assist with intensive training courses	**Research and educational programmes** with a particular aim of assisting developing countries Conduit for global knowledge on sustainable development, based on rigorous research methodology

Notes

1. This chapter benefited from the guidance of an expert who wishes to remain anonymous.
2. http://www.un.org/en/ga/
3. http://www.un.org/Depts/los/general_assembly/general_assembly_resolutions.htm
4. http://www.un.org/en/ga/second/index.shtml
5. http://www.un.org/Depts/los/consultative_process/consultative_process.htm
6. http://www.un.org/Depts/los/biodiversityworkinggroup/biodiversityworkinggroup.htm
7. http://www.un.org/Depts/los/global_reporting/global_reporting.htm
8. http://www.un.org/Docs/sc/
9. http://www.un.org/en/ecosoc/
10. http://sustainabledevelopment.un.org/csd.html
11. http://www.undp.org/content/undp/en/home.html
12. http://www.unep.org/
13. http://www.unep.org/regionalseas/
14. http://www.gpa.unep.org/
15. http://www.unodc.org/
16. http://www.un.org/depts/los/meeting_states_parties/meeting_states_parties.htm
17. http://www.isa.org.jm/en/home
18. http://www.un.org/depts/los/convention_agreements/fish_stocks_agreement_states_parties.htm
19. http://www.un.org/depts/los/convention_agreements/review_conf_fish_stocks.htm
20. http://www.unoceans.org/Index.htm
21. http://www.unemg.org/
22. http://www.fao.org/fishery
23. http://www.codexalimentarius.org/
24. http://www.unesco.org/new/en/
25. http://www.unesco.org/new/en/natural-sciences/ioc-oceans/
26. http://ioc-unesco.org/index.php?option=com_content&view=article&id=364&Itemid=100047
27. http://www.imo.org/Pages/home.aspx
28. http://www.ilo.org/global/lang--en/index.htm
29. http://www.wmo.int/pages/index_en.html
30. http://www.worldbank.org/
31. http://www.cbd.int/
32. http://www.cites.org/
33. http://www.cms.int/
34. http://www.imo.org/OurWork/Environment/SpecialProgrammesAndInitiatives/Pages/London-Convention-and-Protocol.aspx
35. http://unfccc.int/2860.php
36. http://www.ramsar.org/cda/en/ramsar-home/main/ramsar/1_4000_0__
37. http://www.thegef.org/gef/home
38. http://www.wto.org/
39. http://www.iucn.org/
40. http:// http://globaloceanforum.com/
41. http://www.oceancouncil.org/site/
42. http://www.gesamp.org/
43. http://unu.edu/

Index

Note: page numbers in italics refer to figures; page numbers in bold refer to tables.

Governance of Marine Fisheries and Biodiversity Conservation: Interaction and Coevolution, First Edition.
Edited by Serge M. Garcia, Jake Rice and Anthony Charles.